Graduate Texts in Mathematics

55

Arlen Brown
Carl Pearcy

Introduction to Operator Theory I

Elements of Functional Analysis

Springer-Verlag

New York   Heidelberg   Berlin

Arlen Brown
Department of Mathematics
Indiana University
Bloomington, IN 47401
USA

Carl Pearcy
Department of Mathematics
University of Michigan
Ann Arbor, MI 48104
USA

AMS Subject Classifications: 46-02, 47-02

Library of Congress Cataloging in Publication Data

Brown, Arlen, 1926–
    Elements of functional analysis.

    (Their Introduction to operator theory ; v. 1)
(Graduate texts in mathematics ; 55)
    Bibliography: p.
    Includes index.
    1. Functional analysis. I. Pearcy, C.,
1935– joint author. II. Title. III. Series:
Graduate texts in mathematics ; 55.
QA329.B76   vol. 1   [QA320]   515′.72s [515′.7]
                                                        77-23438

© 1977 by Springer-Verlag, New York Inc.
Softcover reprint of the hardcover 1st edition 1977

9 8 7 6 5 4 3 2 1

ISBN 978-1-4612-9928-8
DOI 10.1007/978-1-4612-9926-4

ISBN 978-1-4612-9926-4 (eBook)

For Dodie and Cristina,
who have given constant encouragement
over many years

Preface

This book was written expressly to serve as a textbook for a one- or two-semester introductory graduate course in functional analysis. Its (soon to be published) companion volume, *Operators on Hilbert Space*, is intended to be used as a textbook for a subsequent course in operator theory. In writing these books we have naturally been concerned with the level of preparation of the potential reader, and, roughly speaking, we suppose him to be familiar with the approximate equivalent of a one-semester course in each of the following areas: linear algebra, general topology, complex analysis, and measure theory. Experience has taught us, however, that such a sequence of courses inevitably fails to treat certain topics that are important in the study of functional analysis and operator theory. For example, tensor products are frequently not discussed in a first course in linear algebra. Likewise for the topics of convergence of nets and the Baire category theorem in a course in topology, and the connections between measure and topology in a course in measure theory. For this reason we have chosen to devote the first ten chapters of this volume (entitled Part I) to topics of a preliminary nature. In other words, Part I summarizes in considerable detail what a student should (and eventually must) know in order to study functional analysis and operator theory successfully. The presence of this extensive review of the prerequisite material means that a student who is not familiar with one or more of the four basic courses mentioned above may still successfully read this book by making liberal use of Part I. Indeed, it should be said that perhaps the only critical prerequisite for a profitable reading of this book is a certain mathematical maturity, which, for our purposes, may be taken to mean the ability to follow and construct ε-δ arguments, a level of maturity that any talented student who has had a good course in advanced calculus will have attained.

In keeping with our pedagogical intent in writing this book, we have provided both examples and exercises in copious supply. Indeed, every chapter contains a number of illuminating examples and is followed by a collection of problems. (Some problems appear as simple assertions of fact; in such cases the student is expected to provide a proof of the stated fact.) In this connection we observe that the problem sets constitute an integral part of the book, and that the student must study them along with the text. Working problems is very important in the study of mathematics in general, of course, for that is how mathematics is learned, but in this textbook it is particularly important because many topics of interest are first introduced in the problems. Not infrequently the solution of a problem depends in part on material in one or more preceding problems, a fact that instructors should bear in mind when assigning problems to a class.

While, as noted, this book is intended to serve as a textbook for a course, it is our hope that the wealth of carefully chosen examples and problems, together with the very explicit summary of prerequisite material in Part I, will enable it to be useful as well to the interested student who wishes to study functional analysis individually.

An instructor who plans to use this book as a textbook in a course has several options depending on the time available to him and the level of preparation of his students. He may wish to begin, for example, by devoting some weeks to the study of various chapters in Part I. Whether he does this or not, time limitations may make it impossible for him to treat all of Part II in one semester. With this in mind, we suggest the following abbreviated syllabus for a somewhat shorter course of study.

Chapter 11: Read entire text; omit Problems L–Q and U–Y.
Chapter 12: Read entire text; omit Problems R–Y.
Chapter 13: Read entire text; omit Problems S–T.
Chapter 14: Omit the material on Frechét spaces, viz., Examples H–L and Proposition 14.9; omit Problems Q–W.
Chapter 15: Omit all text after Theorem 15.11; omit Problems O–X.
Chapter 16: Omit the material on dual pairs, viz., everything after Proposition 16.12; omit Problems O–X.
Chapter 17: Read entire text; omit Problems T–Y.
Chapter 18: Omit the material on approximation theory, viz., everything after Example D; omit Problems V–W.
Chapter 19: Omit.

In the writing of this book no systematic effort has been made to attribute results or to assign historical priorities.

The notation and terminology used throughout the book are in essential agreement with those to be found in contemporary (American) textbooks. In particular, the symbols $\mathbb{N}$, $\mathbb{Z}$, $\mathbb{R}$, and $\mathbb{C}$ will consistently represent the systems of positive integers, integers, real numbers, and complex numbers, respectively. We have also found it convenient to reserve the symbol $\mathbb{N}_0$ for the system of nonnegative integers.

Finally, there is one basic convention in force throughout the book: *All vector spaces that appear herein are either real or complex. If nothing is said about the scalar field of a vector space under discussion, it is automatically assumed to be complex.*

ARLEN BROWN
CARL PEARCY

Acknowledgements

Preliminary versions of this book were used several times in courses at Indiana University and the University of Michigan, and we take this opportunity to express our gratitude to the students in those classes for calling to our attention many inaccuracies and oversights. We are indebted to Grahame Bennett, R. G. Douglas, Glen Schober, and Allen Shields for their critical reading of portions of the manuscript and their many valuable suggestions for improvements. We also express our thanks to P. R. Halmos, to whom we are indebted in more ways than can be summarized in this brief space. Finally, we express our gratitude to Donald Deckard, who contributed generously in many ways over a period of several years to the creation of the manuscript.

ARLEN BROWN
CARL PEARCY

Contents

Contents

PART I
PRELIMINARIES

Set theory

1

We shall assume the reader to be familiar with the elements of set theory. Nonetheless, we begin with a review of certain set-theoretic fundamentals, largely to fix notation and terminology. (Readers wishing to improve their acquaintance with set theory, or to pursue in greater depth any of the topics touched on below, might consult [31] or [34]; another excellent source for most topics is [10].) For one thing, at the most elementary level, we reserve certain symbols throughout the book for several important sets. The system of positive integers is denoted by $\mathbb{N}$, the system of nonnegative integers by $\mathbb{N}_0$, the system of all integers by $\mathbb{Z}$, the real number system by $\mathbb{R}$, and the complex number system by $\mathbb{C}$. The empty set is denoted by $\varnothing$, and if X and Y are any two sets, the set-theoretic difference $\{x \in X : x \notin Y\}$ is denoted by $X \backslash Y$ and the symmetric difference $(X \backslash Y) \cup (Y \backslash X)$ by $X \nabla Y$. Moreover, if f is a mapping of X into Y (notation: $f : X \to Y$) and $A \subset X$ and $B \subset Y$, then $f(A)$ will denote the set $\{f(x) : x \in A\}$ and $f^{-1}(B)$ the set $\{x \in X : f(x) \in B\}$.

The reader is also assumed to be familiar with the notion of a partially ordered set. In this context our terminology and notation are quite standard. Thus if $X = (X, \leq)$ is a partially ordered set, then an element x_0 of X is *maximal* [*minimal*] in X if there exists no element x of X such that $x > x_0$ [$x < x_0$]. Likewise, if E is a subset of a partially ordered set X and if x_0 is an element of X such that $x \leq x_0$ for every x in E, then x_0 is an *upper bound* of E. If the set of upper bounds of E in X is nonempty, then E is *bounded above* in X. If, in addition, the set of upper bounds of E possesses a least element, then that *least upper bound* is also called the *supremum* of E and is denoted by sup E. Dually E is *bounded below* if the set of lower bounds of E in X is nonempty; if, in addition, the set of lower bounds of E possesses a greatest element, then that *greatest lower bound* is called the *infimum* of E (notation: inf E). A subset of a

3

partially ordered set is *bounded* if it is bounded both above and below. For finite subsets $\{x_1, \ldots, x_n\}$ of a partially ordered set X we shall also write $x_1 \vee \cdots \vee x_n$ for $\sup\{x_1, \ldots, x_n\}$ and $x_1 \wedge \cdots \wedge x_n$ for $\inf\{x_1, \ldots, x_n\}$. If X has the property that $x \vee y$ and $x \wedge y$ exist for every pair of elements x and y of X, then X is a *lattice*. If, more generally, every subset of X has both a supremum and an infimum, then X is a *complete lattice*. A mapping f of one partially ordered set into another is *monotone increasing* [*decreasing*] if $x \leq y$ implies $f(x) \leq f(y)$ [$f(x) \geq f(y)$] and is *strictly* monotone increasing [decreasing] if $x < y$ implies $f(x) < f(y)$ [$f(x) > f(y)$]. A mapping f of a set X into a partially ordered set Y is *bounded* [*above, below*] if its range $f(X)$ is bounded [above, below] in Y.

Example A. The system $\mathbb{R}$ of real numbers is a lattice (in its usual ordering). Indeed, we have

$$s \vee t = \tfrac{1}{2}[s + t + |s - t|]$$

and

$$s \wedge t = \tfrac{1}{2}[s + t - |s - t|]$$

for every pair of real numbers s and t. If t is a real number the numbers $t \vee 0$ and $-(t \wedge 0)$ are called the *positive* and *negative parts* of t, and are denoted by t^+ and t^-, respectively. Note that t^+ and t^- are nonnegative and satisfy the conditions

$$t^+ + t^- = |t|,$$
$$t^+ - t^- = t,$$

for every real number t.

Example B. Every bounded nonempty subset of $\mathbb{R}$ has a supremum and an infimum in $\mathbb{R}$ (this is, in effect, one formulation of the *Dedekind postulate*; a lattice with this property is said to be *boundedly complete*). It follows that every closed interval $[a, b]$ ($= \{t \in \mathbb{R}: a \leq t \leq b\}$) is a complete lattice. While $\mathbb{R}$ itself is not a complete lattice, it is very useful to imbed $\mathbb{R}$ in a complete lattice. To do this we simply introduce two new "numbers," $+\infty$ and $-\infty$, and define $-\infty < +\infty$ and also $-\infty < t < +\infty$ for every t in $\mathbb{R}$. The enlarged set $\mathbb{R} \cup \{+\infty\} \cup \{-\infty\}$ is called the *extended real number system* and will consistently be denoted by $\mathbb{R}^\natural$. It is clear that $\mathbb{R}^\natural$ is a simply ordered complete lattice, and that if E is a subset of $\mathbb{R}$ that is not bounded above [below] in $\mathbb{R}$, then $\sup E = +\infty$ [$\inf E = -\infty$] in $\mathbb{R}^\natural$. We make a partial extension of the operation of addition to $\mathbb{R}^\natural$ by defining

$$t + (\pm\infty) = (\pm\infty) + t = \pm\infty$$

for every real number t, and

$$(\pm\infty) + (\pm\infty) = \pm\infty.$$

Subtraction is also extended analogously to $\mathbb{R}^2$, but the symbols

$$(\pm\infty) + (\mp\infty) \quad \text{and} \quad (\pm\infty) - (\pm\infty)$$

remain undefined. Similarly we define

$$t(\pm\infty) = (\pm\infty)t = \begin{cases} \pm\infty, & t > 0, \\ 0, & t = 0, \\ \mp\infty, & t < 0, \end{cases}$$

for every real number t, and write

$$(\pm\infty)(\pm\infty) = +\infty \quad \text{and} \quad (\pm\infty)(\mp\infty) = -\infty.$$

Likewise, just as for ordinary real numbers, if t is an extended real number we write $t^+ = t \vee 0, t^- = -(t \wedge 0)$, and $|t| = t^+ + t^-$. When the extended real number system is employed, the ordinary real numbers, that is, the elements of $\mathbb{R}$ itself, are called *finite* (real) numbers.

Example C (The Banach–Knaster–Tarski Lemma). Let X be a complete lattice, and let φ be a monotone increasing mapping of X into itself. If we set $A = \{x \in X : \varphi(x) \leq x\}$, then it is at once clear that $\varphi(A) \subset A$. Let $x_0 = \inf A$, and suppose $x \in A$. Then $x_0 \leq x$ and therefore $\varphi(x_0) \leq \varphi(x) \leq x$. Thus $\varphi(x_0)$ is a lower bound of A, whence it follows that $\varphi(x_0) \leq x_0$, so x_0 is itself an element of A. But then $\varphi(x_0) \in A$, so $x_0 \leq \varphi(x_0)$. Thus we see that $\varphi(x_0) = x_0$, and we have proved the following result: *A monotone increasing mapping of a complete lattice into itself possesses a fixed point.*

In this book the axiom of choice is used without apology or explanation, and is usually employed in the following form.

Zorn's lemma. *Let X be a partially ordered set, and suppose that every simply ordered subset of X is bounded above. Then X possesses a maximal element.*

If $\{X_\gamma\}_{\gamma \in \Gamma}$ is any family of sets indexed by an index set Γ, then the (*Cartesian* or *set-theoretic*) *product* of the family $\{X_\gamma\}$ will be denoted by $\prod_{\gamma \in \Gamma} X_\gamma$. The set $\prod_{\gamma \in \Gamma} X_\gamma$ consists of all indexed families $\{x_\gamma\}_{\gamma \in \Gamma}$ where $x_\gamma \in X_\gamma$ for each γ in Γ. The *projection* π_{γ_0} is the mapping defined by $\pi_{\gamma_0}(\{x_\gamma\}) = x_{\gamma_0}$ for each element $\{x_\gamma\}$ in $\prod_{\gamma \in \Gamma} X_\gamma$.

A nonempty partially ordered set $\Lambda = (\Lambda, \leq)$ is a *directed set* if, for every pair of elements λ_1 and λ_2 of Λ, there exists an element λ of Λ such that $\lambda_1 \leq \lambda$ and $\lambda_2 \leq \lambda$. If Λ is a directed set and f is a mapping of Λ into a set Y, then f is a *net in* Y. A net f will usually (but not always) be written as an indexed family $\{y_\lambda\}_{\lambda \in \Lambda}$, where $y_\lambda = f(\lambda)$, $\lambda \in \Lambda$. As will be seen (Chapter 3), nets play the role of generalized sequences in many situations. If $\{x_\lambda\}_{\lambda \in \Lambda}$ is a net in a set X, Γ is another directed set, and N is a function mapping Γ into Λ, then $\{x_{N(\gamma)}\}_{\gamma \in \Gamma}$ is also a net in X. If N has the property that for every λ_0 in Λ, there exists an index γ_0 in Γ such that $N(\gamma) \geq \lambda_0$ for every $\gamma \geq \gamma_0$, then the net $\{x_{N(\gamma)}\}_{\gamma \in \Gamma}$ is called a *subnet* of $\{x_\lambda\}_{\lambda \in \Lambda}$.

The reader is also assumed to be familiar with the concept of a well-ordered set and with the notions of cardinal and ordinal numbers. The cardinal number of a set E will be denoted by card E. The smallest infinite cardinal number will be denoted by $\aleph_0$, the cardinal number of the continuum by $\aleph$. Thus card $\mathbb{N} = $ card $\mathbb{N}_0 = $ card $\mathbb{Z} = \aleph_0$, while card $\mathbb{R} = $ card $\mathbb{C} = \aleph$. We shall also use the *well-ordering principle*, by which we mean the following fact.

Zermelo's theorem. *For any set X there exists a well-ordering of X. Equivalently, for any cardinal number c there exists an ordinal number α such that $c = $ card α.*

Finally, the reader is assumed to be familiar with the elementary arithmetic properties of cardinal and ordinal numbers. In particular, we shall use the fact that

$$\aleph^{\aleph_0} = \aleph_0^{\aleph_0} = 2^{\aleph_0} = \aleph,$$

as well as the fact that if α is an ordinal number, and if $W(\alpha)$ denotes the *ordinal number segment* consisting of all ordinal numbers ξ such that $\xi < \alpha$, then the ordinal number of the well-ordered set $W(\alpha)$ is α and card $W(\alpha) = $ card α.

PROBLEMS

A. Let X and Y be sets, and suppose given a mapping f of some subset B of X into Y. If A is a subset of B, then the *restriction* of f to A is the mapping $f\,|\,A$ of A into Y defined by $(f\,|\,A)(x) = f(x)$ for every x in A. If f and g are mappings of subsets A and B of X, respectively, into Y, and if f is the restriction of g to A, then g is an *extension* of f. (This requires, of course, that A be a subset of B.) We write $f \subset g$ to indicate that f is a restriction of g. Show that this relation is a partial ordering on the collection $\mathcal{M}$ of all mappings of subsets of X into Y.

B. A partially ordered set X is said to be *simply ordered* or *linearly ordered* if for every pair of elements x, y of X it is the case that either $x \leq y$ or $y \leq x$. Show that a simply ordered set is a lattice, and therefore (if nonempty) a directed set.

C. If Γ is an index set, X is a partially ordered set, and if f and g are mappings of Γ into X, we write $f \leq g$ to mean that $f(\gamma) \leq g(\gamma)$ for every γ in Γ. Show that this relation is a partial ordering on the set of mappings of Γ into X. More generally, the same definition introduces a partial ordering on every Cartesian product $\Pi = \prod_{\gamma \in \Gamma} X_\gamma$ of partially ordered sets. Show that if each X_γ is a [complete] lattice, then Π is a [complete] lattice. In particular, the set of all [extended] real-valued functions on a set Γ is a [complete] lattice.

D. Suppose that $\mathcal{L}$ is a nonempty collection of real-valued functions on a set X with the property that $f + g$, $f - g$, and $f/2$ belong to $\mathcal{L}$ whenever f and g do. Show that $\mathcal{L}$ is a *function lattice* (i.e., that $f \vee g$ and $f \wedge g$ belong to $\mathcal{L}$ whenever f and g do; cf. Problem C) if and only if $|f|$ belongs to $\mathcal{L}$ whenever f does. Show, similarly, that $\mathscr{L}$ is a function lattice if and only if $f^+ = f \vee 0$ and $f^- = -(f \wedge 0)$ belong to $\mathscr{L}$ whenever f does. (The functions f^+ and f^- are the *positive* and *negative parts* of f, respectively.)

E. Let X be a fixed set. For each subset A of X, the *characteristic function* of A is that function χ_A which takes the value 1 at every point of A and the value 0 at every point of $X \backslash A$. If A and B are subsets of X, then $\chi_{A \cap B} = \chi_A \wedge \chi_B = \chi_A \chi_B$ and $\chi_{A \cup B} = \chi_A \vee \chi_B$. Furthermore, $A \subset B$ if and only if $\chi_A \leq \chi_B$, and $\chi_{A \cup B} = \chi_A + \chi_B$ if and only if A and B are disjoint.

F. Let $\{A_\gamma\}$ and $\{B_\gamma\}$ be two similarly indexed families of subsets of a set X. Verify that both

$$\left(\bigcup_\gamma A_\gamma\right) \backslash \left(\bigcup_\gamma B_\gamma\right) \quad \text{and} \quad \left(\bigcap_\gamma A_\gamma\right) \backslash \left(\bigcap_\gamma B_\gamma\right)$$

are subsets of $\bigcup_\gamma (A_\gamma \backslash B_\gamma)$. Verify, likewise, that

$$\left(\bigcup_\gamma A_\gamma\right) \nabla \left(\bigcup_\gamma B_\gamma\right) \quad \text{and} \quad \left(\bigcap_\gamma A_\gamma\right) \nabla \left(\bigcap_\gamma B_\gamma\right)$$

are subsets of $\bigcup_\gamma (A_\gamma \nabla B_\gamma)$.

G. Let a and b be real numbers with $a \leq b$. By a *partition* of the closed interval $[a, b]$ is meant a finite sequence $a = t_0 < t_1 < \cdots < t_n = b$. If $\mathscr{P} = \{s_i\}_{i=0}^m$ and $\mathscr{Q} = \{t_i\}_{i=0}^n$ are two partitions of $[a, b]$, then $\mathscr{Q}$ is *finer than* $\mathscr{P}$ (notation: $\mathscr{P} \leq \mathscr{Q}$) if every number s_i in $\mathscr{P}$ also appears in $\mathscr{Q}$. Show that the set of all partitions of $[a, b]$ forms a directed set under the relation of refinement. Is this directed set a lattice? If $\mathscr{P} = \{s_i\}_{i=0}^m$ is a partition of $[a, b]$, then the *mesh* of $\mathscr{P}$ is by definition the maximum $\max_{i=1,\dots,m} (s_i - s_{i-1})$. Show that the mesh is a monotone decreasing net on the directed set of partitions of $[a, b]$.

H. Let f be a bounded real-valued function on the interval $[a, b]$, let $\mathscr{P} = \{s_i\}_{i=0}^m$ be a partition of $[a, b]$, let M_i be the supremum of f on the ith *subinterval* of $\mathscr{P}$: $M_i = \sup\{f(t): s_{i-1} \leq t \leq s_i\}$, and set $M_\mathscr{P} = \sum_{i=1}^m M_i(s_i - s_{i-1})$. Then $\{M_\mathscr{P}\}$ is a net, called the net of *upper Darboux sums* of the function f. Show that the net $\{M_\mathscr{P}\}$ is monotone decreasing. Dually, if one employs the infimum m_i of f on the ith subinterval instead of M_i, one obtains the net $\{m_\mathscr{P}\}$ of *lower Darboux sums* of f. Show that the net $\{m_\mathscr{P}\}$ is monotone increasing.

I. Let f be a function, real or complex, defined on the real interval $[a, b]$. An interesting and useful net associated with f and indexed by the directed set of all partitions of the interval $[a, b]$ is defined by setting

$$v(\mathscr{P}) = \sum_{i=1}^n |f(s_i) - f(s_{i-1})|,$$

where $\mathscr{P} = \{s_i\}_{i=0}^n$. The number $v(\mathscr{P})$ is called the *variation* of f over $\mathscr{P}$. Show that the net $\{v(\mathscr{P})\}$ is monotone increasing. The function f is said to be of *bounded variation* on $[a, b]$ if the net $\{v(\mathscr{P})\}$ is bounded (above), and in this case the supremum $\sup_\mathscr{P} v(\mathscr{P})$ is called the *total variation* of f over $[a, b]$, and is denoted by $V = V(f; a, b)$. Show that if f and g are both of bounded variation on $[a, b]$ and λ is a complex number, then $V(f + g; a, b) \leq V(f; a, b) + V(g; a, b)$ and $V(\lambda f; a, b) =$

$|\lambda| V(f; a, b)$. (A complex-valued function f on $[a, b]$ may be thought of as a kind of curve in $\mathbb{C}$. Consequently a function f that is of bounded variation on $[a, b]$ is also sometimes said to be *rectifiable*, and the total variation $V(f; a, b)$ is called the *length* of f.)

J. (i) Let f be a real-valued function on a closed interval $[a, b]$. We define two nets of nonnegative real numbers associated with f and indexed by the directed set of partitions of $[a, b]$ by writing

$$v_+(\mathscr{P}) = \sum_{i=1}^{n} [f(s_i) - f(s_{i-1})]^+, \qquad v_-(\mathscr{P}) = \sum_{i=1}^{n} [f(s_i) - f(s_{i-1})]^-$$

(Ex. A), where $\mathscr{P} = \{s_i\}_{i=0}^{n}$. The numbers $v_+(\mathscr{P})$ and $v_-(\mathscr{P})$ are called the *positive* and *negative variations* of f over $\mathscr{P}$, respectively. Show that the nets $\{v_+(\mathscr{P})\}$ and $\{v_-(\mathscr{P})\}$ are both monotone increasing, and that f is of bounded variation on $[a, b]$ if and only if both of the nets $\{v_+(\mathscr{P})\}$ and $\{v_-(\mathscr{P})\}$ are bounded (above). Show also that if f is of bounded variation on $[a, b]$, and if we define $V_+(f; a, b) = \sup_{\mathscr{P}} v_+(\mathscr{P})$ and $V_-(f; a, b) = \sup_{\mathscr{P}} v_-(\mathscr{P})$, then

$$V(f; a, b) = V_+(f; a, b) + V_-(f; a, b)$$

and

$$f(b) - f(a) = V_+(f; a, b) - V_-(f; a, b).$$

The numbers $V_+(f; a, b)$ and $V_-(f; a, b)$ are called the *positive* and *negative variations* of f over $[a, b]$, respectively.

(ii) Verify that a complex-valued function f on $[a, b]$ is of bounded variation on $[a, b]$ when and only when both Re f and Im f are, and that, when this is the case, we have

$$V(\text{Re } f; a, b) \vee V(\text{Im } f; a, b) \leq V(f; a, b) \leq V(\text{Re } f; a, b) + V(\text{Im } f; a, b).$$

K. Let f be a function of bounded variation on the interval $[a, b]$, and let c be a number such that $a \leq c \leq b$. Verify that $V(f; a, b) = V(f; a, c) + V(f; c, b)$, and conclude that if we define $V(t) = V(f; a, t)$, $a \leq t \leq b$, then V is a monotone increasing function of t. Show similarly that, when f is real-valued, the functions $f_+(t) = V_+(f; a, t)$ and $f_-(t) = V_-(f; a, t)$ are also monotone increasing. Use Problem J to show that

$$V = f_+ + f_- \quad \text{and} \quad f = f_+ - (f_- - f(a)). \tag{1}$$

The functions f_+ and f_- are known as the *positive* and *negative variations* of f, respectively. The expression for f in (1) is known as the *Jordan decomposition* of f.

L. If X is an arbitrary set then a collection $\mathscr{P}$ of subsets of X is called a *partition* of X if $\bigcup \mathscr{P} = X$, i.e., if $\mathscr{P}$ covers X, and if the sets in $\mathscr{P}$ are pairwise disjoint. (In the event that X is a closed interval of real numbers there are two distinct notions of partition that have now been introduced, viz., the one in this problem and the one in Problem G. At no time will this slight ambiguity give rise to any misunderstanding.)

(i) If $\mathscr{P}$ and $\mathscr{P}'$ are two partitions of X, then $\mathscr{P}'$ is said to be *finer than* $\mathscr{P}$, or to *refine* $\mathscr{P}$ (and $\mathscr{P}$ is said to be *coarser than* $\mathscr{P}'$; notation: $\mathscr{P} \leq \mathscr{P}'$) if every set in $\mathscr{P}'$ is a subset of some set in $\mathscr{P}$. Show that $\mathscr{P} \leq \mathscr{P}'$ if and only if every set E in $\mathscr{P}$ is partitioned by the subcollection of $\mathscr{P}'$ consisting of the sets in $\mathscr{P}'$ contained in E. Show also that the collection of all partitions of X is a directed set with respect to the partial ordering $\leq$.

(ii) If $\{E_1, \ldots, E_n\}$ is a finite collection of subsets of a set X, then the *partition of X determined by* $\{E_1, \ldots, E_n\}$ consists of the collection of all sets of the form

$$A_1 \cap \cdots \cap A_n$$

where each A_i is either E_i or $X \backslash E_i$. (There are 2^n such sequences $\{A_1, \ldots, A_n\}$, but the number of sets in the partition may be smaller, of course.) Equivalently, the partition determined by $\{E_1, \ldots, E_n\}$ is the coarsest partition of X that partitions each of the sets $E_i, i = 1, \ldots, n$. Show that if s is a function of the form

$$s = \sum_{i=1}^{n} \alpha_i \chi_{E_i},$$

where $E_1, \ldots, E_n$ are subsets of X and $\alpha_1, \ldots, \alpha_n$ are complex numbers (such a function is called a *simple function* on X), then s is constant on each set F in the partition of X determined by $\{E_1, \ldots, E_n\}$ and the value β_F of s on the set F is given by the formula

$$\beta_F = \sum_{F \subset E_i} \alpha_i.$$

M. If Γ is an arbitrary set, then the collection $\mathscr{D}$ of all finite subsets of Γ is a directed set under the inclusion ordering. Let $\{\lambda_\gamma\}_{\gamma \in \Gamma}$ be a family of complex numbers indexed by Γ, and for each D in $\mathscr{D}$ define $\sigma_D = \sum_{\gamma \in D} \lambda_\gamma$. Then $\{\sigma_D\}_{D \in \mathscr{D}}$ is a net directed by $\mathscr{D}$, called the *net of finite sums* of the given family $\{\lambda_\gamma\}_{\gamma \in \Gamma}$. Show that the net $\{\sigma_D\}_{D \in \mathscr{D}}$ is monotone increasing if the numbers λ_γ are all nonnegative.

N. The sets $\mathbb{N}$ and $\mathbb{N}_0$ are directed sets in their usual ordering. Hence infinite sequences $\{x_n\}_{n=1}^{\infty}$ and $\{x_n\}_{n=0}^{\infty}$ of points in a space X are also nets in X. Moreover, every subsequence of a sequence is a subnet of that sequence. Give an example of a subnet of an infinite sequence that is not a subsequence.

O. If Λ_1 and Λ_2 are directed sets, then the Cartesian product $\Lambda_1 \times \Lambda_2$ is also directed when ordered, as in Problem C, by defining $(\lambda_1, \lambda_2) \leq (\lambda'_1, \lambda'_2)$ to mean that $\lambda_1 \leq \lambda'_1$ and $\lambda_2 \leq \lambda'_2$. Show that if $\{x_{\lambda_1}\}_{\lambda_1 \in \Lambda_1}$ is a net indexed by Λ_1, then

$$x_{(\lambda_1, \lambda_2)} = x_{\lambda_1}, \qquad \lambda_1 \in \Lambda_1, \qquad \lambda_2 \in \Lambda_2,$$

defines a subnet $\{x_{(\lambda_1, \lambda_2)}\}$ indexed by $\Lambda_1 \times \Lambda_2$.

P. Let X be a partially ordered set. A subset X' of X is *cofinal* in X if for every x in X there exists x' in X' such that $x \leq x'$. We shall say that X is *countably determined* if it possesses a countable cofinal subset. Show that if Λ is a countably determined directed set, then there exists a monotone increasing sequence $\{\lambda_n\}_{n=1}^{\infty}$ in Λ such that for any λ in Λ there is a positive integer n such that $\lambda_k \geq \lambda$ for all $k \geq n$. Give an example of a directed set that is countably determined and an example of one that is not.

Q. If X and Y are two sets then card X is defined to be less than or equal to card Y if there exists a one-to-one mapping of X into Y. Use Zorn's lemma to show that if $X \neq \emptyset$ then it is also the case that card $X \leq$ card Y if and only if there exists a mapping of Y *onto* X. (Hint: See Problem A.)

R. (Cantor–Bernstein Theorem) Let X and Y be sets, let φ be a one-to-one mapping of X into Y, and let ψ be a one-to-one mapping of Y into X. Use Example C to show that the mapping

$$\Phi(A) = X \backslash \psi(Y \backslash \varphi(A))$$

of the power class $\mathscr{X}$ on X into itself possesses a fixed point. (By the *power class* on X we mean the collection of all subsets of X.) Use this fact to show that there exists a one-to-one mapping of X onto Y, and conclude that if c_1 and c_2 are any two cardinal numbers such that $c_1 \leq c_2$ and $c_2 \leq c_1$, then $c_1 = c_2$.

S. If X is any set and if $\mathscr{X}$ denotes the power class on X, then card $\mathscr{X} >$ card X. (Hint: Suppose there exists a mapping φ of X onto $\mathscr{X}$. Set $A = \{x \in X : x \notin \varphi(x)\}$, and let x_0 be an element of X for which $\varphi(x_0) = A$.)

T. (i) If c is an infinite cardinal number, then $\aleph_0 c = c$. (Hint: It suffices to show that $c = \aleph_0 b$ for *any one* cardinal number b because of the associativity of multiplication of cardinal numbers. Let X be a set with card $X = c$, consider the collection of all disjoint collections of countably infinite subsets of X, and employ Zorn's lemma.) Conclude that if c and d are any two cardinal numbers, one of which at least is infinite, then $c + d = c \vee d$, the larger of the two. More generally, the sum of any finite collection of cardinal numbers, one of which at least is infinite, coincides with the largest number in the collection.

(ii) If c is an infinite cardinal number, then, in fact, $c^2 = c$. (Hint: Let X be a set with card $X = c$, and consider the collection $\mathscr{C}$ of all those mappings f of a subset of X into $X \times X$ with the property that, if A is the domain of f, then f is a one-to-one mapping of A onto $A \times A$. Use Zorn's lemma to show that $\mathscr{C}$ contains a maximal element f_0 with respect to extension (Prob. A). Then use (i) to prove that if A_0 is the domain of f_0, and if the desired conclusion is false, then $X \backslash A_0$ contains a subset A_1 with card $A_1 =$ card A_0. Finally, use (i) again to show that there exists in $\mathscr{C}$ an extension f_1 of f_0 to the domain $A_0 \cup A_1$, a contradiction.) Conclude that if c and d are any two cardinal numbers, one of which at least is infinite, then $cd = c \vee d$. More generally, the product of any finite collection of cardinal numbers, one of which at least is infinite, coincides with the largest number in the collection.

U. (Principle of Transfinite Induction) Let $p(\ \ )$ be a predicate that is either true or false for every ordinal number in some ordinal number segment $W(\alpha)$. Suppose (i) $p(0)$ is true, (ii) if $p(\xi)$ is true and if $\xi + 1 < \alpha$, then $p(\xi + 1)$ is true, and (iii) if λ is a limit ordinal less than α, and if $p(\xi)$ is true for every ξ in $W(\lambda)$, then $p(\lambda)$ is true. (By definition, a *limit ordinal* is a nonzero ordinal number that does not possess an immediate predecessor, i.e., that cannot be written in the form $\alpha + 1$.) Show that $p(\xi)$ is true for every ξ in $W(\alpha)$. (Hint: If this were not the case, then the set consisting of those ordinal numbers ξ in $W(\alpha)$ such that $p(\xi)$ is false would possess a smallest element.) Show also that an alternate formulation of this principle is the following: If Q is a subset of $W(\alpha)$ with the property that $W(\xi) \subset Q$ implies $\xi \in Q$ for every ξ in $W(\alpha)$, then $Q = W(\alpha)$.

V. (Principle of Transfinite Definition) The principle of transfinite induction can also be employed to give definitions. Let λ be a limit ordinal, let G be an arbitrary set, and let g_0 be a fixed element of G. Suppose that for each $\xi \neq 0$ in $W(\lambda)$ there exists a rule R_ξ that associates with each mapping $f : W(\xi) \to G$ a unique element $R_\xi(f)$

of G. Show that there exists a unique mapping $F : W(\lambda) \to G$ such that $F(0) = g_0$ and such that $F(\xi) = R_\xi(F \mid W(\xi))$ for each ξ, $0 < \xi < \lambda$. (Hint: Consider the subset of $W(\lambda)$ consisting of 0 and all of those ordinal numbers η, $0 < \eta < \lambda$, with the property that there exists a unique mapping $F_\eta : W(\eta) \to G$ such that $F_\eta(0) = g_0$ and such that $F_\eta(\xi) = R_\xi(F_\eta \mid W(\xi))$ for all nonzero ξ in $W(\eta)$. We speak here only of defining a function, but since practically everything in mathematics may be construed to be a function in one way or another, this formulation of the principle is adequate.)

W. If c is an infinite cardinal number, then there is a first, or smallest, ordinal number α such that card $\alpha = c$. This ordinal number is called the *initial number* of c. Thus ω, the smallest infinite ordinal number, is the initial number of $\aleph_0$. The first *noncountable* ordinal number, i.e., the initial number of the smallest cardinal number exceeding $\aleph_0$, is customarily denoted by Ω. Thus the initial segment $W(\Omega)$ coincides with the set of all countable ordinal numbers. Show that every countable subset of $W(\Omega)$ is bounded in $W(\Omega)$. In other words, show that if M is any countable set of ordinal numbers such that card $\alpha \leq \aleph_0$ for every α in M, then there exists an ordinal number β such that card $\beta \leq \aleph_0$ and such that $\alpha \leq \beta$ for every α in M. (Hint: If M were cofinal in $W(\Omega)$ (Prob. P), we would have $W(\Omega) = \bigcup_{\alpha \in M} W(\alpha)$.) Thus if $\{\xi_n\}_{n=1}^\infty$ is any infinite sequence of ordinal numbers in $W(\Omega)$, then $\sup_n \xi_n < \Omega$. Show in the converse direction that if λ is an arbitrary limit ordinal in $W(\Omega)$, then there exists an increasing sequence $\{\xi_n\}_{n=1}^\infty$ such that $\lambda = \sup_n \xi_n$. Show finally that if $z(\xi)$ is an arbitrary monotone increasing integer-valued function defined on $W(\Omega)$, then there exists an ordinal number α_0 in $W(\Omega)$ such that $z(\xi)$ is constant on the *tail* $W(\Omega) \setminus W(\alpha_0)$.

2 Linear algebra

We shall assume that the reader is familiar with the rudiments of linear algebra. In particular, he should be acquainted with the notion of a *linear space*, or *vector space*, and the elementary concepts associated with linear spaces. In this chapter we review these ideas, partially to fix terminology and notation. (An exception is our treatment of algebraic tensor products at the end of the chapter; we do not assume any prior knowledge of this subject on the part of the reader.) Readers wishing to improve their acquaintance with any part of linear algebra, or to pursue in greater depth any of the topics touched on below, might consult [38]; another excellent source is [32].

To begin with, all of the linear spaces in this book are either *real* or *complex* (that is, the field of scalars is either $\mathbb{R}$ or $\mathbb{C}$). Furthermore, the following convention will be in force throughout the book: *If nothing is said about the scalar field of a vector space under discussion, the vector space is automatically assumed to be complex.*

If $\mathscr{E}$ is a (real or complex) linear space, and if M_1 and M_2 are arbitrary subsets of $\mathscr{E}$, we shall write $M_1 + M_2$ for the set of sums

$$\{x_1 + x_2 : x_i \in M_i, i = 1, 2\}.$$

More generally, if $\{M_\gamma\}_{\gamma \in \Gamma}$ is an arbitrary indexed family of subsets of $\mathscr{E}$, we write $\sum_{\gamma \in \Gamma} M_\gamma$ for the set of all sums of the form $\sum_{\gamma \in \Gamma} x_\gamma$ where $x_\gamma = 0$ except for some finite set of indices, and $x_\gamma \in M_\gamma$ whenever $x_\gamma \neq 0$. In particular, if $\{M_1, \ldots, M_n\}$ is a finite sequence of subsets of $\mathscr{E}$, then $M_1 + \cdots + M_n$ denotes the set of sums $x_1 + \cdots + x_n$, where $x_i \in M_i, i = 1, \ldots, n$. Similarly, if A denotes a set of scalars and M a set of vectors in $\mathscr{E}$, we shall write AM for the set $\{\alpha x : \alpha \in A, x \in M\}$.

Let $\mathscr{E}$ be a real or complex linear space. An element x of $\mathscr{E}$ is a *linear combination* of vectors $y_1, \ldots, y_n$ in $\mathscr{E}$ if there exist scalars $\alpha_1, \ldots, \alpha_n$ such that

$x = \alpha_1 y_1 + \cdots + \alpha_n y_n$. A nonempty subset $\mathcal{M}$ of $\mathcal{E}$ is a *linear manifold* in $\mathcal{E}$ (or a *linear submanifold* of $\mathcal{E}$) if, for every pair of vectors x, y in $\mathcal{M}$ and every pair of scalars α, β, the linear combination $\alpha x + \beta y$ belongs to $\mathcal{M}$. If $\mathcal{M}$ is a linear submanifold of $\mathcal{E}$ then it is easily seen that every linear combination of vectors in $\mathcal{M}$ belongs to $\mathcal{M}$. Among the linear submanifolds of $\mathcal{E}$ are the space $\mathcal{E}$ itself and the trivial submanifold (0) consisting of the single vector 0. If $\mathcal{M}$ and $\mathcal{N}$ are linear manifolds in $\mathcal{E}$, then the sum $\mathcal{M} + \mathcal{N}$ is also a linear manifold in $\mathcal{E}$; more generally, if $\{\mathcal{M}_\gamma\}_{\gamma \in \Gamma}$ is an indexed family of linear manifolds in $\mathcal{E}$, then $\sum_\gamma \mathcal{M}_\gamma$ is a linear manifold in $\mathcal{E}$. For any set M of vectors in $\mathcal{E}$ there exists a smallest linear manifold $\mathcal{L}$ in $\mathcal{E}$ that contains M. If $M = \varnothing$, then $\mathcal{L} = (0)$; otherwise $\mathcal{L}$ consists of all linear combinations of elements of M. We say that $\mathcal{L}$ is *generated* (*algebraically*) by M, or that M is an (*algebraic*) *system of generators* for $\mathcal{L}$.

Analogously, an element x of $\mathcal{E}$ is a *convex combination* of vectors $y_1, \ldots, y_n$ in $\mathcal{E}$ if there exist nonnegative real numbers $s_1, \ldots, s_n$ such that $s_1 + \cdots + s_n = 1$ and such that $x = s_1 y_1 + \cdots + s_n y_n$. A subset C of $\mathcal{E}$ is *convex* if, for every pair of vectors x and y in C and every pair of nonnegative real numbers s and t such that $s + t = 1$, the convex combination $sx + ty$ belongs to C. It is not difficult to show that if C is convex then every convex combination of vectors in C belongs to C. (A complex linear space becomes a real linear space if one simply declines to multiply by any but real scalars. Clearly a subset C of a complex linear space $\mathcal{E}$ is convex if and only if C is convex in $\mathcal{E}$ when $\mathcal{E}$ is regarded as a real linear space.) For any set M of vectors in $\mathcal{E}$ there exists a smallest convex set C in $\mathcal{E}$ that contains M. This convex set consists of all convex combinations of elements of M, and is called the *convex hull* of M. The convex hull $\sigma = \sigma(x, y)$ of a doubleton $\{x, y\}$ is called the *line segment* joining x and y, and x and y are said to be the *endpoints* of σ. The line segment $\sigma(x, y)$ clearly consists of the set of vectors $\{sx + (1 - s)y : 0 \le s \le 1\}$. According to the above definition, a set C is convex if and only if C contains the line segment joining any two vectors in C. By a *line* in $\mathcal{E}$ is meant any set of the form $L = \{x + sy : s \in \mathbb{R}\}$, where $x, y \in \mathcal{E}$ and $y \ne 0$. (The set L is known, more precisely, as the line *through x along y*.) If x_1 and x_2 are distinct vectors in $\mathcal{E}$ then there exists a unique line joining x_1 and x_2 (that is, containing both x_1 and x_2), and this line coincides with the set $\{sx_1 + (1 - s)x_2 : s \in \mathbb{R}\}$. Thus the line joining x_1 and x_2 contains the line segment joining them.

A nonempty finite set of vectors $J = \{x_1, \ldots, x_n\}$ in a (real or complex) vector space $\mathcal{E}$ is *linearly independent* if the only way in which 0 can be expressed as a linear combination $0 = \alpha_1 x_1 + \cdots + \alpha_n x_n$ is with $\alpha_1 = \cdots = \alpha_n = 0$. An arbitrary subset J of $\mathcal{E}$ is *linearly independent* if every nonempty finite subset of J is linearly independent. A linearly independent set of vectors in $\mathcal{E}$ that is at the same time a system of generators for $\mathcal{E}$ is a *Hamel basis* for $\mathcal{E}$. Every (real or complex) vector space has a Hamel basis (Prob. A). If $\{x_\gamma\}_{\gamma \in \Gamma}$ is an indexed Hamel basis for a (real or complex) linear space $\mathcal{E}$, then for each vector y in $\mathcal{E}$ there exists a uniquely determined indexed family of scalars $\{\lambda_\gamma\}_{\gamma \in \Gamma}$ such that $\lambda_\gamma = 0$ for all but a finite number of indices and such that $y = \sum_{\gamma \in \Gamma} \lambda_\gamma x_\gamma$. The scalars λ_γ are called the *co-ordinates* of y with respect to the basis $\{x_\gamma\}$. If X and Y are any two Hamel bases for $\mathscr{E}$, then card $X = $ card Y (Prob. B). The common cardinal number of all the Hamel bases for $\mathscr{E}$ is the *Hamel dimension of $\mathscr{E}$*. Linear spaces, both real and complex, that have finite Hamel dimension are *finite dimensional*, and those that do not have finite Hamel dimension are *infinite dimensional*. A Hamel basis for a finite dimensional space $\mathscr{E}$ is simply called a *basis* for $\mathscr{E}$, and the Hamel dimension of $\mathscr{E}$ is known as the *dimension* of $\mathscr{E}$ and is denoted by dim $\mathscr{E}$. The vector space (0) consisting of the vector 0 alone is finite dimensional and has dimension 0.

Example A. Let X be an arbitrary set, and let $\mathscr{F}$ denote the linear space of all those mappings f of X into $\mathbb{C}$ with the property that f vanishes everywhere on the complement of some finite subset of X. (The subset of X on which f is nonzero may vary with f; addition, and multiplication by scalars, are defined pointwise on X.) If for each x in X we denote by e_x the element of $\mathscr{F}$ such that $e_x(x) = 1$ while $e_x(y) = 0$ for all $y \neq x$, then $\{e_x\}_{x \in X}$ is a Hamel basis for $\mathscr{F}$. If, as is customary, we simply identify each x in X with the associated vector e_x, then X itself becomes a Hamel basis for $\mathscr{F}$. In this linear space, known as the *free linear space generated by X*, the vectors are *formal linear combinations*

$$\sum_{i=1}^{n} \alpha_i x_i$$

of elements of X.

If $\{\mathscr{E}_\gamma\}_{\gamma \in \Gamma}$ is an indexed family of linear spaces, all over the same scalar field (either $\mathbb{C}$ or $\mathbb{R}$), then the set of all indexed families $\{x_\gamma\}_{\gamma \in \Gamma}$, where $x_\gamma \in \mathscr{E}_\gamma$ for each index γ in Γ, forms a linear space $\mathscr{E}$ under the operations $\{x_\gamma\} + \{y_\gamma\} = \{x_\gamma + y_\gamma\}$ and $\alpha\{x_\gamma\} = \{\alpha x_\gamma\}$. This linear space is called the *full algebraic direct sum* of the family $\{\mathscr{E}_\gamma\}$, and will be denoted by $\sum_{\gamma \in \Gamma} + \mathscr{E}_\gamma$. If all of the spaces $\mathscr{E}_\gamma$ coincide with a single vector space $\mathscr{F}$, then the full algebraic direct sum $\mathscr{E}$ is called the *direct sum of* card Γ *copies of $\mathscr{F}$ indexed by Γ*. If the index set Γ is the finite set $\{1, \ldots, n\}$ we write the elements of $\mathscr{E}$ in the form $(x_1, \ldots, x_n)$ and write $\mathscr{E} = \mathscr{E}_1 + \cdots + \mathscr{E}_n$. In this case $\mathscr{E}$ is called simply the (linear space) *direct sum* of the spaces $\mathscr{E}_i$. The Hamel dimension of the full algebraic direct sum $\sum_\gamma + \mathscr{E}_\gamma$ is $\sum_\gamma d_\gamma$, where d_γ denotes the Hamel dimension of $\mathscr{E}_\gamma$, $\gamma \in \Gamma$ (Prob. F). In particular, if $\mathscr{E}_1, \ldots, \mathscr{E}_n$ are finite dimensional, and if $\mathscr{E} = \mathscr{E}_1 + \cdots + \mathscr{E}_n$, then dim $\mathscr{E} = $ dim $\mathscr{E}_1 + \cdots + $ dim $\mathscr{E}_n$.

Example B. The familiar linear space [real linear space] of all complex [real] n-tuples may be viewed as the direct sum of n copies of $\mathbb{C}[\mathbb{R}]$. Henceforth this space will be denoted by $\mathbb{C}^n[\mathbb{R}^n]$. The n-tuples $e_i = (\delta_{i1}, \ldots, \delta_{in})$, $i = 1, \ldots, n$ (where by definition δ_{ij} is the *Kronecker delta*

$$\delta_{ij} = \begin{cases} 0, & i \neq j, \\ 1, & i = j, \end{cases}$$

for all integers i, j), constitute a basis for $\mathbb{C}^n[\mathbb{R}^n]$ sometimes called the *natural basis*. Clearly $\mathbb{C}^n[\mathbb{R}^n]$ is n-dimensional.

Example C. For any pair of positive integers m and n we shall denote by $\mathbb{C}_{m,n}[\mathbb{R}_{m,n}]$ the collection of all complex [real] $m \times n$ matrices. We assume the reader to be familiar with the rudiments of matrix theory; in particular we take it as known that $\mathbb{C}_{m,n}[\mathbb{R}_{m,n}]$ is a vector space [real vector space] with respect to the usual linear operations. If we think of each $m \times n$ matrix as composed of its n columns, then it is natural to identify $\mathbb{C}_{m,n}$ with the direct sum of n copies of $\mathbb{C}^m$; if we think of each $m \times n$ matrix as composed of its m rows, then it is natural to identify $\mathbb{C}_{m,n}$ with the direct sum of m copies of $\mathbb{C}^n$. The linear space $\mathbb{C}_{m,n}[\mathbb{R}_{m,n}]$ also possesses a natural basis, namely, the system of matrices $\{E_{k,l}\}_{k=1\,l=1}^{m\quad n}$, where $E_{k,l} = (\varepsilon_{ij}^{k,l})$ and $\varepsilon_{ij}^{k,l} = \delta_{ik}\delta_{jl}$ for $i, k = 1, \ldots, m$ and $j, l = 1, \ldots, n$. Clearly $\mathbb{C}_{m,n}[\mathbb{R}_{m,n}]$ is mn-dimensional.

Example D. The direct sum of $\aleph_0$ copies of $\mathbb{C}$ indexed by $\mathbb{N}_0$ is the vector space of all complex sequences $\{\alpha_n\}_{n=0}^{\infty}$. In the sequel this space will be denoted by (ϑ). The subset consisting of all the bounded sequences in (ϑ) is a linear manifold in (ϑ) which we shall denote by (m). Similarly the collections of all convergent sequences and all null sequences (sequences converging to zero) form linear submanifolds of (ϑ) (and of (m)). We denote these spaces by (c) and (c_0), respectively.

Example E. The collection of all complex-valued functions [real-valued functions] on an arbitrary set X is a complex linear space [real linear space] with respect to the pointwise linear operations

$$(f + g)(x) = f(x) + g(x) \quad \text{and} \quad (\alpha f)(x) = \alpha f(x), x \in X, \alpha \in \mathbb{C}[\mathbb{R}].$$

(This space can also be viewed as the full algebraic direct sum of card X copies of $\mathbb{C}[\mathbb{R}]$ indexed by X.) Whenever (as in Example A or in the following example) we refer to a "linear space [real linear space] of functions" on a set X, it is always some linear submanifold of this space that is meant.

Example F. Suppose given a (real or complex) linear space $\mathscr{E}$. A scalar-valued function f defined on $\mathscr{E}$ is a *linear functional* on $\mathscr{E}$ if $f(\alpha x + \beta y) = \alpha f(x) + \beta f(y)$ for all vectors x, y in $\mathscr{E}$ and all scalars α, β. It is a triviality to verify that a linear combination of linear functionals on $\mathscr{E}$ is again a linear functional on $\mathscr{E}$, and hence that the collection of all linear functionals on $\mathscr{E}$ forms a linear submanifold of the space of all scalar-valued functions on $\mathscr{E}$. The linear space of all linear functionals on $\mathscr{E}$ (which is real or complex according as $\mathscr{E}$ is real or complex) will be called the *full algebraic dual* of $\mathscr{E}$.

If $\mathscr{E}$ is a (real or complex) linear space, and if $\mathscr{M}$ is a linear submanifold of $\mathscr{E}$, then the relation $\sim$ on $\mathscr{E}$ defined by setting $x \sim y$ if x is *congruent* to y *modulo* $\mathscr{M}$, that is, if $x - y \in \mathscr{M}$, is an equivalence relation on $\mathscr{E}$. The equivalence class $[x] = x + \mathscr{M}$ of a vector x will be called the *coset* of x modulo $\mathscr{M}$. The set of all cosets $[x]$ modulo $\mathscr{M}$ is turned into a new linear space by the definitions $[x] + [y] = [x + y]$ and $\alpha[x] = [\alpha x]$. This space, denoted by $\mathscr{E}/\mathscr{M}$, is the *quotient space* of $\mathscr{E}$ modulo $\mathscr{M}$, and the transformation π of $\mathscr{E}$ onto $\mathscr{E}/\mathscr{M}$ defined by $\pi(x) = [x]$ is the *natural projection* of $\mathscr{E}$ onto $\mathscr{E}/\mathscr{M}$.

If $\mathscr{E}$ is a real vector space, the *complexification* $\mathscr{E}^+$ of $\mathscr{E}$ is the complex vector space consisting of the Cartesian product of $\mathscr{E}$ with itself with addition defined by $(x_1, y_1) + (x_2, y_2) = (x_1 + x_2, y_1 + y_2)$ and multiplication by a complex scalar $\alpha = s + it$ defined by $\alpha(x, y) = (sx - ty, tx + sy)$. Thus $\mathbb{C}$ is the complexification of $\mathbb{R}$. If the mapping $x \to (x, 0)$ is used to identify $\mathscr{E}$ with a real linear manifold in $\mathscr{E}^+$ regarded as a real space, then, since $i(x, 0) = (0, x)$, every vector in $\mathscr{E}^+$ has a unique expression of the form $x + iy$, where x and y belong to $\mathscr{E}$. (Recall that a complex linear space may always be regarded as a real space simply by refusing to multiply by any but real scalars.)

Example G. If X is a Hamel basis for a real linear space $\mathscr{E}$, and if, as above, we identify $\mathscr{E}$ with the real submanifold $\mathscr{E} \times (0)$ of the complexification $\mathscr{E}^+$, then X is also a basis for $\mathscr{E}^+$. Thus the Hamel dimension of the complex space $\mathscr{E}^+$ is the same as that of the real space $\mathscr{E}$. On the other hand, if $\mathscr{E}^+$ is regarded as a real space, then the union of the two sets

$$X = \{(x, 0) : x \in X\} \quad \text{and} \quad X' = \{(0, x) : x \in X\}$$

is a Hamel basis for $\mathscr{E}^+$. Thus, in particular, if the dimension of $\mathscr{E}$ is n, then the dimension of $\mathscr{E}^+$ regarded as a real space is $2n$. More generally, if $\mathscr{E}$ is any (complex) linear space of dimension n, then $\mathscr{E}$ has dimension $2n$ when regarded as a real space.

Example H. If $\mathscr{F}$ is a complex linear space of complex-valued functions on a set X and if $\mathscr{F}_{\mathbb{R}}$ denotes the set of all real functions in $\mathscr{F}$, then it is clear that $\mathscr{F}_{\mathbb{R}}$ is a real linear space, and that the complexification $(\mathscr{F}_{\mathbb{R}})^+$ may be identified with the linear submanifold of $\mathscr{F}$ consisting of functions of the form $f + ig$, $f, g \in \mathscr{F}_{\mathbb{R}}$. This submanifold, however, does not coincide with $\mathscr{F}$, in general. Indeed it is readily seen that a necessary and sufficient condition for this to be so is that $\mathscr{F}$ contain the complex conjugate $\bar{f}$ of each function f in $\mathscr{F}$, a condition that is customarily expressed by saying that $\mathscr{F}$ is *self-conjugate*. Thus we may, and frequently shall, identify a self-conjugate linear space of functions $\mathscr{F}$ with the complexification of the real linear space of real-valued functions in $\mathscr{F}$.

If $\mathscr{F}$ is a self-conjugate linear space of complex-valued functions on a set X, and if φ is a linear functional defined on $\mathscr{F}$, then φ is said to be *self-conjugate* if $\varphi(\bar{f}) = \overline{\varphi(f)}$ for every function f in $\mathscr{F}$. We observe that φ is self-conjugate if and only if the restriction $\varphi | \mathscr{F}_{\mathbb{R}}$ is a real linear functional on $\mathscr{F}_{\mathbb{R}}$ or, equivalently, if and only if $\varphi(\operatorname{Re} f) = \operatorname{Re} \varphi(f) [\varphi(\operatorname{Im} f) = \operatorname{Im} \varphi(f)]$ for every f in $\mathscr{F}$.

Example I. If a and b are real numbers, $a < b$, we shall denote by $\mathscr{C}((a, b)) = \mathscr{C}^{(0)}((a, b))$ the collection of all continuous complex-valued functions on the open interval $(a, b) (= \{t \in \mathbb{R} : a < t < b\})$. Clearly $\mathscr{C}((a, b))$ is a linear space. Similarly one sees, using the rules of elementary calculus, that the collection $\mathscr{C}^{(n)}((a, b))$ of n times continuously differentiable functions on (a, b), i.e., the collection of those functions f on (a, b) with the property that the nth derivative $f^{(n)}$ exists and is continuous on (a, b), is a linear space. If $0 \le m \le n$, then $\mathscr{C}^{(n)}((a, b))$ is a linear submanifold of $\mathscr{C}^{(m)}((a, b))$. Likewise, if $\mathscr{C}_{\mathbb{R}}^{(n)}((a, b))$ denotes the collection of real-valued functions in $\mathscr{C}^{(n)}((a, b))$, then $\mathscr{C}_{\mathbb{R}}^{(n)}((a, b))$ is a real vector space, and $\mathscr{C}^{(n)}((a, b))$ is the complexification of $\mathscr{C}_{\mathbb{R}}^{(n)}((a, b))$. If $\mathscr{P}[\mathscr{P}_{\mathbb{R}}]$ denotes the space of polynomial functions [real polynomial functions] on (a, b), then $\mathscr{P}[\mathscr{P}_{\mathbb{R}}]$ is a linear submanifold of $\mathscr{C}^{(n)}((a, b))$ $[\mathscr{C}_{\mathbb{R}}^{(n)}((a, b))]$ for every n.

Example J. To define analogs of the spaces of Example I for functions on a *closed* interval, special arrangements must be made regarding the endpoints of the interval. We shall say that a (complex-valued) function f on a closed interval $[a, b] (a < b)$ is continuously differentiable on that interval if (i) f is differentiable on the open interval (a, b), (ii) the one-sided derivatives $f'_+(a)$ and $f'_-(b)$ exist, and (iii) the function

$$f'(t) = \begin{cases} f'_+(a), & t = a, \\ f'(t), & a < t < b, \\ f'_-(b), & t = b, \end{cases}$$

is continuous on $[a, b]$. We then declare $\mathscr{C}([a, b]) = \mathscr{C}^{(0)}([a, b])$ to be the linear space of all continuous functions on $[a, b]$, and define $\mathscr{C}^{(n)}([a, b])$ inductively for positive integers n by setting $\mathscr{C}^{(n)}([a, b])$ equal to the collection of all those continuously differentiable functions f with the property that f' belongs to $\mathscr{C}^{(n-1)}([a, b])$. Here again it is clear that each $\mathscr{C}^{(n)}([a, b])$ is a vector space, that $\mathscr{C}^{(n)}([a, b])$ is a linear submanifold of $\mathscr{C}^{(m)}([a, b])$ when and only when $m \le n$, and that, if $\mathscr{C}_{\mathbb{R}}^{(n)}([a, b])$ denotes the set of real-valued functions in $\mathscr{C}^{(n)}([a, b])$, then $\mathscr{C}_{\mathbb{R}}^{(n)}([a, b])$ is a real vector space and $\mathscr{C}^{(n)}([a, b])$ is the complexification of $\mathscr{C}_{\mathbb{R}}^{(n)}([a, b])$.

If $\mathscr{E}$ and $\mathscr{F}$ are linear spaces over the same scalar field, and if T is a mapping defined on $\mathscr{E}$ and taking its values in $\mathscr{F}$, then T is a *linear transformation* of $\mathscr{E}$ into $\mathscr{F}$ provided $T(\alpha x + \beta y) = \alpha T x + \beta T y$ for all x, y in $\mathscr{E}$ and all scalars α, β. (When $\mathscr{E} = \mathscr{F}$ we refer to T as a linear transformation *on* $\mathscr{E}$. A linear transformation of a linear space $\mathscr{E}$ into its scalar field is a linear functional on $\mathscr{E}$ (Ex. F).) Whether or not T maps $\mathscr{E}$ onto $\mathscr{F}$, the range of T is a linear submanifold of $\mathscr{F}$ which we denote consistently by $\mathscr{R}(T)$. Likewise, the *kernel* or *null space* of T, that is, the set of vectors mapped into 0 by T, is a linear manifold in $\mathscr{E}$ that will be denoted throughout the book by $\mathscr{K}(T)$.

Example K. For any vector space $\mathscr{E}$ and any fixed scalar α the mapping $x \to \alpha x$, $x \in \mathscr{E}$, is a linear transformation on $\mathscr{E}$ which we consistently denote by α or, when necessary in order to avoid confusion, by $\alpha_{\mathscr{E}}$. In particular, the identity mapping 1 and zero mapping 0 are linear transformations.

Example L. Let $\mathscr{E}$ be a linear space, and let $\mathscr{M}$ be a linear submanifold of $\mathscr{E}$. Then the natural projection π of $\mathscr{E}$ onto the quotient space $\mathscr{E}/\mathscr{M}$ is a linear transformation. Moreover, if T is any linear transformation of $\mathscr{E}$ into a linear space $\mathscr{F}$, then there exists a linear transformation $\hat{T}: \mathscr{E}/\mathscr{M} \to \mathscr{F}$ such that $T = \hat{T} \circ \pi$ if and only if $\mathscr{M} \subset \mathscr{K}(T)$. (Briefly: linear transformations T on $\mathscr{E}$ with $T(\mathscr{M}) = (0)$ can be *factored through* $\mathscr{E}/\mathscr{M}$.)

Example M. If $\mathscr{E}$ is a real linear space and T is a linear transformation of $\mathscr{E}$ into a complex linear space $\mathscr{F}$ (regarded as a real space), then

$$T^+(x + iy) = Tx + iTy$$

defines a linear transformation of the complexification $\mathscr{E}^+$ into $\mathscr{F}$. The linear transformation T^+ is called the *complexification* of T.

If $\mathscr{E}$ and $\mathscr{F}$ are linear spaces over the same scalar field, and if T is a one-to-one linear transformation of $\mathscr{E}$ into $\mathscr{F}$ (that is, if $\mathscr{K}(T) = (0)$), then the set-theoretic inverse of T is also a linear transformation (of $\mathscr{R}(T)$ onto $\mathscr{E}$). If in addition $\mathscr{R}(T) = \mathscr{F}$, then T is a *linear space isomorphism* of $\mathscr{E}$ onto $\mathscr{F}$. Two linear spaces are *isomorphic* if there exists a linear space isomorphism of one onto the other.

Example N. If $\mathscr{E}$ is an n-dimensional linear space [real linear space] and $X = \{x_1, \ldots, x_n\}$ is an ordered basis for $\mathscr{E}$, then the mapping

$$\sum_{i=1}^{n} \alpha_i x_i \overset{\eta}{\to} (\alpha_1, \ldots, \alpha_n)$$

that assigns to each vector in $\mathscr{E}$ its n-tuple of coordinates with respect to X is a linear space isomorphism of $\mathscr{E}$ onto $\mathbb{C}^n[\mathbb{R}^n]$. If $Y = \{y_1, \ldots, y_n\}$ is some other ordered basis for $\mathscr{E}$ then there exist unique scalars π_{ij} such that

$$x_j = \sum_{i=1}^{n} \pi_{ij} y_i, \qquad j = 1, \ldots, n.$$

The $n \times n$ matrix $P = (\pi_{ij})$ is the *change of basis matrix* (for changing from the basis Y to the basis X). If η' is the isomorphism of $\mathscr{E}$ onto $\mathbb{C}^n[\mathbb{R}^n]$ that assigns to each vector in $\mathscr{E}$ its n-tuple of coordinates with respect to Y, and if $(\alpha_1, \ldots, \alpha_n)$ denotes an arbitrary element of $\mathbb{C}^n[\mathbb{R}^n]$, then

$$\sum_{i=1}^n \alpha_i x_i = \sum_{i=1}^n \alpha_i \sum_{k=1}^n \pi_{ki} y_k = \sum_{k=1}^n \left(\sum_{i=1}^n \pi_{ki} \alpha_i \right) y_k.$$

Thus $\eta'(x) = (\beta_1, \ldots, \beta_n)$ can be computed in terms of $\eta(x) = (\alpha_1, \ldots, \alpha_n)$ by means of the formula

$$\beta_i = \sum_{j=1}^n \pi_{ij} \alpha_j, \quad i = 1, \ldots, n.$$

If $\mathscr{E}$ and $\mathscr{F}$ are linear spaces over the same scalar field, and if S and T are two linear transformations of $\mathscr{E}$ into $\mathscr{F}$, then the sum $S + T$ is defined by pointwise addition: $(S + T)x = Sx + Tx$ for all x in $\mathscr{E}$. Likewise, for a scalar α, the mapping αS is defined by $(\alpha S)x = \alpha(Sx)$ for all x in $\mathscr{E}$. Clearly $S + T$ and αS are also linear transformations of $\mathscr{E}$ into $\mathscr{F}$. Moreover, these definitions turn the set of all linear transformations of $\mathscr{E}$ into $\mathscr{F}$ into a new linear space—the *full space of linear transformations of $\mathscr{E}$ into $\mathscr{F}$*. The zero element of this linear space is the linear transformation 0 defined by $0x = 0$ for all x in $\mathscr{E}$. (The full space of linear transformations of $\mathscr{E}$ into its scalar field coincides with the full algebraic dual of $\mathscr{E}$ (Ex. F).)

Suppose now that $\mathscr{E}$, $\mathscr{F}$, and $\mathscr{G}$ are all vector spaces over the same scalar field. Let T be a linear transformation of $\mathscr{E}$ into $\mathscr{F}$ and let S be a linear transformation of $\mathscr{F}$ into $\mathscr{G}$. Then the composition $S \circ T$ is a linear transformation of $\mathscr{E}$ into $\mathscr{G}$ called the *product* of S and T and denoted by ST. The multiplication of linear transformations satisfies the following relations whenever the various products are defined:

(i) $R(ST) = (RS)T$,
(ii) $R(S + T) = RS + RT$; $(R + S)T = RT + ST$, (1)
(iii) $\alpha(ST) = (\alpha S)T = S(\alpha T)$.

In particular, if R, S, and T denote linear transformations of a linear space $\mathscr{E}$ into itself, then all of these products are defined, and the relations (1) hold without exception.

Conditions (1) are the main ingredients in the definition of the concept of a *linear algebra*, a notion that is of considerable importance in functional analysis.

Definition. A vector space [real vector space] $\mathscr{A}$ on which is given a product satisfying the conditions

(i) $x(yz) = (xy)z$,
(ii) $x(y + z) = xy + xz$; $(x + y)z = xz + yz$,
(iii) $\alpha(xy) = (\alpha x)y = x(\alpha y)$,

for all elements x, y, z of $\mathscr{A}$ and all scalars α is an (*associative, linear*) *algebra* [*real algebra*]. If x and y are elements of $\mathscr{A}$ such that $xy = yx$, then x and y *commute*. The collection of all elements of $\mathscr{A}$ that commute with every element of $\mathscr{A}$ is the *center* of $\mathscr{A}$, and if the center of $\mathscr{A}$ coincides with $\mathscr{A}$, then $\mathscr{A}$ is a *commutative* or *abelian* algebra. If $\mathscr{A}$ possesses an element 1 such that $1x = x1 = x$ for every x in $\mathscr{A}$, then 1 is the *identity* or *unit* of $\mathscr{A}$ (such an element is obviously unique if it exists, and must belong to the center of $\mathscr{A}$), and $\mathscr{A}$ is said to be a *unital* algebra [real algebra] or an algebra *with identity* or *unit*. If $\mathscr{A}$ is a unital algebra [real algebra] with identity 1, and if λ is a scalar, we shall simply write λ for $\lambda 1$ when no confusion can result. Likewise, if x is an element of $\mathscr{A}$, then an element y of $\mathscr{A}$ is the *inverse* of x in $\mathscr{A}$ if $xy = yx = 1$. (The inverse of an element x is obviously unique if it exists, and is denoted by x^{-1}.) An element of $\mathscr{A}$ that possesses an inverse in $\mathscr{A}$ is said to be *invertible* (in $\mathscr{A}$). If x is an invertible element of $\mathscr{A}$, and if x and y commute, then it is readily seen that x^{-1} and y also commute $(x^{-1}(xy)x^{-1} = yx^{-1}$ while $x^{-1}(yx)x^{-1} = x^{-1}y)$. Thus the inverse x^{-1} not only commutes with x, it also commutes with every element of $\mathscr{A}$ that commutes with x. Such an element of $\mathscr{A}$, that is, one that commutes with every element of $\mathscr{A}$ that commutes with x, is said to *doubly commute* with x.

Thus the full space of linear transformations on a linear space [real linear space] is a unital algebra [real algebra] in which the transformation 1 is the identity element. Another important example of a unital algebra [real algebra] is the system $\mathbb{C}_{n,n}[\mathbb{R}_{n,n}]$ of all complex [real] $n \times n$ matrices. (The product in this algebra is the customary row by column multiplication; see Problem H.) The identity in $\mathbb{C}_{n,n}[\mathbb{R}_{n,n}]$ is the *identity matrix* $1 = (\delta_{ij})$, where, as usual, δ_{ij} denotes the Kronecker delta (Ex. B). If λ is a scalar the *scalar matrix* $\lambda 1$ will be denoted by λ. We recall that if A is an element of the algebra $\mathbb{C}_{n,n}$ and if there exists a matrix B such that either $AB = 1$ or $BA = 1$, then A is invertible and $B = A^{-1}$ (see Problems G and M).

On the algebra $\mathbb{C}_{n,n}$ of $n \times n$ matrices there are two important complex-valued functions. The first of these is the *trace* of an $n \times n$ matrix $A = (\alpha_{ij})$, defined as

$$\operatorname{tr} A = \sum_{i=1}^{n} \alpha_{ii};$$

the second is the *determinant*, defined as

$$\det A = \sum_{\sigma} (\operatorname{sgn} \sigma)\alpha_{1\sigma(1)} \cdots \alpha_{n\sigma(n)},$$

where the sum is taken over all permutations σ of the set $\{1, \ldots, n\}$. The main properties of the trace that we shall need are the readily verified facts that tr is linear and that $\operatorname{tr}(AB) = \operatorname{tr}(BA)$. The central fact concerning determinants is that $A \to \det A$ is a homomorphism (as defined in Problem L) of the algebra $\mathbb{C}_{n,n}$ onto $\mathbb{C}$ with the property that det $A \neq 0$ when and only when A is invertible in $\mathbb{C}_{n,n}$.

If $A = (\alpha_{ij})$ is a matrix in $\mathbb{C}_{m,n}$ (for arbitrary m, n), then the *transpose* of A (denoted by A^t) is the $n \times m$ matrix (β_{ij}) defined by $\beta_{ij} = \alpha_{ji}$, $i = 1, \ldots, n$; $j = 1, \ldots, m$. Furthermore the *adjoint* of A (denoted by A^*) is the $n \times m$ matrix (γ_{ij}) defined by $\gamma_{ij} = \bar{\alpha}_{ji}$, $i = 1, \ldots, n$; $j = 1, \ldots, m$. (If A is a real matrix, then $A^t = A^*$.) A real $n \times n$ matrix A with the property that $A = A^t$ is *symmetric*; a complex $n \times n$ matrix A with the property that $A = A^*$ is *Hermitian* (or *self-adjoint*). A complex $n \times n$ matrix N with the property that $NN^* = N^*N$ is said to be *normal*; such a matrix U with the additional property that $UU^* = U^*U = 1$ is *unitary*.

Suppose now that $\mathscr{E}$ and $\mathscr{F}$ are finite-dimensional linear spaces of dimension n and m, respectively, and let $X = \{x_1, \ldots, x_n\}$ and $Y = \{y_1, \ldots, y_m\}$ be ordered bases in $\mathscr{E}$ and $\mathscr{F}$, respectively. If T is a linear transformation of $\mathscr{E}$ into $\mathscr{F}$, then the equations

$$Tx_j = \sum_{i=1}^{m} \alpha_{ij} y_i, \qquad j = 1, \ldots, n, \tag{2}$$

define an $m \times n$ matrix

$$\begin{pmatrix} \alpha_{11} & \cdots & \alpha_{1n} \\ \vdots & & \vdots \\ \alpha_{m1} & \cdots & \alpha_{mn} \end{pmatrix}$$

called the *matrix of T with respect to X and Y*. (When $\mathscr{E} = \mathscr{F}$ and $X = Y$, this matrix is called the matrix of T with respect to X.) The correspondence $T \leftrightarrow (\alpha_{ij})$ between linear transformations of $\mathscr{E}$ into $\mathscr{F}$ and $m \times n$ complex matrices established by (2) is a linear space isomorphism between the full space of linear transformations of $\mathscr{E}$ into $\mathscr{F}$ and the linear space $\mathbb{C}_{m,n}$ of all complex $m \times n$ matrices. Moreover, when $\mathscr{E} = \mathscr{F}$ and $X = Y$, the correspondence $T \leftrightarrow (\alpha_{ij})$ is an algebra isomorphism as well; see Problems L and M.

Let A be the matrix of a linear transformation $T : \mathscr{E} \to \mathscr{F}$ with respect to the ordered bases X and Y, and suppose that X' and Y' are new ordered bases in the spaces $\mathscr{E}$ and $\mathscr{F}$, respectively. If B denotes the matrix of T with respect to X' and Y', then straightforward calculation shows that

$$B = QAP^{-1}$$

where P and Q denote the change of basis matrices for changing from the bases X' and Y' to the bases X and Y, respectively. (See Example N; we here employ the obvious fact that an $n \times n$ matrix is invertible if and only if it can be viewed as a change of basis matrix.) In particular, if A is the matrix of a linear transformation T on $\mathscr{E}$ with respect to X, and if B is the matrix of T with respect to X', then

$$B = PAP^{-1}. \tag{3}$$

Two $n \times n$ matrices A and B that are related as in (3) for some (invertible) $n \times n$ matrix P are said to be *similar*. From what has been said it is clear that two $n \times n$ matrices are similar if and only if they can be taken to be the matrices of some one linear transformation with respect to suitably chosen ordered bases for an n-dimensional vector space. It is also clear that similarity is an equivalence relation on $\mathbb{C}_{n,n}$.

Thus far we have considered only functions of one variable, but we shall also be interested in certain kinds of functions of two or more variables. Suppose that $\mathscr{E}$, $\mathscr{F}$, and $\mathscr{G}$ are (real or complex) linear spaces over the same scalar field, and let $\varphi = \varphi(x, y)$ be a mapping defined on the direct sum $\mathscr{E} + \mathscr{F}$ and taking values in $\mathscr{G}$. If for each fixed y_0 in $\mathscr{F}$ the function $\varphi(x, y_0)$ is a linear transformation of $\mathscr{E}$ into $\mathscr{G}$, and, for each fixed x_0 in $\mathscr{E}$, $\varphi(x_0, y)$ is a linear transformation of $\mathscr{F}$ into $\mathscr{G}$, then φ is a *bilinear transformation* of $\mathscr{E} + \mathscr{F}$ into $\mathscr{G}$. If the space $\mathscr{G}$ is the scalar field, then φ is a *bilinear functional* on $\mathscr{E} + \mathscr{F}$. If, in addition, $\mathscr{E} = \mathscr{F}$, then φ is a *bilinear functional on $\mathscr{E}$*. The set of all bilinear transformations of $\mathscr{E} + \mathscr{F}$ into $\mathscr{G}$ is a linear space with linear operations defined pointwise. In particular, the set of all bilinear functionals on $\mathscr{E}$ is a linear space.

When $\mathscr{E}$ and $\mathscr{F}$ are complex, there is a notion closely related to that of a bilinear functional on $\mathscr{E} + \mathscr{F}$ that we shall have occasion to use. A mapping $\psi : \mathscr{E} + \mathscr{F} \to \mathbb{C}$ is said to be a *sesquilinear functional* on $\mathscr{E} + \mathscr{F}$ if $\psi(x, y_0)$ is a linear functional on $\mathscr{E}$ for each y_0 in $\mathscr{F}$ and $\bar{\psi}(x_0, y)$ is a linear functional on $\mathscr{F}$ for each x_0 in $\mathscr{E}$. (Another way to state the second of these conditions is to say that $\psi(x_0, \alpha y_1 + \beta y_2)$ is equal to $\bar{\alpha}\psi(x_0, y_1) + \bar{\beta}\psi(x_0, y_2)$ for all complex numbers α, β and all vectors y_1, y_2 in $\mathscr{F}$; such a functional is said to be *conjugate linear*.) When $\mathscr{E} = \mathscr{F}$, ψ is called a sesquilinear functional *on $\mathscr{E}$*. A sesquilinear functional ψ on $\mathscr{E}$ is said to be *symmetric* if $\psi(x, y) = \overline{\psi(y, x)}$ for all x, y in $\mathscr{E}$.

We close this chapter with an account of tensor products of linear spaces. No prior knowledge of this topic is assumed on the part of the reader. If $\mathscr{E}$ and $\mathscr{F}$ are any two linear spaces over the same scalar field, we may form the free linear space $\mathscr{T}$ generated by the set-theoretic product $\mathscr{E} \times \mathscr{F}$, that is, the space of all formal linear combinations $\sum_{i=1}^{n} \lambda_i(x_i, y_i)$ of pairs in $\mathscr{E} \times \mathscr{F}$ (see Example A). Let $\mathscr{R}$ denote the linear manifold in $\mathscr{T}$ generated by all differences of the form

$$(x_1 + x_2, y) - (x_1, y) - (x_2, y) \quad \text{and} \quad (x, y_1 + y_2) - (x, y_1) - (x, y_2)$$

together with all differences of the form

$$(\lambda x, y) - \lambda(x, y) \quad \text{and} \quad (x, \lambda y) - \lambda(x, y)$$

where x, x_1, and x_2 denote arbitrary vectors in $\mathscr{E}$, y, y_1, and y_2 arbitrary vectors in $\mathscr{F}$, and λ an arbitrary scalar. Our interest focuses on the quotient space $\mathscr{T}/\mathscr{R}$, which is called the *algebraic tensor product* of $\mathscr{E}$ and $\mathscr{F}$, and is denoted by $\mathscr{E} \dot{\times} \mathscr{F}$. (If $\mathscr{E}$ and $\mathscr{F}$ are complex [real] vector spaces, then $\mathscr{E} \dot{\times} \mathscr{F}$ is a complex [real] vector space.) If x belongs to $\mathscr{E}$ and y belongs to

$\mathscr{F}$, then the pair (x, y), regarded as an element of $\mathscr{T}$, projects onto an element of $\mathscr{E} \times \mathscr{F}$ which we denote by $x \otimes y$. Elements of $\mathscr{E} \times \mathscr{F}$ of the form $x \otimes y$ are said to be *decomposable*. The mapping of $\mathscr{E} + \mathscr{F}$ into $\mathscr{E} \times \mathscr{F}$ that carries (x, y) to $x \otimes y$ will be denoted by ρ. From the definition of $\mathscr{R}$ we see that the following relations hold for all vectors x, x_1 and x_2 in $\mathscr{E}$, all vectors y, y_1, and y_2 in $\mathscr{F}$, and all scalars λ:

(i) $\qquad\qquad (x_1 + x_2) \otimes y = x_1 \otimes y + x_2 \otimes y,$

(ii) $\qquad\qquad x \otimes (y_1 + y_2) = x \otimes y_1 + x \otimes y_2,$ and $\qquad\qquad$ (4)

(iii) $\qquad\qquad (\lambda x) \otimes y = \lambda(x \otimes y) = x \otimes (\lambda y).$

According to the definition, the decomposable elements of $\mathscr{E} \times \mathscr{F}$ form an algebraic system of generators for $\mathscr{E} \times \mathscr{F}$. By condition (iii) of (4) we see that, in fact, the general element of $\mathscr{E} \times \mathscr{F}$ can be written (in various ways) as a sum of decomposables: $t = \sum_{i=1}^{n} x_i \otimes y_i$. The central facts about algebraic tensor products are readily established.

Proposition 2.1. *The mapping ρ of $\mathscr{E} + \mathscr{F}$ into $\mathscr{E} \times \mathscr{F}$ is bilinear, and the range of ρ generates $\mathscr{E} \times \mathscr{F}$. Moreover, if φ is any bilinear transformation of $\mathscr{E} + \mathscr{F}$ into a linear space $\mathscr{G}$, then there exists a unique linear transformation $\tilde{\varphi}$ of $\mathscr{E} \times \mathscr{F}$ into $\mathscr{G}$ such that $\varphi = \tilde{\varphi} \circ \rho$ (briefly: bilinear mappings on $\mathscr{E} + \mathscr{F}$ can be* factored through $\mathscr{E} \times \mathscr{F}$).

PROOF. It is clear from (4) and the definition that ρ is a bilinear transformation, and it has just been observed that every element of $\mathscr{E} \times \mathscr{F}$ can be written (not uniquely) as a finite sum of decomposable elements. Thus the range of ρ generates $\mathscr{E} \times \mathscr{F}$ algebraically, so $\tilde{\varphi}$ is unique if it exists. With respect to existence we observe that *any* mapping of $\mathscr{E} + \mathscr{F}$ into a linear space $\mathscr{G}$ possesses a unique linear extension to $\mathscr{T}$ (Prob. E). Since φ is bilinear, this extension annihilates the linear submanifold $\mathscr{R}$, and consequently can be factored through $\mathscr{E} \times \mathscr{F}$ (Example L). $\qquad\square$

Definition. Suppose $T_i : \mathscr{E}_i \to \mathscr{F}_i$ is a linear transformation, $i = 1, 2$. Then the mapping $(x_1, x_2) \to T_1 x_1 \otimes T_2 x_2$, $x_i \in \mathscr{E}_i$, $i = 1, 2$, is a bilinear mapping of $\mathscr{E}_1 + \mathscr{E}_2$ into $\mathscr{F}_1 \times \mathscr{F}_2$. Consequently, according to the foregoing result, there exists a unique linear transformation $T : \mathscr{E}_1 \times \mathscr{E}_2 \to \mathscr{F}_1 \times \mathscr{F}_2$ satisfying the condition $T(x_1 \otimes x_2) = T_1 x_1 \otimes T_2 x_2$ for all x_i in $\mathscr{E}_i$, $i = 1, 2$. We shall call T the *algebraic tensor product of T_1 and T_2* and write $T = T_1 \times T_2$.

Proposition 2.1 has an important and useful counterpart that provides a categorical characterization of the algebraic tensor product.

Proposition 2.2. *Let $\mathscr{E}$ and $\mathscr{F}$ be complex [real] linear spaces and suppose given a pair $(\mathscr{G}, \sigma)$, where $\mathscr{G}$ is a complex [real] linear space and σ is a*

23

bilinear transformation of $\mathscr{E} \dotplus \mathscr{F}$ into $\mathscr{G}$. Suppose also that the following conditions are satisfied:

(i) *The range of σ generates $\mathscr{G}$ algebraically,*
(ii) *If φ is any bilinear transformation of $\mathscr{E} \dotplus \mathscr{F}$ into a complex [real] linear space $\mathscr{H}$, then there exists a linear transformation $T : \mathscr{G} \to \mathscr{H}$ such that $\varphi = T \circ \sigma$.*

Then there exists a linear isomorphism Φ of $\mathscr{G}$ onto $\mathscr{E} \dot\times \mathscr{F}$ such that $\Phi \circ \sigma = \rho$. (The isomorphism Φ is unique by virtue of (i) and coincides, in fact, with the result of applying (ii) to ρ.)

PROOF. We simply define Φ to be the result of factoring ρ through $\mathscr{G}$ as in (ii), and denote by Ψ the result of factoring σ through $\mathscr{E} \dot\times \mathscr{F}$. It is clear that Φ and Ψ are mutually inverse mappings on the ranges of σ and ρ, respectively, and since the range of σ generates $\mathscr{G}$ and the range of ρ generates $\mathscr{E} \dot\times \mathscr{F}$, it follows that Φ and Ψ are, in fact, mutually inverse linear isomorphisms between $\mathscr{G}$ and $\mathscr{E} \dot\times \mathscr{F}$. $\qquad\qquad\square$

Example O. Let $\mathscr{E}$ and $\mathscr{F}$ be linear spaces, and let $\mathscr{L}$ and $\mathscr{M}$ be linear manifolds in $\mathscr{E}$ and $\mathscr{F}$, respectively. It is obvious that if we set $\rho_0 = \rho | (\mathscr{L} \dotplus \mathscr{M})$ where ρ is as above, then ρ_0 is a bilinear mapping of $\mathscr{L} \dotplus \mathscr{M}$ into $\mathscr{E} \dot\times \mathscr{F}$. Moreover, if we denote by $\mathscr{G}$ the linear submanifold of $\mathscr{E} \dot\times \mathscr{F}$ generated by $\rho(\mathscr{L} \dotplus \mathscr{M})$, then the pair $(\mathscr{G}, \rho_0)$ satisfies the conditions of Proposition 2.2. Indeed, if φ_0 is a bilinear mapping of $\mathscr{L} \dotplus \mathscr{M}$ into a linear space $\mathscr{H}$, let φ denote a bilinear extension of φ_0 to $\mathscr{E} \dotplus \mathscr{F}$ (Prob. Q). If $\tilde{\varphi}$ denotes the result of factoring φ through $\mathscr{E} \dot\times \mathscr{F}$, and if we set $T = \tilde{\varphi} | \mathscr{G}$, then T is a linear transformation of $\mathscr{G}$ into $\mathscr{H}$ and we have $\varphi_0 = T \circ \rho_0$. Hence the result ι of factoring ρ_0 through $\mathscr{L} \dot\times \mathscr{M}$ is a linear isomorphism of $\mathscr{L} \dot\times \mathscr{M}$ onto the linear manifold $\mathscr{G}$ in $\mathscr{E} \dot\times \mathscr{F}$. It is readily verified that if x and y are vectors in $\mathscr{L}$ and $\mathscr{M}$, respectively, and if $(x \otimes y)_0$ denotes (for the moment) the corresponding element of $\mathscr{L} \dot\times \mathscr{M}$, then $\iota((x \otimes y)_0) = x \otimes y$ in $\mathscr{E} \dot\times \mathscr{F}$. *Throughout this book, whenever tensor products are under discussion, we shall use the canonical isomorphism ι to identify $\mathscr{L} \dot\times \mathscr{M}$ with the linear manifold $\mathscr{G} = \iota(\mathscr{L} \dot\times \mathscr{M})$ in $\mathscr{E} \dot\times \mathscr{F}$.*

Let $\mathscr{E}$ and $\mathscr{F}$ be linear spaces over the same scalar field, and let $t = \sum_{i=1}^{r} x_i \otimes y_i$ be an element of $\mathscr{E} \dot\times \mathscr{F}$. If $\{e_1, \ldots, e_s\}$ is a basis for a linear manifold in $\mathscr{E}$ containing all of the vectors $x_1, \ldots, x_r$, and if

$$x_i = \sum_{j=1}^{s} \lambda_{ij} e_j, \qquad i = 1, \ldots, r,$$

then

$$t = \sum_{i=1}^{r} \sum_{j=1}^{s} \lambda_{ij}(e_j \otimes y_i)$$

$$= \sum_{j=1}^{s} \left[e_j \otimes \left(\sum_{i=1}^{r} \lambda_{ij} y_i \right) \right].$$

This calculation shows that, when representing an element of $\mathscr{E} \dot{\times} \mathscr{F}$ in the form $\sum_i x_i \otimes y_i$, we may always arrange for the x_i's (or the y_j's) to be linearly independent. In this connection the following result is important.

Proposition 2.3. *Let $\mathscr{E}$ and $\mathscr{F}$ be linear spaces over the same scalar field, let X be a subset of $\mathscr{E}$, let Y be a subset of $\mathscr{F}$, and let $T = \{x \otimes y : x \in X, y \in Y\}$. If X and Y generate $\mathscr{E}$ and $\mathscr{F}$, respectively, then T generates $\mathscr{E} \dot{\times} \mathscr{F}$. If X and Y are each linearly independent, then T is linearly independent in $\mathscr{E} \dot{\times} \mathscr{F}$. Finally, if X and Y are Hamel bases for $\mathscr{E}$ and $\mathscr{F}$, respectively, Then T is a Hamel basis for $\mathscr{E} \dot{\times} \mathscr{F}$.*

PROOF. The first assertion is obvious, and is included here only for completeness, while the third assertion is an immediate consequence of the first two. Thus the proof comes down to showing that T is linearly independent when X and Y are, and here it is clearly enough to treat the case in which X and Y are nonempty finite sets. Suppose, then, that $X = \{x_1, \ldots, x_m\}$ and $Y = \{y_1, \ldots, y_n\}$ are linearly independent, and let $\mathscr{L}$ and $\mathscr{M}$ denote the linear manifolds generated by X and Y, respectively. For each vector $x = \sum_{i=1}^{m} \alpha_i x_i$ in $\mathscr{L}$ and $y = \sum_{j=1}^{n} \beta_j y_j$ in $\mathscr{M}$, we define the the $m \times n$ matrix

$$\varphi(x, y) = (\alpha_i \beta_j), \qquad i = 1, \ldots, m, j = 1, \ldots, n,$$

and observe that φ is a bilinear mapping of $\mathscr{L} + \mathscr{M}$ into the linear space $\mathbb{C}_{m,n}$. Moreover, if $\tilde{\varphi}$ denotes the result of factoring φ through $\mathscr{L} \dot{\times} \mathscr{M}$, then

$$\tilde{\varphi}(x_i \otimes y_j) = E_{ij}, \qquad i = 1, \ldots, m, j = 1, \ldots, n,$$

where E_{ij} denotes the matrix with a one in the (i, j) position and all other entries equal to zero (cf. Example C). Since the matrices E_{ij} are linearly independent in $\mathbb{C}_{m,n}$, it is clear that the products $x_i \otimes y_j$ are linearly independent in $\mathscr{E} \dot{\times} \mathscr{F}$, and the proposition follows. (We here use our convention that it is a matter of indifference whether $x_i \otimes y_j$ is regarded as an element of $\mathscr{L} \dot{\times} \mathscr{M}$ or $\mathscr{E} \dot{\times} \mathscr{F}$.) $\qquad \square$

Corollary 2.4. *The Hamel dimension of $\mathscr{E} \dot{\times} \mathscr{F}$ is the product of the Hamel dimensions of $\mathscr{E}$ and $\mathscr{F}$.*

It is easy to see how the above definitions should be modified so as to cover algebraic tensor products of the form $\mathscr{E}_1 \dot{\times} \cdots \dot{\times} \mathscr{E}_r$, and likewise how Propositions 2.1, 2.2, and 2.3 are to be generalized so as to cover this more general case. Details are left to the interested reader.

PROBLEMS

A. The union of a nested collection of linearly independent sets in a (real or complex) linear space $\mathscr{E}$ is again linearly independent. Use this fact and Zorn's lemma to show that there exists a maximal linearly independent set in $\mathscr{E}$. Show also that a maximal linearly independent set in $\mathscr{E}$ is a Hamel basis for $\mathscr{E}$. (The empty set $\varnothing$ is a Hamel basis for (0).) Show similarly that if M is an arbitrary subset of $\mathscr{E}$ and J an arbitrary linearly independent subset of M, and if $\mathscr{L}$ denotes the linear submanifold of $\mathscr{E}$ generated by M, then there exists a Hamel basis X for $\mathscr{L}$ such that $J \subset X \subset M$.

B. (i) Let $X = \{x_1, \ldots, x_m\}$ and $Y = \{y_1, \ldots, y_n\}$ be linearly independent sets of vectors in a (real or complex) vector space $\mathscr{E}$, and suppose X is contained in the submanifold $\mathscr{M}$ of $\mathscr{E}$ that is generated by Y. Show that $m \leq n$ and that it is possible to select a set Z of exactly $n-m$ vectors from Y so that $X \cup Z$ is also a basis for $\mathscr{M}$. (In particular, then, if $m = n$, the set X is itself a basis for $\mathscr{M}$.) Conclude that if a vector space $\mathscr{E}$ possesses a finite basis, then any two bases for $\mathscr{E}$ contain the same number of vectors, and that, if $\dim \mathscr{E} = n$, then $\mathscr{E}$ is itself the only n-dimensional submanifold of $\mathscr{E}$. (Hint: The heart of the matter is that if a vector x belongs to $\mathscr{M}$, and if in the expression $x = \sum_{i=1}^{n} \lambda_i y_i$ some $\lambda_{i_0} \neq 0$, then the set $\tilde{Y}$ consisting of x and $\{y_i : i \neq i_0\}$ is linearly independent, and is therefore another basis for $\mathscr{M}$.)

(ii) Let $\mathscr{E}$ be an infinite dimensional linear space (real or complex), and suppose given two (infinite) Hamel bases X and Y for $\mathscr{E}$. If x is any nonzero vector belonging to X, then there exist (unique) vectors $y_1, \ldots, y_p$ in Y and (unique) nonzero scalars $\lambda_1, \ldots, \lambda_p$ such that $x = \lambda_1 y_1 + \cdots + \lambda_p y_p$. (The vectors y_i and scalars λ_i vary with x, of course.) Let E_x denote the set $\{y_1, \ldots, y_p\}$. Show that $Y = \bigcup_{x \in X} E_x$, and show that this implies that $\operatorname{card} Y \leq \aleph_0 \operatorname{card} X$. Finally, show that $\operatorname{card} X = \operatorname{card} Y$ (and hence that the definition of Hamel dimension makes sense). (Hint: Cf. Problems 1R and 1T.)

C. If $\mathscr{M}$ is a linear manifold in a (real or complex) vector space $\mathscr{E}$ and $\mathscr{N}$ is another linear manifold in $\mathscr{E}$ such that $\mathscr{M} \cap \mathscr{N} = (0)$ and $\mathscr{M} + \mathscr{N} = \mathscr{E}$, then $\mathscr{N}$ is a *complement* of $\mathscr{M}$. Show that every linear manifold $\mathscr{M}$ in $\mathscr{E}$ has a complement. (Hint: Use a Hamel basis.)

D. Let $\mathscr{E}$ and $\mathscr{F}$ be linear spaces, and let T be a linear transformation of some linear submanifold $\mathscr{M}$ of $\mathscr{E}$ into $\mathscr{F}$. Show that there exists a linear transformation $\hat{T}$ of $\mathscr{E}$ into $\mathscr{F}$ such that $T = \hat{T}|\mathscr{M}$. (Hint: Use a Hamel basis or Problem C.)

E. Let $\mathscr{E}$ and $\mathscr{F}$ be linear spaces, and let φ be a mapping of a subset M of $\mathscr{E}$ into $\mathscr{F}$. Show that there exists a linear transformation $T : \mathscr{E} \to \mathscr{F}$ such that $\varphi = T|M$ if and only if φ "respects" all linear dependencies in M, that is, if and only if $\sum_{i=1}^{m} \lambda_i x_i = 0$ with $x_1, \ldots, x_m$ in M implies that $\sum_{i=1}^{m} \lambda_i \varphi(x_i) = 0$. Show also that such a linear transformation T is uniquely determined by φ on the linear submanifold of $\mathscr{E}$ generated by M. (Hint: The stated condition ensures that if

$$\sum_{i=1}^{m} \alpha_i x_i = \sum_{j=1}^{n} \beta_j y_j, \quad \text{then} \quad \sum_{i=1}^{m} \alpha_i \varphi(x_i) = \sum_{j=1}^{n} \beta_j \varphi(y_j)$$

whenever $x_1, \ldots, x_m$ and $y_1, \ldots, y_n$ are vectors in M.) In particular, if X is a Hamel basis for $\mathscr{E}$, and if φ is an *arbitrary* mapping of X into $\mathscr{F}$, then φ possesses a unique linear extension to $\mathscr{E}$. Conclude that two linear spaces are isomorphic if and only if they have the same Hamel dimension.

F. (i) Let $\{\mathscr{E}_\gamma\}_{\gamma \in \Gamma}$ be an indexed family of vector spaces, let $\mathscr{E}$ denote the full algebraic direct sum $\mathscr{E} = \sum_\gamma + \mathscr{E}_\gamma$, and for each index γ_0 let $\mathscr{M}_{\gamma_0}$ denote the set of those elements $\{x_\gamma\}$ of $\mathscr{E}$ such that $x_\gamma = 0$ for all $\gamma \neq \gamma_0$. Verify that $\mathscr{M}_{\gamma_0}$ is a linear manifold in $\mathscr{E}$ and that the mapping τ_{γ_0} obtained by restricting to $\mathscr{M}_{\gamma_0}$ the projection π_{γ_0} of $\mathscr{E}$ onto $\mathscr{E}_{\gamma_0}$ is an isomorphism of $\mathscr{M}_{\gamma_0}$ onto $\mathscr{E}_{\gamma_0}$. (This isomorphism is frequently used to identify $\mathscr{E}_{\gamma_0}$ with its counterpart $\mathscr{M}_{\gamma_0}$.) Show also that the indexed family of submanifolds $\{\mathscr{M}_{\gamma}\}_{\gamma \in \Gamma}$ has the following properties:

(a)
$$\sum_{\gamma} \mathscr{M}_{\gamma} = \mathscr{E},$$

(b)
$$\mathscr{M}_{\gamma} \cap \mathscr{M}_{\gamma'} = (0), \qquad \gamma \neq \gamma'.$$

Use this observation to prove that the Hamel dimension of $\mathscr{E}$ is the sum $\sum_{\gamma} d_{\gamma}$ where d_{γ} denotes the Hamel dimension of $\mathscr{E}_{\gamma}$, $\gamma \in \Gamma$. (Hint: If X_{γ} is a Hamel basis for $\mathscr{M}_{\gamma}$, $\gamma \in \Gamma$, then the sets X_{γ} are pairwise disjoint, and $\bigcup_{\gamma} X_{\gamma}$ is a Hamel basis for $\mathscr{E}$. Recall that the cardinal number of the union of a disjoint collection of sets is the sum of the cardinal numbers of the sets in the collection.)

(ii) An indexed family $\{\mathscr{M}_{\gamma}\}_{\gamma \in \Gamma}$ of linear submanifolds of a linear space $\mathscr{E}$ is said to form an *internal direct sum decomposition* of $\mathscr{E}$ if (a) and (b) of (i) are satisfied. Thus, according to (i), to every (*external*) full algebraic direct sum $\mathscr{E} = \sum_{\gamma} + \mathscr{E}_{\gamma}$ there corresponds a similarly indexed internal direct sum decomposition $\{\mathscr{M}_{\gamma}\}$ of $\mathscr{E}$, where each $\mathscr{M}_{\gamma}$ is isomorphic to $\mathscr{E}_{\gamma}$ in a natural way. Show, conversely, that if $\{\mathscr{M}_{\gamma}\}$ is any internal direct sum decomposition of a linear space $\mathscr{E}$, then there exists a unique isomorphism of $\mathscr{E}$ onto the (external) direct sum $\sum_{\gamma} + \mathscr{M}_{\gamma}$ that agrees with τ_{γ}^{-1} on each linear submanifold $\mathscr{M}_{\gamma}$.

G. If T is a linear transformation of a linear space $\mathscr{E}$ into a linear space $\mathscr{F}$, and if its range $\mathscr{R}(T)$ is finite dimensional, then dim $\mathscr{R}(T)$ is called the *rank* of T (notation: rank T). Likewise, if the null space $\mathscr{K}(T)$ is finite dimensional, them dim $\mathscr{K}(T)$ is the *nullity* of T. Show that if $\mathscr{E}$ is finite dimensional, then for an arbitrary linear transformation $T: \mathscr{E} \to \mathscr{F}$ we have

$$(\text{rank } T) + (\text{nullity of } T) = \dim \mathscr{E}.$$

Conclude that if S is a linear transformation on an n-dimensional linear space $\mathscr{E}$, then S is invertible if either $\mathscr{K}(S) = (0)$ or $\mathscr{R}(S) = \mathscr{E}$, i.e., if either the nullity of S vanishes or the rank of S equals n. Show finally that if there exists a linear transformation R on $\mathscr{E}$ such that $RS = 1$ or $SR = 1$, then S is invertible and $R = S^{-1}$.

H. If $A = (\alpha_{ij})$ is an $m \times n$ matrix and $B = (\beta_{ij})$ an $n \times p$ matrix, then the *product* AB is the $m \times p$ matrix (γ_{ij}), where $\gamma_{ij} = \sum_{k=1}^{n} \alpha_{ik}\beta_{kj}$, $i = 1, \ldots, m$; $j = 1, \ldots, p$. (This operation is, for obvious reasons, known as "row by column" multiplication.) In particular, the product of any two matrices in $\mathbb{C}_{n,n}$ is defined, and this product turns $\mathbb{C}_{n,n}$ into an algebra.

(i) Suppose that $X = \{x_1, \ldots, x_p\}$ and $Y = \{y_1, \ldots, y_n\}$ are ordered bases for the linear spaces $\mathscr{E}$ and $\mathscr{F}$, respectively, and that B is the matrix of a linear transformation $T: \mathscr{E} \to \mathscr{F}$ with respect to X and Y. Verify that if $(\alpha_1, \ldots, \alpha_p)$ is the p-tuple of coordinates of a vector x with respect to the basis X and $(\beta_1, \ldots, \beta_n)$ the n-tuple of coordinates of Tx with respect to Y, then

$$\begin{pmatrix} \beta_1 \\ \vdots \\ \beta_n \end{pmatrix} = B \begin{pmatrix} \alpha_1 \\ \vdots \\ \alpha_p \end{pmatrix}.$$

(ii) Let $Z = \{z_1, \ldots, z_m\}$ be an ordered basis for a third linear space $\mathscr{G}$, and let A be the matrix of a linear transformation $S : \mathscr{F} \to \mathscr{G}$ with respect to Y and Z. Verify that AB is the matrix of ST with respect to the bases X and Z.

(iii) If $A = (\alpha_{ij})$ is an element of $\mathbb{C}_{m,n}$, and if for each $x = (\xi_1, \ldots, \xi_n)$ in $\mathbb{C}^n$ we set $Tx = y = (\eta_1, \ldots, \eta_m)$, where

$$\begin{pmatrix} \eta_1 \\ \vdots \\ \eta_m \end{pmatrix} = A \begin{pmatrix} \xi_1 \\ \vdots \\ \xi_n \end{pmatrix},$$

then the mapping T of $\mathbb{C}^n$ into $\mathbb{C}^m$ is a linear transformation. We shall refer to this as the linear transformation *defined by* the matrix A. Verify that this assignment of a linear transformation to each $m \times n$ matrix is a linear space isomorphism between $\mathbb{C}_{m,n}$ and the full space of linear transformations of $\mathbb{C}^n$ into $\mathbb{C}^m$. In the same spirit, verify that if $\mathscr{E}$ and $\mathscr{F}$ are, respectively, n- and m-dimensional, and if X and Y are any ordered bases for $\mathscr{E}$ and $\mathscr{F}$, respectively, then the correspondence pairing each linear transformation T of $\mathscr{E}$ into $\mathscr{F}$ with the matrix of T with respect to X and Y is a linear space isomorphism between the full space of linear transformations of $\mathscr{E}$ into $\mathscr{F}$ and $\mathbb{C}_{m,n}$. If A is an $m \times n$ matrix, and if T denotes the linear transformation of $\mathbb{C}^n$ into $\mathbb{C}^m$ defined by A, then A is the matrix of T with respect to the natural bases in $\mathbb{C}^n$ and $\mathbb{C}^m$ (Ex. B).

I. By the *rank* of an $m \times n$ matrix A (notation: rank A) is meant the maximal number of linearly independent columns in A (that is, the number of elements in a maximal linearly independent set of columns; the columns of A are here viewed as elements of $\mathbb{C}^m$). Show that if T is a linear transformation of an n-dimensional linear space $\mathscr{E}$ into an m-dimensional linear space $\mathscr{F}$, and if A is the matrix of T with respect to ordered bases for $\mathscr{E}$ and $\mathscr{F}$, then rank A = rank T. Conclude that an $n \times n$ matrix A is invertible if and only if rank $A = n$. (Hint: See Problem G.)

J. Let $\mathscr{A}$ be a (real or complex) algebra. A linear manifold $\mathscr{S}$ in $\mathscr{A}$ that is closed under multiplication is a *subalgebra* of $\mathscr{A}$. If $\mathscr{S}$ has the additional property that for every s in $\mathscr{S}$ and x in $\mathscr{A}$, sx belongs to $\mathscr{S}[xs$ belongs to $\mathscr{S}]$, then $\mathscr{S}$ is a *right [left] ideal* in $\mathscr{A}$. If $\mathscr{S}$ is both a left and a right ideal, then $\mathscr{S}$ is a *two-sided ideal*, or, more simply, an *ideal* in $\mathscr{A}$. Show that if $\mathscr{S}$ is a two-sided ideal in $\mathscr{A}$, then the quotient vector space $\mathscr{A}/\mathscr{S}$ equipped with the product $[x][y] = [xy]$ forms an algebra. This algebra $\mathscr{A}/\mathscr{S}$ is called the *quotient algebra* of $\mathscr{A}$ modulo $\mathscr{S}$.

K. Show that the vector space (s) of Example D forms a unital algebra with respect to the product $\{\alpha_n\}\{\beta_n\} = \{\alpha_n\beta_n\}$. Show that the linear manifolds (m), (c), and (c_0) are subalgebras of (s), and that (c_0) is an ideal in (m).

L. Let $\mathscr{A}$ and $\mathscr{B}$ be algebras over the same scalar field, and suppose that φ is a linear transformation of $\mathscr{A}$ into $\mathscr{B}$ such that φ *preserves products* (i.e., such that $\varphi(xy) = \varphi(x)\varphi(y)$ for every pair x, y of elements of $\mathscr{A}$). Then φ is called an *(algebra) homomorphism* of $\mathscr{A}$ into $\mathscr{B}$. If φ is a vector space isomorphism of $\mathscr{A}$ onto $\mathscr{B}$ that is also an algebra homomorphism, then φ is an *algebra isomorphism* of $\mathscr{A}$ onto $\mathscr{B}$, and if such an isomorphism exists, $\mathscr{A}$ and $\mathscr{B}$ are said to be *isomorphic (as) algebras*.

(i) Let φ be a linear transformation of $\mathscr{A}$ into $\mathscr{B}$, and let $X = \{x_\gamma\}$ be a Hamel basis for $\mathscr{A}$ (regarded as a linear space). Verify that φ is an algebra homomorphism if and only if $\varphi(x_\gamma x_{\gamma'}) = \varphi(x_\gamma)\varphi(x_{\gamma'})$ for every pair x_γ, $x_{\gamma'}$, of elements of the basis X.

(ii) Verify that if $\mathscr{J}$ is an ideal in $\mathscr{A}$, then the natural projection π of $\mathscr{A}$ onto the quotient algebra $\mathscr{A}/\mathscr{J}$ is a homomorphism with kernel $\mathscr{J}$ (see Problem J for definitions). Show in the converse direction that if φ is an arbitrary homomorphism of $\mathscr{A}$ into $\mathscr{B}$, then the kernel $\mathscr{K}(\varphi)$ is an ideal in $\mathscr{A}$, the range $\mathscr{R}(\varphi)$ is a subalgebra of $\mathscr{B}$, and the result of factoring φ through the ideal $\mathscr{K}(\varphi)$ (Ex. L) is an algebra isomorphism of the quotient algebra $\mathscr{A}/\mathscr{K}(\varphi)$ onto $\mathscr{R}(\varphi)$.

(iii) Suppose $\mathscr{A}$ is an algebra with identity 1, and let $\mathscr{P}$ denote the algebra of all complex polynomials $p(\lambda) = \alpha_0 + \alpha_1\lambda + \cdots + \alpha_n\lambda^n$. Show that for each element x of $\mathscr{A}$ there exists a unique homomorphism φ_x of $\mathscr{P}$ into $\mathscr{A}$ satisfying the conditions $\varphi_x(1) = 1$ and $\varphi_x(\lambda) = x$. (Hint: The sequence $\{1, \lambda, \lambda^2, \ldots\}$ constitutes a Hamel basis for $\mathscr{P}$; use (i).) The image $\varphi_x(p)$ of a polynomial p under the homomorphism φ_x is denoted by $p(x)$, and is referred to as the result of *evaluating* p at x. Show that for an arbitrary polynomial p the element $p(x)$ doubly commutes with x, i.e., that $p(x)$ commutes with every element y of $\mathscr{A}$ that commutes with x. Verify that an analogous construction exists, even when $\mathscr{A}$ does not possess a unit, provided the algebra $\mathscr{P}$ is replaced by the ideal $\mathscr{P}_0$ consisting of all polynomials having zero constant term, and discuss the case of a real algebra.

(iv) Show in the same vein that if $\mathscr{P}^{(n)}$ denotes the linear algebra of all polynomials $p(\lambda_1, \ldots, \lambda_n)$ in n indeterminates, and if $x_1, \ldots, x_n$ denote elements of $\mathscr{A}$ that commute in pairs, then there exists a unique homomorphism of $\mathscr{P}^{(n)}$ into $\mathscr{A}$ that assigns the identity 1 in $\mathscr{A}$ to the polynomial 1 and x_i to λ_i, $i = 1, \ldots, n$. The image of $p(\lambda_1, \ldots, \lambda_n)$ under this homomorphism is also referred to as the result of *evaluating* p at $(x_1, \ldots, x_n)$, and is denoted by $p(x_1, \ldots, x_n)$.

(v) Show that if $\mathscr{E}$ and $\mathscr{F}$ are vector spaces and η is a linear space isomorphism of $\mathscr{E}$ onto $\mathscr{F}$, then $\varphi(T) = \eta T\eta^{-1}$ defines an algebra isomorphism φ of the full space of linear transformations on $\mathscr{E}$ onto the full space of linear transformations on $\mathscr{F}$. (The isomorphism φ is said to be *spatially implemented* by η.)

M. If $\mathscr{E}$ is an n-dimensional linear space, then, as noted in Example N, for each ordered basis $X = \{x_1, \ldots, x_n\}$ for $\mathscr{E}$ the mapping

$$\sum_{i=1}^{n} \lambda_i x_i \overset{\eta}{\to} (\lambda_1, \ldots, \lambda_n),$$

which assigns to each vector in $\mathscr{E}$ its n-tuple of coordinates with respect to the basis X, is a linear space isomorphism of $\mathscr{E}$ onto $\mathbb{C}^n$. Show similarly that the mapping ψ that assigns to each linear transformation T on $\mathscr{E}$ its matrix with respect to X is an algebra isomorphism of the algebra of linear transformations on $\mathscr{E}$ onto the matrix algebra $\mathbb{C}_{n,n}$. Show, in the same vein, that if δ denotes the mapping assigning to each matrix A in $\mathbb{C}_{n,n}$ the linear transformation on $\mathbb{C}^n$ defined by A (Prob. H (iii)), then δ is an algebra isomorphism of $\mathbb{C}_{n,n}$ onto the algebra of all linear transformations on $\mathbb{C}^n$. Show, finally, that if $\varphi = \delta \circ \psi$, then $\varphi(T) = \eta T\eta^{-1}$ for every linear transformation T on $\mathscr{E}$, in other words, that φ is the isomorphism spatially implemented by η (Prob. L (v)). (Thus if the isomorphism δ is used to identify the algebra of linear transformations on $\mathbb{C}^n$ with $\mathbb{C}_{n,n}$, then ψ may be said to be spatially implemented.)

N. Show that if A and B are similar $n \times n$ matrices then tr $A =$ tr B and det $A =$ det B. Use this fact to conclude that if T is a linear transformation on an n-dimensional vector space $\mathscr{E}$, then we may define (unambiguously)

$$\text{tr } T = \text{tr } A \quad \text{and} \quad \det T = \det A,$$

where A denotes the matrix of T with respect to any ordered basis for $\mathscr{E}$, and verify that det $T \neq 0$ when and only when T is invertible. (Hint: Recall that det is a homomorphism of $\mathbb{C}_{n,n}$ onto $\mathbb{C}$ with the property that det $A = 0$ if and only if A is not invertible.)

O. If T is a linear transformation on a vector space $\mathscr{E}$, and if there exists a complex number λ and a nonzero vector x in $\mathscr{E}$ such that $Tx = \lambda x$, then λ is an *eigenvalue* of T, and x is an *eigenvector* of T *associated with* the eigenvalue λ. The set $\{y \in \mathscr{E} : Ty = \lambda y\}$ is a linear manifold in $\mathscr{E}$ called the *eigenspace associated with* λ. Show that every linear transformation on a finite dimensional vector space has an eigenvalue. (Hint: Apply the fundamental theorem of algebra to the *characteristic equation* $\det(\lambda - T) = 0$.)

P. Show that if T is a linear transformation on an n-dimensional linear space $\mathscr{E}$, then there is an ordered basis $X = \{x_1, \ldots, x_n\}$ for $\mathscr{E}$ such that the matrix $A = (\alpha_{ij})$ of T with respect to X is in *upper triangular form* (that is, $\alpha_{ij} = 0$ whenever $i > j$). (Hint: According to the preceding problem there is a basis $\{y_1, \ldots, y_n\}$ for $\mathscr{E}$ such that $Ty_1 = \lambda y_1$ for some complex number λ. Let $(\beta_{ij})_{i,j=1}^{n}$ be the matrix of T with respect to this basis, let $\mathscr{M}$ denote the linear manifold generated by the vectors $\{y_2, \ldots, y_n\}$, and let $\mathscr{L}$ denote the linear manifold consisting of the scalar multiples of y_1. The submatrix $(\beta_{ij})_{i,j=2}^{n}$ is the matrix of a linear transformation S on $\mathscr{M}$ with the property that, for every x in $\mathscr{M}$, $Tx - Sx$ belongs to $\mathscr{L}$. Use induction.) Show also that if the matrix A of T is in upper triangular form, then the diagonal entries α_{ii} of A are precisely the eigenvalues of T.

Q. Let $\mathscr{E}$, $\mathscr{F}$, and $\mathscr{G}$ be linear spaces, let $\mathscr{M}$ and $\mathscr{N}$ be linear submanifolds of $\mathscr{E}$ and $\mathscr{F}$, respectively, and let φ be a bilinear transformation of $\mathscr{M} + \mathscr{N}$ into $\mathscr{G}$. Show that φ can be extended to a bilinear transformation of $\mathscr{E} + \mathscr{F}$ into $\mathscr{G}$. (Hint: Recall Problem D.)

R. Let $\mathscr{E}$, $\mathscr{F}$, and $\mathscr{G}$ be linear spaces, let M and N be subsets of $\mathscr{E}$ and $\mathscr{F}$, respectively, and let φ_0 be a mapping of the Cartesian product $M \times N$ into $\mathscr{G}$. Show that necessary and sufficient conditions for the existence of a bilinear transformation φ of $\mathscr{E} + \mathscr{F}$ into $\mathscr{G}$ with the property that $\varphi|(M \times N) = \varphi_0$ are

(i) if $x_1, \ldots, x_n$ is any finite subset of M and if

$$\sum_{i=1}^{n} \alpha_i x_i = 0, \quad \text{then} \quad \sum_{i=1}^{n} \alpha_i \varphi_0(x_i, y) = 0$$

for every y in N, and
(ii) if $\{y_1, \ldots, y_m\}$ is any finite subset of N and if

$$\sum_{i=1}^{m} \beta_i y_i = 0, \quad \text{then} \quad \sum_{i=1}^{m} \beta_i \varphi_0(x, y_i) = 0$$

for every x in M.

(Hint: Use Hamel bases (Prob. A). Note that, according to Problem E, these conditions may be paraphrased as follows: For each y_0 in N the mapping $x \rightarrow \varphi_0(x, y_0)$ admits a linear extension to $\mathscr{E}$, and for each x_0 in M the mapping $y \rightarrow \varphi_0(x_0, y)$ admits a linear extension to $\mathscr{F}$.)

S. Let $\mathscr{E}$ be a linear space, let G be a system of algebraic generators for $\mathscr{E}$, and let ψ_0 be a mapping of $G \times G$ into $\mathbb{C}$. Show that a necessary and sufficient condition for the existence of a sesquilinear functional ψ on $\mathscr{E}$ with the property that $\psi(x, y) = \psi_0(x, y)$ for all x, y in G is that if $\{x_1, \ldots, x_n\}$ is any finite subset of G and if $\sum_{i=1}^{n} \lambda_i x_i = 0$, then $\sum_{i=1}^{n} \lambda_i \psi_0(x_i, y) = 0$ and $\sum_{i=1}^{n} \bar{\lambda}_i \psi_0(y, x_i) = 0$ for every y in G. (This condition can be paraphrased as follows: For each y_0 in G the mappings $x \rightarrow \psi_0(x, y_0)$ and $x \rightarrow \bar{\psi}_0(y_0, x)$ can be extended to linear functionals on $\mathscr{E}$. It would be easy, of course, to state a more general version of this result, along the lines of the preceding problem, but we will have no need for such generality.) Show also that ψ is uniquely determined by ψ_0 when it exists, and that ψ is symmetric if and only if ψ_0 has the property that $\psi_0(x, y) = \overline{\psi_0(y, x)}$ for all x, y in G.

T. Show that there exists a "natural" isomorphism between the algebraic tensor product $\mathbb{C}^m \dot{\times} \mathbb{C}^n$ and the space $\mathbb{C}_{m,n}$ of $m \times n$ matrices. (Hint: For each $x = (\xi_1, \ldots, \xi_m)$ in $\mathbb{C}^m$ and $y = (\eta_1, \ldots, \eta_n)$ in $\mathbb{C}^n$ let $\sigma(x, y)$ be the matrix $(\xi_i \eta_j)$ and set $\mathscr{G} = \mathbb{C}_{m,n}$ in Proposition 2.2.)

U. Let $\mathscr{E}$ and $\mathscr{F}$ be linear spaces, and let $t = \sum_{i=1}^{n} x_i \otimes y_i$ be an element of the algebraic tensor product $\mathscr{E} \dot{\times} \mathscr{F}$. Show that if the vectors x_i are linearly independent, then the vectors y_i are uniquely determined by t (and, similarly, the vectors x_i are uniquely determined by t when the vectors y_i are linearly independent). Conclude that if $x \in \mathscr{E}$ and $y \in \mathscr{F}$, then $x \otimes y = 0$ when and only when one of the vectors x or y is 0, and, likewise, that if x, x' are vectors in $\mathscr{E}$ and y, y' vectors in $\mathscr{F}$, and if $x \otimes y = x' \otimes y' \neq 0$, then there exists a scalar λ such that $x = \lambda x'$ and $y' = \lambda y$.

3 General topology

In this chapter we briefly review the basic facts about general topology that will be used in this book. We omit many of the most basic definitions, and some of the proofs. (The topic of convergence of nets, by contrast, is treated in full detail.) For definitions of terms used below without definition, and for proofs of theorems stated without proof, the reader is referred to any textbook on general topology, e.g., [40], [11], or [20]. To begin with, the terms *topology, topological space, open set, closed set, interior, closure, boundary,* and *neighborhood,* will be used without explanation, and the various relations between these notions will be assumed known. The interior of a set A in a topological space will be denoted by A°, the closure of A by A^-, and the boundary of A by ∂A.

Other terms and concepts with which the reader will be assumed to be familiar are *topological base* and *subbase* and the *first* and *second axioms of countability* (cf. Problem A). In connection with the notion of a base for a topology the following fact, whose verification is routine, is frequently useful.

Proposition 3.1. *A collection $\mathcal{B}$ of subsets of a set X forms a base for a topology on X if and only if every point of X lies in some element of $\mathcal{B}$ (briefly: if $\mathcal{B}$ covers X) and the intersection of every pair of sets belonging to $\mathcal{B}$ is a union of sets belonging to $\mathcal{B}$. (The only property that a collection $\mathcal{S}$ of subsets of X must possess in order to be a subbase for a topology on X is that it cover X.)*

Example A. The collection of all open intervals (a, b), where $a, b \in \mathbb{R}$ and $a < b$, is a base for the *usual* topology on $\mathbb{R}$. Similarly, the collection of all *open cells*, i.e., the collection of all products of open intervals

$$(a_1, b_1) \times \cdots \times (a_n, b_n)$$

is a base for the *usual* topology on $\mathbb{R}^n$. A base for the *usual* topology on the complex plane $\mathbb{C}$ is given by the set of all open discs

$$D_r(\alpha) = \{\lambda \in \mathbb{C} : |\alpha - \lambda| < r\},$$

where α is a point of $\mathbb{C}$ and r is a positive number. Similarly, the collection of all products of open discs

$$D_{r_1}(\alpha_1) \times \cdots \times D_{r_n}(\alpha_n)$$

is a base for the *usual* topology on $\mathbb{C}^n$. Since all of (the real and imaginary parts of) the parameters employed in defining these various bases may be restricted to assume only rational values, the spaces in this example all satisfy the second axiom of countability.

Example B. Let X be a simply ordered set containing at least two elements. For each a in X let

$$L_a = \{x \in X : x < a\} \quad \text{and} \quad R_a = \{x \in X : x > a\}.$$

Then the collection $\mathscr{S} = \{L_a : a \in X\} \cup \{R_a : a \in X\}$ is a subbase for a topology on X called the *order topology*. A base for the order topology is given by $\mathscr{S} \cup \{R_a \cap L_b : a, b \in X, a < b\}$. The set $R_a \cap L_b$, $a < b$, is called an *open interval* and is sometimes denoted by (a, b), just as when X is the system of real numbers. The order topology on $\mathbb{R}$ is the usual topology on $\mathbb{R}$. Another situation in which the order topology is important arises when X is a segment $W(\alpha)$ of ordinal numbers. In particular, if λ is a limit ordinal, then a base for the order topology on $W(\lambda)$ is given by the collection of sets of the form $\{\xi \in W(\lambda) : \alpha \leq \xi < \beta\}$, where $\alpha < \beta < \lambda$.

A subset A of a topological space X is itself a topological space, called a *subspace* of X, when A is given its *relative topology*. This is the topology on A consisting of all sets of the form $A \cap U$, where U is an open set in X. Whenever a subset A of a topological space is regarded as a topological space, the topology on A is always understood to be the relative topology unless the contrary is expressly stipulated.

A set A in topological space X is *dense* in X if $A^- = X$. The space X is *separable* if it possesses a countable dense subset; every topological space that satisfies the second axiom of countability is separable.

Example C. We topologize the extended real number system $\mathbb{R}^\natural$ (Ex. 1B) by giving it its order topology. Note that this has the effect of making $\mathbb{R}$ a dense open subspace of $\mathbb{R}^\natural$.

Recall that if X and Y are topological spaces, if f is a mapping of X into Y, and if $x_0 \in X$, then f is *continuous at* x_0 if for every neighborhood W of $f(x_0)$ in Y there is a neighborhood V of x_0 in X such that $f(V) \subset W$; moreover, f is *continuous* if it is continuous at every point of X. A mapping $f : X \to Y$ is continuous if and only if $f^{-1}(U)$ is open in X for every open set U in Y, or, equivalently, if and only if $f^{-1}(F)$ is closed in X for every closed set F in Y. A mapping $f : X \to Y$ with the property that $f(U)$ is open in Y whenever U is open in X is an *open mapping*. Likewise, f is a *homeomorphism* of X onto Y if f is a one-to-one mapping of X onto Y such that both f and f^{-1} are continuous, i.e., such that f is both continuous and open. Two topological spaces are *homeomorphic* if there exists a homeomorphism of one of them onto the other.

Example D. If X is any topological space, then the collection of continuous complex-valued functions on X will be denoted by $\mathscr{C}(X)$, and the collection of continuous real-valued functions on X by $\mathscr{C}_{\mathbb{R}}(X)$. It is clear that $\mathscr{C}(X)$ $[\mathscr{C}_{\mathbb{R}}(X)]$ is a unital linear algebra [real linear algebra] with respect to the usual operations of pointwise addition and multiplication. The space $\mathscr{C}(X)$ is the complexification of $\mathscr{C}_{\mathbb{R}}(X)$. If A is an arbitrary subset of X, then the set of all functions in $\mathscr{C}(X)$ that vanish on A is an ideal in $\mathscr{C}(X)$. If φ is a homeomorphism of a topological space X onto a topological space Y, then the mapping $f \to f \circ \varphi$ is an algebra isomorphism of $\mathscr{C}(Y)$ onto $\mathscr{C}(X)$ called the *isomorphism induced by* φ.

Example E. Let U denote a nonempty open subset of $\mathbb{R}^n$. If f is a complex-valued function on U, then there are many degrees of smoothness that f may possess beyond mere continuity. Recall that a function f on U is k *times continuously differentiable* on U if all of the kth order partial derivatives $\partial^k f / \partial x_1^{m_1} \cdots \partial x_n^{m_n}$ $(m_1 + \cdots + m_n = k)$ exist and are continuous on U. The collection of all k times continuously differentiable complex-valued functions on U will be denoted by $\mathscr{C}^{(k)}(U)$, the collection of real-valued functions in $\mathscr{C}^{(k)}(U)$ by $\mathscr{C}_{\mathbb{R}}^{(k)}(U)$. (To form the partial derivative $\partial f / \partial x_j$ of a complex-valued function f we simply differentiate the real and imaginary parts:

$$\frac{\partial f}{\partial x_j} = \frac{\partial \operatorname{Re} f}{\partial x_j} + i \frac{\partial \operatorname{Im} f}{\partial x_j}).$$

It is, once again, easily seen that $\mathscr{C}^{(k)}(U)$ $[\mathscr{C}_{\mathbb{R}}^{(k)}(U)]$ is a unital algebra [real algebra], and that $\mathscr{C}^{(k)}(U)$ is the complexification of $\mathscr{C}_{\mathbb{R}}^{(k)}(U)$.

A topological space X is [*countably*] *compact* if from every [countable] open covering of X it is possible to extract a finite subcovering. A subset A of a topological space X is compact if it is compact in its relative topology. Equivalently, A is compact if and only if from every open covering of A by open subsets of X it is possible to extract a finite covering. Countable compactness implies compactness in spaces satisfying the second axiom of countability.

Example F (Heine–Borel Theorem). Every closed interval $\lfloor a, b \rfloor$ in $\mathbb{R}$ is compact. Indeed, if $\mathcal{U}$ is an arbitrary open covering of $[a, b]$, if T denotes the set of those numbers t in $[a, b]$ with the property that the subinterval $[a, t]$ is covered by some finite number of sets in $\mathcal{U}$, and if $t_0 = \sup T$, then it is easy to see that $t_0 > a$ and $t_0 \in T$. But $t_0 < b$ is impossible, so $[a, b] = T$.

More generally, a subset K of $\mathbb{R}^n$ is compact if and only if K is closed and is contained in some *closed cell* $[a_1, b_1] \times \cdots \times [a_n, b_n]$. This theorem, known as the *Heine–Borel theorem*, is an easy consequence of Proposition 3.2, Theorem 3.15 below, and the just established fact that closed intervals are compact.

A topological space X is a *Hausdorff space* if every two distinct points in X have disjoint neighborhoods. A major role is played in modern analysis by compact Hausdorff spaces, and the following two propositions concerning such spaces are frequently useful. (The proofs of both propositions are elementary. Other facts of basic importance pertaining to compact Hausdorff spaces are stated in Propositions 3.4, 3.14, and Theorem 3.15.)

Proposition 3.2. *A subspace A of a compact Hausdorff space X is itself a compact Hausdorff space if and only if A is closed in X.*

Proposition 3.3. *If f is a continuous mapping of a [countably] compact topological space X onto a topological space Y, then Y is [countably] compact. If f is a continuous one-to-one mapping of a compact space X onto a Hausdorff space Y, then f is a homeomorphism.*

A Hausdorff space X is called *regular* if for every closed set F in X and every point x in $X \backslash F$ there are disjoint open sets U and V such that $x \in U$ and $F \subset V$. A Hausdorff space X is *normal* if for each pair of disjoint closed sets E and F in X there are disjoint open sets U and V such that $E \subset U$ and $F \subset V$. Equivalently, a Hausdorff space X is normal if and only if for every closed set F and open set U in X such that $F \subset U$ there exists an open set V such that $F \subset V \subset V^- \subset U$. The facts we shall need concerning normal spaces are the following. (The proof of Proposition 3.4 is quite elementary; see Problem C. Proposition 3.5 is another matter altogether; no really easy proof is known except in special cases; see Problem 4D.)

Proposition 3.4. *Every compact Hausdorff space is normal.*

Proposition 3.5 (Urysohn's Lemma). *If E and F are disjoint closed subsets of a normal space X, then there exists a continuous function $f : X \to [0, 1]$ such that f is identically zero on E and identically one on F, i.e., such that $\chi_F \leq f \leq \chi_{X \backslash E}$.*

Example G. Let X be a normal topological space, and suppose given a finite open covering $\{U_1, \ldots, U_n\}$ of X. If $F = X\backslash(U_1 \cup \cdots \cup U_{n-1})$, then F is a closed set contained in U_n, and there exists an open set V such that $F \subset V \subset V^- \subset U_n$. Using this observation and an obvious induction argument, one verifies without difficulty the following important fact: *If $\{U_1, \ldots, U_n\}$ is any finite open covering of a normal space X, then there exists a corresponding closed covering $\{F_1, \ldots, F_n\}$ of X such that $F_i \subset U_i$, $i = 1, \ldots, n$.*

Suppose now, once again, that $\{U_1, \ldots, U_n\}$ is an open covering of X, and let $\{F_1, \ldots, F_n\}$ be a closed covering of X such that $F_i \subset U_i, i = 1, \ldots, n$. By Urysohn's lemma there exists, for each $i = 1, \ldots, n$, a continuous mapping f_i of X into $[0, 1]$ with the property that $\chi_{F_i} \le f_i \le \chi_{U_i}$. If $f = f_1 + \cdots + f_n$, then $f \ge 1$ on X. Hence if we define $g_i = f_i/f$, then the continuous functions $\{g_i, \ldots, g_n\}$ satisfy the following conditions:

$$g_1 + \cdots + g_n = 1 \quad \text{and} \quad 0 \le g_i \le \chi_{U_i}, \qquad i = 1, \ldots, n.$$

Such a system of functions is customarily called a *partition of unity* on X *subordinate to* the given open covering $\{U_1, \ldots, U_n\}$.

Suppose now that F is a closed set in X and that $\{U_1, \ldots, U_n\}$ is a given open covering of F. Set $U_0 = X\backslash F$, and let $\{g_0, g_1, \ldots, g_n\}$ be a partition of unity of X subordinate to the open covering $\{U_0, U_1, \ldots, U_n\}$. Then the system of functions $\{g_1, \ldots, g_n\}$ possesses the following properties:

(i) $0 \le g_i \le \chi_{U_i}, i = 1, \ldots, n$,
(ii) If $g = g_1 + \cdots + g_n$, then $0 \le g \le 1$ on X,
(iii) $g(x) = 1$ for every x in F.

Such a system is also frequently called a *partition of unity subordinate to* the given sequence $\{U_1, \ldots, U_n\}$.

If for each point x in a Hausdorff space X and for each neighborhood V of x there exists a continuous function $f : X \to [0, 1]$ such that $f(x) = 1$ and f is identically zero on $X\backslash V$, then X is said to be *completely regular*. It is obvious that every completely regular space is regular, and, according to Urysohn's lemma, every normal space is completely regular.

A topological space X is *connected* if it is not possible to express X as the union of two disjoint nonempty open [closed] subsets. Equivalently, a topological space X is connected if and only if the only *closed-open* (i.e., both closed and open) subsets of X are $\varnothing$ and X itself. Still another formulation is the following: X is connected if $A \subset X$ and $\partial A = \varnothing$ imply $A = \varnothing$ or $A = X$. A subset of a topological space is *connected* if it is connected as a subspace. It is easily seen that if C is a connected subset of a topological space X, then C^- is also connected. Another important property of connected sets is given in the following proposition.

Proposition 3.6. *If C is connected subset of a topological space X, and if A is a subset of X such that $C \cap \partial A = \varnothing$, then either $C \subset A$ or $C \cap A = \varnothing$.*

PROOF. If $C \cap \partial A = \varnothing$ and $x \in C \cap A$, then $x \in A^\circ$. Thus $C \cap A = C \cap A^\circ$ is open relative to C. Since A and $X \backslash A$ have the same boundary, this shows that $C \backslash A$ is also open relative to C. Thus $C = (C \cap A) \cup (C \backslash A)$ expresses C as the union of two disjoint relatively open subsets, one of which must be empty. □

Example H. A collection $\mathscr{C}$ of subsets of a set X is said to be *chained* if for any two sets C_0, C_1 in $\mathscr{C}$ there is a finite sequence $\{D_0, \ldots, D_n\}$ of sets in $\mathscr{C}$ such that $D_{i-1} \cap D_i \neq \varnothing, i = 1, \ldots, n$, and such that $D_0 = C_0$ and $D_n = C_1$. Suppose given a chained collection $\mathscr{C}$ of *connected* subsets of a topological space X, let E denote the union of the sets in $\mathscr{C}$, and let A be a closed-open subset of E. Then $\partial A = \varnothing$ in the subspace E. Thus if C is a connected subset of E, then either $C \cap A = \varnothing$ or $C \subset A$. In particular, this is true of each of the sets in $\mathscr{C}$. Thus if $A \neq \varnothing$, then A contains some one of the sets in $\mathscr{C}$. But then, since $\mathscr{C}$ is chained, it is seen at once that A must contain *every* set in $\mathscr{C}$. Thus $A = E$, and we have proved that E is connected.

Example I. Every interval or ray in $\mathbb{R}$, open, closed, or half-open (including the space $\mathbb{R}$ itself), is connected. To see this we note first that if I is either an interval or a ray (or $\mathbb{R}$), then the closed subintervals of I form a chained collection of subsets of I that covers I (Ex. H). Thus it suffices to verify that every closed interval $[a, b]$ is connected. Suppose U is a relatively open subset of $[a, b]$ such that $a \in U$ and such that $V = [a, b] \backslash U$ is also relatively open in $[a, b]$. Let $T = \{t \in [a, b] : [a, t] \subset U\}$ and set $t_0 = \sup T$. Then $a < t_0 \leq b$, and the assumption that $t_0 \in V$ leads at once to a contradiction of the definition of t_0. Hence $t_0 \in U$, and it follows that t_0 is also an element of T. On the other hand, the assumption $t_0 < b$ likewise leads to a contradiction of the definition of t_0. Thus $t_0 = b$ and $U = [a, b]$.

Each point x in a topological space X is contained in a unique largest connected subset of X, called the (*connected*) *component* of x. (It is clear from Example H that the union of all the connected subsets of X that contain x is connected, so this union is the component of x.) The component of any one point is also the component of every other point it contains, and is accordingly also referred to as a (*connected*) *component* of X. The components of a topological space are closed sets, since their closures are also connected.

Proposition 3.7. *If E is an arbitrary subset of the real line $\mathbb{R}$, then the following conditions are equivalent:*

   (i) *E is connected,*
   (ii) *If $a, b \in E$, and if $a < c < b$, then $c \in E$,*
   (iii) *E is either an interval (open, closed, or half-open), or a ray (open or closed), or the entire real line $\mathbb{R}$.*

If U is an open subset of $\mathbb{R}$, then U can be expressed uniquely as a countable union of pairwise disjoint nonempty open intervals, and these intervals are the connected components of U.

PROOF. That (iii) implies (i) was proved in Example I, while the proof that (ii) implies (iii) amounts to nothing more than a careful consideration of cases. Finally, to show that (i) implies (ii) we note that $(-\infty, c) \cup (c, +\infty)$ is a disconnection of the set $\mathbb{R} \backslash \{c\}$ obtained by removing a single point c from $\mathbb{R}$. Hence if C is a connected subset of $\mathbb{R}$ that does not contain c, then either $C \subset (-\infty, c)$ or $C \subset (c, +\infty)$.

To prove the last assertion of the proposition we observe that, since U is open, every nonopen interval that is contained in U is contained in an open interval contained in U. It follows from this that the connected components of U are all open intervals. That no disjoint collection of nonempty open intervals can be uncountable follows from the fact that $\mathbb{R}$ satisfies the second axiom of countability. To complete the proof it suffices to observe that if $\{I_n\}$ is a disjoint (countable) collection of nonempty open intervals such that $U = \bigcup_n I_n$, and if V is a connected component of U that meets some one I_n, then neither endpoint of I_n can belong to V, and therefore $V = I_n$ (Prop. 3.6). $\qquad\square$

If X is a connected space and $f : X \to Y$ is a continuous mapping, then $f(X)$ is a connected set in Y. Thus, in particular, if f is a continuous mapping of a closed interval $[a, b]$ into Y, then the range of f is connected in Y. Such a mapping is called an *arc* in Y, and $[a, b]$ is the *parameter interval* of the arc f. If $f(a) = y_0$ and $f(b) = y_1$, then the arc f is said to *join* y_0 to y_1. If for every pair of points y_0 and y_1 in Y there exists an arc joining y_0 to y_1, then Y is said to be *arcwise connected*. An arcwise connected space is clearly connected.

Proposition 3.8. *If U is a connected open subset of $\mathbb{R}^n$, then U is arcwise connected.*

PROOF. We may assume U to be nonempty. Let x_0 be a point of U, and denote by V the set of those points x in U such that there exists an arc in U joining x_0 to x. If $x \in V$ and if W is an open cell in $\mathbb{R}^n$ containing x and contained in U (Ex. A), then x can clearly be joined to every point of W by an arc in W. Hence $W \subset V$, and it follows that V is an open set.

Suppose now that $y_0 \in U \cap \partial V$, and let W_1 be an open cell containing y_0 and contained in U. Then there exists a point x of V belonging to W_1, and since x can be joined to y_0 by an arc in W_1, it follows that $y_0 \in V$, which is impossible since V is open. Thus $U \cap \partial V = \varnothing$, and therefore $V = U$ by Proposition 3.6. $\qquad\square$

The structure of the most general open subset of $\mathbb{R}$ is set forth in Proposition 3.7. In the topology of the plane matters are not quite so simple, but there are some useful things that can be said. (In this discussion we speak of the complex plane $\mathbb{C}$, but the following facts are equally valid for the real plane $\mathbb{R}^2$, which may be identified with $\mathbb{C}$ via the standard homeomorphism $s + it \leftrightarrow (s, t)$.) We begin by recalling that a *domain* in $\mathbb{C}$ is a nonempty connected open subset of $\mathbb{C}$.

Proposition 3.9. *Every open subset U of $\mathbb{C}$ is uniquely expressible as a countable union of disjoint domains, and these domains are the components of U.*

PROOF. The empty set is the empty union of domains. If U is a nonempty open set in $\mathbb{C}$, and if λ_0 is an element of some component U_0 of U, then (Ex. A) there is an open disc $D_r(\lambda_0)$ about λ_0 ($r > 0$) such that $D_r(\lambda_0) \subset U$. Since discs are (arcwise) connected, it is clear that $D_r(\lambda_0) \subset U_0$, and hence that the components of U are open. That no collection of disjoint open sets in $\mathbb{C}$ can be uncountable is clear from the fact that $\mathbb{C}$ satisfies the second axiom of countability. To complete the proof it suffices to observe that if $\{U_n\}$ is a disjoint (countable) collection of nonempty domains such that $U = \bigcup_n U_n$, and if V is a connected component of U that meets some one U_n, then $V \cap \partial U_n = \varnothing$, and therefore $V = U_n$ (Prop. 3.6). $\qquad\square$

If K is a compact subset of $\mathbb{C}$, then K is *bounded*, that is, there exists a disc $D_R(0)$ large enough so that $K \subset D_R(0)^-$, and since $V = \mathbb{C}\backslash D_R(0)^-$ is (arcwise) connected, it follows that V is entirely contained in some one of the components of $\mathbb{C}\backslash K$. Hence all of the components of $\mathbb{C}\backslash K$ *except this one* are contained in $D_R(0)$, and are therefore bounded.

Definition. If K is a compact subset of $\mathbb{C}$ then there is exactly one *unbounded component* of $\mathbb{C}\backslash K$. The other components of $\mathbb{C}\backslash K$ (if any) are called the *holes* of (or in) K.

Proposition 3.10. *Let K be a compact subset of $\mathbb{C}$, and suppose L is a compact subset of $\mathbb{C}$ such that $K \subset L$ and such that $\partial L \subset K$. Then L consists of the union of K and some of the holes of K. In particular, if K has no holes, then $L = K$.*

PROOF. According to Proposition 3.6 each component of $\mathbb{C}\backslash K$ is either contained in L or disjoint from L. In particular, the unbounded component of $\mathbb{C}\backslash K$ is disjoint from L since L is bounded. $\qquad\square$

In any deep study of plane topology an important role is played by the *Jordan curve theorem*. In this connection we shall employ the following terminology.

Definition. A *Jordan loop* or *Jordan curve* is an arc γ in $\mathbb{C}$ defined on a real parameter interval $[a, b]$ ($a < b$) such that $\gamma(a) = \gamma(b)$ and such that γ is one-to-one and never equal to $\gamma(a)$ on the open interval (a, b). It is easily seen (Prop. 3.3) that the range J of a Jordan loop γ is compact and could equally well be characterized as a homeomorphic image in $\mathbb{C}$ of the unit circle. We shall also refer to the range of a Jordan loop as a *Jordan loop* or *Jordan curve* when that is convenient. In the event that a Jordan curve J is a simple polygon in $\mathbb{C}$ (as defined in elementary geometry) we say that J is a *Jordan polygon*.

A *Jordan domain* in $\mathbb{C}$ is a domain whose (entire) boundary consists of the union of a finite number of pairwise disjoint Jordan curves. The closure of a Jordan domain is, therefore, the union of the domain and the various Jordan curves constituting its boundary. Such a closed set is known as a *Jordan region*.

The following result is the central fact concerning the topology of the plane; an elementary proof can be found in [66; pp. 100–104].

Theorem 3.11 (Jordan Curve Theorem). *If J is a Jordan curve in $\mathbb{C}$, then the open set $\mathbb{C}\backslash J$ is the union of exactly two components, each of which is a Jordan domain having J for its entire boundary.*

Definition. If J is the range of a Jordan loop γ, then the bounded component of $\mathbb{C}\backslash J$ (that is, the hole in J), is called the *interior* domain of J, and is denoted by $\mathrm{Int}(\gamma)$ or $\mathrm{Int}(J)$. The unbounded component of $\mathbb{C}\backslash J$ is the *exterior* domain of J and is denoted by $\mathrm{Ext}(\gamma)$ or $\mathrm{Ext}(J)$.

Proposition 3.12. *Let U be an open subset of $\mathbb{C}$, and let K be a compact subset of U. Then there exists a finite set $\Delta_1, \ldots, \Delta_p$ of Jordan domains such that the corresponding Jordan regions $R_i = \Delta_i^-$ are pairwise disjoint, and such that*

$$K \subset \Delta_1 \cup \cdots \cup \Delta_p \quad \text{and} \quad R_1 \cup \cdots \cup R_p \subset U.$$

Moreover, it is possible to arrange matters so that each of the boundaries $\partial \Delta_i$ is the disjoint union of a finite number of disjoint Jordan polygons.

We shall have no occasion to refer to this fact until Chapter 5. At that time, in connection with some related material, we sketch a proof of Proposition 3.12. (See Problem 5K.)

Using Proposition 3.12 together with the Jordan curve theorem, it is not difficult to determine the structure of the most general Jordan domain.

Proposition 3.13. *If $J_1, \ldots, J_n$ are Jordan loops that are* mutually exterior in pairs, *that is, are so situated that $J_i \subset \mathrm{Ext}(J_j), i \neq j, i, j = 1, \ldots, n$, and if $R_i = (\mathrm{Int}(J_i))^-, i = 1, \ldots, n$, then the complement $\mathbb{C}\backslash(R_1 \cup \cdots \cup R_n)$ is an unbounded Jordan domain with boundary $J_1 \cup \cdots \cup J_n$. Conversely, every unbounded Jordan domain is of this form. If $J_0, J_1, \ldots, J_n$ are Jordan curves such that $J_1, \ldots, J_n$ are mutually exterior in pairs, and such* *that $J_i \subset \mathrm{Int}(J_0)$, $i = 1, \ldots, n$, and if $R_i = (\mathrm{Int}(J_i))^-$, $i = 1, \ldots, n$, then $\mathrm{Int}(J_0) \backslash (R_1 \cup \cdots \cup R_n)$ is a bounded Jordan domain with boundary $J_0 \cup J_1 \cup \cdots \cup J_n$. Conversely, every bounded Jordan domain is of this form.*

A topological space is *totally disconnected* if its connected components are all singletons, i.e., if its one-point subsets are the only nonempty connected subsets that it possesses. The *discrete* topology on a set X is the topology consisting of all the subsets of X; a set X together with the discrete topology on X is a *discrete (topological) space*. Such spaces are obviously totally disconnected. Here is a less trivial example of a totally disconnected space.

Example J. Let $\mathscr{F}_0$ denote the singleton $\{[0, 1]\}$, i.e., the set whose sole element is the unit interval, and, for each positive integer n, let $\mathscr{F}_n$ denote the set of closed intervals obtained by removing the middle open third of each of the intervals in the set $\mathscr{F}_{n-1}$. (Thus $\mathscr{F}_n$ is a set of disjoint closed intervals $I_{n,1}, \ldots, I_{n,2^n}$.) If for each nonnegative integer n we write

$$F_n = \cup \, \mathscr{F}_n = I_{n,1} \cup \cdots \cup I_{n,2^n},$$

then

$$C = \bigcap_{n=0}^{\infty} F_n$$

is the *Cantor set*. For each interval $I_{n,i}$ in $\mathscr{F}_n$ let us write $C_{n,i} = I_{n,i} \cap C$. The set $C_{n,i}$ is closed and is also open relative to C (for an open interval slightly larger than $I_{n,i}$ meets C in the same set). Let c be a point in C, and let a and b be any real numbers such that $a < c < b$. If n is chosen large enough so that $1/3^n < (c - a) \wedge (b - c)$, and if $I_{n,i}$ denotes that interval in $\mathscr{F}_n$ that contains c, then $C_{n,i} \subset (a, b) \cap C$. Thus the system of sets $C_{n,i}$ constitutes a base of closed-open subsets for the (relative) topology on C. From this it is clear that C is totally disconnected.

It is no accident that the Cantor set possesses a base of closed-open sets. In fact, the following result is valid (cf. Problem F).

Proposition 3.14. *If C denotes a connected component of a compact Hausdorff space X, and if U is any open set in X containing C, then there exists a closed-open set E in X such that $C \subset E \subset U$. In particular, if X is a totally disconnected compact Hausdorff space, then the closed-open subsets of X constitute a base for the topology on X.*

It is frequently useful to consider two topologies on the same set X. If $\mathscr{S}$ and $\mathscr{T}$ are two topologies on X, and if every open set in $\mathscr{S}$ is also in $\mathscr{T}$ (briefly: if $\mathscr{S} \subset \mathscr{T}$), then $\mathscr{T}$ is said to be *finer than* $\mathscr{S}$, or to *refine* $\mathscr{S}$, and $\mathscr{S}$ is said to be *coarser than* $\mathscr{T}$. This relation between topologies on X is obviously a partial ordering (as noted, it is just another name for the inclusion ordering), and in this ordering the collection of all topologies on X becomes a complete lattice (Prob. G). In particular, there is a finest topology on X and a coarsest topology. The finest topology on X is clearly the discrete topology introduced above; the coarsest topology on X is the two-element topology $\{X, \varnothing\}$, sometimes known as the *indiscrete* topology.

Example K. If X is a simply ordered set, the collection of all sets $L_a = \{x \in X : x < a\}$, $a \in X$, together with the whole space X, is a base for a topology $\mathscr{T}_l$ on X known as the *left ray topology* on X. Similarly, the collection of all sets $R_a = \{x \in X : x > a\}$, $a \in X$, together with X, is a base for a topology $\mathscr{T}_r$ on X known as the *right ray topology* on X. The infimum $\mathscr{T}_l \wedge \mathscr{T}_r = \mathscr{T}_l \cap \mathscr{T}_r$ is easily seen to coincide with the indiscrete topology on X. The supremum $\mathscr{T}_l \vee \mathscr{T}_r$ contains all rays L_a and R_a, $a \in X$, and therefore coincides with the order topology on X (assuming that X is not a singleton; see Example B).

This construction is of particular importance when $X = \mathbb{R}$. If Z denotes an arbitrary topological space and f is a real-valued function defined on Z, then f is continuous at a point z_0 of Z when $\mathbb{R}$ is equipped with the left-ray topology $\mathscr{T}_l$ if and only if the following condition is satisfied: For every positive number ε there exists a neighborhood V of z_0 such that $f(z) < f(z_0) + \varepsilon$ for every z in V. This situation is expressed by saying that f is *upper semi-continuous at* z_0. Dually, f is *lower semi-continuous* at z_0 if f is continuous at z_0 when $\mathbb{R}$ is equipped with the right-ray topology $\mathscr{T}_r$, which is equivalent to the following: For every positive number ε there exists a neighborhood V of z_0 such that $f(z) > f(z_0) - \varepsilon$ for every z in V. A real-valued function f on a topological space Z is *upper [lower] semi-continuous* if it is upper [lower] semi-continuous at every point z of Z, i.e., if f is continuous as a mapping of Z into $\mathbb{R}$ equipped with the topology $\mathscr{T}_l [\mathscr{T}_r]$. Note that f is upper [lower] semi-continuous on Z if and only if the inverse image under f of every open ray to the left [right] in $\mathbb{R}$ is open in Z, or, equivalently, if and only if the inverse image of every closed ray to the right [left] in $\mathbb{R}$ is closed in Z.

If X is a set and if f is a mapping of X into some topological space Y, then the collection of all inverse images $f^{-1}(U)$ of open sets U in Y forms a topology on X, called the topology *inversely induced by* f (or, when no confusion can result, the topology *induced* by f). The topology inversely induced by f may also be described as the coarsest topology on X making f continuous. If U is merely allowed to run over a base [subbase] for the topology on Y, then the inverse images $f^{-1}(U)$ provide a base [subbase] for the inversely induced topology on X.

Example L. Let X be a topological space, and let A be a subset of X. If we take for f the *inclusion mapping* of A into X ($f(x) = x$ for all x in A), then the topology inversely induced on A by f is the relative topology on A.

A more sophisticated version of the above construction goes as follows.

Definition. Let X be a set, and for each γ in an index set Γ let f_γ be a mapping of X into a topological space Y_γ. Then the coarsest topology on X making all of the mappings f_γ continuous is the topology *inversely induced* on X by the family $\{f_\gamma\}_{\gamma\in\Gamma}$. (That such a topology always exists is an immediate consequence of Problem G. When no confusion is possible, this topology will simply be said to be *induced* by the family $\{f_\gamma\}$.)

If $\{f_\gamma\}$ is a family of mappings of a set X into a family $\{Y_\gamma\}$ of topological spaces, as in the foregoing definition, and if $\mathscr{T}$ denotes the topology induced on X by the family $\{f_\gamma\}$, then it is clear that a subbase for $\mathscr{T}$ is given by the collection of all sets of the form $f_\gamma^{-1}(U)$, where U is an open set in Y_γ and γ runs through Γ. (More generally, if, for each index γ, $\mathscr{S}_\gamma$ is a subbase for the topology on Y_γ, then the set of all inverse images $f_\gamma^{-1}(W)$, $W\in\mathscr{S}_\gamma$, $\gamma\in\Gamma$, constitutes a subbase for $\mathscr{T}$.) Consequently a *base* for $\mathscr{T}$ is given by the collection of all sets of the form

$$f_{\gamma_1}^{-1}(U_1)\cap\cdots\cap f_{\gamma_n}^{-1}(U_n) \tag{1}$$

where each U_i is an open set in Y_{γ_i} (which may be required to belong to a specified base for Y_{γ_i}) and $\{\gamma_1,\ldots,\gamma_n\}$ is an arbitrary finite subset of Γ.

Example M. Let $\{\mathscr{T}_\gamma\}_{\gamma\in\Gamma}$ be an indexed family of topologies on a set X, and for each index γ let Y_γ and f_γ denote, respectively, the topological space consisting of X equipped with the topology $\mathscr{T}_\gamma$, and the identity mapping on X regarded as a mapping of X onto Y_γ. Then the topology inversely induced on X by the family $\{f_\gamma\}_{\gamma\in\Gamma}$ coincides with the supremum $\mathscr{T}$ of the given family of topologies (Prob. G). Consequently a base for $\mathscr{T}$ is given by the collection of all sets of the form $U_1\cap\cdots\cap U_n$, where U_i belongs to $\mathscr{T}_{\gamma_i}$, $i=1,\ldots,n$, and $\{\gamma_1,\ldots,\gamma_n\}$ denotes an arbitrary finite subset of Γ. (Each U_i may also be required to belong to a prescribed base for $\mathscr{T}_{\gamma_i}$.)

Example N. If $\{Y_\gamma\}_{\gamma\in\Gamma}$ is an arbitrary indexed family of topological spaces, and if $X=\prod_\gamma Y_\gamma$, then the *product topology* on X is the topology inversely induced by the family $\{\pi_\gamma\}$ of projections. Thus a base for the product topology on X is given by the collection of all products $\prod_\gamma U_\gamma$, where U_γ is an open subset of Y_γ for all indices γ, and where $U_\gamma=Y_\gamma$ except for a finite number of indices. In the event that the index set Γ is $\{1,\ldots,n\}$, it is customary to write $Y_1\times\cdots\times Y_n$ for the product of the topological spaces $\{Y_i\}_{i=1}^n$ equipped with the product topology. (A base for this topology is given by the collection of all products $U_1\times\cdots\times U_n$, where U_i is an open subset of Y_i, $i=1,\ldots,n$.) Similarly, if X and Y are topological spaces, their product is denoted by $X\times Y$ and is understood to be equipped with the product topology. Indeed, whenever a product of topological spaces is regarded as a topological space, it is the product topology that is understood to be in use unless the contrary is expressly stipulated.

Concerning products of topological spaces the following fact is of primary importance. (See Problem T for a sketch of a proof.)

Theorem 3.15 (Tihonov's Theorem). *An arbitrary product of compact Hausdorff spaces is a compact Hausdorff space.*

In the more classical parts of general topology, in particular in the study of metric spaces, a very considerable role is played by infinite sequences. Unfortunately sequences cannot play this central role in an arbitrary topological space. It turns out, however, that the concept of a *net*, defined in Chapter 1, serves as a natural generalization of the concept of an infinite sequence. The following discussion does not assume any prior knowledge of nets on the part of the reader other than a familiarity with the relevant material in Chapter 1 (see p. 5ff.).

Definition. If X is a topological space, then a net $\{x_\lambda\}_{\lambda \in \Lambda}$ in X is said to *converge* to a point x_0 in X and x_0 is said to be the *limit* of the net $\{x_\lambda\}$ (notation: $\lim_\lambda x_\lambda = x_0$ or $x_\lambda \to x_0$) if for every neighborhood V of x_0 there exists an index λ_V, depending on V, such that $x_\lambda \in V$ for every λ in Λ such that $\lambda \geq \lambda_V$.

Since the system $\mathbb{N}$ of positive integers and the system $\mathbb{N}_0$ of nonnegative integers are directed sets in the natural ordering, infinite sequences may be viewed as a very special, albeit very important, kind of net. Thus the foregoing definition serves to define convergence for an infinite sequence of points in an arbitrary topological space.

Example O. Let X be a simply ordered set containing at least two elements, and let $\{x_\lambda\}_{\lambda \in \Lambda}$ be a monotone increasing net in X. Then $\{x_\lambda\}$ converges in the order topology on X (Ex. B) if and only if $\sup_{\lambda \in \Lambda} x_\lambda$ exists, and, if this supremum does exist, then

$$\lim_\lambda x_\lambda = \sup_\lambda x_\lambda.$$

In particular, a monotone increasing net of real numbers is convergent in $\mathbb{R}$ if and only if it is bounded above in $\mathbb{R}$. Similarly, an arbitrary monotone increasing net of extended real numbers is convergent in $\mathbb{R}^\sharp$. (Analogous remarks apply to decreasing nets.)

Example P. If a and b are real numbers with $a < b$, then the open interval (a, b) is a directed set in the natural ordering of the real numbers. If f is a mapping of (a, b) into a topological space X, and if f converges as a net indexed by the directed set (a, b), then the limit is the *limit of f as t tends to b from below* and is denoted by $\lim_{t \uparrow b} f(t)$ or by $f(b-)$. Similarly, if f converges as a net indexed by the set (a, b) in the *inverse ordering* of $\mathbb{R}$, then the limit is known as the *limit of f as t tends to a from above* and is denoted by $\lim_{t \downarrow a} f(t)$ or by $f(a+)$. (The inverse ordering of $\mathbb{R}$ is the ordering $\prec$ defined by setting $s \prec t$ when and only when $t \leq s$. It is clear that an open interval is directed in this new ordering as well as in the natural ordering.)

Example Q. Let $\{\lambda_\gamma\}_{\gamma \in \Gamma}$ be an indexed family of complex numbers, and let $\{\sigma_D\}_{D \in \mathscr{D}}$ denote the net of finite sums of the given family $\{\lambda_\gamma\}$ (Prob. 1M). If the net $\{\sigma_D\}_{D \in \mathscr{D}}$ is convergent in $\mathbb{C}$ to a limit σ, then the family $\{\lambda_\gamma\}$ is said to be *summable*, σ is called the *sum* of the indexed family $\{\lambda_\gamma\}$, and we write

$$\sigma = \sum_{\gamma \in \Gamma} \lambda_\gamma.$$

Similarly, if $\{t_\gamma\}_{\gamma \in \Gamma}$ is an indexed family of extended real numbers such that $-\infty < t_\gamma \leq +\infty$ $[-\infty \leq t_\gamma < +\infty]$ for every index γ (Ex. 1B), and if the net $\{s_D\}_{D \in \mathscr{D}}$ of finite sums converges to a limit s in the topological space $\mathbb{R}^\sharp$ (Ex. C), then s is called the *sum* of the indexed family $\{t_\gamma\}$ and we write

$$s = \sum_{\gamma \in \Gamma} t_\gamma.$$

(Note that when $s = \pm\infty$ we do *not* say that the family $\{t_\gamma\}$ is summable.) An indexed family $\{t_\gamma\}$ of nonnegative real numbers is summable in $\mathbb{R}$ if and only if the net of finite sums is bounded above in $\mathbb{R}$, and always possesses a sum in $\mathbb{R}^\sharp$ (Ex. O). An indexed family $\{t_\gamma\}_{\gamma \in \Gamma}$ of nonnegative extended real numbers satisfies the condition

$$\sum_{\gamma \in \Gamma} t_\gamma = +\infty$$

if and only if either some $t_\gamma = +\infty$ or the net of finite sums fails to be bounded above in $\mathbb{R}$.

In the event that the index family is $\mathbb{N}$ (or $\mathbb{N}_0$), i.e., when we start with a *sequence* $\{\lambda_n\}_{n=1}^\infty$ of complex numbers, care must be taken to distinguish between the indexed sum $\sum_{n \in \mathbb{N}} \lambda_n$ and the sum of the infinite series $\sum_{n=1}^\infty \lambda_n$. Indeed, it is readily verified that the sequence $\{\lambda_n\}_{n=1}^\infty$ is summable as an indexed family with sum σ if and only if the infinite series $\sum_{n=1}^\infty \lambda_n$ converges unconditionally to σ. (Recall that an infinite series $\sum_{n=1}^\infty \lambda_n$ of complex numbers is unconditionally convergent if it converges to a sum σ that is unchanged by permuting the sequence $\{\lambda_n\}_{n=1}^\infty$ and that a series is unconditionally convergent if and only if it is absolutely convergent.) In the same context, if $\{t_n\}_{n=1}^\infty$ is a sequence of extended real numbers such that $-\infty < t_n \leq +\infty$ $[-\infty \leq t_n < +\infty]$ for every n, then the sequence $\{t_n\}_{n=1}^\infty$ has sum s as an indexed family in $\mathbb{R}^\sharp$ if and only if the series $\sum_{n=1}^\infty t_n$ converges unconditionally to s, i.e., possesses a sum s that is unchanged by permuting the sequence $\{t_n\}_{n=1}^\infty$. (We define s to be the sum of an infinite series $\sum_{n=1}^\infty t_n$ of extended real numbers if the sequence $\{\sum_{i=1}^n t_i\}_{n=1}^\infty$ of partial sums converges to s in $\mathbb{R}^\sharp$.) Like observations apply, of course, when the index family is $\mathbb{N}_0$.

One unpleasant property that a topological space X may have is that a net in X can converge to more than one point. In fact, if X is not a Hausdorff space, then there always exist nets in X that converge to two different points (Prob. K). Fortunately, all of the topological spaces that play any significant role in this book are Hausdorff, and in such spaces this pathology cannot occur.

Proposition 3.16. *If X is a Hausdorff space, then no net in X converges to more than one point.*

PROOF. If x and y are distinct points of X, then there are neighborhoods U and V of x and y, respectively, such that U and V are disjoint. Now suppose that $\{z_\lambda\}_{\lambda \in \Lambda}$ is a net in X that converges to both x and y. Then there exist indices λ_1 and λ_2 in Λ such that $z_\lambda \in U$ for all $\lambda \geq \lambda_1$ and such that $z_\lambda \in V$ for all $\lambda \geq \lambda_2$. Let λ_0 be an index in Λ such that $\lambda_0 \geq \lambda_1, \lambda_2$. Then $z_{\lambda_0} \in U \cap V$, which is impossible. $\qquad\square$

The following propositions, valid in the most general topological space, characterize closure and continuity in terms of nets.

Proposition 3.17. *A point x_0 in a topological space X belongs to the closure of a subset M of X if and only if there exists a net in M that converges to x_0. Consequently a set U in X is open if and only if no net in $X\backslash U$ converges to a limit lying in U.*

Proposition 3.18. *A mapping f of a topological space X into a topological space Y is continuous at a point x_0 of X if and only if, for every net $\{x_\lambda\}_{\lambda \in \Lambda}$ in X converging to x_0, the net $\{f(x_\lambda)\}_{\lambda \in \Lambda}$ converges in Y to $f(x_0)$.*

As has been noted, the preceding results are generalizations of familiar facts about sequences in metric spaces. It is worth pointing out that the possibility of describing closure and continuity in a topological space by means of the special nets that are sequences has nothing to do with the metrizability of that space, but rather with the first axiom of countability (Prob. I).

The proofs of Propositions 3.17 and 3.18 depend upon the following considerations: If the set $\mathcal{N}(x)$ consisting of all the neighborhoods of any one point x in a topological space X is ordered by the *inverse inclusion ordering*, in other words if we write $V \leq V'$ whenever V and V' are neighborhoods of x such that $V \supset V'$, then $\mathcal{N}(x)$ becomes a directed set. (Indeed, if V_1 and V_2 belong to $\mathcal{N}(x)$, then $V_1 \cap V_2$ is a neighborhood of x that is contained in both V_1 and V_2.) Whenever $\mathcal{N}(x)$ is regarded as a directed set, it is this ordering that is understood.

PROOF OF PROPOSITION 3.17. Suppose first that there exists a net $\{x_\lambda\}$ in M that converges to x_0. Then every neighborhood of x_0 contains elements of that net, and therefore meets M. Thus x_0 is in the closure of M. On the other hand, if x_0 is in M^-, then for each neighborhood V of x in $\mathcal{N}(x)$ there exists a point $x = x_V$ such that x_V belongs to $V \cap M$. The indexed family $\{x_V\}_{V \in \mathcal{N}(x)}$ thus obtained is a net lying in M and directed by $\mathcal{N}(x)$. Since this net clearly converges to x_0, the proof is complete. $\square$

PROOF OF PROPOSITION 3.18. Suppose first that f is continuous at x_0, and let $\{x_\lambda\}$ be a net in X that converges to x_0. For any neighborhood V of $f(x_0)$ in Y, there exist a neighborhood W of x_0 such that $f(W) \subset V$ and an index λ_W such that $x_\lambda \in W$ for all $\lambda \geq \lambda_W$. Then, of course, $f(x_\lambda) \in V$ for all $\lambda \geq \lambda_W$, and the net $\{f(x_\lambda)\}$ converges to $f(x_0)$. On the other hand, if f is discontinuous at x_0, then there exists a neighborhood V of $f(x_0)$ such that no member W of the directed set $\mathcal{N}(x)$ satisfies the condition $f(W) \subset V$. Hence for every W in $\mathcal{N}(x)$ there exists a point x_W in W such that $f(x_W) \notin V$. Clearly the net $\{x_W\}_{W \in \mathcal{N}(x)}$ thus obtained converges to x_0, but equally clearly the net $\{f(x_W)\}_{W \in \mathcal{N}(x)}$ does not converge to $f(x_0)$. $\square$

Corollary 3.19. *If X and Y are topological spaces and f is a one-to-one mapping of X onto Y, then f is a homeomorphism of X onto Y if and only if an arbitrary net $\{x_\lambda\}$ in X converges to a limit x_0 when and only when the net $\{f(x_\lambda)\}$ converges in Y to the limit $f(x_0)$. In particular, if $\mathcal{S}$ and $\mathcal{T}$ are two topologies on the same set X, then $\mathcal{S} = \mathcal{T}$ if and only if an arbitrary net $\{x_\lambda\}$ in X converges to a limit x_0 with respect to the topology $\mathcal{S}$ when and only when $\{x_\lambda\}$ converges to x_0 with respect to the topology $\mathcal{T}$.*

The following results are also frequently useful.

Proposition 3.20. *If $\{f_\gamma\}_{\gamma \in \Gamma}$ is a family of mappings $f_\gamma: X \to Y_\gamma$ of a set X into topological spaces Y_γ, then a net $\{x_\lambda\}$ in X converges to a limit x_0 in the topology inversely induced on X by the given family of mappings if and only if the net $\{f_\gamma(x_\lambda)\}$ converges in Y_γ to $f_\gamma(x_0)$ for every index γ. In particular, a net $\{x_\lambda\}$ in the product $\prod_\gamma Y_\gamma$ of an indexed family of topological spaces converges to a limit x_0 if and only if it converges to x_0 "coordinatewise," i.e., if and only if $\{\pi_\gamma(x_\lambda)\}$ converges to $\pi_\gamma(x_0)$ for every index γ. Similarly, if $\{\mathcal{T}_\gamma\}$ is an indexed family of topologies on a set X, then a net $\{x_\lambda\}$ in X converges to a limit x_0 with respect to the supremum $\sup_\gamma \mathcal{T}_\gamma$ of the family $\{\mathcal{T}_\gamma\}$ if and only if $\{x_\lambda\}$ tends to x_0 with respect to each of the topologies $\mathcal{T}_\gamma$.*

PROOF. Since the mappings f_γ are continuous on X with respect to the topology they induce, it follows from Proposition 3.18 that $\lim_\lambda x_\lambda = x_0$ implies $\lim_\lambda f_\gamma(x_\lambda) = f_\gamma(x_0)$ for every γ. Suppose, in the other direction, that the latter condition is satisfied, and let V be a neighborhood of x_0 in the topology inversely induced by the family $\{f_\gamma\}$. Since sets of the form (1) constitute a base for this topology, there exist indices $\gamma_1, \ldots, \gamma_n$ and open subsets U_i of Y_{γ_i}, $i = 1, \ldots, n$, such that

$$x_0 \in f_{\gamma_1}^{-1}(U_1) \cap \cdots \cap f_{\gamma_n}^{-1}(U_n) \subset V.$$

Since $f_{\gamma_i}(x_0) \in U_i$ and $f_{\gamma_i}(x_\lambda) \to f_{\gamma_i}(x_0)$, $i = 1, \ldots, n$, there exist indices λ_i such that $f_{\gamma_i}(x_\lambda) \in U_i$ for all $\lambda \geq \lambda_i$, $i = 1, \ldots, n$. If λ_0 is an index such that $\lambda_i \leq \lambda_0$, $i = 1, \ldots, n$, and if $\lambda \geq \lambda_0$, then $f_{\gamma_i}(x_\lambda) \in U_i$ for every $i = 1, \ldots, n$. Therefore $x_\lambda \in V$ for all $\lambda \geq \lambda_0$. The last two assertions of the proposition follow by virtue of Examples M and N. $\qquad\square$

Corollary 3.21. *If $\{f_\gamma\}_{\gamma \in \Gamma}$ is an indexed family of mappings $f_\gamma : X \to Y_\gamma$ of a set X into topological spaces Y_γ, and if g is a mapping of a topological space Z into X, then g is continuous with respect to the topology inversely induced on X by the family $\{f_\gamma\}$ if and only if $f_\gamma \circ g$ is continuous for each index γ. In particular, a mapping g of a topological space Z into a product space $\prod_\gamma Y_\gamma$ is continuous if and only if $\pi_\gamma \circ g$ is continuous for each projection π_γ. Similarly, if $\{\mathcal{T}_\gamma\}$ is an indexed family of topologies on a set X, then a mapping g of a topological space Z into X is continuous with respect to the topology $\sup_\gamma \mathcal{T}_\gamma$ if and only if g is continuous with respect to each of the topologies $\mathcal{T}_\gamma$, $\gamma \in \Gamma$.*

PROOF. It suffices to prove the first assertion of the corollary. The condition is clearly necessary since the mappings f_γ are all continuous with respect to the topology they induce. To verify the sufficiency of the stated criterion, suppose that all of the compositions $f_\gamma \circ g$ are continuous on Z, and let $\{z_\lambda\}$ be a net in Z converging to a limit z_0. Then $\{f_\gamma(g(z_\lambda))\}$ converges in Y_γ to the limit $f_\gamma(g(z_0))$, and it follows by Proposition 3.20 that $\{g(z_\lambda)\}$ converges to $g(z_0)$ in X with respect to the topology inversely induced by the family $\{f_\gamma\}$. Hence g is continuous by virtue of Proposition 3.18. $\qquad\square$

Example R. Let X be a set, let $\{f_\gamma\}_{\gamma \in \Gamma}$ be a family of mappings $f_\gamma : X \to Y_\gamma$ of X into topological spaces Y_γ, and suppose that for any pair of distinct points x_1, x_2 in X there is some mapping f_γ such that $f_\gamma(x_1) \neq f_\gamma(x_2)$ (such a family of mappings is said to be *separating* on X). If we form the product $\Pi = \prod_\gamma Y_\gamma$ and define

$$F(x) = \{f_\gamma(x)\}_{\gamma \in \Gamma}, \; x \in X,$$

then F is a one-to-one mapping of X into Π. Moreover, if x_0 is a point of X and $\{x_\lambda\}$ a net in X, then, according to Proposition 3.20, $\lim_\lambda x_\lambda = x_0$ in the topology inversely induced by the family $\{f_\gamma\}$ if and only if $\lim_\lambda f_\gamma(x_\lambda) = f_\gamma(x_0)$ for every index γ. On the other hand, the net $\{F(x_\lambda)\}$ tends to $F(x_0)$ if and only if the *very same condition* is satisfied. Hence, by Corollary 3.19, F is a homeomorphism of X in the topology induced by the family $\{f_\gamma\}$ onto the subspace $F(X)$ of Π. Thus we have established the following important fact: Let X be a topological space, and let $\{f_\gamma\}$ be an arbitrary separating family of continuous mappings of X into topological spaces Y_γ such that the topology induced on X by the family $\{f_\gamma\}$ coincides with the given topology on X. Then the mapping

$$F(x) = \{f_\gamma(x)\}_{\gamma \in \Gamma}, \; x \in X,$$

of X into $\Pi = \prod_\gamma Y_\gamma$ is a homeomorphism of X onto the subspace $F(X)$. (Such a mapping is called a *topological embedding* of X in Π.)

Example S. A very important application of the foregoing construction arises when X is a completely regular space. In this case let $\mathscr{F}$ denote the collection of all continuous mappings of X into the unit interval $[0, 1]$, and note that the complete regularity of X ensures that the topology inversely induced on X by $\mathscr{F}$ coincides with the given topology. Moreover, the fact that X is Hausdorff assures that $\mathscr{F}$ is separating on X. Thus we obtain the following result (Tihonov): If we write

$$F(x) = \{f(x)\}_{f \in \mathscr{F}}, x \in X,$$

then F is a topological embedding of X in $\Pi = \prod_f Y_f$, where $Y_f = [0, 1]$ for every f in $\mathscr{F}$. Since Π is a compact Hausdorff space (Th. 3.15), this construction shows that an arbitrary completely regular space can be topologically embedded in a compact Hausdorff space. Conversely, it is clear that every subspace of a compact Hausdorff space is completely regular (Prop. 3.4).

PROBLEMS

A. (i) The *weight* of a topological space X is the minimum cardinal number that a base for the topology on X can have. (Thus the second axiom of countability requires the weight of a space to be a countable cardinal number.) Show that if the weight of X is an infinite cardinal number c and if $\mathscr{B}$ is an arbitrary base for X, then $\mathscr{B}$ contains a base $\mathscr{B}_0$ having cardinal number c. (Hint: Recall Problem 1T.)

(ii) If X is a topological space and x is a point of X, then a *neighborhood base at x* is a collection $\mathscr{V}$ of neighborhoods of x with the property that every neighborhood of x contains some one of the neighborhoods in $\mathscr{V}$. (Briefly: $\mathscr{V}$ is a cofinal subset of $\mathscr{N}(x)$ (Prob. 1P) in the inverse-inclusion ordering.) By the *weight of X at x* is meant the smallest cardinal number a neighborhood base at x can have. (Thus, the first axiom of countability requires that the weight of X be countable at each point of X.) Show that if the weight of X at x is c, and if $\mathscr{V}$ is an arbitrary neighborhood base at x, then $\mathscr{V}$ contains a neighborhood base $\mathscr{V}_0$ at x having cardinal number c.

B. A point x in a topological space X is an *accumulation point* of a subset A of X if x belongs to the closure $(A \backslash \{x\})^-$. (In a space in which all singletons are closed, in particular, in an arbitrary Hausdorff space, it comes to the same thing to require that $V \cap A$ be an infinite set for every neighborhood V of x.) The set A^* consisting of all the points of accumulation of A is called the *derived set* of A.

(i) Show that $A^- = A \cup A^*$ and verify that A^* is closed if X is a space in which all singletons are closed sets.

(ii) A subset D of a topological space X is said to be *dense in itself* if $D \subset D^*$, and to be *perfect* if D is both dense in itself and closed. (Thus, if X is a Hausdorff space, D is a perfect set if and only if $D = D^*$.) An example of a set that is dense in itself is the set Q of rational numbers in the space $\mathbb{R}$. Examples of perfect sets in $\mathbb{R}$ are the closed intervals $[a, b]$, $a < b$. Show that the Cantor set (Ex. J) is perfect.

(iii) A point x in a topological space X is a *condensation point* of a set $A \subset X$ if the intersection $V \cap A$ is *uncountable* for every neighborhood V of x. Show that if X is a Hausdorff space satisfying the second axiom of countability, and if F is a closed subset of X, then F can be expressed as the (disjoint) union of two sets A and B, where A is a countable set while B is a (perfect) set having the property that every point of B is a condensation point of B. In particular, every closed set in $\mathbb{R}^n$ can be so expressed. (Hint: Set B equal to the set of all condensation points of F.)

C. Let K be a compact subset of a Hausdorff space X, and let y be a point of X such that $y \notin K$. Show that there exist disjoint open sets U and V in X such that $K \subset U$ and $y \in V$. (Hint: For each point x of K there exist disjoint open sets U_x and V_x such that $x \in U_x$ and $y \in V_x$, and some finite collection of the sets U_x suffices to cover K.) Use this fact to prove Propositions 3.2 and 3.4.

D. (i) It is clear that an ordinal number segment $W(\beta)$ has a greatest element if and only if β is of the form $\alpha + 1$, in which case $W(\beta) = W(\alpha) \cup \{\alpha\}$. Show that an ordinal number segment $W(\beta)$, $\beta > 0$, is compact in the order topology if and only if $W(\beta)$ contains a greatest element, or, equivalently, if and only if β is not a limit ordinal. Use this fact to show that if Ω denotes the smallest noncountable ordinal number, then $W = W(\Omega)$ is countably compact but not compact in the order topology (Ex. B). (Hint: To show that W is countably compact, let $\{U_n\}_{n=1}^{\infty}$ be a sequence of open sets in W that covers W, and for each countable ordinal number α let $k(\alpha)$ denote the smallest positive integer k such that $\hat{W}(\alpha) = \{\xi \in W : \xi \le \alpha\}$ is covered by $\{U_1, \ldots, U_k\}$ (such a positive integer exists since $\hat{W}(\alpha)$ is compact by the first part of the problem); use Problem 1W.)

(ii) It follows at once from (i) that the topological space W consisting of the ordinal number segment $W(\Omega)$ in the order topology does not satisfy the second axiom of countability. Show that, in fact, if Y is an arbitrary Hausdorff space satisfying the second axiom of countability, and Φ is any continuous mapping of W into Y, then $\Phi(W)$ is a countable, compact subset of Y, and there exists a unique point y_0 in Y and an ordinal number α in W such that $\Phi(\xi) = y_0$ identically on the tail $W \backslash W(\alpha)$. (Hint: Show first that $\Phi(W)$ is compact. Then consider the set of those points y in Y with the property that the set $C = \{\xi \in W : \Phi(\xi) \in V\}$ is uncountable for every neighborhood V of y. If y_1 and y_2 are points of C, then there exist increasing sequences $\{\xi_n\}_{n=1}^{\infty}$ and $\{\eta_n\}_{n=1}^{\infty}$ in W such that

$$\lim_n \Phi(\xi_n) = y_1, \qquad \lim_n \Phi(\eta_n) = y_2,$$

and such that

$$\xi_1 < \eta_1 < \xi_2 < \eta_2 < \cdots .)$$

E. Every real number t, $0 \le t \le 1$, has an expression of the form

$$t = 0. \, \varepsilon_1 \varepsilon_2 \cdots \varepsilon_n \cdots \tag{2}$$

where each of the digits $\varepsilon_1, \varepsilon_2, \ldots$ is either 0, 1, or 2, and where the symbol in (2) represents the number

$$\sum_{j=1}^{\infty} \frac{\varepsilon_j}{3^j}.$$

(The *ternary expansion* (2) is not always unique since each *triadic fraction* $r = m/3^n$, $0 < r < 1$, possesses two ternary expansions, one ending in a sequence of zeros, the other in a sequence of twos. Every other real number in $[0, 1]$ has a unique ternary expansion.) Verify that the set F_1 in Example J consists precisely of those real numbers t in $[0, 1]$ such that $\varepsilon_1 \neq 1$ in the representation (2). (If t is a triadic fraction and therefore has two ternary expansions, then $t \in F_1$ if $\varepsilon_1 \neq 1$ in *either* of its expansions.) Pursue this line of reasoning to show that the Cantor set coincides with the set of all those numbers in the unit interval possessing a ternary expansion in which no digit ε_n is equal to one. (Such expansions are sometimes said to be "one-free.")

F. Let x be a point in a compact Hausdorff space X, and let D denote the intersection of all of the closed-open subsets of X that contain x. Show that D coincides with the connected component of x, and use this fact to prove Proposition 3.14. (Hint: The hard part is to show that D is connected. Suppose that $D = D_1 \cup D_2$, where D_1 and D_2 are closed and disjoint. Let U_1 and U_2 be disjoint open sets such that $D_i \subset U_i$, $i = 1, 2$ (Prop. 3.4). There exists a closed-open set E such that $D \subset E \subset U_1 \cup U_2$.)

G. Show that the intersection of an arbitrary nonempty collection of topologies on a set X is again a topology on X. Use this fact to prove that the collection of all topologies on X is a complete lattice in the inclusion ordering. Show, in other words, that for an arbitrarily given family $\{\mathscr{T}_\gamma\}_{\gamma \in \Gamma}$ of topologies on X there is a coarsest topology on X that refines each $\mathscr{T}_\gamma$ (the supremum $\sup_\gamma \mathscr{T}_\gamma$ of the given family) and a finest topology on X that is refined by each $\mathscr{T}_\gamma$ (the infimum $\inf_\gamma \mathscr{T}_\gamma$ of the given family). (The supremum of the collection of all topologies on X, that is, the finest topology on X, coincides with the power class on X and is called the discrete topology; the supremum of the empty collection of topologies on X, that is, the coarsest topology on X, consists of the two sets $\varnothing$ and X, and is called the indiscrete topology.)

H. Suppose that X is a topological space and that $\sim$ is an equivalence relation on X. Let the *quotient space* of X modulo $\sim$, that is, the family of equivalence classes determined by $\sim$, be denoted by $X/\!\sim$, let π be the projection of X onto $X/\!\sim$, and define a subset U of $X/\!\sim$ to be *open* if $\pi^{-1}(U)$ is open in X. The collection of open sets so obtained forms a topology on $X/\!\sim$ called the *quotient topology*. The mapping π of X onto the space $X/\!\sim$ in the quotient topology is continuous. Let f be a mapping of X into a topological space Y that respects the equivalence relation $\sim$, that is, such that $x \sim y$ implies $f(x) = f(y)$. Show that f can be factored through $X/\!\sim$, so that $f = \tilde{f} \circ \pi$, where $\tilde{f}$ maps $X/\!\sim$ into Y. Show also that if $X/\!\sim$ is equipped with the quotient topology, then f is continuous when and only when $\tilde{f}$ is.

I. Let X be a topological space satisfying the first axiom of countability. Verify that a point x is in the closure of a subset A of X if and only if there exists a sequence in A that converges to x. Show likewise that if X also has the property that all singletons are closed sets, in particular if X is Hausdorff, then x is an accumulation point of A (Prob. B) if and only if there exists a sequence of *distinct* points in A that converges to x. Show, finally, that Proposition 3.18 is valid, with sequences replacing nets, for mappings between topological spaces that satisfy the first axiom of countability.

J. It is well known that an infinite sequence $\{x_n\}$ in a topological space X converges to a limit y if and only if every subsequence of $\{x_n\}$ possesses a subsequence that converges to y. (Proof?) Show that it is also the case that a net $\{x_\lambda\}_{\lambda \in \Lambda}$ in a topological space X converges to a limit y if and only if every subnet of $\{x_\lambda\}$ possesses a subnet that converges to y.

K. Suppose that X is a topological space that is not Hausdorff. Show that there exists a net in X that converges to two distinct points of X. (Hint: Let x and y be distinct points of X that do not possess disjoint neighborhoods, and turn the product $\mathcal{N}(x) \times \mathcal{N}(y)$ into a directed set.)

L. Every convergent sequence in $\mathbb{C}$ is bounded. Give an example of a net in $\mathbb{C}$ that converges but is not bounded.

M. Let a and b be real numbers, $a < b$, and let f be a monotone real-valued function defined on $[a, b]$. Show that $f(c-) = \lim_{t \uparrow c} f(t)$ exists for every $a < c \leq b$, and likewise that $f(c+) = \lim_{t \downarrow c} f(t)$ exists for every $a \leq c < b$ (see Example P). Thus f is continuous except for "jump" discontinuities. At each point $t, a < t < b$, let us define the *jump* in f to be

$$\delta_t = |f(t+) - f(t-)|, \tag{3}$$

and set $\delta_a = |f(a+) - f(a)|$, $\delta_b = |f(b) - f(b-)|$. Show that

$$\sum_{a \leq t \leq b} \delta_t \leq |f(b) - f(a)|.$$

In the same vein verify that if f is an arbitrary function of bounded variation on $[a, b]$ (Prob. 1I), then the one-sided limits $f(t+)$ and $f(t-)$ exist wherever each is defined, so that f is continuous except for jump discontinuities. Moreover, if δ_t is defined as in (3) (and we set $\delta_a = |f(a+) - f(a)|$, $\delta_b = |f(b) - f(b-)|$), then $\sum_{a \leq t \leq b} \delta_t$ is dominated by the total variation of f over $[a, b]$.

N. Let A be a subset of a topological space X, and let f be a mapping of A into a second topological space Y. If x_0 is a point of A^- and y_0 is a point of Y, then we say that f has *limit* y_0 *as* x *tends to* x_0 *along* A (notation: $y_0 = \lim_{\substack{x \to x_0 \\ x \in A}} f(x)$) if for every neighborhood W of y_0 in Y there is a neighborhood V of x_0 in X such that $f(A \cap V) \subset W$.

(i) Show that $\lim_{\substack{x \to x_0 \\ x \in A}} f(x) = y_0$ if and only if $f(x_\lambda) \to y_0$ for every net $\{x_\lambda\}$ in A such that $x_\lambda \to x_0$ (Hint: Cf. the proof of Proposition 3.17.)

(ii) Let $\tilde{A}$ denote the subset of A^- consisting of those points $\tilde{x}$ at which the limit of f exists as x tends to $\tilde{x}$ along A, and for each $\tilde{x}$ in $\tilde{A}$ write

$$\tilde{f}(\tilde{x}) = \lim_{\substack{x \to \tilde{x} \\ x \in A}} f(x).$$

Show that if the space Y is regular, then $\tilde{f}$ is continuous on $\tilde{A}$. Show further that if Y is regular and f is continuous on A, then $A \subset \tilde{A}$ and $\tilde{f}$ is an extension of f.

According to this definition of limit, a function f defined on a subset A of a topological space X possesses a limit at a point x_0 of A when and only when f is *continuous* at x_0 relative to A, in which event

$$\lim_{\substack{x \to x_0 \\ x \in A}} f(x) = f(x_0).$$

In some contexts it is desirable to relax this rather stringent requirement. The slightly different definition of limit that the reader may have encountered elsewhere (notably in his elementary calculus textbook) corresponds in our notation to

$$\lim_{\substack{x \to x_0 \\ x \in A \setminus \{x_0\}}} f(x).$$

O. If X is a topological space and $\{x_\lambda\}_{\lambda \in \Lambda}$ is a net in X, then a point y in X is a *cluster point* of $\{x_\lambda\}$ if for every neighborhood V of y and every λ in Λ there exists an index λ' in Λ such that $\lambda' \geq \lambda$ and $x_{\lambda'} \in V$.

  (i) Let $\{x_\lambda\}_{\lambda \in \Lambda}$ be a net in a topological space X, and for each λ in Λ let $T_\lambda = \{x_{\lambda'} : \lambda' \geq \lambda\}$. Show that y is a cluster point of the net $\{x_\lambda\}$ if and only if $y \in T_\lambda^-$ for every λ in Λ.
  (ii) Show that if X is a Hausdorff space and a net $\{x_\lambda\}$ in X converges to a limit x_0, then x_0 is the unique cluster point of the net $\{x_\lambda\}$.

P. A point y in a topological space X is a cluster point of a net $\{x_\lambda\}_{\lambda \in \Lambda}$ in X if and only if some subnet of $\{x_\lambda\}$ converges to y. (Hint: Turn the product $\Lambda \times \mathcal{N}(y)$ into a directed set.) If X satisfies the first axiom of countability, then a point y in X is a cluster point of a sequence $\{x_n\}$ in X if and only if some subsequence of $\{x_n\}$ converges to y.

Q. A nonempty collection $\mathscr{F}$ of sets is said to have the *finite intersection property* if the intersection of each nonempty finite subcollection of $\mathscr{F}$ is nonempty.

  (i) Show that a topological space X is compact if and only if every collection of closed sets in X that possesses the finite intersection property has nonempty intersection.
  (ii) In the same vein, show that X is countably compact if and only if every decreasing sequence $\{F_n\}_{n=1}^\infty$ of nonempty closed sets in X has nonempty intersection.
  (iii) A point x of a topological space X is said to be an *adherent point* of a collection $\mathscr{C}$ of subsets of X if $x \in E^-$ for every E in $\mathscr{C}$. Verify that X is compact if and only if every collection of sets in X with the finite intersection property possesses an adherent point.

R. Show that a topological space X is compact if and only if every net in X possesses a cluster point. Conclude that X is compact if and only if every net in X has a convergent subnet. (Hint: To go one way, note that if $\{x_\lambda\}$ is an arbitrary net, and if we write $T_\lambda = \{x_{\lambda'} : \lambda' \geq \lambda\}$ for each index λ, then the collection of sets $\{T_\lambda\}$ has the finite intersection property. To go the other way, let $\mathscr{F}$ be a collection of closed sets in X possessing the finite intersection property, and denote by $\mathscr{G}$ the collection consisting of all finite intersections of members of $\mathscr{F}$. Then $\mathscr{G}$ has the finite intersection property and is directed under inverse-inclusion. For each G in $\mathscr{G}$ choose a point x_G in G, and consider the net $\{x_G\}_{G \in \mathscr{G}}$.)

S. A collection $\mathscr{U}$ of subsets of a set X is called an *ultrafilter* on X if (a) $\mathscr{U}$ has the finite intersection property, and (b) $\mathscr{U}$ is maximal with respect to possessing the finite intersection property (in the inclusion ordering on the power class on the power class on X).

   (i) Verify that if $\mathscr{U}$ is an arbitrary ultrafilter on a set X, and if A is a subset of X such that $A \cap E \neq \varnothing$ for every set E in $\mathscr{U}$, then $A \in \mathscr{U}$. Show also that $\mathscr{U}$ is closed with respect to the formation of finite intersections, i.e., show that if $E_1, \dots, E_n$ are sets belonging to $\mathscr{U}$, then $E_1 \cap \cdots \cap E_n$ belongs to $\mathscr{U}$.

   (ii) Prove that a topological space X is compact if and only if every ultrafilter on X possesses an adherent point. (Hint: Use Zorn's lemma and Problem Q.)

T. Let $\mathscr{U}$ be an ultrafilter on a topological space X, and let f be a continuous mapping of X into a compact space Y. Show that there exists a point y_0 in Y such that $f^{-1}(V) \in \mathscr{U}$ for every neighborhood V of y_0 in Y, and use this fact to prove Theorem 3.15. (Hint: Choose y_0 to be an adherent point of the collection $\{f(E) : E \in \mathscr{U}\}$ of subsets of Y, and apply part (i) of Problem S. The hard part of the proof of Tihonov's theorem consists in showing that the product $X = \prod_\gamma Y_\gamma$ of a family $\{Y_\gamma\}$ of compact spaces is compact. Let $\mathscr{U}$ be an ultrafilter on X, apply the stated fact to each projection π_γ, and use Problem S once again.)

U. A topological space X is said to be *sequentially compact* if every infinite sequence in X possesses a convergent subsequence. Show that if X is Hausdorff and satisfies the first axiom of countability, then X is sequentially compact if and only if it is countably compact. (Hint: A necessary and sufficient condition for an arbitrary Hausdorff space to be countably compact is that every infinite sequence in the space possess a cluster point.)

V. A topological space X is said to be *locally compact* if every point of X has a compact neighborhood. Examples of locally compact spaces are $\mathbb{R}^n$, $\mathbb{C}^n$, all compact Hausdorff spaces, and all open subspaces of compact Hausdorff spaces. Show that if x is an arbitrary point of a locally compact Hausdorff space X, then the compact neighborhoods of x form a neighborhood base at x, and conclude that X is completely regular. (Hint: If L is a compact neighborhood of x and V is an arbitrary neighborhood of x, then L is closed (Prob. C) and $W = V^\circ \cap L^\circ$ is an open neighborhood of x such that W^- is contained in L and is therefore compact and normal (Prop. 3.4).) Show also that if K is a compact subset of a locally compact Hausdorff space X, and U is an open subset of X containing K, then there exists a continuous mapping f of X into $[0, 1]$ such that $\chi_K \leq f \leq \chi_U$. (Hint: Construct an open set V such that $K \subset V \subset V^- \subset U$ and such that V^- is compact; apply Urysohn's lemma (Prop. 3.5).)

If for an ordinal number α we write $\hat{W}(\alpha)$ for the *closed* initial segment $W(\alpha) \cup \{\alpha\}$ in the order topology, then

$$\Pi = \hat{W}(\Omega) \times \hat{W}(\omega)$$

is a compact Hausdorff space in which the open subspace $U = \Pi \backslash \{(\Omega, \omega)\}$ fails to be normal. Thus U in its relative topology, known as the *Tihonov plank*, is a locally compact Hausdorff space that is not normal. (The Tihonov plank is thus also an example of a completely regular space that is not normal.)

W. Let X be a locally compact Hausdorff space. Show that there exists a compact Hausdorff space $\hat{X}$ such that $\hat{X}$ contains X as a subspace and such that $\hat{X} \backslash X$ is a singleton $\{\omega\}$. Show also that $\hat{X}$ is unique in the sense that, if $\tilde{X}$ is another compact Hausdorff space such that $\tilde{X}$ contains X as a subspace and such that $\tilde{X} \backslash X$ is a singleton $\{\omega'\}$, then the identity mapping on X extends to a homeomorphism between $\hat{X}$ and $\tilde{X}$ carrying ω to ω'. The (essentially unique) space $\hat{X}$ is called the *one-point compactification* of X, and the point ω is known as the *point at infinity* in $\hat{X}$. Show that the one-point compactification $\hat{\mathbb{C}}$ of the complex plane $\mathbb{C}$ is homeomorphic to the two-dimensional sphere

$$S_2 = \{(x, y, z) \in \mathbb{R}^3 : x^2 + y^2 + (z - 1)^2 = 1\}.$$

(For this reason $\hat{\mathbb{C}}$ is frequently called the *complex* or *Riemann sphere*. The point at infinity in $\hat{\mathbb{C}}$ is denoted by ∞. It is important to note the distinction between the element ∞ of $\hat{\mathbb{C}}$ and the elements $\pm \infty$ of $\mathbb{R}^\natural$.)

4 Metric spaces

We assume the reader to be familiar with the notion of a *metric space*, and with the elementary concepts associated with the theory of metric spaces, In this chapter we briefly review some of these concepts, largely to fix notation and terminology. (An exception is our treatment of the theory of Baire categories; we do not assume any prior knowledge of this topic.) In particular, the reader is assumed to be acquainted with the notion of a *bounded set*, the *diameter* of a bounded set A (notation: diam A), the *distance between* a point x and a set A (notation: $d(x, A)$), and the *distance between* two sets A and B (notation: $d(A, B)$), in a metric space. The reader is also assumed to be familiar with the notion of the *metric topology* on a metric space, and with the elementary properties of a metric space regarded as a topological space. In particular, the following result is assumed known (see Problem 3I).

Proposition 4.1. *Every metric space is normal and satisfies the first axiom of countability. A metric space satisfies the second axiom of countability if and only if it is separable. If E is a subset of a metric space X, then a point x of X belongs to E^- if and only if there exists a sequence $\{x_n\}$ in E such that $x_n \to x$. If f is a mapping of a metric space X into a metric space Y, then f is continuous at a point x of X if and only if, for every sequence $\{x_n\}$ in X, $x_n \to x$ implies $f(x_n) \to f(x)$.*

On occasion, we shall find the more general notion of a *pseudometric* useful. A nonnegative real-valued function σ defined on the Cartesian product of a set X with itself is a *pseudometric* on X if

(i) $\sigma(x, y) = \sigma(y, x)$,
(ii) $\sigma(x, z) \leq \sigma(x, y) + \sigma(y, z)$,
(iii) $\sigma(x, x) = 0$,

for all points x, y, and z in X. A set X equipped with a pseudometric is a *pseudometric space*. If (X, σ) is a pseudometric space and x_0 is a point of X, then for any positive number r the set

$$D_r(x_0) = \{x \in X : \sigma(x, x_0) < r\}$$

is the *open ball* in X with *center* x_0 and *radius* r. If in this definition "$<$" is replaced by "$\leq$", the result is the *closed ball* with the same center and radius. Just as in a metric space, a subset B of a pseudometric space X is *bounded* if there exists a closed ball in X that contains B. The collection of all open balls in X is a base for a topology on X—the *pseudometric topology* or topology *induced* by the pseudometric. Whenever a pseudometric space is regarded as a topological space, it is understood that the topology employed is the pseudometric topology unless the contrary is expressly stipulated. In particular, limits of sequences and nets, as well as accumulation points of sets, are defined in a pseudometric space by regarding it as a topological space in the pseudometric topology. It is important to note that in an arbitrary pseudometric space, just as in the complex plane, a convergent *sequence* is necessarily bounded, but this need not be true of a convergent net (see Problem 3L). We also note that it results from these agreements that if (X, ρ) is a metric space and $\{x_n\}_{n=1}^{\infty}$ is a sequence in X, then $\{x_n\}$ converges to the point x_0 of X if and only if $\lim_n \rho(x_n, x_0) = 0$.

If (X, σ) is a metric [pseudometric] space and if M is a subset of X, then the restriction $\sigma|(M \times M)$ is a metric [pseudometric] on M called the *relative metric* [pseudometric]. The set M equipped with this relative metric [pseudometric] is a *subspace* of the space (X, σ). The relative metric [pseudometric] on M always induces the relative topology on M.

The continuity of a mapping from one pseudometric space into another is defined, in like manner, in terms of the pseudometric topologies of the spaces. In this connection we observe that the topology of a pseudometric space, like that of a metric space, satisfies the first axiom of countability. Consequently, the closure of a subset of a pseudometric space, as well as the continuity of a mapping of one pseudometric space into another, can be characterized in terms of sequences just as in Proposition 4.1.

There is an important concept in the theory of metric spaces that has no exact counterpart in general topology.

Definition. Let (X, ρ) and (Y, ρ') be metric spaces, and let f be a mapping of X into Y. Then f is *uniformly continuous* on X if for every $\varepsilon > 0$ there exists $\delta > 0$ such that $\rho(x', x'') < \delta$ implies $\rho'(f(x'), f(x'')) < \varepsilon$ for every pair of points x', x'' in X.

From this definition it is clear that a uniformly continuous mapping is continuous. As is well known, there is one very important context in which the converse assertion is valid (see Proposition 4.6 below).

Two pseudometrics σ_1 and σ_2 on the same set X are *equivalent* if their pseudometric topologies coincide. Similarly, two metrics ρ_1 and ρ_2 on the same set X are *equivalent* if they give rise to the same metric topology. This amounts, of course, to saying that the identity map on X is a homeomorphism between (X, ρ_1) and (X, ρ_2).

Example A. If ρ is a metric on a set X, then

$$\rho'(x, y) = \frac{\rho(x, y)}{1 + \rho(x, y)}, \qquad x, y \in X,$$

is an equivalent metric on X. (That ρ and ρ' are equivalent is apparent from the fact that the function $f(t) = t/(1 + t)$ is strictly increasing on the ray $[0, +\infty) (= \{u \in \mathbb{R} : u \geq 0\})$. The only hard part is to verify that ρ' is a metric, and this follows at once from the following inequalities:

$$\frac{u + v}{1 + u + v} \leq \frac{u + v + uv}{1 + u + v + uv} \leq \frac{u}{1 + u} + \frac{v}{1 + v}, \qquad u, v \geq 0.)$$

Thus every (nonempty) metric space (X, ρ) admits an equivalent metric ρ' in which diam $X \leq 1$.

If X and Y are metric spaces with metrics ρ and ρ', respectively, then a mapping φ of X into Y is said to be *isometric*, or to be an *isometry*, if $\rho'(\varphi(x), \varphi(x')) = \rho(x, x')$ for every pair of points x, x' in X. If there exists an isometry of X onto Y, then X and Y are *isometric*, or *isomorphic as metric spaces*. Metric spaces that are isometric are indistinguishable for many purposes, and are frequently identified.

A sequence $\{x_n\}$ in a metric space (X, ρ) is a *Cauchy sequence* if $\lim_{m,n} \rho(x_m, x_n) = 0$. More generally, a net $\{x_\lambda\}_{\lambda \in \Lambda}$ in X is a *Cauchy net* if for every $\varepsilon > 0$ there exists an index λ_ε in Λ such that $\rho(x_\lambda, x_{\lambda'}) < \varepsilon$ whenever $\lambda, \lambda' \geq \lambda_\varepsilon$. A metric space in which every Cauchy sequence is convergent is said to be *complete*. As it happens, every Cauchy net in a complete metric space is convergent (Prob. M). The metric spaces $\mathbb{R}$ and $\mathbb{C}$ are complete in the *usual* metric $\rho(\alpha, \beta) = |\alpha - \beta|$. More generally, the metric spaces $\mathbb{R}^n$ and $\mathbb{C}^n$ (as defined in Problem A) are complete metric spaces in their usual metrics.

If X is a metric space with metric ρ, then a *completion* of X is a complete metric space $(\tilde{X}, \tilde{\rho})$ together with an isometry φ of X onto a dense subset of $\tilde{X}$. As is customary, we shall consistently use the embedding φ to identify X with the subset $\varphi(X)$ of $\tilde{X}$. If $(\tilde{X}', \tilde{\rho}')$ is another completion of X, then there exists a unique isometry of $\tilde{X}$ onto $\tilde{X}'$ leaving X pointwise fixed (Prob. I). Every metric space possesses an essentially unique completion (Prob. L). In particular, if X is complete to begin with, then the completion $\tilde{X}$ of X coincides with X. Moreover, if M is a subset of a metric space X, then the completion $\tilde{M}$ of M (in the *relative metric* $\rho|(M \times M)$) can be identified with the closure of M in the completion $\tilde{X}$ of X, and we shall consistently make this identification. In particular, a subset of a complete metric space is itself a complete metric space if and only if it is closed. More generally if M is a subset of an arbitrary metric space X and M is complete as a subspace of X, then M is closed in X (see Problem I).

A topological space X is said to be *metrizable* if there exists a metric ρ on X such that the metric topology induced by ρ coincides with the given topology on X. Clearly the metric that metrizes a given metrizable space is unique only up to equivalence. The following metrization theorem, due to Urysohn, is frequently useful.

Theorem 4.2. *A normal topological space that satisfies the second axiom of countability is metrizable.*

PROOF. Let $\mathcal{B}$ be a countable base for X. There are only countably many pairs (U, V) of sets in $\mathcal{B}$ such that $V^- \subset U$. Let $\{(U_n, V_n)\}_{n=1}^{\infty}$ be an enumeration of the set of all such pairs, and for each positive integer n let f_n be a continuous function on X such that

$$\chi_{V_n} \leq f_n \leq \chi_{U_n}$$

(Prop. 3.5). It is easily seen that the topology inversely induced on X by the sequence of mappings $\{f_n\}$ coincides with the given topology on X. Thus, by Example 3R, the space X can be topologically embedded in the product of a countable number of copies of the unit interval $[0, 1]$. Since this product is metrizable (Prob. C), the result follows. $\qquad\square$

Compact metric spaces are automatically compact Hausdorff spaces, and enjoy all of the properties of the latter. Additional special properties of compact metric spaces are set forth in the following four propositions, proofs of the first three of which are entirely elementary.

Proposition 4.3. *If (X, ρ) is a compact metric space, then for every $\varepsilon > 0$ there exists a finite covering of X consisting of (disjoint) sets of diameter less than ε. (A metric space with this property is said to be* totally bounded.*) Equivalently, X has the property that for every $\varepsilon > 0$ there exists a finite subset N_ε of X with the property that $d(x, N_\varepsilon) < \varepsilon$ for every x in X. (Such a set is called an ε-net in X.)*

Corollary 4.4. *Every compact metric space is separable and satisfies the second axiom of countability. A compact Hausdorff space is metrizable if and only if it satisfies the second axiom of countability.*

Corollary 4.5. *If $f : X \to Y$ is a continuous mapping of a compact topological space X into a metric space Y, then the range $f(X)$ is a bounded set in Y. (A mapping that takes its values in a metric space is said to be* bounded *if its range is bounded. Thus every continuous mapping of a compact space into a metric space is bounded.)*

Proposition 4.6. *Every continuous mapping f of a compact metric space (X, ρ) into a metric space (Y, ρ') is uniformly continuous.*

PROOF. Let ε be a positive number. For each point x of X there exists a $\delta_x > 0$ such that $\rho'(f(x'), f(x)) < \varepsilon/2$ whenever x' is a point of X such that $\rho(x', x) < \delta_x$. The open balls $D_{\delta_x/2}(x)$ cover X, so there exists a finite set of points $x_1, \ldots, x_n$ in X such that the balls

$$D_i = D_{\delta_{x_i}/2}(x_i), \qquad i = 1, \ldots, n,$$

cover X. Let $\delta_i = \delta_{x_i}$, $i = 1, \ldots, n$, set $\delta = \delta_1 \wedge \cdots \wedge \delta_n$, and suppose $\rho(x', x'') < \delta/2$. Then x' belongs to some D_i, and it follows that x' and x'' both belong to $D_{\delta_i}(x_i)$. But then $\rho'(f(x'), f(x_i)) < \varepsilon/2$ and $\rho'(f(x''), f(x_i)) < \varepsilon/2$, so that $\rho'(f(x'), f(x'')) < \varepsilon$. $\qquad\square$

Example B. If φ is a continuous complex-valued function on a closed interval $[a, b]\,(a < b)$, then φ is uniformly continuous on $[a, b]$ by the preceding proposition. It follows that for every $\varepsilon > 0$ there exists a $\delta_\varepsilon > 0$ with the property that if $\{a = t_0 < \cdots < t_N = b\}$ is any partition of $[a, b]$ having mesh less than δ_ε (Prob. 1G), and if t', t'' are two real numbers belonging to any one of the subintervals $[t_{i-1}, t_i]$ then $|\varphi(t') - \varphi(t'')| < \varepsilon$. Consequently, the diameter in $\mathbb{C}$ of the set $\varphi([t_{i-1}, t_i])$ (usually known as the *oscillation* of φ over $[t_{i-1}, t_i]$) is not greater than ε.

Suppose next that φ is a rectifiable (but not necessarily continuous) function on $[a, b]$ (Prob. 1I), and for each t, $a \leq t \leq b$, let $L(t)$ denote the length, or total variation, of φ on the subinterval $[a, t]$. (The function L is monotone increasing and has the property that $L(t) - L(s)$ gives the total variation of φ on the subinterval $[s, t]$ for $a \leq s \leq t \leq b$; see Problem 1K.) Let ε be a positive number, and let $\mathcal{Q}_0 = \{a = s_0 < \cdots < s_M = b\}$ be a partition such that

$$\sum_{i=1}^{M} |\varphi(s_i) - \varphi(s_{i-1})| > L(b) - \varepsilon.$$

If $\{a = t_0 < \cdots < t_N = b\}$ is any refinement of $\mathcal{Q}_0$, then

$$L(t_i) - L(t_{i-1}) \leq |\varphi(t_i) - \varphi(t_{i-1})| + \varepsilon, \qquad i = 1, \ldots, N.$$

From this observation it clearly follows that L is left-continuous [right-continuous] at every point of the interval $[a, b]$ at which the given function φ is left-continuous [right-continuous]. Consequently, when φ is both rectifiable and continuous, the function L is also (uniformly) continuous on $[a, b]$. Hence if $\{\mathcal{P}_n\}_{n=1}^{\infty}$ is any sequence of partitions of $[a, b]$ such that $\lim_n \text{mesh } \mathcal{P}_n = 0$, then the maximum of the lengths of the subarcs of φ determined by the subintervals of each partition $\mathcal{P}_n$ tends to zero as n tends to infinity. (It may also be noted that if $\{\mathcal{P}_n\}$ is any such sequence of partitions, and if V_n denotes the variation of φ over $\mathcal{P}_n$, then $\lim_n V_n = L(b)$, the length of φ. Indeed, whenever mesh $\mathcal{P}_n$ is sufficiently small, the partition points s_i

of $\mathcal{Q}_0$ belong to distinct subintervals of $\mathcal{P}_n$. If mesh $\mathcal{P}_n$ is also so small that the oscillation of φ over each of the subintervals of $\mathcal{P}_n$ does not exceed ε/M, then $V_n > L(b) - 3\varepsilon$.

We turn now to the topic of *Baire categories* in metric spaces. No prior knowledge of this topic is assumed on the part of the reader.

Definition. A subset N of a metric space X is *nowhere dense* in X if the closure N^- contains no nonempty open set. A subset A of X is of *first category* in X if it can be written as a countable union of sets each of which is nowhere dense in X. A subset B is of *second category* in X if it is not of first category in X.

It is clear that if $\{A_n\}$ is any countable collection of sets each of which is of the first category in a metric space X, then $\bigcup_n A_n$ is of first category in X. Likewise, any subset of a set of the first category in X is itself of first category in X. Hence, if X happens to be of first category in itself, then all subsets of X are of first category in X. Thus category distinctions have significance only in spaces that are of second category in themselves, and the most important examples of such spaces are complete metric spaces. The following proposition is central to the proof of this fact.

Proposition 4.7. *A set N is nowhere dense in a metric space X if and only if, for every nonempty open set U in X, there exists a nonempty open set V such that $V \subset U$ and $V \cap N = \varnothing$. Moreover, if N is nowhere dense, it is always possible to arrange for V to be an open ball with arbitrarily small positive radius satisfying the stronger conditions $V^- \subset U$ and $V^- \cap N = \varnothing$.*

PROOF. If N is not nowhere dense, and U is a nonempty open set contained in N^-, then every nonempty open subset V of U clearly meets N. This proves the sufficiency of the given condition. To prove the necessity, set $V = U \backslash N^-$. If $V = \varnothing$, then N is not nowhere dense. Otherwise, V is a nonempty open subset of U that is disjoint from N. Thus the necessity of the given condition is proved. To prove the last assertion of the proposition, observe that if x_0 is any point of V, then there exists a radius $r > 0$ such that $D_r(x_0) \subset V$. But then, if ε is a positive number less than r, the ball $D_\varepsilon(x_0)$ satisfies all the desired conditions. $\square$

Theorem 4.8 (Baire Category Theorem). *Let X be a complete metric space, and let U be a nonempty open subset of X. Then U is of second category in X.*

PROOF. Let $\{N_n\}_{n=1}^\infty$ be an arbitrary sequence of nowhere dense sets in X. It suffices to show that if $A = \bigcup_{n=1}^\infty N_n$, then $U \backslash A \neq \varnothing$. Note first that by the foregoing proposition there exists an open ball $D_1 = D_{r_1}(x_1)$ with radius $r_1 \leq 1$ such that $D_1^- \subset U$ and $D_1^- \cap N_1 = \varnothing$. Continuing via mathematical induction, one easily constructs a nested sequence

$$D_1 \supset D_2 \supset \cdots \supset D_n \supset \cdots$$

of open balls $D_n = D_{r_n}(x_n)$ such that $r_n \leq 1/n$ and such that $D_n^- \cap N_n = \varnothing$ for every n. The sequence of centers $\{x_n\}$ is then obviously a Cauchy sequence, and since X is complete, $\{x_n\}$ must converge to some limit x_0. Since x_0 is also the limit of every tail $\{x_n\}_{n=m}^\infty$, and since this tail lies in D_m, it follows that $x_0 \in D_m^-$ for $m = 1, 2, \ldots$. Thus $x_0 \notin A$, and since $x_0 \in D_1^-$ and $D_1^- \subset U$, we have $x_0 \in U \backslash A$.   □

While the hypothesis of the Baire category theorem is metric in nature, the conclusion is clearly topological. It follows that the theorem holds in any metric space X for which there exists an equivalent metric making it complete. For another, purely topological, setting in which an analog of the Baire category theorem is valid, see Problem W. Readers wishing to learn more about metric spaces could not do better than to consult [42].

PROBLEMS

A.   Establish the *Cauchy inequality*

$$\sum_{k=1}^n |\xi_k \eta_k| \leq \left(\sum_{k=1}^n |\xi_k|^2 \right)^{1/2} \left(\sum_{k=1}^n |\eta_k|^2 \right)^{1/2}$$

for arbitrary n-tuples $(\xi_1, \ldots, \xi_n)$ and $(\eta_1, \ldots, \eta_n)$ in $\mathbb{C}^n$. (Hint: The Cauchy inequality follows from the elementary quadratic inequality

$$ab \leq \frac{a^2 + b^2}{2},$$

valid for all nonnegative a and b; consider first the normalized case

$$\sum_{k=1}^n |\xi_k|^2 = \sum_{k=1}^n |\eta_k|^2 = 1.)$$

Use the Cauchy inequality to prove that

$$d(x, y) = \left(\sum_{k=1}^n |\xi_k - \eta_k|^2 \right)^{1/2} \tag{1}$$

defines a metric on $\mathbb{C}^n$. This metric is called the *usual* metric on $\mathbb{C}^n$, and $\mathbb{C}^n$ equipped with the usual metric is known as n-dimensional *unitary space*. If $x = (x_1, \ldots, x_n)$ and $y = (y_1, \ldots, y_n)$ belong to $\mathbb{R}^n$, then (1) simplifies slightly to

$$d(x, y) = \left(\sum_{k=1}^n (x_k - y_k)^2 \right)^{1/2}.$$

The space $\mathbb{R}^n$ equipped with this metric (also called the *usual* metric) is n-dimensional *Euclidean space*. (The topological space $\mathbb{R}^1$ (Ex. 3C) is homeomorphic to a closed interval of real numbers, and may therefore be metrized whenever that is convenient, but there is no one metric on $\mathbb{R}^1$ that deserves to be called the usual one.)

B. Let X be a metric space with metric ρ. If $\{x_n\}$ and $\{y_n\}$ are sequences in X converging to limits x_0 and y_0, respectively, then $\{\rho(x_n, y_n)\}$ converges to $\rho(x_0, y_0)$. Show that diam $A^- = $ diam A for every (nonempty) set A in X.

C. (i) Let X and Y be metric spaces with metrics ρ and ρ', respectively. Verify that the function ρ_1 defined by

$$\rho_1((x_1, y_1), (x_2, y_2)) = \rho(x_1, x_2) + \rho'(y_1, y_2)$$

is a metric on the product $X \times Y$. Verify likewise that

$$\rho_\infty((x_1, y_1), (x_2, y_2)) = \rho(x_1, x_2) \vee \rho'(y_1, y_2)$$

and

$$\rho_2((x_1, y_1), (x_2, y_2)) = [\rho(x_1, x_2)^2 + \rho'(y_1, y_2)^2]^{1/2}$$

also define metrics on $X \times Y$ (see Problem A). In each of these metrics it is the case that a sequence $\{(x_n, y_n)\}_{n=1}^\infty$ in $X \times Y$ converges to a limit (x_0, y_0) if and only if $x_n \to x_0$ and $y_n \to y_0$. Thus these metrics all metrize the product topology on $X \times Y$ and are therefore all equivalent.

(ii) Show that any finite Cartesian product of metric spaces is metrizable.

(iii) Suppose given an infinite sequence $\{(X_n, \rho_n)\}_{n=1}^\infty$ of metric spaces such that diam $X_n \leq 1$ for every n. Verify that

$$\rho(\{x_n\}, \{y_n\}) = \sum_{n=1}^\infty \frac{1}{2^n} \rho_n(x_n, y_n)$$

defines a metric on the product $\prod_{n=1}^\infty X_n$, and use this fact to show that the product topology on the product of an arbitrary countable collection of metric spaces is metrizable. (Hint: See Example A.)

D. If A is a nonempty subset of a metric space X, the function $d_A(x) = d(x, A)$ is (uniformly) continuous on X. Hence the set $B = \{x \in X : d_A(x) = 0\}$ is closed. Show that, in fact, $B = A^-$. If E and F are nonempty disjoint closed sets, then the sum $d_E(x) + d_F(x)$ is nowhere equal to 0, so that we may define

$$u(x) = \frac{d_E(x)}{d_E(x) + d_F(x)}.$$

The function u is a continuous mapping of X into the unit interval $[0, 1]$ with the properties that $u(x) = 0$ if and only if $x \in E$ and $u(x) = 1$ if and only if $x \in F$. (This proves Urysohn's lemma (Prop. 3.5) for metric spaces.)

E. Let X be a metric space with metric ρ. If $\{x_n\}$ is a Cauchy sequence in X, and if some subsequence of $\{x_n\}$ converges to a limit x_0, then $\{x_n\}$ itself converges to x_0. Show that a Cauchy sequence $\{x_n\}$ always possesses subsequences $\{x_{k_n} = y_n\}$ satisfying the condition $\rho(y_n, y_{n+1}) \leq 1/2^n$, and therefore the condition

$$\sum_{n=1}^\infty \rho(y_n, y_{n+1}) < +\infty. \tag{2}$$

(Conversely, a sequence satisfying (2) is automatically Cauchy.) Show that if every sequence $\{y_n\}$ in X that satisfies (2) is convergent, then X is complete.

F. Suppose given a mapping f defined on a metric space (X, ρ) and taking its values in a second metric space (Y, σ). We say that f is *Lipschitzian* if there exists a positive number M such that $\sigma(f(x), f(y)) \le M\rho(x, y)$ for all points x and y in X. (The number M is then called a *Lipschitz constant* for f.) Show that if f is Lipschitzian on X, then f is uniformly continuous on X. If A is a nonempty subset of X, then the function $d_A(x) = d(x, A)$ is Lipschitzian. So are the projections $(\xi_1, \ldots, \xi_n) \to \xi_i$ of $\mathbb{C}^n[\mathbb{R}^n]$ onto $\mathbb{C}[\mathbb{R}]$, where $\mathbb{C}^n[\mathbb{R}^n]$ is given its usual metric (Prob. A).

G. Let (X, σ) be a pseudometric space. Show that if we define $x \sim y$ to mean $\sigma(x, y) = 0$, then $\sim$ is an equivalence relation on X, and that if $[x]$ denotes the equivalence class of a point x of X, then setting

$$\rho([x], [y]) = \sigma(x, y), \qquad x, y \in X,$$

defines a metric on the quotient space $X/\sim$. Verify also that the metric topology on $X/\sim$ is the quotient topology of the pseudometric topology on X (Prob. 3H). The metric ρ is known as the metric *associated with* σ.

H. Suppose that f is a uniformly continuous mapping whose domain is a dense set A in a metric space (X, ρ) and whose range is a subset of a complete metric space (Y, σ). Show that there exists a unique continuous mapping $\tilde{f}$ of X into Y such that $\tilde{f}|A = f$. The mapping $\tilde{f}$ is said to be obtained by *extending f by continuity*.

I. Suppose A and B are subsets of complete metric spaces X and Y, respectively, and that ψ is an isometry of A onto B. Then there exists a unique isometry $\tilde{\psi}$ of A^- onto B^- that extends ψ. (The isometry $\tilde{\psi}$ is the mapping obtained by extending ψ by continuity.) In particular, if $(\tilde{X}, \tilde{\rho})$ and $(\tilde{X}', \tilde{\rho}')$ are two completions of the same metric space (X, ρ), then there exists a unique isometry ψ of $\tilde{X}$ onto $\tilde{X}'$ leaving every point of X fixed. In this sense the completion of a metric space is unique. Show, in the same vein, that if M is a subset of a metric space X, then there is a unique isometry of the completion $\tilde{M}$ onto the closure of M in the completion $\tilde{X}$ of X that leaves M pointwise fixed. Conclude that if X is complete, then M is complete as a subspace of X if and only if M is closed in X, and show that if M is complete, then M is always closed in X, whether X is complete or not.

J. Two sequences $\{x_n\}$ and $\{y_n\}$ in a metric space X with metric ρ are *equiconvergent* if

$$\lim_n \rho(x_n, y_n) = 0.$$

A necessary and sufficient condition for a sequence $\{x_n\}$ to converge to a limit x_0 is that $\{x_n\}$ be equiconvergent with the *constant sequence* $\{x_0, x_0, \ldots, x_0, \ldots\}$. If either of two equiconvergent sequences is a Cauchy sequence, then both are. If either of two equiconvergent sequences converges, then both converge to the same limit. Let A be a dense subset of X, and suppose every Cauchy sequence in A converges to some limit in X. Show that X is complete.

K. Show that if $\{x_n\}$ and $\{y_n\}$ are Cauchy sequences in a metric space (X, ρ), then $\lim_n \rho(x_n, y_n)$ exists, and that $\sigma(\{x_n\}, \{y_n\}) = \lim_n \rho(x_n, y_n)$ defines a pseudometric σ on the collection of all Cauchy sequences in X. Show also that if $\{x_n\}$ and $\{y_n\}$ are Cauchy sequences and $\{z_n\}$ is any sequence that is equiconvergent with $\{y_n\}$, then $\sigma(\{x_n\}, \{y_n\}) = \sigma(\{x_n\}, \{z_n\})$.

L. Let X be a metric space with metric ρ, and let $\mathscr{C}$ denote the collection of all Cauchy sequences in X. We define a relation on $\mathscr{C}$ by writing $\{x_n\} \sim \{y_n\}$ whenever the

Cauchy sequences $\{x_n\}$ and $\{y_n\}$ are equiconvergent. Verify that $\sim$ is an equivalence relation on $\mathscr{C}$, and that, if σ denotes the pseudometric defined on $\mathscr{C}$ in the preceding problem, then $\sigma(\{x_n\}, \{y_n\}) = 0$ if and only if $\{x_n\} \sim \{y_n\}$. Conclude that if $\tilde{X}$ denotes the quotient space of all equivalence classes $[\{x_n\}]$ of Cauchy sequences under the relation $\sim$, then

$$\tilde{\rho}([\{x_n\}], [\{y_n\}]) = \sigma(\{x_n\}, \{y_n\}) = \lim_n \rho(x_n, y_n)$$

defines a metric $\tilde{\rho}$ on $\tilde{X}$ (Prob. G). Show also that if we define

$$\varphi(x) = [\{x, x, \ldots, x, \ldots\}]$$

for each x in X, then φ is an isometry of X onto a dense subset of $\tilde{X}$. Show, finally, that $\tilde{X}$ is complete. Thus every metric space possesses a completion, and this completion is unique up to a uniquely determined isometry. (Note, in particular, that if X is complete to begin with, then the completion $\tilde{X}$. as constructed here coincides with X.)

M. Let X be a metric space with metric ρ, and let $\{x_\lambda\}_{\lambda \in \Lambda}$ be a Cauchy net in X. Show that there exists an increasing sequence of indices $\{\lambda_n\}_{n=1}^\infty$ such that $\rho(x_{\lambda'}, x_{\lambda''}) < 1/n$ whenever $\lambda', \lambda'' \geq \lambda_n$, and that the corresponding sequence $\{x_n = x_{\lambda_n}\}$ is a Cauchy sequence. Show that if X is complete, then the net $\{x_\lambda\}$ converges to the limit of the sequence $\{x_n\}$.

N. An indexed family $\{\lambda_\gamma\}_{\gamma \in \Gamma}$ of complex numbers is summable in $\mathbb{C}$ (Ex. 3Q) if and only if the corresponding net $\{\sigma_D\}_{D \in \mathscr{D}}$ of finite sums satisfies the following *Cauchy criterion*: For every $\varepsilon > 0$ there exists a finite subset D_ε of Γ such that if D is any finite subset of Γ that is disjoint from D_ε, then $|\sigma_D| < \varepsilon$. Conclude that if $\{\lambda_\gamma\}$ is a summable family of complex numbers, then $\lambda_\gamma = 0$ except for a countable set of indices γ. Show also that $\{\lambda_\gamma\}$ is summable when and only when $\{|\lambda_\gamma|\}$ is, and that, if this is the case, then

$$\left| \sum_\gamma \lambda_\gamma \right| \leq \sum_\gamma |\lambda_\gamma|.$$

(Hint: Do the real case first.)

In the same vein verify that if the index family Γ is partitioned in any way into subsets Γ_1 and Γ_2, and if $\{\lambda_\gamma\}_{\gamma \in \Gamma}$ is a summable family of complex numbers with sum σ, then

$$\sigma = \sum_{\gamma \in \Gamma_1} \lambda_\gamma + \sum_{\gamma \in \Gamma_2} \lambda_\gamma.$$

More generally, if $\{\Gamma_\delta\}_{\delta \in \Delta}$ is an arbitrary indexed partition of the index family, then

$$\sigma = \sum_{\delta \in \Delta} \sum_{\gamma \in \Gamma_\delta} \lambda_\gamma.$$

In particular, if a *doubly indexed* family $\{\lambda_{\gamma, \delta}\}_{\gamma \in \Gamma, \delta \in \Delta}$ of complex numbers is summable over $\Gamma \times \Delta$ and has sum σ, then

$$\sigma = \sum_{\gamma \in \Gamma} \sum_{\delta \in \Delta} \lambda_{\gamma, \delta} = \sum_{\delta \in \Delta} \sum_{\gamma \in \Gamma} \lambda_{\gamma, \delta}.$$

O. Let X be a nonempty set and let (Y, ρ) be a metric space.

   (i) A net $\{f_\lambda\}_{\lambda \in \Lambda}$ of mappings of X into Y is said to be *uniformly convergent* to a limit f (where f is another mapping of X into Y) if for every positive number ε there exists an index λ_0 such that $\sigma(f(x), f_\lambda(x)) < \varepsilon$ for all $\lambda \geq \lambda_0$ and all x in X. Show that if X is a topological space and each f_λ is continuous on X, then the uniform limit f is also continuous on X.

   (ii) Let $\mathscr{B}(X, Y)$ denote the set of all bounded mappings from the set X into the metric space Y. If f and g belong to $\mathscr{B}(X, Y)$, we define

$$\sigma(f, g) = \sup_{x \in X} \rho(f(x), g(x)).$$

Show that σ is a metric on $\mathscr{B}(X, Y)$ and that a net $\{f_\lambda\}$ in $\mathscr{B}(X, Y)$ converges uniformly to a limit f if and only if it converges to f with respect to the metric σ. (For this reason σ is called the *metric of uniform convergence*, and the topology induced by σ the *topology of uniform convergence*, on $\mathscr{B}(X, Y)$.) Show also that $\mathscr{B}(X, Y)$ is complete in the metric of uniform convergence if and only if Y is complete in its metric ρ.

   (iii) When X is a topological space, we denote by $\mathscr{C}_b(X, Y)$ the set of all continuous mappings in $\mathscr{B}(X, Y)$. (Note that $\mathscr{C}_b(X, Y)$ coincides with the collection of *all* continuous mappings of X into Y when the space X is compact; see Corollary 4.5.) Show that $\mathscr{C}_b(X, Y)$ is complete in the metric of uniform convergence when and only when Y is complete.

P. A metric space X is totally bounded (Prop. 4.3) if and only if every sequence $\{x_n\}$ in X possesses a Cauchy subsequence. (Hint: One way is easy; the other way involves the *diagonal process*. Suppose X is totally bounded, and let $\{x_n\}$ be an arbitrary sequence in X. If $\{A_1, \ldots, A_k\}$ is some covering of X, where each A_i is a set of diameter no greater than one, then an infinite number of terms of the sequence $\{x_n\}$ must fall into some one of the sets A_i. Hence $\{x_n\}$ possesses a subsequence $\{x_n^{(1)}\}$ lying in a subset of X of diameter less than or equal to one. If $\{B_1, \ldots, B_q\}$ is a covering of X such that diam $B_i \leq 1/2$, $i = 1, \ldots, q$, then, repeating the above argument, we conclude that $\{x_n^{(1)}\}$ in turn possesses a subsequence $\{x_n^{(2)}\}$ lying in a set of diameter less than or equal to $1/2$. Continuing in this fashion, we obtain by induction a sequence $\{x_n^{(k)}\}_{k=1}^\infty$ of sequences, in which each term (except the first) is a subsequence of the preceding term, all terms are subsequences of the given sequence, and in which the kth term is confined to a set of diameter less than or equal to $1/k$. The "diagonal" sequence $\{x_n^{(n)}\}_{n=1}^\infty$ is Cauchy.)

Q. For any metric space X the following conditions are equivalent:

   (i) X is compact,
   (ii) X is countably compact,
   (iii) X is sequentially compact,
   (iv) X is both totally bounded and complete.

   (Hint: A totally bounded metric space is separable; recall Problem 3U.)

> Note that conditions (i), (ii), and (iii) of Problem Q are topological in nature. Hence these properties are equivalent in any topological space that is metrizable. It follows from this remark that if Ω denotes, as usual, the first noncountable ordinal number, then the ordinal segment $W(\Omega)$ is *not* metrizable in the order topology (see Problem 3D).

R. (Ascoli's Theorem) A collection $\mathscr{F}$ of mappings of a nonempty topological space X into a metric space (Y, ρ) is *equicontinuous* at a point x_0 of X if for every $\varepsilon > 0$ there is a neighborhood V of x_0 such that $\rho(f(x), f(x_0)) < \varepsilon$ for every x in V and every f in $\mathscr{F}$. Likewise, $\mathscr{F}$ is *equicontinuous on* X if $\mathscr{F}$ is equicontinuous at every point x of X.

(i) Show that if $\{f_\lambda\}_{\lambda \in \Lambda}$ is an equicontinuous net of mappings of X into Y, then the set F of those points x in X at which the net $\{f_\lambda(x)\}_{\lambda \in \Lambda}$ is Cauchy in Y is a closed subset of X.

(ii) Show that if X is compact and $\{f_\lambda\}_{\lambda \in \Lambda}$ is an equicontinuous net of mappings of X into Y that is pointwise Cauchy on X [pointwise convergent on X to some limit f], then $\{f_\lambda\}$ is Cauchy in the metric of uniform convergence on $\mathscr{C}_b(X, Y)$ [uniformly convergent to f on X] (Prob. O).

(iii) Prove that if X is a separable compact topological space (in particular, if X is a compact metric space) and Y is a metric space, then a subset $\mathscr{F}$ of the metric space $\mathscr{C}_b(X, Y)$ of (bounded) continuous mappings of X into Y is totally bounded in the metric of uniform convergence on X if and only if (a) it is equicontinuous on X, and (b) the set $\mathscr{F}(x) = \{f(x) \in Y : f \in \mathscr{F}\}$ is totally bounded in Y for every x in X. (Hint: To go one way, show that an arbitrary uniformly continuous mapping carries a totally bounded set onto a totally bounded set, and recall (Prop. 4.3) that a totally bounded set possesses a finite ε-net for every $\varepsilon > 0$. To go the other way, start with a sequence $\{f_n\}$ in $\mathscr{C}_b(X, Y)$ with the property that $\{f_n(x)\}$ is totally bounded in Y for each x in X, and use the diagonal process to obtain a subsequence that is Cauchy at every point of some countable dense subset of X.) This result, usually known as *Ascoli's theorem*, provides an effective criterion for the compactness of subsets of $\mathscr{C}_b(X, Y)$ in the metric of uniform convergence when X is a compact metric space and Y is complete. In particular, if X is a compact metric space, then Ascoli's theorem provides an important criterion for identifying the compact subsets of $\mathscr{C}(X)$ (Ex. 3D) in the topology of uniform convergence: A subset $\mathscr{F}$ of $\mathscr{C}(X)$ that is closed in the topology of uniform convergence on X is compact if and only if $\mathscr{F}$ is equicontinuous on X and $\mathscr{F}(x)$ is bounded in $\mathbb{C}$ for every x in X.

S. Let X be a complete metric space, and suppose given a sequence $\{F_n\}_{n=0}^\infty$ of nonempty closed subsets of X such that

$$F_0 \supset F_1 \supset \cdots \supset F_n \supset \cdots$$

and such that $\lim_n \operatorname{diam} F_n = 0$. Show that the intersection $\bigcap_{n=0}^\infty F_n$ is nonempty and consists of a singleton. Show also (by giving examples) that if the hypothesis $\operatorname{diam} F_n \to 0$ is dropped, then $\bigcap_{n=0}^\infty F_n$ may be either empty or infinite.

T. (i) For each nonnegative integer n, let us write $\mathscr{C}_n$ for the collection of special basic sets $C_{n,i} = C \cap I_{n,i}$, $i = 1, \ldots, 2^n$, introduced in the construction of the Cantor set C (see Example 3J), and let $\mathscr{C}$ denote the topological base $\bigcup_n \mathscr{C}_n$ in C. If φ denotes any continuous mapping of C into an arbitrary metric space X, then the induced mapping $C_{n,i} \xrightarrow{\Phi} \varphi(C_{n,i})$ of $\mathscr{C}$ into the power class on X possesses the following three properties:

(a) The sets $\Phi(C_{n,i})$ are closed and nonempty,

(b) If $C_{n+1,j} \subset C_{n,i}$, then $\Phi(C_{n+1,j}) \subset \Phi(C_{n,i})$,

(c) If $m_n = \sup\{\operatorname{diam} \Phi(C_{n,i}) : i = 1, \ldots, 2^n\}$, then $\lim_n m_n = 0$.

Show conversely that if Φ is a mapping of $\mathscr{C}$ into the power class on a *complete* metric space X, and if Φ satisfies (a), (b), and (c), then there exists a unique· mapping φ of C into X such that $\varphi(C_{n,i}) = \Phi(C_{n,i})$ for every set $C_{n,i}$ in $\mathscr{C}$. Show also that this mapping φ is necessarily continuous. (Hint: For each t in C there exists a uniquely determined sequence $\{i_n\}_{n=0}^{\infty}$ of positive integers such that $t \in C_{n,i_n}$ for every nonnegative integer n. Consequently $\varphi(t)$ must be given by $\{\varphi(t)\} = \bigcap_{n=0}^{\infty} \Phi(C_{n,i_n})$.)

(ii) Prove that if X is an arbitrary nonempty compact metric space, then there exists a continuous mapping of the Cantor set C *onto* X. (Hint: Construct a mapping Φ as in (i) satisfying conditions (a), (b), and (c), along with the added conditions

(d) $\Phi(C_{0,1}) = X$,

(e) If $C_{n,i} = C_{n+1,j} \cup C_{n+1,k}$ then $\Phi(C_{n,i}) = \Phi(C_{n+1,j}) \cup \Phi(C_{n+1,k})$.

Make use of the fact that X is totally bounded (Prop. 4.3).)

(iii) Prove that if P is a nonempty perfect subset of a complete metric space X, then there exists a homeomorphism of the Cantor set C into P. (Hint: Construct a mapping Φ as in (i) satisfying conditions (a), (b), and (c), along with the added condition

(d$'$) If $C_{n,i} \cap C_{n,j} = \varnothing$, then $\Phi(C_{n,i}) \cap \Phi(C_{n,j}) = \varnothing$.)

Since singletons are nowhere dense in a perfect Hausdorff space, it is an obvious consequence of the Baire category theorem that a perfect complete metric space cannot be countable. Problem T (iii) shows more—namely, that a perfect complete metric space must have at least the cardinality of the continuum. Moreover, it is clear on the basis of this last result that a complete metric space satisfying the second axiom of countability must either be countable or have cardinal number $\aleph$ (see Problem 3B).

U. Show that the intersection of a countable collection of dense open subsets of a nonempty complete metric space X is a dense set of the second category in X.

V. Let (X, ρ) be a nonempty complete metric space, and let f be a mapping of X into a metric space (Y, ρ'). Show that if the set C of points of continuity of f is dense in X, then C is of the second category in X. (Hint: For each point x of X and each positive number ε write $\omega(f; x, \varepsilon)$ for the supremum in $\mathbb{R}^{\sharp}$ of the set of distances

$$\{\rho'(f(x), f(y)) : y \in X, \rho(x, y) < \varepsilon\},$$

and set

$$\omega(f; x) = \inf_{\varepsilon > 0} \omega(f; x, \varepsilon), \qquad x \in X.$$

(The extended real number $\omega(f; x)$ is called the *oscillation* of f at x.) Show that f is continuous at a point x if and only if $\omega(f; x) = 0$, and that the set

$$\{x \in X : \omega(f; x) < \eta\}$$

is open in X for every positive number η. Use Problem U.) Conclude from this fact, in particular, that there does not exist any real-valued function f on the unit interval $[0, 1]$ with the property that the set of points at which f is continuous consists exactly of the rational numbers in $[0, 1]$.

W. A subset N of an arbitrary topological space X is said to be *nowhere dense* in X if $(N^-)^\circ = \varnothing$, and a subset A of X is of *first category* in X if A can be expressed as a countable union of sets each of which is nowhere dense in X. Likewise, a subset of X that is not of first category in X is of *second category* in X. Prove that if X is a locally compact Hausdorff space, then every nonempty open subset of X is of second category in X. (Hint: Follow the proof of Theorem 4.8.)

X. Suppose that $\{A_n\}_{n=1}^{\infty}$ and $\{B_n\}_{n=1}^{\infty}$ are two sequences of sets in a topological space X such that for each positive integer n the symmetric difference $A_n \nabla B_n$ is of first category in X. Show that for each pair of positive integers m and n the symmetric difference $(A_n \backslash A_m) \nabla (B_n \backslash B_m)$ is also of first category in X. Show also that $(\bigcup_{n=1}^{\infty} A_n) \nabla (\bigcup_{n=1}^{\infty} B_n)$ is of first category in X. (Hint: See Problem 1F).

5 Complex analysis

We shall have numerous occasions in the sequel to refer to the theory of functions of a complex variable, and we assume the reader to be familiar with the elements of this theory. We present here an outline of the rudiments of complex analysis, chiefly for convenience of reference, but also to fix notation and terminology. To begin at the beginning, we recall that a domain in the complex plane is a nonempty connected open subset of $\mathbb{C}$, and that a (complex-valued) function f defined on a domain Δ is *analytic* (or *holomorphic*) on Δ if its derivative f' exists at each point of Δ. (More generally, if U is an arbitrary nonempty open subset of $\mathbb{C}$ and f is a differentiable complex-valued function defined on U, we shall say that f is *locally analytic* on U. The study of a locally analytic function f on U reduces at once to the study of the analytic functions obtained by restricting f to the components of U; cf. Proposition 3.9.) We also recall that all of the elementary rules of ordinary real differential calculus hold for analytic functions of a complex variable. Thus the sum, product, and quotient rules for computing derivatives all hold for analytic functions, as does the chain rule. In particular, every complex polynomial function is analytic on the entire complex plane, and may be differentiated by means of the same elementary rule learned in calculus. Similarly, every complex rational function is analytic on the complement of the (finite) set of points at which its denominator vanishes.

Example A. If $\{\alpha_n\}_{n=0}^{\infty}$ is any sequence of *coefficients*, then the *power series*

$$\sum_{n=0}^{\infty} \alpha_n (\lambda - \lambda_0)^n \tag{1}$$

has a *radius of convergence* r, $0 \le r \le +\infty$, where r has the property that the series (1) converges for all λ such that $|\lambda - \lambda_0| < r$ and diverges for all λ such that $|\lambda - \lambda_0| > r$. (If $r = +\infty$, then the series (1) converges for every λ and is said to converge *everywhere*; if $r = 0$, then (1) converges only for $\lambda = \lambda_0$. No general statement can be made concerning the convergence of (1) on the *circle of convergence* $C_r(\lambda_0) = \{\lambda \in \mathbb{C} : |\lambda - \lambda_0| = r\}$ when $0 < r < +\infty$; in this connection see Problem A.) The radius of convergence r of the power series (1) is given by the formula

$$\frac{1}{r} = \limsup_n |\alpha_n|^{1/n}. \tag{2}$$

(If $\limsup_n |\alpha_n|^{1/n} = +\infty$, then $r = 0$; if $\limsup_n |\alpha_n|^{1/n} = 0$, then $r = +\infty$.) Moreover, if $r > 0$, the formula

$$f(\lambda) = \sum_{n=0}^{\infty} \alpha_n(\lambda - \lambda_0)^n \tag{3}$$

defines an analytic function f on the *disc of convergence*

$$D_r(\lambda_0) = \{\lambda \in \mathbb{C} : |\lambda - \lambda_0| < r\}$$

of the series, the derivative f' being given by the formally differentiated power series

$$f'(\lambda) = \sum_{n=1}^{\infty} n\alpha_n(\lambda - \lambda_0)^{n-1} \tag{4}$$

for all λ in $D_r(\lambda_0)$. Moreover, the convergence of (3) and (4) is absolute and uniform on compact subsets of $D_r(\lambda_0)$. (When $r = +\infty$, the convergence is absolute everywhere and uniform on arbitrary compact subsets of the plane.)

Thus a power series (1) with a positive radius of convergence defines an analytic function on a disc about its *center* λ_0. To see that the series is in turn determined by its sum f in (3), we note that (4), together with an easy induction argument, shows that each coefficient α_n is given by the familiar formula $\alpha_n = f^{(n)}(\lambda_0)/n!$. Hence the series expansion (3) is automatically and necessarily the *Taylor expansion* of its sum.

Along these lines we also make the following observation: if f is given by the power series expansion (3) in a disc $D_r(\lambda_0)$ with $r > 0$, and if f vanishes on some subset M of $D_r(\lambda_0)$ such that the center λ_0 is an accumulation point of M (Prob. 3B), then (by another induction argument) all of the coefficients α_n must vanish, and therefore f is identically equal to zero. Thus the sum f in (3) is completely determined by its behavior along any set such as M.

Example B. The power series

$$\sum_{n=0}^{\infty} \frac{\lambda^n}{n!} = 1 + \lambda + \frac{\lambda^2}{2!} + \cdots$$

converges everywhere to the exponential function $e^{\lambda} = \exp \lambda$. (We take this to be the definition of e^{λ} in the complex domain.) According to (4) we have

$$\frac{d}{d\lambda} e^{\lambda} = e^{\lambda}, \qquad \lambda \in \mathbb{C}.$$

Similarly, using a straightforward series calculation, it is not difficult to verify that

$$e^{\alpha+\beta} = e^{\alpha}e^{\beta}, \qquad \alpha, \beta \in \mathbb{C}.$$

In this connection we observe that the series

$$\sum_{n=0}^{\infty} \frac{(-1)^n \lambda^{2n}}{(2n!)} \quad \text{and} \quad \sum_{n=0}^{\infty} \frac{(-1)^n \lambda^{2n+1}}{(2n+1)!}$$

also converge everywhere, and we use these expressions to define the circular functions on $\mathbb{C}$, setting

$$\cos \lambda = \sum_{n=0}^{\infty} \frac{(-1)^n \lambda^{2n}}{(2n)!} \quad \text{and} \quad \sin \lambda = \sum_{n=0}^{\infty} \frac{(-1)^n \lambda^{2n+1}}{(2n+1)!}$$

for all complex numbers λ. (These definitions are not capricious, of course; it is readily seen that these are the only analytic functions on $\mathbb{C}$ that extend the real functions e^t, $\cos t$, and $\sin t$, defined on the real axis in $\mathbb{C}$.) It is a simple matter, using these series representations, to show that

$$e^{it} = \cos t + i \sin t$$

for all real numbers t. Hence if we write $\lambda = s + it$, where s and t are real, then

$$e^s = |e^\lambda| \quad \text{while} \quad t = \arg e^\lambda. \tag{5}$$

The facts announced in Example A, all of which may be established by quite elementary methods, principally by means of comparison tests for the convergence of infinite series, show that a function defined by a convergent power series is not only analytic on its disc of convergence but is, in fact, infinitely differentiable there. A converse assertion is the following well-known theorem. (For a sketch of a proof see Example J.)

Theorem 5.1 (Taylor's Theorem). *Let Δ be a domain in $\mathbb{C}$, let f be a complex-valued function defined and analytic on Δ, and let λ_0 be a point of Δ. Then there is a power series $\sum_{n=0}^{\infty} \alpha_n(\lambda - \lambda_0)^n$ with radius of convergence $r \geq d_0 = d(\lambda_0, \mathbb{C}\backslash\Delta)$ such that*

$$f(\lambda) = \sum_{n=0}^{\infty} \alpha_n(\lambda - \lambda_0)^n, \qquad |\lambda - \lambda_0| < d_0, \lambda \in \Delta. \tag{6}$$

(In the event that $\Delta = \mathbb{C}$ the power series (6) converges everywhere.) Moreover, if g is any function analytic on a disc $D_{r_0}(\lambda_0)$ ($r_0 > 0$) that agrees with f on the intersection $D_{r_0}(\lambda_0) \cap \Delta$, then $r_0 \leq r$. (Thus the power series in (6) defines f on the largest disc about λ_0 onto which f can be extended so as to be analytic.)

The following consequence of Taylor's theorem is of the greatest importance.

Theorem 5.2 (Identity Theorem). *Let f and g be analytic functions on the same domain Δ, and suppose $f(\lambda) = g(\lambda)$ for all λ in some subset M of Δ possessing an accumulation point in Δ. Then $f \equiv g$ on Δ.*

PROOF. It clearly suffices to show that $f = 0$ identically on Δ when $f = 0$ on M. Let U denote the subset of Δ consisting of all those points λ in Δ such that f vanishes identically on some open disc of positive radius about λ. Then U is an open subset of Δ which, according to Example A, contains all of the accumulation points of M that lie in Δ, as well as all of the accumulation points of U itself that belong to Δ. Hence U is a nonempty open subset of Δ such that $\partial U \cap \Delta = \varnothing$, whence it follows at once that $U = \Delta$ (Prop. 3.6). □

The proof of Taylor's theorem requires a penetrating analysis of the behavior of analytic functions on discs. In this connection, and in many others, an important role is played by certain *line integrals*. Accordingly, we shall give a brief account of the pertinent concepts. To begin with, we recall that an arc in $\mathbb{C}$ is a continuous complex-valued function defined on some real interval $[a, b]$, called the parameter interval of the arc. In the following discussion we shall consistently identify an arc α defined on an interval $[a, b]$ with the *equivalent* arc $\alpha(t - c)$ defined on the translated interval $[a, b] + c$, $-\infty < c < +\infty$. (The reader may check at each appropriate point as we go along that the concept or construction under discussion is unaffected by thus translating the parameter interval.) If α is an arc defined on the interval $[a, b]$, then the point $\lambda_0 = \alpha(a)$ is the *initial point* of α, the point $\lambda_1 = \alpha(b)$ the *terminal point* of α, and α is said to join λ_0 to λ_1. (If $\lambda_0 = \lambda_1$ the arc α is said to be *closed*.)

If α is an arc in $\mathbb{C}$ defined on the interval $[a, b]$, and if $\{t_0 < \cdots < t_n\}$ is a partition of $[a, b]$, then α is divided by that partition into subarcs $\alpha_1, \ldots, \alpha_n$ where $\alpha_i = \alpha|[t_{i-1}, t_i]$, $i = 1, \ldots, n$. We say that the arcs $\alpha_1, \ldots, \alpha_n$ are *chained*, meaning that the initial point of α_{i+1} coincides with the terminal point of α_i for each $i = 1, \ldots, n - 1$. In this situation it is also customary to say that α is the *sum* of the arcs α_i and to write $\alpha = \alpha_1 + \cdots + \alpha_n$. In the reverse direction, if $\alpha_1, \ldots, \alpha_n$ are any arcs in $\mathbb{C}$ that are chained in this sense, then there exists an arc α (unique up to equivalence) such that $\alpha = \alpha_1 + \cdots + \alpha_n$. In the same spirit we define the arc *opposite to* a given arc α defined on $[a, b]$ to be the arc

$$\tilde{\alpha}(t) = \alpha(-t), \qquad -b \leq t \leq -a.$$

In this connection the following concepts are frequently useful.

Definition. An arc α in $\mathbb{C}$ defined on the parameter interval $[a, b]$ is *smooth* if it belongs to $\mathscr{C}^{(1)}([a, b])$ (recall from Example 2J that this requires α to have a right derivative at a and a left derivative at b), and is *piecewise smooth* if there exists a partition $\{a = t_0 < \cdots < t_n = b\}$ such that each of the subarcs $\alpha \,|\, [t_{i-1}, t_i]$, $i = 1, \ldots, n$, is smooth. If α is a piecewise smooth arc and if the left and right derivatives of α not only exist at every point where they are defined, but are also different from zero at every point, then α is said to be *regular*.

Recall (Prob. 1I) that an arc α defined on the interval $[a, b]$ is *rectifiable* if it is of bounded variation on $[a, b]$, and that, when α is rectifiable, the total variation of α over $[a, b]$ is also called the *length* of α, which we shall here denote by $L(\alpha)$. It is clear that if an arc α is subdivided in any manner into a sum of subarcs, $\alpha = \alpha_1 + \cdots + \alpha_n$, then α is rectifiable if and only if all of the subarcs α_i are, and if this is the case, then $L(\alpha) = L(\alpha_1) + \cdots + L(\alpha_n)$. We take it as known from advanced calculus that a piecewise smooth arc α defined on the parameter interval $[a, b]$ is rectifiable and that $L(\alpha) = \int_a^b |\alpha'(t)| \, dt$.

Suppose now that α is a rectifiable arc defined on a parameter interval $[a, b]$, and let f be a bounded complex-valued function defined on the range of α. If $\mathscr{P} = \{t_0 < \cdots < t_n\}$ is any partition of $[a, b]$, if $\alpha = \alpha_1 + \cdots + \alpha_n$ is the corresponding subdivision of α, and if, for each $i = 1, \ldots, n$, ζ_i denotes a point in the range of α_i, then the sum

$$S = \sum_{i=1}^{n} f(\zeta_i) [\alpha(t_i) - \alpha(t_{i-1})]$$

is called a *Riemann sum* for the function f with respect to α based on the given partition $\mathscr{P}$.

Proposition 5.3. *Let α be a rectifiable arc in $\mathbb{C}$ defined on the parameter interval $[a, b]$, and let f be a continuous complex-valued function defined on the range of α. Then there exists a unique complex number J with the following property: If ε is an arbitrary positive number, then there exists a positive number $\delta = \delta_\varepsilon$ such that if $\mathscr{P}$ is any partition of $[a, b]$ with mesh $\mathscr{P} < \delta$, and if S is an arbitrary Riemann sum for f with respect to α based on $\mathscr{P}$, then $|J - S| < \varepsilon$. This number J is called the* line integral of f along α, *and is denoted by*

$$J = \int_\alpha f(\zeta) d\zeta.$$

From this definition and the foregoing discussion it is clear that if α is a rectifiable arc such that $\alpha = \alpha_1 + \cdots + \alpha_n$ and f is a function defined and continuous on the range of α, then

$$\int_\alpha f(\zeta) d\zeta = \sum_{j=1}^{n} \int_{\alpha_j} f(\zeta) d\zeta$$

and

$$\int_{\tilde{\alpha}} f(\zeta) d\zeta = -\int_\alpha f(\zeta) d\zeta.$$

Moreover, it is readily established that if α is a rectifiable arc in $\mathbb{C}$ with range W, then the mapping

$$f \to \int_\alpha f(\zeta)d\zeta$$

is a linear functional on $\mathscr{C}(W)$ satisfying the condition

$$\left| \int_\alpha f(\zeta)d\zeta \right| \le L(\alpha) \max_{\zeta \in W} |f(\zeta)|, \qquad f \in \mathscr{C}(W). \tag{7}$$

Example C. If λ_0 and λ_1 are complex numbers and $[a, b]$ is a nondegenerate interval, then the *linear parametrization* on $[a, b]$ of the directed line segment $\sigma = \sigma(\lambda_0, \lambda_1)$ joining λ_0 to λ_1 is given by

$$\alpha(t) = \frac{1}{b - a}[(t - a)\lambda_1 + (b - t)\lambda_0], \qquad a \le t \le b.$$

The length of this arc is clearly what it should be, viz., $|\lambda_1 - \lambda_0|$. Moreover, if f is a continuous complex-valued function on σ, that is, on the range of α, then the line integral of f along α is easily seen to be independent of the choice of a and b (cf. Problem E or Problem F). We shall denote this integral by

$$\int_{\sigma(\lambda_0, \lambda_1)} f(\zeta)d\zeta,$$

or, when possible, simply by

$$\int_\sigma f(\zeta)d\zeta.$$

If $\{\lambda_0, \lambda_1, \ldots, \lambda_n\}$ is any finite sequence of complex numbers, then an arc $\pi = \pi(\lambda_0, \ldots, \lambda_n)$ obtained by forming the sum of linear parametrizations of the segments $\sigma(\lambda_{i-1}, \lambda_i)$, $i = 1, \ldots, n$, is a *polygonal arc* joining the given points. The length of such an arc is simply $L(\pi) = \sum_{i=1}^n |\lambda_i - \lambda_{i-1}|$, and if f is a continuous function on the range of π, then $\int_\pi f(\zeta)d\zeta$ exists and is independent of the (piecewise linear) parametrization of π. For future purposes we observe that if α is an arbitrary rectifiable arc in $\mathbb{C}$ defined on an interval $[a, b]$, and if ε is a positive number, then there exists a corresponding positive number δ such that if $\mathscr{P} = \{t_0 < \cdots < t_n\}$ is an arbitrary partition of $[a, b]$ with mesh $\mathscr{P} < \delta$, and if we write $\lambda_i = \alpha(t_i)$, $i = 1, \ldots, n$, then a polygonal arc $\pi = \pi(\lambda_0, \ldots, \lambda_n)$ joining these points has the property that $L(\alpha) < L(\pi) + \varepsilon$. Moreover, reducing δ if necessary, we may also arrange things so that $|\alpha(t) - \pi(t)| < \varepsilon$ for all $a \le t \le b$ simply by choosing the parameter interval of each segment $\sigma(\lambda_{i-1}, \lambda_i)$ to be the corresponding subinterval $[t_{i-1}, t_i]$ of $\mathscr{P}$. (The proofs of these last assertions are not altogether trivial; see Example 4B and Problem D(iii).)

Example D. For each complex number λ_0 and positive number r the arc

$$\gamma(t) = \lambda_0 + re^{it}, \qquad 0 \le t \le 2\pi, \tag{8}$$

is the *standard parametrization* of the circle $C = C_r(\lambda_0)$ of radius r and center λ_0. This arc is rectifiable, of course, with length $2\pi r$. In the sequel, given a continuous complex-valued function f on $C_r(\lambda_0)$, we shall write

$$\int_{C_r(\lambda_0)} f(\zeta)d\zeta,$$

or, when possible, simply

$$\int_C f(\zeta)d\zeta$$

for the integral of f along the arc γ in (8).

While Proposition 5.3 assures us of the existence of the line integral $\int_\alpha f \, d\zeta$ whenever α is a rectifiable arc and f is continuous on the range of α, it gives no hint as to how such an integral is to be evaluated. It is appropriate to recall therefore that when α is piecewise smooth and defined, say, on the parameter interval $[a, b]$, the integral of f along α can be evaluated, at least in principle, by the familiar formula

$$\int_\alpha f(\zeta)d\zeta = \int_a^b f(\alpha(t))\alpha'(t)dt$$

(see Problem F). Thus, for example, if f is continuous on the circle $C = C_r(\lambda_0)$, then

$$\int_C f(\zeta)d\zeta = ri \int_0^{2\pi} f(\lambda_0 + re^{it})e^{it} \, dt.$$

If α, β, and γ are any three complex numbers, then a polygonal arc joining the points $\{\alpha, \beta, \gamma, \alpha\}$ will be denoted by $[\alpha, \beta, \gamma]$. (The symbol $[\alpha, \beta, \gamma]$ thus denotes an object that is unique only up to piecewise linear reparametrization; since we are interested principally in various line integrals along such arcs, this is no drawback.) When α, β, and γ are noncollinear, the range W of such a *triangular arc* is a triangle in $\mathbb{C}$, and the union of W and Int(W) (Ch. 3, p. 40) will be called the *triangular region* determined by α, β, and γ. (In the case of three collinear vertices α, β, and γ, the triangular region determined by them is understood to consist of the line segment constituting the range of $[\alpha, \beta, \gamma]$.)

We here use the fact that a triangular arc is a Jordan loop. As it happens, the Jordan curve theorem (Th. 3.11) for a simple polygon in $\mathbb{C}$ is a theorem in elementary plane geometry, and the notation Int(P) and Ext(P) will be used without further explanation when P is a simple polygon.

The following result is the cornerstone upon which the entire edifice of complex analysis is based.

Theorem 5.4. *Let Δ be a domain in $\mathbb{C}$, let α_0 be a point in Δ, and let f be a function that is continuous on Δ and analytic on $\Delta\backslash\{\alpha_0\}$. If α, β, and γ are complex numbers such that the triangular region determined by them is contained in Δ, then*

$$\int_{[\alpha,\,\beta,\,\gamma]} f(\zeta)d\zeta = 0.$$

For a proof of this fundamental result when f is assumed to be analytic on the entire domain Δ we refer the reader to any standard textbook on complex analysis. For a proof of the theorem as stated here one may consult [57]. (Proofs of the other results stated in this chapter may be constructed by the reader on the basis of the material in the problems.) The principal tool needed to exploit Theorem 5.4 is set forth in the following result, a proof of which is sketched in Problem F.

Proposition 5.5. *Let f be a function defined and continuous on a domain Δ and suppose f possesses a primitive F on Δ. In other words, suppose there is an analytic function F on Δ whose derivative is f. Then for an arbitrary rectifiable arc α in Δ we have*

$$\int_{\alpha} f(\zeta)d\zeta = F(\lambda_1) - F(\lambda_0),$$

where λ_0 and λ_1 denote the initial and terminal points of α, respectively. In particular, if γ is a closed rectifiable arc in Δ, then

$$\int_{\gamma} f(\zeta)d\zeta = 0.$$

Example E. If $p(\lambda) = \alpha_n \lambda^n + \cdots + \alpha_0$ is a polynomial, and if γ is an arbitrary closed rectifiable arc in $\mathbb{C}$, then $\int_{\gamma} p(\lambda)d\lambda = 0$, since p possesses the primitive

$$\frac{\alpha_n}{n+1} \lambda^{n+1} + \cdots + \alpha_0 \lambda$$

on the domain $\mathbb{C}$. Likewise if α is any fixed complex number and γ is a closed rectifiable arc in $\mathbb{C}\backslash\{\alpha\}$, then

$$\int_{\gamma} \frac{d\lambda}{(\lambda-\alpha)^{n+1}} = 0$$

for every positive integer n.

Theorem 5.6. *Let γ be a closed rectifiable arc in $\mathbb{C}$ and let W denote the range of γ. Then the function*

$$w_\gamma(\lambda) = \frac{1}{2\pi i} \int_\gamma \frac{d\zeta}{\zeta - \lambda}$$

is constant and integer-valued on each component of the complement $\mathbb{C} \backslash W$. In particular, w_γ vanishes identically on the unbounded component of $\mathbb{C} \backslash W$. The integer $w_\gamma(\lambda)$ is called the index *or* winding number *of γ at (or about) λ.*

PROOF. Since W is compact it is easily seen that $w_\gamma(\lambda)$ tends to zero as λ tends to infinity. Similarly it is easy to verify that w_γ is continuous on $\mathbb{C} \backslash W$. (If $\lambda_0 \notin W$ and if λ is sufficiently close to λ_0, then $1/(\zeta - \lambda)$ tends uniformly to $1/(\zeta - \lambda_0)$ on W as λ tends to λ_0.) Hence it is enough to prove that w_γ assumes only integral values on $\mathbb{C} \backslash W$. Moreover (Prob. D(iii)) we may, and do, assume that γ is piecewise smooth. Suppose γ is defined on the parameter interval $[a, b]$, so that

$$w_\gamma(\lambda) = \frac{1}{2\pi i} \int_a^b \frac{\gamma'(t)}{\gamma(t) - \lambda} \, dt$$

(Prob. F). If we define

$$\varphi(s) = \exp\left\{ \int_a^s \frac{\gamma'(t)}{\gamma(t) - \lambda} \, dt \right\}, \qquad a \le s \le b,$$

then straightforward calculation shows that

$$\frac{\varphi'(s)}{\varphi(s)} = \frac{\gamma'(s)}{(\gamma(s) - \lambda)},$$

and hence that the function $\psi(s) = \varphi(s)/(\gamma(s) - \lambda)$ has derivative zero whenever it is differentiable, which is at all but a finite number of values of s. Since ψ is continuous on $[a, b]$, this shows that ψ is a constant function, and since $\varphi(a) = 1$, we have $\psi(s) \equiv 1/(\gamma(a) - \lambda)$, and therefore

$$\varphi(s) = \frac{\gamma(s) - \lambda}{\gamma(a) - \lambda}, \qquad a \le s \le b.$$

Since γ is closed, $\gamma(b) = \gamma(a)$, and therefore $\varphi(b) = 1$. But this implies that

$$\int_\gamma \frac{d\zeta}{\zeta - \lambda}$$

is an integral multiple of $2\pi i$ (Ex. B), and the proof is complete. $\qquad \square$

Example F. If γ is the standard parametrization of the circle $C_r(\alpha)$ (Ex. D), then

$$\int_\gamma \frac{d\lambda}{\lambda - \alpha} = ri \int_0^{2\pi} \frac{dt}{r} = 2\pi i.$$

Thus the winding number of γ about every point inside $C_r(\alpha)$ is one, while the winding number of γ about every point outside $C_r(\alpha)$ is zero. For this reason the standard parametrization of a circle is said to be *positive*. Note that in this positive parametrization the *inward tending normal* vector to $C_r(\alpha)$ (i.e., the vector $-(\gamma(t) - \alpha))$ is $\pi/2$ ahead of the tangent vector $(= \gamma'(t))$ (cf. Problem X).

Example G. Let Q be a square region in $\mathbb{C}$ with sides parallel to the real and imaginary axes. In other words, Q is the intersection of a closed strip of some width w in $\mathbb{C}$ parallel to the real axis, and a closed strip of the same width parallel to the imaginary axis. Let λ_0 denote the center of Q, and let C denote the circle with center λ_0 and radius $w/\sqrt{2}$, so that, if γ is the standard parametrization of C, then $\kappa_1 = \gamma(\pi/4)$, $\kappa_2 = \gamma(3\pi/4)$, $\kappa_3 = \gamma(5\pi/4)$, and $\kappa_4 = \gamma(7\pi/4)$, are the four vertices of ∂Q. Then a polygonal arc $\pi = \pi(\kappa_1, \kappa_2, \kappa_3, \kappa_4, \kappa_1)$ is a *standard parametrization* of the square ∂Q. Since λ_0 is in the unbounded component of the complement of the range of each of the four closed arcs formed by adding an edge of ∂Q to the shorter subarc of $\tilde{\gamma}$ joining the end points of that edge, it is clear that the winding number about λ_0 of a standard parametrization π of ∂Q is one. Thus π has winding number one at each point of $Q^\circ = \mathrm{Int}(\partial Q)$ and winding number zero at each point of $\mathbb{C} \backslash Q$. For this reason a standard parametrization of ∂Q is said to be *positive*. Note here again that in such a positive parametrization the inward tending normal vector at each point of ∂Q (other than a vertex) is $\pi/2$ ahead of the tangent vector. Consequently if Q' is another square region in $\mathbb{C}$ that shares *exactly one edge* with Q, then, in standard parametrizations of ∂Q and $\partial Q'$, that common edge is traversed in opposite directions.

Similarly, if α, β, γ are any three noncollinear points in $\mathbb{C}$, then a *positive parametrization of* $[\alpha, \beta, \gamma]$ is a polygonal arc joining the vertices in the order in which they occur in a standard parametrization of the circle passing through them. Suppose $\pi = \pi(\alpha, \beta, \gamma, \alpha)$ is such a positive parametrization, and let T denote the triangular region determined by α, β, and γ. Then w_π is one on T° and zero on $\mathbb{C} \backslash T$. Note here too that if T' is some other triangular region that meets T in exactly one common edge, then that edge is traversed in opposite senses in positive parametrizations of ∂T and $\partial T'$.

Using this last observation it is easy to show that if P is any simple polygon in $\mathbb{C}$, then there exists a closed (essentially unique) polygonal arc π such that P is the range of π and such that w_π is one on $\mathrm{Int}(P)$ and zero on $\mathrm{Ext}(P)$. More generally, it can be shown that if γ is any Jordan loop in $\mathbb{C}$ that is piecewise smooth and regular, then $|w_\gamma| \equiv 1$ on $\mathrm{Int}(\gamma)$, so that one or the other of γ and $\tilde{\gamma}$ has winding number one about every point of $\mathrm{Int}(\gamma)$. This choice of parametrization is said to be *positive*, its opposite to be *negative* (see Problem X).

Theorem 5.7 (Cauchy–Goursat Theorem in a Disc). *Let $D = D_R(\alpha)$ be an open disc in $\mathbb{C}(R > 0)$, let α_0 be a point in D, and let f be a continuous function on D that is analytic on $D\backslash\{\alpha_0\}$. Then f possesses a primitive on D, so that the integral of f about any closed rectifiable arc in D vanishes. In particular, this is true if f is analytic on D.*

PROOF. For each point λ in D we define $F(\lambda) = \int_\sigma f(\zeta)d\zeta$, where σ denotes the directed line segment $\sigma(\alpha, \lambda)$. Then according to Theorem 5.4 we have $F(\lambda) - F(\lambda_0) = \int_\tau f(\zeta)d\zeta$ for any two points λ and λ_0 of D, where τ denotes the segment $\sigma(\lambda_0, \lambda)$. Hold λ_0 fixed, let ε be an arbitrary positive number, and let δ be chosen so that $|f(\lambda) - f(\lambda_0)| < \varepsilon$ whenever $|\lambda - \lambda_0| < \delta$. For each such λ we have

$$F(\lambda) - F(\lambda_0) - f(\lambda_0)(\lambda - \lambda_0) = \int_\tau [f(\zeta) - f(\lambda_0)]d\zeta,$$

and therefore

$$|F(\lambda) - F(\lambda_0) - f(\lambda_0)(\lambda - \lambda_0)| < \varepsilon|\lambda - \lambda_0|$$

by virtue of Proposition 5.5 and the estimate (7). Thus

$$\left|\frac{F(\lambda) - F(\lambda_0)}{\lambda - \lambda_0} - f(\lambda_0)\right| < \varepsilon$$

whenever $0 < |\lambda - \lambda_0| < \delta$, which shows that $F'(\lambda_0) = f(\lambda_0)$, and hence that F is a primitive of f on D. $\square$

Theorem 5.8 (Cauchy Integral Formula in a Disc). *Let $D = D_R(\alpha)$ be a disc in $\mathbb{C}(R > 0)$, let f be a function defined and analytic on D, and let γ be a closed rectifiable arc in D. Then*

$$w_\gamma(\lambda)f(\lambda) = \frac{1}{2\pi i}\int_\gamma \frac{f(\zeta)}{\zeta - \lambda}\,d\zeta$$

at every point λ of D that is not in the range of γ, where $w_\gamma(\lambda)$ denotes the winding number of γ at λ. In particular, if $C = C_r(\alpha)$ is a circle about α with radius r, $0 < r < R$, then

$$f(\lambda) = \frac{1}{2\pi i}\int_C \frac{f(\zeta)}{\zeta - \lambda}\,d\zeta$$

for every λ inside C.

PROOF. The function

$$g(\zeta) = \begin{cases} \dfrac{f(\zeta) - f(\lambda)}{\zeta - \lambda}, & \zeta \neq \lambda, \\[2mm] f'(\lambda), & \zeta = \lambda, \end{cases}$$

is continuous on D and analytic on $D\backslash\{\lambda\}$ for any one fixed λ in D. Hence

$$\int_\gamma g(\zeta)d\zeta = \int_\gamma \frac{f(\zeta)}{\zeta - \lambda}\,d\zeta - f(\lambda)\int_\gamma \frac{d\zeta}{\zeta - \lambda} = 0$$

by Theorem 5.7. $\square$

The consequences of these two theorems are very numerous and of enormous importance. Here are a few that we shall find useful.

Example H. Let K denote a compact subset of a disc $D = D_R(\alpha)$, let r be a positive radius such that $r < R$ and such that $K \subset D_r(\alpha)$, and let $d_0 = d(K, C_r(\alpha))$. Then for any analytic function f on D, and for arbitrary λ in K and ξ such that $|\xi| < d_0$, we have

$$f(\lambda + \xi) - f(\lambda) = \frac{1}{2\pi i} \int_{C_r(\alpha)} f(\zeta) \left[\frac{1}{\zeta - \lambda - \xi} - \frac{1}{\zeta - \lambda} \right] d\zeta$$

$$= \frac{\xi}{2\pi i} \int_{C_r(\alpha)} \frac{f(\zeta) d\zeta}{(\zeta - \lambda)(\zeta - \lambda - \xi)}.$$

Hence there exist positive constants ε and M (e.g., $\varepsilon = d_0/2$ and $M = 2r/d_0^2$) such that

$$|f(\lambda + \xi) - f(\lambda)| \le M|\xi| \max_{\zeta \in C_r(\alpha)} |f(\zeta)|$$

whenever $\lambda \in K$ and $|\xi| < \varepsilon$.

This discussion only pertains directly to compact subsets of a disc and functions analytic on that disc. However, every compact subset K of a domain Δ can be covered by a finite number of open discs each as small as desired (Heine–Borel theorem; Example 3F), and K can then be decomposed into the union of compact subsets each of which is contained in one of the covering discs (see Example 3G). Hence the foregoing observations lead directly to the following conclusion. Let K be a compact subset of a domain Δ, and let U be any bounded open subset of Δ such that $K \subset U$ and $U^- \subset \Delta$. Then there exist positive constants ε and M (which depend on the geometry of K and U but on nothing else) such that if $\lambda \in K$ and $|\xi| < \varepsilon$, and if f is any analytic function on Δ, then

$$|f(\lambda + \xi) - f(\lambda)| \le M|\xi| \sup_{\zeta \in U} |f(\zeta)|.$$

In particular, any collection of analytic functions on Δ that is uniformly bounded on U is uniformly equicontinuous on K, and is therefore totally bounded in the metric of uniform convergence on K (see Problem 4R).

Example I. Let D, K, and r be as in the preceding example, and let f be an analytic function on D. Then, as we just saw, if $\lambda \in K$ and $|\xi|$ is small enough ($\xi \ne 0$),

$$\frac{f(\lambda + \xi) - f(\lambda)}{\xi} = \frac{1}{2\pi i} \int_{C_r(\alpha)} \frac{f(\zeta) d\zeta}{(\zeta - \lambda)(\zeta - \lambda - \xi)}.$$

Similarly, if η is another nonzero complex number with $|\eta|$ sufficiently small,

$$\frac{f(\lambda + \eta) - f(\lambda)}{\eta} = \frac{1}{2\pi i} \int_{C_r(\alpha)} \frac{f(\zeta) d\zeta}{(\zeta - \lambda)(\zeta - \lambda - \eta)},$$

whence it follows that

$$\frac{f(\lambda + \xi) - f(\lambda)}{\xi} - \frac{f(\lambda + \eta) - f(\lambda)}{\eta}$$

$$= \frac{\xi - \eta}{2\pi i} \int_{C_r(\alpha)} \frac{f(\zeta)d\zeta}{(\zeta - \lambda)(\zeta - \lambda - \xi)(\zeta - \lambda - \eta)}.$$

Hence, using the Heine–Borel theorem just as in the preceding example, we conclude that if K is a compact subset of an arbitrary domain Δ, and if U is any bounded open subset of Δ such that $K \subset U$ and $U^- \subset \Delta$, then there exist positive constants ε and M such that

$$\left| \frac{f(\lambda + \xi) - f(\lambda)}{\xi} - \frac{f(\lambda + \eta) - f(\lambda)}{\eta} \right| \le M|\xi - \eta| \sup_{\zeta \in U} |f(\zeta)| \qquad (9)$$

for all λ in K, all ξ and η such that $0 < |\xi|, |\eta| < \varepsilon$, and all analytic functions f on Δ. Letting η tend to zero in (9) we find that

$$\left| \frac{f(\lambda + \xi) - f(\lambda)}{\xi} - f'(\lambda) \right| \le M|\xi| \sup_{\zeta \in U} |f(\zeta)|$$

for all λ in K and all ξ such that $0 < |\xi| < \varepsilon$. Thus the difference quotient of f tends uniformly to f' on K.

Example J (Taylor's Theorem and the Cauchy Estimates). Let r, r', and R be positive numbers such that $r' < r < R$, and let α be a point in the complex plane. Straightforward calculation discloses that for each ζ on $C_r(\alpha)$ we have

$$\frac{1}{\zeta - \lambda} = \sum_{n=0}^{\infty} \frac{(\lambda - \alpha)^n}{(\zeta - \alpha)^{n+1}} \qquad (10)$$

for every λ inside $C_r(\alpha)$. Moreover, the convergence in (10) is uniform in the variable ζ on $C_r(\alpha)$ for each λ inside $C_{r'}(\alpha)$. Hence if f is an arbitrary analytic function on $D = D_R(\alpha)$, we have

$$f(\lambda) = \frac{1}{2\pi i} \int_{C_r(\alpha)} \frac{f(\zeta)}{\zeta - \lambda} d\zeta$$

$$= \sum_{n=0}^{\infty} (\lambda - \alpha)^n \frac{1}{2\pi i} \int_{C_r(\alpha)} \frac{f(\zeta)}{(\zeta - \alpha)^{n+1}} d\zeta$$

$$= \sum_{n=0}^{\infty} \alpha_n (\lambda - \alpha)^n, \qquad |\lambda - \alpha| < r', \qquad (11)$$

where

$$\alpha_n = \frac{1}{2\pi i} \int_{C_r(\alpha)} \frac{f(\zeta)}{(\zeta - \alpha)^{n+1}} d\zeta, \qquad n \in \mathbb{N}_0. \qquad (12)$$

(The coefficients α_n are also given by the familiar formula $\alpha_n = f^{(n)}(\alpha)/n!$ as noted in Example A, cf. Problem G.) Since r' can be taken to be any positive number smaller than r, it follows that the representation (11) is valid for all λ in $D_r(\alpha)$, and since r can be taken to be any positive number smaller than R, we see that, in fact, the representation (11) holds everywhere in D. Thus Taylor's theorem is proved. Moreover, from the expression (12) for the Taylor coefficients of f we learn that

$$|\alpha_n| \leq \frac{2\pi r}{2\pi} \frac{M_r}{r^{n+1}} = \frac{M_r}{r^n}, \qquad n \in \mathbb{N}_0,$$

where M_r denotes the maximum modulus of f on the circle $C_r(\alpha): M_r = \max\{|f(\zeta)| : |\zeta - \alpha| = r\}$. These inequalities are known as the *Cauchy estimates* for f about the point α.

Example K (Liouville's Theorem on Entire Functions). A function f that is defined and analytic on the entire complex plane $\mathbb{C}$ is called an *entire* function. Examples of entire functions are polynomial functions, the exponential function, and the circular functions. If f is an entire function and α an arbitrary complex number, then the Taylor series expansion

$$f(\lambda) = \sum_{n=0}^{\infty} \alpha_n(\lambda - \alpha)^n$$

of f about α converges everywhere. Hence the Cauchy estimates hold for every $r > 0$. Thus $|\alpha_n| \leq M_r/r^n$ for every positive integer n and every positive number r. From this we conclude immediately that *if f is a bounded entire function, then f is a constant function.* This result is known as *Liouville's theorem*.

Example L (Morera's Theorem). Taylor's theorem shows that an analytic function (that is, a differentiable function) on a domain Δ is actually infinitely differentiable on Δ. Hence any function that possesses a primitive on Δ is automatically analytic on Δ. Suppose f is a continuous function on Δ such that

$$\int_{[\alpha, \beta, \gamma]} f(\zeta)d\zeta = 0$$

for every triangular arc $[\alpha, \beta, \gamma]$ in Δ, and let λ_0 be a point of Δ. If $D_r(\lambda_0)$ is any disc of positive radius that is contained in Δ, then, just as in the proof of Theorem 5.7, the function

$$F(\lambda) = \int_{\sigma(\lambda_0, \lambda)} f(\zeta)d\zeta$$

may be seen to be a primitive of f, so f is analytic on $D_r(\lambda_0)$. In particular f is differentiable at λ_0, and since λ_0 is an arbitrary point of Δ, we see that f is analytic on Δ. Thus we have proved that *if a function f is continuous on a domain Δ and if the integral of f around every triangular arc in Δ is zero, then f is analytic on Δ.* This result is known as *Morera's theorem.* (As the argument shows, it suffices to consider the integral of f about arbitrarily small triangles each of which is contained in some disc contained in Δ.)

Example M (The Maximum Modulus Principle). Let f be an analytic function on a disc $D = D_R(\alpha)$ $(R > 0)$, and for each radius r, $0 \leq r < R$, let M_r denote the maximum modulus of f on the circle $C_r(\alpha)$ as defined in Example J. Since

$$f(\alpha) = \frac{1}{2\pi i} \int_{C_r(\alpha)} \frac{f(\zeta)}{\zeta - \alpha} d\zeta, \tag{13}$$

it follows from (7) that $|f(\alpha)| = M_0 \leq M_r$ for every r, $0 < r < R$. More is true however. Let us suppose, as we may without loss of generality, that $f(\alpha)$ is positive, so that $f(\alpha) = |f(\alpha)|$, and evaluate the integral in (13) according to Problem F. Taking real parts we obtain

$$f(\alpha) = \frac{1}{2\pi} \int_0^{2\pi} \text{Re } f(\alpha + re^{it}) dt,$$

and therefore

$$\frac{1}{2\pi} \int_0^{2\pi} [f(\alpha) - \text{Re } f(\alpha + re^{it})] dt = 0. \tag{14}$$

Suppose now that $f(\alpha) = M_r$. Then the integrand $[f(\alpha) - \text{Re } f(\alpha + re^{it})]$ in (14) is nonnegative and continuous, and must therefore vanish identically, whence it follows at once that f is constantly equal to $f(\alpha)$ on $C_r(\alpha)$, and therefore, by the identity theorem, constantly equal to $f(\alpha)$ on the entire domain on which f is defined (and analytic). It follows that if f is an arbitrary analytic function on a domain Δ, and if K is any compact subset of Δ, then the maximum of $|f(\lambda)|$ on K is assumed on the boundary ∂K. In particular, the maximum modulus function M_r itself is seen to yield the maximum of $|f(\lambda)|$ over the entire closed disc $D_r(\alpha)^-$, not just over the boundary $C_r(\alpha)$. Hence for any nonconstant analytic function on a domain Δ, and for any point α in Δ, the maximum modulus function M_r is a *strictly increasing* function of r wherever it is defined.

More general versions of the Cauchy integral formula may be formulated and employed to obtain other important results. The version given in Problem O is appropriate for the following result, the proof of which may be patterned after the proof of Taylor's theorem.

Proposition 5.9. *Let r_0 and r_1 be nonnegative numbers such that $r_0 < r_1$, let α be a complex number, and suppose f is an analytic function on the* *annular domain* $A = \{\lambda \in \mathbb{C} : r_0 < |\lambda - \alpha| < r_1\}$. *Then there exist coefficients* $\{\alpha_n\}_{n=-\infty}^{+\infty}$ *such that*

$$f(\lambda) = \sum_{n=-\infty}^{+\infty} \alpha_n(\lambda - \alpha)^n, \qquad \lambda \in A. \tag{15}$$

The coefficients α_n *in* (15) *are uniquely determined by* f, *and the convergence in* (15) *is absolute in* A *and uniform on compact subsets of* A. *The series* (15) *is known as the* Laurent expansion *of* f *in* A.

A special case of particular importance arises when the inner radius r_0 is zero.

Definition. If f is an analytic function on a domain Δ, and if for some complex number α not belonging to Δ there exists a positive number r such that Δ contains the punctured disc $\{\lambda \in \mathbb{C} : 0 < |\lambda - \alpha| < r\}$, or, in other words, if α is an isolated point of $\partial\Delta$, then α is an *isolated singularity of* f.

If α is an isolated singularity of an analytic function f on a domain Δ, then f possesses a Laurent expansion

$$f(\lambda) = \sum_{n=-\infty}^{+\infty} \alpha_n(\lambda - \alpha)^n$$

in which the power series

$$\sum_{n=0}^{\infty} \alpha_n(\lambda - \alpha)^n$$

has some positive radius of convergence r, while the series

$$\sum_{n=1}^{\infty} \alpha_{-n}(\lambda - \alpha)^{-n}$$

converges for every $\lambda \neq \alpha$. (The sum of this latter series is known as the *principal part* of f at α.) Thus if we write $\tau = 1/(\lambda - \alpha)$, then the function $\sum_{n=1}^{\infty} \alpha_{-n}\tau^n$ is an entire function of τ.

There are three possible cases to be distinguished. To begin with, the function f is bounded on some punctured disc about α if and only if the principal part of f at α is so bounded, and by Liouville's theorem this can only happen when $\alpha_{-n} = 0$ for every positive integer n. In this case f can be extended to α by continuity by setting $f(\alpha) = \alpha_0$, and if this is done, the enlarged function thus obtained is analytic on $\Delta \cup \{\alpha\}$. In this situation we say that the singularity α is *removable* and that the analytic function on $\Delta \cup \{\alpha\}$ obtained by setting $f(\alpha) = \alpha_0$ is obtained by *removing* the singularity at α. (Thus, for example, the apparent singularity allowed for in the statements of Theorems 5.4 and 5.7 is already removed, the function f in each case being analytic on Δ.)

If the isolated singularity is not removable there are some nonzero coefficients in the sequence $\{\alpha_{-n}\}_{n=1}^{\infty}$, and two sharply different cases arise. If there are infinitely many nonzero, negatively indexed coefficients, then α is an *essential singularity* of f. If there are only finitely many nonzero, negatively indexed coefficients, and if N denotes the largest positive integer n for which $\alpha_{-n} \neq 0$, then the function $(\lambda - \alpha)^N f(\lambda)$ has a removable singularity at α. If g denotes the function obtained by removing the singularity at α, then g satisfies the condition $g(\alpha) \neq 0$, and we have

$$f(\lambda) = \frac{g(\lambda)}{(\lambda - \alpha)^N}.$$

In this situation it is said that f has a *pole* of *order N* at α.

In this connection we note that the classification of singularities is extended to the point at infinity in a systematic manner (see Problem 3W). If f is a function defined and analytic on the complement in $\mathbb{C}$ of some closed disc $D_r(0)^-$, then the point at infinity is an *isolated singularity* of f, and is said to be an *essential singularity* of f or a *pole* of *order N* if the function $h(\tau) = f(1/\tau)$ has an essential singularity or pole of order N at $\tau = 0$. On the other hand, if the function h has a removable singularity at $\tau = 0$, then f is said to have a *removable singularity at infinity*, and when this is the case, the function obtained by defining f at ∞ by continuity is said to be *analytic at infinity*.

In the same vein we also recall that if f is an analytic function on a domain Δ and if $f(\alpha) = 0$, then α is called a *zero* of f. If α is a zero of f then by Taylor's theorem (Th. 5.1) there exist a unique positive integer N and an analytic function g on Δ such that $g(\alpha) \neq 0$ and $f(\lambda) = (\lambda - \alpha)^N g(\lambda)$. In this situation it is said that α is a zero of *order N* of f. If f has a zero of order N at α then $1/f$ has a pole of order N at α. Conversely, if f has a pole of order N at α, then the result of removing the singularity in $1/f$ at α has a zero of order N there.

Example N. An entire function f with a removable singularity at infinity is bounded, and is therefore a constant by Liouville's theorem (Ex. K). An entire function f with a pole at infinity is bounded away from zero outside some disc centered at the origin, and can therefore have only a finite number of zeros. If $\alpha_1, \ldots, \alpha_n$ is a complete list of the zeros of f, and k_i denotes the order of the zero α_i, $i = 1, \ldots, n$, then the function g obtained by removing the removable singularities of $f/(\lambda - \alpha_1)^{k_1} \cdots (\lambda - \alpha_n)^{k_n}$ at the zeros $\alpha_1, \ldots, \alpha_n$ is entire, has no zeros, and also has at worst a pole at infinity. Hence $1/g$ is a bounded entire function, and is therefore a nonzero constant. If we denote this constant by $1/\alpha_0$, then f is seen to coincide with the polynomial $\alpha_0(\lambda - \alpha_1)^{k_1} \cdots (\lambda - \alpha_n)^{k_n}$. Thus the polynomial functions on $\mathbb{C}$ are characterized as those entire functions with (at worst) a pole at infinity. Similarly, it is easy to see that the rational functions on $\mathbb{C}$ may be characterized as those functions that are holomorphic at every point of $\mathbb{C}$ except for a finite number of poles and that possess (at worst) a pole at infinity.

We conclude this chapter with a brief remark concerning the relations between analytic functions and harmonic functions. To facilitate this discussion we note first that there is an obvious one-to-one correspondence between domains Δ in $\mathbb{C}$ and domains (i.e., nonempty connected open sets) Δ' in $\mathbb{R}^2$ obtained via the correspondence $x + iy \leftrightarrow (x, y)$. Moreover, for each complex-valued function f on a complex domain Δ, there exist two real-valued functions u and v on the corresponding real domain Δ' such that

$$f(x + iy) = u(x, y) + iv(x, y), \qquad (x, y) \in \Delta'.$$

The functions u and v are known customarily (if not with total accuracy) as the *real* and *imaginary parts* of f.

Definition. Let Δ' be a domain in $\mathbb{R}^2$, and let f be a function in $\mathscr{C}^{(2)}(\Delta')$
(Ex. 3E). Then f is *harmonic* on Δ' if the differential equation

$$f_{xx} + f_{yy} = 0$$

is satisfied everywhere on Δ'. Similarly, a function f on a complex domain Δ is *harmonic* on Δ if the function $f_0(x, y) = f(x + iy)$ is harmonic on the corresponding real domain Δ'.

Proposition 5.10. *Let f be a complex-valued function on a complex domain Δ, and let u and v be the real and imaginary parts of f on the corresponding real domain Δ'. If f is analytic on Δ then u and v satisfy the equations*

$$u_x = v_y, \qquad u_y = -v_x, \tag{16}$$

known as the Cauchy–Riemann equations, *identically on Δ'. Conversely, if u and v belong to $\mathscr{C}^{(1)}(\Delta')$ and satisfy the Cauchy–Riemann equations identically on Δ', then f is analytic on Δ.*

PROOF. Let $\lambda_0 = x_0 + iy_0$ be a point of Δ (x_0, y_0 real), let R be a positive radius small enough so that the disc $D_R(\lambda_0) \subset \Delta$, and suppose that f is differentiable at λ_0. If λ is restricted so that $0 < |\lambda - \lambda_0| < R$ and also so that the difference $\lambda - \lambda_0 = h$ is a real number, and if the limit of the difference quotient $(f(\lambda) - f(\lambda_0))/(\lambda - \lambda_0) = (f(\lambda) - f(\lambda_0))/h$ is then taken as h tends to zero, a simple calculation shows that

$$f'(\lambda_0) = u_x(x_0, y_0) + iv_x(x_0, y_0).$$

On the other hand, if $f'(\lambda_0)$ is computed using values of λ such that $\lambda - \lambda_0 = ih$ where h is real, a parallel calculation discloses that

$$f'(\lambda_0) = v_y(x_0, y_0) - iu_y(x_0, y_0).$$

Thus the Cauchy–Riemann equations are satisfied on Δ' whenever f is analytic on Δ.

Suppose now, to go the other way, that $u, v \in \mathscr{C}^{(1)}(\Delta')$. Then, as is shown in advanced calculus (theorem on the total differential), there exist real-valued functions ε', ε'', η', and η'', defined on the disc

$$D' = \{(x, y): (x - x_0)^2 + (y - y_0)^2 < R\}$$

and tending to zero as $(x, y) \to (x_0, y_0)$, such that

$$u(x, y) - u(x_0, y_0) = [u_x(x_0, y_0) + \varepsilon'(x, y)](x - x_0)$$
$$+ [u_y(x_0, y_0) + \varepsilon''(x, y)](y - y_0)$$

and

$$v(x, y) - v(x_0, y_0) = [v_x(x_0, y_0) + \eta'(x, y)](x - x_0)$$
$$+ [v_y(x_0, y_0) + \eta''(x, y)](y - y_0)$$

identically on D'. If, in addition, the Cauchy–Riemann equations are satisfied at (x_0, y_0) then straightforward calculation shows that

$$f(\lambda) - f(\lambda_0) = u_x(x_0, y_0)(\lambda - \lambda_0) + iv_x(x_0, y_0)(\lambda - \lambda_0)$$
$$+ [\varepsilon'(x, y) + i\eta'(x, y)](x - x_0)$$
$$+ [\varepsilon''(x, y) + i\eta''(x, y)](y - y_0).$$

Hence $f'(\lambda_0)$ exists (and equals $u_x(x_0, y_0) + iv_x(x_0, y_0)$), and the result follows. $\qquad \square$

Corollary 5.11. *If f is an analytic function on a complex domain Δ, then the real and imaginary parts of f are harmonic functions on the corresponding real domain Δ'.*

Proposition 5.12. *If u is a real harmonic function on a disc*

$$D' = \{(x, y): (x - x_0)^2 + (y - y_0)^2 < r^2\} \qquad (r > 0)$$

in $\mathbb{R}^2$, then u possesses a harmonic conjugate $\tilde{u}$ on D', that is, there exists an analytic function f on the complex disc $D_r(\alpha)$, where $\alpha = x_0 + iy_0$, such that u and $\tilde{u}$ are the real and imaginary parts of f, respectively. The harmonic conjugate $\tilde{u}$ is unique up to an additive constant.

PROOF. The uniqueness of the harmonic conjugate follows from the fact that the Cauchy–Riemann equations must be satisfied (see also Problem B). To see that $\tilde{u}$ exists, we observe that for any function u in $\mathscr{C}_{\mathbb{R}}^{(1)}(D')$ the functions

$$U_1(x, y) = -\int_{x_0}^{x} u_y(s, y_0)ds + \int_{y_0}^{y} u_x(x, t)dt, \qquad (x, y) \in D',$$

and

$$U_2(x, y) = -\int_{x_0}^{x} u_y(s, y)ds + \int_{y_0}^{y} u_x(x_0, t)dt, \qquad (x, y) \in D',$$

satisfy the conditions

$$\frac{\partial U_1}{\partial y} = u_x \quad \text{and} \quad \frac{\partial U_2}{\partial x} = -u_y$$

on D'. Moreover if u belongs to $\mathscr{C}^{(2)}_{\mathbb{R}}(D')$ we have

$$U_1 - U_2 = \int_{x_0}^{x} \int_{y_0}^{y} [u_{xx}(s, t) + u_{yy}(s, t)] ds \, dt.$$

Thus $U_1 \equiv U_2$ when u is harmonic, and if we set

$$\tilde{u} = U_1 = U_2,$$

then $u + i\tilde{u}$ satisfies the Cauchy–Riemann equations and is therefore analytic by Proposition 5.10.   □

PROBLEMS

A. (Abel's Theorem) Let

$$f(\lambda) = \sum_{n=0}^{\infty} \alpha_n \lambda^n, \tag{17}$$

where the power series (17) has radius of convergence r_0, $0 < r_0 < +\infty$, and suppose that for some point $\lambda_0 = r_0 e^{it_0}$ on the circle of convergence the series $\sum_n \alpha_n \lambda_0^n$ converges, say to σ. Show that

$$\lim_{r \uparrow r_0} f(re^{it_0}) = \sigma.$$

Thus f converges "radially" to the sum of the power series representing f at each point on the circle of convergence at which the series converges. (Hint: It is no loss of generality to assume $r_0 = 1$ and $t_0 = 0$. Multiply the given series by the geometric series $\sum_n \lambda^n$ to show that

$$f(\lambda) = (1 - \lambda) \sum_{n=0}^{\infty} \sigma_n \lambda^n,$$

where $\sigma_n = \sum_{i=0}^{n} \alpha_i$, $n \in \mathbb{N}_0$, and hence that

$$\sigma - f(\lambda) = (1 - \lambda) \sum_{n=0}^{\infty} \rho_n \lambda^n, \qquad |\lambda| < 1,$$

where the sequence $\{\rho_n = \sigma - \sigma_n\}$ tends to zero. Write this last series as the sum of a partial sum and a remainder $\sum_{n=N+1}^{\infty} \rho_n \lambda^n$.)

B. (i) Let f be an analytic function on a domain Δ. Show that the complex conjugate $\bar{f}$ is differentiable at precisely those points λ of Δ at which $f'(\lambda) = 0$. Hence $\bar{f}$ is not analytic on Δ (or on any subdomain of Δ) unless f is a constant. Conclude that if an analytic function f maps any nonempty open set onto a subset of a straight line or onto a subset of a circle, then f must be constant. In particular, a nonconstant analytic function cannot have constant modulus or be real-valued. (Hint: $\lim_{\lambda \to \lambda_0} (\bar{\lambda} - \bar{\lambda}_0)/(\lambda - \lambda_0)$ does not exist at any point λ_0.)

(ii) Let f be an analytic function on a domain Δ, and let $\bar{\Delta}$ denote the reflection of Δ in the real axis—$\bar{\Delta} = \{\lambda \in \mathbb{C} : \bar{\lambda} \in \Delta\}$. Verify that the function

$$f^{\wedge}(\lambda) = \overline{f(\bar{\lambda})}$$

is analytic on $\bar{\Delta}$ and that, in fact, $(f^{\wedge})' = (f')^{\wedge}$ on $\bar{\Delta}$. If $\Delta = \bar{\Delta}$, then $f = f^{\wedge}$ if and only if f is real-valued along the intersection of Δ with the real axis.

As it turns out, if f is an analytic function on a domain Δ, and if $f'(\lambda_0) \neq 0$ at some point λ_0 of Δ, then there exists an open neighborhood U of λ_0 such that $f | U$ is a homeomorphism of U onto an open set in $\mathbb{C}$. Thus Problem B(i) reveals only a fragment of the truth concerning the range of an analytic function.

C. In real analysis it is apparent that the hyperbolic functions $\sinh t$ and $\cosh t$ are closely related to the circular functions. This relation becomes particularly lucid in complex analysis. Show that, in fact,

$$\cosh t = \cos(it) \quad \text{and} \quad \sinh t = -i \sin(it)$$

for every real number t. (These relations are used to define the entire functions $\cosh \zeta$ and $\sinh \zeta$ on $\mathbb{C}$.)

D. (i) Let α be a rectifiable arc in $\mathbb{C}$, and let f be a bounded complex-valued function defined on the range of α. Show that the line integral $\int_\alpha f(\zeta)d\zeta$ exists if and only if the following Cauchy criterion is satisfied: Given $\varepsilon > 0$ there exists $\delta > 0$ such that if $\mathscr{P}$ and $\mathscr{P}'$ are arbitrary partitions of the parameter interval of α with mesh $\mathscr{P}$, $\mathscr{P}' < \delta$ and if S and S' are arbitrary Riemann sums for f with respect to α based on $\mathscr{P}$ and $\mathscr{P}'$, respectively, then $|S - S'| < \varepsilon$. Verify also that if the line integral $\int_\alpha f(\zeta)d\zeta$ does exist, and if ε and δ are as stated, then

$$\left| S - \int_\alpha f(\zeta)d\zeta \right| \leq \varepsilon$$

for any Riemann sum S for f with respect to α based on a partition $\mathscr{P}$ of the parameter interval of α such that mesh $\mathscr{P} < \delta$.

(ii) Let α be a rectifiable arc in $\mathbb{C}$ with parameter interval $[a, b]$, and let f and $\hat{f}$ be bounded complex-valued functions defined on the range of α such that for some fixed nonnegative number η, $|f(\zeta) - \hat{f}(\zeta)| \leq \eta$ for every ζ in that range. Suppose that for a given positive number ε there exists a positive δ with the property that $|f(\alpha(t)) - f(\alpha(t'))| < \varepsilon$ for all t, t' in $[a, b]$ such that $|t - t'| < \delta$. Show that if $\mathscr{P}$ and $\mathscr{P}'$ are partitions of $[a, b]$ such that mesh $\mathscr{P}$, $\mathscr{P}' < \delta$, and S and S' are Riemann sums for f and $\hat{f}$ with respect to α based on $\mathscr{P}$ and $\mathscr{P}'$, respectively, then $|S - S'| < L(2\varepsilon + \eta)$ where L denotes the length of α. (Hint: Consider first the case in which $\mathscr{P}'$ is a refinement of $\mathscr{P}$ and $f = \hat{f}$.) In particular, if S and S' are both Riemann sums for f with respect to α based on $\mathscr{P}$ and $\mathscr{P}'$ respectively, then $|S - S'| \leq 2L\varepsilon$. Complete the proof of Proposition 5.3, and verify that if f is continuous on the range of α, and if ε and δ are related as above, then

$$\left| S - \int_\alpha f(\zeta)d\zeta \right| \leq 2L\varepsilon \qquad \left[\left| S' - \int_\alpha f(\zeta)d\zeta \right| \leq L(2\varepsilon + \eta) \right]$$

for any Riemann sum S for $f[S'$ for $\tilde{f}]$ with respect to α based on any partition $\mathcal{P}$ of $[a, b]$ with mesh $\mathcal{P} < \delta$.

(iii) Let f be a continuous complex-valued function on a domain Δ in $\mathbb{C}$, and let α be a rectifiable arc in Δ. Show that for any arbitrarily given positive number ε there exists a positive number δ such that if $\mathcal{P} = \{t_0 < \cdots < t_n\}$ is any partition of the parameter interval of α such that mesh $\mathcal{P} < \delta$, and if we write $\lambda_i = \alpha(t_i)$, $i = 0, \ldots, n$, then the polygonal arc $\pi = \pi(\lambda_0, \ldots, \lambda_n)$ lies entirely in Δ, and we have

$$\left| \int_\alpha f(\zeta)d\zeta - \int_\pi f(\zeta)d\zeta \right| < \varepsilon.$$

Thus, in particular, every line integral can be approximated as closely as desired by integrals along (regular) piecewise smooth arcs. (Hint: Choose a bounded open set U such that $W \subset U \subset U^- \subset \Delta$, where W denotes the range of α. If $d = d(W, \mathbb{C}\backslash U)$ and if δ is small enough so that mesh $\mathcal{P} < \delta$ implies $|\lambda_i - \lambda_{i-1}| < d, i = 1, \ldots, n$, then π lies in U. Use Example C and part (ii).)

E. Let φ be an arbitrary *strictly increasing* continuous mapping of a real interval $[a, b]$ onto another real interval $[c, d]$. Show that if α is a rectifiable arc in $\mathbb{C}$ defined on the parameter interval $[c, d]$, then $\alpha \circ \varphi$ is also a rectifiable arc having the same length as α, and that if f is any function defined and continuous on the range of α, then

$$\int_\alpha f(\zeta)d\zeta = \int_{\alpha \circ \varphi} f(\zeta)d\zeta.$$

F. Let α be a piecewise smooth arc in $\mathbb{C}$ defined on the parameter interval $[a, b]$, and let f be a continuous function on the range of α. Show that

$$\int_\alpha f(\zeta)d\zeta = \int_a^b f(\alpha(t))\alpha'(t)dt,$$

and use this formula along with Problem D(iii) to prove Proposition 5.5. (Hint: One may assume that α is smooth. If $\mathcal{P} = \{t_0 < \cdots < t_n\}$ is a partition of $[a, b]$, and if τ_i is chosen so that $t_{i-1} \le \tau_i \le t_i, i = 1, \ldots, n$, then

$$S = \sum_{i=1}^n f(\alpha(\tau_i))[\alpha(t_i) - \alpha(t_{i-1})] = \int_a^b s(t)\alpha'(t)dt$$

where s denotes the step function that takes the value $f(\alpha(\tau_i))$ on the interval (t_{i-1}, t_i).)

G. Let α be a rectifiable arc in $\mathbb{C}$ and let g be an arbitrary continuous function on the range W of α. Show that the function

$$f(\lambda) = \int_\alpha \frac{g(\zeta)}{\zeta - \lambda} d\zeta$$

is infinitely differentiable on the complement $\mathbb{C}\backslash W$ by verifying the formula

$$f^{(n)}(\lambda) = n! \int_\alpha \frac{g(\zeta)}{(\zeta - \lambda)^{n+1}} d\zeta, \qquad \lambda \notin W, n \in \mathbb{N}_0.$$

(Hint: If λ and λ_0 are two points in $\mathbb{C} \setminus W$, then

$$\frac{f(\lambda) - f(\lambda_0)}{\lambda - \lambda_0} = \int_\alpha g(\zeta) \frac{d\zeta}{(\zeta - \lambda)(\zeta - \lambda_0)},$$

and the integrand in this last integral tends uniformly in ζ on W to $g(\zeta)/(\zeta - \lambda_0)^2$ as λ tends to λ_0. Use an induction argument.)

H. Thus far we have employed the notion of the sum of a sequence of arcs $\{\alpha_1, \ldots, \alpha_n\}$ only when these arcs are chained. It turns out that a more general concept is also useful. If $\alpha_1, \ldots, \alpha_n$ are arbitrary arcs in $\mathbb{C}$ we introduce the *formal sum* $\alpha = \alpha_1 + \cdots + \alpha_n$ with the understanding that two formal sums $\alpha = \alpha_1 + \cdots + \alpha_n$ and $\beta = \beta_1 + \cdots + \beta_m$ are equal if (and only if) each α_i and β_i can be partitioned into a *chained* sum $\alpha_i = \sum_{j=1}^{n_i} \alpha_{ij}$, $\beta_i = \sum_{j=1}^{m_i} \beta_{ij}$, in such a way that the two systems of arcs $\{\alpha_{ij}\}$ and $\{\beta_{ij}\}$ are pairwise equivalent except for order. Show that if $\alpha = \alpha_1 + \cdots + \alpha_n$ is such a formal sum of rectifiable arcs, and if for each continuous function f on W, where W denotes the union of the ranges of the arcs α_i, we define

$$\int_\alpha f(\zeta) d\zeta = \sum_{i=1}^n \int_{\alpha_i} f(\zeta) d\zeta,$$

then the mapping $f \to \int_\alpha f(\zeta) d\zeta$ is a linear functional on $\mathscr{C}(W)$. In particular, if $\gamma = \gamma_1 + \cdots + \gamma_n$ is a formal sum of closed rectifiable arcs, and if W denotes the union of the ranges of the various arcs γ_i, we define the *winding number* w_γ of γ on the open complement $\mathbb{C} \setminus W$ to be

$$w_\gamma(\lambda) = \frac{1}{2\pi i} \int_\gamma \frac{d\zeta}{\zeta - \lambda},$$

so that $w_\gamma = w_{\gamma_1} + \cdots + w_{\gamma_n}$ at every point of $\mathbb{C} \setminus W$.

I. Suppose given a finite collection $\mathscr{S} = \{\sigma_i\}$ of directed line segments in $\mathbb{C}$, each equipped with a linear parametrization, and suppose that

(i) No two of the segments in $\mathscr{S}$ meet, except possibly at a common endpoint,
(ii) For any endpoint λ_0 of any segment in $\mathscr{S}$ there are exactly as many segments in $\mathscr{S}$ having λ_0 for terminal point as there are having λ_0 for initial point.

(Such a set of directed line segments is called *balanced*.) Show that the formal sum $\sum_{\sigma_i \in \mathscr{S}} \sigma_i$ can also be written as a formal sum $\pi_1 + \cdots + \pi_m$ of closed polygonal arcs whose ranges intersect only at common vertices. (If $\mathscr{S}$ satisfies (i) and the stronger condition

(ii′) Each endpoint of a segment in $\mathscr{S}$ is the initial point, and likewise the terminal point, of *exactly one* segment in $\mathscr{S}$,

then the polygons in the formal sum $\pi_1 + \cdots + \pi_m$ will be simple and pairwise disjoint.)

J. Let m be a positive number. By the *grid* $\mathscr{G}_m$ in $\mathbb{C}$ of *mesh* m we shall mean the system of all the horizontal lines in $\mathbb{C}$ obtained by translating the real axis by amounts kmi, $k \in \mathbb{Z}$, together with the system of all the vertical lines obtained by translating the imaginary axis by amounts km, $k \in \mathbb{Z}$. The closed strip bounded by two adjacent horizontal lines, $\mathbb{R} + kmi$ and $\mathbb{R} + (k + 1)mi$, in the grid is a *horizontal strip of $\mathscr{G}_m$*. Likewise, a *vertical strip of $\mathscr{G}_m$* is a closed strip bounded by two adjacent vertical lines in $\mathscr{G}_m$. Finally, a square region Q is a *square region of $\mathscr{G}_m$* if it is the intersection of a horizontal strip of $\mathscr{G}_m$ and a vertical strip of $\mathscr{G}_m$, and a complex number λ_0 is a *vertex of $\mathscr{G}_m$* if it is a vertex of some square region of $\mathscr{G}_m$.

Now let $\{Q_1, \ldots, Q_n\}$ be an arbitrary finite, nonempty collection of square regions of a grid $\mathscr{G}_m$, and set $L = Q_1 \cup \cdots \cup Q_n$. Show that the boundary ∂L can be parametrized as a formal sum $\pi = \pi_1 + \cdots + \pi_m$ of closed polygonal arcs in such a way that the winding number of π is *one* at each point of L° and zero on the complement of L. (Hint: Let γ_i denote a standard parametrization of ∂Q_i, $i = 1, \ldots, n$, set $\gamma = \sum_{i=1}^{n} \gamma_i$, and verify that the winding number of γ is one in the interior of each Q_i and zero outside L. Then write γ as a formal sum of all of the edges of all of the squares ∂Q_i, and thus as a formal sum of directed line segments. Remove all matched pairs of oppositely directed line segments, and show that this removes all line segments that meet L°, so that what is left is really a parametrization of ∂L given as a formal sum of directed line segments. Then apply Problem I.)

K. Let U be an open subset of $\mathbb{C}$ and let K be a compact subset of U.

(i) Show that there exists a formal sum $\gamma = \gamma_1 + \cdots + \gamma_n$ of closed rectifiable arcs in $U \backslash K$ with the following properties:

(a) If V denotes the (open) set of points λ at which $w_\gamma(\lambda) = 1$, then $K \subset V$ and V^- is a (compact) subset of U,

(b) The range of γ constitutes the topological boundary of V, and

(c) The winding number of γ is zero at every point of $\mathbb{C} \backslash V^-$.

Such a formal sum γ will be called an *oriented envelope of K in U*. (Hint: If $U = \mathbb{C}$ or $K = \varnothing$ the result is trivial; otherwise let $d = d(K, \mathbb{C} \backslash U)$, choose a mesh m such that $m < d/\sqrt{2}$, and let $Q_1, \ldots, Q_p$ be the set of closed squares of the grid $\mathscr{G}_m$ that touch K. Set $L = Q_7 \cup \cdots \cup Q_p$, and parametrize ∂L as in the preceding problem.)

(ii) Show further that in the foregoing construction it is possible to arrange matters so that the individual arcs γ_i, $i = 1, \ldots, n$, are Jordan loops with pairwise disjoint ranges, and hence so that V^- is a finite union of disjoint Jordan regions (see Chapter 3 for definitions). (Hint: The only way that the set L in (i) can fail to be a disjoint union of polygonal Jordan regions is for one or more vertices of the grid $\mathscr{G}_m$ to be an endpoint of all four of the line segments joining it to adjacent vertices of $\mathscr{G}_m$—let us call such a vertex a *quadruple point* in ∂L. Verify that a quadruple point in ∂L cannot belong to K, and show that L can be modified by subtracting from it small open squares centered at each quadruple point in ∂L so as to obtain a closed set M such that $K \subset M^\circ$ and such that ∂M is the union of a set $\mathscr{S}$ of line segments with the property that each endpoint of a line segment in $\mathscr{S}$ is an endpoint of *exactly two* line segments in $\mathscr{S}$. Orient the line segments in $\mathscr{S}$ suitably, and use Problem I.)

The reader should note that Problem K(ii) constitutes a proof of Proposition 3.12. The device employed there to avoid quadruple points would be completely unnecessary if, as has been suggested by W. Gustin, we had paved the plane with congruent hexagons instead of squares.

L. Let U and V be open subsets of $\mathbb{C}$, and let α and β be rectifiable arcs lying in U and V, respectively. Let W_α and W_β denote the ranges of α and β, and let $g(\zeta, \lambda)$ be a continuous complex-valued function (of two variables) on $W_\alpha \times W_\beta$. Show that the function $f_1(\zeta) = \int_\beta g(\zeta, \lambda) d\lambda$ is continuous on W_α, and likewise that the function $f_2(\lambda) = \int_\alpha g(\zeta, \lambda) d\zeta$ is continuous on W_β. (Hint: If $[a, b]$ and $[c, d]$ denote, respectively, the parameter intervals of α and β, then the function $g(\alpha(s), \beta(t))$ is (uniformly) continuous on the closed rectangle $Q = [a, b] \times [c, d]$. Hence for a given positive ε there exists a positive δ such that

$$|g(\alpha(s), \beta(t)) - g(\alpha(s'), \beta(t'))| < \varepsilon$$

for all (s, t), (s', t') in Q such that $|s - s'|, |t - t'| < \delta$. Verify that if $a \leq s, s' \leq b$ and $|s - s'| < \delta$, then

$$\left| \int_\beta g(\alpha(s), \lambda) d\lambda - \int_\beta g(\alpha(s'), \lambda) d\lambda \right| \leq L_\beta \varepsilon,$$

where L_β denotes the length of β; see Problem D(ii).) Show further that

$$\int_\alpha \left[\int_\beta g(\zeta, \lambda) d\lambda \right] d\zeta = \int_\beta \left[\int_\alpha g(\zeta, \lambda) d\zeta \right] d\lambda.$$

(Hint: Let ε and δ be as above, and let $\mathscr{P}_1 = \{s_0 < \cdots < s_M\}$ and $\mathscr{P}_2 = \{t_0 < \cdots < t_N\}$ be partitions of $[a, b]$ and $[c, d]$, respectively, such that mesh $\mathscr{P}_1, \mathscr{P}_2 < \delta$. Show first that

$$\left| f_1(\zeta) - \sum_{j=1}^{N} g(\zeta, \beta(t_j)) [\beta(t_j) - \beta(t_{j-1})] \right| \leq 2 L_\beta \varepsilon$$

for all ζ in W_α, and conclude from this that the *double* sum

$$\sigma = \sum_{i=1}^{M} \sum_{j=1}^{N} g(\alpha(s_i), \beta(t_j)) [\beta(t_j) - \beta(t_{j-1})] [\alpha(s_i) - \alpha(s_{i-1})] \tag{18}$$

satisfies the condition

$$\left| \sigma - \int_\alpha f_1(\zeta) d\zeta \right| \leq L_\alpha (2 L_\beta \varepsilon + 2 L_\beta \varepsilon) = 4 L_\alpha L_\beta \varepsilon,$$

where L_α denotes the length of α. Then reverse the order of summation in (18).)

M. Let f be a locally analytic function on an open subset U of $\mathbb{C}$, and let

$$g(\zeta, \lambda) = \begin{cases} \dfrac{f(\zeta) - f(\lambda)}{\zeta - \lambda}, & \zeta \neq \lambda, \\[2mm] f'(\lambda), & \zeta = \lambda, \end{cases} \qquad \zeta, \lambda \in U.$$

Show that g is continuous on $U \times U$. (Hint: One may either express the difference $g(\zeta, \lambda) - g(\alpha, \alpha)$ as the integral of $[f'(\xi) - f'(\alpha)]$ over $\sigma(\lambda, \zeta)$ (for ζ, λ sufficiently close to α), or refer to Example I.)

N. (Dixon [19]) Let U be an open set in $\mathbb{C}$, and let γ be a formal sum of closed rectifiable arcs in U (Prob. H). Suppose given a locally analytic function f on U and set

$$g(\zeta, \lambda) = \begin{cases} \dfrac{f(\zeta) - f(\lambda)}{\zeta - \lambda}, & \zeta \neq \lambda, \\ f'(\lambda), & \zeta = \lambda, \end{cases} \qquad \zeta, \lambda \in U.$$

Show that if V denotes the (open) subset of $\mathbb{C}$ at which w_γ vanishes, then

$$h(\lambda) = \begin{cases} \dfrac{1}{2\pi i} \displaystyle\int_\gamma g(\zeta, \lambda)d\zeta, & \lambda \in U, \\ \dfrac{1}{2\pi i} \displaystyle\int_\gamma \dfrac{f(\zeta)}{\zeta - \lambda} d\zeta, & \lambda \in V, \end{cases} \tag{19}$$

defines a locally analytic function on $U \cup V$. (Hint: It is obvious that the two definitions of h agree on $U \cap V$, so (19) really defines a function on $U \cup V$. Moreover it is also clear (Prob. G) that h is differentiable at each point of V. To complete the proof, use Problems L and M and Morera's theorem (Ex. L).)

O. (The Cauchy–Goursat Theorem and Cauchy Integral Formula: The General Case)

(i) Let U be an open set in $\mathbb{C}$, let f be a locally analytic function on U, and let γ be a formal sum of closed rectifiable arcs in U with the property that if $\lambda \notin U$ then $w_\gamma(\lambda) = 0$. (This condition says, intuitively, that γ does not wind around any point of the complement of U, and is frequently indicated by writing $\gamma \sim 0$ in U.) Show that

$$w_\gamma(\lambda)f(\lambda) = \frac{1}{2\pi i} \int_\gamma \frac{f(\zeta)}{\zeta - \lambda} d\zeta, \qquad \lambda \in U \backslash W,$$

where W denotes the union of the ranges of the arcs constituting γ. (Hint: If V denotes the set of points at which w_γ vanishes, then $U \cup V = \mathbb{C}$; use Liouville's theorem (Ex. K).) Verify in particular that if C_1 and C_2 are any two circles such that $C_2 \subset \text{Int}(C_1)$, and if f is analytic on some domain containing A^-, where A denotes the annular domain $\text{Int}(C_1) \cap \text{Ext}(C_2)$, then

$$f(\lambda) = \frac{1}{2\pi i} \left(\int_{C_1} - \int_{C_2} \right) \frac{f(\zeta)}{\zeta - \lambda} d\zeta, \qquad \lambda \in A,$$

(cf. Example F).

(ii) Let U be an open subset of $\mathbb{C}$, let f be a locally analytic function on U, and let γ be a formal sum of closed rectifiable arcs in U such that $\gamma \sim 0$ in U as in (i). Show that

$$\int_\gamma f(\zeta)d\zeta = 0.$$

(Hint: Apply the Cauchy integral formula to the function $g(\zeta) = (\zeta - \alpha)f(\zeta)$ at the point α, where α is some point of U not in the union W of the ranges of the arcs constituting γ. (That such a point α exists is clear from the fact that U is open and W is compact.))

P. A domain Δ in $\mathbb{C}$ is said to be *simply connected* if every closed rectifiable arc γ in Δ satisfies the condition $\gamma \sim 0$ in Δ.

(i) Use the results of Problem K to prove that a domain Δ in $\mathbb{C}$ is simply connected if and only if the complement of Δ in the *Riemann sphere* $\hat{\mathbb{C}}$ is connected (cf. Problem 3W).

(ii) Show that every analytic function on a simply connected domain Δ possesses a primitive on Δ. (Hint: Fix a point λ_0 in Δ. If f is analytic on Δ and if α_1 and α_2 are any two rectifiable arcs in Δ joining λ_0 to a point λ, then $\int_{\alpha_1} f(\zeta)d\zeta = \int_{\alpha_2} f(\zeta)d\zeta$ by the Cauchy-Goursat theorem.)

Q. Let Δ be a simply connected domain, and let f be an analytic function on Δ with no zeros in Δ. Show that there exists an analytic determination of $\log f(\lambda)$ on Δ, that is, an analytic function $\log f(\lambda)$ such that $\exp\{\log f(\lambda)\} \equiv f(\lambda)$ on Δ. In particular, if $0 \notin \Delta$, then there is an analytic determination of $\log \lambda$ on Δ, unique up to an additive constant of the form $2k\pi i, k \in \mathbb{Z}$. (For example, if C denotes an arbitrary closed, connected, unbounded subset of $\mathbb{C}$ that contains 0, then there is an analytic determination of $\log \lambda$ on each component of $\mathbb{C}\backslash C$.) Conclude that if Δ is a simply connected domain not containing 0, then there exists a real harmonic function $\arg \lambda$ on Δ (unique up to an additive constant of the form $2k\pi, k \in \mathbb{Z}$) satisfying the condition $\lambda \equiv |\lambda| \exp\{i \arg \lambda\}$ in Δ. In the event that the domain Δ contains the positive ray of the real axis there is precisely one of these determinations of the *argument function* that vanishes on that ray; this uniquely determined function is called the *principal branch* of the argument on Δ, and is denoted by $\mathrm{Arg}\ \lambda$. Show, analogously, that in a simply connected domain Δ that contains the positive ray of the real axis but does not contain 0 there exists, for each complex number α, a unique determination of the power function $f_\alpha(\lambda) = \lambda^\alpha, \lambda \in \Delta$, satisfying the usual exponential laws identically on Δ and having the property that $f_t(s) = s^t$ when s is positive and t is real. (Hint: If F denotes any primitive of the function f'/f, then fe^{-F} is a constant β; hence $f = \beta e^F$. Choose any complex number α such that $e^\alpha = \beta$, and set $\log f(\lambda) = F(\lambda) + \alpha$.)

R. A pair (D, f) consisting of an open disc D in $\mathbb{C}$ and an analytic function f defined on D is called a *function element*. Let $(D_1, f_1), \ldots, (D_n, f_n)$ be a finite set of function elements, and suppose given an analytic function f_0 on a domain Δ such that

(i) f_i agrees with f_0 on $D_i \cap \Delta \neq \varnothing$ for all $i = 1, \ldots, n$,
(ii) $D_i \cap D_j \neq \varnothing$ implies $D_i \cap D_j \cap \Delta \neq \varnothing$ for all $i, j = 1, \ldots, n$.

(Condition (ii) will hold, for example, if Δ is convex and the centers of the discs D_i all lie in Δ^-.) Show that there exists a (unique) analytic function f on the union $\Delta \cup D_1 \cup \cdots \cup D_n$ that extends all of the functions $f_0, f_1, \ldots, f_n$.

S. A *circle chain* in $\mathbb{C}$ is a finite sequence $\mathcal{K} = \{D_0, \ldots, D_n\}$ of open discs in $\mathbb{C}$ such that $D_i \cap D_{i-1} \neq \varnothing$ for every $i = 1, \ldots, n$. If $\mathcal{K} = \{D_0, \ldots, D_n\}$ is a circle chain, and if f_i is an analytic function on $D_i, i = 0, \ldots, n$, then the sequence $\{(D_0, f_0), \ldots, (D_n, f_n)\}$ of function elements constitutes an *analytic continuation* along $\mathcal{K}$ if f_i agrees with f_{i-1} on $D_i \cap D_{i-1}, i = 1, \ldots, n$. It is clear from the identity theorem (Th. 5.2) that an analytic continuation along $\mathcal{K}$ is uniquely determined by any one of its function elements. Customarily it is said that f_n is obtained by *continuing f_0 analytically along $\mathcal{K}$*.

(i) Let l be a line in $\mathbb{C}$, and let $\mathscr{K} = \{D_0, \dots, D_n\}$ be a circle chain such that the center α_i of each disc D_i lies on l. Show that if $\{(D_0, f_0), \dots, (D_n, f_n)\}$ is an analytic continuation along $\mathscr{K}$, then there exists a (unique) analytic function f on the union $\Delta = D_0 \cup \dots \cup D_n$ that extends each f_i, $i = 0, \dots, n$. (If the centers α_i all lie on the line segment $\sigma(\alpha_0, \alpha_n)$, such an analytic continuation is known as an *analytic continuation along the directed line segment* $\sigma(\alpha_0, \alpha_n)$.) (Hint: Deal first with the case in which the centers α_i, $i = 0, \dots, n$, are arranged in order in one direction or the other along l. Begin by showing that if a disc D_i in $\mathscr{K}$ is wholly contained in some other disc in $\mathscr{K}$, then the function element (D_i, f_i) can be removed from the analytic continuation without prejudice to any hypothesis. Hence one may assume without loss of generality that no disc in $\mathscr{K}$ is contained in any other. Show that if this is the case, and if $D_i \cap D_k \neq \varnothing$ for some i and k with $i < k - 1$, then $D_i \cap D_{k-1} \neq \varnothing$.)

(ii) If $\mathscr{K}_1, \dots, \mathscr{K}_p$ is a sequence of circle chains, and if there is given an analytic continuation along each of these circle chains, we say that this sequence of analytic continuations is *chained* if the last function element in the analytic continuation along $\mathscr{K}_i$ coincides with the first function element in the analytic continuation along $\mathscr{K}_{i+1}$, $i = 1, \dots, p - 1$. In this case we can form a *single* analytic continuation simply by juxtaposing the given ones in the given order (and then, if desired, suppressing consecutive repetitions). We may call this the result of *chaining together* the given analytic continuations. If $\pi = \pi(\lambda_0, \dots, \lambda_n)$ is a polygonal arc, then by an *analytic continuation along* π we mean an analytic continuation obtained by chaining together analytic continuations along the individual edges $\sigma(\lambda_{i-1}, \lambda_i)$, $i = 1, \dots, n$. Let Δ be a domain and let f be an analytic function on Δ. Show that if $\pi = \pi(\lambda_0, \dots, \lambda_n)$ is a polygonal arc in Δ, and if $(D_r(\lambda_0), f_0)$ is a function element centered at λ_0 such that f_0 agrees with f on $D_r(\lambda_0)$, then there exists an analytic continuation of f_0 along π such that every function element in that continuation is obtained simply by restricting f to the appropriate disc.

T.   (i) Let α, β, γ be noncollinear complex numbers, and let T denote the triangular region determined by them. Suppose given a function element (D_0, f_0) centered at one of the vertices of T, say at α, and suppose this function element can be continued along every polygonal arc in T. Show that if f_0 is continued analytically around the triangular arc $[\alpha, \beta, \gamma]$ then the last function element in the continuation agrees with f_0 in a neighborhood of α. (Hint: Let $\beta_t = (1 - t)\alpha + t\beta$ and $\gamma_t = (1 - t)\alpha + t\gamma$, $0 \leq t \leq 1$, and consider the set E of those points t in the unit interval with the property that analytic continuation of f_0 along $[\alpha, \beta_t, \gamma_t]$ has the stated property that the last function element agrees with f_0 in a neighborhood of α. Use Problem S(i) to show that E is a nonempty, closed, relatively open subset of $[0, 1]$.)

(ii) Let P be a simple polygon in $\mathbb{C}$ and suppose given a function element (D_0, f_0) centered at a vertex λ_0 of P with the property that f_0 can be continued analytically along every polygonal arc beginning at λ_0 and lying wholly in $P \cup \text{Int}(P)$. Show that if $\pi = \pi(\lambda_0, \dots, \lambda_n, \lambda_0)$ is a polygonal arc parametrizing P (such an arc exists by Problem I), then the result of continuing f_0 along π has the property that the last function element in the analytic continuation, which is also centered at λ_0, agrees with f_0 in a neighborhood of λ_0. (Hint: It is a theorem in elementary plane geometry that the region $R = P \cup \text{Int}(P)$ can be dissected into triangular regions $T_1, \dots, T_k$ in such a way that the vertices of all these triangular regions are also vertices of P, and such that any two triangular regions T_i and T_j, $i \neq j$, that are not disjoint meet only in a common edge (which must then be a line segment lying entirely in Int(P) except for its end-points, which are vertices of P). Use (i) and an induction on k.)

U. (Monodromy Principles) Let Δ be a simply connected domain, let D_0 be an open disc contained in Δ, and let f_0 be an analytic function defined on D_0. Suppose f_0 can be continued analytically along every polygonal arc in Δ. Prove that if π_1 and π_2 are any polygonal arcs in Δ joining the center λ_0 to the same point λ, then the last function elements in the analytic continuations of f_0 along π_1 and π_2, respectively, agree in a neighborhood of λ. Conclude that there exists a (unique) analytic function f on Δ that agrees with f_0 on D_0. (Hint: It comes to the same thing to show that the analytic continuation of f_0 along the arc $\pi_1 + \tilde{\pi}_2$ returns to the function f_0 in a neighborhood of λ_0. By interpolating new vertices (and new function elements) as needed along $\pi_1 + \tilde{\pi}_2$, one may arrange things so that the resulting closed polygonal arc π intersects itself only at its own vertices, in which case it is the formal sum of various polygonal arcs of the form $\rho + \tilde{\rho}$ and of polygonal Jordan loops. Use Problem T and an induction on the number of edges in π.)

V. Let u be a real harmonic function on a simply connected domain Δ' in $\mathbb{R}^2$. Show that there exists an analytic function f on the corresponding complex domain Δ having u for its real part, and conclude that u possesses a (real) harmonic conjugate in Δ' that is unique up to an additive constant.

W. If f is an analytic function on a domain Δ, then f and $\bar{f}$ are both (complex-valued) harmonic functions on Δ. Hence if f and g are any two analytic functions on Δ, then $f + \bar{g}$ is harmonic on Δ. Show, in the converse direction, that if Δ is simply connected and h is a complex-valued harmonic function on Δ, then h can be written as $h = f + \bar{g}$, where f and g are analytic on Δ. Show also that this decomposition of h is unique up to an additive constant. (Hint: A complex-valued function is harmonic if and only if its real and imaginary parts are. Let u and v be the real and imaginary parts of h, let $\tilde{u}$ and $\tilde{v}$ denote harmonic conjugates of u and v, respectively, and set $w_1 = (u - \tilde{v})/2$, $w_2 = (u + \tilde{v})/2$. If f and g denote analytic functions on Δ having w_1 and w_2, respectively, for their real parts, then $f + \bar{g}$ differs from h by an additive constant.)

X. Let γ be a piecewise smooth, regular, Jordan loop in $\mathbb{C}$, and let $\Delta = \mathrm{Int}(\gamma)$. The purpose of this problem is to show that either $w_\gamma \equiv 1$ or $w_\gamma \equiv -1$ on Δ.

(i) Let $[a, b]$ be the parameter interval of γ, let $t_0 < \cdots < t_n$ be a partition of $[a, b]$ such that the subarcs $\alpha_i = \gamma|[t_{i-1}, t_i]$ are all smooth, and let t be a point of $[a, b]$ strictly between some pair of consecutive partition points—say $t_{i-1} < t < t_i$. Then at the point $\lambda_0 = \alpha_i(t)$ there are unique tangent and normal lines to α_i, and if λ_1 is any other point on the normal n to α_i at λ_0, then $\rho(s) = s\lambda_1 + (1 - s)\lambda_0$, $-\infty < s < +\infty$, is a parametrization of n. Show that for sufficiently small positive ε one of the two segments $\sigma_1 = \{\rho(t): -\varepsilon < t < 0\}$ and $\sigma_2 = \{\rho(t): 0 < t < +\varepsilon\}$ lies entirely in Δ, while the other lies entirely in Ext(γ). If λ_1 is chosen so that it is σ_2 that lies in Δ, then ρ is called an *inward tending* parametrization of n. (Hint: If a line segment σ has the property that λ_0 is the midpoint of σ, and if σ lies (except for its midpoint) entirely in one complementary component of γ, then σ is an edge of a polygonal Jordan loop P such that λ_0 is the only point of γ on P (see Proposition 3.8). Hence the range of γ lies (except for the one point λ_0) either in Int(P) or Ext(P).)

(ii) Let α_i be one of the smooth arcs of which γ is the chained sum, and for each t, $t_{i-1} < t < t_i$, let $\lambda_0 = \lambda_0(t) = \alpha_i(t)$, let $\lambda_1 = \lambda_1(t)$ denote the point of the normal to α_i at $\lambda_0(t)$ such that the parametrization $\rho(s) = s\lambda_1 + (1 - s)\lambda_0$ is inward tending and such that $|\lambda_1 - \lambda_0| = 1$, and let $\lambda_2 = \lambda_2(t)$ denote the point on the tangent line to α_i at λ_0 such that $\lambda_2 - \lambda_0$ is a positive multiple of $\alpha'(t)$ and such that $|\lambda_2 - \lambda_0| = 1$. Show that if $\lambda_j - \lambda_0 = x_j + iy_j$, where x_j, y_j are real, $j = 1, 2$, then the determinant

$$\delta(t) = \begin{vmatrix} x_2 & y_2 \\ x_1 & y_1 \end{vmatrix}$$

is either $+1$ for every t or -1 for every t, $t_{i-1} < t < t_i$. When $\delta(t) \equiv 1$, we say that α_i is oriented *positively* with respect to Δ. (Hint: For arbitrary t', $t_{i-1} \leq t' < t_i$, let α_0 and α' denote, respectively, $\alpha_i(t')$ and the (right) derivative of α_i at t'. Set $\beta_1 = \omega\alpha'$, $\beta_2 = \bar{\omega}\alpha'$, where $\omega = \exp(iu_0)$, $0 < u_0 < \pi/2$, and define $\rho_j(s) = \alpha_0 + s\beta_j$, $0 \leq s < +\infty$, $j = 1, 2$. Thus ρ_1, ρ_2 are parametrizations of two rays emanating from α_0 that straddle the (right) tangent vector to α_i at α_0, and make an angle less than π (Figure 1). Show that there exists a number $\delta > 0$ such that if $0 < s_0 < \delta$, and if $\xi_j = \rho_j(s_0)$, $j = 1, 2$, then the set of values of the parameter t, $t' \leq t \leq b$, for which $\gamma(t)$ lies in or on the triangle $[\alpha_0, \xi_1, \xi_2]$ consists of a single interval $[t', t' + \varepsilon]$ for some positive number ε, where $\gamma(t)$ is strictly inside $[\alpha_0, \xi_1, \xi_2]$ on this interval except for $t = t'$, $t' + \varepsilon$. Use this observation to prove that for every t', $a < t' < b$, there exists a positive number r_0 such that if $0 < r < r_0$, then the set of values of the parameter t, $a \leq t \leq b$, for which $\gamma(t)$ lies in or on the circle $C = C_r(\gamma(t'))$ consists of a single closed interval $[c, d]$, $a \leq c < t' < d \leq b$, where $\gamma(t)$ is strictly inside C on this interval except for $t = c, d$. (A similar assertion can also be made in a neighborhood of the point $\gamma(a) = \gamma(b)$, of course, but it must be worded differently.))

(iii) Show that if *one* subarc α_i of γ is oriented positively with respect to Δ, then *all* the subarcs α_i are. When this is the case, we say that γ is *oriented positively* with respect to Δ. (Hint: It suffices to show that if the subarc α_i is positively oriented

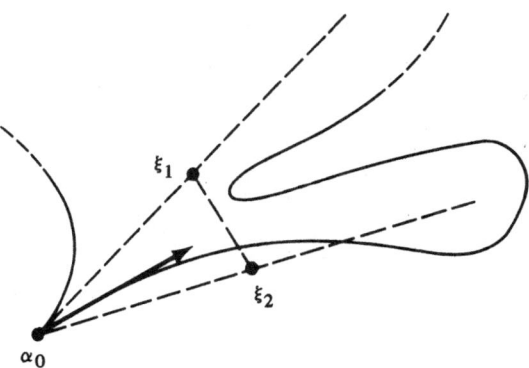

Figure 1.

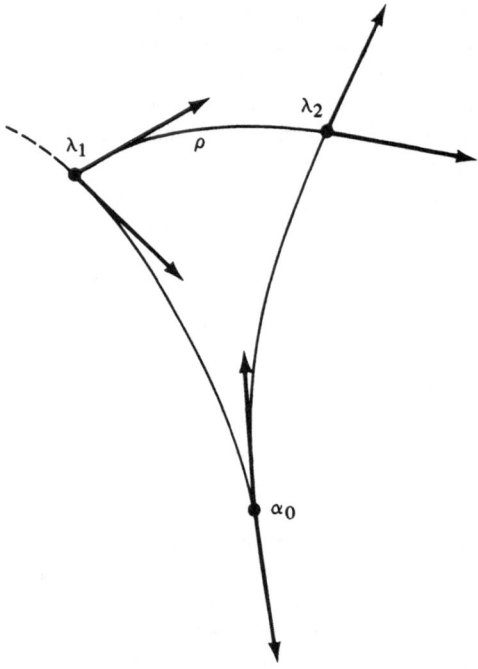

Figure 2.

with respect to Δ, then α_{i+1} is too. Let $\alpha_0 = \gamma(t_i)$, and let $C = C_r(\alpha_0)$ be a circle with the property set forth in the hint in (ii) and small enough so that α'_1 changes very little on the subinterval $[s_1, t_i]$ on which $\alpha_i(t)$ lies in $\mathrm{Int}(C) \cup C$, while α'_{i+1} also changes very little on the subinterval $[t_i, s_2]$ on which α_{i+1} lies in $\mathrm{Int}(C) \cup C$. Then the points $\lambda_j = \gamma(s_j)$, $j = 1, 2$, divide C into two arcs, one lying entirely in Δ except for its endpoints, and the other lying in $\mathrm{Ext}(\gamma)$ except for its endpoints, and if ρ denotes a parametrization of the subarc of C (or of $\tilde{C}$) that is in Δ and has λ_1 and λ_2 for initial and terminal points, respectively, then the tangent vector to ρ lies (roughly) along the inward tending normal to γ at λ_1, and (roughly) opposite to the inward tending normal to γ at λ_2; see Figure 2.)

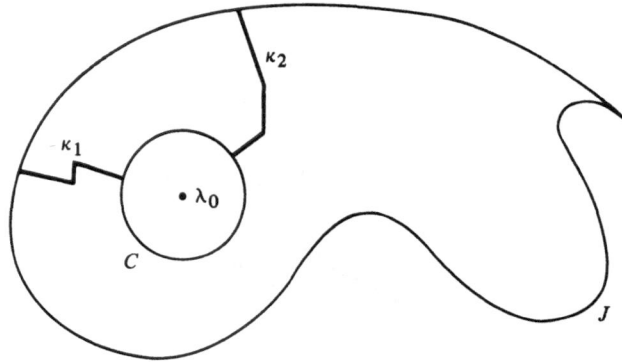

Figure 3.

(iv) Verify that if γ is positively oriented with respect to Δ, then $w_\gamma \equiv 1$ on Δ. (Hint: Let $C = C_r(\lambda_0)$ be a circle in Δ. Show first that there exist two arcs κ_1 and κ_2, lying in $\Delta \cap \text{Ext}(C)$ except for their endpoints, such that the initial points of κ_1 and κ_2 lie on C, while the terminal points of κ_1 and κ_2 belong to the range J of γ, and such that the ranges of κ_1 and κ_2 are disjoint. Chain sub-arcs of γ and $\tilde{C}$ together with κ_1 and $\tilde{\kappa}_2$ (and $\tilde{\kappa}_1$ and κ_2) to form two simply closed arcs γ_1 and γ_2 such that $w_{\gamma_1} = w_{\gamma_2} = 0$ at λ_0, while the winding number of $\gamma_1 + \gamma_2$ at λ_0 is the same as that of $\gamma + \tilde{C}$; see Figure 3.)

6

Measurability

In this chapter and the following four we review some needed facts from the theory of measure and integration. Most of the proofs of those propositions that are stated below without proof are outlined in the problems. The reader may consult [30], [58], or [36] for more detail.

A nonempty collection **R** of subsets of a set X is a *ring (of sets)* if **R** is closed with respect to the formation of unions and differences (i.e., if for every pair of sets A and B in **R**, $A \cup B$ and $A \backslash B$ belong to **R**). Every ring of sets contains the empty set, and is closed under the formation of intersections. A ring of sets that is closed with respect to the formation of countable unions is a *σ-ring* of sets. Every σ-ring **S** of sets is closed under the formation of countable intersections. Moreover, **S** is closed under the formation of complements if and only if the entire space X belongs to **S**. The intersection of a nonempty family of rings [σ-rings] is itself a ring [σ-ring]. If $\mathscr{C}$ is any collection of subsets of a set X, then there exists a smallest ring [σ-ring] of subsets of X that contains $\mathscr{C}$, namely, the intersection of the collection of all those rings [σ-rings] of subsets of X that contain $\mathscr{C}$. This ring [σ-ring] is called the ring [σ-ring] *generated by* $\mathscr{C}$ and is denoted by $\mathbf{R}(\mathscr{C})[\mathbf{S}(\mathscr{C})]$. The ring $\mathbf{R}(\mathscr{C})$ can be described directly in a fairly straightforward manner (see Problems E and F); the general description of the σ-ring $\mathbf{S}(\mathscr{C})$ requires a transfinite procedure (Prob. G).

A set X together with a σ-ring **S** of subsets of X such that X belongs to **S** is a *measurable space*. Thus a measurable space is a pair $(X, \mathbf{S})$ such that **S** is a σ-ring of subsets of X that is closed with respect to the formation of complements. However we shall frequently denote the measurable space $(X, \mathbf{S})$ by the single symbol X, and refer to X as *measurable with respect to* **S** or as *measurable* [**S**]. The elements of **S** are called the *measurable* ([**S**]) subsets of X. If $(X, \mathbf{S})$ is a measurable space, and A is an arbitrary subset of X, then the *trace* $S_A = \{E \cap A : E \in S\}$ of S on A is a σ-ring of sets (Prob. D), and (A, S_A) is a measurable space. In case $A \in S$, so that the elements of S_A are exactly the measurable subsets of A, (A, S_A) is called a *(measurable) subspace* of (X, S).

Example A. Let X be an arbitrary set, and let $\mathscr{C}$ be an arbitrary collection of subsets of X. Then there exists a smallest σ-ring S of subsets of X that contains $\mathscr{C}$ and turns X into a measurable space, viz., the σ-ring $S(\mathscr{C} \cup \{X\})$. If $\mathscr{C}$ is itself a σ-ring, then S consists of the union of $\mathscr{C}$ and the collection of all the complements of the sets in $\mathscr{C}$.

Example B. Let X be a set and let $\mathscr{D}$ be a nonempty collection of subsets of X with the property that $\mathscr{D}$ is closed with respect to the formation of countable intersections. If S denotes the collection of all those subsets A of X such that either A or its complement $X \backslash A$ contains some set D belonging to $\mathscr{D}$, then it is readily verified that S is closed with respect to the formation of complements and countable intersections. Hence S is a σ-ring turning X into a measurable space (Prob. E(ii)). We note for future reference that the sub-collection of S consisting of sets disjoint from some one fixed set D in $\mathscr{D}$ is a σ-ideal in S (Prob. I).

Example C. If S and T are two σ-rings of subsets of X, then we say that S is *refined by* T (or that S is *coarser than* T, or that T is *finer than* S) if $S \subset T$. For an arbitrary set X the power class on X (Problem 1R) is a σ-ring on X that turns X into a measurable space, and it is clear that this is the finest σ-ring of subsets of X. The coarsest σ-ring of subsets of X turning X into a measurable space is the pair $\{\varnothing, X\}$. Every other σ-ring that turns X into a measurable space lies between these two extremes. The coarsest σ-ring that turns X into a measurable space in which every singleton is a measurable set consists of the collection of all countable subsets of X together with the collection of all those subsets of X that have countable complements.

Example D. In $\mathbb{R}^n$ consider the collection of all *half-open cells* of the form

$$H = (a_1, b_1] \times (a_2, b_2] \times \cdots \times (a_n, b_n] \tag{1}$$

(where $(a, b]$ denotes the half-open interval $\{t \in \mathbb{R} : a < t \leq b\}$). If H_0 and H_1 are two such cells, then $H_0 \cap H_1$ is another, and $H_1 \backslash H_0$ can be expressed (not uniquely) as a finite disjoint union of still other half-open cells. In fact, every set in the ring $\mathbf{H}_r$ generated by the collection of all half-open cells of the form (1) can be expressed as a finite disjoint union of such cells. Similarly, the ring $\mathbf{H}_l$ generated by the collection of all half-open cells of the form

$$\tilde{H} = [a_1, b_1) \times [a_2, b_2) \times \cdots \times [a_n, b_n)$$

(where $[a, b) = \{t \in \mathbb{R} : a \leq t < b\}$) consists of the collection of all finite disjoint unions of such cells. Since any σ-ring containing either $\mathbf{H}_r$ or $\mathbf{H}_l$ must also contain all the *degenerate* closed cells $[a_1, b_1] \times [a_2, b_2] \times \cdots \times [a_n, b_n]$ in which $a_i = b_i$ for at least one i, $1 \leq i \leq n$, it is clear that $\mathbf{H}_r$ and $\mathbf{H}_l$ generate the same σ-ring.

Example E. Let $\mathscr{F}$ denote the collection of all finite unions of closed cells

$$Z = [a_1, b_1] \times \cdots \times [a_n, b_n]$$

in Euclidean space $\mathbb{R}^n$. The collection $\mathscr{F}$ is closed with respect to the formation of finite unions and intersections, since the intersection of two closed cells is again a closed cell. (Such a collection of subsets of a set X is called a *lattice* of subsets of X.) It is easily seen that $\mathscr{F}$ also generates the same σ-ring as that generated by $\mathbf{H}_r$ and $\mathbf{H}_l$ in Example D. The elements of $\mathscr{F}$ are known as *elementary figures* in $\mathbb{R}^n$.

Suppose now that X is a topological space. Then the σ-ring $\mathbf{B}$ generated by the collection of all open subsets of X is the σ-ring of *Borel sets* in X. The σ-ring $\mathbf{B}$ is also the σ-ring generated by the collection of all closed sets in X. Every set in X that can be written as the intersection of a countable collection of open sets is called a G_δ. Likewise, every set in X that can be written as the union of a countable collection of closed sets is called an F_σ. The union of a countable collection of G_δ's is a $G_{\delta\sigma}$, and so forth. All of these sets are Borel sets.

If $(X, \mathbf{S})$ and $(Y, \mathbf{T})$ are measurable spaces, and if Φ is a mapping of X into Y, then Φ is *measurable with respect to* $\mathbf{S}$ *and* $\mathbf{T}$, or *measurable* $[\mathbf{S}, \mathbf{T}]$ (or, if no confusion is possible, simply *measurable*) if $\Phi^{-1}(E) \in \mathbf{S}$ for every set E in $\mathbf{T}$. More generally, if Φ is defined only on some subset of X that contains a measurable set A, then Φ is *measurable on* A if $\Phi | A$ is measurable on the subspace $(A, \mathbf{S}_A)$. In the special case that Y is a topological space, a mapping Φ of X into Y will be called *measurable with respect to* $\mathbf{S}$, or *measurable* $[\mathbf{S}]$ if it is measurable $[\mathbf{S}, \mathbf{B}_Y]$ where $\mathbf{B}_Y$ denotes the ring of Borel sets in Y. In particular, a complex-valued function f defined on a measurable space X is measurable if and only if $f^{-1}(U)$ is a measurable set in X for every open set of complex numbers U (see Problem B). If f is a real-valued function and all of the sets $\{x \in X : f(x) < t\}$, $-\infty < t < +\infty$, are measurable sets in X, then f is measurable. (Other similar sufficient conditions for a real-valued function to be measurable are mentioned in Problem C.) We observe that a measurable real-valued function is measurable when regarded as a complex-valued function (because $\mathbb{R}$ is topologically embedded in $\mathbb{C}$ as the real axis), and that a complex-valued function is measurable if and only if its real and imaginary parts are (because $\mathbb{C}$ is the topological product of the real and imaginary axes). In the same vein we recall that the extended real number system $\mathbb{R}^\sharp$ is a topological space in its order topology (Ex. 3C), and we observe that if f is an extended real-valued function on a measurable space $(X, \mathbf{S})$, then f is measurable if and only if all of the sets $\{x \in X : f(x) < t\}$, $-\infty < t \leq +\infty$, are measurable. Equivalently, f is measurable when and only when the sets $E_{+\infty} = \{x \in X : f(x) = +\infty\}$ and $E_{-\infty} = \{x \in X : f(x) = -\infty\}$ are both measurable and the finite real-valued function $f_0 = f\,|\,X_0$ is measurable on X_0 where $X_0 = X \backslash (E_{+\infty} \cup E_{-\infty})$.

If both X and Y are topological spaces, then a mapping of X into Y is *Borel measurable* if it is measurable $[\mathbf{B}_X, \mathbf{B}_Y]$, where $\mathbf{B}_X$ and $\mathbf{B}_Y$ denote the σ-rings of Borel sets in X and Y, respectively. In particular, every continuous mapping of X into Y is Borel measurable.

Example F. Let X be a metric space, and let F be a nonempty closed set in X. Then the function $f(x) = d(x, F)$ is continuous on X and satisfies the condition

$$F = \{x \in X : f(x) = 0\}$$

(Prob. 4D). It follows that if $\mathbf{S}$ is any σ-ring of subsets of X such that all the continuous real-valued functions on X are measurable $[\mathbf{S}]$, then $\mathbf{S}$ must contain the entire ring $\mathbf{B}$ of Borel sets in X. Thus $\mathbf{B}$ is the smallest σ-ring of subsets of the metric space X with respect to which all complex (or even all real) continuous functions are measurable.

The following proposition is an easy consequence of the various definitions.

Proposition 6.1. *Let $(X, \mathbf{S})$, $(Y, \mathbf{T})$, and $(Z, \mathbf{U})$ be measurable spaces, and let $\Phi : X \to Y$ and $\Psi : Y \to Z$ be measurable mappings. Then the composition $\Psi \circ \Phi$ is measurable. In particular, if Y and Z are topological spaces, and if Ψ is a Borel measurable mapping of Y into Z, then $\Psi \circ \Phi$ is measurable $[\mathbf{S}]$ whenever Φ is.*

It follows that if f is a measurable complex-valued function, then so are the functions $\bar{f}$ and f^k, $k = 2, 3, \ldots$, as well as the functions $|f|^r$, $r > 0$. Likewise, the functions f^{-k} and $|f|^{-r}$ are measurable on the measurable set $\{x : f(x) \neq 0\}$. If f and g are measurable complex-valued functions on the same measurable space, then fg and all linear combinations $\alpha f + \beta g$ are also measurable. If f and g are real functions, then $f \vee g$ and $g \wedge f$ are measurable. (To verify that these last named facts are consequences of Proposition 6.1 it suffices to note that if f and g are measurable complex-valued functions on $(X, \mathbf{S})$, then $x \to (f(x), g(x))$ is a measurable mapping of X into $\mathbb{C}^2$, and the functions $\lambda \mu$, and $\alpha \lambda + \beta \mu$ are continuous on $\mathbb{C}^2$. Likewise, if f and g are measurable real-valued functions, then $x \to (f(x), g(x))$ is a measurable mapping of X into $\mathbb{R}^2$, and $s \vee t$ and $s \wedge t$ are continuous functions on $\mathbb{R}^2$; alternatively, see Problem C.) With respect to sequences of measurable functions, we recall the following two results (cf. Problems R and C).

Proposition 6.2. *Let X be a measurable space, and let $\{f_n\}$ be a sequence of measurable complex-valued functions on X. Denote by E the set of all those points x at which the numerical sequence $\{f_n(x)\}$ is convergent. Then E is a measurable set, and the function f defined on E by $f(x) = \lim_n f_n(x)$ is measurable on the subspace E.*

Proposition 6.3. *Let X be a measurable space, and let $\{f_n\}$ be a sequence of measurable [extended] real-valued functions on X. Then the extended real-valued functions $\sup_n f_n$, $\inf_n f_n$, $\limsup_n f_n$, and $\liminf_n f_n$, are all measurable on X. In particular, the set of those points x at which the numerical sequence $\{f_n(x)\}$ is bounded above in $\mathbb{R}$ is measurable, as is the set of those point x at which $\{f_n(x)\}$ is bounded below in $\mathbb{R}$.*

The following theorem is a convenient summary of some of the preceding facts.

Theorem 6.4. *The collection $\mathscr{A}$ of all measurable complex-valued functions on a measurable space X is a unital algebra that contains all constant functions, is self-conjugate (Ex. 2H), and is closed with respect to the formation of limits of pointwise convergent sequences. The collection of real-valued functions in $\mathscr{A}$ forms a unital real algebra that is also a function lattice (Prob. 1D).*

This theorem has a converse which is sometimes useful, and we record it here for completeness. The proof is sketched in Problems T and U.

Theorem 6.5. Let X be a set and let $\mathscr{A}$ be a real algebra of real-valued functions on X that contains the constant functions and is closed with respect to the formation of limits of pointwise convergent sequences. Then there is a unique σ-ring $\mathbf{S}$ of subsets of X such that X is a measurable space with respect to $\mathbf{S}$ and such that $\mathscr{A}$ is precisely the set of all real-valued functions on X that are measurable [$\mathbf{S}$]. Similarly, if $\mathscr{A}$ is a self-conjugate algebra of complex-valued functions on X that contains the constant functions and is closed with respect to the formation of limits of pointwise convergent sequences, then there exists a unique σ-ring $\mathbf{S}$ of subsets of X such that $(X, \mathbf{S})$ is a measurable space and such that $\mathscr{A}$ is precisely the set of all complex-valued functions on X that are measurable [$\mathbf{S}$].

Recall that a complex-valued function defined on a set X is said to be simple if it assumes only a finite number of distinct values (Prob. 1L). Equivalently, a function on X is simple if and only if it can be expressed as a finite linear combination of characteristic functions of subsets of X. Among such representations of a simple function s there is precisely one,

$$s = \sum_{i=1}^{m} \alpha_i \chi_{E_i}, \tag{2}$$

in which the sets E_i are disjoint and nonempty, and the coefficients α_i are distinct and different from zero. (If $s = 0$, the sum in (2) is empty.) If X is a measurable space, then a simple function s on X is measurable if and only if it assumes each of its values on a measurable set, or, equivalently, if and only if the sets E_i in the representation (2) are measurable. The following useful result shows how measurable real-valued functions can be approximated by measurable simple functions; for a sketch of a proof as well as a discussion of the complex case see Problem S.

Proposition 6.6. *Let X be a measurable space and let f be a real-valued function defined on X. Then the following three conditions are equivalent:*

(i) *f is measurable,*

(ii) *There exists a sequence $\{s_n\}_{n=1}^{\infty}$ of measurable simple real-valued functions converging pointwise to f,*

(iii) *There exists a sequence $\{s_n\}_{n=1}^{\infty}$ of measurable simple real-valued functions converging pointwise to f and satisfying the following additional conditions:*

   (a) *$|s_n(x)| \leq |f(x)|$ for all n in $\mathbb{N}$ and x in X,*

   (b) *At each point x of X, either $0 \leq s_1(x) \leq \cdots \leq s_n(x) \leq \cdots$, or $0 \geq s_1(x) \geq \cdots \geq s_n(x) \geq \cdots$,*

   (c) *If N is a positive integer and if $|f(x)| \leq N$, then $|f(x) - s_n(x)| \leq 1/2^n$ for every $n \geq N$. In particular, the convergence of $\{s_n\}$ to f is uniform on every set on which f is bounded.*

PROBLEMS

A. Let X and Y be sets, and let $\Phi : X \to Y$ be a mapping of X into Y. Then the induced mapping Φ^{-1} of the power class on Y into the power class on X preserves arbitrary unions and intersections as well as complements, differences, and symmetric differences. (Thus, for example, $\Phi^{-1}(A \nabla B) = \Phi^{-1}(A) \nabla \Phi^{-1}(B)$ for all subsets A and B of Y.) This is sometimes expressed by saying that Φ^{-1} is a *complete Boolean homomorphism*. Verify that if S is a σ-ring in Y, then $\Phi^{-1}(S) = \{\Phi^{-1}(A) : A \in S\}$ is a σ-ring in X. Similarly, if S is a σ-ring in X, then $\{B \subset Y : \Phi^{-1}(B) \in S\}$ is a σ-ring in Y.

B. Let $(X, \mathbf{S})$ be a measurable space, and let Φ be a mapping of $(X, \mathbf{S})$ into a measurable space $(Y, \mathbf{T})$. If $\Phi^{-1}(C)$ is measurable for every C in some collection of sets $\mathscr{C}$ in Y, then $\Phi^{-1}(E)$ is measurable for every E in $S(\mathscr{C})$. Thus, to show that Φ is measurable [S, T] it suffices to verify that $\Phi^{-1}(C)$ is measurable [S] for every set C in some subcollection $\mathscr{C}$ of $\mathbf{T}$ sufficiently large so that $\mathbf{T} \subset S(\mathscr{C})$. In particular, if Y is a topological space satisfying the second axiom of countability, and if $\Phi^{-1}(W)$ is measurable for every open set W in some base (or subbase) for the topology on Y, then Φ is measurable [S].

C. Let M denote any one fixed set of real numbers that is dense in $\mathbb{R}$, e.g., the set of rationals, or the set of dyadic fractions $m/2^n$. Then the set of rays $\{[t, +\infty) : t \in M\}$ generates the ring of Borel sets in $\mathbb{R}$, as does each of the sets of rays $\{(t, +\infty) : t \in M\}$, $\{(-\infty, t] : t \in M\}$, and $\{(-\infty, t) : t \in M\}$. Hence a real-valued function on a measurable space X is measurable if and only if the inverse image of each of the sets of any one of these four special forms is measurable. Devise analogous criteria for the measurability of extended real-valued functions on X, and use your findings to show that if $\{f_n\}$ is an arbitrary sequence of measurable extended real-valued functions defined on X then $\sup_n f_n$ and $\inf_n f_n$ are measurable. Show also that an arbitrary monotone real-valued function defined on an arbitrary subset A of $\mathbb{R}$ is Borel measurable on A.

D. If $\mathscr{C}$ is a collection of subsets of X, and if M is a fixed subset of X, then the collection $\mathscr{C}_M = \{C \cap M : C \in \mathscr{C}\}$ is called the *trace* of $\mathscr{C}$ on M. Show that the σ-ring generated by the trace $\mathscr{C}_M$ coincides with the trace on M of the σ-ring $\mathbf{S}(\mathscr{C})$ generated by $\mathscr{C}$. (Hint: It is readily verified that the trace of $\mathbf{S}(\mathscr{C})$ on M is a σ-ring of subsets of M that contains $\mathscr{C}_M$ and therefore contains $\mathbf{S}(\mathscr{C}_M)$; to go the other way, let $\mathbf{T}$ denote the collection of subsets A of X such that $A \cap M \in \mathbf{S}(\mathscr{C}_M)$, and show that $\mathbf{T}$ is a σ-ring.)

E. (i) Let $\{E_1, \ldots, E_n\}$ be a finite collection of subsets of a set X, and let $\mathscr{P}_0$ denote the collection of sets obtained by deleting from the partition of X determined by $\{E_1, \ldots, E_n\}$ the one set $X \backslash (E_1 \cup \cdots \cup E_n)$. (See Problem 1L for an explanation of this terminology; if the sets $E_1, \ldots, E_n$ cover X, the set $X \backslash (E_1 \cup \cdots \cup E_n) = \varnothing$ need not be excluded.) Show that the ring of sets generated by $\{E_1, \ldots, E_n\}$ consists of all (finite) unions of sets belonging to $\mathscr{P}_0$. Show also that the ring $\mathbf{R}(\mathscr{C})$ generated by an arbitrary nonempty collection $\mathscr{C}$ of subsets of X is the union of the rings generated by the various finite subsets of $\mathscr{C}$. (Hint: This union forms a ring.) Show in a like vein that the σ-ring $\mathbf{S}(\mathscr{C})$ is the union of the σ-rings generated by the various countable subsets of $\mathscr{C}$.

(ii) A ring of sets is closed with respect to the formation of both finite unions and finite intersections. Likewise, a σ-ring is closed with respect to the formation of both countable unions and countable intersections. Show that if $\mathbf{S}$ is a collection of subsets of a set X that is closed with respect to the formation of complements, then $\mathbf{S}$ is a ring [σ-ring] if $\mathbf{S}$ is closed with respect to *either* the formation of finite [countable] unions or intersections.

F. (i) Show that the binary set operation ∇ is *associative*, that is, that $A \nabla (B \nabla C) = (A \nabla B) \nabla C$ for any three sets A, B, C, and conclude that the symmetric difference $E_1 \nabla E_2 \nabla \cdots \nabla E_n$ of an arbitrary finite sequence of sets $\{E_1, \ldots, E_n\}$ can be defined without regard to either order or bracketing. (Hint: If $E_1, \ldots, E_n$ are subsets of a set X, then an element x of X belongs to $E_1 \nabla \cdots \nabla E_n$ if and only if $x \in E_i$ for an odd number of indices i.) ·

(ii) Let $\mathscr{C}$ be a nonempty collection of subsets of a set X, and let $\mathscr{C}_d$ denote the collection of all finite intersections $C_1 \cap \cdots \cap C_k$ of sets belonging to $\mathscr{C}$. Then the ring $\mathbf{R} = \mathbf{R}(\mathscr{C})$ generated by $\mathscr{C}$ coincides with the collection of all finite symmetric differences $P_1 \nabla \cdots \nabla P_m$ of sets in $\mathscr{C}_d$.

> If $\mathscr{X}$ denotes the power class on a set X, and if the binary operations ∇ and $\cap$ are interpreted as "addition" and "multiplication," respectively, then $\mathscr{X}$ becomes a ring in the sense of abstract algebra. Moreover, the *rings of sets* discussed in this section are precisely the *subrings* of $\mathscr{X}$ in the algebraic sense (hence the terminology). Viewed in this light, Problem F can be seen to be simply the translation into set-theoretic terms of the standard description in ring theory of the subring generated by a subset of a ring.
>
> Viewed algebraically, rings of sets have the curious properties that every nonzero element has additive order two ($E \nabla E = \varnothing$) and every element is idempotent ($E \cap E = E$). Such rings have been called *Boolean rings* by M. H. Stone [63].

G. For an arbitrary nonempty collection $\mathscr{C}$ of subsets of a set X let $\mathscr{C} \# \mathscr{C}$ denote the collection of all differences $E \backslash F$, where E and F belong to $\mathscr{C}$. Furthermore, let $\mathscr{C}_\sigma$ denote the collection of all unions of countable subcollections of $\mathscr{C}$, and let $\mathscr{C}^* = (\mathscr{C} \# \mathscr{C})_\sigma$. In terms of this operation it is possible to describe the σ-ring $S(\mathscr{C})$ generated by $\mathscr{C}$. We suppose, without loss of generality, that $\varnothing \in \mathscr{C}$. Let $\mathscr{C}^{(0)} = \mathscr{C}$, and, for a given countable ordinal number α, suppose $\mathscr{C}^{(\xi)}$ already defined for every ordinal number ξ less than α. Then if α is not a limit ordinal we set $\mathscr{C}^{(\alpha)} = (\mathscr{C}^{(\xi)})^*$, where $\alpha = \xi + 1$, while if α is a limit ordinal we set $\mathscr{C}^{(\alpha)} = \bigcup_{\xi < \alpha} \mathscr{C}^{(\xi)}$. (See Problem 1V for a discussion of transfinite definitions.) Show that if Ω denotes, as usual, the smallest uncountable ordinal number (Prob. 1W), then $S(\mathscr{C}) = \bigcup_{\xi < \Omega} \mathscr{C}^{(\xi)}$.

H. Let $\mathscr{C}$ be a collection of subsets of a set X, and let $\mathbf{R}$ and $\mathbf{S}$ denote, respectively, the ring and σ-ring generated by $\mathscr{C}$. If $\mathscr{C}$ is finite, then so is $\mathbf{R}$. What about $\mathbf{S}$? If $\mathscr{C}$ is countable, then so is $\mathbf{R}$, while for $\mathbf{S}$ one has the estimate card $\mathbf{S} \leq \aleph$ (where $\aleph$ denotes the cardinal number of the continuum). Indeed, card $\mathscr{C} \leq \aleph$ implies card $\mathbf{S} \leq \aleph$. (Hint: If $\mathscr{C}$ consists of k sets, then $\mathbf{R}(\mathscr{C})$ cannot contain more than 2^{2^k} sets; if card $\mathscr{C} \leq \aleph$, then card $\mathscr{C}_\sigma \leq \aleph$, where $\mathscr{C}_\sigma$ is defined as in Problem G.)

I. If $\mathbf{R}$ is a ring of sets then a nonempty subset $\mathbf{J}$ of $\mathbf{R}$ is an *ideal* in $\mathbf{R}$ if $\mathbf{J}$ is closed with respect to the formation of finite unions and if $A \in \mathbf{J}$ whenever $A \subset B$ with A in $\mathbf{R}$ and B in $\mathbf{J}$. If $\mathbf{S}$ is a σ-ring of sets, then an ideal in $\mathbf{S}$ is a *σ-ideal* if it is closed with respect to the formation of countable unions. Verify that an ideal [σ-ideal] in a ring $\mathbf{R}$ [σ-ring $\mathbf{S}$] is itself a ring [σ-ring]. If $\mathbf{S}$ is a ring [σ-ring] of sets, and if $\mathscr{C}$ is an arbitrary subcollection of $\mathbf{S}$, then $\mathscr{C}$ is contained in a smallest ideal [σ-ideal] $\mathbf{J}$, called the ideal [σ-ideal] *generated by* $\mathscr{C}$. Show that $\mathbf{J}$ consists of all the sets A belonging to $\mathbf{S}$ such that there exists a finite [countable] collection of sets in $\mathscr{C}$ that covers A. In particular, if $\mathscr{C}$ is an arbitrary nonempty collection of subsets of a set X, then every set in $\mathbf{R}(\mathscr{C})[S(\mathscr{C})]$ is covered by some finite [countable] subcollection of $\mathscr{C}$.

> This terminology is readily explained. If the rings of subsets of a set X are viewed as the subrings of the Boolean ring consisting of the power class on X equipped with the operations $\triangledown$ and $\cap$ (cf. the remark following Problem F), then a subset of a ring $\mathbf{R}$ is an ideal in $\mathbf{R}$ in the above sense if and only if it is an ideal in $\mathbf{R}$ in the sense of abstract ring theory.

J. A nonempty collection $\mathbf{Q}$ of subsets of a set X is a *quasiring* in X if (i) $P \backslash Q$ belongs to $\mathbf{Q}$ whenever P and Q belong to $\mathbf{Q}$ and $Q \subset P$, and (ii) $P \cup Q$ belongs to $\mathbf{Q}$ whenever P and Q are *disjoint* sets belonging to $\mathbf{Q}$. A quasiring $\mathbf{Q}$ is a *σ-quasiring* if it is closed with respect to the formation of countable disjoint unions, that is, if $\bigcup_n Q_n$ belongs to $\mathbf{Q}$ for every sequence $\{Q_n\}_{n=1}^\infty$ of pairwise disjoint sets belonging to $\mathbf{Q}$. Show that a σ-quasiring is a σ-ring if and only if it is closed with respect to the formation of finite intersections. Give an example of a σ-quasiring that is not a σ-ring. Show that the intersection of any nonempty collection of σ-quasirings in X is again a σ-quasiring in X, and conclude that an arbitrary collection $\mathscr{C}$ of subsets of X is contained in a unique smallest σ-quasiring. This σ-quasiring is called the σ-quasiring *generated* by $\mathscr{C}$ and will be denoted by $\mathbf{Q}(\mathscr{C})$.

K. Show that if $\mathscr{D}$ is a nonempty collection of subsets of a set X that is closed with respect to the formation of finite intersections, then the σ-quasiring $\mathbf{Q}(\mathscr{D})$ generated by $\mathscr{D}$ coincides with the σ-ring $S(\mathscr{D})$. (Hint: Let $\mathbf{A}$ denote the collection of all subsets

A of *X* with the property that if $D \in \mathcal{D}$, then $A \cap D \in \mathbf{Q}(\mathcal{D})$, and let **B** denote the collection of all subsets *B* of *X* with the property that if $Q \in \mathbf{Q}(\mathcal{D})$, then $B \cap Q \in \mathbf{Q}(\mathcal{D})$. Verify that **A** and **B** are both σ-quasirings containing $\mathcal{D}$, and use Problem J.)

L. Let *X* be a topological space and let *A* be an arbitrary subset of *X*. If $\mathcal{T}$ denotes the topology of *X*, then the trace $\mathcal{T}_A$ of $\mathcal{T}$ on *A* is the relative topology on the subspace *A*. Use this fact to show that the ring of Borel sets in the subspace *A* is the trace $\mathbf{B}_A$ of **B** on *A*, where **B** denotes the ring of Borel sets in *X* (see Problem D).

M. Let *F* be a nonempty closed set in a metric space *X*, and for each positive integer *n* let G_n denote the set $G_n = \{x \in X : d(x, F) < 1/n\}$. Show that each G_n is open and that $F = \bigcap_{n=1}^{\infty} G_n$. Thus *F* is a G_δ. Dually, every open set in *X* is the union of an increasing sequence of closed subsets. In particular, every open set is an F_σ. Show that the characteristic functions of open sets and closed sets are pointwise limits of sequences of continuous functions. (Hint: See Problem 4D.)

> The above construction uses the metric on *X*, and cannot be duplicated in an arbitrary topological space. In fact, the results of Problem M are not valid in a general topological space.

N. If *X* is a metric space there is a transfinite construction of the σ-ring of Borel sets in *X* that goes as follows. Set $\mathcal{B}^{(0)}$ equal to the collection of all open sets in *X*, then define $\mathcal{B}^{(1)}$ to consist of the collection of all G_δ's, $\mathcal{B}^{(2)}$ to consist of the collection of all $G_{\delta\sigma}$'s, etc. In general, let β be an ordinal number, $0 < \beta < \Omega$, and suppose $\mathcal{B}^{(\xi)}$ has been defined for all $0 \leq \xi < \beta$. If β is a limit number, we define $\mathcal{B}^{(\beta)} = \bigcup_{\xi < \beta} \mathcal{B}^{(\xi)}$. If β is not a limit number, and if, say, $\beta = \alpha + 1$, we set $\mathcal{B}^{(\beta)}$ equal to the collection $(\mathcal{B}^{(\alpha)})_\delta$ of all countable intersections of sets in $\mathcal{B}^{(\alpha)}$ if β is odd, and to the collection $(\mathcal{B}^{(\alpha)})_\sigma$ of all countable unions of sets in $\mathcal{B}^{(\alpha)}$ if β is even. (An ordinal β can be written in exactly one way as $\beta = \lambda + n$, where λ is either a limit number or 0 and *n* is an ordinary nonnegative integer; we say that α is even or odd according as *n* is.) Show that the union $\bigcup_{\xi < \Omega} \mathcal{B}^{(\xi)}$ coincides with the ring of Borel sets in *X*.

There is likewise a construction, dual to the one set forth here, that begins with the collection of *closed* sets, continues with the F_σ, the $F_{\sigma\delta}$'s and so on. This construction also produces the ring of all Borel sets in *X*.

> Except for fairly trivial metric spaces *X* the ordinal number Ω cannot be replaced in the above construction by any smaller number. If, for example, *X* is a subset of $\mathbb{R}^n$ that contains a nonempty perfect set, then $\mathcal{B}^{(\eta)}$ is strictly larger than $\mathcal{B}^{(\xi)}$ whenever $\xi < \eta < \Omega$ [34: pp. 182–184].

O. The ring **B** of Borel sets in Euclidean space $\mathbb{R}^n$ coincides with the σ-ring generated by the ring $\mathbf{H}_r[\mathbf{H}_l]$ of Example D. The ring **B** is also generated by the collection of all closed cells as well as by the collection of all open cells (see Examples 3A and 3F). Hence the lattice of elementary figures of Example E also generates **B** as a σ-ring.

P. Let *X* be a metric space, and let $\mathcal{C}$ be a countable collection of subsets of *X* with the property that for every point *x* in *X* and every $\varepsilon > 0$ there exists a set *C* in $\mathcal{C}$ such that $x \in C$ and diam $C < \varepsilon$. Show that the σ-ring generated by $\mathcal{C}$ contains the σ-ring of Borel sets in *X*.

Q. Let *W* denote the topological space $W(\Omega)$ of countable ordinal numbers in the order topology (see Problem 1W and Example 3B). If *E* and *F* are closed and unbounded subsets of W, then $E \cap F$ is also closed and unbounded. Show, more generally, that if $\{F_n\}_{n=1}^{\infty}$ is an arbitrary sequence of closed subsets of W each of which is unbounded, then $\bigcap_{n=1}^{\infty} F_n$ is also unbounded, and use this fact to verify that every Borel set A in W has the property that either A or its complement $W \setminus A$ contains a closed unbounded subset of W. (Hint: Given a countable ordinal number α, there are various ways to construct a strictly increasing sequence $\{\xi_n\}_{n=1}^{\infty}$ in $\bigcup_n F_n$ such that $\xi_n > \alpha$ for every n and such that, for each $m = 1, 2, \ldots, \{\xi_n\}$ possesses a subsequence lying in F_m. The supremum of any such sequence belongs to $\bigcap_n F_n$; use Example B.)

R. Let $(X, \mathbf{S})$ be a measurable space, let (Y, ρ) be a complete metric space, and let $\{\varphi_n\}_{n=1}^{\infty}$ be a sequence of measurable mappings of X into Y. Show that if E denotes the subset of X consisting of the points x at which the sequence $\{\varphi_n(x)\}$ is convergent in Y, then E is measurable, and setting

$$\Phi(x) = \lim_n \varphi_n(x), \qquad x \in E,$$

defines a measurable mapping Φ of E into Y. (Hint: For each triple (p, q, r) of positive integers write $E_{p,q,r} = \{x \in X : \rho(\varphi_p(x), \varphi_q(x)) \leq 1/r\}$, and for each pair (N, r) of positive integers set $F_{N,r} = \bigcap_{p,q=N}^{\infty} E_{p,q,r}$. Then $\bigcap_r \bigcup_N F_{N,r}$ is precisely the set of points x at which the sequence $\{\varphi_n(x)\}$ is Cauchy. To see that the pointwise limit Φ is measurable on E, let U be any open set in Y different from Y itself, and set $F = Y \setminus U$. For each positive integer k let $V_k = \{y \in Y : d(y, F) > 1/k\}$, and for each pair (N, k) of positive integers set $G_{N,k} = \bigcap_{n=N}^{\infty} \varphi_n^{-1}(V_k)$. Then a point x of E is mapped into U by Φ if and only if $x \in \bigcup_{k,N} G_{N,k}$.)

S. (i) Provide a proof for Proposition 6.6. (Hint: It suffices to show that (i) implies (iii). For each positive integer N partition the interval $[-N, +N]$ into $N2^{N+1}$ subintervals by means of the points $-N + i/2^N$, $i = 0, 1, \ldots, N2^{N+1}$. If $|f(x)| \leq N$, set $s_N(x)$ equal to the closest endpoint to zero of any subinterval of the partition that contains $f(x)$. If $|f(x)| > N$, set $s_N(x) = 0$.)

(ii) Proposition 6.6 shows, in particular, that every measurable real-valued function f is the pointwise limit of a sequence $\{s_n\}$ of measurable, simple, real-valued functions such that $|s_n| \leq |f|$ for all n, and such that the convergence of $\{s_n\}$ to f is uniform on any set on which f is bounded. Show that the analogous assertion for measurable complex-valued functions is also valid. (Hint: For each positive integer N dissect the complex plane into squares of edge $1/2^N$, and adapt the hint given in part (i).)

The following two problems provide a sketch of a proof of Theorem 6.5. In this argument we employ the Weierstrass approximation theorem, as well as the Stone–Weierstrass theorem, at least in a special setting. Since these matters are not discussed in this book until Chapter 18, it is appropriate to point out that Theorem 6.5 is not used anywhere in the sequel, so no even apparent danger of circularity is to be feared.

T. Let $\mathscr{A}$ denote a real algebra of real-valued functions on a set X that contains the constant functions and is closed with respect to the formation of limits of pointwise convergent sequences. Set $\mathbf{S} = \{E \subset X : \chi_E \in \mathscr{A}\}$ and verify the following: (i) $\mathbf{S}$ is a σ-ring, (ii) X is a measurable space with respect to $\mathbf{S}$, (iii) the functions in $\mathscr{A}$ are all measurable [$\mathbf{S}$], and (iv) every real-valued function on X that is measurable [$\mathbf{S}$] belongs to $\mathscr{A}$. (Hint: The proofs of (i) and (ii) are straightforward, and (iv) follows at once from Proposition 6.6. To establish (iii) let f be a function in $\mathscr{A}$ and let U be an open subset of $\mathbb{R}$. Use Problem M and the Weierstrass approximation theorem (see Example 18C) to show that there exists a sequence $\{p_n(t)\}$ of real polynomials that converges pointwise to χ_U, and observe that the function $h_n(t) = p_n(f(t))$ belongs to $\mathscr{A}$ for every positive integer n.)

U. Use Problem M and the Stone–Weierstrass theorem (see Example 18D) to show that if U is an open subset of the complex plane. then there exists a sequence $\{p_n(\lambda, \bar{\lambda})\}$ of complex polynomials in λ and $\bar{\lambda}$ that converges pointwise to χ_U. Then follow the argument sketched in Problem T to prove the complex version of Theorem 6.5.

Integrals and measures 7

In the language of modern integration theory the term *integral* refers to a number of somewhat different concepts arrived at through a variety of constructions and definitions. About the only thing that can be said about integration in full generality is that an integral on a space X is a linear transformation that is defined on a linear space of functions on X and satisfies certain continuity requirements. As regards the *Lebesgue* integral, however, matters are in a much less chaotic state. Indeed, while here too a considerable number of different definitions and constructions can be found in the literature, there is unanimous agreement on what a Lebesgue integral is. Fortunately, the concept is easy to characterize axiomatically.

Definition. Let $X = (X, \mathbf{S})$ be a measurable space. A linear functional L defined on a linear space $\mathscr{L}$ of complex-valued measurable functions on X is a *(complex) Lebesgue integral* on X provided the following two conditions are satisfied.

(L_1) If f belongs to $\mathscr{L}$ and if g is a measurable function on X such that $|g| \leq |f|$, then

(a) g belongs to $\mathscr{L}$, and
(b) $|L(g)| \leq L(|f|)$.

(L_2) If the sequence $\{f_n\}_{n=1}^{\infty}$ belongs to $\mathscr{L}$, and if

(a) $\lim_{m,n} L(|f_m - f_n|) = 0$, and
(b) $\{f_n\}$ converges pointwise to g,

then g belongs to $\mathscr{L}$ and $\lim_n L(f_n) = L(g)$.

Similarly a real linear functional L on a real linear space of real-valued measurable functions on X is a *real Lebesgue integral* on X if conditions (L_1) and (L_2) are satisfied.

113

The reader will note that, strictly speaking, a [real] Lebesgue integral is a triple $(X, \mathscr{L}, L)$, where X is a measurable space, $\mathscr{L}$ a [real] linear space of measurable functions on X, and L a [real] linear functional defined on $\mathscr{L}$. In line with the convention enunciated in Chapter 2 (and in the preface), *in any discussion concerning a Lebesgue integral, the integral is understood to be complex if no reference is made to the scalar field*. The functions belonging to the domain $\mathscr{L}$ of a (real or complex) Lebesgue integral L are said to be *integrable* with respect to L. Note that the right member of (L_1b) is defined. Indeed, the first conclusion we draw from (L_1a) is that $|f|$ is integrable along with f. Likewise, $\bar{f}$, Re f, and Im f are all integrable along with f, and one has

$$|L(f)| \leq L(|f|).$$

Note also that (L_2b) implies that g is measurable (Prop. 6.2). A sequence of functions integrable with respect to L that satisfies condition (L_2a) will be said to be *Cauchy in the mean*. In the same vein, if the functions f and f_n are integrable with respect to L, and if $L(|f - f_n|) \to 0$, we shall say that the sequence $\{f_n\}$ *converges to f in the mean*.

Example A. Let $(X, \mathbf{S})$ be a measurable space, and let $\mathscr{M}$ denote the linear space of all complex-valued measurable functions on X. Then the zero functional on $\mathscr{M}$ is a Lebesgue integral on X. Similarly, the zero functional on the trivial submanifold $(0) \subset \mathscr{M}$ is a Lebesgue integral on X. More generally, let $\mathbf{J}$ be an arbitrary σ-ideal in $\mathbf{S}$ (Prob. 6I) and let $\mathscr{L}_{\mathbf{J}}$ denote the collection of all those functions f in $\mathscr{M}$ such that f vanishes outside of some set belonging to $\mathbf{J}$. Then the zero functional on $\mathscr{L}_{\mathbf{J}}$ is a Lebesgue integral on X.

Example B. If L is a Lebesgue integral on a set X, if $\mathscr{L}$ denotes the linear space of integrable functions with respect to L, and if a is a positive real number, then the functional aL is also a Lebesgue integral on X having $\mathscr{L}$ for its space of integrable functions. (For $a = 0$ this assertion is not valid in general.) Likewise, if L' is another Lebesgue integral on X and $\mathscr{L}'$ is the linear space of functions integrable with respect to L', then $L + L'$ acting on $\mathscr{L} \cap \mathscr{L}'$ is a Lebesgue integral on X.

Example C. Let x_0 be a point in a measurable space $(X, \mathbf{S})$ and let $\mathscr{M}$ denote the linear space of all measurable functions on X. The linear functional D defined on $\mathscr{M}$ by setting $D(f) = f(x_0)$ for all f in $\mathscr{M}$ is a Lebesgue integral on X. So is the functional $\tilde{D}$ defined by

$$\tilde{D}(f) = f(x_0) + \cdots + f(x_n)$$

where $\{x_0, \ldots, x_n\}$ is any fixed set of $n + 1$ points in X.

Example D. There is an interesting infinite analog of the foregoing construction. Suppose $\{x_n\}_{n=0}^{\infty}$ is a fixed infinite sequence of distinct points in a measurable space $(X, \mathbf{S})$. Consider the collection $\mathscr{L}$ of all those measurable functions f on X with the property that the infinite series $\sum_n |f(x_n)|$ is convergent, and for each f in $\mathscr{L}$ set $S(f) = \sum_n f(x_n)$. Then S is a Lebesgue integral on X. The nontrivial part of the proof of this fact, of course, is the verification of axiom (L_2). Let $\{f_k\}$ be a sequence of functions in $\mathscr{L}$ that converges pointwise on X to a function g and is Cauchy in the mean with respect to S—$\lim_{k,k'} S(|f_k - f_{k'}|) = 0$. Then for any given $\varepsilon > 0$, there exists a positive integer K such that $S(|f_k - f_{k'}|) < \varepsilon$ for all integers $k, k' \geq K$. Hence, in particular,

$$\sum_{n=0}^{N} |f_k(x_n) - f_{k'}(x_n)| < \varepsilon$$

for every positive integer N and all $k, k' \geq K$. Taking the limit as k' tends to infinity, we obtain

$$\sum_{n=0}^{N} |f_k(x_n) - g(x_n)| \leq \varepsilon, \qquad k \geq K,$$

for all N, and therefore

$$\sum_{n=0}^{\infty} |f_k(x_n) - g(x_n)| \leq \varepsilon, \qquad k \geq K. \tag{1}$$

This shows, in the first place, that $f_k - g$ belongs to $\mathscr{L}$, and hence that g does so too. In the second place, (1) shows that

$$|S(f_k) - S(g)| \leq S(|f_k - g|) \leq \varepsilon$$

for all $k \geq K$, and the verification is complete. The simplest instance of this construction is obtained by taking for X the set $\mathbb{N}_0$ of nonnegative integers, for $\mathbf{S}$ the power class on $\mathbb{N}_0$, and setting $x_n = n$.

Example E. There is also a generalization of the construction in Example D that is sometimes useful. Suppose given an arbitrary indexed family $\{L_\gamma\}_{\gamma \in \Gamma}$ of Lebesgue integrals, all on the same measurable space $(X, \mathbf{S})$, but with each integral L_γ having its own space $\mathscr{L}_\gamma$ of integrable functions. We denote by $\mathscr{L}$ the collection of all those functions f in the intersection $\bigcap_\gamma \mathscr{L}_\gamma$ with the property that $\sum_{\gamma \in \Gamma} L_\gamma(|f|) < +\infty$, and for each f in $\mathscr{L}$ we define

$$L(f) = \sum_{\gamma \in \Gamma} L_\gamma(f).$$

(Since $|L_\gamma(f)| \leq L_\gamma(|f|)$ for each index γ, the indexed family $\{L_\gamma(f)\}_{\gamma \in \Gamma}$ is summable for each f in $\mathscr{L}$; see Problem 4N.) It is clear that $\mathscr{L}$ is a linear space of measurable functions on X and that L is a linear functional on $\mathscr{L}$. Here again it turns out that L is a Lebesgue integral on X. Just as in Example D, the only difficulty encountered in proving this assertion comes in verifying axiom (L_2). Suppose $\{f_k\}$ is a sequence of functions in $\mathscr{L}$ that converges pointwise on X to a function g and that is Cauchy in the mean with respect

115

to $L-\lim_{k,k'} L(|f_k - f_{k'}|) = 0$. Then the sequence $\{f_k\}$ is also Cauchy in the mean with respect to each of the integrals L_γ, so $g \in \mathscr{L}_\gamma$ and $\lim_k L_\gamma(f_k) = L_\gamma(g)$ for each index γ. For any given $\varepsilon > 0$ there exists a positive integer K such that $L(|f_k - f_{k'}|) < \varepsilon$ for all integers $k, k' \geq K$. Hence, in particular, we have

$$\sum_{\gamma \in D} L_\gamma(|f_k - f_{k'}|) < \varepsilon$$

for all integers $k, k' \geq K$, where D denotes an arbitrary finite subset of Γ. Taking the limit as k' tends to infinity, we obtain

$$\sum_{\gamma \in D} L_\gamma(|f_k - g|) \leq \varepsilon, \qquad k \geq K,$$

for an arbitrary finite set D of indices, whence it follows that

$$\sum_{\gamma \in \Gamma} L_\gamma(|f_k - g|) \leq \varepsilon, \qquad k \geq K. \tag{2}$$

This shows, in the first place, that $f_k - g$ belongs to $\mathscr{L}$, and hence that g does so too. In the second place, (2) shows that

$$|L(f_k) - L(g)| \leq L(|f_k - g|) \leq \varepsilon$$

for all $k \geq K$, and the verification is complete.

Example F. The Cauchy integral of elementary calculus (say, on the unit interval) is a linear functional defined on the real linear space $\mathscr{C}_{\mathbb{R}}([0, 1])$ (Ex. 2J), but it is obviously not a real Lebesgue integral. Likewise, the Riemann integral R of advanced calculus defined on the real linear space $\mathscr{R}$ of Riemann integrable functions on $[0, 1]$ fails to be a real Lebesgue integral. Indeed, if R were a real Lebesgue integral, then the functions in $\mathscr{R}$ would all be measurable with respect to some σ-ring $\mathbf{S}$ of subsets of $[0, 1]$, and according to Example 6F, $\mathbf{S}$ would contain all Borel sets. But then the characteristic function of the set of rational numbers in $[0, 1]$ would be Riemann integrable, contrary to fact.

Proposition 7.1. *Let L be a Lebesgue integral on a measurable space X and let $\mathscr{L}$ denote the linear space of functions integrable with respect to L. If f is a real-valued [nonnegative] function in $\mathscr{L}$, then $L(f)$ is also real [nonnegative]. Hence L is monotone increasing on the real-valued functions in $\mathscr{L}$, and is self-conjugate on $\mathscr{L}$, so that*

$$L(\bar{f}) = \overline{L(f)}, L(\mathrm{Re}\ f) = \mathrm{Re}\ L(f), L(\mathrm{Im}\ f) = \mathrm{Im}\ L(f), f \in \mathscr{L}$$

(Ex. 2H). Moreover, if f and g are real functions in $\mathscr{L}$, then so are $f \vee g$ and $f \wedge g$. (In other words, the real-valued functions in $\mathscr{L}$ form a function lattice (Prob. 1D).) Finally,

$$\sigma(f, g) = L(|f - g|), \qquad f, g \in \mathscr{L},$$

defines a pseudometric σ (Ch. 4, p. 56) on the vector space $\mathscr{L}$ with the property that convergence in the mean with respect to L is convergence with respect to the pseudometric σ.

PROOF. If f is a nonnegative function that belongs to $\mathscr{L}$, then $f = |f|$, and it follows that $L(f) \geq 0$ (set $g = 0$ in axiom (L_1)). If f is real-valued and belongs to $\mathscr{L}$, then the positive and negative parts of f also belong to $\mathscr{L}$ ($0 \leq f^\pm \leq |f|$) and $L(f) = L(f^+) - L(f^-)$, so $L(f)$ is real. The rest of the proof is routine. □

Example G. Let L be a (complex) Lebesgue integral on a measurable space X, and let $\mathscr{L}$ denote the vector space of integrable functions with respect to L. Then $\mathscr{L}$ is self-conjugate and is therefore (identifiable with) the complex-ification $(\mathscr{L}_{\mathbb{R}})^+$ of the real vector space $\mathscr{L}_{\mathbb{R}}$ of real-valued functions in $\mathscr{L}$ (see Example 2H). Moreover, it is clear that if this identification is made, then L itself coincides with the complexification $(L|\mathscr{L}_{\mathbb{R}})^+$ (Example 2M). Con-versely, if one starts with a real Lebesgue integral L defined on a real linear space $\mathscr{L}$ of real-valued functions on a measurable space X, then the complexification L^+ (defined on the space $\mathscr{L}^+$ of complex-valued functions) is a complex Lebesgue integral; see Problem C.

Proposition 7.2. *Let L be a (real or complex) Lebesgue integral on a measurable space X, and let $\{f_n\}$ be a monotone increasing sequence of nonnegative functions on X, each of which is integrable with respect to L. Suppose also that $\{f_n\}$ converges pointwise to a real function g. Then g is integrable with respect to L if and only if the numerical sequence $\{L(f_n)\}$ is bounded, and in this case we have $L(f_n) \to L(g)$.*

PROOF. According to Proposition 7.1 the sequence of integrals $\{L(f_n)\}$ is nonnegative and monotone increasing. Moreover, if g is integrable, then $0 \leq L(f_n) \leq L(g)$ for every n. On the other hand, if the sequence $\{L(f_n)\}$ is bounded in $\mathbb{R}$, then it is convergent, and is therefore a Cauchy sequence:

$$\lim_{m,n} |L(f_m) - L(f_n)| = \lim_{m,n} |L(f_m - f_n)| = 0.$$

But, since the sequence $\{f_n\}$ is monotone, we have $|L(f_m - f_n)| = L(|f_m - f_n|)$ for all positive integers m and n, so the sequence $\{f_n\}$ is Cauchy in the mean, and the result follows at once from axiom (L_2). □

The most important characteristic of a Lebesgue integral is its close connection with a *measure*, a concept which we now review. Recall that a function φ whose domain of definition is a collection $\mathscr{C}$ of subsets of some set X is a *set function* on X. We shall be concerned with set functions that are either complex-valued or extended real-valued. In either of these two cases we say of a set function φ that it is *finitely additive* on $\mathscr{C}$ if for every finite sequence $\{A_k\}_{k=1}^n$ of pairwise disjoint sets in $\mathscr{C}$ such that the union $A_1 \cup \cdots \cup A_n$ also belongs to $\mathscr{C}$ we have

$$\varphi\left(\bigcup_{k=1}^n A_k\right) = \sum_{k=1}^n \varphi(A_k). \tag{3}$$

Similarly, we say that φ is *countably additive* on $\mathscr{C}$ if for every infinite sequence $\{A_n\}_{n=1}^\infty$ of pairwise disjoint sets in $\mathscr{C}$ such that the union $\bigcup_{n=1}^\infty A_n$ also belongs to $\mathscr{C}$ we have

$$\varphi\left(\bigcup_{n=1}^\infty A_n\right) = \sum_{n=1}^\infty \varphi(A_n). \tag{4}$$

If φ takes on only ordinary complex values the sum appearing in (4) is simply a convergent infinite series of complex numbers. Note that since the left member of (4) is not changed if an arbitrary permutation is applied to the sequence $\{A_n\}_{n=1}^\infty$, the series appearing in (4) is unconditionally convergent, and therefore

$$\varphi\left(\bigcup_{n=1}^\infty A_n\right) = \sum_{n=1}^\infty \varphi(A_n) = \sum_{n \in \mathbb{N}} \varphi(A_n), \tag{5}$$

where the latter expression represents, of course, the sum of the indexed family $\{\varphi(A_n)\}_{n \in \mathbb{N}}$ of complex numbers (see Example 3Q). If φ is an extended real-valued set function, then the sums appearing in (3) and (4) must be interpreted in $\mathbb{R}^\sharp$. In particular, the validity of (3) implies that there do not exist disjoint sets A and B in $\mathscr{C}$ such that $\varphi(A) = -\infty$, $\varphi(B) = +\infty$, and $A \cup B \in \mathscr{C}$. Moreover, the sum appearing in (4) is an infinite series of extended real numbers convergent in $\mathbb{R}^\sharp$ as discussed in Example 3Q. In this case also, the fact that the series $\sum_{n=1}^\infty \varphi(A_n)$ is unconditionally convergent in $\mathbb{R}^\sharp$ implies that (5) is valid. Thus, if φ is either a complex-valued or extended real-valued set function that is countably additive, we may write $\sum_n \varphi(A_n)$ for the sum in (5) without fear of ambiguity.

Another important concept with which we shall be concerned is defined only in the real case; a nonnegative, extended real-valued set function φ is said to be *countably subadditive* on $\mathscr{C}$ if for every set B in $\mathscr{C}$ and every sequence of sets $\{A_n\}_{n=1}^\infty$ in $\mathscr{C}$ such that $B \subset \bigcup_n A_n$, we have $\varphi(B) \le \sum_{n=1}^\infty \varphi(A_n)$.

Definition. If $(X, \mathbf{S})$ is a measurable space, then a *measure* on $(X, \mathbf{S})$ is a nonnegative, extended real-valued, countably additive set function μ defined on the σ-ring $\mathbf{S}$ of measurable sets and satisfying the additional requirement $\mu(\varnothing) = 0$. The value $\mu(E)$ of μ at a set E is the *measure of E*. A measurable space $(X, \mathbf{S})$ together with a measure μ on $(X, \mathbf{S})$ will be called a *measure space* and will be denoted by $(X, \mathbf{S}, \mu)$, or, when no confusion is possible, simply by X. If $\mu(X) < +\infty$, then the measure μ and the measure space $(X, \mathbf{S}, \mu)$ are said to be *finite*.

If μ is a measure on a measure space X, and if $\{E_\gamma\}_{\gamma \in \Gamma}$ is any countable indexed family of disjoint measurable sets in X, then

$$\mu\left(\bigcup_{\gamma \in \Gamma} E_\gamma\right) = \sum_{\gamma \in \Gamma} \mu(E_\gamma)$$

in the sense of the sum of an indexed family of extended real numbers (Example 3Q). In particular, measures are finitely additive. Certain elementary properties of measures are shared by any finitely additive, nonnegative, extended real-value set function defined on a ring of sets and vanishing at $\varnothing$. Such a set function is known as a *finitely additive measure*. The proofs of the following two results are routine and are therefore omitted.

Proposition 7.3. *Let v be a finitely additive measure defined on a ring of sets* **R**. *Then v is* monotone *and* subtractive *in the sense that, if A and B belong to* **R** *and if $A \subset B$, then $v(A) \le v(B)$ and, provided $v(A) < +\infty$, $v(B \backslash A) = v(B) - v(A)$.*

Proposition 7.4. *Every measure μ is countably subadditive and also semicontinuous in the sense that*

$$\mu\left(\bigcup_n E_n\right) = \lim_n \mu(E_n)$$

for every monotone increasing sequence $\{E_n\}$ of measurable sets. Conversely, if v is a finitely additive measure defined on the σ-ring of measurable sets in a measurable space $(X, \mathbf{S})$, and if v is either countably subadditive or semicontinuous in the above-named sense, then v is countably additive and is therefore a measure.

The following result concerning the convergence of a sequence of functions is useful in many situations.

Theorem 7.5 (Egorov's Theorem). *Let $(X, \mathbf{S}, \mu)$ be a measure space, let E be a measurable subset of X having finite measure, and let $\{f_n\}$ be a sequence of complex-valued functions defined and measurable on E and converging pointwise on E to a function f. Then for any positive number ε, there exists a subset F of E such that $\mu(F) < \varepsilon$ and such that $\{f_n\}$ converges uniformly to f on $E \backslash F$.*

Egorov's theorem is actually valid for measurable mappings taking their values in an arbitrary metric space; see Problem G. The following fundamental theorem sets forth the intimate connection between measures and Lebesgue integrals.

Theorem 7.6. *Let* $(X, \mathbf{S})$ *be a measurable space, and suppose given a Lebesgue integral [real Lebesgue integral]* $(X, \mathscr{L}, L)$ *on* X*. For each measurable set* E *in* X*, define*

$$\mu(E) = \begin{cases} L(\chi_E), & \chi_E \in \mathscr{L}, \\ +\infty, & \chi_E \notin \mathscr{L}. \end{cases} \tag{6}$$

Then μ *is a measure on* $(X, \mathbf{S})$*. Conversely, if* μ *is any measure on* $(X, \mathbf{S})$*, then there exists a unique Lebesgue integral [real Lebesgue integral]* $(X, \mathscr{L}, L)$ *on* X *that is related to* μ *as in* (6)*. Thus* (6) *gives rise to a one-to-one correspondence between the set of all Lebesgue integrals [real Lebesgue integrals] on* $(X, \mathbf{S})$ *and the set of all measures on* $(X, \mathbf{S})$*.*

The proof that (6) defines a measure on $(X, \mathbf{S})$ is an easy exercise (Prob. A). Furthermore, the uniqueness of the Lebesgue integral [real Lebesgue integral] L satisfying (6) with respect to a given measure μ is a consequence of the fact that (6) determines which characteristic functions, and therefore which simple functions, are integrable with respect to L, as well as the value of L at all integrable simple functions (see Problem B). To prove the existence assertion of the theorem is a nontrivial task. The reader who wishes to provide a proof may do so by following the procedure for constructing the Lebesgue integral with respect to a given measure as set forth in any textbook on integration theory. (Programs for such a construction can be derived from Problems O and P; see also Example 9D and the proof of Theorem 9.10.)

If L is a (real or complex) Lebesgue integral on a measurable space X, and μ is the measure determined by L via (6), then μ is called the measure *associated with* L. Likewise we say that L is *integration with respect to* the measure μ. In the same spirit, the value $L(f)$ of L at an integrable function f is called the *integral of* f *with respect to* μ, and is denoted by

$$\int_X f \, d\mu \quad \text{or} \quad \int f \, d\mu,$$

or, when necessary to avoid confusion, by

$$\int_X f(x) d\mu(x).$$

Similarly, the functions that have heretofore been spoken of as integrable with respect to L are ordinarily said to be *integrable (over* X*) with respect to* the measure μ associated with L, or, more briefly, to be *integrable* $[\mu]$. More generally, let E be any measurable subset of X, and let f be a scalar-valued function defined on a subset of X that contains E. If the function $\tilde{f}$ defined by

$$\tilde{f}(x) = \begin{cases} f(x), & x \in E, \\ 0, & x \in X \backslash E, \end{cases}$$

is integrable [μ], then f is said to be *integrable* [μ] *over* E, $L(\tilde{f})$ is declared to be the *integral of f over E with respect to* μ, and we write $L(\tilde{f}) = \int_E f \, d\mu$. Thus, if f is integrable [μ] over X, then

$$L(\chi_E f) = \int_E f \, d\mu$$

for every measurable set E (for in this case $\tilde{f} = \chi_E f$ is integrable [μ] along with f; cf. axiom (L_1)).

Let us pause to restate the defining properties of a Lebesgue integral in this more familiar terminology and notation. To begin with, we now suppose given a measure space (X, S, μ). Axiom (L_1) states that if f and g are two complex-valued measurable functions on X such that $|g| \le |f|$, and if f is integrable [μ], then g is also integrable [μ] and $|\int g \, d\mu| \le \int |f| \, d\mu$. In particular, $|f|$ is integrable [μ] and

$$\left| \int f \, d\mu \right| \le \int |f| \, d\mu.$$

Axiom (L_2) states that if a sequence $\{f_n\}$ of functions integrable [μ] is *Cauchy in the mean with respect to* μ, i.e., if

$$\lim_{m,n} \int |f_m - f_n| \, d\mu = 0,$$

and if $\{f_n\}$ also converges pointwise to g, then g is integrable [μ] and $\int g \, d\mu = \lim_n \int f_n \, d\mu$. (It should also be remarked, of course, that the collection $\mathscr{L} = \mathscr{L}_\mu$ of functions integrable [μ] is a linear space, and that integration with respect to μ is a linear functional on $\mathscr{L}$.) The following proposition helps to clarify the relations between measures and their integrals. (All parts of Proposition 7.7 are easy consequences of the definitions; cf. Problem B.)

Proposition 7.7. *Let (X, S, μ) be a measure space, let E be a measurable subset of X, and let f be a real-valued function that is integrable over E. If a is a real number such that $a \le f(x)[f(x) \le a]$ for all x in E, then*

$$a\mu(E) \le \int_E f \, d\mu \quad \left[\int_E f \, d\mu \le a\mu(E) \right].$$

In particular, if a is positive [negative], then $\mu(E) < +\infty$. If s is a measurable simple function, then s is integrable if and only if it vanishes outside a set of finite measure. If this is the case, and if $s = \sum_{i=1}^n \alpha_i \chi_{E_i}$ is a representation of s with $\sum_{i=1}^n \mu(E_i) < +\infty$, then

$$\int s \, d\mu = \sum_{i=1}^n \alpha_i \mu(E_i).$$

Example H. Let (X, S) be a measurable space, and let J be a σ-ideal in S. The measure associated with the Lebesgue integral of Example A based on the σ-ideal J assigns measure 0 to every set E in J and measure $+\infty$ to every set E in $S \backslash J$. In particular, the zero linear functional on the space of all measurable functions on (X, S) is Lebesgue integration with respect to the zero measure on S.

Example I. Recall the Lebesgue integral D constructed in Example C. The measure δ_{x_0} associated with this integral is given by

$$\delta_{x_0}(E) = \chi_E(x_0), \qquad E \in S.$$

(The measure δ_{x_0} is known as the *Dirac measure* or *Dirac mass* concentrated at x_0.) More generally, if $\{x_n\}$ is a sequence of distinct points in a measurable space (X, S), and if S denotes the Lebesgue integral constructed in Example D, then the measure μ associated with S may be described as follows: If E is any measurable set, and if the number of points of the sequence $\{x_n\}$ belonging to E is finite, then $\mu(E)$ is the number of those points; if the number of points of the sequence $\{x_n\}$ belonging to E is infinite, then $\mu(E) = +\infty$. It should be noted that if the σ-ring of measurable sets S in this example is insufficiently rich, then μ and its integral S may not be very interesting. Suppose for example, to take the worst possible case, that the sequence $\{x_n\}$ were so chosen that every measurable set either contained all of the points of the sequence or no point of it. Then the only functions integrable with respect to μ would have integral zero, and S would be one of the trivial integrals of Example A.

Example J. If S is the Lebesgue integral on $\mathbb{N}_0$ described in Example D, then the associated measure κ is characterized by the fact that $\kappa(\{n\}) = 1$ for every nonnegative integer n. The measure κ is called the *counting measure* on $\mathbb{N}_0$. Quite generally, if X is an arbitrary set, then the *counting measure* on (X, S), where S is any σ-ring that contains X and all of the singletons in X, is the (unique) measure on (X, S) that assigns measure one to each singleton in X.

Example K. Let μ and μ' be measures on the same measurable space (X, S), and let $\mathscr{L}$ and $\mathscr{L}'$ denote the linear spaces of functions integrable with respect to μ and μ', respectively. If a and b are positive real numbers, then $a\mu + b\mu'$ is also a measure on (X, S), and the space of functions integrable $[a\mu + b\mu']$ is precisely the linear space $\mathscr{L} \cap \mathscr{L}'$. On the other hand, if either a or b is allowed to be zero, then the measure $a\mu + b\mu'$ will, in general, have a larger space of integrable functions. (Compare this example with Example B.)

Example L. More generally, suppose given an indexed family $\{\mu_\gamma\}_{\gamma \in \Gamma}$ of measures on a measurable space (X, S). If we form the sum L of the indexed family of Lebesgue integrals with respect to these measures as in Example E, then it is a simple matter to determine the measure μ associated with L. Indeed, direct calculation shows that

$$\mu(E) = \sum_\gamma \mu_\gamma(E), \qquad E \in \mathbf{S},$$

where the sum of the indexed family $\{\mu_\gamma(E)\}$ is formed in $\mathbb{R}^\sharp$ (see Example 3Q). Thus the sum of an indexed family of measures is a measure. (This can also be verified directly, of course.)

Example M. Let $(X, \mathbf{S}, \mu)$ be a measure space, and let E denote a fixed measurable subset of X. If for every measurable function f on E that is integrable $[\mu]$ over E we write

$$L_0(f) = \int_E f \, d\mu,$$

then L_0 is a Lebesgue integral on the subspace $(E, \mathbf{S}_E)$. The measure associated with L_0, which we shall refer to as the *restriction* of μ to E and denote by $\mu|E$, is given by $(\mu|E)(F) = \mu(F)$ for every measurable subset F of E. If f is any function defined and measurable on E, then

$$\int_E (f \,|\, E) d(\mu \,|\, E) = \int_E f \, d\mu \tag{7}$$

in the sense that equality holds if either member of (7) is defined. A measure space of the form $(E, \mathbf{S}_E, \mu|E)$ is said to be a *subspace* of the measure space $(X, \mathbf{S}, \mu)$.

An important element in the theory of Lebesgue integration, and one that helps to give that theory its characteristic flavor, is the prominent role played by sets of measure zero, or *null sets*, as they are frequently called. Given a property $p(\ )$ that is predicable of the points of a measure space $(X, \mathbf{S}, \mu)$, one says that the property holds *almost everywhere* with respect to μ, or *almost everywhere* $[\mu]$, if there exists a null set Z such that $p(x)$ holds for every x in $X \backslash Z$. If there can be no doubt as to which measure is intended, we simply say that the property holds *almost everywhere*. We shall also employ the abbreviations a.e. and a.e. $[\mu]$. In particular, two functions f and g on X are *equal almost everywhere* $[\mu]$ ($f = g$ a.e. $[\mu]$) if there exists a null set Z such that $\{x \in X : f(x) \neq g(x)\} \subset Z$, and two subsets E and F of X are *almost equal* $[\mu]$ if $\chi_E = \chi_F$ a.e. $[\mu]$. Likewise, a sequence of functions $\{f_n\}$ converges a.e. to a limit g if there exists a null set Z such that $\{f_n\}$ converges pointwise to g on $X \backslash Z$. If each of a countable sequence $\{p_n(\ )\}$ of properties holds a.e. $[\mu]$, then all of the properties $p_n(\ )$ hold simultaneously a.e. $[\mu]$, since the union of a countable collection of null sets is again a null set. The following proposition summarizes some important facts pertinent to this circle of ideas.

Proposition 7.8. *Let $(X, \mathbf{S}, \mu)$ be a measure space. If f is a nonnegative measurable function on X, then $\int f\, d\mu = 0$ if and only if $f = 0$ a.e. $[\mu]$. If f is an arbitrary complex-valued measurable function on X such that $f = 0$ a.e. $[\mu]$, then f is integrable and $\int f\, d\mu = 0$. If f and g are measurable functions on X such that $f = g$ a.e. $[\mu]$, then f is integrable $[\mu]$ if and only if g is, and in this case $\int f\, d\mu = \int g\, d\mu$.*

PROOF. Let f be a nonnegative measurable function on X, and let $\{s_n\}$ be a monotone increasing sequence of nonnegative measurable simple functions converging pointwise to f (Prop. 6.6). If $\int f\, d\mu = 0$, then $\int s_n\, d\mu = 0$ and therefore $s_n = 0$ a.e. $[\mu]$ for every positive integer n by Proposition 7.7. But then $f = 0$ a.e. $[\mu]$. On the other hand, if $f = 0$ a.e. $[\mu]$, then $s_n = 0$ a.e. $[\mu]$, and therefore $\int s_n\, d\mu = 0$, for every n, whence it follows by Proposition 7.2 that f is integrable and that $\int f\, d\mu = 0$.

Suppose now that f is a complex-valued, measurable function on X and that $f = 0$ a.e. $[\mu]$. Then $(\operatorname{Re} f)^{\pm}$ and $(\operatorname{Im} f)^{\pm}$ all vanish almost everywhere, and it follows at once from what has just been proved that f is integrable and that $\int f\, d\mu = 0$. Moreover, if f and g are measurable functions such that $f = g$ a.e. $[\mu]$, then since $|f| \le |g| + |f - g|$ and $|g| \le |f| + |f - g|$, f and g are integrable together. Finally, if $f = g$ a.e. $[\mu]$ and f and g are integrable $[\mu]$, then it is clear that $\int f\, d\mu = \int g\, d\mu$. $\qquad\square$

It is illuminating to paraphrase Proposition 7.8 in the language of linear spaces. Let $(X, \mathbf{S}, \mu)$ be a measure space and let $\mathscr{L}_\mu$ and $\mathscr{M}$ denote, respectively, the linear space of all integrable functions and the linear space of all measurable functions on X. Then according to Proposition 7.8 the set $\mathscr{M}_0$ of those functions in $\mathscr{M}$ that vanish a.e. $[\mu]$, which is clearly a linear submanifold of $\mathscr{M}$, is also a linear submanifold of $\mathscr{L}_\mu$ and is, in fact, the kernel of the integral with respect to μ. Let f be a function in $\mathscr{M}$ and let $[f]$ denote the coset of f in the quotient space $\mathscr{M}/\mathscr{M}_0$. Then a function g in $\mathscr{M}$ belongs to $[f]$ if and only if $g = f$ a.e. $[\mu]$. In particular, if f is integrable, and if g belongs to $[f]$, then $\int g\, d\mu = \int f\, d\mu$. Thus it makes sense to define $\int_X [f]\, d\mu = \int_X f\, d\mu$. In other words, if we denote by $\dot{\mathscr{L}}_\mu$ the quotient space $\mathscr{L}_\mu/\mathscr{M}_0$, then the linear functional consisting of integration with respect to μ can be factored through $\dot{\mathscr{L}}_\mu$ (cf. Example 2L). Indeed, once a measure μ is fixed, it is frequently advantageous to think of integration with respect to μ as a linear functional on the quotient space $\dot{\mathscr{L}}_\mu$ rather than on $\mathscr{L}_\mu$ itself. (This is standard practice, in fact, in the study of Lebesgue spaces (Chapter 17).) In this same context we observe that

$$\rho([f], [g]) = \int_X [|f - g|]\, d\mu, \qquad [f], [g] \in \dot{\mathscr{L}}_\mu,$$

defines a metric on $\dot{\mathscr{L}}_\mu$, and that this is precisely the metric associated with the pseudometric $\sigma(f, g) = \int_X |f - g|\, d\mu$ (Prob. 4G).

Proposition 7.8 also suggests a modest generalization of the concepts of measurability and integrability.

Definition. Let $(X, \mathbf{S}, \mu)$ be a measure space, and let f be a complex-valued [extended real-valued] function whose domain of definition is a subset A of X. Then f is said to be *defined almost everywhere* $[\mu]$, or *defined* a.e. $[\mu]$, on X if there exists a null set $Z \subset X$ such that $(X \backslash A) \subset Z$. Two functions f_1 and f_2 that are defined a.e. $[\mu]$ are *equal almost everywhere* $[\mu]$ (notation: $f_1 = f_2$ a.e. $[\mu]$) if there exists a null set $Z \subset X$ such that $f_1(x) = f_2(x)$ for every x in $X \backslash Z$. Let f be a complex-valued [extended real-valued] function that is defined a.e. $[\mu]$ on X. Then f is said to be *measurable* $[\mu]$ on X if there exists a measurable complex-valued [extended real-valued] function g that is defined everywhere on X such that $f = g$ a.e. $[\mu]$. (There may well also be nonmeasurable functions h on X such that $f = h$ a.e. $[\mu]$.) Similarly, f is said to be *integrable* $[\mu]$ if there exists a (finite-valued, everywhere defined) function g that is integrable $[\mu]$ such that $f = g$ a.e. $[\mu]$. Moreover, in this situation, the *integral of* f with respect to μ (notation: $\int_X f \, d\mu$ or $\int f \, d\mu$) is declared to be the integral $\int_X g \, d\mu$. (If one such integrable function g exists, and if h is any (finite-valued) measurable function defined on all of X such that $f = h$ a.e. $[\mu]$, then $g = h$ a.e., so $\int h \, d\mu = \int g \, d\mu$ by Proposition 7.8.)

Note that it follows from this definition that an extended real-valued function f that is everywhere defined on X is integrable if and only if there exists an integrable, finite real-valued function g on X such that $f = g$ a.e. $[\mu]$. In particular, an integrable extended real-valued function must be finite-valued almost everywhere. It should also be noted that the collection of functions integrable $[\mu]$ in this sense does not any longer constitute a vector space in general, but it is the case, of course, that if f is an arbitrary function that is integrable $[\mu]$, then there exists a unique element of the vector space $\mathscr{L}_\mu$ consisting exclusively of functions each of which is equal to f a.e. $[\mu]$. Another concept that should be mentioned in connection with this general circle of ideas is the following.

Definition. Let $(X, \mathbf{S}, \mu)$ be a measure space and let f be a complex-valued function that is measurable $[\mu]$ on X. Then f is *essentially bounded* with respect to μ, or *essentially bounded* $[\mu]$, if there exists a real constant M such that $|f| \leq M$ a.e. $[\mu]$. More generally, if E is a measurable set, then f is *essentially bounded on* E if $f\chi_E$ is essentially bounded on X.

If f is a real-valued function (or extended real-valued function) that is measurable $[\mu]$ on a measure space $(X, \mathbf{S}, \mu)$, and if E is a measurable subset of X having positive measure, then it is easily seen that there is a smallest extended real number M with the property that $f \leq M$ a.e. $[\mu]$ on E. This number is called the *essential supremum of* f *over* E with respect to μ and is denoted by ess $\sup_E f$. The *essential infimum of f over E* with respect to μ is defined analogously, and is denoted by ess $\inf_E f$. If E is a measurable set of positive measure, and if f is a real-valued function that is defined and measurable $[\mu]$ on E, then f is essentially bounded on E if and only if both ess $\sup_E f$ and ess $\inf_E f$ are finite, in which case ess $\sup_E |f| = |\text{ess sup} f_E| \vee |\text{ess inf}_E f|$. If E is a measurable set of finite measure, and if f is defined and measurable $[\mu]$ and essentially bounded on E with respect to μ, then f is integrable $[\mu]$ over E, and we have

$$\int_E |f| \, d\mu \leq (\text{ess} \sup_E |f|)\mu(E).$$

It should be noted that if f is an arbitrary measurable real-valued function on a measure space (X, S, μ) with $\mu(X) > 0$, then $M = \text{ess sup}_E f$ may also be characterized as the largest extended real number with the property that the set $\{x \in X : f(x) > t\}$ has positive measure for every $t < M$. In this same general connection the following notion is frequently useful.

Definition. Let (X, S, μ) be a measure space and let f be a complex-valued function that is defined and measurable $[\mu]$ on X. Then the *essential range* of f with respect to μ is the set of all those complex numbers λ with the property that the set $\{x \in X : |f(x) - \lambda| < \varepsilon\}$ has positive measure for every $\varepsilon > 0$.

We conclude this chapter with a sequence of important theorems having to do with passing to the limit under the integral sign. In stating these theorems, and elsewhere, it is convenient to employ the following convention: If f is an extended real-valued function that is defined a.e. and measurable $[\mu]$ and *nonnegative* a.e. $[\mu]$ on a measure space (X, S, μ), then we write

$$\int_X f \, d\mu = +\infty$$

to indicate that f is *not* integrable $[\mu]$ over X. (According to the above definition of integrability for extended real-valued functions this amounts to saying that if g is a measurable nonnegative extended real-valued function on X that is equal to f a.e. $[\mu]$, then g is either not integrable over the set on which g is finite-valued, or the set $E_{+\infty}$ on which g takes the value $+\infty$ has positive measure. Justification for this notational convention is to be found in Problem O. Note that this is consonant with the practice of writing $\sum_\gamma x_\gamma = +\infty$ for an indexed family $\{x_\gamma\}_{\gamma \in \Gamma}$ of extended real numbers when the net of finite sums converges to $+\infty$ in $\mathbb{R}^\natural$; cf. Example 3Q.)

Theorem 7.9 (Monotone Convergence Theorem). *Let (X, S, μ) be a measure space, and let $\{f_n\}$ be a sequence of extended real-valued functions defined*

126

and nonnegative a.e. [μ] and measurable [μ] on X. Suppose also that the sequence $\{f_n\}$ is monotone increasing a.e. [μ]. Then

$$\int_X (\lim_n f_n)\, d\mu = \lim_n \int_X f_n\, d\mu.$$

(Note that $\lim_n f_n$ is defined a.e. as a nonnegative extended real-valued function.) In particular, $\lim_n f_n$ is integrable if and only if the numerical sequence $\{\int f_n\, d\mu\}$ is bounded above in $\mathbb{R}$.

PROOF. By hypothesis there exists a measurable set X_0 such that $\mu(X \backslash X_0) = 0$ and such that the sequence $\{f_n\}$ is defined, nonnegative, and monotone increasing pointwise on X_0. Clearly it suffices to prove the theorem with X replaced by X_0, so that we may, without loss of generality, assume that $X = X_0$. Let us write f for the extended real-valued function $f(x) = \lim_n f_n(x)$, $x \in X$. Denote by E the set of those points x in X at which $f(x) < +\infty$, by Y the complementary set of points at which $f(x) = +\infty$, and let $\tilde{f}$ be the function

$$\tilde{f}(x) = \begin{cases} f(x), & x \in E, \\ 0, & x \in Y. \end{cases}$$

Note that E, Y, and $\tilde{f}$ are all measurable (Prop. 6.2). Suppose first that f is integrable [μ]. Then, by definition, Y is a null set and $\int f\, d\mu = \int \tilde{f}\, d\mu$. Moreover, the sequence $\{f_n \chi_E\}$ converges pointwise to $\tilde{f}$. Therefore by Proposition 7.2 the sequence $\{\int_E f_n\, d\mu\}$ is bounded, and we conclude that

$$\int_X f_n\, d\mu = \int_E f_n\, d\mu \to \int_X \tilde{f}\, d\mu = \int_X f\, d\mu.$$

To complete the proof we must show that if f is not integrable, then the sequence $\{\int f_n\, d\mu\}$ is not bounded above in $\mathbb{R}$, and thus converges to $+\infty$ in $\mathbb{R}^\natural$. Suppose on the contrary that $\{\int f_n\, d\mu\}$ is bounded in $\mathbb{R}$—say by the constant M. It follows immediately from Proposition 7.2 that the function $\tilde{f}$ is integrable [μ] over E and that $\int_E \tilde{f}\, d\mu = \lim_n \int_E f_n\, d\mu$. Thus the proof will be complete if we show that Y is a null set (for this will show that the measurable extended real-valued function f is, in fact, integrable, contrary to hypothesis).

Our first step in this direction is a modest one but essential. The set Y is at least σ-finite with respect to μ. Indeed, Y is contained in the union of the supports of the functions f_n, and each f_n has σ-finite support (see Problems H and J). Hence it suffices to show that every measurable subset Z of Y that has finite measure actually has measure zero (Prob. K). Suppose, on the contrary, that there exists a subset Z of Y such that $\mu(Z) = a$, where $0 < a < +\infty$. Then there also exists a subset W of Z such that $\mu(W) < a/2$ and such that the convergence of $\{f_n\}$ to $+\infty$ is uniform on $Z \backslash W$. (This is no more than a special version of Egorov's theorem; see Problem G.) Thus

127

there exists a positive integer n_0 such that $f_{n_0}(x) \geq 2M/a$ for all x in $Z \setminus W$. But then, since $\mu(Z \setminus W) > a/2$, we obtain (Prop. 7.7)

$$\int_X f_{n_0} \, d\mu \geq \int_{Z \setminus W} f_{n_0} \, d\mu \geq (2M/a)\mu(Z \setminus W) > M,$$

which is impossible. □

Corollary 7.10 (Theorem of Beppo–Levi). *Let* $(X, \mathbf{S}, \mu)$ *be a measure space, let* $\{p_n\}$ *be a sequence of extended real-valued functions defined and non-negative a.e.* $[\mu]$ *and measurable* $[\mu]$ *on* X, *and let* p *denote the (almost everywhere defined) pointwise sum* $p(x) = \sum_n p_n(x)$. *Then*

$$\int_X p \, d\mu = \sum_n \int_X p_n \, d\mu.$$

In particular, p *is integrable* $[\mu]$ *if and only if all of the functions* p_n *are integrable and the numerical series* $\sum_n \int p_n \, d\mu$ *is convergent in* $\mathbb{R}$. *Hence* $\sum_n \int p_n \, d\mu < +\infty$ *implies* $\sum_n p_n(x) < +\infty$ *at almost every point* x.

PROOF. Just as in the preceding proof it is easy to see that it is enough to treat the case in which the functions p_n are defined and nonnegative everywhere on X. Set $f_n = p_1 + p_2 + \cdots + p_n, n = 1, 2, \ldots$, and apply the monotone convergence theorem. □

Theorem 7.11 (Fatou's Lemma). *Let* $(X, \mathbf{S}, \mu)$ *be a measure space, and let* $\{f_n\}$ *be a sequence of functions defined and nonnegative a.e.* $[\mu]$ *and integrable* $[\mu]$ *on* X. *Suppose that there exists a real number* M *such that* $\int f_n \, d\mu \leq M$ *for every* n, *and suppose also that* $\{f_n\}$ *converges a.e. to the function* f. *Then* f *is integrable* $[\mu]$ *and* $\int f \, d\mu \leq M$.

PROOF. For each positive integer n the function $g_n = \inf_{k \geq n} f_k$ is defined and nonnegative almost everywhere. Moreover, g_n is measurable and, since $0 \leq g_n \leq f_n$ a.e., it follows that g_n is integrable $[\mu]$. Finally, the sequence $\{g_n\}$ is monotone increasing and convergent a.e. to f. Hence the desired conclusion follows at once from the monotone convergence theorem (Th. 7.9). □

Example N. Let $\{f_n\}_{n=1}^{\infty}$ be a sequence of integrable complex-valued functions on a measure space $(X, \mathbf{S}, \mu)$ such that

$$\sum_{n=1}^{\infty} \int |f_n - f_{n+1}| \, d\mu < +\infty.$$

Then by the theorem of Beppo–Levi, the numerical series

$$\sum_{n=1}^{\infty} (f_n(x) - f_{n+1}(x))$$

converges absolutely almost everywhere $[\mu]$. Since this series telescopes, we see that the sequence $\{f_n\}$ converges a.e. to some limit—say f. Since $\{f_n\}$ is Cauchy in the mean, one sees, by applying Fatou's lemma to the sequence $\{f_n - f_m\}_{m=1}^{\infty}$, that $\{f_n\}$ also tends to f in the mean with respect to μ.

The three convergence theorems studied thus far (Th. 7.9, Cor. 7.10, Th. 7.11) all suffer from the limitation that they apply only to functions that are nonnegative almost everywhere. We turn now to the standard convergence theorems for complex-valued functions.

Theorem 7.12 (Dominated Convergence Theorem). *Let (X, S, μ) be a measure space, let $\{f_n\}$ be a sequence of complex-valued functions measurable $[\mu]$ on X, and suppose that $\{f_n\}$ converges a.e. to a function f. Suppose also that there exists a function g on X such that g is integrable $[\mu]$ and such that $|f_n| \le g$ a.e. $[\mu]$ for every n. Then f is integrable $[\mu]$,*

$$\lim_n \int_X |f_n - f| \, d\mu = 0,$$

and therefore

$$\int_X f \, d\mu = \lim_n \int_X f_n \, d\mu.$$

Theorem 7.13 (Bounded Convergence Theorem). *Let (X, S, μ) be a finite measure space, and let $\{f_n\}$ be a sequence of complex-valued functions measurable $[\mu]$ on X. Suppose that $\{f_n\}$ converges a.e. to a function f, and that there exists a real number M such that $|f_n| \le M$ a.e. for every n. Then the functions f_n and f are integrable $[\mu]$,*

$$\lim_n \int_X |f_n - f| \, d\mu = 0,$$

and therefore

$$\int_X f \, d\mu = \lim_n \int_X f_n \, d\mu.$$

The bounded convergence theorem is an easy consequence of the dominated convergence theorem, which can, in turn, be derived from the monotone convergence theorem and a version of Fatou's lemma; see Problem S. A simpler, more direct proof of the dominated convergence theorem is sketched in Problem T.

PROBLEMS

A. Show that the set function μ defined in (6) is indeed a measure. (Hint: The finite additivity of μ follows from the linearity of L. To establish the countable additivity of μ use Propositions 7.2 and 7.4.)

B. Let $(X, \mathbf{S})$ be a measurable space, let $(X, \mathscr{L}, L)$ be a Lebesgue integral on $(X, \mathbf{S})$, and let μ be the measure associated with L as in Theorem 7.6. Verify that if E is a set in $\mathbf{S}$ and α is a complex number such that $\alpha \neq 0$ then $\alpha \chi_E$ belongs to $\mathscr{L}$ if and only if $\mu(E) < +\infty$, and use this fact to show that a measurable simple function s on X belongs to $\mathscr{L}$ if and only if it vanishes on the complement of a set of finite measure. Conclude that for each simple function s in $\mathscr{L}$ there exist representations of the form $s = \sum_{i=1}^{n} \alpha_i \chi_{E_i}$, where $\mu(E_i) < +\infty$, $i = 1, \ldots, n$, and that, for any such representation, $L(s) = \sum_{i=1}^{n} \alpha_i \mu(E_i)$.

   (i) Let $(X, \mathscr{L}', L')$ be a second Lebesgue integral on $(X, \mathbf{S})$, let μ' be the measure associated with L' as in Theorem 7.6, and suppose $\mu(E) \leq \mu'(E)$, $E \in \mathbf{S}$. Prove that $\mathscr{L}' \subset \mathscr{L}$ and that $L(f) \leq L'(f)$ for every nonnegative function f in $\mathscr{L}'$. (Hint: Use Propositions 6.6 and 7.2.)

   (ii) Verify the uniqueness assertion of Theorem 7.6 by showing that if $(X, \mathbf{S}, \mu)$ is a measure space and $(X, \mathscr{L}_1, L_1)$ and $(X, \mathscr{L}_2, L_2)$ are Lebesgue integrals on X, each of which satisfies (6) with respect to μ, then $\mathscr{L}_1 = \mathscr{L}_2$ and $L_1 = L_2$.

C. If $(X, \mathscr{L}, L)$ is a complex Lebesgue integral on a measurable space X, then, as noted in Example G, the restriction of L to the real linear space $\mathscr{L}_{\mathbb{R}}$ of real integrable functions in $\mathscr{L}$ is a real Lebesgue integral on X. Show, conversely, that if $(X, \mathscr{L}, L)$ is a real Lebesgue integral, if $\mathscr{L}^+$ is the complexification of $\mathscr{L}$ (Ex. 2H), and L^+ is the complexification of L (Ex. 2M), then L^+ is a complex Lebesgue integral on X. Briefly, the complexification of a real Lebesgue integral is a complex Lebesgue integral, and every complex Lebesgue integral is the complexification of its real part. (Hint: This is not an altogether trivial exercise; one approach is to employ Theorem 7.6.)

D. Let $\mathscr{C}$ be a collection of subsets of a set X such that the empty set $\varnothing$ belongs to $\mathscr{C}$, and let φ be an extended real-valued set function defined on $\mathscr{C}$ that is either finitely or countably additive. Show that $\varphi(\varnothing)$ must be either 0 or $\pm\infty$. Conclude that every countably additive, nonnegative, extended real-valued set function defined on $\mathscr{C}$ is also finitely additive. Show also that $\psi(\varnothing) = 0$ if ψ is a complex-valued set function defined on $\mathscr{C}$ that is either finitely or countably additive.

E. Propositions 7.3 and 7.4 are valid for complex-valued set functions insofar as they make sense. Thus, a finitely additive complex-valued set function defined on a ring of sets is subtractive. Likewise, a finitely additive complex-valued set function φ defined on the σ-ring of measurable subsets of a measurable space $(X, \mathbf{S})$ is countably additive if and only if φ is *semicontinuous* in the sense that $\varphi(\bigcup_n E_n) = \lim_n \varphi(E_n)$ for every monotone increasing sequence of sets $\{E_n\}$ in $\mathbf{S}$.

F. Let μ be a measure on a measurable space X. It is stated in the text (Proposition 7.4) that if $\{E_n\}$ is an arbitrary increasing sequence of measurable sets, then the numerical sequence $\{\mu(E_n)\}$ tends upward to $\mu(\bigcup_n E_n)$. On the other hand, if $\{E_n\}$ is a decreasing sequence, then $\{\mu(E_n)\}$ need not tend to $\mu(\bigcap_n E_n)$ (example?). Show, on the other hand, that the equation $\mu(\bigcap_n E_n) = \lim_n \mu(E_n)$ is valid for a decreasing sequence of measurable sets $\{E_n\}$ if $\mu(E_n) < +\infty$ for any one set E_n.

G. Let $(X, \mathbf{S}, \mu)$ be a finite measure space, let (Y, ρ) be a metric space, and let $\{\Phi_n\}$ be a sequence of measurable mappings of X into Y that converges pointwise on X to a limit Φ.

   (i) Show that the sets $E_{m,n} = \{x \in X : \rho(\Phi_n(x), \Phi(x)) < 1/m\}$, and with them the sets $G_{m,k} = \bigcap_{n=k}^{\infty} E_{m,n}$ and $F_{m,k} = X \backslash G_{m,k}$, are measurable for all k, m, n

in $\mathbb{N}$ (recall Problem 6R). Show also that $\bigcup_{k=1}^{\infty} G_{m,k} = X$ for each m, and conclude that for each m and for arbitrary positive ε there exists a positive integer $k(m)$ such that $\mu(F_{m,k(m)}) < \varepsilon/2^m$. Set $F = \bigcup_{m=1}^{\infty} F_{m,k(m)}$, and show that $\mu(F) < \varepsilon$ and that $\{\Phi_n\}$ converges uniformly to Φ on $X \backslash F$.

(ii) Use these observations to give a proof of Theorem 7.5. Conclude also that if $\{f_n\}$ is a sequence of measurable real-valued functions on X converging point-wise to $+\infty$ in $\mathbb{R}^\sharp$ on a set E of finite measure, then for arbitrary positive ε there exists a subset F of E such that $\mu(F) < \varepsilon$ and such that $\{f_n\}$ converges uniformly to $+\infty$ on $E \backslash F$.

H. Show that an essentially bounded measurable function f on a measure space $(X, \mathbf{S}, \mu)$ is integrable $[\mu]$ if its *support* $N_f = \{x \in X : f(x) \neq 0\}$ has finite measure. If f is bounded away from zero on N_f, the condition is both necessary and sufficient. (Thus, for example, as was noted in Problem B, a measurable simple function s is integrable $[\mu]$ if and only if $\mu(N_s) < +\infty$.)

I. Let $(X, \mathbf{S}, \mu)$ be a measure space and let f be an arbitrary complex-valued function defined on an arbitrary subset of X. Show that if f is integrable $[\mu]$ over some measurable subset E of X, and if F is a measurable subset of E, then f is also integrable $[\mu]$ over F. Let $\mathbf{R}$ denote the collection of measurable sets over which f is integrable $[\mu]$. Show that $\mathbf{R}$ is a ring of sets (and therefore an ideal in $\mathbf{S}$; Problem 6I), and verify that if $v(E) = \int_E f \, d\mu$ for every set E in $\mathbf{R}$, then v is a finitely additive set function on $\mathbf{R}$. (If f is nonnegative and integrable over X, then v is actually a finite measure on $(X, \mathbf{S})$; we shall return to this idea in Chapter 9.)

J. If $(X, \mathbf{S}, \mu)$ is a measure space, then a measurable set E is said to be σ-*finite* with respect to μ if there exists a countable sequence $\{E_n\}$ of sets of finite measure with respect to μ such that $E = \bigcup_n E_n$. In particular, if X is σ-finite with respect to μ, then μ is σ-*finite* and $(X, \mathbf{S}, \mu)$ is a σ-*finite measure space*. The collection of all sets that are σ-finite with respect to μ is a σ-ideal in $\mathbf{S}$ (Prob. 6I). So is the collection $\mathbf{Z}$ of all sets of measure zero with respect to μ. (The sets of finite measure form an ideal in $\mathbf{S}$ between these two σ-ideals.) Show that if f is a function that is integrable $[\mu]$, then the support of f is σ-finite with respect to μ.

K. If E is a σ-finite set in a measure space $(X, \mathbf{S}, \mu)$, and if $\mu(F) \leq a$ for every measurable subset F of E having finite measure, then $\mu(E) \leq a$. Show by an example that the assumption that E is σ-finite cannot be dropped.

L. Let $(X, \mathbf{S}, \mu)$ be a measure space and let A be a measurable subset of X such that $\mu(A) = +\infty$ and such that if E is any measurable subset of A, then either $\mu(E) = +\infty$ or $\mu(E) = 0$. (Such a set A is called an *infinite atom*; we shall return to this notion in Chapter 8.) Show that if B is any σ-finite subset of X, then $\mu(A \cap B) = 0$. Conclude that if f is any function integrable $[\mu]$, then $f = 0$ a.e. on A and hence that $\int_X f \, d\mu = \int_{X \backslash A} f \, d\mu$. In particular, if $\{x_0\}$ is a measurable singleton such that $\mu(\{x_0\}) = +\infty$, then $f(x_0) = 0$.

M. Suppose $(X, \mathbf{S})$ and $(X, \mathbf{S}_0)$ are two measurable spaces (with the same carrier X) such that $\mathbf{S}_0 \subset \mathbf{S}$, and suppose given a measure μ on $(X, \mathbf{S})$. Then the restriction $\mu_0 = \mu | \mathbf{S}_0$ is clearly a measure on $(X, \mathbf{S}_0)$. Show that a function f on X that is measurable $[\mathbf{S}_0]$ is integrable $[\mu_0]$ if and only if it is integrable $[\mu]$, and that, if this is the case, then $\int f \, d\mu_0 = \int f \, d\mu$.

N. Let $(X, \mathbf{S}, \mu)$ be a measure space, and suppose given a σ-ring $\mathbf{S}_0$ contained in $\mathbf{S}$ with the property that every set in $\mathbf{S}$ is almost equal $[\mu]$ to some set in $\mathbf{S}_0$. (In other words, for each set E in $\mathbf{S}$ there is a set E_0 in $\mathbf{S}_0$ such that $\mu(E \vee E_0) = 0$.) Show that every function measurable $[\mathbf{S}]$ on X is equal a.e. $[\mu]$ to a function that is measurable $[\mathbf{S}_0]$. Hence for every integrable function f on $(X, \mathbf{S}, \mu)$ there is a function f_0 on X that is equal to f a.e. $[\mu]$ such that f_0 is integrable $[\mu|\mathbf{S}_0]$ on $(X, \mathbf{S}_0)$ and such that

$$\int_X f \, d\mu = \int_X f_0 \, d(\mu|\mathbf{S}_0).$$

(Hint: Show that if f is measurable $[\mathbf{S}]$ and if $\varepsilon > 0$, then there exists a function f_ε that is measurable $[\mathbf{S}_0]$ and possesses the property that $|f - f_\varepsilon| \le \varepsilon$ a.e. $[\mu]$. Construct a sequence of functions.)

O. Let $(X, \mathbf{S}, \mu)$ be a measure space, and let f be nonnegative function defined and measurable on X and having σ-finite support (Prob. J). Consider the supremum M in $\mathbb{R}^\sharp$ of the set of all integrals $\int s \, d\mu$ of integrable simple functions s such that $0 \le s \le f$. Show that f is integrable if and only if $M < +\infty$ and that in this case $M = \int f \, d\mu$. (This fact is exploited in many textbooks on integration theory to construct the Lebesgue integral with respect to a given measure μ.) Show also that the supremum M can be finite even if the support of f is not σ-finite. (Hint: Use Propositions 6.6 and 7.2. To obtain an example as asked for in the last part of the problem, consider Example H.)

These last results show clearly that if f is a nonnegative measurable function on a measure space $(X, \mathbf{S}, \mu)$, then the only way for f *not* to be integrable is for f to be too large. Indeed, if the support of f is not σ-finite, then f is a *fortiori* too large to be integrable (Prob. J). On the other hand, if f has σ-finite support, then there exists a monotone increasing sequence of nonnegative integrable functions $\{f_n\}$ that converges pointwise to f, and the sequence $\{\int f_n \, d\mu\}$ is either bounded, in which case f is integrable, or else $\int f_n \, d\mu \to +\infty$, in which case f is not integrable.

P. Let $(X, \mathbf{S}, \mu)$ be a finite measure space and let f be a real-valued measurable function on X.

(i) Suppose first that f is bounded and that the range of f is contained in the half-open interval $(a, b]$. Let

$$a = t_0 < t_1 < \cdots < t_n = b$$

be a partition of $[a, b]$, and for each $i = 1, \ldots, n$, let $E_i = \{x : t_{i-1} < f(x) \le t_i\}$. Let ε be a positive number and suppose that $t_i - t_{i-1} \le \varepsilon$ for all i. Show that

$$0 \le \sum_{i=1}^{n} t_i \mu(E_i) - \int_X f \, d\mu \le \varepsilon \mu(X).$$

(ii) In the general case let ε be a positive number, and let $\{t_n\}_{n=-\infty}^{+\infty}$ be a two-way infinite sequence such that $0 < t_n - t_{n-1} \le \varepsilon$ for every integer n and such that $t_n \to \pm\infty$ as $n \to \pm\infty$. Let $E_n = \{x : t_{n-1} < f(x) \le t_n\}$ for every integer n.

Then f is integrable $[\mu]$ if and only if the series

$$\sum_{n=-\infty}^{n=+\infty} t_n \mu(E_n)$$

is absolutely convergent, and in this event

$$0 \le \sum_{n=-\infty}^{+\infty} t_n \mu(E_n) - \int_X f \, d\mu \le \varepsilon \mu(X).$$

(These observations have also been made the basis for a construction of the Lebesgue integral with respect to a given finite measure μ.)

Q. Show that an indexed family $\{\lambda_\gamma\}_{\gamma \in \Gamma}$ of complex numbers is summable (Ex. 3Q) when and only when the function $f(\gamma) = \lambda_\gamma$ is an integrable function on the index set Γ with respect to the counting measure κ on Γ (Ex. J), and that if this is the case, then $\int_\Gamma f \, d\kappa = \sum_{\gamma \in \Gamma} \lambda_\gamma$. Thus the summation of (summable) indexed families of complex numbers is a special case of Lebesgue integration. (Hint: It suffices to treat the case $\lambda_\gamma \ge 0$; use Proposition 7.2.)

R. Let $(X, \mathbf{S}, \mu)$ be a measure space and let f be a measurable complex-valued function on X.

(i) Show that there exists a radius $R \ge 0$ such that $\mu(f^{-1}(\mathbb{C} \backslash D_R(0)^-)) = 0$ if and only if f is essentially bounded, and that if f *is* essentially bounded, then ess $\sup_X |f|$ is the smallest such radius. Show, in general, that the essential range of f is the smallest closed set F in $\mathbb{C}$ such that $\mu(f^{-1}(\mathbb{C} \backslash F)) = 0$.

(ii) Let f be essentially bounded, and let $\dot{\mathscr{L}}_\mu$ denote the linear space of all equivalence classes $[g]$ of integrable functions g on X (where $[g]$ consists of all the measurable functions h on X such that $h = g$ a.e. $[\mu]$). Show that the mapping M_f defined by

$$M_f([g]) = [fg], \qquad [g] \in \dot{\mathscr{L}}_\mu,$$

is a linear transformation on $\dot{\mathscr{L}}_\mu$. The collection of all such linear transformations forms a unital commutative algebra, and M_f is invertible in this algebra if and only if 0 does not belong to the essential range of f.

S. (i) The assumption that the functions f_n in Fatou's lemma (Th. 7.11) are nonnegative a.e. cannot be dropped (example?) but it can be relaxed, and the conclusion can also be strengthened slightly. Here is an alternate version of Fatou's lemma. Let $(X, \mathbf{S}, \mu)$ be a measure space and let $\{f_n\}$ be a sequence of real-valued functions on X such that each f_n is integrable $[\mu]$. Suppose that there exists a real function φ, integrable $[\mu]$ over X, such that $\varphi \le f_n$ a.e. for all n. Then the pointwise inferior limit of $\{f_n\}$ is integrable $[\mu]$ and we have

$$\int (\liminf_n f_n) \, d\mu \le \liminf_n \int f_n \, d\mu.$$

(Hint: In the proof of Fatou's lemma given in the text we actually have $\int g_n \, d\mu \le \inf_{k \ge n} \int f_k \, d\mu$.)

(ii) Let $(X, \mathbf{S}, \mu)$ be a measure space, let $\{f_n\}$ be a sequence of measurable real-valued functions on X, and suppose that there exist real functions φ and Φ, both integrable $[\mu]$ over X, such that $\varphi \le f_n \le \Phi$ a.e. $[\mu]$ for all n. Use (i) to show that $\lim \inf_n f_n$ and $\lim \sup_n f_n$ are integrable $[\mu]$ and that

$$\int (\lim_n \inf f_n) \, d\mu \le \lim_n \inf \int f_n \, d\mu$$

$$\le \lim_n \sup \int f_n \, d\mu \le \int (\lim_n \sup f_n) \, d\mu.$$

Use this fact to give a proof of the dominated convergence theorem (Th. 7.12).

T. Let (X, S, μ) be a measure space and let g be a nonnegative function on X that is integrable $[\mu]$.

(i) Show that for an arbitrarily given positive number ε there exists a measurable set W such that $\int_W g \, d\mu < \varepsilon$ and such that g is bounded on $X \backslash W$, and use this fact to conclude that for an arbitrarily given positive number ε there exists a positive number δ such that $\mu(E) < \delta$ implies $\int_E g \, d\mu < \varepsilon$. (Hint: For each positive integer N set

$$g_N(x) = \begin{cases} g(x), & g(x) \le N, \\ 0, & g(x) > N, \end{cases}$$

and invoke the monotone convergence theorem.)

(ii) Use these observations to give another proof of the dominated convergence theorem (Th. 7.12). (Hint: First treat the case $\mu(X) < +\infty$ using Egorov's theorem; then employ the fact that the support of g is σ-finite (Prob. J).)

U. Let (X, S, μ) be a measure space and let $\mathscr{L}_\mu$ denote, as usual, the linear space of all complex-valued functions on X that are integrable with respect to μ. Let $\mathscr{M}_0$ be the linear submanifold of $\mathscr{L}_\mu$ consisting of those functions that vanish a.e. $[\mu]$, let $\mathscr{L}_\mu = \mathscr{L}_\mu/\mathscr{M}_0$, and let ρ be the metric

$$\rho([f], [g]) = \int_X [|f - g|] d\mu = \int_X |f - g| d\mu$$

on $\mathscr{L}_\mu$. Show that if $\{f_n\}$ is an arbitrary sequence in $\mathscr{L}_\mu$ such that the sequence $\{[f_n]\}$ is Cauchy with respect to the metric ρ, then $\{f_n\}$ possesses a subsequence $\{f_{n_k}\}$ that is convergent to some limit f a.e. $[\mu]$ and possesses the further property that there exists a single function φ in $\mathscr{L}_\mu$ such that $|f_{n_k}| \le \varphi$ a.e. $[\mu]$ for every index k. Show, finally, that the metric space $(\mathscr{L}_\mu, \rho)$ is complete. (Hint: See Example N and Problem 4E.)

V. Let (X, S, μ) be a finite measure space, and let $\{f_n\}$ be a sequence of integrable functions on X converging uniformly to a function f. Show that f is integrable and that $\{f_n\}$ converges to f in the mean. Show by examples that if the requirement that μ be finite is dropped, then it is possible for the limit f not to be integrable and that, even if f is integrable, the sequence $\{\int f_n \, d\mu\}$ may fail to converge, or may converge to some limit other than $\int f \, d\mu$.

W. (i) Let (X, S, μ) be a measure space, and let $\{f_\lambda\}_{\lambda \in \Lambda}$ be a net of real-valued functions defined everywhere on X and integrable $[\mu]$. Suppose that the net $\{f_\lambda\}$ is monotone increasing in the sense that $f_\lambda \le f_{\lambda'}$ pointwise on X whenever $\lambda \le \lambda'$, and suppose finally that the directed set Λ is countably determined

(Prob. 1P). Then the pointwise limit $\lim_\lambda f_\lambda$ is integrable if and only if the numerical net $\{\int f_\lambda \, d\mu\}_{\lambda \in \Lambda}$ is bounded in $\mathbb{R}$, and in any case we have $\int (\lim_\lambda f_\lambda) d\mu = \lim_\lambda \int f_\lambda \, d\mu$. Show by example that the assumption that Λ is countably determined cannot be dropped.

(ii) Let $(X, \mathbf{S}, \mu)$ be a measure space, let $\{f_\lambda\}_{\lambda \in \Lambda}$ be a net of complex-valued measurable functions on X, and suppose $\{f_\lambda\}$ converges a.e. $[\mu]$ to the function f. Suppose also that Λ is countably determined and that there exists a function g on X such that g is integrable $[\mu]$ and such that $|f_\lambda| \leq g$ a.e. $[\mu]$ for every index λ. Show that $\lim_\lambda \int_X |f_\lambda - f| d\mu = 0$. (Hint: Let $\{\lambda_n\}_{n=1}^\infty$ be an increasing cofinal sequence in Λ, and write $f_n = f_{\lambda_n}, n \in \mathbb{N}$. Suppose the desired conclusion is false. Then there exist a positive number ε_0, a second sequence $\{\lambda'_n\}$ in Λ, and a strictly increasing sequence $\{k_n\}$ of positive integers such that $\lambda_{k_n} \leq \lambda'_n \leq \lambda_{k_n+1}$ and such that $\int |f_{\lambda'_n} - f| d\mu \geq \varepsilon_0$ for every n.)

X. Let $(X, \mathbf{S}, \mu)$ be a measure space, let f be a function that is measurable $[\mathbf{S}]$, and let $\{f_n\}$ be a sequence of functions each of which is measurable $[\mathbf{S}]$. Let us write $E_{\varepsilon, n} = \{x \in X : |f(x) - f_n(x)| \geq \varepsilon\}$ for every positive number ε and positive integer n. Then $\{f_n\}$ is said to converge to f in *measure* if $\lim_n \mu(E_{\varepsilon, n}) = 0$ for every $\varepsilon > 0$. Show that if $\mu(X) < +\infty$ and if $\{f_n\}$ converges to f a.e. $[\mu]$, then $\{f_n\}$ also converges to f in measure. Show likewise that if the functions f_n and f are all integrable $[\mu]$, and if $\{f_n\}$ tends to f in the mean, then $\{f_n\}$ also tends to f in measure. Show, in addition, that if the sequence $\{f_n\}$ converges to f in measure, then a subsequence $\{f_{n_k}\}$ converges to f a.e. $[\mu]$. (Hint: Choose f_{n_k} so that

$$\mu\left(\left\{x \in X : |f(x) - f_{n_k}(x)| \geq \frac{1}{k}\right\}\right) \leq \frac{1}{2^k}.)$$

Show, finally, that Theorem 7.12 is valid with convergence almost everywhere replaced by convergence in measure. (Hint: One possibility is to use Problem T(i).)

8 Measure theory

As we have seen (Th. 7.6), there is a one-to-one correspondence between the set of measures on a measurable space $(X, \mathbf{S})$ and the set of Lebesgue integrals on $(X, \mathbf{S})$. In this section we discuss the properties of some specific measures that play an important role in this book and, more generally, in all of mathematical analysis, and we also review some of the standard theory concerning these and other measures.

With the exception of various uncomplicated measures such as those discussed in Examples 7H, 7I and 7J, which are simple enough to be described directly, the measures of interest in mathematical analysis are obtained via some indirect procedure. Frequently this procedure consists of extensions or approximations, sometimes encompassing several stages, starting from some more primitive set function, not itself a measure. The theory of such processes lies largely outside the scope of this book, though some fragments of it are developed in the problems. The reader interested in a more detailed treatment should consult a treatise on measure theory such as [30] or [26].

Example A. The most familiar and important measure of all is *Lebesgue measure on the real line.* This is the measure μ, constructed by Lebesgue on the σ-ring of *Lebesgue measurable* subsets of $\mathbb{R}$, with the property that

$$\mu([a, b]) = b - a, \qquad a, b \in \mathbb{R}, \qquad a \le b \tag{1}$$

(see Problems E and F). In particular, all singletons have Lebesgue measure zero, so that if $a \le b$, then all of the intervals (a, b), $(a, b]$, $[a, b)$, and $[a, b]$ have the same Lebesgue measure $b - a$. If I denotes any such interval, we shall write $|I|$ for its length, so that $\mu(I) = |I|$ for all bounded intervals I.

Lebesgue measure μ is the unique measure on the σ-ring of Lebesgue measurable subsets of $\mathbb{R}$ that satisfies (1). Furthermore, μ is *translation invariant* in the sense that if E is any Lebesgue measurable set and r is a real number, then $\mu(E + r) = \mu(E)$. The Lebesgue integral associated with the measure μ (according to Theorem 7.6) is the *Lebesgue integral* of elementary real analysis, and possesses the property that if f is any continuous function on a real interval $[a, b]$, then the value of the ordinary Cauchy–Riemann integral of f, as studied in calculus, is identical with the value of the Lebesgue integral of f with respect to μ. In view of this fact, if E is a Lebesgue measurable subset of $\mathbb{R}$ and f is a function that is integrable $[\mu]$ over E, we shall write

$$\int_E f(t)dt$$

for the Lebesgue integral of f over E and, if E is an interval with endpoints a and b, we shall also use the standard notation

$$\int_a^b f(t)dt.$$

Example B. The σ-ring **S** of Lebesgue measurable subsets of $\mathbb{R}$ contains all closed intervals, and therefore contains the σ-ring **B** of Borel subsets of $\mathbb{R}$ (Ch. 6, p. 104). Furthermore, as it happens (Prob. H), every Lebesgue measurable set is almost equal $[\mu]$ to some Borel set. It follows (Prob. 7N) that for purposes of integration theory it does not matter whether one does business with the measure space $(\mathbb{R}, \mathbf{S}, \mu)$ or with $(\mathbb{R}, \mathbf{B}, \mu|\mathbf{B})$. In this book we shall usually prefer to employ the measure $\mu|\mathbf{B}$, called *Lebesgue–Borel measure* on $\mathbb{R}$. (See Problem F. As a matter of historical fact, Lebesgue–Borel measure was originally constructed by E. Borel.)

If n is a positive integer, and if $I_1, \ldots, I_n$ are arbitrary bounded intervals of real numbers—open, closed, or half-open—then the product

$$Z = I_1 \times \cdots \times I_n \tag{2}$$

is called a *cell* in $\mathbb{R}^n$, and for any such cell we define $|Z|$ to be the product

$$|Z| = |I_1| \cdots |I_n|.$$

(It is clear that the intervals $I_1, \ldots, I_n$ in (2) are uniquely determined by Z unless Z is empty, which happens when and only when one of the intervals $I_1, \ldots, I_n$ is empty, and hence that $|Z|$ is really a function of Z itself, and not of the representation (2). The reader will note that the open cells [closed cells] in Chapter 3 (see Examples 3A and 3F) are simply those cells in the present sense that happen to be open [closed].)

Example C. For each positive integer n there exists a unique measure μ_n defined on the σ-ring **S** of *Lebesgue measurable* subsets of $\mathbb{R}^n$ (see Problem E) with the property that

$$\mu_n(Z) = |Z| \tag{3}$$

for every cell Z. (Thus μ_1 is Lebesgue measure on $\mathbb{R}$.) The measure μ_n, which is *Lebesgue measure* on $\mathbb{R}^n$, is also translation invariant in the obvious sense. Once again, the σ-ring $\mathbf{S}$ contains the σ-ring $\mathbf{B}$ of Borel subsets of $\mathbb{R}^n$, and every set in $\mathbf{S}$ is almost equal to a Borel set (Prob. H), so that for purposes of integration theory one may use Lebesgue measure μ_n and *Lebesgue–Borel measure* $\mu_n | \mathbf{B}$ interchangeably (Prob. 7N). In this book we shall usually choose to do business with the measure space $(\mathbb{R}^n, \mathbf{B}, \mu_n | \mathbf{B})$.

Lebesgue measure is a natural extension of other, older, schemes for assigning a "content" to various subsets of $\mathbb{R}^n$. In particular, for $n = 3$ the Lebesgue measure of any geometrical solid is readily seen to agree with its ordinary (Euclidean) volume. Likewise, the Lebesgue measure of a geometrical figure in $\mathbb{R}^2$ coincides with its area. Thus, for example, the Lebesgue measure of a rectangle in the plane is the product of its length and width, whether or not the edges are parallel to the coordinate axes. Since the measure of an arbitrary Lebesgue measurable set in $\mathbb{R}^2$ may be defined in terms of the measures of various rectangles (Prob. F), this implies, in turn, that Lebesgue measure on $\mathbb{R}^2$ is invariant under rigid motions. More generally, Lebesgue measure on $\mathbb{R}^n$ is invariant under rigid motions for every positive integer n.

If f is a bounded complex-valued function on some set $Q \subset \mathbb{R}^n$ on which the Cauchy–Riemann integral, as studied in advanced calculus, is defined, then, just as in the case $n = 1$, the Lebesgue integral $\int_Q f \, d\mu_n$ coincides with that integral. Accordingly we shall use the familiar notation

$$\int \cdots \int_E f(x_1, \ldots, x_n) dx_1 \cdots dx_n$$

for the Lebesgue integral of f over a Lebesgue measurable set E. In the same spirit, if Z is the cell $I_1 \times \cdots \times I_n$, where the interval I_j has endpoints a_j and $b_j, j = 1, \ldots, n$, we shall also write

$$\int_{a_n}^{b_n} \cdots \int_{a_1}^{b_1} f(x_1, \ldots, x_n) dx_1 \cdots dx_n$$

for $\int_Z f \, d\mu_n$.

Example D. There exists a unique measure ν on the σ-ring of Borel subsets of the complex plane $\mathbb{C}$ satisfying the condition that if a, b and c, d are any real numbers such that $a \leq b$ and $c \leq d$, then $\nu(R) = (b - a)(d - c)$, where R denotes the rectangle

$$\{\lambda \in \mathbb{C} : a \leq \operatorname{Re} \lambda \leq b, c \leq \operatorname{Im} \lambda \leq d\}.$$

The measure ν, known as *Lebesgue–Borel* measure on $\mathbb{C}$, is also characterized by the fact that in the standard identification $s + it \leftrightarrow (s, t)$ of $\mathbb{C}$ with $\mathbb{R}^2$, if E' denotes the Borel set in $\mathbb{R}^2$ corresponding to a Borel set E in $\mathbb{C}$, then $\nu(E) = \mu_2(E')$. Similarly, there is a unique measure ν_n on the Borel subsets of $\mathbb{C}^n$, called *Lebesgue–Borel* measure on $\mathbb{C}^n$, satisfying the condition that if E

is a Borel set in $\mathbb{C}^n$, and if E' denotes the corresponding subset of $\mathbb{R}^{2n}$ under the identification

$$(s_1 + it_1, \ldots, s_n + it_n) \leftrightarrow (s_1, t_1, \ldots, s_n, t_n),$$

then $v_n(E) = \mu_{2n}(E')$.

Example E. Let f be a monotone increasing real-valued function defined on $\mathbb{R}$. Then there exists a unique measure μ_f defined on the σ-ring of Borel subsets of $\mathbb{R}$ such that for all a, b in $\mathbb{R}$, $a < b$, we have

$$\mu_f((a, b)) = f(b-) - f(a+) = \lim_{t \uparrow b} f(t) - \lim_{t \downarrow a} f(t). \tag{4}$$

(For an explanation of the notation employed in (4) see Example 3P; for a sketch of a proof of the asserted fact see Problem I.) The measure μ_f is called the *Stieltjes–Borel* measure associated with f, and integration with respect to μ_f is known as *Lebesgue–Stieltjes integration with respect to f*. Clearly when $f(t) \equiv t$, the measure μ_f coincides with Lebesgue–Borel measure on $\mathbb{R}$.

Starting from (4) it is easy to verify that $\mu_f(\{t\}) = \delta_t = f(t+) - f(t-)$, so that $\mu_f(\{t\}) = 0$ except for those real numbers t at which f is discontinuous, where $\mu_f(\{t\})$ is equal to the jump in f (cf. Problem 3M for notation and terminology). More generally, we have $\mu_f([a, b]) = f(b+) - f(a-)$ for all $a \leq b$. Likewise, $\mu_f((a, b]) = f(b+) - f(a+)$ for all $a < b$.

If $\tilde{f}$ is another monotone increasing function that agrees with f at every point of continuity of f, then $\tilde{f}(t) = f(t)$ except for at most countably many values of t, whence it follows that $\tilde{f}(t+) = f(t+)$ and $\tilde{f}(t-) = f(t-)$ for *every* t, and therefore that $\mu_{\tilde{f}} = \mu_f$. Thus, starting from a given monotone increasing function f, we may change the value of f at any point t at which f is discontinuous to any value in the interval $[f(t-), f(t+)]$ without changing μ_f in any way. In particular, we may choose $\tilde{f}$ such that $\tilde{f}(t) = f(t+)$ for all t (so that f is right-continuous on $\mathbb{R}$), and if this is done we have

$$\mu_f((a, b]) = \tilde{f}(b) - \tilde{f}(a), \qquad a, b \in \mathbb{R}, \qquad a \leq b.$$

Because of this fact we may assume without loss of generality, when dealing with a Stieltjes–Borel measure μ_f on $\mathbb{R}$, that the function f is right-continuous.

Suppose now that f is a monotone increasing function defined on a closed interval $[a, b]$. If we extend f by setting

$$\hat{f}(t) = \begin{cases} f(a), & t \leq a, \\ f(t), & a \leq t \leq b, \\ f(b), & t \geq b, \end{cases}$$

then $\hat{f}$ is a monotone increasing function on $\mathbb{R}$ and $\mu_{\hat{f}}$ has the property that $\mu_{\hat{f}}((-\infty, a)) = \mu_{\hat{f}}((b, +\infty)) = 0$. Accordingly, in this situation we shall write μ_f for the restriction of μ_f to $[a, b]$. It is readily seen that μ_f is uniquely characterized by the two conditions

$$\mu_f((s, t)) = f(t-) - f(s+), \qquad a \le s < t \le b,$$

and

$$\mu_f(\{a\}) = f(a+) - f(a), \qquad \mu_f(\{b\}) = f(b) - f(b-)$$

provided $a < b$. (The measure μ_f is clearly the zero measure if the interval $[a, b]$ is degenerate.) Here again an analysis exactly like the foregoing one shows that there is no loss of generality in assuming f to be right-continuous at every point of the open interval (a, b) (since changing f so as to make it right-continuous on (a, b) has no effect on μ_f), and that, if this is done, then μ_f is also characterized by the formulas

$$\mu_f((s, t]) = f(t) - f(s), \qquad a < s \le t \le b,$$
$$\mu_f([a, t]) = f(t) - f(a), \qquad a < t \le b.$$

We note that the second of these two formulas actually implies the first, and also serves to show that a right-continuous monotone increasing function f is uniquely determined up to an additive constant by its associated Lebesgue–Stieltjes measure μ_f.

In the preceding example, if f has a jump discontinuity at some real number t, so that $\delta_t > 0$, then the set $\{t\}$ is a singleton with positive measure with respect to μ_f. This is an instance of a phenomenon that has no parallel in the theory of Lebesgue measure on $\mathbb{R}^n$ and one that deserves special attention.

Definition. A measurable set A in a measure space $(X, \mathbf{S}, \mu)$ is an *atom* of μ (or with respect to μ) if $\mu(A) > 0$ and if for every measurable subset E of A either $\mu(E) = 0$ or $\mu(E) = \mu(A)$. An atom A is *finite* or *infinite* according as $\mu(A) < +\infty$ or $\mu(A) = +\infty$. If μ possesses no atoms, then μ and $(X, \mathbf{S}, \mu)$ are said to be *atom-free*.

Thus, if the function f in Example E has a discontinuity at t, then $\{t\}$ is an atom with respect to μ_f, and μ_f is atom-free if and only if f is continuous. It is clear that any (measurable) singleton with positive measure is an atom, but atoms need not consist of a single point. Indeed, in the measure space of Example 7I, in which a single point x_0 of a space X is fixed and δ_{x_0} is defined by $\delta_{x_0}(E) = \chi_E(x_0)$, every subset of X containing x_0 is an atom.

If $0 < \mu(A) < +\infty$, then to say that A is an atom with respect to μ is the same as saying that every measurable subset of A either is a null set or is almost equal to A. For an infinite atom, however, things are rather different, since it is quite possible for an atom A with $\mu(A) = +\infty$ to split into the disjoint union of sets E and F with $\mu(E) = \mu(F) = +\infty$ (cf. Example 7H). In this connection, and in others as well, the following concept is of interest.

Definition. If $(X, \mathbf{S}, \mu)$ is a measure space, and if μ has the property that

$$\mu(E) = \sup\{\mu(F) : F \in \mathbf{S}, F \subset E, \mu(F) < +\infty\}$$

for every set E in $\mathbf{S}$, then μ and $(X, \mathbf{S}, \mu)$ are said to be *locally finite*.

If a measure is locally finite, then all of its values are determined by the values it assumes on sets of finite measure. In this context the following result is important.

Proposition 8.1. *A necessary and sufficient condition for a measure space $(X, \mathbf{S}, \mu)$ to be locally finite is that it possess no infinite atoms.*

PROOF. If $\mu(A) = +\infty$ and A is an atom with respect to μ, then $\mu(F) < +\infty$ and $F \subset A$ imply $\mu(F) = 0$, so μ is certainly not locally finite. Suppose, in the converse direction, that μ fails to be locally finite. Then there exists a measurable set E_0 such that $\mu(E_0) = +\infty$ and such that

$$M = \sup\{\mu(F) : F \in \mathbf{S}, F \subset E_0, \mu(F) < +\infty\} < +\infty.$$

Let $\{F_n\}_{n=1}^{\infty}$ be a sequence of measurable subsets of E_0 of finite measure such that $\mu(F_n) \to M$, and set $E_n = F_1 \cup \cdots \cup F_n$. Then $\mu(F_n) \leq \mu(E_n) < +\infty$ for each n, whence it follows at once that $\{\mu(E_n)\}$ also tends to M. But then, if $F_0 = \bigcup_{n=1}^{\infty} F_n = \bigcup_{n=1}^{\infty} E_n$, we have $\mu(F_0) = M$. Set $A = E_0 \backslash F_0$, and suppose F is a measurable subset of A such that $0 < \mu(F) < +\infty$. Then for sufficiently large n we have $\mu(E_n) > M - \mu(F)$, and therefore

$$\mu(E_n \cup F) = \mu(E_n) + \mu(F) > M.$$

Since $E_n \cup F$ is a subset of E_0 of finite measure, this contradicts the definition of M. Thus every measurable subset of A either has measure zero or $+\infty$, and since $\mu(A) = +\infty$, A is an infinite atom. $\qquad\square$

The proof of Proposition 8.1 is a typical illustration of a line of argument that occurs frequently in measure theory and is known generally as the "method of exhaustion." It is appropriate to recall that, as was seen in Problem 7L, if f is an integrable function on a measure space $(X, \mathbf{S}, \mu)$, and if A is an infinite atom with respect to μ, then f must vanish a.e. $[\mu]$ on A. Thus, while it is of importance measure-theoretically whether or not a given measure possesses infinite atoms, the distinction is of little consequence in matters concerning integration with respect to that measure.

We turn now to a useful scheme for defining a new measure in terms of a given one. Let $(X, \mathbf{S})$ and $(Y, \mathbf{T})$ be measurable spaces, let Φ be a mapping of X into Y that is measurable $[\mathbf{S}, \mathbf{T}]$, and suppose given a measure μ on $(X, \mathbf{S})$. Let $\mathscr{L}_0$ denote the collection of all those measurable functions f on Y with the property that $f \circ \Phi$ is integrable $[\mu]$. It is clear that $\mathscr{L}_0$ is a linear space and that the equation

$$L_0(f) = \int_X (f \circ \Phi)d\mu$$

defines a linear functional L_0 on $\mathscr{L}_0$. Moreover, it is a routine chore to verify that L_0 is a Lebesgue integral on $(Y, \mathbf{T})$. The measure v associated with the integral L_0 is called the measure *induced on Y by* Φ *and* μ.

Proposition 8.2. *Let $(Y, \mathbf{T})$ be a measurable space, let $(X, \mathbf{S}, \mu)$ be a measure space, and let Φ be a mapping of X into Y that is measurable* [$\mathbf{S}, \mathbf{T}$]. *If v denotes the measure induced on Y by Φ and μ, and if f is a measurable function on Y that is either integrable* [v] *or nonnegative, then*

$$\int_Y f \, dv = \int_X (f \circ \Phi) d\mu.$$

In particular, f is integrable [v] *if and only if $f \circ \Phi$ is integrable* [μ], *and*

$$v(E) = \mu(\Phi^{-1}(E)), \qquad E \in \mathbf{T}.$$

Example F. Let $f(t) = e^{it}$, $t \in \mathbb{R}$. Then f maps $\mathbb{R}$ onto the unit circle Z in a continuous fashion, and it follows that if A belongs to the σ-ring $\mathbf{B}$ of Borel subsets of Z, then $B = f^{-1}(A)$ is a Borel subset of $\mathbb{R}$. Moreover, B has the property that $B + 2\pi = B$, from which it follows at once that, if we write $I_a = [a, a + 2\pi)$ for each real number a, then the Lebesgue measure of $B \cap I_a$ is independent of a. Hence, if $f_a = f|I_a$, then the measure induced on $(Z, \mathbf{B})$ by f_a and Lebesgue–Borel measure on I_a does not depend on the choice of a. This measure on Z is known as *arc-length measure* and throughout this book will consistently be denoted by θ. From the foregoing discussion it is clear that θ is *rotation invariant* in the sense that, if A is a Borel subset of Z and γ is a complex number of modulus one, then $\theta(\gamma A) = \theta(A)$. It should also be remarked that θ can be characterized as the unique measure on $(Z, \mathbf{B})$ that assigns to every simple arc on Z its arc-length as measure (hence the name, of course). If for each function g on Z we define $\hat{g}$ by setting

$$\hat{g}(t) = g(e^{it}), \qquad t \in \mathbb{R},$$

then g is Borel measurable if and only if $\hat{g}$ is, and g is integrable [θ] if and only if $\hat{g}$ is Lebesgue integrable over each period interval I_a. Moreover, if g is integrable [θ] then

$$\int_Z g \, d\theta = \int_a^{a+2\pi} \hat{g}(t)dt, \qquad a \in \mathbb{R}. \tag{5}$$

Example G. Let A denote the affine transformation on $\mathbb{R}^n$ defined by

$$A(x) = x_0 + ax, \qquad x \in \mathbb{R}^n,$$

where a denotes a fixed positive number and x_0 a fixed n-tuple in $\mathbb{R}^n$. Let E_0 be a Borel subset of $\mathbb{R}^n$, and let $E_1 = A(E_0)$. If v denotes the measure induced on E_1 by A and $\mu_n|E_0$, then $\nu = \mu_n/a^n$ on E_1 (Prob. F). Hence if f is an arbitrary Lebesgue integrable function on E_1, then

$$\int_{E_1} f \, d\mu_n = a^n \int_{E_0} f \circ A \, d\mu_n$$

In particular, if f is Lebesgue integrable over $\mathbb{R}^n$, then

$$\int f \, d\mu_n = a^n \int f \circ A \, d\mu_n.$$

A useful way of making one large measure space out of several smaller ones is the formation of direct sums.

Definition. Let $\{(X_\gamma, S_\gamma)\}_{\gamma \in \Gamma}$ be an indexed family of measurable spaces with the property that the sets X_γ are pairwise disjoint. Then by the *direct sum* $\sum_{\gamma \in \Gamma} \oplus (X_\gamma, S_\gamma)$ is meant the measurable space (X, S), where $X = \bigcup_\gamma X_\gamma$ and S denotes the σ-ring consisting of all those subsets E of X with the property that $E \cap X_\gamma \in S_\gamma$ for every index γ. (It is obvious that S is indeed a σ-ring and that $X \in S$.) If, in addition, there is given a measure μ_γ on (X_γ, S_γ) for each index γ, then it is an easy consequence of Example 7L that the set function μ defined by

$$\mu(E) = \sum_{\gamma \in \Gamma} \mu_\gamma(E \cap X_\gamma), \qquad E \in S, \tag{6}$$

is a measure on the measurable space (X, S). The measure space (X, S, μ) is called the *direct sum* of the indexed family $\{(X_\gamma, S_\gamma, \mu_\gamma)\}_{\gamma \in \Gamma}$ of measure spaces, and is denoted by $\sum_{\gamma \in \Gamma} \oplus (X_\gamma, S_\gamma, \mu_\gamma)$. The σ-ring S is likewise called the *direct sum* of the σ-rings S_γ, and μ is called the *direct sum* of the measures μ_γ. We write $\mu = \sum_\gamma \oplus \mu_\gamma$, or when the index family is the finite set $\{1, \ldots, n\}$, $\mu = \mu_1 \oplus \cdots \oplus \mu_n$.

Observe that the σ-ring S in this definition can equally well be described as the collection of all unions $\bigcup_\gamma E_\gamma$, where E_γ is a measurable subset of X_γ for each index γ. The following proposition, a proof of which can easily be given on the basis of Example 7L, summarizes the main facts about direct sums of measure spaces.

Proposition 8.3. *Let $\{(X_\gamma, S_\gamma, \mu_\gamma)\}_{\gamma \in \Gamma}$ be an indexed family of measure spaces with the property that the sets X_γ are pairwise disjoint, and let (X, S, μ) denote the direct sum of the given family. Then X contains each of the measure spaces X_γ as a subspace (Ex. 7M), and a function f on X is measurable [S] if and only if $f|X_\gamma$ is measurable on X_γ for every index γ. Furthermore, if f is measurable on X, then f is integrable [μ] if and only if the indexed family $\{\int_{X_\gamma} |f| d\mu_\gamma\}_{\gamma \in \Gamma}$ is summable, in which case*

$$\int_X f \, d\mu = \sum_{\gamma \in \Gamma} \left(\int_{X_\gamma} f \, d\mu_\gamma \right).$$

Example H. Let $\{t_n\}_{n=0}^{\infty}$ be a strictly increasing sequence of nonnegative numbers such that

$$t_0 = 0 \quad \text{and} \quad \lim_n t_n = +\infty.$$

Let **B** denote the σ-ring of Borel subsets of the half-line $[0, +\infty)$, and for each positive integer n let $\mathbf{B}_n$ denote the trace of **B** on the interval $[t_{n-1}, t_n)$ (Prob. 6D). Then

$$([0, +\infty), \mathbf{B}, \mu) = \sum_{n=1}^{\infty} \oplus ([t_{n-1}, t_n), \mathbf{B}_n, \mu|[t_{n-1}, t_n)),$$

where μ denotes Lebesgue–Borel measure on $[0, +\infty)$. More generally, if $(X, \mathbf{S}, \mu)$ is an arbitrary measure space and if $\{E_n\}$ is any partition of X into a countable collection of measurable sets, then

$$(X, \mathbf{S}, \mu) = \sum_n \oplus (E_n, \mathbf{S}_{E_n}, \mu|E_n).$$

Example I. Let X be an arbitrary set, let **S** denote the power class on X, and let κ be the counting measure on $(X, \mathbf{S})$ (Ex. 7J). Then $(X, \mathbf{S}, \kappa)$ is, in an obvious sense, the direct sum of the (self-indexed) family of all the singletons in X.

The subject of measure theory is ordinarily understood to deal with various set functions other than measures. Two useful generalizations of the notion of a measure are given in the following definition.

Definition. Let $(X, \mathbf{S})$ be a measurable space. An extended real-valued set function μ on $(X, \mathbf{S})$ is a *signed measure* if $\mu(\varnothing) = 0$ and if μ is countably additive on **S**. Similarly, a complex-valued set function ζ defined on **S** is a *complex measure* on $(X, \mathbf{S})$ if it is countably additive. (There is no need to assume $\zeta(\varnothing) = 0$ in the case of a complex measure since this follows from countable additivity; see Problem 7D). Thus if μ is either a signed measure or a complex measure, and if $\{E_n\}_{n=1}^{\infty}$ is an infinite sequence of pairwise disjoint measurable sets, then

$$\mu\left(\bigcup_{n=1}^{\infty} E_n\right) = \sum_{n=1}^{\infty} \mu(E_n) = \sum_{n \in \mathbb{N}} \mu(E_n).$$

(Recall the discussion of countably additive set functions in Chapter 7.) A signed measure μ on $(X, \mathbf{S})$ is *finite* if $|\mu(X)| < +\infty$, and is *σ-finite* if X is a countable union of measurable sets E_n such that $|\mu(E_n)| < +\infty$ for every n.

The above terminology, while fairly standard in the literature, is less logical than it might be. Thus a complex measure is not a measure, in general, since it need not be non-negative. Likewise, a measure is not, in general, a complex measure, since it may take on the value $+\infty$, and a signed measure may be neither a measure nor a complex measure. On the other hand, a finite signed measure is a complex measure, and a finite measure is both a signed measure and a complex measure.

If ζ is a complex measure on (X, S), then the real and imaginary parts of ζ are real-valued complex measures, i.e., finite-valued signed measures on (X, S). Accordingly we turn first to a discussion of signed measures. Our initial observation is a repetition of one made earlier: If μ is a signed measure, and if E and F are disjoint measurable sets, then $\mu(E) + \mu(F)$ must be defined, so $\mu(E)$ and $\mu(F)$ cannot be infinities of opposite sign. It follows that if E and F are measurable sets such that $E \subset F$, and if $\mu(E) = \pm\infty$, then $\mu(F) = \pm\infty$. This proves half of the following elementary lemma.

Lemma 8.4. *If μ is a signed measure on a measurable space (X, S), and if $\mu(E) < +\infty$ for some measurable set E, then $\mu(F) < +\infty$ for every measurable set F contained in E. Similarly, if $\mu(E) > -\infty$, then $\mu(F) > -\infty$ for every measurable set F contained in E. In particular, if $\mu(E)$ is finite, then μ is finite-valued on the measurable subsets of E. If $\mu(E) = +\infty[\mu(E) = -\infty]$ for any one measurable set E, then $\mu(F) > -\infty[\mu(F) < +\infty]$ for every measurable set F.*

COMPLETION OF PROOF. Suppose $\mu(E) = +\infty$ and $\mu(F) = -\infty$. Then $-\infty < \mu(E \cap F) < +\infty$ by what has already been said, so $\mu(E \backslash F) = +\infty$ and $\mu(F \backslash E) = -\infty$, which is impossible. $\qquad\square$

One way to build a signed measure on a measurable space (X, S) is to start with two measures μ and ν on (X, S), one of which is finite, and form the difference $\mu - \nu$. The central fact concerning signed measures is that this simple construction produces the most general signed measure.

Definition. If μ is a signed measure on a measurable space (X, S), then a measurable set A is said to be *positive* with respect to μ if $\mu(F) \geq 0$ for every measurable set F contained in A. Similarly, a measurable set B is *negative* with respect to μ if $\mu(F) \leq 0$ for every measurable set F contained in B.

The importance of positive and negative sets with respect to a signed measure is indicated by the following theorem, a proof of which is outlined in Problem T.

Theorem 8.5. *If μ is a signed measure on a measurable space (X, S), then there exists a partition of X into two (disjoint) sets A and B such that A is positive and B is negative with respect to μ.*

Two sets A and B with the properties set forth in Theorem 8.5 are said to constitute a *Hahn decomposition* of X with respect to μ. A Hahn decomposition is not (usually) uniquely determined, but it is easy to see that

145

if $\{A, B\}$ and $\{A', B'\}$ are two Hahn decompositions of a space X with respect to the same signed measure μ, then

$$\mu(E \cap A) = \mu(E \cap A') \quad \text{and} \quad \mu(E \cap B) = \mu(E \cap B')$$

for every measurable set E.

Definition. If μ is a signed measure on a measurable space $(X, \mathbf{S})$, and if $X = A \cup B$ is a Hahn decomposition of X with respect to μ, then the set functions defined (unambiguously) on $\mathbf{S}$ by

$$\mu_+(E) = \mu(E \cap A) \quad \text{and} \quad \mu_-(E) = -\mu(E \cap B), \qquad E \in \mathbf{S},$$

are called the *positive* and *negative variations* of μ, respectively.

The following basic result is an easy consequence of the definitions, Lemma 8.4 and Theorem 8.5.

Theorem 8.6. *The positive and negative variations μ_+ and μ_- of a signed measure μ on a measurable space $(X, \mathbf{S})$ are measures on $(X, \mathbf{S})$, one of which at least is finite, and $\mu = \mu_+ - \mu_-$ setwise on $\mathbf{S}$. Moreover, μ_+ and μ_- are minimal in this respect in the sense that if v_1 and v_2 are any two measures on $(X, \mathbf{S})$ such that $\mu = v_1 - v_2$, then there exists a finite measure δ on $(X, \mathbf{S})$ such that $v_1 = \mu_+ + \delta$ and $v_2 = \mu_- + \delta$. If μ is finite or σ-finite, then so are μ_+ and μ_-.*

PROOF. It is clear that μ_+ and μ_- are measures, one of which is necessarily finite, that both are finite (or σ-finite) if μ is, and that $\mu = \mu_+ - \mu_-$. Suppose v_1 and v_2 are measures on $(X, \mathbf{S})$ such that $\mu = v_1 - v_2$, and let $X = A \cup B$ be a Hahn decomposition of X, where A is positive and B negative with respect to μ. Then

$$\delta(E) = v_2(E \cap A) + v_1(E \cap B), \qquad E \in \mathbf{S},$$

defines a finite measure on $(X, \mathbf{S})$ satisfying the required conditions.   □

The representation of a signed measure μ as the difference of its positive and negative variations is called the *Jordan decomposition* of μ, in analogy with the older notion of the Jordan decomposition of a real-valued function of bounded variation (see Problems W and 1K). The fact that a signed measure has a Jordan decomposition enables us to define the integral of a function with respect to it.

Definition. Let μ be a signed measure on a measurable space $(X, \mathbf{S})$ and let $\mu = \mu_+ - \mu_-$ be the Jordan decomposition of μ. A complex-valued function f on X is *integrable* with respect to μ (or *integrable* $[\mu]$) if it is integrable both $[\mu_+]$ and $[\mu_-]$. If f is integrable $[\mu]$, then the *integral* of f with respect to μ is defined to be

$$\int_X f \, d\mu = \int_X f \, d\mu_+ - \int_X f \, d\mu_-.$$

Example J. Let g be a real-valued function of bounded variation on a closed interval of real numbers $[a, b]$, $a \le b$. (See Problems 1I, 1J, and 1K for definitions and terminology.) Then there exists a unique signed measure μ_g on the σ-ring of Borel subsets of $[a, b]$ satisfying the conditions

$$\mu_g((s, t)) = g(t-) - g(s+), \qquad a \le s < t \le b,$$

and

$$\mu_g(\{a\}) = g(a+) - g(a), \qquad \mu_g(\{b\}) = g(b) - g(b-).$$

Indeed, if we write g_+ and g_- for the positive and negative variations, respectively, of g on the interval $[a, b]$, then g_+ and g_- are monotone increasing functions on $[a, b]$ satisfying the condition

$$g(t) - g(a) = g_+(t) - g_-(t), \qquad a \le t \le b.$$

Hence, if v_1 and v_2 denote the Stieltjes–Borel measures associated with g_+ and g_-, respectively, as in Example E, then the signed measure $\mu_g = v_1 - v_2$ has the prescribed properties. Just as in Example E, if g is normalized so as to make $g(t+) = g(t)$ for all $a < t < b$, then the formulas

$$\mu_g((s, t]) = g(t) - g(s), \qquad a < s \le t \le b,$$
$$\mu_g([a, t]) = g(t) - g(a), \qquad a < t \le b,$$

also characterize μ_g provided $a < b$. (If $a = b$, μ_g is the zero measure. Here, as before, the second of these formulas implies the first, and also shows that g is uniquely determined up to an additive constant by μ_g when g is right-continuous.) It is also an easily verifiable fact that $\mu_g = v_1 - v_2$ is the Jordan decomposition of μ_g when g is right-continuous, so that $\mu_{g+} = (\mu_g)_+$ and $\mu_{g-} = (\mu_g)_-$ in this case; cf. Problem W. Integration with respect to the signed measure μ_g is known as *Lebesgue–Stieltjes* integration with respect to g.

Suppose now that $(X, \mathbf{S})$ is a measurable space and ζ is a given complex measure on $(X, \mathbf{S})$. If we write

$$\mu(E) = \operatorname{Re} \zeta(E) \quad \text{and} \quad v(E) = \operatorname{Im} \zeta(E), \qquad E \in \mathbf{S},$$

then, as has been noted, μ and v are finite-valued signed measures on $(X, \mathbf{S})$ known as the *real* and *imaginary* parts of ζ, respectively. It follows of course that if $\{E_1, \ldots, E_n\}$ is any finite partition of X into disjoint measurable sets, then the sum

$$\sum_{i=1}^{n} |\zeta(E_i)| \qquad (7)$$

is dominated by the finite number $\mu_+(X) + \mu_-(X) + \nu_+(X) + \nu_-(X)$. The supremum of the set of sums of the form (7), taken over the set of all finite partitions of X into measurable subsets, is therefore a finite real number called the *total variation* of ζ. More generally, we define the *total variation* of ζ *over* each measurable set E to be the total variation of the complex measure $\zeta | E$ obtained by restricting ζ to the measurable subsets of E (Ex. 7M). The total variation of ζ over E will be denoted by $|\zeta|(E)$, so that the total variation of ζ is given by $|\zeta|(X)$. The reader will note that this defines the total variation of a finite signed measure, since a finite signed measure is a complex measure; a discussion of the total variation of an arbitrary signed measure may be found in Problem U(i). The following proposition, whose proof is left as an exercise (Prob. U(ii)), summarizes the elementary facts concerning the total variation of a complex measure.

Proposition 8.7. *Let ζ be a complex measure on a measurable space (X, S), and let μ and ν denote the real and imaginary parts of ζ, respectively. Then $|\mu| = \mu_+ + \mu_-$, $|\nu| = \nu_+ + \nu_-$, and $|\zeta|$ is a finite measure on (X, S) such that*

$$|\mu|(E) \vee |\nu|(E) \leq |\zeta|(E) \leq |\mu|(E) + |\nu|(E), \qquad E \in S.$$

It is an immediate consequence of Proposition 8.7 and Problem 7B that a complex-valued function on (X, S) is integrable with respect to the total variation of a complex measure ζ if and only if it is integrable with respect to both the real and imaginary parts of ζ. We are thus led to the following definition.

Definition. Let ζ be a complex measure on a measurable space (X, S), and let μ and ν denote the real and imaginary parts of ζ, respectively. A complex-valued function f on X is *integrable* with respect to ζ (or *integrable* [ζ]) if it is integrable with respect to the total variation $|\zeta|$. If f is integrable [ζ], then the *integral* of f with respect to ζ is defined to be

$$\int_X f \, d\zeta = \int_X f \, d\mu + i \int_X f \, d\nu.$$

Proposition 8.8. *Let ζ be a complex measure on a measurable space (X, S), and let $\mathscr{L}$ denote the set of all those complex-valued measurable functions* *on X that are integrable* [ζ]. *Then* $\mathscr{L}$ *is a linear space of functions and the mapping*

$$f \to \int_X f \, d\zeta$$

is a linear functional on $\mathscr{L}$ *satisfying the inequality*

$$\left| \int_X f \, d\zeta \right| \leq \int_X |f| \, d|\zeta|, \qquad f \in \mathscr{L}. \tag{8}$$

PROOF. It is obvious that $\mathscr{L}$ is a linear space and that integration with respect to ζ is a linear functional on $\mathscr{L}$. In order to verify (8) we note that this inequality is clearly valid when f is a simple function. Indeed, if $s = \sum_{i=1}^{n} \alpha_i \chi_{E_i}$, where the sets E_i are disjoint and measurable, then

$$\left| \int_X s \, d\zeta \right| = \left| \sum_{i=1}^{n} \alpha_i \zeta(E_i) \right| \leq \sum_{i=1}^{n} |\alpha_i| |\zeta(E_i)| \leq \sum_{i=1}^{n} |\alpha_i| |\zeta|(E_i) = \int_{X.} |s| \, d|\zeta|.$$

Since an arbitrary integrable function f is the pointwise limit of a sequence $\{s_n\}$ of integrable simple functions such that $|s_n| \leq |f|$ for all n (Prob. 6S), the proposition follows by the dominated convergence theorems (see Theorem 7.12 and Problem U). $\qquad \square$

Example K. Let α be a complex-valued function of bounded variation on the interval $[a, b]$, $a \leq b$ (Prob. 1I). Then there exists a unique complex measure ζ_α on the σ-ring of Borel subsets of $[a, b]$ satisfying the conditions

$$\zeta_\alpha((s, t)) = \alpha(t-) - \alpha(s+), \qquad a \leq s < t \leq b,$$

and

$$\zeta_\alpha(\{a\}) = \alpha(a+) - \alpha(a), \qquad \zeta_\alpha(\{b\}) = \alpha(b) - \alpha(b-).$$

Moreover, if α is modified at its points of discontinuity in the open interval (a, b) so as to make it right-continuous at every such point, then ζ_α is unchanged and is characterized by the formulas

$$\zeta_\alpha((s, t]) = \alpha(t) - \alpha(s), \qquad a < s \leq t \leq b,$$
$$\zeta_\alpha([a, t]) = \alpha(t) - \alpha(a), \qquad a < t \leq b,$$

provided $a < b$. (If $a = b$, ζ_α is the zero measure. Here again the second of these formulas implies the first and shows that α is uniquely determined up to an additive constant by ζ_α when α is right-continuous.) Integration with respect to the complex measure ζ_α is known as *Lebesgue–Stieltjes* integration with respect to α. If α is a rectifiable arc, i.e., if α is a continuous function of bounded variation on $[a, b]$, and if f is a continuous complex-valued function defined on the range of α, then it is readily seen that the Lebesgue–Stieltjes integral of the composition $f \circ \alpha$ with respect to α coincides with the *line integral of f along* α introduced in Chapter 5:

$$\int_a^b f \circ \alpha \, d\zeta_\alpha = \int_\alpha f(\zeta)d\zeta.$$

Cf. Problem X and Problem 5D.

In Example A, at the very beginning of this chapter, we introduced the notion of Lebesgue measure on the real line $\mathbb{R}$. This historically important concept constitutes the root of the development of all modern real analysis. Up to this point we have concerned ourselves primarily with that portion of real analysis directly associated with the theory of abstract Lebesgue integration and, as a matter of fact, little of the rest of real analysis is needed for our purposes in this book. There are, however, some extremely useful results in the theory of differentiation that we shall want to use later and we conclude this chapter with a brief scrutiny of them. The attentive reader will observe that the arguments and constructions that follow consist largely of unsophisticated calculations depending on little more than elementary arithmetic. Indeed, the bulk of the following discussion might very well have been presented earlier, say in Chapter 3, where the structure of open subsets of $\mathbb{R}$ is set forth. It has been delayed until now because Lebesgue measure is also an essential ingredient in the discussion.

We shall be concerned primarily with continuous real-valued functions on real intervals and their derivatives and difference quotients. Let f be a continuous real-valued function on the closed interval $[a, b]$. If m is a real number and (c, d) an open subinterval of $[a, b]$, then we define the set $R_m = R_m(f; (c, d))$ to consist of all those points t of (c, d) for which there exists a point u in (t, d) such that

$$\frac{f(u) - f(t)}{u - t} > m.$$

Similarly, the set $L_m = L_m(f; (c, d))$ is defined to consist of those points t in (c, d) for which there exists a point s of (c, t) such that

$$\frac{f(s) - f(t)}{s - t} < m.$$

More generally, if U is any open subset of $[a, b]$, then the set

$$R_m = R_m(f; U) \qquad [L_m = L_m(f; U)]$$

is defined to be the union of all sets of the form

$$R_m(f; (c, d)) \qquad [L_m(f; (c, d))]$$

where (c, d) is an open interval that is a connected component of U. (There may be either finitely many or countably infinitely many such intervals; see Proposition 3.7.)

Proposition 8.9. *Let f be a continuous real-valued function defined on a closed interval* $[a, b]$, *let* U *be an open subset of* $[a, b]$, *and let* m *be a real number. Then the set* $R_m(f; U)$ $[L_m(f; U)]$ *is open, and if* (c_k, d_k) *is any one of its components, then*

$$\frac{f(d_k) - f(c_k)}{d_k - c_k} \geq m \qquad \left[\frac{f(d_k) - f(c_k)}{d_k - c_k} \leq m\right].$$

PROOF. It clearly suffices to treat the case in which U is a single open interval, which we may as well assume to be (a, b). Let $g(t) = f(t) - mt$, $a \leq t \leq b$. Then the difference quotient of g across any subinterval of $[a, b]$ is less than that of f across the same subinterval by exactly m, whence it follows that $R_m(f; (a, b)) = R_0(g; (a, b))$ $[L_m(f; (a, b)) = L_0(g; (a, b))]$. Thus it suffices to prove the proposition for $m = 0$. Moreover, the left-hand version follows at once from the right-hand version by applying the latter to the function $\tilde{f}(t) = f(-t)$ (defined on the interval $[-b, -a]$). Hence it is enough to prove the following lemma.

Lemma 8.10 (Riesz [54]). *For any continuous real-valued function* f *defined on a closed interval* $[a, b]$, *the set* $R_0 = R_0(f; (a, b))$ *is open, and if* (c, d) *is any one of the components of* R_0, *then* $f(c) \leq f(d)$.

PROOF. A point t of (a, b) belongs to R_0 if and only if there exists a point u in (t, b) such that $f(t) < f(u)$. From this and the fact that f is continuous it is obvious that R_0 is open. Let (c, d) be a component of R_0, let t be a point in (c, d), and suppose $f(t) > f(d)$. Since f is continuous there is a largest number t_1 in $[t, d)$ such that $f(t_1) = f(t)$, and it is clear that $f(u) < f(t)$ for all u in the interval (t_1, d). Since $t_1 \in R_0$ there is a real number u_1 in (t_1, b) such that $f(u_1) > f(t_1)$, and it is clear that u_1 cannot belong to $(t_1, d]$. But then $f(d) < f(t_1) < f(u_1)$ and $d < u_1 < b$, which implies that $d \in R_0$, contrary to fact. Thus we see that $f(t) \leq f(d)$ for every t in (c, d), and hence that $f(c) \leq f(d)$. $\qquad\square$

The application we wish to make of Proposition 8.9 depends on the notion of the *derivates* of a function.

Definition. If f is a real-valued function defined on a subset of $\mathbb{R}$ containing some interval $[a, b)$, then the extreme limits

$$\limsup_{t \downarrow a} \frac{f(t) - f(a)}{t - a} \quad \text{and} \quad \liminf_{t \downarrow a} \frac{f(t) - f(a)}{t - a}$$

are known, respectively, as the *upper* and *lower right-hand derivates* of f at a, and will be denoted by $D_+(a) = D_+(a; f)$ and $d_+(a) = d_+(a; f)$. Similarly, if f is defined on a set containing some interval $(a, b]$, then the *upper* and *lower left-hand derivates* of f at b, denoted by $D_-(b) = D_-(b; f)$ and $d_-(b) = d_-(b; f)$, are defined as

$$\limsup_{t \uparrow b} \frac{f(t) - f(b)}{t - b} \quad \text{and} \quad \liminf_{t \uparrow b} \frac{f(t) - f(b)}{t - b},$$

respectively.

The basic significance of these derivates in differentiation theory is clear. A real-valued function f is differentiable to the left [right] at a point t if and only if $D_-(t) = d_-(t) [D_+(t) = d_+(t)]$. Hence f is differentiable at a point t in the interior of its domain of definition if and only if $D_-(t) = d_-(t) = D_+(t) = d_+(t)$. (We are here admitting $\pm\infty$ as possible values of the derivative.)

Suppose now that f is a continuous real-valued function defined on an interval $[a, b]$ and that the upper right-hand derivate of f at a point t, $a < t < b$, exceeds some real number m. Then for any open subset U of $[a, b]$ such that $t \in U$ we have $t \in R_m(f; U)$. Similarly, if $d_-(t; f) < m$, then $t \in L_m(f; U)$ whenever $t \in U \subset [a, b]$. These observations are exploited in the following lemma.

Lemma 8.11. *Let f be a continuous monotone increasing function defined on an interval $[a, b]$ and let m and M be positive numbers such that $m < M$. Let*

$$E = \{t \in (a, b) : D_+(t) > M \quad \text{and} \quad d_-(t) < m\},$$

and let U be an open subset of $[a, b]$ containing E. Then there exists a second open subset W such that $E \subset W \subset U$ and such that $\mu_1(W) \le (m/M)\mu_1(U)$, where μ_1 denotes Lebesgue measure on $\mathbb{R}$.

PROOF. Let $V = L_m(f; U)$, and set $W = R_M(f; V)$. According to the above remarks we have $E \subset W \subset V \subset U$. Moreover, if (c, d) denotes any one component of V, then, according to Proposition 8.9, $f(d) - f(c) \le m(d - c)$. Hence if $\{(c_k, d_k)\}$ is an enumeration of the components of V, then

$$\sum_k (f(d_k) - f(c_k)) \le m\mu_1(V).$$

(The sum on the left in this inequality exists and is finite, even if it represents an infinite series, since f is monotone.) Likewise, for exactly analogous reasons, if $\{(p_i, q_i)\}$ is an enumeration of the components of W, then

$$\sum_i (f(q_i) - f(p_i)) \ge M\mu_1(W).$$

But W is contained in V, and since f is increasing, it follows that

$$\sum_i (f(q_i) - f(p_i)) \le \sum_k (f(d_k) - f(c_k)).$$

Hence $M\mu_1(W) \le m\mu_1(V)$, and the result follows.

Theorem 8.12 (Lebesgue). *If f is a continuous monotone function defined on an interval [a, b], then the derivative f' exists at every point of (a, b)\Z, where Z is a set of Lebesgue measure zero (briefly: f is differentiable a.e. $[\mu_1]$).*

PROOF. We assume, as we clearly may, that f is increasing. Let m and M be positive numbers such that $m < M$ and consider the set $E_{m, M}$ of those points t in (a, b) at which $d_-(t) < m < M < D_+(t)$. (Since f is increasing, all four of its derivates are nonnegative at every point, so it will suffice to use positive numbers for m and M.) According to the preceding lemma (with $U = (a, b)$), there exists an open subset W_1 of (a, b) such that $E_{m, M} \subset W_1$ and $\mu_1(W_1) \le (m/M)(b - a)$. Then, applying the lemma again with $U = W_1$, we obtain a second open set W_2 such that $E_{m, M} \subset W_2 \subset W_1$ and such that $\mu_1(W_2) \le (m/M)\mu_1(W_1) \le (m/M)^2(b - a)$. Continuing in this manner, we obtain by mathematical induction a nested sequence of open sets

$$W_1 \supset W_2 \supset \cdots \supset W_n \supset \cdots \supset E_{m, M}$$

such that $\mu_1(W_n) \downarrow 0$. Hence $\mu_1(E_{m, M}) = 0$. (See Problems C and E; note that it is not asserted here that $E_{m, M}$ is a Borel set, only that it is Lebesgue measurable by virtue of having measure zero.)

Next we consider the set $E = \{t \in (a, b) : d_-(t) < D_+(t)\}$. For each point t of E there exists a pair of positive rational numbers m and M such that $d_-(t) < m < M < D_+(t)$. Hence E is a countable union of subsets of the form $E_{m, M}$, whence it follows that $\mu_1(E) = 0$, and therefore that $D_+(t) \le d_-(t)$ a.e. $[\mu_1]$.

Finally, let us consider the function $\hat{f}(t) = -f(-t)$ on the interval $[-b, -a]$. Since $\hat{f}$ is also continuous and monotone increasing, we conclude by what has just been shown that $D_+(t; \hat{f}) \le d_-(t; \hat{f})$ a.e. on $[-b, -a]$. But a straightforward calculation shows that

$$D_\pm(t; \hat{f}) = D_\mp(-t; f) \quad \text{and} \quad d_\pm(t; \hat{f}) = d_\mp(-t; f)$$

on $(-b, -a)$. Hence $D_-(t; f) \le d_+(t; f)$ a.e. $[\mu_1]$ on (a, b). But also $d_\pm(t; f) \le D_\pm(t; f)$ at *every* point of (a, b). Hence

$$D_+(t; f) \le d_-(t; f) \le D_-(t; f) \le d_+(t; f) \le D_+(t; f)$$

a.e. $[\mu_1]$ on $[a, b]$, and the theorem is proved. $\square$

The foregoing remarkable theorem is complemented by the following result.

Theorem 8.13 (Lebesgue). *Let f be a continuous monotone increasing function defined on an interval [a, b], and let f' denote the derivative of f (existent a.e. $[\mu_1]$ by the preceding theorem). Then f' is Lebesgue measurable and integrable $[\mu_1]$ over [a, b], and*

$$\int_a^b f' \, d\mu_1 \le f(b) - f(a).$$

PROOF. Let $\{\mathscr{P}_n\}_{n=1}^{\infty}$ be a nested sequence of partitions of the interval $[a, b]$ with the property that mesh $\mathscr{P}_n \downarrow 0$ (see Problem 1G for basic definitions), let $\mathscr{P}_n = \{a = t_0^{(n)} < \cdots < t_{N_n}^{(n)} = b\}$, and for each n let

$$q_n(t) = \frac{f(t_i^{(n)}) - f(t_{i-1}^{(n)})}{t_i^{(n)} - t_{i-1}^{(n)}}, \qquad t \in (t_{i-1}^{(n)}, t_i^{(n)}),$$

for $i = 1, \ldots, N_n$. Then each of the step functions q_n (undefined on the finite set of points appearing in the partition $\mathscr{P}_n$) satisfies the condition

$$\int_a^b q_n(t)dt = f(b) - f(a).$$

On the other hand, if $t_{i-1}^{(n)} < t < t_i^{(n)}$, then it is a matter of elementary arithmetic to verify that $q_n(t)$ is straddled by the two difference quotients

$$\frac{f(t_{i-1}^{(n)}) - f(t)}{t_{i-1}^{(n)} - t} \quad \text{and} \quad \frac{f(t_i^{(n)}) - f(t)}{t_i^{(n)} - t}.$$

Hence at every point t of (a, b) at which $f'(t)$ exists, and which is not one of the countable set of points appearing in the various partitions $\mathscr{P}_n$, the sequence $\{q_n(t)\}_{n=1}^{\infty}$ converges to $f'(t)$. This shows that f' is Lebesgue measurable $[\mu_1]$, and the asserted inequality follows by Fatou's lemma (Th. 7.11). □

The following extension of Theorems 8.12 and 8.13 is little more than a corollary.

Proposition 8.14. *If g is any continuous complex-valued function of bounded variation defined on an interval $[a, b]$, then the derivative g' exists a.e. $[\mu_1]$ and satisfies the condition*

$$\int_a^b |g'| d\mu_1 \leq V(g; a, b) \tag{9}$$

(see Problem 1I for notation).

PROOF. That the derivative of g exists a.e. $[\mu_1]$ and is Lebesgue integrable on $[a, b]$ follows at once from consideration of Jordan decompositions of the real and imaginary parts of g (Prob. 1K). To verify (9), let $\{\mathscr{P}_n\}_{n=1}^{\infty}$ be a nested sequence of partitions of $[a, b]$ with the property that mesh $\mathscr{P}_n \downarrow 0$, and define the sequence $\{q_n\}$ of step functions as in the proof of the preceding theorem. Then, just as before, the sequence $\{q_n\}$ tends pointwise to g' a.e. $[\mu_1]$, and it is clear from the definition that $\int_a^b |q_n| d\mu_1$ is dominated by the total variation of g over $[a, b]$ for every n. Hence (9) follows by Fatou's lemma. □

It is only fair to say that the above discussion develops but a very small fragment of a vast theory of differentiation. Much more could be said. For one thing, the hypothesis of continuity can be dropped outright in Theorem 8.12 (and therefore in Theorem 8.13 and Proposition 8.14 as well). Secondly, the theory of differentiation can be extended to much more general functions with much more general domains of definition. The reader who desires to study these topics more fully should consult a treatise on real analysis, such as [58]. Alternatively, he might consult the encyclopedic [26].

PROBLEMS

A. Let $(X, \mathbf{S}_1)$ and $(X, \mathbf{S}_2)$ be two measurable spaces with the same carrier X, and let $\mathscr{D}$ be a collection of sets belonging to both $\mathbf{S}_1$ and $\mathbf{S}_2$ that is closed with respect to the formation of finite intersections. Suppose φ_1 and φ_2 are signed [complex] measures on $(X, \mathbf{S}_1)$ and $(X, \mathbf{S}_2)$, respectively, such that $\varphi_1(E)$ and $\varphi_2(E)$ are finite and equal for every set E in $\mathscr{D}$. Show that $\varphi_1 = \varphi_2$ on the entire σ-ring $\mathbf{S}(\mathscr{D})$. (Hint: If E_0 is any one fixed set in $\mathscr{D}$, and if $\mathbf{Q}$ denotes the collection of all those subsets F of E_0 such that $\varphi_1(F) = \varphi_2(F)$, then $\mathbf{Q}$ is a σ-quasiring containing the trace $\mathscr{D}_{E_0}$. Recall Problems 6D and 6K.)

> Problem A provides the basic uniqueness theorem used to establish all of the assertions of uniqueness made throughout Chapter 8. Neither of the principal hypotheses in this theorem can be omitted, even in the presence of the other. Thus, if κ denotes the counting measure on $\mathbb{R}$, then, for example, 2κ agrees with κ on the ring $\mathbf{H}_r$ of half-open intervals (Ex. 6D) since both measures are infinite on all nonempty sets in $\mathbf{H}_r$, but $\kappa \neq 2\kappa$ on the σ-ring $\mathbf{B}$ of Borel subsets of $\mathbb{R}$ generated by $\mathbf{H}_r$. That the requirement that $\mathscr{D}$ be closed with respect to the formation of finite intersections cannot be dropped may be seen by experimenting with a three point space.

B. A nonnegative, extended real-valued set function μ^* defined on the power class on a set X is an *outer measure* on X if μ^* is countably subadditive and satisfies the condition $\mu^*(\varnothing) = 0$. Show that an outer measure μ^* on X is *finitely subadditive* in the sense that if $A_1, \ldots, A_n$ and B are subsets of X such that $B \subset A_1 \cup \cdots \cup A_n$, then

$$\mu^*(B) \leq \mu^*(A_1) + \cdots + \mu^*(A_n).$$

In particular, if $A \subset B \subset X$, then $\mu^*(A) \leq \mu^*(B)$.

C. (Theorem of Carathéodory) Let μ^* be an outer measure on a set X. A subset E of X is said to be *measurable (in the sense of Carathéodory) with respect to μ^** if for every subset A of X we have

$$\mu^*(A) \geq \mu^*(A \cap E) + \mu^*(A \backslash E).$$

and therefore

$$\mu^*(A) = \mu^*(A \cap E) + \mu^*(A \backslash E).$$

Show that every set Z with the property that $\mu^*(Z) = 0$ is measurable with respect to μ^*, and that $X \backslash E$ is measurable with respect to μ^* whenever E is. Let E_1 and E_2 be two sets that are measurable with respect to μ^*, and write $E_{11} = E_1 \cap E_2$, $E_{12} = E_1 \backslash E_2$, $E_{21} = E_2 \backslash E_1$, and $E_{22} = X \backslash (E_1 \cup E_2)$, for the four sets into which E_1 and E_2 partition X. Show that if A denotes an arbitrary subset of X, and if we write $A_{ij} = A \cap E_{ij}$, $i, j = 1, 2$, then

$$\mu^*(A) = \mu^*(A_{11}) + \mu^*(A_{12}) + \mu^*(A_{21}) + \mu^*(A_{22}).$$

Use this fact to show that the collection **M** of all those subsets of X that are measurable with respect to μ^* is a ring of sets and that $\mu^*|\mathbf{M}$ is countably additive on **M**. (Hint: A finitely additive, countably subadditive, nonnegative set function defined on a ring of sets is automatically countably additive.) Show finally that **M** is a σ-ring, and hence that $(X, \mathbf{M}, \mu^*|\mathbf{M})$ is a measure space. The measure $\mu^*|\mathbf{M}$ is called the measure *defined by* μ^*. (Hint: It suffices to verify that **M** is a σ-quasiring (Prob. 6J). If $\{E_n\}_{n=1}^{\infty}$ is a disjoint sequence of sets in **M** with $E = \bigcup_n E_n$, and A is an arbitrary subset of X, then it follows from what has already been shown that

$$\mu^*(A) = \mu^*(A \setminus (E_1 \cup \cdots \cup E_n)) + \sum_{i=1}^{n} \mu^*(A \cap E_i)$$

$$\geq \mu^*(A \setminus E) + \sum_{i=1}^{n} \mu^*(A \cap E_i)$$

for every positive integer n.)

D. Let us say that a collection $\mathscr{C}$ of subsets of a set X is *admissible* if (i) the empty set $\varnothing$ belongs to $\mathscr{C}$ and (ii) X is the union of some countable collection of sets belonging to $\mathscr{C}$. Let $\mathscr{C}$ be an admissible collection of subsets of X, and let φ be a nonnegative, extended real-valued set function defined on $\mathscr{C}$ and satisfying the condition $\varphi(\varnothing) = 0$. (Such a set function is sometimes called a *gauge*.) For an arbitrary subset A of X let us define

$$\mu^*(A) = \inf \left\{ \sum_{n=1}^{\infty} \varphi(C_n) : A \subset \bigcup_n C_n \right\},$$

where the infimum is taken over all countable coverings of A by means of sets belonging to $\mathscr{C}$. Show that μ^* is an outer measure on X, that $\mu^*(C) \leq \varphi(C)$ for every set C in $\mathscr{C}$, and that $\mu^* = \varphi$ on $\mathscr{C}$ if and only if φ is countably subadditive. The outer measure μ^* is called the outer measure *generated* by φ. (Hint: If ε is a positive number and A_m is the mth set in some sequence of subsets of X, then there is a sequence $\{C_n\}$ of sets in $\mathscr{C}$ such that $A_m \subset \bigcup_n C_n$ and such that

$$\sum_{n=1}^{\infty} \varphi(C_n) < \mu^*(A_m) + \frac{\varepsilon}{2^m}.)$$

Suppose that φ is a finite-valued gauge defined on an admissible collection $\mathscr{C}$ of subsets of X and that μ^* is the outer measure generated by φ. Verify that if $\mu^*(A) > 0$ and if r satisfies $0 < r < 1$, then there exist sets C in $\mathscr{C}$ such that $\mu^*(A \cap C) > r\varphi(C)$.

E. For each positive integer n the collection $\mathscr{Z}_n$ of all cells in $\mathbb{R}^n$ is admissible, and the set function $\varphi_n(Z) = |Z|$, $Z \in \mathscr{Z}_n$, satisfies the condition $\varphi_n(\varnothing) = 0$. Hence φ_n generates an outer measure on $\mathbb{R}^n$ according to the procedure set forth in the preceding problem. This outer measure is *Lebesgue outer measure* on $\mathbb{R}^n$, and will be denoted by μ_n^*. A subset E of $\mathbb{R}^n$ is said to be *Lebesgue measurable* if it is measurable with respect to μ_n^* (see Problem C). Show that every cell in $\mathbb{R}^n$ is Lebesgue measurable, and hence that the σ-ring $\mathbf{S}$ of Lebesgue measurable sets contains all Borel subsets of $\mathbb{R}^n$. (Hint: Show first that if a is an arbitrary real number and i_0 is an integer, $i_0 = 1, \ldots, n$, then the half-space $\{(r_1, \ldots, r_n) \in \mathbb{R}^n : r_{i_0} \le a\}$ is Lebesgue measurable.)

F. The measure defined by Lebesgue outer measure μ_n^* introduced in the preceding problem agrees with *Lebesgue–Borel measure* μ_n on the Borel subsets of $\mathbb{R}^n$ (Examples B and C). Verify this by showing that if Z is a cell in $\mathbb{R}^n$, then $\mu_n^*(Z) = |Z|$ (recall Problem A). Show also that if for any one fixed point $a = (a_1, \ldots, a_n)$ we write T_a for the translation carrying each point $x = (x_1, \ldots, x_n)$ to $x + a = (x_1 + a_1, \ldots, x_n + a_n)$, then $\mu_n^*(T_a(A)) = \mu_n^*(A)$ for every subset A of $\mathbb{R}^n$. Show, finally, that if Φ denotes the mapping of $\mathbb{R}^n$ onto itself defined for each point x in $\mathbb{R}^n$ by $\Phi((x_1, \ldots, x_n)) = (a_1 x_1, \ldots, a_n x_n)$, where $(a_1, \ldots, a_n)$ is some fixed n-tuple of positive numbers, then $\mu_n^*(\Phi(A)) = a_1 \cdots a_n \mu_n^*(A)$ for every subset A of $\mathbb{R}^n$. (Hint: It is necessary to show that if Z_0 is a cell and $\{Z_n\}$ is a countable covering of Z_0 by cells, then $|Z_0| \le \sum_n |Z_n|$. The gauge $\varphi(Z) = |Z|$ is continuous on $\mathscr{Z}_n$ in the sense that if a cell Z is replaced by a slightly smaller compact cell contained in Z, or by a slightly larger open cell containing Z, the value of φ is changed as little as desired. Use this fact to reduce the problem to the case of a finite covering. The finite case may be disposed of by direct calculation.) Verify that if A is a subset of $\mathbb{R}^n$ such that $\mu_n^*(A) > 0$, then for any ratio r satisfying $0 < r < 1$ there exist cells Z such that $\mu_n^*(A \cap Z) > r|Z|$. Show also that such cells exist having edges as short as desired, and having, for example, rational points for vertices. (Hint: If a cell Z has the property that $\mu_n(A \cap Z) > r|Z|$, and if Z is partitioned in any manner into subcells Z_i, $i = 1, \ldots, p$, then $\mu_n^*(A \cap Z_i) > r|Z_i|$ for at least one index i.)

G. If U is an open subset of $\mathbb{R}$, then U is uniquely expressible as a countable union of disjoint open intervals (the connected components of U; Proposition 3.7). Hence if μ denotes Lebesgue measure on $\mathbb{R}$, then $\mu(U)$ is given by the sum of the lengths of these constituent intervals. Using this fact, show that the Cantor set (Ex. 3J) has Lebesgue measure zero. Conclude that there exist (many) Lebesgue measurable sets in $\mathbb{R}$ that are not Borel sets. (Hint: All of the subsets of the Cantor set are Lebesgue measurable; use Problems 6H and 1S.)

H. If E is a Lebesgue measurable subset of $\mathbb{R}^n$ and if ε is a positive number, then there exists an open set U containing E such that $\mu_n(U \backslash E) < \varepsilon$. Likewise there exists a closed set F contained in E such that $\mu_n(E \backslash F) < \varepsilon$. Conclude that there exist sets G and H such that $G \subset E \subset H$ and such that $\mu_n(H \backslash E) = \mu_n(E \backslash G) = 0$, where G is an F_σ and H is a G_δ. (Hint: See Problem E.)

I. Let f be a monotone increasing real-valued function on $\mathbb{R}$, and for each open interval $(a, b), a < b$, set $\varphi_f((a, b)) = f(b-) - f(a+)$. If we also define $\varphi_f(\varnothing) = 0$, then φ_f is a gauge defined on an admissible collection of subsets of $\mathbb{R}$. Show that $\mu_f^*((a, b)) = \varphi_f((a, b))$ for all open intervals (a, b), where μ_f^* denotes the outer measure on $\mathbb{R}$ generated by φ_f (i.e., show that φ_f is countably subadditive), and show also that all intervals are measurable with respect to μ_f^*. Conclude that the σ-ring of subsets of $\mathbb{R}$ measurable with respect to μ_f^* contains the σ-ring $\mathbf{B}$ of Borel subsets of $\mathbb{R}$ and that $\mu_f^* | \mathbf{B}$ is the Stieltjes–Borel measure of Example E. (Hint: Show first that the ray $\{r \in \mathbb{R} : r \le t_0\}$ is measurable with respect to μ_f^* whenever t_0 is a point of continuity of f.)

J. Let **B** denote the σ-ring of Borel subsets of $\mathbb{R}$, and suppose given a measure v on $(\mathbb{R}, \mathbf{B})$ which is finite on the bounded sets in **B**. Let

$$f(t) = \begin{cases} v((0, t]), & t \geq 0, \\ -v((t, 0]), & t < 0. \end{cases}$$

Show that f is a monotone increasing right-continuous function on $\mathbb{R}$ and that v is the Stieltjes–Borel measure associated with f.

K. Let f be a monotone increasing function defined on a real interval $[a, b]$. If g is a bounded, real-valued function defined on $[a, b]$ and $\mathscr{P} = \{t_i\}_{i=0}^{N}$ is a partition of $[a, b]$, then the *upper Darboux sum* (for g with respect to f) based on $\mathscr{P}$ is the sum

$$D_{\mathscr{P}} = \sum_{i=1}^{N} M_i[f(t_i) - f(t_{i-1})],$$

where $M_i = \sup\{g(t) : t_{i-1} \leq t \leq t_i\}$, $i = 1, \ldots, N$ (see Problem 1G for basic definitions). Similarly, the *lower Darboux sum* (for g with respect to f) based on $\mathscr{P}$ is the sum

$$d_{\mathscr{P}} = \sum_{i=1}^{N} m_i[f(t_i) - f(t_{i-1})],$$

where $m_i = \inf\{g(t) : t_{i-1} \leq t \leq t_i\}$, $i = 1, \ldots, N$. Verify that for fixed f and g the nets $\{D_{\mathscr{P}}\}$ and $\{d_{\mathscr{P}}\}$, indexed by the directed set of partitions of $[a, b]$, are monotone decreasing and monotone increasing, respectively. (This is a straightforward generalization of the construction introduced in Problem 1H where $f(t) \equiv t$.) Show also that the limits

$$\underline{J} = \lim_{\mathscr{P}} d_{\mathscr{P}} \quad \text{and} \quad \bar{J} = \lim_{\mathscr{P}} D_{\mathscr{P}}$$

exist and satisfy the condition $\underline{J} \leq \bar{J}$. The limits $\underline{J}$ and $\bar{J}$ are known as the *lower* and *upper Darboux integrals* over $[a, b]$ of g with respect to f, respectively. In the event that $\underline{J} = \bar{J}$, the common value is the *Darboux integral* over $[a, b]$ of g with respect to f. Show that if g is a bounded real-valued Borel measurable function on $[a, b]$ and μ_f is the Stieltjes–Borel measure associated with f (Ex. E), then the Lebesgue–Stieltjes integral $\int_{[a,b]} g \, d\mu_f$ is straddled by the upper and lower Darboux integrals over $[a, b]$ of g with respect to f, so that the Darboux integral over $[a, b]$ of g with respect to f coincides with $\int_{[a,b]} g \, d\mu_f$ whenever the former exists. (Hint: The upper and lower Darboux sums based on an arbitrary partition of $[a, b]$ are the integrals with respect to μ_f of a pair of simple functions that straddle g. When f is continuous this is easily seen; if f is discontinuous at any of the internal partition points, extra care must be taken.)

L. Let φ and α be bounded complex-valued functions defined on an interval $[a, b]$ $(a \leq b)$. If $\mathscr{P} = \{t_i\}_{i=0}^{N}$ is a partition of $[a, b]$, then by a *Riemann–Stieltjes sum* (for φ with respect to α) based on $\mathscr{P}$ is meant a sum of the form

$$S = \sum_{i=1}^{N} \varphi(\tau_i)[\alpha(t_i) - \alpha(t_{i-1})],$$

where $t_{i-1} \leq \tau_i \leq t_i$, $i = 1, \ldots, N$. The number J is defined to be the *Riemann–Stieltjes integral* of φ over $[a, b]$ with respect to α if for each given positive number ε there exists a positive number δ such that $|J - S| < \varepsilon$ for every Riemann–Stieltjes sum S (for φ with respect to α) based on any partition $\mathscr{P}$ of $[a, b]$ such that the mesh of $\mathscr{P}$ is less than δ. If J is the Riemann–Stieltjes integral over $[a, b]$ of φ with respect to α, we write

$$J = \int_a^b \varphi(t) d\alpha(t).$$

(In particular, when $\alpha(t) \equiv t$, the Riemann–Stieltjes integral of φ with respect to α is the *Riemann integral* of φ over $[a, b]$, denoted by $\int_a^b \varphi(t) dt$.) If the Riemann–Stieltjes integral of φ over $[a, b]$ with respect to α exists, then φ is said to be *(Riemann–Stieltjes) integrable* over $[a, b]$ with respect to α. In the Riemann–Stieltjes integral $\int_a^b \varphi(t) d\alpha(t)$ the function φ is known as the *integrand*, the function α as the *integrator*.

(i) Show that a bounded function φ is integrable over an interval $[a, b]$ with respect to a bounded function α if and only if the following Cauchy criterion is satisfied: Given $\varepsilon > 0$ there exists $\delta > 0$ such that if $\mathscr{P}$ and $\mathscr{P}'$ are arbitrary partitions of $[a, b]$ with mesh $\mathscr{P}$, $\mathscr{P}' < \delta$, and if S and S' are arbitrary Riemann–Stieltjes sums for φ with respect to α based on $\mathscr{P}$ and $\mathscr{P}'$, respectively, then $|S - S'| < \varepsilon$. Verify also that if φ is integrable with respect to α over $[a, b]$, and if ε and δ are as stated, then

$$\left| S - \int_a^b \varphi(t) d\alpha(t) \right| \leq \varepsilon$$

for any Riemann–Stieltjes sum S for φ with respect to α based on a partition $\mathscr{P}$ of $[a, b]$ such that mesh $\mathscr{P} < \delta$.

(ii) Let φ and α be complex-valued functions on the real interval $[a, b]$ and suppose φ is bounded and α is of bounded variation over $[a, b]$. Verify that if for a given positive number ε there exists a positive number δ with the property that $|\varphi(t) - \varphi(t')| < \varepsilon$ for all t, t' in $[a, b]$ such that $|t - t'| < \delta$, and if $\mathscr{P}$ and $\mathscr{P}'$ are partitions of $[a, b]$ such that mesh $\mathscr{P}$, $\mathscr{P}' < \delta$, and S and S' are Riemann–Stieltjes sums for φ with respect to α based on $\mathscr{P}$ and $\mathscr{P}'$, respectively, then $|S - S'| < 2V\varepsilon$, where $V = V(\alpha; a, b)$ denotes the total variation of α. (Hint: Consider first the case in which $\mathscr{P}'$ is a refinement of $\mathscr{P}$.) Conclude that if φ is continuous, then φ is integrable over $[a, b]$ with respect to α, and if ε and δ are as above, then $|S - \int_a^b \varphi(t) d\alpha(t)| \leq 2V\varepsilon$ for any Riemann–Stieltjes sum S for φ with respect to α based on a partition $\mathscr{P}$ of $[a, b]$ with mesh $\mathscr{P} < \delta$.

(iii) Suppose f is a monotone increasing real-valued function on $[a, b]$ and φ is a Borel measurable function that is Riemann–Stieltjes integrable over $[a, b]$ with respect to f. Show that

$$\int_a^b \varphi(t) df(t) = \int_{[a, b]} \varphi \, d\mu_f,$$

where μ_f denotes the Stieltjes–Borel measure associated with f (Ex. E). Show likewise that if φ is real-valued, then $\int_a^b \varphi(t) df(t)$ is also equal to the Darboux integral over $[a, b]$ of φ with respect to f. (Hint: First treat the case in which φ

is real. If ε is a positive number, if δ is chosen as in the definition of the Riemann–Stieltjes integral, and if $\mathscr{P}$ is a partition of $[a, b]$ with mesh $\mathscr{P} < \delta$, then the upper and lower Darboux sums for φ with respect to f based on $\mathscr{P}$ cannot differ by more than 2ε. Hence $\bar{J} - \underline{J} \leq 2\varepsilon$.)

The condition of continuity in the result obtained in Problem L(ii) is only a convenient sufficient condition for the integrability of φ with respect to α. A closer study of the question discloses that if φ is any bounded complex-valued function and α is a function of bounded variation, both defined on $[a, b]$, then a necessary and sufficient condition for the existence of the Riemann–Stieltjes integral of φ with respect to α over $[a, b]$ is that φ be continuous almost everywhere with respect to the measure μ_V, where $V(t)$ denotes the total variation of α over the interval $[a, t]$. It is worth remarking that this condition, in turn, implies that φ is measurable with respect to the *completion* of μ_V (Prob. Q). Thus, for example, a Riemann integrable function on an interval $[a, b]$ is always Lebesgue measurable, though it need not be Borel measurable.

M. Let **R** be an admissible ring of subsets of a set X, let λ be a countably additive, nonnegative, extended real-valued set function defined on **R** such that $\lambda(\varnothing) = 0$, and let μ^* be the outer measure generated by λ (Prob. D). Show that every set in **R** is measurable with respect to μ^* and that $\mu^* | \mathbf{R} = \lambda$. Conclude that if $(X, \mathbf{S}, \mu)$ is a measure space and **R** is a ring of measurable sets of finite measure such that **R** generates **S** as a σ-ring, then for any set E of finite measure and any positive number ε there exists a set R in **R** such that $\mu(E \nabla R) < \varepsilon$. Thus, for example, if f is a monotone increasing function on an interval $[a, b]$ and if μ_f is the associated Stieltjes–Borel measure on $[a, b]$ (Ex. E), then for any Borel subset E of $[a, b]$ and any $\varepsilon > 0$ there exist half-open intervals $(a_i, b_i]$, $i = 1, \ldots, n$, such that $\mu_f(E \nabla H) < \varepsilon$, where $H = \bigcup_{i=1}^{n} (a_i, b_i]$.

N. Let $(X, \mathbf{S}, \mu)$ be a finite measure space, and let f be a real-valued integrable function on X. For each real number t let $E_t = \{x \in X : f(x) \leq t\}$ and define $m(t) = \mu(E_t)$, $t \in \mathbb{R}$. Then m is a monotone increasing, right-continuous function on $\mathbb{R}$. (The function m is known as the *distribution function* of f.) Show that the Stieltjes–Borel measure μ_m (Ex. E) is the measure induced on $(\mathbb{R}, \mathbf{B})$ by the function f and μ (Prop. 8.2). Conclude that if φ is a continuous function on $\mathbb{R}$ such that $\varphi \circ f$ is integrable $[\mu]$, then

$$\int_X \varphi \circ f \, d\mu = \int_{-\infty}^{+\infty} \varphi(t) dm(t) \qquad (10)$$

where the integral in the right member of (10) is the improper Riemann–Stieltjes integral

$$\lim_{\substack{a \to -\infty \\ b \to +\infty}} \int_a^b \varphi(t) dm(t).$$

In particular,

$$\int_X f \, d\mu = \int_{-\infty}^{+\infty} t \, dm(t).$$

O. Let θ denote arc-length measure on the unit circle Z (Ex. F). Show that $\int_Z \lambda^n \, d\theta(\lambda) = 0$ for all integers $n \neq 0$.

P. A measure space (X, S, μ) is said to be *complete* if every subset of an arbitrary set of measure zero with respect to μ is measurable and is, therefore, also a set of measure zero. (Another way to say this is to say that the collection **Z** of null sets is a σ-ideal (Prob. 6I) not only in **S**, but in the power class on X.) If (X, S, μ) is complete, the measure μ is also said to be *complete*. Show that if (X, S, μ) is a complete measure space, and if f is a measurable function on (X, S), then every function g on X such that $g = f$ a.e. $[\mu]$ is also measurable $[S]$. Similarly, if (X, S, μ) is complete and E belongs to **S**, then every set F that is almost equal $[\mu]$ to E also belongs to **S** and we have $\mu(E) = \mu(F)$. If Z is a null set in a complete measure space (X, S, μ), then an arbitrary complex-valued function defined on Z is measurable on Z. Show that if μ^* is an outer measure on a set X, and if **M** denotes the σ-ring of sets that are measurable with respect to μ^* (Prob. C), then $(X, M, \mu^*|M)$ is complete.

Q. Let (X, S, μ) be a measure space. Consider the collection $\hat{S}$ of those subsets of X that are almost equal $[\mu]$ to some set E in **S**. If F belongs to $\hat{S}$ and if E_1 and E_2 are sets in **S** such that F and E_i, $i = 1, 2$, are almost equal $[\mu]$, then E_1 and E_2 are almost equal $[\mu]$ and therefore $\mu(E_1) = \mu(E_2)$. Thus we may and do define $\hat{\mu}(F) = \mu(E_1) = \mu(E_2)$. Show that $\hat{S}$ is a σ-ring of subsets of X and that $\hat{\mu}$ is a measure on the measurable space $(X, \hat{S})$. Show also that $(X, \hat{S}, \hat{\mu})$ is complete, and if (X, T, ν) is any complete measure space such that $S \subset T$ and $\mu = \nu|S$, then $\hat{S} \subset T$ and $\hat{\mu} = \nu|\hat{S}$. The space $(X, \hat{S}, \hat{\mu})$ is the *completion* of (X, S, μ) and $\hat{\mu}$ is the *completion* of μ. Thus Lebesgue measure on $\mathbb{R}^n$ is the completion of Lebesgue–Borel measure on $\mathbb{R}^n$. (Hint: See Problem H.)

R. Show that a measure space is σ-finite (Prob. 7J) if and only if it can be expressed as the direct sum of a countable collection of finite measure spaces. Show also that an arbitrary direct sum of finite measure spaces is locally finite.

S. A measure space (X, S, μ) is said to be *purely atomic* if every measurable subset of X is the union of a set of measure zero and a disjoint collection of atoms. Show that an arbitrary direct sum of finite measure spaces can be expressed (essentially uniquely) as the direct sum of two subspaces A and B, where A is purely atomic, while B is atom-free. In particular, this is true of every σ-finite measure space. (Hint: First treat the case in which $\mu(X) < +\infty$. Show that a disjoint collection of atoms in X is necessarily countable, and use the method of exhaustion.)

T. (i) Let μ be a signed measure on a measurable space (X, S), and suppose $\mu(X) < 0$. Show that there is a negative set B_0 with respect to μ such that $\mu(B_0) < 0$. (Hint: Suppose the contrary. Then for any set E in **S**, $\mu(E) < 0$ implies the existence of a subset F of E such that $\mu(F) > 0$. For each E in **S** having negative measure let $k(E)$ denote the smallest positive integer k with the property that there exists a subset F of E such that $\mu(F) \geq 1/k$. Note that if $\mu(E) < 0$ and $\mu(F) > 0$, where $F \subset E$, then $\mu(E\backslash F) \leq \mu(E) < 0$ and $k(E\backslash F) \geq k(E)$. Set $k_1 = k(X)$ and construct by induction a disjoint sequence $\{F_n\}$ of measurable sets and a corresponding sequence $\{k_n\}$ of positive integers such that

$$k_{n+1} = k(X\backslash(F_1 \cup \cdots \cup F_n)) \quad \text{and} \quad \mu(F_n) \geq \frac{1}{k_n}$$

for every positive integer n. Set $F = \bigcup_n F_n$ and show that $B_0 = X\backslash F$ is negative with respect to μ and that $\mu(B_0) \leq \mu(X)$.)

(ii) Use (i) to give a proof of Theorem 8.5. (Hint: Suppose, without loss of generality, that $-\infty < \mu(E) \leq +\infty$ for every measurable set E, and exhaust the negative measure from X.)

U. (i) Let μ be a signed measure on a measurable space $(X, \mathbf{S})$, and let E be a measurable set. Verify that

$$\mu_+(E) + \mu_-(E) = \sup \sum_{i=1}^{n} |\mu(E_i)|,$$

where the supremum is taken in $\mathbb{R}^\sharp$ over all finite partitions $\{E_1, \ldots, E_n\}$ of E into disjoint measurable subsets. Thus the *total variation* $|\mu|(E)$ of μ over E, defined to be the indicated supremum, is given by $\mu_+(E) + \mu_-(E)$. Verify that a measurable function f on $(X, \mathbf{S})$ is integrable $[\mu]$ if and only if f is integrable $[|\mu|]$, and that if f is integrable $[\mu]$, then $|\int_X f \, d\mu| \leq \int_X |f| \, d|\mu|$. Show also that if $\{f_n\}$ is a sequence of measurable functions on $(X, \mathbf{S})$ that converges to a limit f a.e. $[|\mu|]$, and if there exists a function φ that is integrable $[\mu]$ such that $|f_n| \leq \varphi$ a.e. $[|\mu|]$ for every index n, then $\lim_n \int |f - f_n| d\mu = 0$ and $\lim_n \int f_n \, d\mu = \int f \, d\mu$. (In other words, the dominated convergence theorem is valid for signed measures.)

(ii) Prove Proposition 8.7 for a complex measure ζ on a measurable space $(X, \mathbf{S})$. (Hint: Show first that $|\zeta|$ is finitely additive.) Show also that if $\{f_n\}$ is a sequence of measurable complex-valued functions on $(X, \mathbf{S})$ that converges to a limit f a.e. $[|\zeta|]$, and if there exists a function φ that is integrable $[\zeta]$ such that $|f_n| \leq \varphi$ a.e. $[|\zeta|]$ for every index n, then $\lim_n \int |f - f_n| d\zeta = 0$ and $\lim_n \int f_n \, d\zeta = \int f \, d\zeta$. (In other words, the dominated convergence theorem is valid for complex measures.)

V. Let ζ be a complex measure on a measurable space $(X, \mathbf{S})$. Show that the total variation of ζ is given by

$$\sup \left| \int_X f \, d\zeta \right|,$$

where the supremum is taken over the collection of all measurable functions f on X such that $|f| \leq 1$ a.e. $[|\zeta|]$.

W. (i) Suppose given two monotone increasing real-valued functions f_1 and f_2 on the same real interval $[a, b]$. Show that $f = f_1 - f_2$ is of bounded variation on $[a, b]$. Show also that $f_1 = f_+ + d$ and $f_2 = (f_- - f(a)) + d$, where d is another monotone increasing function on $[a, b]$ and f_+ and f_- denote the positive and negative variations, respectively, of f (Prob. 1K). Use this minimality of the positive and negative variations of a given real-valued function f of bounded variation to verify that in the Jordan decomposition $f = f_+ - (f_- - f(a))$ of such a function we have $\mu_{f_+} = (\mu_f)_+$ and $\mu_{f_-} = (\mu_f)_-$ whenever f is right-continuous on (a, b). (Hint: Show first that the positive [negative] variation of f on any subinterval of $[a, b]$ (Prob. 1J) is dominated by the increment of $f_1[f_2]$ on that subinterval; then use Theorem 8.6.)

(ii) Let α be a complex-valued function of bounded variation on $[a, b]$ that is right-continuous on (a, b), let ζ_α denote the complex Stieltjes–Borel measure on $[a, b]$ associated with α (Ex. K), and for each t, $a \leq t \leq b$, let $L(t)$ denote the length, or total variation, of α on the interval $[a, t]$. Show that the total variation

$|\zeta_\alpha|$ is the measure μ_L associated with the monotone increasing function L. (Hint: It suffices to show that $|\zeta_\alpha|$ agrees with the measure associated with L on subintervals of the form $[a, t]$, $a < t \leq b$ (Prob. A). This amounts to verifying that $|\zeta_\alpha|([a, t]) = L(t)$. Inequality one way is obvious; to obtain inequality the other way use Problem M.)

X. Let α be a complex-valued function of bounded variation on a real interval $[a, b]$, and let φ be a bounded, complex-valued, Borel measurable function on $[a, b]$. Show that if the Riemann–Stieltjes integral $\int_a^b \varphi(t)d\alpha(t)$ exists (see Problem L), then

$$\int_a^b \varphi(t)d\alpha(t) = \int_{[a,b]} \varphi \, d\zeta_\alpha,$$

where ζ_α is the Stieltjes–Borel measure associated with α (Ex. K). Using this fact, show that if the Riemann–Stieltjes integral of φ with respect to α over $[a, b]$ exists, then

$$\left| \int_a^b \varphi(t)d\alpha(t) \right| \leq \int_{[a,b]} |\varphi(t)|d\mu_L \leq L(b) \max_{a \leq t \leq b} |\varphi(t)|,$$

where $L(t)$ denotes the total variation of α on the interval $[a, t]$.

Y. Let α be a complex-valued function defined on a real interval $[a, b]$. Show that α is Lipschitzian on $[a, b]$ (Prob. 4F) if and only if there exists a bounded, Lebesgue measurable, complex-valued function β on $[a, b]$ such that

$$\alpha(t) - \alpha(a) = \int_a^t \beta(u)du, \qquad a \leq t \leq b, \tag{11}$$

and that, when α is given in this manner, then $\beta = \alpha'$ a.e. $[\mu_1]$. (Hint: If α is Lipschitzian on $[a, b]$, then α is continuous and of bounded variation on $[a, b]$. Moreover, if α is Lipschitzian, then the sequence $\{q_n\}$ of step functions defined in the proof of Theorem 8.13 tends to α' a.e. $[\mu_1]$ and is uniformly bounded, so

$$\int_a^b \alpha'(t)dt = \alpha(b) - \alpha(a)$$

by the bounded convergence theorem (Th. 7.13). Conclude that $\alpha(t) - \alpha(a) = \int_a^t \alpha'(u)du$, $a \leq t \leq b$, and also that if α is related to β as in (11), then $\beta - \alpha'$ is a Lebesgue integrable function whose integral with respect to Lebesgue measure is zero over every Lebesgue measurable subset of $[a, b]$.)

9 More integration theory

In this chapter we treat some additional important topics in the theory of measure and integration. The first of these concerns another way that new measures can be constructed from old ones. If $(X, \mathbf{S})$ and $(Y, \mathbf{T})$ are measurable spaces, then a set of the form $E \times F$, where $E \in \mathbf{S}$ and $F \in \mathbf{T}$, is called a *measurable rectangle* in $X \times Y$. The σ-ring of subsets of $X \times Y$ generated by the collection of all measurable rectangles is denoted by $\mathbf{S} \times \mathbf{T}$, and the space $X \times Y$ equipped with the *product* σ-ring $\mathbf{S} \times \mathbf{T}$ is a measurable space called the *product* of $(X, \mathbf{S})$ and $(Y, \mathbf{T})$.

Example A. If X and Y are topological spaces, and if $\mathbf{B}_X$ and $\mathbf{B}_Y$ denote the σ-rings of Borel subsets of X and Y, respectively, then $\mathbf{B}_X \times \mathbf{B}_Y$ is, in general, a subring of the σ-ring $\mathbf{B}$ of Borel subsets of $X \times Y$. If X and Y both satisfy the second axiom of countability, then $\mathbf{B}_X \times \mathbf{B}_Y = \mathbf{B}$. (In this connection see Problem A.)

Lemma 9.1. *Let* $(X, \mathbf{S}, \mu)$ *and* $(Y, \mathbf{T}, v)$ *be measure spaces, and let* f *be a complex-valued function on* $X \times Y$ *that is measurable* $[\mathbf{S} \times \mathbf{T}]$. *Then for each* x *in* X *the function* $f(x, y)$, *considered as a function of the single variable* y, *is measurable* $[\mathbf{T}]$. *Moreover, if* v *is* σ-*finite, and if* A *denotes the set of those points* x *in* X *such that* $f(x, y)$ *is integrable* $[v]$ *as a function of* y, *then* A *is measurable and the function*

$$g(x) = \int_Y f(x, y) dv(y), \qquad x \in A,$$

is a measurable function on A. *The analogous assertions obtained by interchanging the roles of* X *and* Y *are also valid.*

If X and Y are arbitrary sets, and if E is a subset of $X \times Y$, then for each point x of X we denote by E_x the set

$$E_x = \{y \in Y : (x, y) \in E\}.$$

This set is called the *X-section* of E determined by x; *Y-sections* of E are defined analogously, and are denoted by E^y. It is an immediate consequence of Lemma 9.1 that if (X, S, μ) and (Y, T, v) are any two measure spaces, and if E is a subset of $X \times Y$ belonging to $S \times T$, then the section E_x is a measurable subset of Y for each x in X, and, in the event that v is σ-finite, the function $g(x) = v(E_x)$ is a measurable function on X. In this connection see Problems B and C.

Example B. Let us take for both (X, S) and (Y, T) the unit interval $[0, 1]$ equipped with its ring of Borel sets, so that $(X \times Y, S \times T)$ is simply the unit square Q equipped with its ring of Borel sets $\mathbf{B}$ (see Example A). Let μ be ordinary Lebesgue–Borel measure on (X, S), and let κ be the counting measure on (Y, T) (Ex. 7J). According to Lemma 9.1, if A is any Borel set in Q, then the function $h(y) = \mu(A^y)$ is a Borel measurable function on $[0, 1]$. Irrespective of that fact, we may certainly define

$$\xi(A) = \int_Y h(y) d\kappa(y) = \sum_{0 \le y \le 1} \mu(A^y),$$

and it follows by the theorem of Beppo–Levi (Cor. 7.10) that the set function ξ is a measure on $(Q, \mathbf{B})$. On the other hand, if A is a given Borel subset of Q, and if we define

$$g(x) = \kappa(A_x), \qquad 0 \le x \le 1,$$

then the function g *need not be* Borel measurable on $[0, 1]$. (This is one of those set-theoretic mysteries, an explanation of which would take us far afield from the proper concerns of this book; the interested reader is referred to [34; §37, Satz IV].)

Example B shows clearly that the hypothesis of σ-finiteness is of great importance in Lemma 9.1. Note that the measure space (Y, T, κ), while not σ-finite, is locally finite. The following theorem, the proof of which is sketched in Problem E, is of basic importance.

Theorem 9.2 (Fubini Theorem). *If (X, S, μ) and (Y, T, v) are σ-finite measure spaces, then there exists a unique measure $\mu \times v$ (called the* product *of μ and v) on the measurable space $(X \times Y, S \times T)$ satisfying the condition $(\mu \times v)(E \times F) = \mu(E)v(F)$ for every measurable rectangle $E \times F$ in $X \times Y$. (The measurable space $(X \times Y, S \times T)$ equipped with the measure $\mu \times v$ is the* product *of the measure spaces (X, S, μ) and (Y, T, μ).) If f is* *a complex-valued function on* $X \times Y$ *that is integrable* $[\mu \times v]$, *then the function* $g(x) = \int_Y f(x, y)dv(y)$ *is measurable and integrable* $[\mu]$ *and we have*

$$\int_X g \, d\mu = \int_X \left[\int_Y f(x, y)dv(y) \right] d\mu(x) = \int_{X \times Y} f \, d(\mu \times v).$$

Similarly, the iterated integral

$$\int_Y \left[\int_X f(x, y)d\mu(x) \right] dv(y)$$

exists and is equal to the product (*or* double) *integral* $\int_{X \times Y} f \, d(\mu \times v)$.

In applications of the Fubini theorem it is of great importance to be able to determine when a measurable function on a product space $X \times Y$ is integrable with respect to the product of two measures. In this connection the following result, the proof of which is sketched in Problems C and D, is very useful.

Theorem 9.3 (Tonelli Theorem). *Let* $(X, \mathbf{S}, \mu)$ *and* $(Y, \mathbf{T}, v)$ *be* σ-*finite measure spaces. If* f *is a function that is measurable* $[\mathbf{S} \times \mathbf{T}]$ *and nonnegative on* $X \times Y$, *then*

$$\int_X \left[\int_Y f(x, y)dv(y) \right] d\mu = \int_Y \left[\int_X f(x, y)d\mu(x) \right] dv$$

$$= \int_{X \times Y} f \, d(\mu \times v).$$

Thus if φ *is an arbitrary measurable complex-valued function on* $X \times Y$, *and if the function* $g(x) = \int_Y |\varphi(x, y)|dv(y)$ *satisfies the condition* $\int_X g \, d\mu < +\infty$ (*or if* $h(y) = \int_X |\varphi(x, y)|d\mu(x)$ *satisfies the condition* $\int_Y h \, dv < +\infty$), *then* φ *is integrable* $[\mu \times v]$.

Corollary 9.4. *Let* $(X, \mathbf{S}, \mu)$ *and* $(Y, \mathbf{T}, v)$ *be* σ-*finite measure spaces. If* E *is a measurable subset of* $X \times Y$, *then* $(\mu \times v)(E) < +\infty$ *if and only if* $\int_X v(E_x)d\mu < +\infty$. *Furthermore,* $(\mu \times v)(E) = 0$ *if and only if* $v(E_x) = 0$ *for almost every* x *in* X *or, equivalently, if and only if* $\mu(E^y) = 0$ *for almost every* y *in* Y.

The Fubini and Tonelli theorems extend, of course, to arbitrary finite products $X_1 \times \cdots \times X_n$ of σ-finite measure spaces (see Problem J).

The assumption that the measure spaces $(X, \mathbf{S}, \mu)$ and $(Y, \mathbf{T}, v)$ are both σ-finite is certainly sufficient to assure the validity of the Fubini and Tonelli theorems, but this relatively strong condition has seemed to many to be unjustifiably restrictive. In this connection it is appropriate to remark that if $(X, \mathbf{S}, \mu)$ and $(Y, \mathbf{T}, v)$ are any two, absolutely arbitrary, measure spaces, then there exists a measure $\hat{\lambda}$ on $\mathbf{S} \times \mathbf{T}$ satisfying the condition

$$\hat{\lambda}(E \times F) = \mu(E)v(F) \tag{1}$$

on measurable rectangles, and this measure, which is constructed in a completely standard fashion, possesses certain rather appealing properties. For example, the Fubini theorem is valid for the measure $\hat{\lambda}$, again without any restrictions whatever on X or Y (Prob. H). Thus it might seem natural simply to define the "product measure" on $(X \times Y, \mathbf{S} \times \mathbf{T})$ to be the measure $\hat{\lambda}$. In the most general case, however, the measure $\hat{\lambda}$ is not uniquely determined by (1), which makes the terminology "product measure" somewhat suspect. (Indeed, at least one serious student of this circle of ideas has proposed another candidate altogether for the "product" of two arbitrary measures [5].) Another (more serious) difficulty is that the Tonelli theorem is *not* valid, in general, for the measure $\hat{\lambda}$, and since this is ordinarily the only feasible way of determining when the Fubini theorem is applicable, we have, in this book, relegated the measure $\hat{\lambda}$ to the problems (see Problems G and H). Readers who wish to probe more deeply into these matters may consult the articles [49], [50], [6], and [45].

Example C. If we identify $\mathbb{R}^m \times \mathbb{R}^n$ with $\mathbb{R}^{m+n}$ in the usual manner, then the σ-ring of Borel subsets of $\mathbb{R}^{m+n}$ is identified with the product of the σ-rings of Borel subsets of $\mathbb{R}^m$ and $\mathbb{R}^n$. Thus Lebesgue–Borel measure on $\mathbb{R}^{m+n}$ is also identified with the product of the Lebesgue–Borel measures on $\mathbb{R}^m$ and $\mathbb{R}^n$. In the same way, Lebesgue–Borel measure on $\mathbb{R}^n$ may be thought of as the product of n copies of Lebesgue–Borel measure on the real line, and Lebesgue–Borel measure on $\mathbb{C}$ as the product of the Lebesgue–Borel measures on the real and imaginary axes in $\mathbb{C}$.

Example D. Let $(Y, \mathbf{T}, v)$ be a σ-finite measure space, let $(\mathbb{R}, \mathbf{B}, \mu)$ denote the measure space consisting of Lebesgue–Borel measure on $\mathbb{R}$, and let f be a nonnegative real-valued function defined on Y. If R_f denotes the set $\{(t, y) : 0 \le t < f(y)\}$, then f is measurable [$\mathbf{T}$] when and only when R_f is measurable [$\mathbf{B} \times \mathbf{T}$]. Likewise, f is integrable [v] when and only when $(\mu \times v)(R_f) < +\infty$, and, when f is integrable [v], we have $(\mu \times v)(R_f) = \int_Y f \, dv$. (It is easy to see that these observations can be used to construct the real Lebesgue integral with respect to the measure v, and thus to give a proof of Theorem 7.6 in the σ-finite case.)

For any given measure μ there are other measures associated with and, in a sense, subordinate to μ. To make this more precise, let $(X, \mathbf{S}, \mu)$ be a measure space, let f be a measurable, nonnegative, extended real-valued function defined on X, and let $F = \{x \in X : f(x) < +\infty\}$. Clearly the mapping $g \to gf$ is a linear transformation of the space $\mathcal{M}_0$ into itself, where $\mathcal{M}_0$ denotes the linear space of all those measurable complex-valued functions on X that vanish on $X \backslash F$. Let $\mathcal{L}_f$ denote the linear space of all those measurable functions g in $\mathcal{M}_0$ with the property that gf is integrable [μ], and set

$$L_f(g) = \int gf \, d\mu, \qquad g \in \mathcal{L}_f.$$

Then L_f is a linear functional on $\mathcal{L}_f$, and it is a routine chore to verify that L_f is, in fact, a Lebesgue integral on $(X, \mathbf{S})$.

Proposition 9.5. *If f is a measurable extended real-valued function defined and nonnegative a.e.* [μ] *on a measure space* (X, S, μ), *then the set function*

$$v_f(E) = \int_E f \, d\mu, \qquad E \in S,$$

is a measure on (X, S) *called the* indefinite integral of f with respect to μ. *If g is an arbitrary measurable function that is either integrable with respect to* v_f *or nonnegative, then*

$$\int_X g \, dv_f = \int_X gf \, d\mu.$$

Example E. Let T denote the strip $\{(r, t) : r \geq 0, 0 \leq t < 2\pi\}$ in the plane $\mathbb{R}^2$, and consider the mapping

$$\varphi(r, t) = re^{it}, \qquad (r, t) \in T,$$

of T onto the complex plane $\mathbb{C}$. The mapping φ is clearly Borel measurable; indeed, if $0 \leq r_0 < r_1 < +\infty$ and $0 \leq t_0 < t_1 < 2\pi$, then the half-open rectangle $H = [r_0, r_1) \times [t_0, t_1)$ is the inverse image under φ of the segment S of the annulus $\{\lambda \in \mathbb{C} : r_0 \leq |\lambda| < r_1\}$ cut out by the angle $\{\lambda \in \mathbb{C} : t_0 \leq \operatorname{Arg} \lambda < t_1\}$, and sets of the form S generate the σ-ring of Borel subsets of $\mathbb{C}$. Moreover, it is easily seen that the Lebesgue–Borel measure of S is given by

$$v(S) = \frac{(t_1 - t_0)(r_1^2 - r_0^2)}{2}.$$

(Indeed, if t_0 and t_1 are both rational multiples of 2π, this follows at once from the fact that Lebesgue–Borel measure on $\mathbb{C}$ is invariant under rotations, and the general case then follows by continuity.) But this number is also the value of the integral

$$\int_H r \, dr \, dt = \int_{r_0}^{r_1} r \, dr \int_{t_0}^{t_1} dt.$$

Thus the measure induced on $\mathbb{C}$ (Prop. 8.2) by the mapping φ and the indefinite integral v_r of the coordinate function r on T agrees with Lebesgue–Borel measure v on $\mathbb{C}$ on all segments S of the above form. Since, as has been noted, these sets suffice to generate the σ-ring of Borel subsets of $\mathbb{C}$ (just as the half-open rectangles of the form H generate the σ-ring of Borel subsets of the strip T), it follows (Prob. 8A) that Lebesgue–Borel measure on $\mathbb{C}$ is identical with the measure induced by φ and v_r. Hence if E is an arbitrary Borel set in $\mathbb{C}$, then

$$v(E) = \iint_{\varphi^{-1}(E)} r \, dr \, dt.$$

Thus, for example, if A denotes the annulus $A = \{\lambda \in \mathbb{C} : r_0 < |\lambda| < r_1\}$, and if f is any Lebesgue integrable function on A, then

$$\int_A f \, dv = \int_0^{2\pi} \int_{r_0}^{r_1} f(re^{it}) r \, dr \, dt.$$

In particular, if f is Lebesgue integrable over a disc $D_R(0)$, then

$$\int_{D_R(0)} f \, dv = \int_0^{2\pi} \int_0^R f(re^{it}) r \, dr \, dt.$$

The properties of the indefinite integral v_f of a function f with respect to a measure μ as defined in Proposition 9.5 reflect, quite naturally, the properties of the underlying measure space (X, S, μ) as well as those of the function f. Thus if f is finite-valued (a.e. $[\mu]$), then v_f is easily seen to be σ-finite whenever μ is σ-finite, and locally finite whenever μ is locally finite; cf. Proposition 8.1. On the other hand, it is possible for $v_f(E)$ to be equal to $+\infty$ even when $\mu(E)$ is finite and f is finite-valued. An important observation concerning indefinite integrals that holds quite generally is the following: For any measurable nonnegative extended real-valued function f, v_f vanishes wherever μ does (Prop. 7.8).

Definition. Let (X, S, μ) be a measure space, and suppose that v is a measure on (X, S) with the property that $\mu(E) = 0$ implies $v(E) = 0$ for every set E in S. Then we say that v is *absolutely continuous* with respect to μ, or *absolutely continuous* $[\mu]$, and we write $v \ll \mu$.

Example F. Let (X, S) be a measurable space, let μ and v be measures on (X, S), and suppose the following condition is satisfied:

> For any given positive number ε there exists a positive number δ such that $\mu(E) < \delta$ implies $v(E) < \varepsilon$ for every set E in S. $\qquad$ (2)

Then $\mu(Z) = 0$ implies $v(Z) < \varepsilon$ for every positive number ε and therefore $v(Z) = 0$. Hence v is absolutely continuous with respect to μ. On the other hand, if condition (2) is not satisfied, then there exists a positive number ε_0 and a sequence $\{E_n\}_{n=1}^\infty$ of measurable sets such that $v(E_n) \geq \varepsilon_0$ and $\mu(E_n) < 1/2^n$ for every positive integer n. Set

$$F_n = E_{n+1} \cup E_{n+2} \cup \cdots.$$

Then it is clear that $v(F_n) \geq \varepsilon_0$ and $\mu(F_n) < 1/2^n$ for every positive integer n, but the sequence $\{F_n\}$ is decreasing and has empty intersection. Since $\{v(F_n)\}$ does not converge to zero, this shows that v *cannot be a finite measure* (see Problem 7F). Thus if μ and v are two measures on the same measurable space (X, S), and if v is *finite*, then v is absolutely continuous with respect to μ if and only if (2) is satisfied. (The finiteness restriction on v cannot be dropped, or even relaxed; consider the indefinite integral of the function $1/t$ with respect to Lebesgue measure on the unit interval $(0, 1)$.)

According to the discussion preceding the above definition the indefinite integral of an arbitrary nonnegative measurable function f with respect to a measure μ is absolutely continuous with respect to μ. It is somewhat astonishing that the converse of this elementary assertion is also valid when the measure μ is σ-finite. (Indeed, even weaker conditions on μ suffice to make the converse valid; see Problem T.)

Theorem 9.6 (Radon–Nikodym Theorem). *Let $(X, \mathbf{S}, \mu)$ be a σ-finite measure space, and suppose given a measure v on $(X, \mathbf{S})$ that is absolutely continuous with respect to μ. Then there exists a measurable nonnegative extended real-valued function f on X such that*

$$v(E) = \int_E f \, d\mu, \qquad E \in \mathbf{S}. \tag{3}$$

In other words, $v = v_f$ is the indefinite integral of f with respect to μ. The function f, which is unique in the sense that if f_0 is any other function satisfying (3), then $f = f_0$ a.e. $[\mu]$, is called the Radon–Nikodym derivative *of v with respect to μ, and is frequently denoted by $dv/d\mu$.*

PROOF. Let $\{E_n\}_{n=1}^{\infty}$ be a disjoint sequence of sets of finite measure with respect to μ such that $X = \bigcup_n E_n$, and suppose the theorem valid for each of the subspaces E_n. If, for each n, f_n is a Radon–Nikodym derivative of $v|E_n$ with respect to $\mu|E_n$, and if we define

$$f(x) = f_n(x), \qquad x \in E_n,$$

for every n in $\mathbb{N}$, then it is clear that f is a Radon–Nikodym derivative of v with respect to μ. Moreover, if f_0 is any other Radon–Nikodym derivative of v with respect to μ, then $f_0|E_n$ must agree with f_n a.e. $[\mu]$, so the set $Z = \{x \in X : f(x) \neq f_0(x)\}$ meets each set E_n in a set of measure zero with respect to μ. But then $\mu(Z) = 0$. Thus it suffices to prove the theorem when $\mu(X) < +\infty$, and we henceforth assume this to be the case.

Suppose first that f and f_0 are two Radon–Nikodym derivatives of v with respect to μ. For each positive integer n let $F_n = \{x \in X : f_0(x) \leq n\}$, and set $S = \{x \in X : f(x) > f_0(x)\}$. Then for each n

$$\int_{S \cap F_n} f \, d\mu = \int_{S \cap F_n} f_0 \, d\mu < +\infty,$$

whence it follows (Ch. 7, p. 125) that f is finite-valued almost everywhere on $S \cap F_n$ and that $\int_{S \cap F_n} (f - f_0) d\mu = 0$. But this implies that $\mu(S \cap F_n) = 0$ for every n (Prop. 7.8), and it follows that $\mu(S) = 0$. Similarly, the set $\{x \in X : f(x) < f_0(x)\}$ has measure zero with respect to μ, and it follows that

$f = f_0$ a.e. $[\mu]$. This proves the uniqueness assertion of the theorem. It remains to establish the existence of a measurable nonnegative extended real-valued function f on X satisfying (3).

For each positive rational number r let $\{A_r, X\backslash A_r\}$ be a Hahn decomposition of X with respect to the signed measure $r\mu - \nu$ (Theorem 8.5), so that $\nu(E) \leq r\mu(E)$ for every measurable subset E of A_r, while $\nu(E) \geq r\mu(E)$ for every measurable subset E of $X\backslash A_r$. The countable collection of sets $\{A_r\}$ thus obtained need not be nested, but if r and s are any two rational numbers such that $0 < r < s$, then

$$s\mu(A_r\backslash A_s) \leq \nu(A_r\backslash A_s) \leq r\mu(A_r\backslash A_s),$$

which implies that $\mu(A_r\backslash A_s) = 0$, and hence that $\nu(A_r\backslash A_s) = 0$ too, since ν is absolutely continuous with respect to μ. Using this fact it is not difficult to verify that if we define

$$B_t = \bigcup_{0 < r < t} A_r$$

for each positive real number t, where the union is extended over all those rational numbers r such that $0 < r < t$, then the new family $\{B_t\}$ retains the property that $\{B_t, X\backslash B_t\}$ is a Hahn decomposition of X for the signed measure $t\mu - \nu, 0 < t < +\infty$, and also satisfies the condition

$$B_u = \bigcup_{0 < t < u} B_t,$$

for every positive real number u.

We now define the function f by setting

$$f(x) = \begin{cases} \inf\{t : x \in B_t\}, & x \in B_\infty, \\ +\infty, & x \in X\backslash B_\infty, \end{cases}$$

where B_∞ denotes the union $\bigcup_{0 < t < +\infty} B_t$. Clearly f is defined and nonnegative everywhere on X. Furthermore, it is easy to see that if t is a positive number, then

$$\{x \in X : f(x) < t\} = B_t,$$

which shows that f is measurable (Prob. 6C). We complete the proof by verifying the validity of (3). Suppose first that E is a measurable subset of B_t for some positive number t, and let ε be a given positive number. If $0 = t_0 < t_1 < \cdots < t_n = t$ is any partition of the interval $[0, t]$ such that each $\Delta t_i = t_i - t_{i-1}, i = 1, \ldots, n$, is less than ε, and if we write $E_i = E \cap (B_{t_i}\backslash B_{t_{i-1}})$, $i = 1, \ldots, n$, then it is easy to see that both $\nu(E_i)$ and $\nu_f(E_i)$ are straddled by the numbers $t_{i-1}\mu(E_i)$ and $t_i\mu(E_i)$. Hence

$$|\nu(E_i) - \nu_f(E_i)| \leq \Delta t_i\mu(E_i), \qquad i = 1, \ldots, n,$$

and therefore

$$|\nu(E) - \nu_f(E)| \leq \varepsilon\mu(E).$$

Since μ is finite and ε is arbitrary, this implies that $v(E) = v_f(E)$. Next, if E is any measurable subset of B_∞, and if $\{t_n\}$ is an increasing sequence of positive real numbers tending to $+\infty$, then

$$v(E) = \lim_n v(E \cap B_{t_n}) = v_f(E \cap B_{t_n}) = v_f(E)$$

by what has just been shown and the semicontinuity of measures (Prop. 7.4).

Suppose, finally, that E is a measurable subset of $X \backslash B_\infty$. If $\mu(E) = 0$, then $v(E) = v_f(E) = 0$ because v and v_f are both absolutely continuous $[\mu]$. On the other hand, if $\mu(E) > 0$, then $v(E) \geq t\mu(E)$ for every positive real number t, and therefore $v(E) = +\infty = v_f(E)$. Thus v and v_f agree on the measurable subsets of both B_∞ and $X \backslash B_\infty$, and the theorem follows. $\square$

We note that if μ and v are as in the statement of the Radon–Nikodym theorem, and if g is either integrable $[v]$ or nonnegative and measurable $[S]$, then

$$\int g \, dv = \int g \frac{dv}{d\mu} \, d\mu \tag{4}$$

by virtue of Proposition 9.5.

Definition. Let (X, S) be a measurable space, and suppose that μ and v are measures on (X, S) with the property that $v \ll \mu$ and $\mu \ll v$. Then we say that μ and v are *equivalent* measures (notation: $\mu \equiv v$).

It is clear that two measures on the same measurable space (X, S) are equivalent if and only if they possess exactly the same null sets, and that the relation of being equivalent is an equivalence relation on the set of all measures on (X, S).

Proposition 9.7. *Let λ, μ, and v be σ-finite measures on a measurable space (X, S), and suppose that $\lambda \ll \mu \ll v$. Then*

$$\frac{d\lambda}{dv} = \left(\frac{d\lambda}{d\mu}\right)\left(\frac{d\mu}{dv}\right)$$

almost everywhere with respect to all three measures. In particular, if μ and v are equivalent measures, then

$$\frac{d\mu}{dv} = \frac{1}{dv/d\mu}$$

a.e. with respect to both μ and v.

PROOF. If E is a set belonging to S, then

$$\lambda(E) = \int_E \left(\frac{d\lambda}{d\mu}\right) d\mu = \int_E \left(\frac{d\lambda}{d\mu}\right)\left(\frac{d\mu}{dv}\right) dv. \qquad \square$$

The antithesis of the concept of equivalence for measures is that of singularity.

Definition. Let $(X, \mathbf{S})$ be a measurable space, and suppose μ and ν are measures on $(X, \mathbf{S})$ with the property that there exists a partition of X into disjoint measurable sets A and B such that $\mu(A) = 0$ and $\nu(B) = 0$. Then μ and ν are *(mutually) singular* and we write $\mu \perp \nu$ or, equivalently, $\nu \perp \mu$. We also say that ν is *singular* with respect to μ, and that μ is *singular* with respect to ν.

A sketch of the proof of the following proposition is to be found in Problem K.

Proposition 9.8. *If $(X, \mathbf{S})$ is a measurable space, and if μ and ν are σ-finite measures on $(X, \mathbf{S})$, then there exists a unique decomposition of ν into the sum of two measures ν_1 and ν_2 on $(X, \mathbf{S})$ such that $\nu_1 \ll \mu$ and $\nu_2 \perp \mu$. The decomposition $\nu = \nu_1 + \nu_2$ is called the* Lebesgue decomposition *of ν with respect to μ.*

Let $(X, \mathbf{S}, \mu)$ be a measure space, and for each pair of sets E_1, E_2 in $\mathbf{S}$ let us write $E_1 \sim E_2$ to indicate that E_1 and E_2 are almost equal $[\mu]$, i.e., that $\mu(E_1 \,\triangledown\, E_2) = 0$. It is clear that $\sim$ is an equivalence relation on $\mathbf{S}$. We shall denote by $[E]$ the equivalence class of a measurable set E with respect to the relation $\sim$, and denote by $\dot{\mathbf{S}}$ the collection of all equivalence classes into which $\mathbf{S}$ is thus partitioned. Note that if $E_1 \sim E_2$, then $\mu(E_1) = \mu(E_2)$. Thus a function $\dot{\mu}$ is unambiguously defined on $\dot{\mathbf{S}}$ by setting $\dot{\mu}([E]) = \mu(E)$, $E \in \mathbf{S}$. Similarly, if we define $[E] \leq [F]$ to mean that $\mu(E \backslash F) = 0$, then $\leq$ is an unambiguously defined order relation on $\dot{\mathbf{S}}$ turning it into a partially ordered set. In this ordering $\dot{\mathbf{S}}$ is a lattice in which $[E] \vee [F] = [E \cup F]$ and $[E] \wedge [F] = [E \cap F]$. More generally (since the union of a countable sequence of null sets is again a null set), every sequence $\{[E_n]\}_{n=1}^{\infty}$ in $\dot{\mathbf{S}}$ possesses a *supremum* $\bigvee_{n=1}^{\infty} [E_n]$ and an *infimum* $\bigwedge_{n=1}^{\infty} [E_n]$ given by

$$\bigvee_{n=1}^{\infty} [E_n] = \left[\bigcup_{n=1}^{\infty} E_n \right] \quad \text{and} \quad \bigwedge_{n=1}^{\infty} [E_n] = \left[\bigcap_{n=1}^{\infty} E_n \right],$$

respectively. (Such a system is called a *σ-complete lattice*.) In this same vein we set

$$[E] \backslash [F] = [E \backslash F] \quad \text{and} \quad [E] \,\triangledown\, [F] = [E \,\triangledown\, F].$$

The pair $(\dot{\mathbf{S}}, \dot{\mu})$, equipped with these *Boolean operations*, is called the *measure ring* of the measure space $(X, \mathbf{S}, \mu)$. Likewise, if $\dot{\mathbf{S}}_{\mathscr{F}}$ denotes the subset of $\dot{\mathbf{S}}$ consisting of those elements $[E]$ of $\dot{\mathbf{S}}$ such that $\dot{\mu}([E])$ is finite, then the pair $(\dot{\mathbf{S}}_{\mathscr{F}}, \dot{\mu}|\dot{\mathbf{S}}_{\mathscr{F}})$ is the *finite measure ring* of $(X, \mathbf{S}, \mu)$. It is easy to see that the function ρ defined by $\rho([E], [F]) = \dot{\mu}([E \,\triangledown\, F])$ for all $[E]$, $[F]$ in $\dot{\mathbf{S}}_{\mathscr{F}}$ is a metric on $\dot{\mathbf{S}}_{\mathscr{F}}$. The metric space $(\dot{\mathbf{S}}_{\mathscr{F}}, \rho)$ is the *metric space associated with*

$(X, \mathbf{S}, \mu)$. A measure space is said to be *separable* if its associated metric space is separable. (The completeness of the measure space $(X, \mathbf{S}, \mu)$ has nothing to do with the completeness of its associated metric space; see Problem U.)

> The terminology "measure ring" has an easily explained origin. The system $\dot{\mathbf{S}}$ associated with a measure space $(X, \mathbf{S}, \mu)$ is, in fact, a Boolean ring if $\wedge$ is interpreted as multiplication and ∇ as addition. As was remarked in Chapter 6, such rings were studied initially in [63].

Example G. Let $(\mathbb{R}^n, \mathbf{S}, \mu_n)$ be the measure space consisting of Lebesgue measure on the Lebesgue measurable sets in $\mathbb{R}^n$, and let $(\mathbb{R}^n, \mathbf{B}, \mu_n | \mathbf{B})$ be Lebesgue–Borel measure on $\mathbb{R}^n$. Then each equivalence class in $(\dot{\mathbf{S}}, \dot{\mu}_n)$ contains exactly one equivalence class in the measure ring $(\dot{\mathbf{B}}, (\mu_n | \mathbf{B}))$ (Prob. 8H). Thus the measure rings, as well as the associated metric spaces, of these two measures spaces are virtually identical. More generally, the measure ring, and the associated metric space of any measure space $(X, \mathbf{S}, \mu)$ may be identified with the measure ring and associated metric space, respectively, of the completion of $(X, \mathbf{S}, \mu)$ (Prob. 8Q).

Example H. Let $(X, \mathbf{S}, \mu)$ be a measure space, and let $\mathbf{R}$ be any ring of measurable sets of finite measure with respect to μ that generates $\mathbf{S}$ as a σ-ring. Then, according to Problem 8M, $\dot{\mathbf{R}} = \{[A] \in \dot{\mathbf{S}} : A \in \mathbf{R}\}$ is dense in the metric space $(\dot{\mathbf{S}}_{\mathscr{F}}, \rho)$ associated with $(X, \mathbf{S}, \mu)$. Thus, for instance, the set of equivalence classes of all finite unions of (half-open) cells in $\mathbb{R}^n$ is dense in the metric space associated with Lebesgue measure on $\mathbb{R}^n$. From this observation it is easily seen that Lebesgue measure on $\mathbb{R}^n$ is separable.

Proposition 9.9. *The function $\dot{\mu}$ on the metric space $(\dot{\mathbf{S}}_{\mathscr{F}}, \rho)$ associated with an arbitrary measure space $(X, \mathbf{S}, \mu)$ satisfies the condition*

$$|\dot{\mu}([E]) - \dot{\mu}([F])| \leq \rho([E], [F]), \qquad [E], [F] \in \dot{\mathbf{S}}_{\mathscr{F}}.$$

In particular, $\dot{\mu}$ is (uniformly) continuous on $\dot{\mathbf{S}}_{\mathscr{F}}$. Similarly, the Boolean operations $\vee, \wedge, \backslash, and \nabla$ are all continuous mappings of $\dot{\mathbf{S}}_{\mathscr{F}} \times \dot{\mathbf{S}}_{\mathscr{F}}$ into $\dot{\mathbf{S}}_{\mathscr{F}}$.

PROOF. For any two sets E and F of finite measure in X it is easy to verify that

$$|\mu(E) - \mu(F)| \leq \mu(E \nabla F),$$

whence the first part of the proposition follows at once. Similarly, if Φ denotes any one of the four Boolean operations listed above, and if (E_1, F_1) and (E_2, F_2) are any two pairs of sets of finite measure with respect to μ, then a routine computation shows that

$$\Phi([E_1], [F_1]) \nabla \Phi([E_2], [F_2]) \leq ([E_1] \nabla [E_2]) \vee ([F_1] \nabla [F_2])$$

(cf. Problem 1F), and hence that

$$\rho(\Phi([E_1], [F_1]), \Phi([E_2], [F_2])) \leq \rho([E_1], [E_2]) + \rho([F_1], [F_2]). \qquad \square$$

Example I. For each fixed point a in $\mathbb{R}^n$ let us write T_a for the translation that carries each x in $\mathbb{R}^n$ to $T_a(x) = x + a$, and if E is a subset of $\mathbb{R}^n$ let us also write $E + a$ for $T_a(E)$. Since

$$(E + a) \nabla (F + a) = (E \nabla F) + a$$

for all subsets E and F and every point a, it is clear that the mapping $\dot{T}_a$ defined by $\dot{T}_a([E]) = [E + a]$ for each Lebesgue measurable set E is an isometry on the metric space $(\dot{S}_{\mathscr{F}}, \rho)$ associated with Lebesgue measure on $\mathbb{R}^n$.

Suppose now that H is a half-open cell in $\mathbb{R}^n$:

$$H = [a_1, b_1) \times \cdots \times [a_n, b_n). \tag{5}$$

If U is an open cell slightly larger than H, and K a closed cell slightly smaller than H, then there is a neighborhood V of the origin in $\mathbb{R}^n$ such that if $x \in V$, then $K \subset H + x \subset U$, and therefore such that $H \nabla (H + x) \subset U \backslash K$ for every x in V. Thus for any given $\varepsilon > 0$ there exists a neighborhood V of the origin such that $\rho([H], \dot{T}_x([H]) < \varepsilon$ for every x in V. Consider next the ring $\mathbf{H}_l$ of sets generated by all half-open cells of the form (5). If A is an element of $\mathbf{H}_l$, then A is the union $H_1 \cup \cdots \cup H_k$ of such cells (Ex. 6D), and since

$$\rho([A], [A + x]) \le \rho([H_1], [H_1 + x]) + \cdots + \rho([H_k], [H_k + x])$$

(Prob. 1F), we see again that for any given $\varepsilon > 0$ there exists a neighborhood V of the origin in $\mathbb{R}^n$ such that

$$\rho([A], \dot{T}_x([A])) < \varepsilon, \qquad x \in V.$$

Consider now the real-valued function

$$f_A(x) = \rho([A], \dot{T}_x([A])) = \mu(A \nabla T_x(A)), \qquad x \in \mathbb{R}^n.$$

Since

$$\begin{aligned}
|f_A(x) - f_A(y)| &= |\mu(A \nabla T_x(A)) - \mu(A \nabla T_y(A))| \\
&\le \rho([A \nabla T_x(A)], [A \nabla T_y(A)]) \\
&= \mu(A \nabla T_x(A) \nabla A \nabla T_y(A)),
\end{aligned}$$

and $A \nabla T_x(A) \nabla A \nabla T_y(A) = T_x(A) \nabla T_y(A)$, we see that

$$\begin{aligned}
|f_A(x) - f_A(y)| &\le \rho(\dot{T}_x([A]), \dot{T}_y([A])) \\
&= \rho([A], \dot{T}_{y-x}([A])).
\end{aligned}$$

Hence f_A is uniformly continuous on $\mathbb{R}^n$. Moreover, if E denotes any Borel set in $\mathbb{R}^n$ of finite Lebesgue measure, then, as noted in Example H, there exists a set A in $\mathbf{H}_l$ such that $\rho([E], [A]) < \varepsilon$. But then, once again, it is easily seen that

$$|\rho([E], [E + x]) - \rho([A], [A + x])| < 2\varepsilon$$

for every x in $\mathbb{R}^n$. Thus the function $f_E(x) = \rho([E], \dot{T}_x([E]))$ is uniformly approximable by uniformly continuous functions of the form f_A, $A \in \mathbf{H}_l$, and it follows that f_E is also uniformly continuous on $\mathbb{R}^n$.

It is in terms of measure rings that we define the notion of isomorphism appropriate to measure spaces.

Definition. Let $(X, \mathbf{S}, \mu)$ and $(Y, \mathbf{T}, \nu)$ be measure spaces, and let $(\dot{\mathbf{S}}, \dot{\mu})$ and $(\dot{\mathbf{T}}, \dot{\nu})$, respectively, be their measure rings. A one-to-one mapping Φ of $\dot{\mathbf{S}}$ onto $\dot{\mathbf{T}}$ is said to be a *weak measure ring isomorphism* if

   (i) $\Phi([E] \backslash [F]) = \Phi([E]) \backslash \Phi([F])$ for every pair $[E]$, $[F]$ in $\dot{\mathbf{S}}$, and
   (ii) $\Phi(\bigvee_{n=1}^{\infty} [E_n]) = \bigvee_{n=1}^{\infty} \Phi([E_n])$ for every sequence $\{[E_n]\}_{n=1}^{\infty}$ in $\dot{\mathbf{S}}$.

(It is easy to see that a weak measure ring isomorphism preserves *all* Boolean operations; cf. the proof of Theorem 9.10.) If Φ is a weak measure ring isomorphism of $(\dot{\mathbf{S}}, \dot{\mu})$ onto $(\dot{\mathbf{T}}, \dot{\nu})$ satisfying the further condition

   (iii) $\dot{\nu}(\Phi([E])) = \dot{\mu}([E])$, $[E] \in \dot{\mathbf{S}}$,

then Φ is said to be a *measure ring isomorphism*. If there exists a [weak] measure ring isomorphism of $(\dot{\mathbf{S}}, \dot{\mu})$ onto $(\dot{\mathbf{T}}, \dot{\nu})$, then the measure spaces $(X, \mathbf{S}, \mu)$ and $(Y, \mathbf{T}, \nu)$, as well as the measure rings $(\dot{\mathbf{S}}, \dot{\mu})$ and $(\dot{\mathbf{T}}, \dot{\nu})$, are said to be [*weakly*] *isomorphic*.

Example J. Let $(X, \mathbf{S}, \mu)$ and $(Y, \mathbf{T}, \nu)$ be measure spaces, and let X_0 and Y_0 be measurable subsets of X and Y, respectively, such that $\mu(X \backslash X_0) = \nu(Y \backslash Y_0) = 0$. Suppose given a one-to-one mapping φ of X_0 onto Y_0 with the property that a subset E of X_0 belongs to $\mathbf{S}$ if and only if $\varphi(E)$ belongs to $\mathbf{T}$. (This is equivalent to requiring that φ be measurable $[\mathbf{S}, \mathbf{T}]$ on X_0 and that φ^{-1} be measurable $[\mathbf{T}, \mathbf{S}]$ on Y_0; such a mapping is said to be *measurability preserving*.) Suppose also that φ possesses the further property that $\mu(E) = 0$ if and only if $\nu(\varphi(E)) = 0$. (This condition can also be stated in another way; it says that ν is equivalent to the measure induced on $(Y_0, \mathbf{T}_{Y_0})$ by φ and the measure $\mu | X_0$.) Then

$$\Phi([E]) = [\varphi(E \cap X_0)],   E \in \mathbf{S},$$

defines a weak isomorphism Φ of $(\dot{\mathbf{S}}, \dot{\mu})$ onto $(\dot{\mathbf{T}}, \dot{\nu})$. The weak isomorphism Φ that arises in this manner is said to be *induced* by the *point transformation* φ. The weak isomorphism Φ induced by φ is an isomorphism if and only if φ possesses the added property that

$$\mu(E) = \nu(\varphi(E)),   E \in \mathbf{S}, E \subset X_0.$$

(Such a mapping φ is said to be *measure preserving*.) An isomorphism of $(\dot{\mathbf{S}}, \dot{\mu})$ onto $(\dot{\mathbf{T}}, \dot{\nu})$ that is induced by a measure preserving point transformation φ is called a *strong isomorphism*. If such a strong isomorphism exists, then the measure spaces $(X, \mathbf{S}, \mu)$ and $(Y, \mathbf{T}, \nu)$, as well as the measure rings $(\dot{\mathbf{S}}, \dot{\mu})$ and $(\dot{\mathbf{T}}, \dot{\nu})$, are said to be *strongly isomorphic*. (The point mapping φ itself is also frequently called a *strong isomorphism*.)

Example K. Let $\{(X_\gamma, \mathbf{S}_\gamma, \mu_\gamma)\}$ and $\{(\tilde{X}_{.,}, \tilde{\mathbf{S}}_{.,}, \tilde{\mu}_{.,})\}$ be two similarly indexed families of measure spaces such that $X_\gamma \cap X_{\gamma'} = \emptyset = \tilde{X}_\gamma \cap \tilde{X}_{\gamma'}$ for all indices $\gamma \neq \gamma'$, and let us form the direct sums

$$(X, \mathbf{S}, \mu) = \sum_\gamma \oplus (X_\gamma, \mathbf{S}_\gamma, \mu_\gamma)$$

and

$$(\tilde{X}, \tilde{\mathbf{S}}, \tilde{\mu}) = \sum_\gamma \oplus (\tilde{X}_\gamma, \tilde{\mathbf{S}}_\gamma, \tilde{\mu}_\gamma).$$

If for each index γ the measure spaces X_γ and $\tilde{X}_\gamma$ are isomorphic, then it is easy to see that X and $\tilde{X}$ are also isomorphic. Likewise, if for each index γ the spaces X_γ and $\tilde{X}_\gamma$ are *strongly* isomorphic, and if the index family is countable, then it is readily seen that X and $\tilde{X}$ are also strongly isomorphic. On the other hand, if for each index γ there exists a strong isomorphism φ_γ of *all* of X_γ onto *all* of $\tilde{X}_\gamma$, then the isomorphisms φ_γ can be combined to construct a strong isomorphism of (all of) X onto (all of) $\tilde{X}$, no matter what the cardinality of the index family may be. Using this remark it is possible to introduce a notion of weak direct sum of an indexed family of measure spaces $\{(X_\gamma, \mathbf{S}_\gamma, \mu_\gamma)\}$ even when the carriers $X_{.,}$ are not pairwise disjoint. The trick is simply to construct a new, pairwise disjoint, family of sets $\tilde{X}_\gamma$ such that $\tilde{X}_\gamma$ is in one-to-one correspondence with X_γ for each index γ, to use the one-to-one correspondence to import the structure of the measure space X_γ into $\tilde{X}_\gamma$, and then to form the direct sum of the new indexed family $\{\tilde{X}_\gamma\}$. This procedure does not result in a uniquely defined measure space, to be sure, but, as we have just seen, any two such spaces are strongly isomorphic.

Example L. In a *binary expansion*

$$t = .\varepsilon_1\varepsilon_2 \cdots \varepsilon_n \cdots$$

of a number t in the unit interval (here $\{\varepsilon_n\}_{n=1}^\infty$ is a sequence of zeros and ones and $t = \sum_{n=1}^\infty \varepsilon_n/2^n$), it is well known that the number t is rational if and only if the sequence $\{\varepsilon_n\}$ is ultimately periodic. On the other hand, it is clear that two sequences $\{\varepsilon_n\}$ and $\{\eta_n\}$ of zeros and ones are both ultimately periodic if and only if the "perfect shuffle"

$$\{\varepsilon_1, \eta_1, \varepsilon_2, \eta_2, \ldots, \varepsilon_n, \eta_n, \ldots\}$$

is ultimately periodic. It follows that if X_0 denotes the subset of the unit square $X = [0, 1] \times [0, 1]$ consisting of pairs (r, s) in which either r or s is irrational, then

$$\varphi(.\varepsilon_1\varepsilon_2 \cdots, .\eta_1\eta_2 \cdots) = .\varepsilon_1\eta_1\varepsilon_2\eta_2 \cdots$$

defines (unambiguously) a one-to-one mapping of X_0 onto the set Y_0 of irrational numbers in the unit interval Y. Note that if μ_1 denotes Lebesgue–Borel measure on the real line, and μ_2 denotes Lebesgue–Borel measure on the plane $\mathbb{R}^2$, then $\mu_1(Y \setminus Y_0) = \mu_2(X \setminus X_0) = 0$. Moreover, if r and s are any two dyadic fractions of the form $k/2^n$, $0 \leq k < 2^n$, and if Q denotes the square $[r, r + 1/2^n] \times [s, s + 1/2^n]$, then direct calculation discloses that φ maps $Q \cap X_0$ onto the set of irrational numbers in an interval of length $1/2^{2n}$. Since

$$\mu_2(Q \cap X_0) = |Q| = \left(\frac{1}{2^n}\right)^2 = \frac{1}{2^{2n}} = \mu_1(\varphi(Q \cap X_0)),$$

and since squares of the special form here considered suffice to generate the σ-ring of Borel subsets of X, while subintervals of the form $[k/2^{2n}, (k + 1)/2^{2n}]$ likewise suffice to generate the σ-ring of Borel subsets of Y, it follows readily that φ is a measure preserving mapping of X_0 onto Y_0 (see Problem 8A). Thus the unit square equipped with Lebesgue–Borel measure and the unit interval equipped with Lebesgue–Borel measure are strongly isomorphic measure spaces. Similarly the real line $\mathbb{R}$ equipped with Lebesgue–Borel measure is strongly isomorphic to the plane $\mathbb{R}^2$ equipped with Lebesgue–Borel measure. More generally, $(\mathbb{R}, \mathbf{B}, \mu_1)$ is strongly isomorphic to $\mathbb{R}^n$ equipped with Lebesgue–Borel measure.

The following theorem shows that the Lebesgue integrals on two isomorphic measure spaces are not very different. In order to formulate the desired result we employ the notation $\dot{\mathscr{L}}_\mu$, as in Chapter 7, for the linear space of equivalence classes of complex-valued functions defined and integrable $[\mu]$ on a measure space $(X, \mathbf{S}, \mu)$. If $[f]$ and $[g]$ are elements of $\dot{\mathscr{L}}_\mu$, then $[f] = [g]$ if and only if $f = g$ a.e. $[\mu]$, and if $E \in \mathbf{S}$, then $\int_E [f] d\mu = \int_E f \, d\mu$ by definition.

Theorem 9.10. *If Φ is a measure ring isomorphism of the measure ring $(\dot{\mathbf{S}}, \mu)$ of a measure space $(X, \mathbf{S}, \mu)$ onto the measure ring $(\dot{\mathbf{T}}, \nu)$ of a measure space $(Y, \mathbf{T}, \nu)$, then there exists a unique linear isomorphism T_Φ of the linear space $\dot{\mathscr{L}}_\mu$ onto the linear space $\dot{\mathscr{L}}_\nu$ satisfying the following conditions:*

(i) *If E is a set in $\mathbf{S}$ such that $\mu(E) < +\infty$, and if $\Phi([E]) = [F]$, then*
$$T_\Phi([\chi_E]) = [\chi_F],$$
(ii) *If $[f] \in \dot{\mathscr{L}}_\mu$ and $[g] = T_\Phi([f])$, then $[|g|] = T_\Phi([|f|])$ and*

$$\int_X [f] d\mu = \int_Y [g] d\nu.$$

The isomorphism T_Φ is also an isometry between the space $\dot{\mathscr{L}}_\mu$ equipped with the metric

$$\rho_\mu([f_1], [f_2]) = \int_X |f_1 - f_2| d\mu, \quad f_1, \quad f_2 \in \dot{\mathscr{L}}_\mu,$$

and the space $\dot{\mathscr{L}}_\nu$ equipped with the metric

$$\rho_\nu([g_1],[g_2]) = \int_Y |g_1 - g_2| dv, \qquad g_1, g_2 \in \mathscr{L}_\nu.$$

PROOF. It is easily seen that $\Phi([\varnothing]) = [\varnothing]$ and $\Phi([X]) = [Y]$, so that $\Phi([X\backslash E]) = [Y]\backslash\Phi([E])$ for every $[E]$ in $\dot{S}$, and likewise

$$\Phi([E_1] \wedge [E_2]) = \Phi([E_1]) \wedge \Phi([E_2]), \qquad [E_1], [E_2] \in \dot{S}.$$

It follows that if $\{E_1, \ldots, E_n\}$ is an arbitrary finite collection of measurable subsets of X, and if $F_1, \ldots, F_n$ are chosen so that $[F_i] = \Phi([E_i])$, $i = 1, \ldots, n$, then Φ carries the equivalence classes of the sets belonging to the partition of X determined by $\{E_1, \ldots, E_n\}$ (Prob. 1L) onto the equivalence classes of the sets belonging to the partition determined by $\{F_1, \ldots, F_n\}$ in such a manner that if G is in the former partition and $\Phi([G]) = [H]$, then $[G] \leq [E_i]$ if and only if $[H] \leq [F_i]$, $i = 1, \ldots, n$. (This much is valid even for weak isomorphisms, of course.)

Suppose now that we denote by $\mathscr{S}_\mu$ the linear submanifold of $\dot{\mathscr{L}}_\mu$ consisting of all equivalence classes of the form $[s]$, where s is an integrable *simple* function, and that $\mathscr{S}_\nu$ is defined analogously. It is clear that if $[s] \in \mathscr{S}_\mu$, where $s = \sum_{i=1}^n \alpha_i \chi_{E_i}$ and $\mu(E_i) < +\infty$, $i = 1, \ldots, n$, and if T is any linear transformation of $\dot{\mathscr{L}}_\mu$ into $\dot{\mathscr{L}}_\nu$ satisfying (i), then we must have $T([s]) = [t]$ where $t = \sum_{i=1}^n \alpha_i \chi_{F_i}$ and $[F_i] = \Phi([E_i])$, $i = 1, \ldots, n$. On the other hand, on the basis of what has already been said, it is easy to verify that setting

$$T_0\left(\left[\sum_{i=1}^n \alpha_i \chi_{E_i}\right]\right) = \left[\sum_{i=1}^n \alpha_i \chi_{F_i}\right],$$

where $[F_i] = \Phi([E_i])$, $i = 1, \ldots, n$, does indeed define a linear transformation T_0 of $\mathscr{S}_\mu$ onto $\mathscr{S}_\nu$ satisfying condition (i) (see Problem 1L). Moreover, it is clear that if $[t] = T_0([s])$, then $[|t|] = T_0([|s|])$ and $\int [t] dv = \int [s] d\mu$. Hence if s_1 and s_2 are any two integrable simple functions on X, and if $[t_i] = T_0([s_i])$, $i = 1, 2$, then

$$\int_X [|s_1 - s_2|] d\mu = \int_Y [|t_1 - t_2|] dv.$$

This says that the mapping T_0 is an isometry of $\mathscr{S}_\mu$ onto $\mathscr{S}_\nu$ with respect to the metrics ρ_μ and ρ_ν (see Problem 7U). Moreover, $\mathscr{S}_\mu$ is dense in $\dot{\mathscr{L}}_\mu$. (Proof: If f belongs to $\mathscr{L}_\mu$ and $\{s_n\}$ is any sequence of measurable simple functions on X that converges pointwise to f and satisfies the condition $|s_n| \leq |f|$ for every n (Prob. 6S), then $\{s_n\}$ converges to f in the mean by the dominated convergence theorem (Th. 7.12), so the sequence $\{[s_n]\}$ tends to $[f]$ in the metric ρ_μ.) Likewise $\mathscr{S}_\nu$ is dense in $\dot{\mathscr{L}}_\nu$. Since $\dot{\mathscr{L}}_\mu$ and $\dot{\mathscr{L}}_\nu$ are complete metric spaces with respect to the metrics ρ_μ and ρ_ν, respectively, it follows that T_0 admits a unique isometric extension T_Φ mapping $\dot{\mathscr{L}}_\mu$

onto $\dot{\mathscr{L}}_v$ (see Problem 4H), and it is a simple matter to verify that T_Φ is a linear transformation satisfying (i) and (ii).

To complete the proof it suffices to show that T_Φ is uniquely determined by (i) and (ii). But, as has already been observed, if T is any linear transformation of $\mathscr{L}_\mu$ into $\dot{\mathscr{L}}_v$ satisfying (i), then T must agree with T_0 on $\mathscr{S}_\mu$. If T also shares condition (ii) with T_Φ, if $[f]$ and $\{[f_n]\}$ belong to $\mathscr{L}_\mu$ and are such that $\{f_n\}$ converges to f in the mean, and if we write $[g_n] = T([f_n])$ and $[g] = T([f])$, then $\{g_n\}$ converges to g in the mean. Hence $T = T_\Phi$. $\square$

The version of Theorem 9.10 appropriate for weak isomorphisms goes as follows.

Theorem 9.11. *Let Φ be a weak measure ring isomorphism of the measure ring $(\dot{S}, \dot{\mu})$ of a σ-finite measure space (X, S, μ) onto the measure ring $(\dot{T}, \dot{v})$ of a σ-finite measure space (Y, T, v). Then there exists a nonnegative measurable function h on Y and a linear isomorphism T_Φ of $\dot{\mathscr{L}}_\mu$ onto $\dot{\mathscr{L}}_v$ satisfying the following conditions:*

(i) *If E is a set in S such that $\mu(E) < +\infty$, and if $\Phi([E]) = [F]$, then $T_\Phi([\chi_E]) = [h\chi_F]$,*

(ii) *If $[f] \in \dot{\mathscr{L}}_\mu$ and $T_\Phi([f]) = [g]$, then $T_\Phi([|f|]) = ([|g|]$ and*

$$\int_X [f]d\mu = \int_Y [g]dv.$$

Both the function h and the linear transformation T_Φ are uniquely determined by Φ and conditions (i) and (ii), and the isomorphism T_Φ is an isometry between the metric spaces $\dot{\mathscr{L}}_\mu$ and $\dot{\mathscr{L}}_v$ of Theorem 9.10.

SKETCH OF PROOF. If we define $v_0(F) = \dot{\mu}(\Phi^{-1}([F]))$, $F \in T$, it is easy to verify that v_0 is a σ-finite measure on (Y, T) such that v_0 is equivalent to v, and that Φ is a measure ring isomorphism of $(\dot{S}, \dot{\mu})$ onto the measure ring $(\dot{T}, \dot{v}_0)$ of the measure space (Y, T, v_0). Since Theorem 9.10 applies to this latter isomorphism, a moment's reflection discloses that it suffices to treat the case in which μ and v are equivalent measures on the same measurable space (Y, T) and the isomorphism Φ is the identity mapping on $\dot{T}$. In this case there exists a nonnegative measurable function $h = d\mu/dv$ on Y such that

$$\int_Y f\, d\mu = \int_Y hf\, dv, \qquad f \in \mathscr{L}_\mu$$

(Th. 9.6), and if we simply define $T_\Phi([f]) = [hf]$ for each element $[f]$ of $\mathscr{L}_\mu$, then it is obvious that T_Φ is a linear transformation satisfying (i) and (ii). The fact that T_Φ is also an isometry follows from condition (ii), just as before. The uniqueness of the function h follows from the fact that h must coincide with $d\mu/dv$ on sets of finite measure with respect to μ, along with the fact that μ is σ-finite; the uniqueness of T_Φ is then also clear.

The scope of Theorem 9.10 is considerably enhanced by the fact that, up to isomorphism, there is only one separable atom-free measure space of a given size.

Theorem 9.12 (Halmos–von Neumann Theorem [33]). *If $(X, \mathbf{S}, \mu)$ is a separable atom-free measure space with $0 \leq \mu(X) \leq +\infty$, then $(X, \mathbf{S}, \mu)$ is isomorphic to the measure space consisting of Lebesgue–Borel measure on the interval $[0, \mu(X))$.*

PROOF. If $(\dot{\mathbf{S}}_{\mathscr{F}}, \rho)$ is the metric space associated with $(X, \mathbf{S}, \mu)$, then by hypothesis there exists a dense sequence $\{[E_n]\}$ in $\dot{\mathbf{S}}_{\mathscr{F}}$. Set $X_0 = \bigcup_n E_n$. If E is a measurable subset of $X \setminus X_0$, then $\mu(E)$ must be either 0 or $+\infty$. Since μ is atom-free, this implies that $\mu(X \setminus X_0) = 0$, and hence that μ is σ-finite. Thus it suffices to treat the case $0 < \mu(X) < +\infty$ (cf. Example 8H, Example K, and Problem 8R). Accordingly we assume henceforth that $\{[E_n]\}_{n=1}^{\infty}$ is a fixed dense sequence in the metric space associated with the finite measure space $(X, \mathbf{S}, \mu)$ (with $\mu(X) > 0$), and we write $(Y, \mathbf{B}, \nu)$ for the measure space consisting of the interval $[0, \mu(X)]$ equipped with Lebesgue–Borel measure. For each positive integer n let $\mathscr{P}_n$ be the partition of X determined by the sets $E_1, E_2, \ldots, E_n$, that is, the partition consisting of all the nonempty subsets of X of the form $A_1 \cap \cdots \cap A_n$, where, for each $i = 1, \ldots, n$, A_i denotes either E_i or $X \setminus E_i$ (Prob. 1L). Let $\mathbf{R}_n$ denote the ring of subsets of X generated by $\mathscr{P}_n$ (thus $\mathbf{R}_n$ consists of the unions of the various sets belonging to $\mathscr{P}_n$), and set $\dot{\mathbf{R}}_n = \{[A] : A \in \mathbf{R}_n\}$. Note that $\mathscr{P}_{n+1}$ refines $\mathscr{P}_n$ and that $\mathbf{R}_n \subset \mathbf{R}_{n+1}$ for each positive integer n. Hence, using Problem X repeatedly and starting with $n = 1$, we may construct by induction a sequence $\{\varphi_n\}_{n=1}^{\infty}$ of mappings and an accompanying sequence $\{\mathscr{Q}_n\}_{n=1}^{\infty}$ of partitions of the interval $[0, \mu(X)]$ possessing the following properties for each positive integer n:

(i) The partition $\mathscr{Q}_{n+1}$ refines $\mathscr{Q}_n$,
(ii) If $\mathscr{Q}_n = \{t_i\}_{i=0}^{N_n}$, then φ_n maps $\dot{\mathbf{R}}_n$ onto the subset of $\dot{\mathbf{B}}$ consisting of equivalence classes $[B]$, where B runs over the ring of unions of the subintervals $[t_{i-1}, t_i)$, $i = 1, \ldots, N_n$, of $\mathscr{Q}_n$,
(iii) $\varphi_{n+1} | \mathbf{R}_n = \varphi_n$,
(iv) If $A \in \mathbf{R}_n$, then $\dot{\nu}(\varphi_n([A])) = \dot{\mu}([A])$,
(v) If $A_1, A_2 \in \mathbf{R}_n$, then $\varphi_n([A_1] \vee [A_2]) = \varphi_n([A_1]) \vee \varphi_n([A_2])$ and $\varphi_n([A_1] \setminus [A_2]) = \varphi_n([A_1]) \setminus \varphi_n([A_2])$.

Let $\dot{\mathbf{R}} = \bigcup_n \dot{\mathbf{R}}_n$ and define Φ_0 on $\dot{\mathbf{R}}$ by setting $\Phi_0([A]) = \varphi_n([A])$ for any one n such that $[A] \in \dot{\mathbf{R}}_n$. The mapping Φ_0 is unambiguously defined by virtue of (iii), and it is clear from (iv) and (v) that if $[A] \in \dot{\mathbf{R}}$ then $\dot{\nu}(\Phi_0([A])) = \dot{\mu}([A])$, and likewise that Φ_0 preserves the operations $\vee$ and $\setminus$. Hence Φ_0 is an isometry of $\dot{\mathbf{R}}$, regarded as a subset of $(\dot{\mathbf{S}}, \rho)$, into the metric space associated with $(Y, \mathbf{B}, \nu)$. Moreover, $\dot{\mathbf{R}}$ is dense in $\dot{\mathbf{S}}$, since $\dot{\mathbf{R}}$ contains the entire sequence $\{[E_n]\}$, and the range of Φ_0 is also dense in $\dot{\mathbf{B}}$, since the sequence $\{\text{mesh } \mathscr{Q}_n\}$ tends to zero as may be seen by consulting Problem W. Hence, extending Φ_0

by continuity, we obtain an isometry Φ of $\dot{S}$ onto $\dot{B}$. Since Φ is isometric, we have $\dot{v}(\Phi([E])) = \dot{\mu}([E])$ for every E in S. Likewise, $\Phi | \dot{R} = \Phi_0$ preserves $\vee$ and $\setminus$, and since $\dot{R}$ is dense in $\dot{S}$, and since the operations $\vee$ and $\setminus$ are continuous on $\dot{S}$ and on $\dot{B}$ (Prop. 9.9), we see that Φ also preserves $\vee$ and $\setminus$. Finally, this shows that Φ is order preserving, from which it follows at once that

$$\Phi\left(\bigvee_{n=1}^{\infty} [F_n]\right) = \bigvee_{n=1}^{\infty} \Phi([F_n])$$

for every sequence $\{[F_n]\}$ in $\dot{S}$. Thus Φ is a measure ring isomorphism of $(\dot{S}, \dot{\mu})$ onto $(\dot{B}, \dot{v})$. $\square$

PROBLEMS

A. Let W denote the topological space $W(\Omega)$ of countable ordinal numbers in the order topology (Prob. 1W, Ex. 3B), and let **B** denote the σ-ring of Borel subsets of W. Show that the diagonal $\Delta = \{(\alpha, \alpha) : \alpha \in W\}$ is a Borel set in the topological space $W \times W$, but that Δ does *not* belong to the σ-ring **B** $\times$ **B**. (Hint: Every set A in the σ-ring **B** $\times$ **B** has the property that either A or its complement $(W \times W) \setminus A$ contains a set of the form $F \times F$, where F is a closed unbounded subset of W; cf. Problem 6Q.)

> It is easy to show that if (X, S) and (Y, T) are arbitrary measurable spaces, and if E is any one set in $S \times T$, then there exist σ-rings $S_0 \subset S$ and $T_0 \subset T$ such that $E \in S_0 \times T_0$ and such that both S_0 and T_0 are generated by countable collections of sets (see Problem 6E). According to Problem 6H, the σ-rings S_0 and T_0 both have cardinal number no greater than $\aleph$, and according to Lemma 9.1, all of the X-sections E_x, $x \in X$, belong to T_0, and all of the Y-sections E^y, $y \in Y$, belong to S_0. Thus an *arbitrary* set E belonging to a product σ-ring $S \times T$ can have no more than $\aleph$ distinct sections in either direction. In particular, if card $X > \aleph$, then the diagonal $\Delta = \{(x, x) : x \in X\}$ does not belong to *any* σ-ring of the form $S \times S$. On the other hand, Δ is a closed set, *a fortiori* a Borel set, whenever X has a Hausdorff topology and $X \times X$ is given its product topology. (We are indebted to Professor J. C. Oxtoby for calling our attention to this rather startling counterexample.)

B. Show that if (X, S) and (Y, T) are measurable spaces, and if E is a subset of $X \times Y$ that belongs to the product $S \times T$, then every X-section E_x of E is measurable [**T**], and every Y-section E^y is measurable [**S**]. Show also that if $f(x, y)$ is a function on $X \times Y$ that is measurable [$S \times T$], then, for each x_0 in X, $f(x_0, y)$ is measurable [**T**] as a function of the single variable y, and likewise that, for each y_0 in Y, $f(x, y_0)$ is measurable [**S**] as a function of the single variable x. (Hint: The collection of subsets E of $X \times Y$ with the property that $E_x \in T$ for every x in X is a σ-ring.)

C. Let (X, S, μ) and (Y, T, v) be measure spaces, and suppose that v is σ-finite.

(i) Show that if G is a set belonging to $S \times T$, then $v(G_x)$ is a measurable function on X. (That this function is defined everywhere on X follows from Problem B.) Show also that if f is a nonnegative, measurable, extended real-valued function on $(X \times Y, \mathbf{S} \times \mathbf{T})$, then $\int_Y f(x, y)dv(y)$ is a measurable extended real-valued function of x on X, and complete the proof of Lemma 9.1. (Hint: If $v(Y) < +\infty$, then the collection of all measurable sets G in $X \times Y$ with the property that $v(G_x)$ is a measurable function of x is a σ-quasiring (Prob. 6J). Verify first that $\int_Y s(x, y)dv(y)$ is a measurable function of x for a nonnegative finite-valued measurable simple function s, and use the monotone convergence theorem.)

(ii) Show that if we write

$$\lambda(G) = \int_X v(G_x)d\mu, \qquad G \in \mathbf{S} \times \mathbf{T},$$

then λ is a measure on $(X \times Y, \mathbf{S} \times \mathbf{T})$ satisfying the condition $\lambda(E \times F) = \mu(E)v(F)$ on measurable rectangles. Show also that if f is a nonnegative, measurable, extended real-valued function on $(X \times Y, \mathbf{S} \times \mathbf{T})$, then

$$\int_{X \times Y} f\, d\lambda = \int_X \left[\int_Y f(x, y)dv(y) \right] d\mu(x). \tag{6}$$

(Hint: Use the theorem of Beppo–Levi (Cor. 7.10) to prove that λ is a measure. Verify (6) first for a nonnegative measurable simple function, and use the monotone convergence theorem.)

D. Show that if $(X, \mathbf{S}, \mu)$ and $(Y, \mathbf{T}, v)$ are both σ-finite measure spaces, then there exists a unique measure $\mu \times v$ on $(X \times Y, \mathbf{S} \times \mathbf{T})$ such that $(\mu \times v)(E \times F) = \mu(E)v(F)$ for every measurable rectangle $E \times F$, and use this fact to prove the Tonelli theorem (Th. 9.3). (Hint: Use Problem C and Problem 8A.)

E. Complete the proof of the Fubini theorem (Th. 9.2). (Hint: Suppose f is measurable $[\mathbf{S} \times \mathbf{T}]$ and integrable $[\mu \times v]$. Apply the Tonelli theorem to $|f|$ to conclude that $f(x, y)$ is integrable $[v]$ as a function of the single variable y for almost every x. Write f as a linear combination of four nonnegative integrable functions, and use Problem C.)

F. Let ξ and ζ be complex measures on measurable spaces $(X, \mathbf{S})$ and $(Y, \mathbf{T})$, respectively.

(i) Show that if G is a set belonging to $\mathbf{S} \times \mathbf{T}$, then the function $\zeta(G_x)$ is measurable $[\mathbf{S}]$ on X, and likewise that the function $\xi(G^y)$ is measurable $[\mathbf{T}]$ on Y. Show also that if f is a measurable complex-valued function on $(X \times Y, \mathbf{S} \times \mathbf{T})$, then the set of all points x of X such that $\int_Y f(x, y)d\zeta(y)$ exists is a measurable subset of X and

$$g(x) = \int_Y f(x, y)d\zeta(y)$$

is a measurable function on that set. Similarly, $h(y) = \int_X f(x, y)d\xi(x)$ defines a measurable function on a measurable subset of Y. (Hint: Just as in Problem C, the collection of all sets G in $\mathbf{S} \times \mathbf{T}$ with the property that $\zeta(G_x)$ is a measurable function of x is a σ-quasiring containing all measurable rectangles. Since $f(x, y)$ is integrable $[\zeta]$ with respect to y if and only if $|f(x, y)|$ is integrable $[|\zeta|]$ with respect to y (Ch. 8, p. 148), the set on which this holds coincides with

$$A = \left\{ x \in X : \int_Y |f(x, y)|d|\zeta|(y) < +\infty \right\},$$

a measurable set by Problem C. If $\{s_n\}$ is a sequence of measurable simple functions on $X \times Y$ such that $s_n \to f$ and $|s_n| \leq |f|$ pointwise on X (Prob. 6S), then

$$\lim_n \int_Y s_n(x, y)d\zeta(y) = \int_Y f(x, y)d\zeta(y)$$

for every x in A by the dominated convergence theorem for complex measures (Prob. 8U).)

(ii) Verify that there exists a unique complex measure $\xi \times \zeta$ on $(X \times Y, \mathbf{S} \times \mathbf{T})$ satisfying the condition

$$(\xi \times \zeta)(E \times F) = \xi(E)\zeta(F) \tag{7}$$

on measurable rectangles (the measure $\xi \times \zeta$ is called the *product* of ξ and ζ), and show that the total variation $|\xi \times \zeta|$ of $\xi \times \zeta$ coincides with $|\xi| \times |\zeta|$. (Hint: If $G \in \mathbf{S} \times \mathbf{T}$ then $\zeta(G_x)$ is not only a measurable function of x, it is also bounded (by $|\zeta|(Y)$). Hence one may define $(\xi \times \zeta)(G) = \int_X \zeta(G_x)d\xi$, and it is clear that this set function satisfies (7). To show that $\xi \times \zeta$ is a complex measure, use the dominated convergence theorem. To obtain the identity $|\xi \times \zeta| = |\xi| \times |\zeta|$, verify it first for measurable rectangles and use Problem 8A.)

(iii) If f is a complex-valued function on $X \times Y$ that is integrable $[\xi \times \zeta]$, then the function $g(x) = \int_Y f(x, y)d\zeta(y)$ is a measurable and integrable function $[|\xi|]$, and

$$\int_X g\, d\xi = \int_X \left[\int_Y f(x, y)d\zeta(y)\right]d\xi(x) = \int_{X \times Y} f\, d(\xi \times \zeta).$$

Similarly, the iterated integral

$$\int_Y \left[\int_X f(x, y)d\xi(x)\right]d\zeta(y)$$

exists and is equal to the *product* (or *double*) integral $\int_{X \times Y} f\, d(\xi \times \zeta)$. (Hint: According to (ii) and Chapter 8 a function f is integrable $\xi \times \zeta$ if and only if $|f|$ is integrable $[|\xi| \times |\zeta|]$. Hence if a sequence $\{s_n\}$ of measurable simple functions is constructed as in (i), then the sequence $\{\int_Y s_n(x, y)d\zeta(y)\}$ converges to $\int_Y f(x, y)d\zeta(y)$ a.e. $[|\xi|]$ by the dominated convergence theorem. Two further similar applications of the dominated convergence theorem reduce the problem to the case of a single integrable simple function.)

G. Let $(X, \mathbf{S}, \mu)$ and $(Y, \mathbf{T}, \nu)$ be arbitrary measure spaces, and let $\mathbf{R}$ denote the ring of sets in $X \times Y$ generated by the measurable rectangles $E \times F$, so that the σ-ring $\mathbf{S} \times \mathbf{T}$ coincides with $\mathbf{S}(\mathbf{R})$.

(i) Verify that every set R in $\mathbf{R}$ can be expressed (not necessarily uniquely) as a disjoint union of a finite number of measurable rectangles.

(ii) Show that if $\{E_n \times F_n\}$ is any countable disjoint collection of measurable rectangles such that

$$E \times F = \bigcup_n (E_n \times F_n),$$

then

$$\mu(E)\nu(F) = \sum_n \mu(E_n)\nu(F_n).$$

(Hint: For any one fixed x in E the collection of rectangles $E_n \times F_n$ that meet the set $\{x\} \times F$ is exactly the same as the collection of rectangles $E_n \times F_n$ such that $x \in E_n$, and the corresponding collection of sets F_n forms a disjoint covering of F. Use this observation to show that

$$v(F)\chi_E(x) = \sum_n v(F_n)\chi_{E_n}(x), \qquad x \in X;$$

then apply the theorem of Beppo–Levi (Cor. 7.10).)

(iii) Conclude that there exists a unique finitely additive set function λ on $\mathbf{R}$ such that $\lambda(E \times F) = \mu(E)v(F)$ for every measurable rectangle $E \times F$, and that λ is, in fact, *countably* additive on $\mathbf{R}$. Conclude further that if λ^* denotes the outer measure on $X \times Y$ generated by λ (which is the same as the outer measure generated by the gauge $\mu(E)v(F)$ on the set of all measurable rectangles; cf. Problem 8D), then every set in $\mathbf{S} \times \mathbf{T}$ is measurable with respect to λ^*, so that $\hat{\lambda} = \lambda^* | (\mathbf{S} \times \mathbf{T})$ is a measure on $(X \times Y, \mathbf{S} \times \mathbf{T})$ with the property that

$$\hat{\lambda}(E \times F) = \mu(E)v(F), \qquad E \in \mathbf{S}, F \in \mathbf{T}. \tag{8}$$

(Hint: See Problem 8M.)

(iv) Show that a set A in $\mathbf{S} \times \mathbf{T}$ is σ-finite with respect to $\hat{\lambda}$ when and only when there exist sets E and F, σ-finite with respect to μ and v, respectively, such that $A \subset E \times F$. Use this fact to show that any measure on $(X \times Y, \mathbf{S} \times \mathbf{T})$ that satisfies condition (8) must agree with $\hat{\lambda}$ on every set that is σ-finite with respect to $\hat{\lambda}$. Conclude from this that any measure on $(X \times Y, \mathbf{S} \times \mathbf{T})$ that satisfies condition (8) is necessarily dominated (setwise) by $\hat{\lambda}$ and agrees with $\hat{\lambda}$ on any set in $\mathbf{S} \times \mathbf{T}$ that is not an infinite atom with respect to $\hat{\lambda}$. Conclude, finally, that if $\hat{\lambda}$ is locally finite, then $\hat{\lambda}$ is the *unique* measure on $\mathbf{S} \times \mathbf{T}$ satisfying condition (8).

H. Let $(X, \mathbf{S}, \mu)$ and $(Y, \mathbf{T}, v)$ be measure spaces, and let $\hat{\lambda}$ be the measure on $(X \times Y, \mathbf{S} \times \mathbf{T})$ introduced in the preceding problem.

(i) Verify that the Fubini theorem is valid for every function on $X \times Y$ that is measurable $[\mathbf{S} \times \mathbf{T}]$ and integrable $[\hat{\lambda}]$. (Hint: An integrable function always vanishes outside some σ-finite set; see Problem 7J.)

(ii) (Mukherjea [49]) Show that if $\hat{\lambda}$ is locally finite, then Tonelli's theorem is also valid for an arbitrary nonnegative function f on $X \times Y$ that is measurable $[\mathbf{S} \times \mathbf{T}]$, in the sense that if either of the iterated integrals

$$\int_X \left[\int_Y f(x, y)dv(y) \right] d\mu \quad \text{or} \quad \int_Y \left[\int_X f(x, y)d\mu(x) \right] dv$$

exists and is finite, then f is integrable $[\hat{\lambda}]$ on $X \times Y$. (Hint: It clearly suffices to prove that f vanishes outside some set that is σ-finite with respect to $\hat{\lambda}$. Thus one may assume that

$$\int_X \left[\int_Y f(x, y)dv(y) \right] d\mu = M < +\infty,$$

and also that $f(x, y) \geq k > 0$ on some set A such that $\hat{\lambda}(A) = +\infty$. If $\hat{\lambda}$ is locally finite then A contains a subset B such that $\hat{\lambda}(B)$ is finite and as large as desired; use part (i) to obtain a contradiction.)

The last two problems show that if (X, S, μ) and (Y, T, ν) are any two measure spaces such that the measure $\hat{\lambda}$ constructed in Problem G is locally finite, then both the Fubini and Tonelli theorems hold for measurable functions on $(X \times Y, S \times T, \hat{\lambda})$ without any further restriction on X or Y. It should not go unmentioned, however, that in this more general version of the Tonelli theorem it is a part of the hypothesis of the theorem that the function $g(x) = \int f(x, y)d\nu(y)$ be measurable $[\mu]$ (or that the function $h(y) = \int f(x, y)d\mu(x)$ be measurable $[\nu]$).

I. Let $(X, S) = (Y, T)$ be the unit interval $[0, 1]$ equipped with its σ-ring of Borel sets, so that $(X \times Y, S \times T)$ is the unit square Q equipped with its σ-ring $\mathbf{B}$ of Borel sets (see Example A). Let μ denote Lebesgue–Borel measure on (X, S) and let κ denote the counting measure on (Y, T). Let $\hat{\lambda}$ be the measure defined on $(Q, \mathbf{B})$ as in Problem G (with $\nu = \kappa$), and let ξ be the measure on $(Q, \mathbf{B})$ introduced in Example B. Verify that ξ satisfies (8), and that if A is an arbitrary Borel subset of the diagonal $\Delta = \{(t, t) : 0 \leq t \leq 1\}$, then $\xi(A) = 0$. Show also that $\hat{\lambda}(\Delta) = +\infty$ and that, in fact, Δ is an infinite atom for $\hat{\lambda}$, so that $\hat{\lambda}$ is not locally finite. (Note that in this construction the measure space X is finite, while Y is locally finite but not σ-finite.) (Hint: Recall Problem G(iv).)

J. (i) Suppose given an indexed family $\{(X_\gamma, S_\gamma)\}_{\gamma \in \Gamma}$ of measurable spaces. Form the product $\Pi = \prod_{\gamma \in \Gamma} X_\gamma$, and call a product $Z = \prod_{\gamma \in \Gamma} E_\gamma$ a *measurable cell* in Π if $E_\gamma \in S_\gamma$ for every γ and if, in addition (when the index family Γ is infinite), $E_\gamma = X_\gamma$ for all but a finite number of indices. If S denotes the σ-ring generated by the collection of all measurable cells, then (Π, S) is a measurable space known as the *product* of the given family $\{(X_\gamma, S_\gamma)\}$. Show that if the index set Γ is partitioned into nonempty subsets Γ_1 and Γ_2, if we write (Π_i, S_i) for the product of the indexed family $\{(X_\gamma, S_\gamma)\}_{\gamma \in \Gamma_i}$, $i = 1, 2$, and if we identify the set Π with $\Pi_1 \times \Pi_2$ in the usual manner, then (Π, S) is identified with the product of (Π_1, S_1) and (Π_2, S_2). Show further that if A is any fixed subset of Π_1, then $\{B \subset \Pi_2 : A \times B \in S\}$ is a σ-ring.

(ii) Let $\{(X_\gamma, S_\gamma, \mu_\gamma)\}_{\gamma \in \Gamma}$ be a *finite* indexed family of σ-finite measure spaces, and let (Π, S) denote the product of the corresponding family $\{(X_\gamma, S_\gamma)\}_{\gamma \in \Gamma}$ in the sense of (i). Show that there exists a unique measure μ on (Π, S) satisfying the condition

$$\mu\left(\prod_{\gamma \in \Gamma} E_\gamma\right) = \prod_{\gamma \in \Gamma} \mu_\gamma(E_\gamma), \qquad E_\gamma \in S_\gamma, \gamma \in \Gamma.$$

The measure μ is called the *product* of the measures μ_γ and is denoted by $\prod_{\gamma \in \Gamma} \mu_\gamma$. (When the index family is $\{1, \ldots, n\}$, we shall also write $\mu_1 \times \cdots \times \mu_n$ for the product measure.) Show further that if $\Gamma = \Gamma_1 \cup \Gamma_2$ is any partition of Γ into nonempty subsets, and if Π is identified with the product

$$\left(\prod_{\gamma \in \Gamma_1} X_\gamma\right) \times \left(\prod_{\gamma \in \Gamma_2} X_\gamma\right)$$

in the usual manner, then the product measure μ on $(\Pi, \mathbf{S})$ coincides with the measure

$$\left(\prod_{\gamma \in \Gamma_1} \mu_\gamma \right) \times \left(\prod_{\gamma \in \Gamma_2} \mu_\gamma \right).$$

The construction of product measures extends in a natural way to the product of infinite, even noncountably infinite, indexed families $\{(X_\gamma, \mathbf{S}_\gamma, \mu_\gamma)\}$ provided all but a finite number of the factors X satisfy the condition $\mu_\gamma(X_\gamma) = 1$ (see [30; §38]).

K. Let μ and ν be finite measures on the same measurable space $(X, \mathbf{S})$. Show that there exists a function f that is measurable [$\mathbf{S}$] such that $0 \le f \le 1$ on X and such that

$$\int_E (1 - f) d\nu = \int_E f \, d\mu, \qquad E \in \mathbf{S}. \tag{9}$$

Show also that if f is as in (9), and if $A = \{x \in X : f(x) < 1\}$ and $B = \{x \in X : f(x) = 1\}$, then $\mu(B) = 0$, while if $\mu(E) = 0$ for some measurable subset E of A, then $\nu(E) = 0$. Complete the proof of Proposition 9.8. (Hint: Observe that ν is absolutely continuous with respect to $\mu + \nu$, and also that if two finite measures ν_2 and $\tilde{\nu}_2$ are both singular with respect to μ, then there exists a single set C such that $\mu(C) = \nu_2(X \backslash C) = \tilde{\nu}_2(X \backslash C) = 0$.)

L. (i) If $(X, \mathbf{S}, \mu)$ is a σ-finite measure space, and if ζ is a complex measure on $(X, \mathbf{S})$, then ζ is said to be *absolutely continuous* with respect to μ (notation: $\zeta \ll \mu$) if $\mu(E) = 0$ implies $\zeta(E) = 0$. Show that ζ is absolutely continuous with respect to μ if and only if $|\zeta|$ is. Show also that if $\zeta \ll \mu$ then there exists an integrable function φ on X such that

$$\zeta(E) = \int_E \varphi \, d\mu, \qquad E \in \mathbf{S}. \tag{10}$$

The function φ, which is uniquely determined by (10) up to a set of measure zero with respect to μ, is called the *Radon–Nikodym derivative* of ζ with respect to μ, and is denoted by $d\zeta/d\mu$. (The complex measure ζ is also called the *indefinite integral* of φ with respect to μ.) Show that ζ is real-valued [a measure] if and only if $d\zeta/d\mu$ is real-valued [nonnegative] a.e. [μ]. Show finally that if a function g is integrable [ζ], then $\int g \, d\zeta = \int g(d\zeta/d\mu)d\mu$.

   (ii) Let $(X, \mathbf{S}, \mu)$ be a measure space, and let f be a measurable real-valued function on X. If either f^+ or f^- is integrable [μ] (even if the other fails to be), then the difference

$$\int_E f^+ \, d\mu - \int_E f^- \, d\mu$$

can be formed in $\mathbb{R}^\sharp$ for every set E in $\mathbf{S}$, and this set function is a signed measure ν_f on $(X, \mathbf{S})$ (called the *indefinite integral* of f with respect to μ) which is *absolutely continuous* with respect to μ (notation: $\nu_f \ll \mu$) in the sense that $\mu(E) = 0$ implies $\nu_f(E) = 0$. Show conversely that if μ is σ-finite and ν is a signed measure that is absolutely continuous with respect to μ, then there exists a measurable extended real-valued function f on X such that

$$v(E) = \int_E f^+ \, d\mu - \int_E f^- \, d\mu, \qquad E \in \mathbf{S}. \tag{11}$$

The function f, which is uniquely determined by (11) up to a set of measure zero with respect to μ, is called the *Radon–Nikodym derivative* of v with respect to μ, and is denoted by $dv/d\mu$. Show that if a function g is integrable $[v]$, then $\int g \, dv = \int gf \, d\mu$.

M. Let μ be a σ-finite signed measure on a measurable space $(X, \mathbf{S})$, and let ζ be a complex measure on $(X, \mathbf{S})$. Show that $\zeta \ll |\mu|$ if and only if there exists an integrable complex-valued function φ on X such that

$$\zeta(E) = \int_E \varphi \, d\mu, \qquad E \in \mathbf{S}, \tag{12}$$

and show that the function φ in (12) is uniquely determined up to a set of measure zero with respect to $|\mu|$. (Here again the complex measure ζ is called the *indefinite integral* of φ with respect to μ, and φ is the *Radon–Nikodym derivative* $d\zeta/d\mu$.) State and prove an analogous theorem for two complex measures. In both cases state and prove the appropriate counterpart of formula (4). (Hint: Use a Hahn decomposition of X with respect to μ.)

N. A complex-valued function α defined on a real interval $[a, b]$ is said to be *absolutely continuous* there if for every $\varepsilon > 0$ there exists a $\delta > 0$ such that if $\{(c_k, d_k)\}_{k=1}^p$ is any finite disjoint system of open subintervals of $[a, b]$ with $\sum_{k=1}^p (d_k - c_k) < \delta$, then $\sum_{k=1}^p |\alpha(d_k) - \alpha(c_k)| < \varepsilon$.

(i) Verify that an absolutely continuous function α on an interval $[a, b]$ is both continuous and of bounded variation there, and that α is absolutely continuous if and only if the associated function $L(t) = V(\alpha; a, t)$ is. (Hint: Show first that α is continuous if it is absolutely continuous, and recall (Ex. 4B) that L is continuous along with α.) Show further that if α is a complex-valued function of bounded variation on $[a, b]$, and if ζ_α denotes the Stieltjes–Borel measure on $[a, b]$ associated with α (Ex. 8K), then α is absolutely continuous on $[a, b]$ if and only if ζ_α is absolutely continuous with respect to Lebesgue–Borel measure on $[a, b]$. (Hint: Recall (Prob. 8W) that $|\zeta_\alpha|$ is the Stieltjes–Borel measure associated with the function L. Use Example F.)

(ii) Let α be a complex-valued function defined on an interval $[a, b]$. Show that α is absolutely continuous on $[a, b]$ if and only if there exists a Lebesgue integrable function β on $[a, b]$ such that

$$\alpha(t) - \alpha(a) = \int_a^t \beta(u)du, \qquad a \le t \le b, \tag{13}$$

and that, when this is the case, $\beta = \alpha'$ almost everywhere. (Hint: Let α and β be related as in (13) with $\alpha(a) = 0$ and $\beta \ge 0$. Set $\beta_n = \beta \wedge n$ and

$$\alpha_n(t) = \int_a^t \beta_n(u)du, \qquad a \le t \le b,$$

for every positive integer n. Then $\int_a^t \alpha_n'(u)du = \alpha_n(t)$, $a \le t \le b$, for all n (Prob. 8Y), and the sequence $\{\alpha_n(t)\}$ tends upward to $\alpha(t)$. Show that $\alpha_n' \le \alpha'$ for all n, and conclude that $\alpha(t) = \int_a^t \alpha'(u)du$ for $a \le t \le b$, and hence that $\beta - \alpha'$ is a Lebesgue integrable function whose indefinite integral vanishes identically on the σ-ring of Lebesgue measurable subsets of $[a, b]$.)

(iii) (Integration by parts) Prove that

$$\int_a^b \alpha(t)\beta'(t)dt = [\alpha(b)\beta(b) - \alpha(a)\beta(a)] - \int_a^b \beta(t)\alpha'(t)dt$$

for any two absolutely continuous functions α and β on the real interval $[a, b]$. (Hint: Verify that the product $\alpha\beta$ is absolutely continuous.)

O. If μ and ν are two measures on the same measurable space (X, S), let us say that ν is *accessible* with respect to μ if for every set E in S we have

$$\nu(E) = \sup\{\nu(F) : F \in S, F \subset E, \mu(F) < +\infty\}.$$

Thus a measure μ is accessible with respect to itself if and only if it is locally finite.

(i) Show that ν is accessible with respect to μ if and only if for each set E in S such that $\nu(E) > 0$ there exists a subset F of E such that $\nu(F) > 0$ and $\mu(F) < +\infty$. (Hint: Use the method of exhaustion.)

(ii) Let X be a set, let S denote the power class on X, and let κ be the counting measure on (X, S). Describe the measures on (X, S) that are accessible with respect to κ. (Whether there exists a set X and a finite measure ν on (X, S) such that ν is *not* accessible with respect to κ is a version of an old problem known as the *problem of the existence of a measurable cardinal*, a question of some consequence in the study of the foundations of mathematics.)

P. Let (X, S, μ) be a locally finite measure space, let f be an arbitrary measurable, nonnegative, extended real-valued function on X, and let ν_f denote the indefinite integral of f. Show that ν_f is accessible with respect to μ. Show also that ν_f is locally finite if and only if f is finite-valued a.e. $[\mu]$. Can the requirement that μ be locally finite be dropped?

Q. A measure space (X, S, μ) is said to be *localizable* if its measure ring $(\dot{S}, \dot{\mu})$ is complete as a lattice, that is, if given any subset $\dot{\mathscr{C}}$ of $\dot{S}$, the supremum $\sup \dot{\mathscr{C}}$ exists in $\dot{S}$. In terms of measurable sets this comes out as follows: Given an arbitrary subset $\mathscr{C}$ of S, there exists a set S in S such that

(i) If $C \in \mathscr{C}$, then $\mu(C \backslash S) = 0$,
(ii) If T is any other set in S with property (i), then $\mu(S \backslash T) = 0$.

(Such a set S will be called a *supremum* $[\mu]$ of $\mathscr{C}$.) Show that any finite measure space is localizable, and that an arbitrary direct sum of localizable spaces is localizable. (Hint: Use the method of exhaustion.)

R. Show that a localizable measure space (X, S, μ) is the direct sum of a locally finite subspace A and an infinite atom B. (Hint: Consider indexed collections $\{D_\gamma\}$ of measurable sets of positive finite measure with the property that $\mu(D_{\gamma_1} \cap D_{\gamma_2}) = 0$ for $\gamma_1 \ne \gamma_2$, and use Zorn's lemma.)

S. According to Problem Q and Problem 8R, an arbitrary direct sum of finite measure spaces is locally finite and localizable. (Thus, in particular, every σ-finite measure space is localizable.) Show, in the converse direction, that if $(X, \mathbf{S}, \mu)$ is a locally finite and localizable measure space, then $(X, \mathbf{S}, \mu)$ is *isomorphic* to a direct sum of finite measure spaces. (Hint: Using the definition of localizability and Zorn's lemma show that there exists a family $\{D_\gamma\}_{\gamma \in \Gamma}$ of pairwise almost disjoint measurable sets such that $0 < \mu(D_\gamma) < +\infty$ for every index γ and such that X is a supremum $[\mu]$ of the family $\{D_\gamma\}$. Show that if E is an arbitrary measurable subset of X, then E is a supremum $[\mu]$ of the family $\{E \cap D_\gamma\}$, and that if $\mu(E) > 0$, then $\mu(E \cap D_\gamma) > 0$ for at least one γ. Conclude that, in general, $\mu(E) = \sum_\gamma \mu(E \cap D_\gamma)$, and hence that $(X, \mathbf{S}, \mu)$ is isomorphic to the direct sum

$$\sum_{\gamma \in \Gamma} \oplus (D_\gamma, \mathbf{S}_{D_\gamma}, \mu | D_\gamma).)$$

T. The following is an outline of what is substantially the most general version of the Radon–Nikodym theorem for two ordinary measures. (The appropriately formulated versions of the Radon–Nikodym theorem suggested in Problems L and M continue to hold in the present context. Moreover, it is easy to see that appropriate versions of Propositions 9.7 and 9.8 are also valid.)

(i) Let μ and ν be measures on a measurable space $(X, \mathbf{S})$, and suppose that μ is locally finite and localizable and that ν is absolutely continuous with respect to μ. Show that there exists a partition of X into the union of two disjoint measurable sets A and B such that $\nu(E) \le \mu(E)$ for every measurable subset E of A, while $\nu(E) \ge \mu(E)$ for every measurable subset E of B. (Hint: Let us call a set N a ν-*set* if $\nu(E) \ge \mu(E)$ for every measurable subset E of N. Let B denote a supremum $[\mu]$ of the collection of all ν-sets, and set $A = X \backslash B$. If $\nu(E) > \mu(E)$ for some set $E \subset A$, then Theorem 8.5 may be applied to the signed measure $(\nu | E) - (\mu | E)$ to obtain a ν-set $N \subset A$ such that $\mu(N) > 0$, in contradiction to the definition of B. Thus $\nu(E) \le \mu(E)$ for every measurable subset E of A. To show that B itself is a ν-set use the method of exhaustion to show that if $E \subset B$ and $0 < \mu(E) < +\infty$, then E contains a ν-set N such that $\mu(E \backslash N) = 0$, and hence that E is itself a ν-set. Then invoke the local finiteness of μ.)

(ii) Let μ be a locally finite and localizable measure on the measurable space $(X, \mathbf{S})$, and let ν be a measure on $(X, \mathbf{S})$ that is absolutely continuous and accessible with respect to μ. Show that there exists a measurable, nonnegative, extended real-valued function f on X such that ν is the indefinite integral of f with respect to μ. Show that the function f (to be called the *Radon–Nikodym derivative* of ν with respect to μ and denoted by $d\nu/d\mu$) is unique up to a set of measure zero with respect to μ, and that if φ is any function that is either integrable $[\nu]$ or nonnegative and measurable $[\mathbf{S}]$, then $\int \varphi \, d\nu = \int \varphi f \, d\mu$. (Hint: Follow the same procedure used to prove Theorem 9.6 to construct f, using (i) instead of Theorem 8.5 to obtain the necessary sets A_r. The same argument used in proving Theorem 9.6 also shows that $\nu(E) = \int_E f \, d\mu$ whenever $\mu(E) < +\infty$. Use Problem P.)

U. Prove that the metric space $(\dot{\mathbf{S}}_{\mathscr{F}}, \rho)$ associated with an arbitrary measure space $(X, \mathbf{S}, \mu)$ is complete. (Hint: If $\{[E_n]\}$ is a sequence in $(\dot{\mathbf{S}}_{\mathscr{F}}, \rho)$ such that

$$\sum_n \rho([E_n], [E_{n+1}]) < +\infty,$$

then the sequence $\{\chi_{E_n}\}$ is convergent both in the mean and almost everywhere with respect to μ. See Example 7N and Problem 4E.)

V. Let $(X, \mathbf{S}, \mu)$ and $(Y, \mathbf{T}, v)$ be measure spaces, let $\mathscr{L}_\mu$ and $\mathscr{L}_v$ denote the linear spaces of functions integrable $[\mu]$ and $[v]$, respectively, and suppose φ is a strong isomorphism between $(X, \mathbf{S}, \mu)$ and $(Y, \mathbf{T}, v)$ as in Example J. Show that there exists a linear transformation $T_\varphi : \mathscr{L}_\mu \to \mathscr{L}_v$ such that

$$[T_\varphi f] = T_\Phi[f], \qquad f \in \mathscr{L}_\mu,$$

where Φ denotes the measure ring isomorphism of $(\dot{\mathbf{S}}, \dot\mu)$ onto $(\dot{\mathbf{T}}, \dot v)$ induced by the point mapping φ and T_Φ is the isomorphism of Theorem 9.10. (Thus the isomorphism T_Φ induced by a strong isomorphism can be *lifted* from $\dot{\mathscr{L}}_\mu$ to $\mathscr{L}_\mu$.)

W. Suppose given a finite measure space $(X, \mathbf{S}, \mu)$ with associated metric space $(\dot{\mathbf{S}}, \rho) = (\dot{\mathbf{S}}_{\mathscr{F}}, \rho)$, and let $\{\mathscr{P}_n\}_{n=1}^\infty$ be a sequence of partitions of X such that each $\mathscr{P}_n$ consists of a finite number of measurable sets and such that $\mathscr{P}_{n+1}$ refines $\mathscr{P}_n$ for each n. Suppose also that $\{E_n\}_{n=1}^\infty$ is a nested sequence of sets such that $E_n \in \mathscr{P}_n$ for all n and such that $\mu(\bigcap_n E_n) = \delta > 0$. Show that if F is a measurable subset of $\bigcap_n E_n$ such that $0 < \mu(F) < \delta$, and if ε denotes the smaller of the two numbers $\mu(F)$ and $\delta - \mu(F)$, then $\rho([F], [A]) \geq \varepsilon$ for every set A in the ring of sets $\mathbf{R}$ generated by all of the sets in the various partitions $\mathscr{P}_n$. Conclude that if μ is atom-free, if $\{\mathscr{P}_n\}$ is such a sequence of partitions, and if the ring $\mathbf{R}$ generated by all of the sets in the partitions has the property that $\dot{\mathbf{R}} = \{[A] \in \dot{\mathbf{S}} : A \in \mathbf{R}\}$ is dense in $(\dot{\mathbf{S}}, \rho)$, then

$$\lim \max\{\mu(E) : E \in \mathscr{P}_n\} = 0.$$

X. Let $(X, \mathbf{S}, \mu)$ be a measure space with $0 < \mu(X) < +\infty$, let $(\dot{\mathbf{S}}, \dot\mu)$ be its measure ring, and let $(Y, \mathbf{B}, v)$ denote the measure space consisting of the interval $[0, \mu(X)]$ equipped with Lebesgue–Borel measure. Suppose given a partition $\mathscr{P}$ of X into a finite number of measurable subsets, let $\mathbf{R}$ denote the ring of sets generated by $\mathscr{P}$, and let $\dot{\mathbf{R}} = \{[A] \in \dot{\mathbf{S}} : A \in \mathbf{R}\}$. Show that there exists a partition $\mathscr{Q} = \{t_i\}_{i=1}^N$ of $[0, \mu(X)]$ and a mapping φ of $\dot{\mathbf{R}}$ into $(\dot{\mathbf{B}}, \dot v)$ such that (i) the range of φ is the subset of $\dot{\mathbf{B}}$ consisting of equivalence classes $[B]$ of the elements B of the ring generated by the subintervals of $\mathscr{Q}$, (ii) φ preserves measures in the sense that $\dot v(\varphi([A])) = \dot\mu([A])$ for every A in $\mathbf{R}$, and (iii) φ preserves the Boolean operations $\vee$, $\wedge$ and $\setminus$. Show also that if $\mathscr{P}'$ is another such partition of X, and if $\mathscr{P}'$ refines $\mathscr{P}$, so that the ring $\mathbf{R}'$ generated by $\mathscr{P}'$ contains $\mathbf{R}$, then the corresponding partition $\mathscr{Q}'$ of $[0, \mu(X)]$ may be chosen to refine $\mathscr{Q}$ and the mapping φ' constructed so as to extend φ. (Hint: If $E_1, \ldots, E_N$ is any enumeration of the sets of positive measure in $\mathscr{P}$, then each element $[A]$ of $\dot{\mathbf{R}}$ has a unique representation as a union of certain of the sets E_i. Define $t_i = \mu(E_1) + \cdots + \mu(E_i)$, $i = 1, \ldots, N$. In dealing with the refinement $\mathscr{P}'$, take care in choosing an enumeration of the sets of positive measure in $\mathscr{P}'$ so that it fits nicely with the enumeration $E_1, \ldots, E_N$.)

Y. Show that every separable, locally finite measure space is isomorphic to the direct sum of two measure spaces, the first of which is an interval (possibly infinite) of the real line $\mathbb{R}$ equipped with Lebesgue–Borel measure, and the second of which is a purely atomic measure space consisting of a countable disjoint collection of finite atoms. (Hint: A separable, locally finite measure space is σ-finite; use the Halmos–von Neumann theorem (Th. 9.12) and Problem 8S.)

As it turns out, there exist *finite* measure spaces that are not separable. (Indeed, the product of a noncountable number of copies of the unit interval, each equipped with Lebesgue measure, fails to be separable.) Thus the converse of the implication stated in Problem Y is false.

Measure and topology

10

If X is an arbitrary topological space, then, as defined in Chapter 6, the σ-ring **B** of Borel sets in X is the σ-ring generated by the open [closed] subsets of X. There are important cases, however, in which **B** is generated by smaller collections of sets satisfying additional conditions.

Definition. A subset A of a topological space X is said to be σ-*compact* if there exists a countable covering of A consisting of compact subsets of X. In particular, the space X itself is σ-compact if it is a countable union of compact subsets.

Proposition 10.1. *If X is a σ-compact Hausdorff space, then the collection $\mathscr{K}$ of all compact subsets of X generates the σ-ring* **B** *of Borel subsets of X.*

PROOF. It suffices to show that every closed set F in X belongs to $\mathbf{S}(\mathscr{K})$. If $X = \bigcup_{n=1}^{\infty} K_n$, where each K_n is compact, then $F = \bigcup_{n=1}^{\infty} (F \cap K_n)$, and each set $F \cap K_n$ is compact. $\qquad\square$

Definition. If X is a locally compact Hausdorff space (cf. Problem 3V), a subset S of X is said to be *topologically bounded* if its closure S^- is compact. (Note that a topologically bounded subset of a locally compact metric space is a bounded set, but that the converse need not hold.)

Example A. Let U be an open set in $\mathbb{R}^n$ such that both U and its complement $\mathbb{R}^n \backslash U$ are nonempty. If we write $F_n = \{x \in \mathbb{R}^n : d(x, \mathbb{R}^n \backslash U) \geq 1/n\}$ for each positive integer n, then the sets

$$K_n = \{x = (x_1, \ldots, x_n) \in F_n : x_1^2 + \cdots + x_n^2 \leq n^2\}$$

are compact, nested, and have union U. This shows that every open subset of $\mathbb{R}^n$ is a σ-compact locally compact Hausdorff space in its relative topology. We also observe that, with the sequence $\{K_n\}$ defined in this way, the sets K_n° are nested, topologically bounded, open subsets of U such that $U = \bigcup_{n=1}^{\infty} K_n^{\circ}$.

Proposition 10.2. *If X is a σ-compact, locally compact, Hausdorff space, then the collection of all topologically bounded open sets in X generates the σ-ring $\mathbf{B}$ of Borel sets in X.*

PROOF. It suffices to show that every open subset U of X is a countable union of topologically bounded open subsets of X. We write $X = \bigcup_{n=1}^{\infty} K_n$, where each K_n is compact. According to Problem A, there exists for each positive integer n a topologically bounded open set V_n containing K_n, so that $X = \bigcup_{n=1}^{\infty} V_n$. But then $U = \bigcup_{n=1}^{\infty} (U \cap V_n)$, and each $U \cap V_n$ is a topologically bounded open set. $\qquad\square$

If μ is a measure on a measurable space $(X, \mathbf{B})$, where X is a topological space and $\mathbf{B}$ is its σ-ring of Borel sets, then it is natural to require that the values of μ on arbitrary Borel sets be related to the values of μ on the topologically important open sets and compact sets. Furthermore, it is sufficient for our purposes to deal with topological spaces that are locally compact Hausdorff spaces.

Definition. Let X be a locally compact Hausdorff space, and let $\mathbf{B}$ denote its σ-ring of Borel subsets. A *Borel measure* on X is a measure μ on the measurable space $(X, \mathbf{B})$ with the property that $\mu(K)$ is finite for every compact subset K of X. A Borel set E is said to be *outer regular* with respect to the Borel measure μ if

$$\mu(E) = \inf\{\mu(U) : U \supset E, U \text{ open}\},$$

and to be *inner regular* with respect to μ if

$$\mu(E) = \sup\{\mu(K) : K \subset E, K \text{ compact}\}.$$

A Borel set E is *regular* with respect to μ if E is both inner and outer regular with respect to μ. If every Borel set in X is regular with respect to μ, then μ is a *regular* Borel measure.

The following proposition, whose proof is left as an exercise (Problems F and G), is frequently useful in connection with the notion of regularity.

Proposition 10.3. *Let μ be a Borel measure on a locally compact Hausdorff space X, and let $\mathscr{C}$ be a collection of regular sets of finite measure with respect to μ. Then every set in the σ-ring $\mathbf{S}(\mathscr{C})$ generated by $\mathscr{C}$ is regular with respect to μ.*

This proposition has several useful consequences.

Corollary 10.4. *Let μ be a Borel measure on a locally compact Hausdorff space X, and let $\mathcal{K}$ denote the collection of all compact subsets of X. If every set K in $\mathcal{K}$ is outer regular with respect to μ, or if every topologically bounded open set in X is inner regular with respect to μ, then every set in the σ-ring $S(\mathcal{K})$ is regular with respect to μ. In particular, if X is σ-compact, then a Borel measure μ on X is regular if and only if every compact subset of X is outer regular with respect to μ (or, equivalently, if and only if every topologically bounded open subset of X is inner regular with respect to μ).*

PROOF. It is an immediate consequence of Problem A that the σ-ring generated by the collection of topologically bounded open sets coincides with $S(\mathcal{K})$. If X is σ-compact, this σ-ring coincides with the σ-ring of all Borel subsets of X (Prop. 10.1). □

Corollary 10.5. *If X is a locally compact Hausdorff space, and if $\mathcal{K}_0$ denotes the collection of all compact G_δ's in X, then the sets belonging to the σ-ring $S(\mathcal{K}_0)$ are all regular with respect to every Borel measure on X.*

PROOF. Every set K_0 in $\mathcal{K}_0$ can be written as the intersection of a decreasing sequence of open sets, each of which may be taken to be topologically bounded (Prob. A). Hence K_0 is outer regular by Problem 7F. □

Corollary 10.6. *If X is a locally compact Hausdorff space that is metrizable, then the sets belonging to the σ-ring $S(\mathcal{K})$ generated by the collection $\mathcal{K}$ of compact sets in X are all regular with respect to every Borel measure on X. If X is a locally compact Hausdorff space that satisfies the second axiom of countability, then every Borel measure on X is regular.*

PROOF. If X is metrizable, then every compact set in X is a G_δ (Prob. 6M). If X is a locally compact Hausdorff space satisfying the second axiom of countability, then X is metrizable and σ-compact (Prob. B). □

Example B. A Borel measure whose carrier is an arbitrary Borel subset of any of the familiar spaces $\mathbb{R}^n$ and $\mathbb{C}^n$ is regular. In particular, the Stieltjes-Borel measures introduced in Example 8E are all regular, as is arc-length measure on the unit circle Z (Ex. 8F).

The preceding remarks show that in order to find a Borel measure that is *not* regular one must be prepared to examine some fairly large locally compact Hausdorff spaces. Here is a standard example of a Borel measure that is not regular. (This example is due to Dieudonné; see [30, p. vii]. Another example is outlined in Problem L.)

Example C. Let W denote the space $W(\Omega)$ of all countable ordinal numbers in the order topology (Problem 1W and Example 3B). Then, for every

Borel set E in W, either E or $W \backslash E$ contains some closed unbounded subset of W (Prob. 6Q), and the collection $\mathbf{J}$ of those Borel sets E in W such that $W \backslash E$ contains a closed unbounded subset of W forms a σ-ideal in the σ-ring $\mathbf{B}$ of Borel subsets of W. Hence, if a denotes an arbitrary nonnegative extended real number, and if we define

$$\mu_a(E) = \begin{cases} 0, & E \in \mathbf{J}, \\ a, & E \in \mathbf{B} \backslash \mathbf{J}, \end{cases}$$

then μ_a is a measure on $(W, \mathbf{B})$ with the property that $\mu_a(K) = 0$ for every compact set K in W. Thus μ_a is a Borel measure on W, but μ_a is not regular unless $a = 0$ (in which case $\mu_a = 0$).

Whether a Borel measure is regular or not has profound consequences for the Lebesgue integral with respect to that measure.

Proposition 10.7. *Let μ be a regular Borel measure on a locally compact Hausdorff space X. Then for any Borel measurable function f on X that is integrable $[\mu]$ there exists a sequence $\{g_n\}$ of integrable continuous functions on X that converges to f in the mean.*

PROOF. It suffices to show that if f is integrable $[\mu]$, then there exists a continuous function g on X such that $\int |f - g| d\mu$ is as small as desired. To this end, suppose first that E is a Borel set such that $\mu(E) < +\infty$ and that ε is a positive number. Since μ is regular, there exist a compact set K and an open set U such that $K \subset E \subset U$ and such that $\mu(U \backslash K) < \varepsilon$. By an easy extension of Urysohn's lemma (Prob. 3V), there exists a continuous function g on X such that $\chi_K \le g \le \chi_U$, whence it clearly follows that $\int |g - \chi_E| d\mu \le \mu(U \backslash K) < \varepsilon$. From this observation we conclude at once that if s is an integrable simple function on X, then there exists a continuous function g on X such that $\int |g - s| d\mu < \varepsilon$. But if f is integrable $[\mu]$, then, as we have noted before (cf. the proof of Theorem 9.10), there are always integrable simple functions s such that $\int |s - f| d\mu < \varepsilon$, and the result follows. $\square$

Proposition 10.8. *Let μ and v be regular Borel measures on the same locally compact Hausdorff space X, and suppose the integrals with respect to μ and v agree on continuous functions; that is, suppose $\mathscr{L}_\mu$ and $\mathscr{L}_v$ contain the same set $\mathscr{C}$ of continuous functions on X and*

$$\int_X g\, d\mu = \int_X g\, dv, \qquad g \in \mathscr{C}.$$

Then $\mu = v$.

PROOF. Let K be a compact set in X, and let

$$U_1 \supset U_2 \supset \cdots \supset U_n \supset \cdots \supset K$$

be a nested sequence of topologically bounded open sets such that $\mu(U_n) \to \mu(K)$ and $\nu(U_n) \to \nu(K)$ (see Problem D). By Problem 3V there exists for each n a continuous function g_n such that $\chi_K \leq g_n \leq \chi_{U_n}$, and therefore such that

$$\mu(K) \leq \int g_n \, d\mu \leq \mu(U_n) \quad \text{and} \quad \nu(K) \leq \int g_n \, d\nu \leq \nu(U_n).$$

Thus the sequence $\{\int g_n \, d\mu = \int g_n \, d\nu\}$ converges to both $\mu(K)$ and $\nu(K)$. Hence $\mu(K) = \nu(K)$, and it follows at once that $\mu = \nu$ since both measures are regular. □

We close this chapter by deriving a preliminary version of an important representation theorem due to F. Riesz. To this end we introduce the following definition.

Definition. Let $\mathcal{E}$ be a self-conjugate linear space of complex-valued functions on a set X with the property that the real-valued functions in $\mathcal{E}$ form a function lattice (see Example 2H and Problem 1D). Then a linear functional φ on $\mathcal{E}$ is *positive* if $\varphi(f) \geq 0$ for every nonnegative function f in $\mathcal{E}$. Likewise, if $\mathcal{E}$ is a real linear space of real-valued functions that is a function lattice, then a linear functional φ on $\mathcal{E}$ is *positive* if $\varphi(f) \geq 0$ for every nonnegative function f in $\mathcal{E}$.

Example D. If (X, S, μ) is an arbitrary measure space, then the Lebesgue integral with respect to μ is a positive linear functional on the linear space $\mathcal{L}$ of all integrable functions on X (Prop. 7.1).

If $\mathcal{E}$ is a linear space of functions as specified in the foregoing definition, and if φ is a positive linear functional on $\mathcal{E}$, then $\varphi(f)$ is real whenever f is a real-valued function in $\mathcal{E}$ (in other words, φ is self-conjugate). If f and g are real functions in $\mathcal{E}$ such that $f \leq g$, then $\varphi(f) \leq \varphi(g)$. In particular, if $\mathcal{E}$ contains the constant functions, and if f is a function in $\mathcal{E}$ such that $0 \leq f \leq 1$, then $0 \leq \varphi(f) \leq \varphi(1)$. Hence if $\varphi(1) = 0$, then φ annihilates all of the bounded functions in $\mathcal{E}$.

Theorem 10.9. *Let φ be a positive linear functional on the linear space $\mathscr{C}(X)$ of continuous complex-valued functions on a compact Hausdorff space X. Then there exists a unique regular Borel measure μ on X such that*

$$\varphi(f) = \int_X f \, d\mu, \qquad f \in \mathscr{C}(X). \tag{1}$$

Example E. If X is a compact Hausdorff space and x_0 a point of X, then point-evaluation at x_0, that is, the linear functional mapping each function f in $\mathscr{C}(X)$ to its value $f(x_0)$, is a positive linear functional on $\mathscr{C}(X)$. As we already know (Ex. 7I), a measure satisfying condition (1) for this linear functional is the Dirac mass δ_{x_0} concentrated at the point x_0, and since Dirac masses are obviously regular as Borel measures, δ_{x_0} is in this case the measure referred to in Theorem 10.9.

We preface the proof of Theorem 10.9 with a sequence of lemmas.

Lemma 10.10. *Let φ_0 be a positive linear functional on $\mathscr{C}(X)$, where X is a compact Hausdorff space, and suppose in addition that $\varphi_0(1) = 1$. For each open set U in X let*

$$\rho(U) = \sup\{\varphi_0(f) : 0 \leq f \leq \chi_U, f \in \mathscr{C}(X)\}.$$

Then ρ is a countably subadditive gauge (Prob. 8D) on the collection $\mathscr{T}$ of all open sets in X. In particular, ρ is monotone increasing on $\mathscr{T}$.

PROOF. Clearly $\rho(\varnothing) = 0$. Let V be an open subset of X, let $\{U_n\}$ be a countable open covering of V, let ε be a positive number, and let f be a continuous function on X such that $0 \leq f \leq \chi_V$. We set $K = \{x \in X : f(x) \geq \varepsilon\}$. If f_ε denotes the function $f_\varepsilon = (f - \varepsilon) \vee 0$, then $0 \leq f_\varepsilon \leq \chi_K$ and $f \leq f_\varepsilon + \varepsilon$. Since K is compact, there exists a positive integer N such that

$$K \subset U_1 \cup \cdots \cup U_N,$$

and there exist corresponding nonnegative continuous functions $g_1, \ldots, g_N$ on X such that $g_i \leq \chi_{U_i}$, $i = 1, \ldots, N$, and such that if $g = g_1 + \cdots + g_N$, then $0 \leq g \leq 1$ on X, while $g(x) = 1$ for all x in K (this is just a partition of unity subordinate to $\{U_1, \ldots, U_N\}$; see Example 3G). Thus $f_\varepsilon = f_\varepsilon g = f_\varepsilon g_1 + \cdots + f_\varepsilon g_N$, whence it follows that $\varphi_0(f_\varepsilon) \leq \rho(U_1) + \cdots + \rho(U_N)$. Since $f \leq f_\varepsilon + \varepsilon$, we conclude that

$$\varphi_0(f) \leq \rho(U_1) + \cdots + \rho(U_N) + \varepsilon \leq \sum_{n=1}^{\infty} \rho(U_n) + \varepsilon.$$

By the definition of ρ this shows that $\rho(V) \leq \sum_{n=1}^{\infty} \rho(U_n) + \varepsilon$, and since ε is arbitrary, the result follows. $\qquad\square$

Lemma 10.11. *Let X, φ_0 and ρ be as in the preceding lemma, let U be an open set in X, and let ε be a positive number. If f is a continuous function on X such that $0 \leq f \leq \chi_U$ and $\varphi_0(f) > \rho(U) - \varepsilon$, and if K denotes the set $K = \{x \in X : f(x) \geq \varepsilon\}$, then $\rho(U \backslash K) \leq 2\varepsilon$.*

PROOF. Suppose g is a continuous function on X such that $0 \leq g \leq \chi_{U \backslash K}$. Then $f + g \leq \chi_U + \varepsilon$. If we set $h = (f + g - \varepsilon) \vee 0$, then $0 \leq h \leq \chi_U$ and $f + g \leq h + \varepsilon$, so we have $\varphi_0(f) + \varphi_0(g) \leq \varphi_0(h) + \varepsilon \leq \rho(U) + \varepsilon$. Hence

$$\varphi_0(g) \leq \rho(U) - \varphi_0(f) + \varepsilon < 2\varepsilon,$$

and the lemma follows. $\qquad\square$

Lemma 10.12. *Let X, φ_0 and ρ be as in Lemma 10.10, and let U and V be open sets in X. Then*

$$\rho(U) + \rho(V) \le \rho(U \cup V) + \rho(U \cap V).$$

PROOF. Let f and g be arbitrary continuous functions on X such that $0 \le f \le \chi_U$ and $0 \le g \le \chi_V$, and set

$$h = (f + g) \wedge 1 \quad \text{and} \quad k = f + g - h.$$

Then it is easy to verify that $0 \le h \le \chi_{U \cup V}$ and $0 \le k \le \chi_{U \cap V}$. Since $f + g = h + k$, we have

$$\varphi_0(f) + \varphi_0(g) = \varphi_0(h) + \varphi_0(k) \le \rho(U \cup V) + \rho(U \cap V).$$

and the lemma follows. □

Lemma 10.13. *Let X, φ_0 and ρ be as in Lemma 10.10, and let μ^* denote the outer measure on X generated by ρ (see Problem 8D for definitions). Then μ^* agrees with ρ on open sets, and*

$$\mu^*(V \setminus U) + \mu^*(V \cap U) = \mu^*(V)$$

for all open sets U and V.

PROOF. That μ^* agrees with ρ on open sets follows at once from Lemma 10.10. Let U and V be open sets, and let ε be any positive number. Let f be any continuous function on X such that $0 \le f \le \chi_U$ and $\varphi_0(f) > \rho(U) - \varepsilon$, and let $K = \{x \in X : f(x) \ge \varepsilon\}$. If we write $W = V \setminus K$, then $W \cup (U \cap V) = V$ and $W \cap (U \cap V) \subset U \setminus K$. Thus, applying Lemmas 10.11 and 10.12, we see that

$$\rho(W) + \rho(U \cap V) \le \rho(V) + \rho(U \setminus K) \le \rho(V) + 2\varepsilon.$$

Since $V \setminus U \subset W$, this implies that

$$\mu^*(V \setminus U) + \mu^*(V \cap U) \le \mu^*(V) + 2\varepsilon.$$

Since ε is arbitrary and μ^* is subadditive, the lemma follows. □

PROOF OF THEOREM 10.9. If $\varphi(1) = 0$, then $\varphi = 0$, and (1) is satisfied by $\mu = 0$. If $\varphi(1) = a_0 > 0$, then $\varphi_0 = \varphi/a_0$ is a positive linear functional on $\mathscr{C}(X)$ such that $\varphi_0(1) = 1$, and if μ_0 satisfies the appropriate modification of (1) for φ_0, then $\mu = a_0 \mu_0$ satisfies (1) for φ. Thus we may and do assume that φ satisfies the extra condition $\varphi(1) = 1$. Let ρ and μ^* be defined as in Lemmas 10.10 and 10.13, so that μ^* is an outer measure on X such that $0 \le \mu^* \le 1$ and such that if U is an open set in X then

$$\mu^*(U) = \rho(U) = \sup\{\varphi(f) : 0 \le f \le \chi_U, f \in \mathscr{C}(X)\}.$$

We first show that every open set is measurable with respect to μ^*. To this end, let U be an open subset of X and let A be an arbitrary subset of X. If

ε is a positive number, then by the definition of μ^* there exists a sequence $\{V_n\}_{n=1}^{\infty}$ of open sets in X such that

$$A \subset \bigcup_{n=1}^{\infty} V_n \quad \text{and} \quad \sum_{n=1}^{\infty} \rho(V_n) < \mu^*(A) + \varepsilon. \tag{2}$$

By the countable subadditivity of μ^* we have

$$\mu^*(A \cap U) \leq \sum_{n=1}^{\infty} \mu^*(V_n \cap U) \quad \text{and} \quad \mu^*(A \setminus U) \leq \sum_{n=1}^{\infty} \mu^*(V_n \setminus U).$$

But then, by Lemma 10.13 and (2), we obtain the inequality

$$\mu^*(A \cap U) + \mu^*(A \setminus U) \leq \mu^*(A) + \varepsilon,$$

and since ε is arbitrary, this implies that U is measurable with respect to μ^*. From this it follows, of course, that the σ-ring of sets measurable with respect to μ^* contains the σ-ring $\mathbf{B}$ of Borel sets in X (see Problem 8C), and hence that $\mu = \mu^* | \mathbf{B}$ is a Borel measure on X with the property that $\mu(U) = \rho(U)$ for every open set U. Since Lemma 10.11 clearly shows that the measure μ is regular, and since the uniqueness of μ was settled in Proposition 10.8, the proof will be complete if we verify (1) for the measure μ.

To do this, we note first that if f is a continuous function on X and V is an open set in X such that $f \geq \chi_V$, then $\varphi(f) \geq \varphi(g)$ for every continuous function g on X such that $0 \leq g \leq \chi_V$, and therefore $\varphi(f) \geq \rho(V) = \mu(V)$. Thus if U and V are open sets in X such that $V \subset U$, and if f is a continuous function on X such that $\chi_V \leq f \leq \chi_U$, then the numbers $\varphi(f)$ and $\int f \, d\mu$ are both straddled by the pair of numbers $\mu(V)$ and $\mu(U)$. To put this observation to use, suppose that f is a nonnegative function in $\mathscr{C}(X)$ and that a is a positive number such that the interval $[0, a]$ contains the range of f. Let $0 = t_0 < t_1 < \cdots < t_N = a$ be a partition of $[0, a]$. For each $i = 1, \ldots, N$, let us write $\Delta t_i = t_i - t_{i-1}$ and denote by f_i the continuous function

$$f_i = [(f \vee t_{i-1}) \wedge t_i] - t_{i-1} = [(f \wedge t_i) \vee t_{i-1}] - t_{i-1}.$$

Likewise, for $i = 0, \ldots, N$, let U_i denote the open set $\{x \in X : f(x) > t_i\}$. (Note that U_N is the empty set.) Then, on the one hand, we have

$$\Delta t_i \chi_{U_i} \leq f_i \leq \Delta t_i \chi_{U_{i-1}}, \quad i = 1, \ldots, N,$$

while, on the other hand, as is readily verified, $f = f_1 + \cdots + f_N$. Thus

$$\left| \varphi(f_i) - \int_X f_i \, d\mu \right| \leq \Delta t_i \mu(U_{i-1} \setminus U_i), \quad i = 1, \ldots, N,$$

and therefore

$$\left| \varphi(f) - \int_X f \, d\mu \right| \leq \left(\max_i \Delta t_i \right) \mu(U_0) \leq \max_i \Delta t_i.$$

Since the mesh of the partition may be taken to be as small as desired, this shows that (1) holds for all nonnegative functions f in $\mathscr{C}(X)$, and the theorem follows. $\square$

Thus far we have considered only Borel *measures*. Appropriate generalizations of this concept are also frequently useful.

Definition. Let X be a locally compact Hausdorff space, and let **B** denote, as usual, the σ-ring of Borel sets in X. A complex measure ξ on the measurable space $(X, \mathbf{B})$ is called a *complex Borel measure*. Similarly, a signed measure μ on $(X, \mathbf{B})$ is called a *signed Borel measure* provided $\mu(K)$ is finite for every compact set K in X.

Proposition 10.14. *Let X be a locally compact Hausdorff space. If μ is a signed Borel measure on X, then μ_+, μ_- and $|\mu|$ are Borel measures on X. If ξ is a complex Borel measure on X, then* Re ξ *and* Im ξ *are signed Borel measures on X, and $|\xi|$ is a finite Borel measure.*

PROOF. If μ is a signed Borel measure on X, and if K is a compact subset of X, then both $\mu_+(K)$ and $\mu_-(K)$ must be finite, since $\mu(K) = \mu_+(K) - \mu_-(K)$ is. But then $|\mu|(K) = \mu_+(K) + \mu_-(K) < +\infty$ (Prob. 8U). Hence μ_+, μ_- and $|\mu|$ are Borel measures on X. If ξ is a complex Borel measure on X, then $v_1 = \text{Re } \xi$ and $v_2 = \text{Im } \xi$ are clearly signed Borel measures such that both $v_1(X)$ and $v_2(X)$ are finite (since $\xi(X) = v_1(X) + iv_2(X)$). But then

$$|\xi|(X) \le |v_1|(X) + |v_2|(X) < +\infty. \qquad \square$$

Example F. Let X be a locally compact Hausdorff space, let ξ be a nonzero complex Borel measure on X, and let φ be a measurable complex-valued function defined on X. Then

$$f(\lambda) = \int_X \frac{d\xi(x)}{\varphi(x) - \lambda}$$

defines a locally analytic function on the complement U in $\mathbb{C}$ of the essential range of φ with respect to $|\xi|$ (Ch. 7, p. 126). Indeed, if $\lambda_0 \in U$ and if $D_R(\lambda_0)$ is a disc such that $Z = \varphi^{-1}(D_R(\lambda_0))$ has measure zero $[|\xi|]$, then for each $0 < r < R$ we have

$$\left| \frac{\lambda - \lambda_0}{\varphi(x) - \lambda_0} \right| < \frac{r}{R} < 1$$

for all λ in the disc $D_r(\lambda_0)$ and all x in $X \backslash Z$. Consequently the series

$$\sum_{n=0}^{\infty} \frac{(\lambda - \lambda_0)^n}{(\varphi(x) - \lambda_0)^{n+1}} \qquad (3)$$

converges uniformly on $X \backslash Z$ to the sum $1/(\varphi(x) - \lambda)$ for each λ in $D_r(\lambda_0)$, and we may therefore integrate the series (3) term by term, thus obtaining the power series expansion

$$\int_X \frac{d\xi(x)}{\varphi(x) - \lambda} = \sum_{n=0}^{\infty} \alpha_n (\lambda - \lambda_0)^n$$

on $D_r(\lambda_0)$, where $\alpha_n = \int_X d\xi(x)/(\varphi(x) - \lambda_0)^{n+1}$, $n \in \mathbb{N}_0$.

In particular, if ξ is a complex Borel measure on $\mathbb{C}$ itself, then

$$h(\lambda) = \int_{\mathbb{C}} \frac{d\xi(\zeta)}{\zeta - \lambda}$$

is a locally analytic function on the complement U of the support of ξ (see Problem P). The function h is usually called the *Cauchy transform* of the measure ξ.

Definition. A signed Borel measure μ on a locally compact Hausdorff space X is *regular* if its total variation $|\mu|$ is regular. Similarly, a complex Borel measure ξ on X is *regular* if $|\xi|$ is regular.

The proof of the following result is an easy exercise which we leave to the reader.

Proposition 10.15. *A signed Borel measure μ on a locally compact Hausdorff space X is regular if and only if its positive and negative variations μ_+ and μ_- are both regular. A complex Borel measure ξ on X is regular if and only if $\operatorname{Re} \xi$ and $\operatorname{Im} \xi$ are regular.*

Thus a regular complex Borel measure is simply a linear combination of (four) ordinary regular Borel measures. It should be noted that if ξ is a regular complex Borel measure on a locally compact Hausdorff space X, if E is a Borel set in X, and if ε is a positive number, then there exist a compact set K and an open set U in X such that $K \subset E \subset U$ and $|\xi|(U \backslash K) < \varepsilon$. In particular, $|\xi(E) - \xi(K)| = |\xi(E \backslash K)|$ and $|\xi(U) - \xi(E)| = |\xi(U \backslash E)|$ are both dominated by ε.

It follows from the above definition and Corollary 10.6 that all complex Borel measures and all signed Borel measures on a locally compact Hausdorff space satisfying the second axiom of countability are regular. Thus, in particular, a complex Borel measure on a locally compact subset of $\mathbb{R}^n$ or $\mathbb{C}^n$ is regular.

Example G. If α is an arbitrary complex-valued function of bounded variation on a real interval $[a, b]$ $(a \leq b)$, then the Stieltjes–Borel measure ζ_α associated with α (Ex. 8K) is a regular complex Borel measure on $[a, b]$. (Conversely, every complex Borel measure on $[a, b]$ is of this form; see Problem R.)

Finally we observe the following fact, an immediate consequence of Proposition 8.8.

Proposition 10.16. *Let X be a compact Hausdorff space and let ξ be a complex Borel measure on X. Then the mapping*

$$f \to \int_X f \, d\xi, \qquad f \in \mathscr{C}(X),$$

is a linear functional on the space $\mathscr{C}(X)$ of continuous complex-valued functions on X such that

$$\left| \int_X f \, d\xi \right| \le \int_X |f| \, d|\xi|, \qquad f \in \mathscr{C}(X).$$

PROBLEMS

A. Show that a closed subset F of a locally compact Hausdorff space X is compact if and only if there exists a topologically bounded open set U in X such that $F \subset U$.

B. Let X be a locally compact Hausdorff space satisfying the second axiom of countability, and let $\hat{X}$ denote the one-point compactification of X (Prob. 3W). Show that X is σ-compact, and use this fact to show that $\hat{X}$ also satisfies the second axiom of countability. Conclude that $\hat{X}$ and X are both metrizable. Finally, in the converse direction, show that a locally compact metrizable space X satisfies the second axiom of countability if and only if X is σ-compact.

C. (Berberian and Jakobsen [8].) Let X be a locally compact Hausdorff space, and let $\mathscr{H}$ denote the collection of all compact subsets of X. Show that the σ-ring $S(\mathscr{H})$ is a σ-ideal (Prob. 6I) in the σ-ring $\mathbf{B}$ of Borel subsets of X, and use this fact to conclude that $S(\mathscr{H})$ consists precisely of those Borel sets in X that are σ-compact. (Hint: Let $\mathbf{T}$ denote the collection of subsets T of X with the property that $E \in S(\mathscr{H})$ implies $E \cap T \in S(\mathscr{H})$, and show that $\mathbf{T}$ contains all closed subsets of X.)

D. Let μ be a Borel measure on a locally compact Hausdorff space X, and let E be a Borel set in X that is outer regular with respect to μ. Show that there exists a decreasing sequence $\{U_n\}$ of open sets in X such that $E \subset U_n$ for every n and such that $\mu(U_n) \downarrow \mu(E)$. Show also that if ν is another Borel measure on X and if E is also outer regular with respect to ν, then the sequence $\{U_n\}$ may be chosen so that $\nu(U_n) \downarrow \nu(E)$ as well. Formulate and prove the dual assertions for an inner regular Borel set.

E. If μ is a Borel measure on a locally compact Hausdorff space X, and if $\{E_n\}_{n=1}^{\infty}$ is a sequence of inner regular sets of finite measure with respect to μ, then $\bigcap_{n=1}^{\infty} E_n$ is inner regular with respect to μ. On the other hand, if $\{F_n\}_{n=1}^{\infty}$ is a *decreasing* sequence of outer regular sets of finite measure with respect to μ, then $\bigcap_{n=1}^{\infty} F_n$ is also outer regular with respect to μ.

F. Let X be a locally compact Hausdorff space, and let μ be a Borel measure on X.
   (i) Verify that a Borel set E in X is regular and of finite measure with respect to μ when and only when, for arbitrarily given $\varepsilon > 0$, there exist a compact set K and an open set U such that $K \subset E \subset U$ and such that $\mu(U \backslash K) < \varepsilon$.
   (ii) Show that if E and F are regular Borel sets of finite measure with respect to μ, then $E \backslash F$ is also regular (and of finite measure) with respect to μ.

(iii) Show similarly that if each set E_n of a countable collection $\{E_n\}$ of Borel sets is regular with respect to μ, then the union $\bigcup_n E_n$ is also regular with respect to μ. (Hint: Sets of infinite measure are trivially outer regular.)

(iv) Conclude that the collection **R** of all regular Borel sets of finite measure with respect to μ is a ring of sets, and that **R** is, in fact, a σ-ring if $\mu(X) < +\infty$. (In particular, then, this is true if X is compact.)

G. Provide a proof for Proposition 10.3. (Hint: Use the results of the preceding problem to show that if E denotes any one set in $\mathscr{C}$, then the collection of all Borel sets M such that $M \subset E$ and M is regular with respect to μ is a σ-ring containing the trace $\mathscr{C}_E$ and therefore containing the σ-ring $S(\mathscr{C}_E) = (S(\mathscr{C}))_E$ (Prob. 6D). Recall that every set in $S(\mathscr{C})$ is contained in the union of some countable collection of sets in $\mathscr{C}$ (Prob. 6I).)

H. Derive Proposition 10.8 from Proposition 10.7. (Hint: See Example 7N.)

I. Let ξ be a regular complex Borel measure on a compact Hausdorff space X. Prove that the total variation of ξ is given by

$$\sup \left| \int_X f \, d\xi \right|,$$

where the supremum is taken over the collection of all continuous functions f on X such that $|f| \leq 1$. (The reader should compare this result with Problem 8V.) (Hint: Use a partition of unity; see Example 3G.)

J. Suppose we set $a = 1$ in Example C. Determine which Borel subsets of W are regular with respect to the measure μ_1 defined there, and which are not. Show that the Borel sets that are regular with respect to μ_1 constitute a σ-ring.

K. Let $\{X_\gamma\}_{\gamma \in \Gamma}$ be an indexed family of pairwise disjoint topological spaces, and let X denote the union $X = \bigcup_\gamma X_\gamma$. Verify that there exists a unique topology on X with respect to which each space X_γ is an open subspace of X, and that a subset U of X is open in this topology if and only if $U \cap X_\gamma$ is open in X_γ for each index γ. The space X equipped with this topology is called the (*topological*) *direct sum* of the given family $\{X_\gamma\}$. (If each space X_γ is homeomorphic to some one fixed space Y, and if card $\Gamma = c$, then X is sometimes said to be the direct sum of c *copies of Y indexed by* Γ. It is clear that the resulting topological space is uniquely determined up to homeomorphism by Y and c.) Verify that the direct sum X is a locally compact Hausdorff space [a metrizable space] when and only when each of the subspaces X_γ is a locally compact Hausdorff space [a metrizable space]. Show further that if $\mathbf{B}_\gamma$ denotes the σ-ring of Borel subsets of X_γ for each index γ and $(X, \mathbf{S})$ denotes the direct sum of the family of measurable spaces $\{(X_\gamma, \mathbf{B}_\gamma)\}$, then $\mathbf{S} \supset \mathbf{B}$ where $\mathbf{B}$ denotes the σ-ring of Borel subsets of X. Show also that if μ_γ is a Borel measure on X_γ for each index γ, and if $\mu = \sum_\gamma \oplus \mu_\gamma$, then the restriction μ_0 of μ to $\mathbf{B}$ is a Borel measure on X (cf. Problem 7M). Conclude that an arbitrary countable direct sum of Borel measures is a Borel measure. (For a discussion of direct sums of measures, see Chapter 8.)

L. Let Γ be an index family with card $\Gamma = c$, where $c > \aleph_0$, and let X be the topological direct sum of an indexed family $\{X_\gamma\}_{\gamma \in \Gamma}$ of open subspaces, each of which is homeomorphic to the unit interval $[0, 1]$ via a homeomorphism φ_γ. (In other words, X is the direct sum of c copies of $[0, 1]$ indexed by Γ.) For each index γ let μ_γ denote the Borel measure on X_γ corresponding to Lebesgue-Borel measure on $[0, 1]$ under the mapping φ_γ, and let μ denote the direct sum of the family of measures $\{\mu_\gamma\}$. According to Problem K the space X is a metrizable locally compact Hausdorff space, and the restriction μ_0 of μ to the σ-ring of Borel subsets of X is a Borel measure on X. Determine which Borel sets in X are regular with respect to μ_0 and which are not. (Hint: Every open subset of X that meets uncountably many of the subspaces X_γ has infinite measure.) Find a closed set F and an open set U in X, both of which are regular with respect to μ_0 such that $U \subset F$ but such that $F \backslash U$ fails to be regular. Thus the regular sets with respect to a Borel measure need not form a ring, or even a quasiring (Prob. 6J).

> This construction, which is essentially due to S. K. Berberian [7], is really much more illuminating than the one given in Example C. Here we have a measure space that is simple and natural in every way, except of course that it is quite large, and yet there are many Borel sets that fail to be regular. The message is clear: For very large locally compact Hausdorff spaces (even metrizable ones), the requirement of regularity for a Borel measure, as we have defined it, is not only quite restrictive, it may even be unnatural. The following six problems introduce a notion that is less restrictive than regularity, but is still strong enough to yield useful results.

M. A Borel measure μ on a locally compact Hausdorff space X will be said to be *semiregular* if every Borel set in X is *inner* regular with respect to μ. Verify that if μ is a semiregular Borel measure on X, then every Borel subset of X that is contained in some topologically bounded open set is regular with respect to μ, and use this fact to conclude that every σ-compact Borel subset of X is regular with respect to μ. Thus a semiregular Borel measure on X is automatically regular if X is σ-compact. Show also that the restriction to Borel sets of an arbitrary direct sum of semiregular Borel measures is semiregular. (Thus the restriction to Borel sets of a direct sum of regular Borel measures is always at least semiregular; in particular, the measure μ_0 in Problem L is semiregular.) Show, finally, that if E is an arbitrary Borel set of finite [σ-finite] measure with respect to a semiregular measure μ, then E can be expressed (not necessarily uniquely) as the union of a null set and a σ-compact set.

N. Verify that a Borel measure μ on a locally compact Hausdorff space X is semiregular if and only if for each Borel set E in X there exists a Borel set F that is regular with respect to μ such that $F \subset E$ and $\mu(F) = \mu(E)$.

O. (i) Verify Proposition 10.7 for a semiregular Borel measure μ. (Hint: According to the last part of Problem M, if E is a Borel set in X such that $\mu(E) < +\infty$, then there exists a regular Borel set $F \subset E$ such that $\mu(E \backslash F) = 0$).
(ii) Verify Proposition 10.8 for semiregular measures μ and ν.

P. Let μ be a semiregular Borel measure on a locally compact Hausdorff space X. Show that for each Borel set E in X there exists a unique smallest closed set F such that $\mu(E \backslash F) = 0$. (In other words, the set F is closed, $\mu(E \backslash F) = 0$, and if F_1 is any other closed set such that $\mu(E \backslash F_1) = 0$, then $F \subset F_1$.) The set F is called the *kernel* of E with respect to μ. The kernel of the entire space X is also frequently called the *support* of μ. Show that the kernel of an atom with respect to a semiregular Borel measure μ is necessarily a singleton.

Q. (Lusin's Lemma) Let μ be a semiregular Borel measure on a locally compact Hausdorff space X, let E be a Borel subset of X that is σ-finite with respect to μ, let f be a complex-valued function that is defined and Borel measurable on E, and let ε be a positive number. Show that there exists a Borel subset G of E such that $\mu(E \setminus G) < \varepsilon$ and such that $f\,|\,G$ is continuous on the subspace G. Show also that G can always be taken to be the union of a countable collection of compact sets, and that G can even be taken to be compact when $\mu(E) < +\infty$. (Hint: First take care of the case of a Borel measurable simple function on a set of finite measure, then the case of a bounded Borel measurable function on a set of finite measure, recalling that such a function is the uniform limit of a sequence of simple functions (Prob. 6S). Finally, to take care of the general case, construct an increasing sequence $\{U_n\}_{n=1}^{\infty}$ of topologically bounded open sets such that E is almost covered by $\{U_n\}$, and set $E_n = E \cap (U_n \setminus U_{n-1})$, $n \in \mathbb{N}$, where U_0 is taken to be the empty set.)

R. Let X be a locally compact Hausdorff space, and let $\mathbf{L}$ denote the ring of all topologically bounded Borel sets in X. A countably additive, nonnegative, real-valued set function μ_0 defined on $\mathbf{L}$ will be called a *local Borel measure* on X. Similarly, a countably additive complex-valued set function defined on $\mathbf{L}$ is called a *local complex Borel measure* on X.

(i) If ξ_0 is a local complex Borel measure, and if K is a compact subset of X, then the restriction ξ_K of ξ_0 to K (i.e., the restriction of ξ_0 to the Borel subsets of K) is a complex Borel measure on K, and the system $\{\xi_K\}$ of Borel measures thus obtained is *coherent* in the sense that if K and L are two compact subsets of X, and if E is a Borel set such that $E \subset K \cap L$, then $\xi_K(E) = \xi_L(E)$. Show, conversely, that if such a coherent system $\{\xi_K\}$ of complex Borel measures is given, then there exists a unique local complex Borel measure ξ_0 on X such that ξ_K coincides with the restriction of ξ_0 to K for every compact set K in X.

(ii) By the *support* of a continuous complex-valued function f on X is meant the closure of the set $\{x \in X : f(x) \neq 0\}$. (If there ever is any chance of confusion between this concept of support and the one introduced in Chapter 7 for a measurable function defined on a measure space, we shall refer to the present notion as the *closed support*.) Let $\mathscr{C}_0(X)$ denote the collection of all continuous complex-valued functions on X having compact support. Verify that $\mathscr{C}_0(X)$ is a linear space. Show also that if ξ_0 is a local complex Borel measure on X, then there exists a unique linear functional φ on $\mathscr{C}_0(X)$ satisfying the condition

$$\varphi(f) = \int_K f\, d\xi_K, \qquad f \in \mathscr{C}_0(X),$$

where K denotes an arbitrary compact subset of X that contains the support of f.

(iii) Let us take for X an open interval (a, b) $(-\infty \le a < b \le +\infty)$ of the (extended) real line. Show that if ζ_0 is an arbitrary local complex Borel measure on (a, b), then there exists a function α on (a, b) (unique up to an additive constant) such that α is right-continuous and of bounded variation on each (bounded) closed subinterval of (a, b) (Prob. 1I), and such that $\zeta_{[c, d]}$ coincides on each such interval $[c, d]$ with the Stieltjes-Borel measure ζ_α associated with $\alpha\,|\,[c, d]$ (Ex. 8K). Show conversely that if α is a function on (a, b) that is *locally* of bounded variation there, i.e., has the property that it is of bounded variation on each compact subinterval of (a, b), then there exists a unique local complex Borel measure on (a, b) that coincides with ζ_α on every compact subinterval of (a, b).

Thus the local complex Borel measures on a real interval may be satisfactorily identified as deriving from those functions on the interval that are locally of bounded variation there.

(iv) Let μ_0 be a local Borel measure on a locally compact Hausdorff space X. Show that there exists a unique semiregular Borel measure μ on X that extends μ_0.

S. Let X be a locally compact Hausdorff space, and let $\mathscr{C}_0(X)$ denote the linear space of continuous complex-valued functions on X having compact support, as defined in the preceding problem. Show that if φ is a positive linear functional on $\mathscr{C}_0(X)$, then there exists a unique semiregular Borel measure μ on X such that

$$\varphi(f) = \int f \, d\mu, \qquad f \in \mathscr{C}_0(X).$$

Thus Theorem 10.9 is valid, more generally, on locally compact Hausdorff spaces if we replace $\mathscr{C}(X)$ by $\mathscr{C}_0(X)$ and admit semiregular measures in place of regular ones.

The construction in Problem L shows that the measure μ in Problem S need not be regular in general.

T. If X is an arbitrary topological space, then the σ-ring $\mathbf{B}_0$ of subsets of X generated by the collection of all those closed sets F such that F is a G_δ in X is the σ-ring of *Baire sets* in X (or *Baire subsets* of X). Note that $\mathbf{B}_0 \subset \mathbf{B}$, where $\mathbf{B}$ denotes the σ-ring of Borel subsets of X. Verify that $\mathbf{B}_0$ turns X into a measurable space in such a way that every continuous complex-valued function on X is measurable [$\mathbf{B}_0$]. Verify also that the σ-ring $\mathbf{B}_0$ is generated by the collection of all open F_σ's in X. Show, finally, that $\mathbf{B}_0 = \mathbf{B}$ when the space X is metrizable (cf. Example 6F).

U. (i) If $\mathscr{C}$ is a collection of complex-valued functions on an arbitrary set X, then we say that $\mathscr{C}$ is *closed in the sense of Baire* if every function f that is the pointwise limit of a sequence $\{f_n\}$ of functions belonging to $\mathscr{C}$ is itself in $\mathscr{C}$. (Thus, according to Theorem 6.4, the collection of all measurable complex-valued functions on an arbitrary measurable space $(X, \mathbf{S})$ is closed in the sense of Baire.) Show that for an arbitrary collection of functions $\mathscr{C}$ on X there exists a unique smallest collection $\tilde{\mathscr{C}}$ that contains $\mathscr{C}$ and is closed in the sense of Baire. (The collection $\tilde{\mathscr{C}}$ will be referred to as the *Baire hull* of $\mathscr{C}$.) Verify that if $\mathscr{C}$ is a linear space [algebra] of functions, then $\tilde{\mathscr{C}}$ is also a linear space [algebra].

(ii) For an arbitrary collection $\mathscr{C}$ of functions on a set X denote by $\mathscr{C}_\lambda$ the collection of all functions g such that g is the pointwise limit of some sequence $\{f_n\}$ in $\mathscr{C}$. (Thus $\mathscr{C} = \mathscr{C}_\lambda$ when and only when $\mathscr{C}$ is closed in the sense of Baire.) Consider the following construction. Set $\mathscr{C}^{(0)} = \mathscr{C}$, and suppose $\mathscr{C}^{(\xi)}$ already defined for every ordinal number ξ less than some countable ordinal number α. Then, if α is not a limit ordinal, we set $\mathscr{C}^{(\alpha)} = (\mathscr{C}^{(\xi)})_\lambda$, where $\alpha = \xi + 1$, while if α is a limit ordinal, we set $\mathscr{C}^{(\alpha)} = \bigcup_{\xi < \alpha} \mathscr{C}^{(\xi)}$. (See Problem 1V for a discussion of such transfinite definitions.) Show that if Ω denotes, as usual, the smallest noncountable ordinal number, then

$$\mathscr{C}^{(\Omega)} = \bigcup_{\xi < \Omega} \mathscr{C}^{(\xi)}$$

coincides with the Baire hull of $\mathscr{C}$.

V. Let X be a locally compact Hausdorff space. Verify in general that the Baire hull $\widetilde{\mathscr{C}}(X)$ of the collection $\mathscr{C}(X)$ of continuous functions on X is contained in the collection $\mathscr{B}_0$ of Baire measurable functions on X. Show also that if X is either metrizable or σ-compact, then $\widetilde{\mathscr{C}}(X)$ and $\mathscr{B}_0$ coincide. (Hint: One way is easy; the Baire measurable functions form a class that contains $\mathscr{C}(X)$ and is closed in the sense of Baire. To go the other way, it suffices to show that the characteristic function of each Baire set belongs to $\widetilde{\mathscr{C}}(X)$. Show that the sets whose characteristic functions belong to $\widetilde{\mathscr{C}}(X)$ form a σ-ring.)

W. Let X be a locally compact Hausdorff space, let $\mathscr{C}$ denote a collection of Borel subsets of X, and let λ be a nonnegative extended real-valued set function defined on $\mathscr{C}$. A set E in $\mathscr{C}$ will be said to be *regular* with respect to λ and $\mathscr{C}$ if

$$\lambda(E) = \sup_{K \subset E} \lambda(K) = \inf_{E \subset V} \lambda(V),$$

where the supremum is taken over just those compact sets K *in* $\mathscr{C}$ that are contained in E, and the infimum is taken over just the open sets V *in* $\mathscr{C}$ that contain E. In the same vein we say that λ is *regular on* $\mathscr{C}$ if every set in $\mathscr{C}$ is regular with respect to λ and $\mathscr{C}$. Show that if $\mathscr{K}_0$ denotes the collection of all those compact subsets of X that are G_δ's, and if λ is an arbitrary nonnegative, extended real-valued, countably additive set function defined on $\mathbf{S}(\mathscr{K}_0)$ with the property that $\lambda(K) < +\infty$ for every K in $\mathscr{K}_0$, then λ is automatically regular on $\mathbf{S}(\mathscr{K}_0)$. (Hint: Show that the collection of sets that are regular with respect to λ and $\mathbf{S}(\mathscr{K}_0)$ is a σ-ring.)

> A set function such as λ is sometimes called a *Baire measure* on X, and in this terminology the foregoing result may be paraphrased by saying that every Baire measure on $\mathbf{S}(\mathscr{K}_0)$ is regular. This usage will not be employed in this book, however, since $(X, \mathbf{S}(\mathscr{K}_0))$ is not a measurable space except when X is σ-compact.

X. Let X be a locally compact Hausdorff space, let $\mathscr{K}_0$ denote the collection of compact G_δ's in X, and let λ be a countably additive, nonnegative, extended real-valued set function defined on $\mathbf{S}(\mathscr{K}_0)$ with the property that λ is finite on all of the sets in $\mathscr{K}_0$. Show that there exists a unique semiregular Borel measure on X that agrees with λ on $\mathbf{S}(\mathscr{K}_0)$.

PART II
BANACH SPACES

Normed linear spaces **11**

We here commence our study of the fundamentals of the branch of mathematics known as functional analysis. In the course of this study and, indeed, throughout the balance of this book, we shall constantly be dealing with various linear spaces that are at the same time equipped with a topology. Frequently the topology will be induced by a metric. In this connection the following concept is useful.

Definition. A *value* on a (real or complex) linear space $\mathscr{E}$ is a nonnegative real-valued function v on $\mathscr{E}$ having the following properties:

  (i) $v(x) = 0$ if and only if $x = 0$,
  (ii) $v(-x) = v(x)$ for all x in $\mathscr{E}$,
  (iii) $v(x + y) \le v(x) + v(y)$ for all x, y in $\mathscr{E}$.
  (The inequality (iii) is known as the *triangle inequality* for v.)

It follows from properties (ii) and (iii) in the above definition that

$$v(x - y) = v(y - x) \quad \text{and} \quad v(x - z) \le v(x - y) + v(y - z)$$

for all vectors x, y, z in a linear space $\mathscr{E}$ equipped with a value v. Hence the function $\rho(x, y) = v(x - y)$ is a metric on $\mathscr{E}$, called the metric *defined* by v. This metric is *invariant* in the sense that $\rho(x + z, y + z) = \rho(x, y)$ for all vectors x, y, z. In the converse direction we have the following elementary proposition.

Proposition 11.1. *If ρ is an invariant metric on a (real or complex) linear space $\mathscr{E}$, then*

$$v(x) = \rho(x, 0), \tag{1}$$

defines a value on $\mathscr{E}$, and the given metric ρ is the metric defined by v.

PROOF. Suppose ρ is a given invariant metric on $\mathscr{E}$, and let v be defined as in (1). Then $v(x) > 0$ for all $x \neq 0$, and $v(-x) = \rho(-x, 0) = \rho(0, x) = \rho(x, 0) = v(x)$ for all x in $\mathscr{E}$, because ρ is invariant and symmetric. Likewise, if x and y are any two vectors in $\mathscr{E}$, then

$$v(x + y) = \rho(x + y, 0) \leq \rho(x + y, y) + \rho(y, 0) = v(x) + v(y).$$

This shows that v is a value on $\mathscr{E}$. That the metric defined by v on $\mathscr{E}$ coincides with ρ follows at once from (1). $\qquad\square$

The topology induced by the metric defined by a value v on a (real or complex) linear space $\mathscr{E}$ will also be said to be *induced* by v. The following elementary but important results concern this topology on $\mathscr{E}$ and the corresponding product topology on $\mathscr{E} \times \mathscr{E}$. (The reader must take care to distinguish between this notion of induced topology and the *inversely* induced topology introduced in Chapter 3 (see page 43).)

Proposition 11.2. *Let $\mathscr{E}$ be a real or complex linear space and let v be a value on $\mathscr{E}$. Then $(x, y) \to x + y$ is a continuous $\mathscr{E}$-valued mapping on $\mathscr{E} \times \mathscr{E}$. Likewise, the real-valued function $x \to v(x)$ is continuous on $\mathscr{E}$.*

PROOF. The proposition results immediately from the following inequalities, which are themselves easy consequences of the defining properties of a value:

$$\rho(x_1 + y_1, x_2 + y_2) = v((x_1 + y_1) - (x_2 + y_2)) \leq v(x_1 - x_2) + v(y_1 - y_2),$$

$$|v(x_1) - v(x_2)| \leq v(x_1 - x_2). \qquad\square$$

Corollary 11.3. *If $\mathscr{E}$ is any linear space equipped with a value v, then the mapping $x \to x + x_0$ (translation by x_0) is a homeomorphism of $\mathscr{E}$ onto itself in the topology induced on $\mathscr{E}$ by v.*

We shall be principally interested in a special type of value satisfying a very strong homogeneity requirement.

Definition. A *norm* on a linear space $\mathscr{E}$ is a nonnegative real-valued function $\| \ \|$ on $\mathscr{E}$ having the following properties:

(i) $\|x\| = 0$ if and only if $x = 0$,
(ii′) $\|\lambda x\| = |\lambda| \|x\|$ for all λ in $\mathbb{C}$ and x in $\mathscr{E}$,
(iii) $\|x + y\| \leq \|x\| + \|y\|$ for all x, y in $\mathscr{E}$.

A linear space possessing a norm is called a *normed linear space* or, more simply, a *normed space*. If $\mathscr{E}$ is a real linear space, then a *norm* on $\mathscr{E}$ is a nonnegative function $\| \ \|$ on $\mathscr{E}$ that satisfies (i) and (iii), and (ii′) for all real scalars. A real linear space possessing a norm is a *real normed space*. A complex linear space becomes a real linear space if one refuses to multiply by any but real scalars, and it is evident that a complex normed space, when regarded as a real linear space, is a real normed space with respect to the given norm. (According to the agreement adopted at the beginning of the book (see Chapter 2 or the preface), a normed space is understood to be complex if nothing is said concerning its scalar field.)

A norm is clearly a value, so the foregoing observations concerning values apply equally well to norms. In particular, if $\mathscr{E}$ is a normed space, then $\rho(x, y) = \|x - y\|$, $x, y \in \mathscr{E}$, defines an invariant metric on $\mathscr{E}$ with respect to which $\mathscr{E}$ becomes a metric space, and consequently a topological space as well. Whenever, in the sequel, a normed space is regarded as a metric space, it is the metric $\|x - y\|$ that is understood, and the topology induced by this metric will be referred to as the *norm topology* on $\mathscr{E}$ (if it is necessary to make a distinction). The metrics defined by norms are easily characterized.

Proposition 11.4. *If $\mathscr{E}$ is a normed space, then the metric ρ defined on $\mathscr{E}$ by the norm is an invariant metric satisfying the condition*

$$\rho(\lambda x, \lambda y) = |\lambda| \rho(x, y), \qquad x, y \in \mathscr{E}, \qquad \lambda \in \mathbb{C}. \qquad (2)$$

Conversely, if ρ is an invariant metric on a linear space $\mathscr{E}$ that satisfies (2), then there exists a unique norm $\| \ \|_\rho$ on $\mathscr{E}$ such that ρ is defined by $\| \ \|_\rho$.

PROOF. One way is clear enough; if ρ is defined by a norm $\| \ \|$ on $\mathscr{E}$, then $\rho(\lambda x, \lambda y) = \|\lambda x - \lambda y\| = |\lambda| \|x - y\| = |\lambda| \rho(x, y)$ for all vectors x, y and complex numbers λ. Moreover, if ρ is defined by $\| \ \|$, then $\|x\| = \rho(x, 0)$ for every vector x in $\mathscr{E}$, so that $\| \ \|$ is uniquely determined by ρ. On the other hand, if ρ is an invariant metric on $\mathscr{E}$, then, as we know, $\|x\|_\rho = \rho(x, 0)$ defines a value on $\mathscr{E}$. If ρ also satisfies (2), then $\| \ \|_\rho$ is a norm. $\square$

Example A. The usual metric on n-dimensional unitary space $\mathbb{C}^n$ (Prob. 4A) is defined by the norm

$$\|(\alpha_1, \ldots, \alpha_n)\| = \left[\sum_{i=1}^n |\alpha_i|^2 \right]^{1/2}.$$

Similarly, Euclidean space $\mathbb{R}^n$ becomes a real normed space if we set

$$\|(t_1, \ldots, t_n)\| = \left[\sum_{i=1}^n t_i^2 \right]^{1/2},$$

and the usual metric on $\mathbb{R}^n$ is the one defined by this norm. In particular, $\mathbb{C}$ itself is a normed space (and $\mathbb{R}$ a real normed space) with respect to the norm $\|\lambda\| = |\lambda|$.

It is not hard to see that the space $\mathbb{C}$ equipped with the norm $\|\lambda\| = |\lambda|$ more or less exhausts the class of one-dimensional normed spaces (Prob. J). Examples of normed spaces abound, however, in higher dimensions. Thus on the space $\mathbb{C}^n$ of complex n-tuples $a = (\alpha_1, \ldots, \alpha_n)$ the formula $\|a\|_1 = \sum_{i=1}^n |\alpha_i|$ clearly defines a norm distinct from the one given in Example A, as does the formula $\|a\|_\infty = \max_i |\alpha_i|$. We next consider an interesting and important family of norms on $\mathbb{C}^n$ of which $\|\ \|_1$ and $\|\ \|_\infty$ are special cases.

Example B. For an arbitrary n-tuple $a = (\alpha_1, \ldots, \alpha_n)$ in $\mathbb{C}^n$ and an arbitrary positive real number p we write

$$\|a\|_p = \left[\sum_{i=1}^n |\alpha_i|^p \right]^{1/p}$$

(The reader will note that this is consistent with the notation $\|\ \|_1$ used above, and likewise that $\|\ \|_2$ coincides with the quadratic norm on $\mathbb{C}^n$ introduced in Example A.) We shall see that the function $\|\ \|_p$ is a norm on $\mathbb{C}^n$ for every $p \geq 1$ (but not for $0 < p < 1$; cf. Example C below). To show that this is so it clearly suffices to treat the case $p > 1$, and to each such number p there corresponds a unique positive number q satisfying the condition $1/p + 1/q = 1$ or, equivalently, the condition $p + q = pq$. The number q is called the *Hölder conjugate* of p, and p and q are said to be *Hölder conjugates*. As it turns out, if p and q are any two Hölder conjugates, and if u and v denote arbitrary nonnegative numbers, then

$$uv \leq \frac{u^p}{p} + \frac{v^q}{q}. \tag{3}$$

(The proof of (3) is an exercise in elementary calculus; see Problem A.) Using this fact it is an easy matter to verify the *Hölder inequality*

$$\sum_{i=1}^n |\alpha_i \beta_i| \leq \|a\|_p \|b\|_q, \tag{4}$$

valid for all n-tuples $a = (\alpha_1, \ldots, \alpha_n)$ and $b = (\beta_1, \ldots, \beta_n)$ and every pair p, q of Hölder conjugates. Indeed, suppose $\|a\|_p = \|b\|_q = 1$. Then by (3) we have

$$|\alpha_i \beta_i| = |\alpha_i||\beta_i| \leq \frac{|\alpha_i|^p}{p} + \frac{|\beta_i|^q}{q}$$

for each index i, whence we obtain

$$\sum_{i=1}^n |\alpha_i \beta_i| \leq \frac{\|a\|_p^p}{p} + \frac{\|b\|_q^q}{q} = \frac{1}{p} + \frac{1}{q} = 1.$$

But then (4) follows in the general case upon replacing a and b by $a/\|a\|_p$ and $b/\|b\|_q$, respectively. (If either $a = 0$ or $b = 0$, then (4) is trivially valid.) Finally, the Hölder inequality implies the triangle inequality for $\|\ \|_p$ ($p > 1$), which here assumes the form

$$\left[\sum_{i=1}^n |\alpha_i + \beta_i|^p \right]^{1/p} \leq \left[\sum_{i=1}^n |\alpha_i|^p \right]^{1/p} + \left[\sum_{i=1}^n |\beta_i|^p \right]^{1/p}, \tag{5}$$

known as the *Minkowski inequality*, valid for all n-tuples $a = (\alpha_1, \ldots, \alpha_n)$ and $b = (\beta_1, \ldots, \beta_n)$ (cf. Problem B). Since it is obvious that $\| \ \|_p$ satisfies the other two defining conditions for a norm, we see that $\| \ \|_p$ is indeed a norm on $\mathbb{C}^n$ for all $p \geq 1$. (The reader may note that this argument is a straightforward generalization of the one used to derive the triangle inequality for $\| \ \|_2$ from the Cauchy inequality, to which the Hölder inequality reduces when $p = q = 2$; see Problem 4A.) It should be observed that a sequence $\{x_k\} = \{(\xi_1^{(k)}, \ldots, \xi_n^{(k)})\}$ of n-tuples converges to a limit $a = (\alpha_1, \ldots, \alpha_n)$ with respect to the (metric defined by the) norm $\| \ \|_p$, $1 \leq p \leq +\infty$, if and only if it converges *termwise* or *coordinatewise* to a, i.e., if and only if $\lim_k \xi_i^{(k)} = \alpha_i$, $i = 1, \ldots, n$. Thus, while these norms differ from one another in numerical value, they all induce the same norm topology on $\mathbb{C}^n$, namely, the usual product topology (Cor. 3.19, Prob. 3I). Two norms on the same linear space that induce the same norm topology are said to be *equivalent*. Thus the norms $\| \ \|_p$, $1 \leq p \leq +\infty$, are all equivalent norms on $\mathbb{C}^n$.

Example C. The function $\| \ \|_p$ is *not* a norm on $\mathbb{C}^n$ for any p such that $0 < p < 1$ (unless $n = 1$). Consider the case of the two pairs $(1, 0)$ and $(0, 1)$. Clearly $\|(1, 0)\|_p = \|(0, 1)\|_p = 1$ for every p, while $\|(1, 0) + (0, 1)\|_p = 2^{1/p}$, and $2^{1/p} > 2$ whenever $0 < p < 1$. Thus the triangle inequality fails to hold even on $\mathbb{C}^2$ except when $p \geq 1$.

Example D. For each number p, $0 < p < +\infty$, let (ℓ_p) denote the collection of all infinite sequences $a = \{\alpha_n\}_{n=0}^\infty$ with the property that $\sum_{n=0}^\infty |\alpha_n|^p < +\infty$, and for each such a in (ℓ_p) set

$$\|a\|_p = \left[\sum_{n=0}^\infty |\alpha_n|^p \right]^{1/p}.$$

By virtue of the elementary inequality $(u + v)^p \leq 2^p(u^p + v^p)$, valid for all $u, v \geq 0$, it is easily seen that, with linear operations defined termwise, each (ℓ_p) is a linear submanifold of the linear space (m) of all bounded sequences (Ex. 2D). In the same vein we denote by (ℓ_∞) the vector space (m) itself equipped with the norm

$$\|a\|_\infty = \sup_n |\alpha_n|.$$

Just as in the preceding examples, it is clear that $\| \ \|_1$ and $\| \ \|_\infty$ are norms on the spaces (ℓ_1) and (ℓ_∞), respectively, and that $\| \ \|_p$ is not a norm on (ℓ_p) for $0 < p < 1$. Furthermore, the only difficulty that arises in verifying that $\| \ \|_p$ is a norm on (ℓ_p) for $1 < p < +\infty$ comes in checking the triangle inequality. This inequality, exactly like (5) but stated for infinite series instead of finite sums, and still known as the *Minkowski inequality*, may be obtained simply by passing to the limit in (5) as n tends to infinity. It is rather more instructive, however, first to verify the *Hölder inequality* (4) for infinite series (the same proof works), and then to use that inequality to obtain the Minkowski inequality for elements of (ℓ_p) (see Problem B).

The justification for the notation $\| \ \|_\alpha$ is to be found in the relation

$$\|a\|_\alpha = \lim_{p \to +\alpha} \|a\|_p,$$

valid for every sequence $a = \{\alpha_n\}$ belonging to any of the spaces (ℓ_p), $p < +\infty$. The space $\mathbb{C}^n$ equipped with the quadratic norm $\| \ \|_2$ and its generalization (ℓ_2) are examples, indeed prototypes, of a class of normed spaces called *Hilbert spaces*, the study of which we shall take up in earnest in Volume II.

Notation and Terminology. The closed ball of radius r $(r > 0)$ centered at the origin in a normed space $\mathcal{E}$ will be denoted by $\mathcal{E}_r$. It is easily seen that the open ball about 0 with the same radius r is the topological interior of $\mathcal{E}_r$, and will accordingly be denoted by $\mathcal{E}_r^\circ$. In this connection it should be observed that the balls $\mathcal{E}_r^\circ$, $r > 0$, along with their various translates $x + \mathcal{E}_r^\circ$, $x \in \mathcal{E}$, form a base for the norm topology on $\mathcal{E}$. By the *sphere* of radius r centered at the origin is meant the set $\{x \in \mathcal{E} : \|x\| = r\} = \mathcal{E}_r \backslash \mathcal{E}_r^\circ$. In particular, the *unit sphere* consists of the set of all *unit vectors*, i.e., vectors x such that $\|x\| = 1$.

Proposition 11.5. *If $\mathcal{E}$ is a normed space with norm $\| \ \|$, then the $\mathcal{E}$-valued mappings $(\alpha, x) \to \alpha x$ and $(x, y) \to x + y$ are continuous. Likewise, the real function $x \to \|x\|$ is continuous. Indeed, $| \|x\| - \|y\| | \le \|x - y\|$ for any vectors x, y in $\mathcal{E}$.*

PROOF. The continuity of the mapping $(x, y) \to x + y$ and the facts concerning the function $x \to \|x\|$ have already been established, more generally, for vector spaces equipped with a value (Prop. 11.2), and these assertions are included here purely for convenience of reference. The fact that the $\mathcal{E}$-valued mapping $(\alpha, x) \to \alpha x$ is continuous on $\mathbb{C} \times \mathcal{E}$ follows at once from the inequality

$$\|\alpha x - \beta y\| \le |\alpha - \beta| \|x\| + |\beta| \|x - y\|,$$

valid for all scalars α, β and all vectors x, y in $\mathcal{E}$, and this inequality is, in turn, an easy consequence of the defining properties of a norm. $\square$

Corollary 11.6. *If x_0 is an arbitrary vector in a normed space $\mathcal{E}$, and α_0 is an arbitrary nonzero complex number, then the mappings $x \to x + x_0$ (translation by x_0) and $x \to \alpha_0 x$ (dilatation by α_0) are both homeomorphisms of $\mathcal{E}$ onto itself.*

As has been noted (Ch. 4, p. 58) an important property that a metric space may or may not possess is that of completeness.

Definition. A [real] normed space that is complete as a metric space is called a [real] *Banach space*.

Example E. The space $\mathbb{C}^n$ equipped with any one of the norms $\| \ \|_p$ ($1 \le p \le +\infty$) is complete, and is therefore a Banach space. This is so for the simple reason that a sequence in $\mathbb{C}^n$ is Cauchy [convergent] with respect to any one of the norms $\| \ \|_p$ if and only if it is Cauchy [convergent] termwise, and the field $\mathbb{C}$ of complex numbers is complete in the usual metric $|\alpha - \beta|$. (As a matter of fact, *every* finite dimensional normed space is complete (Prob. J).) The spaces (ℓ_p), $1 \le p \le +\infty$, introduced in Example D are also Banach spaces, but here the verification of completeness is not so simple (except for $p = +\infty$; the verification that (ℓ_∞) is complete is an easy exercise which we leave to the reader). The argument goes as follows. Fix p, $1 \le p < \infty$, and suppose $\{x_n\}_{n=1}^\infty$ is a Cauchy sequence in (ℓ_p), where $x_n = \{\zeta_m^{(n)}\}_{m=0}^\infty$, $n \in \mathbb{N}$. Then $\{x_n\}$ is certainly Cauchy termwise, and therefore convergent termwise to some sequence $x = \{\xi_m\}_{m=0}^\infty$. Let ε be an arbitrary positive number, and let N be a positive integer such that $\| x_n - x_m \|_p < \varepsilon$ for all $m, n \ge N$. Then for every positive integer k it is the case that

$$\sum_{i=0}^{k} |\xi_i^{(n)} - \xi_i^{(m)}|^p < \varepsilon^p$$

for all $m, n \ge N$. Hence, letting m tend to infinity, we see that

$$\sum_{i=0}^{k} |\xi_i^{(n)} - \xi_i|^p \le \varepsilon^p$$

for every positive integer k and all positive integers n such that $n \ge N$. But then, letting k tend to infinity, we see that

$$\sum_{i=0}^{\infty} |\xi_i^{(n)} - \xi_i|^p \le \varepsilon^p \tag{6}$$

for every positive integer n such that $n \ge N$. This shows, in the first place, that $x_n - x$ belongs to (ℓ_p) for $n \ge N$, and hence that x belongs to (ℓ_p) as well. In the second place (6) shows that $\| x_n - x \|_p \le \varepsilon$ whenever $n \ge N$, and hence that $\{x_n\}$ converges to x in the metric of (ℓ_p). (For an entirely similar argument in the special case $p = 1$, see Example 7D.)

Example F. Let X be a nonempty compact Hausdorff space, and let $\mathscr{C}(X)$ denote the linear space of all continuous complex-valued functions defined on X (Ex. 3D). The space $\mathscr{C}(X)$ becomes a Banach space when it is equipped with the norm $\| f \| = \sup_{x \in X} |f(x)|$. (This supremum norm is known, affectionately, if somewhat inelegantly, as the "sup norm"; by an obvious analogy with the norm on (ℓ_∞) it is also frequently denoted by $\| \ \|_\infty$. That a continuous function f on X is bounded, so that $\| f \|$ is finite, and that there exists a point x_0 in X such that $\| f \| = |f(x_0)|$, so that we may equally well write $\| f \| = \max_{x \in X} |f(x)|$, are routine consequences of the compactness of X; see Proposition 3.3 and Corollary 4.5.) Indeed, it is clear that the function $\| \ \|$ is a norm. Moreover, a moment's thought shows that the metric it defines on $\mathscr{C}(X)$ is simply the metric of uniform convergence, and $\mathscr{C}(X)$ is complete in this metric (cf. Problem 4O). Similarly, the collection $\mathscr{C}_{\mathbb{R}}(X)$ of all real-valued functions in $\mathscr{C}(X)$ is a real Banach space in the sup norm.

(The requirement that X be nonempty in this construction is necessitated by the reference to the sup norm. In the event that X is empty the space $\mathscr{C}(X)$ reduces to the trivial linear space (0) consisting of the empty function. Hence there is but one possible norm on $\mathscr{C}(\varnothing)$, viz., $\|0\| = 0$, and in this norm $\mathscr{C}(\varnothing)$ becomes the trivial Banach space. Thus we may and do regard $\mathscr{C}(X)$ as a well-defined Banach space even when X is empty, and, by a slight abuse of language, also refer to the "sup norm" in this trivial case.)

A broad and frequently useful generalization of the above construction suggests itself at once. Suppose X is a compact Hausdorff space and $\mathscr{E}$ is a given normed space. The collection $\mathscr{C}(X;\mathscr{E})$ of all continuous $\mathscr{E}$-valued mappings on X is a linear space with respect to pointwise linear operations, and a normed space in the sup norm

$$\|\Phi\| = \sup_{x \in X} \|\Phi(x)\|.$$

It is easily seen that if X is nonempty then $\mathscr{C}(X;\mathscr{E})$ is a Banach space when and only when $\mathscr{E}$ is (cf. Problem D). The case $X = \varnothing$ is covered as in the preceding paragraph.

In the study of linear algebra one becomes accustomed to thinking of the continuous scalar-valued functions on an abstract topological space as elements of various linear spaces of functions. As Example F shows, the continuous complex-valued functions on a compact Hausdorff space are, in a natural way, elements of a *Banach* space. The following example shows that this point of view can lead to valuable economies in both thought and notation.

Example G (Tietze Extension Theorem). Let X be a compact Hausdorff space, let F be a nonempty closed subset of X, and let f be a continuous real-valued function defined on F such that $\|f\|_\infty \leq 1$ (computed in $\mathscr{C}(F)$). Set $E_{1/3} = \{x \in X : f(x) \geq 1/3\}$, $E_{-1/3} = \{x \in X : f(x) \leq -1/3\}$, and let g_1 be a continuous real-valued function on X such that $g_1 = 1/3$ on $E_{1/3}$, $g_1 = -1/3$ on $E_{-1/3}$, and $\|g_1\|_\infty = 1/3$. (That such a function g_1 exists is an immediate consequence of Urysohn's lemma; cf. Proposition 3.5.) Then, as a consideration of cases shows, $\|f - g_1\|_\infty \leq 2/3$ (computed on F). Next set

$$E_{2/9} = \{x \in X : f(x) - g_1(x) \geq 2/9\}$$

and

$$E_{-2/9} = \{x \in X : f(x) - g_1(x) \leq -2/9\}.$$

If g_2 is a continuous real-valued function on X such that $g_2 = 2/9$ on $E_{2/9}$, $g_2 = -2/9$ on $E_{-2/9}$, and $\|g_2\|_\infty = 2/9$, then $\|f - g_1 - g_2\|_\infty \leq 4/9$

(computed on F). Continuing in this fashion we obtain by mathematical induction an infinite sequence $\{g_n\}_{n=1}^{\infty}$ in $\mathscr{C}_{\mathbb{R}}(X)$ such that

$$\|g_n\|_{\infty} \le \frac{2^{n-1}}{3^n} \quad \text{in} \quad \mathscr{C}_{\mathbb{R}}(X), \qquad n \in \mathbb{N},$$

and such that

$$\left\| f - \sum_{k=1}^{n} g_k \right\|_{\infty} \le \left(\frac{2}{3} \right)^n \quad \text{in} \quad C_{\mathbb{R}}(F), \qquad n \in \mathbb{N}.$$

From the first of these conditions it follows readily that the infinite series $\sum_{n=1}^{\infty} g_n$ converges in $\mathscr{C}_{\mathbb{R}}(X)$ and that, if we set $g = \sum_{n=1}^{\infty} g_n$, then

$$\|g\|_{\infty} \le \sum_{n=1}^{\infty} \|g_n\| \le \frac{1}{3} \frac{1}{1 - 2/3} = 1$$

in $\mathscr{C}_{\mathbb{R}}(X)$ (cf. Problem G). From the second condition we conclude that $g = f$ on F. Thus, except for a trivial normalization, we have proved the following theorem: If X is a compact Hausdorff space, F is a closed subset of X, and f is an element of the real Banach space $\mathscr{C}_{\mathbb{R}}(F)$, then there exists an element g of $\mathscr{C}_{\mathbb{R}}(X)$ such that $\|g\|_{\infty}$ (computed on X) is equal to $\|f\|_{\infty}$ (computed on F), and such that $g|F = f$.

In order to extend this result to the case of complex-valued functions, we observe that for any radius $R \ge 0$ the complex plane $\mathbb{C}$ admits a retraction onto the closed disc $D_R^{-} = \{\lambda \in \mathbb{C} : |\lambda| \le R\}$. (That is, there exists a continuous mapping ρ of $\mathbb{C}$ onto D_R^{-} such that $\rho(\lambda) = \lambda$ for each λ in D_R^{-}. Indeed, the mapping

$$\rho(\lambda) = \begin{cases} \dfrac{R\lambda}{|\lambda|}, & |\lambda| > R, \\ \lambda, & |\lambda| \le R, \end{cases}$$

is readily seen to possess both of the desired properties.)

Now let X and F be as above, and let f be an element of $\mathscr{C}(F)$. Then there exists an extension g of f in $\mathscr{C}(X)$, since the real and imaginary parts of f may simply be extended individually by virtue of what has already been proved. Set $R = \|f\|_{\infty}$ (computed on F, of course) and let ρ denote a retraction of $\mathbb{C}$ onto the disc D_R^{-}. Then composing ρ with g yields the following result, known as the *Tietze extension theorem*: *If X is a compact Hausdorff space and F is a closed subset of X, then every element f of $\mathscr{C}(F)$ possesses a continuous extension g to X such that $\|g\|_{\infty} = \|f\|_{\infty}$.*

If $\mathscr{E}$ is a normed space and $\mathscr{M}$ is an arbitrary linear manifold in $\mathscr{E}$, then it is clear that the restriction to $\mathscr{M}$ of the norm $\| \; \|$ on $\mathscr{E}$ is a norm on $\mathscr{M}$ turning it into a normed space. It is also clear that the metric defined on $\mathscr{M}$ by this restricted norm coincides with the restriction to $\mathscr{M}$ of the metric defined on

$\mathscr{E}$ by $\|\ \|$, so that $\mathscr{M}$ is automatically a subspace of $\mathscr{E}$ as a metric space and as a topological space as well. Whenever a linear manifold in a normed space is regarded as a normed space in its own right, it is this restricted norm that is understood unless some other norm is stipulated. Since a subset of a complete metric space is complete if and only if it is closed (Prob. 4I), a linear sub-manifold of a Banach space $\mathscr{E}$ is itself a Banach space if and only if it is closed in $\mathscr{E}$. For this reason the closed linear manifolds are of particular importance. To emphasize this distinction we introduce the following terminology.

Definition. A linear submanifold $\mathscr{M}$ of a normed space $\mathscr{E}$ is called a *subspace* of $\mathscr{E}$ (sometimes, for special emphasis, a *closed subspace* of $\mathscr{E}$) if and only if $\mathscr{M}$ is closed in $\mathscr{E}$.

In any normed space $\mathscr{E}$ the linear submanifolds (0) and $\mathscr{E}$ are subspaces. Moreover, as will be seen below (Prob. P), unless $\mathscr{E}$ is one-dimensional, there are many other subspaces as well. The following results, while easily proved, are nonetheless of basic importance.

Proposition 11.7. *The intersection of an arbitrary nonempty collection of subspaces of a normed space $\mathscr{E}$ is again a subspace of $\mathscr{E}$. Consequently, given an arbitrary subset M of $\mathscr{E}$, there exists a smallest subspace $\mathscr{M}$ of $\mathscr{E}$ that contains M.*

PROOF. An intersection of linear manifolds in a linear space is a linear mani-fold, and an intersection of closed sets in a topological space is a closed set. Hence an intersection of subspaces of $\mathscr{E}$ is a subspace of $\mathscr{E}$. The smallest sub-space of $\mathscr{E}$ containing a given subset M of $\mathscr{E}$ is clearly the intersection of the collection of all those subspaces of $\mathscr{E}$ that contain M. (This collection is never empty since, as noted above, $\mathscr{E}$ is a subspace of itself.) □

Corollary 11.8. *The collection of all subspaces of a normed space is a complete lattice in the inclusion ordering.*

Definition and Notation. If M is a subset of a normed space $\mathscr{E}$, then the smallest subspace of $\mathscr{E}$ that contains M is called the subspace *spanned by M*, and is denoted by $\vee M$. (Thus a sharp distinction is drawn between the *linear submanifold* of $\mathscr{E}$ generated (algebraically) by M (which need not be closed) and the *subspace* $\vee M$ *spanned* by M (which *is* closed by definition).) A subset M of a normed space $\mathscr{E}$ is a *spanning set* for $\mathscr{E}$, or *spans* $\mathscr{E}$, if $\vee M = \mathscr{E}$. If $\{\mathscr{M}_\gamma\}_{\gamma \in \Gamma}$ is an indexed family of subspaces of $\mathscr{E}$, the supremum of the family in the complete lattice of subspaces of $\mathscr{E}$ will be denoted by $\vee_{\gamma \in \Gamma} \mathscr{M}_\gamma$ or, when no confusion can result, simply by $\vee_\gamma \mathscr{M}_\gamma$. (Thus $\vee_\gamma \mathscr{M}_\gamma = \vee (\bigcup_\gamma \mathscr{M}_\gamma)$.) It would also be appropriate, of course, to write $\wedge_\gamma \mathscr{M}_\gamma$ for the infimum of the family $\{\mathscr{M}_\gamma\}$, but since $\wedge_{\gamma \in \Gamma} \mathscr{M}_\gamma = \bigcap_{\gamma \in \Gamma} \mathscr{M}_\gamma$ whenever Γ is nonempty, this added notational complexity will be avoided.

Proposition 11.9. *The topological closure $\mathscr{L}^-$ of a linear manifold $\mathscr{L}$ in a normed space $\mathscr{E}$ is again a linear manifold, and is therefore a subspace of $\mathscr{E}$. Hence the subspace spanned by a subset M of $\mathscr{E}$ coincides with the closure of the linear submanifold generated (algebraically) by M. In particular, the supremum $\bigvee_\gamma \mathscr{M}_\gamma$ of an indexed family $\{\mathscr{M}_\gamma\}$ of subspaces of $\mathscr{E}$ is the closure of the sum $\sum_\gamma \mathscr{M}_\gamma$ (as defined in Chapter 2).*

PROOF. The second and third assertions are immediate consequences of the first. To prove the first, let x and y belong to $\mathscr{L}^-$ and let $\{x_n\}$ and $\{y_n\}$ be sequences in $\mathscr{L}$ such that $x_n \to x$ and $y_n \to y$ (Prop. 4.1). If α and β are arbitrary complex numbers, then $\alpha x_n + \beta y_n \in \mathscr{L}$ for all n, and $\alpha x_n + \beta y_n \to \alpha x + \beta y$. Hence $\alpha x + \beta y \in \mathscr{L}^-$. $\qquad\square$

Example H. Each of the Banach spaces (ℓ_p), $1 \le p \le +\infty$, contains the sequence of vectors $\{e_n\}_{n=0}^\infty$, where $e_n = \{\delta_{0n}, \delta_{1n}, \ldots\}$ and δ_{mn} denotes, as usual, the Kronecker delta. (Cf. Example 2B; the sequence $\{e_n\}$ of unit vectors will appear often in the sequel, and will consistently be denoted in this way.) The linear submanifold of each space (ℓ_p) generated algebraically by this sequence is the same, namely, the linear space (φ) consisting of all those sequences $\{\xi_n\}_{n=0}^\infty$ with the property that there exists a nonnegative integer N such that $\xi_n = 0$ for $n > N$. When $1 \le p < +\infty$, this linear submanifold is dense in (ℓ_p), so the subspace of (ℓ_p) spanned by the sequence $\{e_n\}_{n=0}^\infty$ is (ℓ_p) itself. (Hence (ℓ_p) is a separable metric space for $1 < p < +\infty$; see Problem K.)

In the space (ℓ_∞) the story is quite different; the subspace of (ℓ_∞) spanned by the sequence $\{e_n\}$ (which is also the closure in (ℓ_∞) of the submanifold (φ)) coincides with the space (c_0) of null sequences equipped with the norm $\|\ \|_\infty$; cf. Example 2D. Thus the sequence $\{e_n\}_{n=0}^\infty$ is a spanning set in (ℓ_p) for $1 \le p < +\infty$, but not in (ℓ_∞). Indeed, if E and F are any two distinct subsets of $\mathbb{N}_0$, then $\|\chi_E - \chi_F\|_\infty = 1$, whence it follows that (ℓ_∞) is *not* separable as a metric space. On the other hand, the collection $\{\chi_E\}$ of *all* characteristic functions of subsets E of $\mathbb{N}_0$ is readily seen to span (ℓ_∞). (We also observe, for future reference, that the collection $\{\chi_J\}$ of all characteristic functions of *infinite* subsets J of $\mathbb{N}_0$ suffices to span (ℓ_∞).)

Suppose now that $\mathscr{E}$ is a normed space and that $\mathscr{M}$ is a subspace of $\mathscr{E}$. Writing $[x]$ for the coset $x + \mathscr{M}$ of the vector x in the quotient space $\mathscr{E}/\mathscr{M}$, we define

$$\|[x]\| = \inf_{z \in \mathscr{M}} \|x + z\| = \inf_{z \in \mathscr{M}} \|x - z\|. \tag{7}$$

(See Chapter 2 for basic definitions; that these two expressions for $\|[x]\|$ are really equal follows at once from the fact that $\mathscr{M}$ is a linear manifold.) Then it is easy to see that (7) defines a norm on $\mathscr{E}/\mathscr{M}$. Indeed, since $\{x + z; z \in \mathscr{M}\}$ and $\{x + z_0 + z; z \in \mathscr{M}\}$ are the same set of vectors for any

z_0 in $\mathcal{M}$, it is clear that the expressions in (7) depend only on the coset $[x]$ and not on the representative x. Likewise, the second of the two expressions for $\|[x]\|$ in (7) shows that $\|[x]\|$ may also be described as the distance $d(x, \mathcal{M})$ from x to the closed set $\mathcal{M}$, so that $\|[x]\| = 0$ only when $x \in \mathcal{M}$, i.e., when $[x] = 0$ (Prob. 4D). To check that the function defined in (7) has the other properties required of a norm is equally easy, and will be left as an exercise. In the sequel it is this *quotient norm* that is understood whenever a quotient space is regarded as a normed space. Observe that

$$\|[x]\| \leq \|x\| \tag{8}$$

for every x in $\mathcal{E}$. The quotient norm topology on $\mathcal{E}/\mathcal{M}$ coincides with the quotient topology (see Problem 3H), and the natural projection $x \overset{\pi}{\to} [x]$ is a continuous open mapping of $\mathcal{E}$ in the norm topology onto $\mathcal{E}/\mathcal{M}$ in the quotient norm topology (see Problem H). The following fact concerning the quotient space $\mathcal{E}/\mathcal{M}$ is considerably less elementary.

Theorem 11.10. *If $\mathcal{E}$ is a Banach space and $\mathcal{M}$ is a subspace of $\mathcal{E}$, then $\mathcal{E}/\mathcal{M}$ is also a Banach space.*

PROOF. We must show that $\mathcal{E}/\mathcal{M}$ is complete. Accordingly, we suppose given a Cauchy sequence $\{[x_n]\}_{n=1}^{\infty}$ of cosets in $\mathcal{E}/\mathcal{M}$. As noted (Prob. H) the natural projection of $\mathcal{E}$ onto $\mathcal{E}/\mathcal{M}$ is continuous, so it suffices to construct a convergent sequence $\{x'_n\}$ in $\mathcal{E}$ such that $[x'_n] = [x_n]$ for every n, for then $\{[x_n]\}$ will converge to $[\lim_n x'_n]$. Moreover, as always, one may assume that the given sequence is not merely Cauchy, but satisfies the stronger condition

$$\sum_{n=1}^{\infty} \|[x_{n+1}] - [x_n]\| = \sum_{n=1}^{\infty} \|[x_{n+1} - x_n]\| < +\infty$$

(Prob. 4E). But then for each n in $\mathbb{N}$ we may select a vector z_n in $\mathcal{M}$ such that $\|x_{n+1} - x_n + z_n\| < \|[x_{n+1} - x_n]\| + 1/2^n$, and it follows that the series

$$x_1 + (x_2 - x_1 + z_1) + \cdots + (x_{n+1} - x_n + z_n) + \cdots$$

is absolutely convergent, and therefore convergent (Prob. G). This means that the sequence $\{s_n = x_n + \sum_{i=1}^{n-1} z_i\}$ of partial sums is convergent, and to complete the proof we simply define $x'_n = s_n$. $\qquad\square$

If $\mathcal{E}$ and $\mathcal{F}$ are two normed spaces, then it is easy to see that

$$\|(x, y)\| = \|x\| + \|y\|, \qquad (x, y) \in \mathcal{E} \dotplus \mathcal{F},$$

defines a norm on the algebraic direct sum $\mathcal{E} \dotplus \mathcal{F}$ (see Chapter 2). The space $\mathcal{E} \dotplus \mathcal{F}$ equipped with this norm will be called the (*normed space*) *direct sum* of $\mathcal{E}$ and $\mathcal{F}$ and will be denoted by $\mathcal{E} \oplus_1 \mathcal{F}$. More generally, if $\{\mathcal{E}_1, \ldots, \mathcal{E}_n\}$ is any finite sequence of normed spaces, then

$$\|(x_1, \ldots, x_n)\| = \sum_{i=1}^{n} \|x_i\|, \qquad x_i \in \mathcal{E}_i, i = 1, \ldots, n,$$

defines a norm on the algebraic direct sum $\mathscr{E} = \mathscr{E}_1 \dotplus \cdots \dotplus \mathscr{E}_n$, and the result of equipping $\mathscr{E}$ with this norm will be denoted by $\mathscr{E}_1 \oplus_1 \cdots \oplus_1 \mathscr{E}_n$. (For a further exploration of this idea, along with some closely related concepts, see Problems R and T.)

Proposition 11.11. *The direct sum $\mathscr{E}_1 \oplus_1 \cdots \oplus_1 \mathscr{E}_n$ of a finite sequence $\{\mathscr{E}_1, \ldots, \mathscr{E}_n\}$ of normed spaces is complete if and only if each space $\mathscr{E}_i$ is complete, $i = 1, \ldots, n$.*

PROOF. A sequence $\{(x_1^{(k)}, \ldots, x_n^{(k)})\}_{k=1}^{\infty}$ is Cauchy [convergent] in $\mathscr{E}_1 \oplus_1 \cdots \oplus_1 \mathscr{E}_n$ if and only if each sequence $\{x_i^{(k)}\}_{k=1}^{\infty}$ is Cauchy [convergent] in its respective space $\mathscr{E}_i$. $\quad\square$

There is an interesting and frequently useful generalization of the concept of a normed space. The idea is to weaken radically the homogeneity requirement in the definition of a norm. Consider a nonnegative function $\mathbf{\|} \ \mathbf{\|}$ on a linear space $\mathscr{E}$ satisfying the following conditions:

(i) $\|x\| = 0$ if and only if $x = 0$,
(ii″) $\|\lambda x\| \le \|x\|$ for every x in $\mathscr{E}$ and all complex numbers λ such that $|\lambda| \le 1$,
(iii) $\|x + y\| \le \|x\| + \|y\|$ for all x, y in $\mathscr{E}$.

It follows at once from (ii″) that $\|\lambda x\| = \|x\|$ whenever $|\lambda| = 1$, and therefore that the function $\mathbf{\|} \ \mathbf{\|}$ is a value on $\mathscr{E}$. Hence $\rho(x, y) = \|x - y\|$ defines an invariant metric on $\mathscr{E}$ with respect to which the functions $x + y$ and $\|x\|$ are continuous functions of their arguments (Proposition 11.2). There is no reason to suppose, however, that the mapping $(\alpha, x) \to \alpha x$ is continuous, and it is therefore appropriate to impose a further requirement. We shall say that $\mathbf{\|} \ \mathbf{\|}$ is a *quasinorm* on $\mathscr{E}$ provided it also satisfies the condition

(iv) $\alpha_n \to 0$ implies $\|\alpha_n x\| \to 0$

for every null sequence $\{\alpha_n\}$ of complex numbers and every x in $\mathscr{E}$. A linear space $\mathscr{E}$ equipped with a quasinorm will be called a *quasinormed space*. (A quasinorm on a real vector space is defined by requiring (i), (ii″), (iii), and (iv) to hold, with real scalars instead of complex ones in (ii″) and (iv). A real linear space equipped with a quasinorm is a *real quasinormed space*.) The topology induced on $\mathscr{E}$ by (the invariant metric defined by) a quasinorm on $\mathscr{E}$ is called the *quasinorm topology*. If $\mathscr{E}$ is a quasinormed space, we shall continue to use the notation $\mathscr{E}_n[\mathscr{E}_n^\circ]$ for the closed [open] ball in $\mathscr{E}$ with center 0 and radius r. The following characterization of the invariant metric defined by a quasinorm is easily verified and we omit the proof (cf. Proposition 11.4).

Proposition 11.12. *If $\mathscr{E}$ is a quasinormed space, then the metric ρ defined by the quasinorm is an invariant metric on ρ satisfying the condition*

$$\rho(\lambda x, \lambda y) \le \rho(x, y), \qquad x, y \in \mathscr{E}, |\lambda| \le 1. \tag{9}$$

Conversely, if ρ is an invariant metric on a linear space $\mathscr{E}$ that satisfies (9), and if $\alpha_n \to 0$ implies $\rho(\alpha_n x, 0) \to 0$ for every vector x in $\mathscr{E}$, then there exists a unique quasinorm $|\ |_\rho$ on $\mathscr{E}$ such that ρ is defined by $|\ |_\rho$.

It is an immediate consequence of condition (iv) in the definition of a quasinorm that the $\mathscr{E}$-valued mapping $\alpha \to \alpha x$ is continuous on $\mathbb{C}$ for each fixed vector x in $\mathscr{E}$. Much more is true however.

Lemma 11.13. *If $\mathscr{E}$ is a quasinormed linear space and α is a fixed complex number, then the dilatation $x \to \alpha x$ is a continuous mapping of $\mathscr{E}$ into itself.*

PROOF. It suffices to show that if ε is a given positive number, then there exists a positive number δ such that $|x| < \delta$ implies $|\alpha x| < \varepsilon$. Moreover, for given $\varepsilon > 0$ we know that there exists $\delta > 0$ such that $|x| < \delta$ and $|y| < \delta$ imply $|x + y| < \varepsilon$ (since the mapping $(x, y) \to x + y$ is continuous at $(0, 0)$). Hence, in particular, if $|x| < \delta$, then $|2x| < \varepsilon$. In other words, the very special dilatation $x \to 2x$ is continuous. But then (by induction) all of the dilatations $x \to 2^n x$ are continuous. Choose n so that $|\alpha| \le 2^n$, and for given positive ε choose a positive number δ such that $|x| < \delta$ implies $|2^n x| < \varepsilon$. Then $|\alpha x| = |(\alpha/2^n)2^n x| \le |2^n x| < \varepsilon$ whenever $|x| < \delta$, and the lemma is proved. $\qquad\square$

Proposition 11.14. *If $\mathscr{E}$ is a quasinormed linear space with quasinorm $|\ |$, then the $\mathscr{E}$-valued mappings $(\alpha, x) \to \alpha x$ and $(x, y) \to x + y$ are continuous. Likewise, the real function $x \to |x|$ is continuous. Indeed, $||x| - |y|| \le |x - y|$ for any vectors x, y in $\mathscr{E}$.*

PROOF. Since $|\ |$ is a value, the continuity of the mapping $(x, y) \to x + y$ and the facts concerning the function $|\ |$ have already been established (Prop. 11.2), and these assertions are included here purely for convenience of reference. In order to see that the mapping $(\alpha, x) \to \alpha x$ is continuous on $\mathbb{C} \times \mathscr{E}$, let us write

$$\alpha x - \beta y = (\alpha - \beta)y + (\alpha - \beta)(x - y) + \beta(x - y). \tag{10}$$

With β and y held fixed, it is clear that we can make the first and third summands in the right member of (10) as close to 0 as desired by taking α sufficiently close to β and x sufficiently close to y (since αx is known to be continuous in either variable separately by condition (iv) in the definition of a quasinorm and the preceding lemma). Moreover, if we simply require $|\alpha - \beta| \le 1$, then $|(\alpha - \beta)(x - y)| \le |x - y|$, so the second summand can also be made as close to 0 as desired. Thus the mapping $(\alpha, x) \to \alpha x$ is continuous at (β, y), and the proposition is proved. $\qquad\square$

Corollary 11.15. *In a quasinormed linear space every translation $x \to x + x_0$ and every dilatation $x \to \alpha_0 x$, $\alpha_0 \ne 0$, is a homeomorphism of the space onto itself.*

Example I. The function $f(t) = t/(1 + t)$ is a strictly increasing, continuous function on the ray $[0, +\infty)$ that vanishes at $t = 0$ and satisfies the condition

$$f(s + t) \leq f(s) + f(t).$$

From these observations it follows at once that if $|\ \ |$ is a given quasinorm on a linear space $\mathscr{E}$, then

$$|x|_0 = \frac{|x|}{1 + |x|}, \qquad x \in \mathscr{E},$$

defines a new quasinorm on $\mathscr{E}$. The metric defined on $\mathscr{E}$ by $|\ |_0$ is easily seen to be equivalent to the metric defined by the original quasinorm $|\ |$, so the topologies induced by these two quasinorms coincide. Two quasinorms on the same linear space that induce the same topology are said to be *equivalent*. Thus the quasinorm $|\ |_0$ is equivalent to the given one. (The reader may wish to consult Example 4A in connection with this example.)

Example J. If $\mathscr{E}_1, \ldots, \mathscr{E}_n$ are quasinormed spaces, then it is easily seen that

$$|(x_1, \ldots, x_n)| = \sum_{i=1}^{n} |x_i| \tag{11}$$

defines a quasinorm on $\mathscr{E}_1 + \cdots + \mathscr{E}_n$. Likewise, if $\{\mathscr{E}_n\}_{n=0}^{\infty}$ is an infinite sequence of quasinormed spaces, then

$$|\{x_n\}_{n=0}^{\infty}| = \sum_{n=0}^{\infty} \frac{1}{2^n} \frac{|x_n|}{1 + |x_n|} \tag{12}$$

defines a quasinorm on the full algebraic direct sum $\sum_{n \in \mathbb{N}_0} + \mathscr{E}_n$. Indeed, in both the finite case and the infinite case, conditions (i), (ii″) and (iii) in the definition of a quasinorm are trivial to verify directly (using the construction in Example I). To show that condition (iv) holds, one may either argue directly, or observe that the invariant metrics defined by the values in the left members of (11) and (12) metrize the product topology on $\mathscr{E}_1 + \cdots + \mathscr{E}_n$ and $\sum_{n \in \mathbb{N}_0} + \mathscr{E}_n$, respectively (see Problem 4C and Example 4A). Thus the product topology on any countable product of quasinormed spaces is induced by a quasinorm. Similarly, if $\mathscr{M}$ is a linear submanifold of a quasinormed space $\mathscr{E}$, then the relative topology on $\mathscr{M}$ is induced by a quasinorm (simply restrict the quasinorm on $\mathscr{E}$ to $\mathscr{M}$), and so is the quotient topology on $\mathscr{E}/\mathscr{M}$ provided $\mathscr{M}$ is closed (see Problem I).

It is worth remarking that the construction in this example applies, in particular, when the spaces $\mathscr{E}_n$ are normed spaces. Thus, for instance, the formula

$$|\{\xi_n\}_{n=0}^{\infty}| = \sum_{n=0}^{\infty} \frac{|\xi_n|}{1 + |\xi_n|}$$

defines a quasinorm on the linear space (∂) of *all* complex sequences (Ex. 2D). The quasinorm topology on (∂) induced by $|\ |$ coincides with the product topology, that is, with the topology of termwise convergence.

Example K. Let $(X, \mathbf{S}, \mu)$ be a finite measure space (Ch. 7, p.118), let $\mathcal{M}$ denote the linear space of all complex-valued measurable functions on X, and let $\dot{\mathcal{M}}$ be the quotient space of equivalence classes $[f]$ of functions in $\mathcal{M}$, where f and g are equivalent if and only if $f = g$ a.e. $[\mu]$ (see Chapter 7 for definitions). Using the same calculations as in Example I, it is easy to see that

$$\mathbf{|}[f]\mathbf{|} = \int_X \frac{|f|}{1 + |f|} \, d\mu, \qquad f \in \mathcal{M},$$

defines a quasinorm on $\dot{\mathcal{M}}$. If f and g are two functions in $\mathcal{M}$, and if we write $E_\varepsilon = \{x \in X : |f(x) - g(x)| > \varepsilon\}$, then straightforward computation shows that

$$\frac{\varepsilon \mu(E_\varepsilon)}{1 + \varepsilon} \le \int_{E_\varepsilon} \frac{|f - g|}{1 + |f - g|} \, d\mu$$

$$\le \int_X \frac{|f - g|}{1 + |f - g|} \, d\mu \le \varepsilon \mu(X \backslash E_\varepsilon) + \mu(E_\varepsilon),$$

and hence that

$$\frac{\varepsilon \mu(E_\varepsilon)}{1 + \varepsilon} \le \mathbf{|}[f - g]\mathbf{|} \le \varepsilon \mu(X \backslash E_\varepsilon) + \mu(E_\varepsilon).$$

It follows that if f is a function in $\mathcal{M}$ and $\{f_n\}$ a sequence of functions in $\mathcal{M}$, then $\mathbf{|}[f] - [f_n]\mathbf{|} \to 0$ if and only if $\{f_n\}$ converges to f in measure (Prob. 7X). For this reason the invariant metric defined on $\dot{\mathcal{M}}$ by the quasinorm $\mathbf{|} \; \mathbf{|}$ is known as the *metric of convergence in measure*.

Just as in the case of a normed space, completeness is an important property for a quasinormed space to possess.

Definition. A [real] quasinormed space that is complete as a metric space is called a [*real*] *F-space*.

Example L. If $\{x_n\}$ is a sequence in the space $\mathscr{E}$ of Example I, then $\{x_n\}$ is Cauchy [convergent] with respect to the quasinorm $\mathbf{|} \; \mathbf{|}_0$ if and only if it is Cauchy [convergent] with respect to the quasinorm $\mathbf{|} \; \mathbf{|}$. Thus $\mathscr{E}$ is an F-space in the quasinorm $\mathbf{|} \; \mathbf{|}_0$ if and only if it was an F-space to begin with. Similarly, a sequence $\{X_k\}$ in either of the direct sums of Example J is Cauchy if and only if its is termwise Cauchy. Indeed, it is clear that this is so for the quasinorm (11). Likewise, if $\{X_k\}_{k=1}^\infty$ is a Cauchy sequence with respect to the quasinorm (12), and if $X_k = \{x_0^{(k)}, x_1^{(k)}, \ldots\}$ for each k, then the coordinate sequence $\{x_n^{(k)}\}_{k=1}^\infty$ is Cauchy in $\mathscr{E}_n$ for each n. To complete the argument, suppose that the sequence $\{x_n^{(k)}\}_{k=1}^\infty$ is Cauchy in $\mathscr{E}_n$, $n \in \mathbb{N}_0$, and let ε be a positive number. If N is chosen large enough so that

$$\frac{1}{2^N} = \sum_{n=N+1}^\infty \frac{1}{2^n} < \frac{\varepsilon}{2},$$

then

$$\mid X_k - X_m \mid < \sum_{n=0}^{N} \mid x_n^{(k)} - x_n^{(m)} \mid + \frac{\varepsilon}{2}$$

for all k, m in $\mathbb{N}$, from which it follows at once that there exists a positive integer K such that $\mid X_k - X_m \mid < \varepsilon$ whenever k, $m \geq K$. It results from this analysis (and that in Example J) that if each of the spaces $\mathcal{E}_n$ is complete, and if $\{X_k\}$ is a Cauchy sequence with respect to the quasinorm (12), then $\{X_k\}$ is termwise convergent, and therefore convergent with respect to the quasinorm (12); consequently $\sum_{n \in \mathbb{N}_0} + \mathcal{E}_n$ is complete. Thus the direct sum of a finite or countably infinite number of F-spaces, as constructed in Example J, is an F-space. Finally, the space $\dot{\mathcal{M}}$ in Example K is complete in the quasinorm defined there by virtue of the properties of convergence in measure (see Problem 7X).

It is important to know that every normed space can be embedded isometrically and isomorphically as a dense linear manifold in a Banach space, and likewise that every quasinormed space can be similarly embedded in an F-space.

Theorem 11.16. *If $\mathcal{E}$ is a quasinormed space, then there exists an F-space $\hat{\mathcal{E}}$ and a linear isomorphism φ of $\mathcal{E}$ onto a dense linear manifold $\varphi(\mathcal{E})$ in $\hat{\mathcal{E}}$ such that $\mid \varphi(x) \mid = \mid x \mid$ for every x in $\mathcal{E}$. Furthermore, the pair $(\hat{\mathcal{E}}, \varphi)$ is unique in the following sense: If $\hat{\mathcal{E}}_1$ is another F-space and φ_1 a linear isomorphism of $\mathcal{E}$ into $\hat{\mathcal{E}}_1$ possessing the properties that $\varphi_1(\mathcal{E})^- = \hat{\mathcal{E}}_1$ and $\mid \varphi_1(x) \mid = \mid x \mid$ for every x in $\mathcal{E}$, then there exists a unique linear isomorphism Φ of $\hat{\mathcal{E}}$ onto $\hat{\mathcal{E}}_1$ such that $\mid \Phi(y) \mid = \mid y \mid$ for every y in $\hat{\mathcal{E}}$ and such that $\Phi \circ \varphi = \varphi_1$. Finally, if the quasinorm on $\mathcal{E}$ is a norm (so that $\mathcal{E}$ is a normed space), then $\hat{\mathcal{E}}$ is a Banach space.*

Proof. Suppose first that $(\hat{\mathcal{E}}, \varphi)$ and $(\hat{\mathcal{E}}_1, \varphi_1)$ are two pairs satisfying the stated conditions. Setting $\Phi_0(\varphi(x)) = \varphi_1(x)$, $x \in \mathcal{E}$, defines a linear isomorphism Φ_0 of $\varphi(\mathcal{E})$ onto $\varphi_1(\mathcal{E})$ which is simultaneously an isometry of $\varphi(\mathcal{E})$ onto $\varphi_1(\mathcal{E})$. Hence Φ_0 admits a unique isometric extension Φ mapping $\hat{\mathcal{E}}$ onto $\hat{\mathcal{E}}_1$ (Prob. 4H), and Φ is also a linear transformation (and therefore a linear space isomorphism) because both Φ and the linear operations in $\hat{\mathcal{E}}$ and $\hat{\mathcal{E}}_1$ are continuous. Thus Φ exists and is uniquely determined by $(\hat{\mathcal{E}}, \varphi)$ and $(\hat{\mathcal{E}}_1, \varphi_1)$.

To complete the proof, consider the linear space $\mathscr{C}$ obtained by defining linear operations termwise on the set of all Cauchy sequences $\{x_n\}_{n=1}^{\infty}$ of vectors in $\mathcal{E}$. If $\{x_n\}$ is an element of $\mathscr{C}$, then it is readily verified that the numerical sequence $\{\mid x_n \mid\}$ is Cauchy and therefore convergent in $\mathbb{R}$, and we define

$$v_0(\{x_n\}) = \lim_n \mid x_n \mid$$

for every sequence $\{x_n\}$ in $\mathscr{C}$. A brief check shows that v_0 is nonnegative and satisfies conditions (ii'') and (iii) in the definition of a quasinorm (and that

v_0 also satisfies condition (ii′) in the definition of a norm when $\| \ \|$ is a norm on $\mathscr{E}$). It follows from these observations that the set

$$\mathscr{M} = \{\{x_n\} \in \mathscr{C} : v_0(\{x_n\}) = 0\}$$

is a linear manifold in $\mathscr{E}$. Moreover, if two sequences $\{x_n\}$ and $\{y_n\}$ in $\mathscr{C}$ are congruent modulo $\mathscr{M}$, that is, if there exists a sequence $\{z_n\}$ in $\mathscr{M}$ such that $\{x_n\} = \{y_n + z_n\}$, then $v_0(\{x_n\}) \leq v_0(\{y_n\}) + v_0(\{z_n\}) = v_0(\{y_n\})$. Similarly $v_0(\{y_n\}) \leq v_0(\{x_n\})$, of course, so $v_0(\{x_n\}) = v_0(\{y_n\})$. Hence we may, and do, define

$$\| [\{x_n\}] \|_0 = v_0(\{x_n\}), \qquad \{x_n\} \in \mathscr{C},$$

where $[\{x_n\}]$ denotes the coset $\{x_n\} + \mathscr{M}$ of $\{x_n\}$ in the quotient vector space $\mathscr{C}/\mathscr{M}$. Clearly $\| \ \|_0$ is a value on $\mathscr{C}/\mathscr{M}$ that satisfies condition (ii″) in the definition of a quasinorm. Hence it is only necessary to verify condition (iv) in that definition to show that $\| \ \|_0$ is a quasinorm on $\mathscr{C}/\mathscr{M}$. This we shall do by showing that if $\{\alpha_k\}_{k=1}^{\infty}$ is a null sequence in $\mathbb{C}$, if $\{x_n\}_{n=1}^{\infty}$ is a Cauchy sequence in $\mathscr{E}$, and if $\varepsilon > 0$, then there exists a positive integer K such that $\lim_n \| \alpha_k x_n \| \leq \varepsilon$ for all $k \geq K$. To this end we first choose a positive integer N such that $\| x_n - x_N \| < \varepsilon/2$ for all $n \geq N$, and then a second positive integer K such that $\| \alpha_k x_N \| < \varepsilon/2$ and $|\alpha_k| \leq 1$ for all $k \geq K$. Then for $k \geq K$ we have

$$\| \alpha_k x_n \| \leq \| \alpha_k x_N \| + \| x_n - x_N \| < \frac{\varepsilon}{2} + \frac{\varepsilon}{2} = \varepsilon$$

for all $n \geq N$, so that $\lim_n \| \alpha_k x_n \| \leq \varepsilon$. (In the event that $\| \ \|$ is a norm on $\mathscr{E}$ it is immediate that $\| \ \|_0$ is also a norm on $\mathscr{C}/\mathscr{M}$, and the last portion of the proof is unnecessary.)

For each x in $\mathscr{E}$ we next define $\varphi(x)$ to be the coset $[\{x, x, \ldots\}]$ of the constant sequence $\{x, x, \ldots\}$. It is clear that φ is a linear transformation of $\mathscr{E}$ into $\mathscr{C}/\mathscr{M}$ and that $\|\varphi(x)\|_0 = \|x\|$ for all vectors x. Hence it suffices to prove that $\mathscr{C}/\mathscr{M}$ is complete with respect to the quasinorm $\| \ \|_0$ and that $\varphi(\mathscr{E})$ is dense in $\mathscr{C}/\mathscr{M}$ (for then the proof may be completed by defining $\hat{\mathscr{E}}$ to be $\mathscr{C}/\mathscr{M}$). To see this we note that if $\{x_n\} \in \mathscr{C}$ and if $\varepsilon > 0$, then there exists a positive integer N such that $\| x_n - x_m \| < \varepsilon$ for all $m, n \geq N$, from which it follows that $\| [\{x_n\}] - \varphi(x_m) \|_0 \leq \varepsilon$ for all $m \geq N$. Thus $\lim_m \varphi(x_m) = [\{x_n\}]$ in $\mathscr{C}/\mathscr{M}$. This shows, in particular, that $\varphi(\mathscr{E})$ is indeed dense in $\mathscr{C}/\mathscr{M}$. But also, if $\{x_n\}$ is any sequence in $\mathscr{E}$ such that the sequence $\{\varphi(x_n)\}$ is Cauchy in $\mathscr{C}/\mathscr{M}$, then $\{x_n\}$ is itself Cauchy in $\mathscr{E}$ (since φ is an isometry), and this in turn implies that $\{\varphi(x_n)\}$ converges in $\mathscr{C}/\mathscr{M}$ (to $[\{x_n\}]$). Since $\varphi(\mathscr{E})$ is dense in $\mathscr{C}/\mathscr{M}$, it follows that $\mathscr{C}/\mathscr{M}$ is complete (Prob. 4J), and the theorem is proved. $\qquad\square$

We observe that this result can be paraphrased by saying that the completion of a quasinormed [normed] space *as a metric space* can be taken to be a quasinormed [normed] space, and therefore an F-space [Banach space]. In any event, the (essentially unique) space $\hat{\mathscr{E}}$ of Theorem 11.16 is called the *completion* of $\mathscr{E}$. We shall regularly adopt the common practice of using the isometric isomorphism φ to identify $\mathscr{E}$ with $\varphi(\mathscr{E})$. Thus we shall ordinarily regard $\mathscr{E}$ as a dense linear submanifold of its completion $\hat{\mathscr{E}}$. (The reader may wish to compare the above argument with that sketched in Problems 4H–4L.)

Up to this point we have concerned ourselves exclusively with linear spaces that are equipped with some sort of invariant metric. Consideration of the role played by the function v_0 in the proof of Theorem 11.16 suggests the desirability of considering yet other functions on linear spaces that, like quasinorms, are similar to but more general than norms, and experience shows that this is, indeed, frequently useful.

Definition. A nonnegative real-valued function σ on a [real] linear space $\mathscr{E}$ is a *pseudonorm* on $\mathscr{E}$ if it satisfies the following conditions:

(i′) $\sigma(0) = 0$,
(ii′) $\sigma(\lambda x) = |\lambda|\sigma(x)$ for every x in $\mathscr{E}$ and every [real] scalar λ,
(iii) $\sigma(x + y) \leq \sigma(x) + \sigma(y)$ for all x, y in $\mathscr{E}$.

It is readily seen that if σ is a pseudonorm on a linear space $\mathscr{E}$, then

$$\rho_0(x, y) = \sigma(x - y), \qquad x, y \in \mathscr{E},$$

defines a pseudometric on $\mathscr{E}$ (Ch. 4, p. 56) that is *invariant*, i.e., satisfies the condition

$$\rho_0(x + z, y + z) = \rho_0(x, y)$$

for all vectors x, y, z in $\mathscr{E}$. Conversely, it is clear that if ρ_0 is an invariant pseudometric on $\mathscr{E}$, then $\sigma(x) = \rho_0(x, 0)$ defines a pseudonorm on $\mathscr{E}$ (that in turn defines the given pseudometric) if and only if ρ_0 satisfies the condition

$$\rho_0(\lambda x, \lambda y) = |\lambda|\rho_0(x, y), \qquad x, y \in \mathscr{E}, \qquad \lambda \in \mathbb{C}$$

(cf. Proposition 11.4). Furthermore, it is also easy to see that the topology induced by the pseudometric defined by a pseudonorm σ on $\mathscr{E}$ (this topology is said to be *induced* by σ) turns $\mathscr{E}$ into a topological space in such a way that the $\mathscr{E}$-valued mappings $(x, y) \to x + y$ and $(\alpha, x) \to \alpha x$ are continuous on $\mathscr{E} \times \mathscr{E}$ and $\mathbb{C} \times \mathscr{E}$, respectively. Likewise the function σ satisfies the inequality $|\sigma(x) - \sigma(y)| \leq \sigma(x - y)$ for all vectors x, y in $\mathscr{E}$, and is therefore continuous with respect to this topology (cf. Proposition 11.5). All of these facts work out exactly as in the earlier situation in which we dealt with norms, and to derive them in detail would be repetitious. The novel feature of a pseudonorm σ is the possible existence of vectors $x \neq 0$ such that $\sigma(x) = 0$. The set of vectors x with the property that $\sigma(x) = 0$ is called the *zero space* with respect to σ. Concerning this space we have the following elementary but useful result.

Proposition 11.17. *Let σ be a pseudonorm on a linear space $\mathscr{E}$, and let $\mathscr{Z}$ denote the zero space with respect to σ. Then, in the topology induced on $\mathscr{E}$ by σ, $\mathscr{Z}$ is a closed linear manifold in $\mathscr{E}$ coinciding with the closure $(0)^-$ of the trivial linear manifold (0). If x and y are congruent vectors modulo $\mathscr{Z}$ (that is, if $x - y \in \mathscr{Z}$), then $\sigma(x) = \sigma(y)$. Hence, if for each vector x in $\mathscr{E}$ we denote by $\dot{x}$ the coset $x + \mathscr{Z}$, then $\|\dot{x}\| = \sigma(x)$ defines unambiguously a nonnegative function on the quotient space $\dot{\mathscr{E}} = \mathscr{E}/\mathscr{Z}$, and this function is a norm on $\dot{\mathscr{E}}$. (The norm $\|\ \|$ is called the norm associated with σ, and $\dot{\mathscr{E}}$ equipped with the associated norm $\|\ \|$ is called the normed space associated with the given space $\mathscr{E}$ and pseudonorm σ.) Finally, the metric defined by the associated norm $\|\ \|$ coincides with the metric associated with the invariant pseudo-metric defined by σ (Prob. 4G).*

PROOF. That $\mathscr{Z}$ is a linear manifold in $\mathscr{E}$ is clear from the defining properties of a pseudonorm. If $x \notin \mathscr{Z}$, then $d = \sigma(x) > 0$, and if $\sigma(x - y) < d/2$, then $\sigma(y) > d/2$ by the triangle inequality. Thus the set $\{y \in \mathscr{E} : \sigma(x - y) < d/2\}$ is a neighborhood of x (in the topology induced by σ) that is disjoint from $\mathscr{Z}$. Hence the complement $\mathscr{E} \backslash \mathscr{Z}$ is open, and $\mathscr{Z}$ is closed, in that topology. Moreover, if $z \in \mathscr{Z}$ and if $\varepsilon > 0$, then $\sigma(z - 0) = \sigma(z) = 0 < \varepsilon$, so every neighborhood of z in the topology induced by σ contains 0. Thus $\mathscr{Z}$ is contained in $(0)^-$ and, being closed, must equal $(0)^-$. The rest of the proof is completely routine, and is omitted. $\qquad \square$

Example M. Let $\mathscr{F}$ be a normed space, and let T be a linear transformation of a linear space $\mathscr{E}$ into $\mathscr{F}$. If we define

$$\sigma(x) = \|Tx\|, \qquad x \in \mathscr{E},$$

then it is a simple matter to verify that σ is a pseudonorm on $\mathscr{E}$. The zero space of σ is just the null space $\mathscr{K}(T)$ (Ch. 2, p. 18). Thus σ is a norm if and only if T is one-to-one. (This example turns out to be universal. If σ is a given pseudonorm on a linear space $\mathscr{E}$, if $\|\ \|$ is the associated norm on $\dot{\mathscr{E}}$, and if π denotes the natural projection of $\mathscr{E}$ onto $\dot{\mathscr{E}}$, then $\sigma(x) = \|\pi(x)\|$ for every x in $\mathscr{E}$.)

There are still other generalizations of the foregoing concepts that merit investigation. One idea is to omit any reference, direct or indirect, to any sort of metric or pseudometric, and simply consider linear spaces that are equipped with a useful, compatible topology. The resulting notion of a *topological linear space* will not play a large or central role in this book, but it is certainly important in its own right, and is valuable for us because it unifies and clarifies the relations between those more special types of linear spaces in which we shall be principally interested.

Definition. A topology on a [real] linear space $\mathscr{E}$ is called a *linear topology* on $\mathscr{E}$ if the mappings $(\alpha, x) \rightarrow \alpha x$ and $(x, y) \rightarrow x + y$ of $[\mathbb{R} \times \mathscr{E}]$ $\mathbb{C} \times \mathscr{E}$ and $\mathscr{E} \times \mathscr{E}$, respectively, into $\mathscr{E}$ are both continuous. A [real] linear space equipped with a linear topology is a [*real*] *topological linear space* or [*real*] *topological vector space*. (If a (complex) topological linear space is regarded as a real linear space by refusing to multiply by any but real scalars, then it is clear that it is also a real topological linear space with respect to the given topology.)

In this terminology it is the substance of Proposition 11.5 that a normed space is a topological linear space in its norm topology, and likewise the substance of Proposition 11.14 that a quasinormed space is a topological linear space in its quasinorm topology. The following proposition is an immediate consequence of the definition of a topological linear space.

Proposition 11.18. *If $\mathscr{E}$ is a topological linear space, then all translations $x \to x + x_0$ and all dilatations $x \to \alpha_0 x$ ($\alpha_0 \neq 0$) are homeomorphisms of $\mathscr{E}$ onto itself. Thus, in particular, $x_0 + V$ is a neighborhood of x_0 in $\mathscr{E}$ if and only if V is a neighborhood of 0.*

Corollary 11.19. *The weight of a topological linear space (Prob. 3A) is the same at every one of its points. In particular, a topological linear space $\mathscr{E}$ satisfies the first axiom of a countability if and only if there exists a countable neighborhood base at the origin 0 in $\mathscr{E}$.*

PROOF. According to Proposition 11.18, if x_0 is any vector in $\mathscr{E}$, then $\mathscr{V}$ is a neighborhood base at the origin 0 if and only if $x_0 + \mathscr{V} = \{x_0 + V : V \in \mathscr{V}\}$ is a neighborhood base at x_0.   □

Definition. A topological linear space $\mathscr{E}$ is said to be *separated* if $\mathscr{E}$ is Hausdorff as a topological space.

Proposition 11.20. *The following conditions are equivalent for an arbitrary topological linear space $\mathscr{E}$:*

  (i) *$\mathscr{E}$ is separated,*
  (ii) *Every singleton in $\mathscr{E}$ is a closed set,*
  (iii) *At least one singleton in $\mathscr{E}$ is a closed set,*
  (iv) *The origin 0 is not in the closure $\{x\}^-$ of any singleton $\{x\}$, $x \neq 0$.*

PROOF. It is clear that (i) implies (ii) and that (ii) implies (iv). Moreover, Proposition 11.18 shows that (ii) and (iii) are equivalent in every topological linear space. To complete the proof, suppose (iv) holds and let x be a vector in $\mathscr{E}$ such that $x \neq 0$. Then there exists a neighborhood V of 0 such that $x \notin V$. Let W be a neighborhood of 0 such that $W - W$ ($= \{u - v : u, v \in W\}$) is contained in V. (Such a neighborhood exists because the mapping $(u, v) \to u - v$ is continuous at $(0, 0)$; cf. Problem L.) Then W and $x + W$ are disjoint neighborhoods of 0 and x respectively, and it follows from Proposition 11.18 that $\mathscr{E}$ is separated.   □

If $\mathscr{E}$ is a topological linear space and $\mathscr{L}$ is a linear manifold in $\mathscr{E}$, then it is easily seen that $\mathscr{L}$ is itself a topological linear space in the relative topology. In this context the following elementary result is valid.

Proposition 11.21. *If $\mathscr{L}$ is a linear manifold in a topological linear space $\mathscr{E}$, then $\mathscr{L}^-$ is also a linear manifold in $\mathscr{E}$.*

PROOF. Let x and y be vectors in $\mathscr{L}^-$ and suppose V is a neighborhood of $x + y$. There exist neighborhoods W_1 and W_2 of x and y, respectively, such that $W_1 + W_2 \subset V$. Since both W_1 and W_2 contain vectors belonging to $\mathscr{L}$, this shows that V contains a vector belonging to $\mathscr{L}$, and it follows that $x + y \in \mathscr{L}^-$. Similarly, one shows that $\mathscr{L}^-$ is closed with respect to multiplication by scalars. $\qquad\square$

Proposition 11.22. *Let $\mathscr{E}$ be a topological linear space and let $\mathscr{M}$ be a linear submanifold of $\mathscr{E}$. Then the quotient topology on $\mathscr{E}/\mathscr{M}$ (Prob. 3H) is a linear topology with the property that the natural projection π of $\mathscr{E}$ onto $\mathscr{E}/\mathscr{M}$ is both open and continuous. The quotient space $\mathscr{E}/\mathscr{M}$ (consisting of the quotient linear space $\mathscr{E}/\mathscr{M}$ equipped with the quotient topology) is separated if and only if $\mathscr{M}$ is closed in $\mathscr{E}$.*

PROOF. It is clear in general from the definition that the natural projection π is continuous. (Recall that a set U in $\mathscr{E}/\mathscr{M}$ is only said to be open in the quotient topology when $\pi^{-1}(U)$ is open in $\mathscr{E}$.) In the case at hand, if A is an arbitrary subset of $\mathscr{E}$, then $\pi^{-1}(\pi(A))$ coincides with the set $\{x + z : x \in A,\ z \in \mathscr{M}\}$, and this set can be written either as a union of cosets ($\pi^{-1}(\pi(A)) = \bigcup_{x \in A} x + \mathscr{M}$) or as a union of translates of A ($\pi^{-1}(\pi(A)) = \bigcup_{z \in \mathscr{M}} z + A$). The latter representation shows that $\pi^{-1}(\pi(U))$ is open in $\mathscr{E}$ along with U, and hence that π is an open mapping.

To see that the quotient topology is also a linear topology, let x and y be vectors in $\mathscr{E}$, and suppose V is a neighborhood in $\mathscr{E}/\mathscr{M}$ of $[x] + [y] = [x + y]$. Then $\pi^{-1}(V)$ is a neighborhood of $x + y$ in $\mathscr{E}$, so there exist neighborhoods W_1 and W_2 of x and y, respectively, such that $W_1 + W_2 \subset \pi^{-1}(V)$. Since π is open, it follows that $\pi(W_1)$ and $\pi(W_2)$ are neighborhoods of $[x]$ and $[y]$, respectively, in $\mathscr{E}/\mathscr{M}$, and

$$\pi(W_1) + \pi(W_2) = \pi(W_1 + W_2) \subset \pi(\pi^{-1}(V)) = V.$$

Thus vector addition is continuous with respect to the quotient topology on $\mathscr{E}/\mathscr{M}$, and an entirely similar argument shows that the mapping $(\alpha, x) \to \alpha x$ is continuous too.

To complete the proof, we note first that if $\mathscr{E}/\mathscr{M}$ is separated, then the singleton $\{[0]\}$ is closed in $\mathscr{E}/\mathscr{M}$ and $\pi^{-1}(\{[0]\}) = \mathscr{M}$. Thus the condition is necessary. On the other hand, if $\mathscr{M}$ is closed in $\mathscr{E}$, then the complement $\mathscr{E}\backslash\mathscr{M}$ is open, and this set projects onto the complement of $\{[0]\}$. Since π is open, it follows that $\mathscr{E}/\mathscr{M}$ is separated. Thus the condition is also sufficient. $\qquad\square$

Corollary 11.23. *If $\mathscr{E}$ is an arbitrary topological linear space, then $\mathscr{Z} = (0)^-$ is a closed linear submanifold of $\mathscr{E}$ with the property that the quotient space $\mathscr{E}/\mathscr{Z}$ is separated. (The linear manifold $\mathscr{Z}$ is called the* zero space *of $\mathscr{E}$; the space $\mathscr{E}/\mathscr{Z}$ is called the* separated space *associated with $\mathscr{E}$, and is customarily denoted by $\mathring{\mathscr{E}}$. Cf. Proposition 11.17.)*

Regarding products of topological linear spaces, we have the following sweeping, but not very fruitful, result. The proof is elementary, and is left as an exercise (Prob. R).

Proposition 11.24. *If $\{\mathscr{E}_\gamma\}_{\gamma \in \Gamma}$ is an arbitrary indexed family of topological vector spaces, then the product topology (Ex. 3N) is a linear topology on the full algebraic direct sum of the given family. (The full algebraic direct sum of the indexed family $\{\mathscr{E}_\gamma\}$ equipped with the product topology is usually called the* product *of the family.)*

It is frequently necessary to deal simultaneously with several linear topologies on a space. In this connection the following easy result is of basic importance.

Proposition 11.25. *If $\mathscr{E}$ is a linear space and $\{\mathscr{T}_\gamma\}_{\gamma \in \Gamma}$ is a family of linear topologies on $\mathscr{E}$, then the topology $\mathscr{T} = \sup_\gamma \mathscr{T}_\gamma$ (Prob. 3G) is also a linear topology on $\mathscr{E}$.*

PROOF. Let x and y be vectors in $\mathscr{E}$, and let

$$U = U_1 \cap \cdots \cap U_n$$

be a typical basic open set in $\mathscr{T}$ containing the vector $x + y$, where $U_i \in \mathscr{T}_{\gamma_i}$, $i = 1, \ldots, n$ (Ex. 3M). Then for each index i there are open sets V_i and W_i in $\mathscr{T}_{\gamma_i}$ such that $x \in V_i, y \in W_i$, and $V_i + W_i \subset U_i$. But then $V = V_1 \cap \cdots \cap V_n$ and $W = W_1 \cap \cdots \cap W_n$ are open sets in the topology $\mathscr{T}$ containing x and y, respectively, and $V + W$ is contained in U. This shows that addition is continuous as a mapping of $\mathscr{E} \times \mathscr{E}$ into $\mathscr{E}$ with respect to $\mathscr{T}$. An entirely similar argument shows that the mapping $(\alpha, x) \to x$ is also continuous, and the result follows, at least when the index set Γ is nonempty. The supremum of the empty family of topologies is the indiscrete topology on $\mathscr{E}$, clearly a linear topology. $\square$

The collections of linear topologies with which we shall be concerned will frequently be induced by some corresponding collection of pseudonorms, and the linear topologies that are obtained in this way are of particular importance. We conclude this chapter with a study of some of their properties. (A second, independent, characterization of this class of topological vector spaces will be given in Chapter 14.) Our first result in this direction facilitates the comparison of a pseudonorm topology with a given linear topology.

Proposition 11.26. *Let $\mathscr{E}$ be a topological vector space with linear topology $\mathscr{T}$ and let σ be a pseudonorm on $\mathscr{E}$. The following conditions are equivalent:*

(i) *The topology $\mathscr{T}$ refines the topology induced by σ,*

(ii) *For an arbitrary positive number ε there exists in $\mathscr{E}$ a neighborhood V of 0 with respect to $\mathscr{T}$ such that $V \subset \{x \in \mathscr{E} : \sigma(x) < \varepsilon\}$,*

(iii) *The pseudonorm σ is continuous on $\mathscr{E}$ with respect to $\mathscr{T}$.*

PROOF. Since the ball $\{x \in \mathscr{E} : \sigma(x) < \varepsilon\}$ is a neighborhood of 0 in the topology induced by σ, it is clear that (i) implies (ii). To show that (ii) implies (iii) we observe that if x and x_0 are vectors in $\mathscr{E}$, then $|\sigma(x) - \sigma(x_0)| \leq \sigma(x - x_0)$. Hence if V is a neighborhood of 0 with respect to $\mathscr{T}$ that is contained in $\{x \in \mathscr{E} : \sigma(x) < \varepsilon\}$, then $x \in x_0 + V$ implies that $|\sigma(x) - \sigma(x_0)| < \varepsilon$. Since ε is arbitrary, this shows that σ is continuous at the arbitrary vector x_0, and hence that σ is continuous on $\mathscr{E}$ with respect to $\mathscr{T}$. Finally, to see that (iii) implies (i), let ε be a positive number, and suppose $y \in \{x \in \mathscr{E} : \sigma(x) < \varepsilon\}$. If (iii) holds, then there exists a neighborhood V of 0 with respect to $\mathscr{T}$ such that if $z \in y + V$ then $|\sigma(y) - \sigma(z)| < \varepsilon - \sigma(y)$, and therefore $\sigma(z) < \varepsilon$. Thus $\{x \in \mathscr{E} : \sigma(x) < \varepsilon\}$ belongs to $\mathscr{T}$, and since both that topology and the one induced by σ are linear, the result follows. $\qquad\square$

Suppose now that there is given an indexed family $\{\sigma_\gamma\}_{\gamma \in \Gamma}$ of pseudonorms on a linear space $\mathscr{E}$. According to Proposition 11.25, the supremum of the family of linear topologies induced by the various pseudonorms σ_γ is a linear topology on $\mathscr{E}$, and according to Proposition 11.26, this topology may also be characterized as the coarsest linear topology on $\mathscr{E}$ with respect to which all of the pseudonorms σ_γ are continuous.

Definition. If $\{\sigma_\gamma\}_{\gamma \in \Gamma}$ is a family of pseudonorms on a linear space $\mathscr{E}$, then the supremum of the collection of linear topologies induced by them is called the topology *induced* by the given family $\{\sigma_\gamma\}$.

The following facts are easy consequences of the foregoing definition, Proposition 3.20, and Example 3M.

Proposition 11.27. *If $\{\sigma_\gamma\}_{\gamma \in \Gamma}$ is an indexed family of pseudonorms on a linear space $\mathscr{E}$, then the collection of all sets of the form*

$$U(\gamma_1, \ldots, \gamma_n; \varepsilon) = \{x \in \mathscr{E} : \sigma_{\gamma_i}(x) < \varepsilon, \ i = 1, \ldots, n\},$$

where $\{\gamma_1, \ldots, \gamma_n\}$ runs over all finite subsets of Γ and ε runs independently over all positive numbers, forms a base of open neighborhoods of 0 in the topology induced on $\mathscr{E}$ by the family $\{\sigma_\gamma\}$, and the collection of all translates $x_0 + U(\gamma_1, \ldots, \gamma_n; \varepsilon)$, $x_0 \in \mathscr{E}$, of these sets forms a base for that topology. Moreover, if $\{x_\lambda\}$ is a net in $\mathscr{E}$, then $\{x_\lambda\}$ converges to a limit x in the topology induced on $\mathscr{E}$ by the family $\{\sigma_\gamma\}$ if and only if $\lim_\lambda \sigma_\gamma(x - x_\lambda) = 0$ for every index γ.

PROOF. According to Example 3M, the collection of all finite intersections of the form

$$V = \{x \in \mathscr{E}: \sigma_{\gamma_1}(x - x_1) < \varepsilon_1\} \cap \cdots \cap \{x \in \mathscr{E}: \sigma_{\gamma_n}(x - x_n) < \varepsilon_n\}$$

constitutes a base for the topology induced on $\mathscr{E}$ by the family $\{\sigma_\gamma\}$. To prove that the collection of all translates $x_0 + U(\gamma_1, \ldots, \gamma_n; \varepsilon)$ also forms a base for this topology, it suffices to show that if $x_0 \in V$, then there exists a positive number ε such that the set

$$W_\varepsilon = \{x \in \mathscr{E}: \sigma_{\gamma_i}(x - x_0) < \varepsilon, \, i = 1, \ldots, n\}$$

is contained in V (since $W_\varepsilon = x_0 + U(\gamma_1, \ldots, \gamma_n; \varepsilon)$). To this end, suppose $x_0 \in V$ and let $\delta_i = \sigma_{\gamma_i}(x_0 - x_i)$, $i = 1, \ldots, n$. Then $\delta_i < \varepsilon_i$, so $\varepsilon = \inf_{1 \le i \le n} \{\varepsilon_i - \delta_i\}$ is positive, and if $x \in W_\varepsilon$, then

$$\sigma_{\gamma_i}(x - x_i) < \sigma_{\gamma_i}(x - x_0) + \sigma_{\gamma_i}(x_0 - x_i) < \varepsilon + \delta_i < \varepsilon_i, \qquad i = 1, \ldots, n,$$

so $x \in V$ and therefore $W_\varepsilon \subset V$. The assertion concerning a neighborhood base at 0 follows from this same calculation, and the assertion concerning convergence of nets is a consequence of Proposition 3.20. $\square$

Definition. A family $\{\sigma_\gamma\}$ of pseudonorms on a linear space $\mathscr{E}$ is said to be *separating* on $\mathscr{E}$ if for each vector $x \neq 0$ in $\mathscr{E}$ there is a pseudonorm $\sigma_{..}$ in the family such that $\sigma_\gamma(x) \neq 0$.

Example N. Let $\mathscr{E}$ be a linear space, let $\mathscr{F}$ be a normed space, and let $\{T_\gamma\}$ be an indexed family of linear transformations of $\mathscr{E}$ into $\mathscr{F}$. If for each index γ we write σ_γ for the pseudonorm $\sigma_\gamma(x) = \| T_\gamma x \|$, $x \in \mathscr{E}$ (Ex. M), then the family $\{\sigma_\gamma\}$ is separating on $\mathscr{E}$ when and only when the family $\{T_\gamma\}$ of linear transformations is separating (Ex. 3R).

Proposition 11.28. *Let $\mathscr{E}$ be a vector space, and let $\{\sigma_\gamma\}$ be an indexed family of pseudonorms on $\mathscr{E}$. Then the intersection*

$$\mathscr{N} = \bigcap_\gamma \{x \in \mathscr{E}: \sigma_\gamma(x) = 0\}$$

is the zero space for the topology induced on $\mathscr{E}$ by the family $\{\sigma_\gamma\}$. In particular, the latter topology is separated if and only if the family $\{\sigma_\gamma\}$ is separating.

PROOF. It is clear that $\mathscr{N}$ is a linear manifold in $\mathscr{E}$ and that $\mathscr{N}$ is contained in the zero space $\mathscr{Z} = (0)^-$ (see Proposition 11.17 and Corollary 11.23). If $x_0 \notin \mathscr{N}$, and if, say, $\sigma_\gamma(x_0) = d > 0$, then $x_0 + \{x \in \mathscr{E}: \sigma_\gamma(x) < d/2\}$ is a neighborhood of x_0 in the topology induced on $\mathscr{E}$ by $\{\sigma_\gamma\}$ that does not contain 0, so $x_0 \notin \mathscr{Z}$. $\square$

It is important to note that the collection of all pseudonorms on a given linear space is a partially ordered set in the ordinary ordering of real-valued functions. In this connection we make the following elementary observation, an immediate consequence of Proposition 11.26.

Proposition 11.29. *If σ and τ are pseudonorms on a linear space $\mathscr{E}$, and if $\sigma \leq \tau$, then the topology induced on $\mathscr{E}$ by τ refines the one induced by σ.*

If σ and τ are two pseudonorms on a linear space $\mathscr{E}$, then a trivial calculation discloses that $\sigma + \tau$ and $\sigma \vee \tau$ are also pseudonorms. This observation clarifies the first assertion of the next result.

Proposition 11.30. *The collection of all pseudonorms on a given linear space $\mathscr{E}$ is a directed set in the usual ordering of real-valued functions. If $\sigma_1, \ldots, \sigma_n$ are pseudonorms on $\mathscr{E}$, then both $\sigma_1 + \cdots + \sigma_n$ and $\sigma_1 \vee \cdots \vee \sigma_n$ are pseudonorms that dominate $\sigma_1, \ldots, \sigma_n$ in that ordering. If $\mathscr{T}_i$ denotes the topology induced on $\mathscr{E}$ by $\sigma_i, i = 1, \ldots, n$, then both $\sigma_1 + \cdots + \sigma_n$ and $\sigma_1 \vee \cdots \vee \sigma_n$ induce the topology $\mathscr{T} = \mathscr{T}_1 \vee \cdots \vee \mathscr{T}_n$.*

PROOF. As noted above, it is trivial to verify that $\sigma_1 + \cdots + \sigma_n$ and $\sigma_1 \vee \cdots \vee \sigma_n$ are pseudonorms, and it is visible that they both dominate the given pseudonorms. Hence, by the foregoing proposition, both $\sigma_1 + \cdots + \sigma_n$ and $\sigma_1 \vee \cdots \vee \sigma_n$ induce topologies on $\mathscr{E}$ that refine $\mathscr{T}$. Moreover, if ε is a positive number, and if $\sigma_i(x) < \varepsilon/n, i = 1, \ldots, n$ (equivalently, if x belongs to the intersection of the balls $\{x \in \mathscr{E} : \sigma_i(x) < \varepsilon/n\}$), then $\sigma_1(x) \vee \cdots \vee \sigma_n(x) \leq \sigma_1(x) + \cdots + \sigma_n(x) < \varepsilon$. Thus both pseudonorms are continuous with respect to $\mathscr{T}$, and the proof is complete by virtue of Proposition 11.26. ☐

Definition. An indexed family $\{\sigma_\gamma\}$ of pseudonorms on a vector space $\mathscr{E}$ is said to be *saturated* if it is a *directed* subset of the directed set of all pseudonorms on $\mathscr{E}$, that is, if for any two pseudonorms σ_{γ_1} and σ_{γ_2} in the family, there is a σ_γ in the family such that $\sigma_{\gamma_i} \leq \sigma_\gamma, i = 1, 2$.

Example O. Let $\mathscr{E}$ be a topological linear space whose topology is induced by an indexed family $\{\sigma_\gamma\}$ of pseudonorms, and let $\mathscr{M}$ be a linear submanifold of $\mathscr{E}$. Each of the pseudonorms σ'_γ obtained by setting

$$\sigma'_\gamma(x) = \inf_{z \in \mathscr{M}} \sigma_\gamma(x + z), \qquad x \in \mathscr{E},$$

is continuous on $\mathscr{E}/\mathscr{M}$ in the quotient topology (see Problem H), so the topology $\mathscr{T}'$ induced on $\mathscr{E}/\mathscr{M}$ by the family $\{\sigma'_\gamma\}$ is refined by the quotient topology according to Proposition 11.26. If the family $\{\sigma_\gamma\}$ is saturated, then it is readily seen that $\mathscr{T}'$ coincides with the quotient topology. (If the requirement of saturation is not imposed, these two topologies may well be different. Consider the pair of pseudonorms

$$\sigma(\xi, \eta) = |\xi| \quad \text{and} \quad \tau(\xi, \eta) = |\eta|$$

on $\mathbb{C}^2$ with $\mathscr{M} = \{(\xi, \eta): \xi = \eta\}$.) It should be remarked that the restrictions to $\mathscr{M}$ of the pseudonorms σ_γ always induce the relative topology on $\mathscr{M}$, no matter whether the family $\{\sigma_\gamma\}$ is saturated or not.

The following result is an immediate consequence of Proposition 11.27.

Proposition 11.31. *Let $\{\sigma_\gamma\}_{\gamma \in \Gamma}$ be a saturated family of pseudonorms on a linear space $\mathscr{E}$. Then a base of neighborhoods of 0 in the topology induced on $\mathscr{E}$ by the family is provided by the collection of open balls $\{x \in \mathscr{E} : \sigma_\gamma(x) < \varepsilon\}$ (or by the collection of all closed balls $\{x \in \mathscr{E} : \sigma_\gamma(x) \leq \varepsilon\}$), $\gamma \in \Gamma$, $\varepsilon > 0$.*

Note. If, starting from an arbitrary family $\{\sigma_\gamma\}$ of pseudonorms on a vector space $\mathscr{E}$, we simply adjoin to that family the sums (or suprema) of all of its finite subsets, we obtain a larger saturated family that induces the same topology on $\mathscr{E}$ as did the given family $\{\sigma_\gamma\}$. Thus one may always assume, without loss of generality, that a given family of pseudonorms is saturated whenever that is convenient.

We close this discussion with a useful criterion for the metrizability of a topology induced by a family of pseudonorms. (For a more general criterion for the metrizability of a linear topology see Problem Y.)

Proposition 11.32. *Let $\{\sigma_\gamma\}$ be a separating indexed family of pseudonorms on a linear space $\mathscr{E}$ and let $\mathscr{T}$ denote the topology induced on $\mathscr{E}$ by $\{\sigma_\gamma\}$. If the family $\{\sigma_\gamma\}$ is countably determined (Prob. 1P), then $\mathscr{T}$ is metrizable, and is, in fact, induced by a (suitably constructed) quasinorm. (A topological vector space whose topology is induced by a quasinorm is said to be quasi-normable.) Conversely, if $\mathscr{T}$ is metrizable, then a countable subfamily of the family $\{\sigma_\gamma\}$ suffices to induce $\mathscr{T}$.*

PROOF. Suppose the family $\{\sigma_\gamma\}$ is countably determined, and let the sequence $\{\sigma_n = \sigma_{\gamma_n}\}_{n=1}^\infty$ be cofinal in $\{\sigma_\gamma\}$, so that the topology induced on $\mathscr{E}$ by the sequence $\{\sigma_n\}$ coincides with $\mathscr{T}$ (Prop. 11.29). Define

$$|x| = \sum_{n=1}^\infty \frac{1}{2^n} \frac{\sigma_n(x)}{1 + \sigma_n(x)}, \qquad x \in \mathscr{E}.$$

If $x \neq 0$ in $\mathscr{E}$ then $\sigma_\gamma(x) > 0$ for some index γ, and if $\sigma_\gamma \leq \sigma_{\gamma_m}$, then $\sigma_m(x) > 0$ and therefore $|x| > 0$. Hence $|\ |$ is a value on $\mathscr{E}$. Moreover, if $\{x_p\}_{p=1}^\infty$ is a sequence in $\mathscr{E}$ such that $\lim_p \sigma_n(x_p) = 0$ for every positive integer n, and if ε is a positive number, then we can choose a positive integer K such that $1/2^K = \sum_{n=K+1}^\infty 1/2^n < \varepsilon/2$, and then a second positive integer P such that

$$\sum_{n=1}^K \frac{1}{2^n} \frac{\sigma_n(x_p)}{1 + \sigma_n(x_p)} < \frac{\varepsilon}{2}$$

for every $p > P$. Hence $|x_p| < \varepsilon/2 + \varepsilon/2 = \varepsilon$ for all $p > P$, and it follows that $\lim_p |x_p| = 0$. In particular, if $\{\alpha_p\}_{p=1}^\infty$ is a null sequence in $\mathbb{C}$, then $|\alpha_p x| \to 0$ for each fixed vector x in $\mathscr{E}$, which shows that $|\ |$ is a quasinorm. To see that the linear topology induced by $|\ |$ coincides with that induced by the sequence $\{\sigma_n\}$ (and therefore with $\mathscr{T}$), we note first that if $|x| < \varepsilon/2^n(1 + \varepsilon)$, then $\sigma_n(x) < \varepsilon$. From this it follows at once that the topology induced by $\|\ \|$ refines $\mathscr{T}$ (see Proposition 11.26). Next, let ε be a positive number, and, as above, choose a positive integer K such that $1/2^K < \varepsilon/2$. If $\sigma_n(x) < \varepsilon/2K$, $n = 1, \ldots, K$ (i.e., if x belongs to the intersection of the balls $\{x \in \mathscr{E} : \sigma_n(x) < \varepsilon/2K\}$, $n = 1, \ldots, K$), then $\|x\| < \varepsilon/2 + \varepsilon/2 = \varepsilon$. Thus every neighborhood of 0 with respect to $\|\ \|$ contains a neighborhood of 0 in the topology $\mathscr{T}$, and we have proved that $\mathscr{T}$ is the topology induced by $\|\ \|$.

To go the other way, suppose $\mathscr{T}$ is metrizable, and let $\{U_n\}$ be a countable neighborhood base at 0 with respect to $\mathscr{T}$. For each n there exists a finite subset D_n of Γ and a positive number ε_n such that

$$\{x \in \mathscr{E} : \sigma_\gamma(x) < \varepsilon_n, \gamma \in D_n\} \subset U_n,$$

and if we set $\Gamma_0 = \bigcup_n D_n$, then the countable subfamily $\{\sigma_\gamma\}_{\gamma \in \Gamma_0}$ induces $\mathscr{T}$. $\qquad\square$

PROBLEMS

A. Verify inequality (3) of Example B. (Hint: One may use the techniques of elementary calculus to show that if $c > 0$, then the function $cu - u^p/p$ is bounded above on the ray $u \geq 0$, and has maximum c^q/q there. Alternatively, (3) may be verified by means of Lagrange multipliers.)

B. Verify the Hölder inequality for the case in which a and b belong to (ℓ_p) and (ℓ_q), respectively (cf. (4)), and use that result to derive the Minkowski inequality. (Hint: Since p exceeds one, we may write

$$\sum_i |\alpha_i + \beta_i|^p \leq \sum_i |\alpha_i| |\alpha_i + \beta_i|^{p-1} + \sum_i |\beta_i| |\alpha_i + \beta_i|^{p-1}$$

where, of course, $p - 1 > 0$.)

C. Show that if $1 \leq p < p' < +\infty$, then (ℓ_p) is a dense linear submanifold of $(\ell_{p'})$. Conclude from this fact that the normed space obtained by equipping (ℓ_p) with the norm $\|\ \|_{p'}$ is not complete.

D. Let X be an arbitrary nonempty topological space, let $\mathscr{C}_b(X)$ denote the linear space of all bounded, continuous, complex-valued functions on X, and for each f in $\mathscr{C}_b(X)$ set $\|f\|_\infty = \sup_{x \in X} |f(x)|$. Show that with this definition $\mathscr{C}_b(X)$ becomes a Banach space. (Just as in the special case of a compact Hausdorff space X, where $\mathscr{C}_b(X) = \mathscr{C}(X)$ (Ex. F), this norm is known as the "sup norm".) More generally, if $\mathscr{E}$ is a given normed space, we write $\mathscr{C}_b(X; \mathscr{E})$ for the linear space of all bounded, continuous, $\mathscr{E}$-valued mappings on X, and for each Φ in $\mathscr{C}_b(X; \mathscr{E})$ we set $\|\Phi\|_\infty = \sup_{x \in X} \|\Phi(x)\|$. Show that $\mathscr{C}_b(X; \mathscr{E})$ is a normed space, and that $\mathscr{C}_b(X; \mathscr{E})$ is a Banach space when and only when $\mathscr{E}$ is complete. Show also that if X_0 is a topological space obtained by replacing the given topology on X by some coarser topology, then $\mathscr{C}_b(X_0; \mathscr{E})$ is a subspace of $\mathscr{C}_b(X; \mathscr{E})$.

It is easily seen that the construction employed in Example G goes through without change in the Banach space $\mathscr{C}_b(X)$ provided only that the topological space X is normal so that Urysohn's lemma may be invoked. Hence the Tietze extension theorem is valid for *bounded* functions defined and continuous on closed subsets of a normal space X, and, indeed, it is this more general result that is commonly known in the literature as the *Tietze extension theorem.*

E. If X is an arbitrary nonempty set, then the space $\mathscr{B}(X)$ of all bounded complex-valued functions on X is a Banach space in the sup norm. (cf. Problem 4O). If X is a topological space, then the space $\mathscr{C}_b(X)$ defined in Problem D is a subspace of $\mathscr{B}(X)$. Suppose now that X is a metric space. Show that the uniformly continuous functions in $\mathscr{B}(X)$ form a linear manifold in $\mathscr{B}(X)$. Show, likewise, that the same is true of the Lipschitzian functions in $\mathscr{B}(X)$ (cf. Problem 4F). Are these linear manifolds closed?

F. Consider the collection $\mathscr{C}^{(k)} = \mathscr{C}^{(k)}([a, b])$ of all those complex-valued functions f on the closed interval $[a, b]$, $a < b$, that possess k continuous derivatives $f', f'', \ldots, f^{(k)}$. (See Example 2J; recall that we use right derivatives at a and left derivatives at b. By convention $\mathscr{C}^{(0)}$ coincides with $\mathscr{C}([a, b])$.) Show that for every positive integer k the space $\mathscr{C}^{(k)}$ is a dense linear submanifold of $\mathscr{C}^{(0)}$. Show also that in the norm

$$\| f \|_{(k)} = \| f \|_\infty + \| f' \|_\infty + \cdots + \| f^{(k)} \|_\infty$$

$\mathscr{C}^{(k)}$ becomes a Banach space in its own right.

G. (i) If $\{x_n\}_{n=1}^\infty$ is an infinite sequence of elements of a normed space $\mathscr{E}$, then we say that the infinite series $\sum_{n=1}^\infty x_n$ has *sum* s, or *converges to* s (notation: $s = \sum_{n=1}^\infty x_n$), if s is the limit of the sequence $\{s_n = \sum_{k=1}^n x_k\}_{n=1}^\infty$ of *partial sums*. Show that in a Banach space every *absolutely convergent* series is convergent. That is, show that if $\{x_n\}_{n=1}^\infty$ is a sequence of vectors in a Banach space $\mathscr{E}$ with the property that $\sum_n \| x_n \| < +\infty$, then the series $\sum_{n=1}^\infty x_n$ is convergent in $\mathscr{E}$. Show also that if the series $\sum_{n=1}^\infty x_n$ is absolutely convergent, then

$$\left\| \sum_{n=1}^\infty x_n \right\| \leq \sum_{n=1}^\infty \| x_n \|.$$

(ii) More generally, suppose that $\mathscr{E}$ is a linear space that is complete with respect to an arbitrary invariant metric ρ, and let v denote the associated value on $\mathscr{E}$ ($v(x) = \rho(x, 0)$, $x \in \mathscr{E}$; see Proposition 11.1). Show that if $\{x_n\}_{n=1}^\infty$ is a sequence of vectors in $\mathscr{E}$ such that $\sum_n v(x_n) < +\infty$, then the infinite series $\sum_{n=1}^\infty x_n$ is convergent in $\mathscr{E}$, that is, show that the sequence of partial sums is convergent in $\mathscr{E}$ to some sum s (notation: $s = \sum_{n=1}^\infty x_n$). Show also that in this case

$$v(\textstyle\sum_{n=1}^\infty x_n) \leq \sum_{n=1}^\infty v(x_n).$$

H. (i) Let σ be a pseudonorm on a linear space $\mathscr{E}$, let $\mathscr{M}$ be an arbitrary linear manifold in $\mathscr{E}$, and let $[x]$ denote the coset of x modulo $\mathscr{M}$. Show that

$$\sigma'([x]) = \inf_{z \in \mathscr{M}} \sigma(x + z), \qquad x \in \mathscr{E},$$

defines a pseudonorm on the quotient space $\mathscr{E}/\mathscr{M}$, and that

$$\pi(\{x \in \mathscr{E} : \sigma(x) < r\}) = \{[x] \in \mathscr{E}/\mathscr{M} : \sigma'([x]) < r\}$$

for every positive real number r (where π denotes the natural projection of $\mathscr{E}$ onto $\mathscr{E}/\mathscr{M}$). Conclude that π is both open and continuous with respect to the topologies on $\mathscr{E}$ and $\mathscr{E}/\mathscr{M}$ induced by σ and σ', respectively, and that the latter topology coincides with the quotient topology on $\mathscr{E}/\mathscr{M}$ (Prob. 3H). Show, finally, that σ' is a norm on $\mathscr{E}/\mathscr{M}$ when and only when $\mathscr{M}$ is closed in the topology induced on $\mathscr{E}$ by σ.

(ii) Conclude, in particular, that if $\mathscr{E}$ is a normed space and $\mathscr{M}$ is a subspace of $\mathscr{E}$, then the function defined on $\mathscr{E}/\mathscr{M}$ in (7) is a norm on $\mathscr{E}/\mathscr{M}$.

I. (Generalization and converse of Theorem 11.10) Let v be a value on a linear space $\mathscr{E}$, let $\mathscr{M}$ be a linear submanifold of $\mathscr{E}$ that is closed with respect to the invariant metric defined by v, and let $[x]$ denote the coset in $\mathscr{E}/\mathscr{M}$ of a vector x in $\mathscr{E}$. Show that

$$v'([x]) = \inf_{z \in \mathscr{M}} v(x + z), \qquad x \in \mathscr{E}, \tag{13}$$

defines a value on $\mathscr{E}/\mathscr{M}$, and verify that $\mathscr{E}$ is complete if and only if $\mathscr{M}$ and $\mathscr{E}/\mathscr{M}$ are both complete (with respect to the metrics defined by v and v', respectively). (Hint: To go one way, just follow the proof of Theorem 11.10. To go the other way, show first that if $\{x_n\}$ is a Cauchy sequence in $\mathscr{E}$, then $\{[x_n]\}$ is also a Cauchy sequence in $\mathscr{E}/\mathscr{M}$, and hence that there exists a vector y in $\mathscr{E}$ such that $[x_n] \to [y]$ in $\mathscr{E}/\mathscr{M}$, assuming that the latter is complete. The balance of the argument consists in employing the completeness of $\mathscr{M}$ to find a vector z in $\mathscr{M}$ such that $x_n \to y + z$; use (13).) Verify, in addition, that if v is a quasinorm on $\mathscr{E}$, then v' is a quasinorm on $\mathscr{E}/\mathscr{M}$.

J. Show that if $\mathscr{L}$ is a one-dimensional linear space, then any two norms on $\mathscr{L}$ are simply multiples of one another, and use this fact to prove that every one-dimensional normed space is complete and that every linear functional on a one-dimensional normed space is continuous. Conclude that every finite dimensional linear submanifold of a normed space is complete, and therefore closed in that space. Prove also that every linear functional on a finite dimensional normed space is continuous. (Hint: Use the two preceding problems, and recall that if f is a linear functional on a linear space $\mathscr{E}$, then $\dim(\mathscr{E}/\mathscr{K}(f)) \leq 1$.) Show, finally, that any two norms on the same finite dimensional linear space are equivalent. (Hint: If $X = \{x_1, \ldots, x_n\}$ is an ordered basis for a normed space $\mathscr{E}$, then the mapping that assigns to every vector in $\mathscr{E}$ its n-tuple of coordinates with respect to X is a homeomorphism of $\mathscr{E}$ onto $\mathbb{C}^n$.)

K. An important property that a metric space may or may not possess is that of *separability* (Ch. 3, p. 33). Show that for a normed space $\mathscr{E}$ the following conditions are equivalent:

(i) $\mathscr{E}$ is separable as a metric space,

(ii) There exists a dense linear submanifold of $\mathscr{E}$ having Hamel dimension no greater than $\aleph_0$,

(iii) There exists a countable set of vectors in $\mathscr{E}$ that spans $\mathscr{E}$.

Show also that if $\mathscr{E}$ is a separable normed space and $\hat{\mathscr{E}}$ denotes the completion of $\mathscr{E}$, then $\hat{\mathscr{E}}$ is separable. Show finally that a separable normed space satisfies the second axiom of countability (Prob. 3A). Which of the foregoing assertions are valid for quasinormed spaces?

L. (i) Let $\mathscr{E}$ be a topological linear space, and let $\mathscr{E}^n$ denote the direct sum of n copies of $\mathscr{E}$ (indexed by $\{1, \ldots, n\}$ and equipped with the product topology). Show that

$$((\lambda_1, \ldots, \lambda_n), (x_1, \ldots, x_n)) \rightarrow \lambda_1 x_1 + \cdots + \lambda_n x_n$$

is a continuous mapping of $\mathbb{C}^n \times \mathscr{E}^n$ into $\mathscr{E}$. In particular, the mapping

$$(x_1, \ldots, x_n) \rightarrow x_1 + \cdots + x_n$$

is a continuous mapping of $\mathscr{E}^n$ into $\mathscr{E}$. Conclude that if V is an arbitrary neighborhood of 0 in $\mathscr{E}$, then there exists a neighborhood W of 0 in $\mathscr{E}$ such that

$$\overbrace{W + \cdots + W}^{n} \subset V.$$

(Recall that $\overbrace{W + \cdots + W}^{n}$ denotes the set of all sums $x_1 + \cdots + x_n$, where $x_i \in W, i = 1, \ldots, n$.)

(ii) Show that if M is a dense set in a topological linear space $\mathscr{E}$ and $\mathscr{V}$ is a base of open neighborhoods of 0 in $\mathscr{E}$, then the collection of sets $\{x + V : x \in M, V \in \mathscr{V}\}$ is a base for the topology on $\mathscr{E}$, and conclude that if $\mathscr{E}$ is separable, then $\mathscr{E}$ satisfies the second axiom of countability if and only if it satisfies the first.

This last fact is but the first of a number of ways in which topological linear spaces resemble metric spaces. Not surprisingly, there exists a theory that englobes these two concepts, namely, the theory of *uniform spaces*. For an account thereof the reader may consult [40] or [65].

M. A subset G of a linear space $\mathscr{E}$ is said to be *balanced* if $\lambda x \in G$ whenever $x \in G$ and $|\lambda| \le 1$. Show that if D^- denotes the closed unit disc in $\mathbb{C}$, then a subset G of $\mathscr{E}$ is balanced if and only if $D^- G = G$. (Recall that if A is a set of scalars and M a set of vectors, then AM denotes the set $\{\alpha x : \alpha \in A, x \in M\}$. A set G in a real linear space is *balanced* if $x \in G$ implies $tx \in G$, $-1 \le t \le +1$.) Show that if $\mathscr{E}$ is a topological linear space and V is a neighborhood of 0 in $\mathscr{E}$, then there exists a balanced neighborhood W of 0 such that $W \subset V$, and conclude that in every topological linear space there exists a neighborhood base at 0 consisting exclusively of balanced open [closed] sets. (Hint: There exist a neighborhood V_1 of 0 and a positive radius r such that $D_r^- V_1 \subset V$, where $D_r = \{\lambda \in \mathbb{C} : |\lambda| < r\}$. To show that W can be taken to be open, verify that the interior of a balanced set is balanced. To show that W can be taken to be closed, verify that the closure of a balanced set is balanced, and choose a balanced neighborhood W' of 0 such that $W' + W' \subset W$.)

N. Show that there exists a unique separated linear topology on a one-dimensional linear space $\mathscr{L}$ by showing that if $\mathscr{L}$ is equipped with such a topology, and if x_0 is a nonzero vector in $\mathscr{L}$, so that the mapping $\lambda \overset{\tau}{\rightarrow} \lambda x_0$ is an algebraic isomorphism of the linear space $\mathbb{C}$ onto $\mathscr{L}$, then τ is also a homeomorphism of $\mathbb{C}$ onto $\mathscr{L}$. Conclude that every linear functional on a separated one-dimensional topological linear space is continuous. (Hint: The mapping τ is surely continuous, so it suffices to verify that τ is open or, equivalently, that τ^{-1} is continuous. Choose a balanced open neighborhood V of 0 in $\mathscr{L}$ such that $x_0 \notin V$.) Does $\mathscr{L}$ admit any nonseparated linear topologies?

O. Let $\mathscr{E}$ be a separated topological linear space, and let $\mathscr{L}$ be a one-dimensional linear manifold in $\mathscr{E}$. Show that $\mathscr{L}$ is automatically closed in $\mathscr{E}$. Show also that if the topology on $\mathscr{E}$ is induced by some invariant metric, then $\mathscr{L}$ is automatically complete with respect to that metric. (Hint: If x_0 is a nonzero vector in $\mathscr{L}$ and a net $\{\alpha_\lambda x_0\}_{\lambda \in \Lambda}$ converges in $\mathscr{E}$, then for an arbitrary neighborhood V of 0 in $\mathscr{E}$ there exists an index λ_0 such that $(\alpha_\lambda - \alpha_{\lambda'})x_0 \in V$ for all $\lambda, \lambda' \geq \lambda_0$. Hence the net $\{\alpha_\lambda\}$ is Cauchy in $\mathbb{C}$ and is therefore convergent (Prob. 4M).)

P. (i) Let $\mathscr{M}$ be a linear manifold in a topological vector space $\mathscr{E}$, let x_0 be a vector in $\mathscr{E}$ that does not belong to $\mathscr{M}$, and denote by $\mathscr{M}_0$ the linear submanifold of $\mathscr{E}$ generated (algebraically) by $\mathscr{M}$ and x_0. Show that $\mathscr{M}_0$ is closed whenever $\mathscr{M}$ is, and use this observation to prove that every finite dimensional linear submanifold of a separated topological linear space is closed. In the same spirit, show that every finite dimensional topological linear space whose topology is induced by an invariant metric is complete with respect to that metric.

(ii) Show that there exists a unique separated linear topology on each finite dimensional linear space, and hence that every linear space isomorphism between two finite dimensional separated topological linear spaces is automatically a homeomorphism. (Hint: Suppose $\{x_1, \ldots, x_n\}$ is a basis for the separated topological vector space $\mathscr{E}$. Then $\varphi(\lambda_1, \ldots, \lambda_n) = \lambda_1 x_1 + \cdots + \lambda_n x_n$ defines a linear space isomorphism φ of $\mathbb{C}^n$ onto $\mathscr{E}$, and it is easy to see that φ is continuous (if $\mathbb{C}^n$ is given the usual topology). Show that φ is also open; see Problem J.) Conclude that every pseudonorm on a finite dimensional separated topological linear space is continuous.

(iii) Let $\mathscr{E}$ be a separated topological vector space and suppose given a balanced compact neighborhood V_0 of 0 in $\mathscr{E}$, a closed linear manifold $\mathscr{M}$ in $\mathscr{E}$, and a vector z_0 in $\mathscr{E} \backslash \mathscr{M}$. Prove that the set $R = \{t > 0 : (z_0 + tV_0) \cap \mathscr{M} \neq \varnothing\}$ is a ray to the right in $\mathbb{R}$ with $d = \inf R > 0$. (Hint: The sets of the form $tV_0, t > 0$, form a neighborhood base at 0.) Show also that if $m_0 = z_0 + (3d/2)v_0$ with m_0 in $\mathscr{M}$ and v_0 in V_0 (such vectors exist because of the definition of d), then $v_0 \notin \mathscr{M} + \frac{1}{2}V_0$, and use this observation to prove that if $y_1, \ldots, y_k$ are so chosen that

$$V_0 \subset (y_1 + \tfrac{1}{2}V_0) \cup \cdots \cup (y_k + \tfrac{1}{2}V_0),$$

then the vectors $y_1, \ldots, y_k$ generate $\mathscr{E}$ algebraically. (Hint: Set $\mathscr{M}$ equal to the linear manifold generated by $y_1, \ldots, y_k$.) Conclude that a separated topological linear space is locally compact (Prob. 3V) if and only if it is finite dimensional. In particular, a Banach space $\mathscr{E}$ is finite dimensional if and only if $\mathscr{E}_1$ is compact.

Q. Every linear submanifold $\mathscr{M}$ of a topological linear space $\mathscr{E}$ is either dense in $\mathscr{E}$ or nowhere dense in $\mathscr{E}$. (Hint: If $\mathscr{M}^-$ has nonempty interior, then $\mathscr{M}^-$ contains a neighborhood of 0 and therefore coincides with $\mathscr{E}$.) In particular, if $\mathscr{E}$ is separated and not finite dimensional, then every finite dimensional linear submanifold of $\mathscr{E}$ is a closed nowhere dense subset of $\mathscr{E}$. Show that no linear space of Hamel dimension $\aleph_0$ can be normed in such a way as to make it complete. Are there any cardinal numbers c such that if $\mathscr{E}$ is a linear space of Hamel dimension c, then $\mathscr{E}$ cannot be normed at all?

R. (i) Show that if $\mathscr{E}_1, \ldots, \mathscr{E}_n$ are normed spaces, then the norm topology on the direct sum $\mathscr{E}_1 \oplus_1 \cdots \oplus_1 \mathscr{E}_n$ coincides with the product topology. Show, more generally, that if $\mathscr{E}_1, \ldots, \mathscr{E}_n$ are arbitrary topological linear spaces, then the product topology is a linear topology on the direct sum $\mathscr{E}_1 + \cdots + \mathscr{E}_n$.

(ii) Show likewise that if $\mathscr{F}$ is a topological vector space and T is a linear transformation of a linear space $\mathscr{E}$ into $\mathscr{F}$, then the coarsest topology on $\mathscr{E}$ that makes T continuous (that is, the topology inversely induced on $\mathscr{E}$ by T; see Chapter 3) is a linear topology. Develop these ideas to give a proof of Proposition 11.24.

S. Let $\{\mathscr{E}_\gamma\}_{\gamma \in \Gamma}$ be a nonempty indexed family of normed spaces. The full algebraic direct sum $\mathscr{E} = \sum_{\gamma \in \Gamma} + \mathscr{E}_\gamma$ is a topological linear space in the product topology (Prop. 11.24), and can even be quasinormed if the index family is countable (Ex. J), but this space is too large in most cases to be of much interest. Various linear submanifolds of it are frequently of importance, however. Thus, for example, the *bounded* families $\{x_\gamma\}$ in $\mathscr{E}$, i.e., those such that $\sup_\gamma \|x_\gamma\| < +\infty$, form a linear manifold $\mathscr{B}$ in $\mathscr{E}$, and $\|\{x_\gamma\}\|_\infty = \sup_\gamma \|x_\gamma\|$ defines a norm on $\mathscr{B}$. Show that $\mathscr{B}$ is complete in the norm $\|\ \|_\infty$ if and only if all the spaces $\mathscr{E}_\gamma$ are. (If $\Gamma = \aleph_0$ and all spaces $\mathscr{E}_n$ coincide with the scalar field $\mathbb{C}$, then $\mathscr{B}$ coincides with the space (ℓ_∞).)

T. Continuing in the vein of the preceding problem, let $\{\mathscr{E}_\gamma\}_{\gamma \in \Gamma}$ be an indexed family of normed spaces, let $\mathscr{E}$ denote the full algebraic direct sum of the family, and let p be a positive real number. Show that the collection of elements $\{x_\gamma\}_{\gamma \in \Gamma}$ of $\mathscr{E}$ such that $\sum_\gamma \|x_\gamma\|^p < +\infty$ (Ex. 3Q) is a linear manifold $\mathscr{K}_p$ in $\mathscr{E}$, and that, if $p \geq 1$, then $\|\{x_\gamma\}\|_p = [\sum_\gamma \|x_\gamma\|^p]^{1/p}$ is a norm on $\mathscr{K}_p$. Verify that $\mathscr{K}_p$ is complete if and only if each $\mathscr{E}_\gamma$ is complete. (Note, once again, that if $\Gamma = \aleph_0$ and all of the spaces coincide with $\mathbb{C}$, then $\mathscr{K}_p$ coincides with the Banach space (ℓ_p).) When Γ is the finite set $\{1, \ldots, n\}$, then all of the norms $\|\ \|_p$, $1 \leq p \leq +\infty$, are defined on the full algebraic direct sum $\mathscr{E}$. In this special case we shall employ the notation $\mathscr{E}_1 \oplus_p \cdots \oplus_p \mathscr{E}_n$ for $\mathscr{E}$ equipped with the norm $\|\ \|_p$. (In particular, $\mathscr{E}$ equipped with the norm $\|\ \|_1$ is what we have referred to earlier as the *direct sum* of the spaces $\mathscr{E}_1, \ldots, \mathscr{E}_n$.)

U. (i) Consider the linear space $\mathscr{C}(X)$ of all continuous complex-valued functions on a locally compact Hausdorff space X (Prob. 3V). If K is a nonempty compact subset of X then

$$\sigma_K(f) = \sup_{x \in K} |f(x)|, \qquad f \in \mathscr{C}(X),$$

defines a pseudonorm on $\mathscr{C}(X)$. Show that the family of pseudonorms $\{\sigma_K\}_{K \in \mathscr{K}}$, where $\mathscr{K}$ denotes the collection of all nonempty compact subsets of X, is separating and saturated. Show also that the family $\{\sigma_K\}_{K \in \mathscr{K}}$ is countably determined (Prob. 1P) if and only if X is σ-compact (Ch. 10, p. 193). Verify that a net $\{f_\lambda\}$ in $\mathscr{C}(X)$ converges to a limit f in the topology induced on $\mathscr{C}(X)$ by the family $\{\sigma_K\}$ if and only if $\{f_\lambda\}$ converges to f uniformly on every compact subset K of X (cf. Problem 4O). For that reason this topology is known as the topology of *uniform convergence on compact subsets* of X. According to Proposition 11.32, the topology of uniform convergence on compact subsets of X is induced by a quasinorm when X is σ-compact. Construct such a quasinorm and show that the topology of uniform convergence on compact subsets of X is not induced by any equivalent *norm* on $\mathscr{C}(X)$ (unless, of course, X is compact). (Hint: If $\|\ \|$ is a norm on $\mathscr{C}(X)$ that induces the topology of uniform convergence on compact subsets of X, then the open unit ball with respect to $\| \ \|$ contains some neighborhood $\{f \in \mathscr{C}(X) : \sigma_K(f) < \varepsilon\}$ where K is a compact subset of X and ε is some positive number.)

(ii) Let Δ be a domain in the complex plane, and let $\mathscr{A}$ denote the linear space of all analytic functions on Δ. Show that $\mathscr{A}$ is closed in $\mathscr{C}(\Delta)$ in the topology of uniform convergence on compact subsets of Δ.

V. Let U be a nonempty open subset of Euclidean space $\mathbb{R}^n$ and let m be a nonnegative integer. We denote by $\mathscr{C}^{(m)}(U)$ the linear space of all m times continuously differentiable complex-valued functions on U (Ex. 3E) and by $\mathscr{C}^{(\infty)}(U)$ the space of infinitely differentiable functions on $U - \mathscr{C}^{(\infty)}(U) = \bigcap_{m=0}^{\infty} \mathscr{C}^{(m)}(U)$. If $k_1, \ldots, k_n$ are nonnegative integers such that $k_1 + \cdots + k_n \leq m$, and if $f \in \mathscr{C}^{(m)}(U)$, we write $D^{k_1, \ldots, k_n}f$ for the (continuous) partial derivative of f

$$\frac{\partial^{k_1 + \cdots + k_n}f}{\partial x_1^{k_1} \cdots \partial x_n^{k_n}}$$

obtained by differentiating k_i times with respect to the variable x_i, $i = 1, \ldots, n$. If K is a nonempty compact subset of the open set U, we shall also write, as in the preceding problem,

$$\sigma_K(f) = \sup_{x \in K}|f(x)|, \qquad f \in \mathscr{C}^{(0)}(U),$$

and, combining these two notational conventions,

$$\sigma_{K,m}(f) = \sup_{k_1 + \cdots + k_n \leq m} \sigma_K(D^{k_1, \ldots, k_n}f), \qquad f \in \mathscr{C}^{(m)}(U),$$

for each nonnegative integer m. (Thus $\sigma_{K,0}$ coincides with σ_K.)

(i) Verify that each $\sigma_{K,m}$ is a pseudonorm on each of the spaces $\mathscr{C}^{(p)}(U)$, $p \geq m$, and conclude that *all* of the functions $\sigma_{K,m}$ (K a nonempty compact subset of U, m a nonnegative integer) are pseudonorms on $\mathscr{C}^{(\infty)}(U)$. Show also that the family of pseudonorms $\{\sigma_{K,m}\}$ is separating, saturated, and countably determined on $\mathscr{C}^{(\infty)}(U)$ (cf. Example 10A), so that the topology this family induces on $\mathscr{C}^{(\infty)}(U)$ is separated and quasinormable. Show, finally, that a sequence $\{f_p\}_{p=1}^{\infty}$ in $\mathscr{C}^{(\infty)}(U)$ converges in this topology to a limit f if and only if the sequence $\{D^{k_1, \ldots, k_n}f_p\}_{p=1}^{\infty}$ converges to $D^{k_1, \ldots, k_n}f$ uniformly on compact subsets of U for every sequence $k_1, \ldots, k_n$ of nonnegative integers. (When this is the case, we shall say of the sequence $\{f_p\}$ that it is *D-convergent* to f, and refer generally to the topology induced on $\mathscr{C}^{(\infty)}(U)$ by the family of pseudonorms $\{\sigma_{K,m}\}$ as the *topology of D-convergence on* $\mathscr{C}^{(\infty)}(U)$.)

(ii) The *support* of a continuous complex-valued function f on U is by definition the closure of the set $\{x \in U : f(x) \neq 0\}$; see Problem 10R. A function f in $\mathscr{C}^{(\infty)}(U)$ is said to be a *test function* on U if the support of f is compact, or, equivalently, if there exists a compact subset K of U such that f vanishes on $U \backslash K$. We denote by $\mathscr{C}_0^{(\infty)}(U)$ the collection of all test functions on U. While it is clear that there are many functions in $\mathscr{C}^{(\infty)}(U)$ (all polynomials belong to $\mathscr{C}^{(\infty)}(U)$, for example), and it is likewise clear that $\mathscr{C}_0^{(\infty)}(U)$ is a linear manifold in $\mathscr{C}^{(\infty)}(U)$, it requires some ingenuity to show that there *are* any nontrivial test functions. To this end let r be a positive number, and let ψ_r denote the function defined on $\mathbb{R}$ by

$$\psi_r(t) = \begin{cases} e^{r/t}, & t < 0, \\ 0, & t \geq 0. \end{cases}$$

Show that ψ_r is infinitely differentiable on $\mathbb{R}$. (Hint: Show first by mathematical induction that $\psi_r^{(k)}(t) = R_k(t)e^{r/t}$, $t < 0$, for all positive integers k, where R_k is a polynomial in $1/t$. Use this fact to show that the limits

$$\lim_{t\uparrow 0} \psi_r^{(k)}(t) \quad \text{and} \quad \lim_{t\uparrow 0} \psi_r^{(k)}(t)/t$$

are both equal to zero for every nonnegative integer k.) Next show that for each positive real number a the function

$$\varphi_a(t) = \begin{cases} \exp 1/(t^2 - a^2), & |t| < a, \\ 0, & |t| \geq a, \end{cases}$$

is a test function on $\mathbb{R}$ which is positive on the interval $(-a, +a)$, has support $[-a, +a]$, and also possesses the property that $(\varphi_a)^r$ is a test function for every positive number r. Conclude that for every closed cell

$$Z = [a_1, b_1] \times \cdots \times [a_n, b_n]$$

in $\mathbb{R}^n$ (where $a_i < b_i$, $i = 1, \ldots, n$) there exist test functions on $\mathbb{R}^n$ that are positive on Z°, have support Z, and possess the property that all of their positive powers are also test functions. Show similarly that if x_0 is a point in $\mathbb{R}^n$ and r is a positive radius, then there exist test functions φ_r on $\mathbb{R}^n$ having for support the closed ball $D_r(x_0)^-$ and possessing the property that $\int_{\mathbb{R}^n} \varphi_r \, d\mu_n = 1$, where μ_n denotes Lebesgue-Borel measure on $\mathbb{R}^n$.

W. There is a concept that relates to that of a quasinorm as pseudonorms relate to norms, and this notion plays a role of some importance in the general theory of topological vector spaces. A *deminorm* on a linear space $\mathscr{E}$ is a nonnegative real-valued function δ on $\mathscr{E}$ having the following properties:

(i′) $\delta(0) = 0$,
(ii″) $\delta(\lambda x) \leq \delta(x)$ for every x in $\mathscr{E}$ and all complex numbers λ such that $|\lambda| \leq 1$,
(iii) $\delta(x + y) \leq \delta(x) + \delta(y)$ for all x, y in $\mathscr{E}$,
(iv) $\alpha_n \to 0$ implies $\delta(\alpha_n x) \to 0$

for every null sequence $\{\alpha_n\}$ of complex numbers and every x in $\mathscr{E}$.

(i) If δ is a deminorm on a linear space $\mathscr{E}$, then

$$\rho_0(x, y) = \delta(x - y), \qquad x, y \in \mathscr{E},$$

defines an invariant pseudometric on $\mathscr{E}$. Show that $\mathscr{E}$ is a topological linear space in the topology induced by ρ_0. (This topology is also said to be *induced* by the deminorm δ.) Show also that if ρ_0 is a given invariant pseudometric on $\mathscr{E}$ and if we define

$$\delta(x) = \rho_0(x, 0), \qquad x \in \mathscr{E},$$

then δ is a deminorm on $\mathscr{E}$ if and only if the following conditions are satisfied:

(a) The ball $\{x \in \mathscr{E} : \rho_0(x, 0) < r\}$ is a balanced set (Prob. M) for every positive number r,
(b) The mapping $\alpha \to \alpha x$ is continuous at $\alpha = 0$ for every x in $\mathscr{E}$ with respect to the topology induced by ρ_0.

(ii) Let δ be a deminorm on a linear space $\mathscr{E}$ and let $\mathscr{Z}$ denote the zero space of $\mathscr{E}$ in the topology induced by δ. Verify that $\mathscr{Z} = \{x \in \mathscr{E} : \delta(x) = 0\}$, and that if for each vector x in $\mathscr{E}$ we denote by $\dot{x}$ the coset $x + \mathscr{Z}$, then

$$\| \dot{x} \| = \delta(x), \qquad x \in \mathscr{E}, \tag{14}$$

defines unambiguously a quasinorm $\| \ \|$ on $\mathscr{E}/\mathscr{Z}$. Show also that this quasinorm induces the topology of the associated space $\dot{\mathscr{E}}$. (The quasinorm defined in (14) is said to be *associated* with δ. See Corollary 11.23.)

(iii) If $\{\delta_\gamma\}_{\gamma \in \Gamma}$ is a family of deminorms on a linear space $\mathscr{E}$, then the supremum of the collection of linear topologies induced by them is called the topology *induced* by the given family. Similarly, $\{\delta_\gamma\}$ is said to be *separating* on $\mathscr{E}$ if for each vector $x \neq 0$ in $\mathscr{E}$ there is a deminorm δ_γ in the family such that $\delta_\gamma(x) \neq 0$. Verify that Propositions 11.26–11.32 are all valid for families of deminorms in place of families of pseudonorms. (Hint: The property (ii′) of pseudonorms is not used in any of the proofs.)

X. (i) Suppose given a nonnegative function w on a vector space $\mathscr{E}$ satisfying the two conditions $w(0) = 0$ and $w(-x) = w(x)$ for every x in $\mathscr{E}$. Using the function w, define ρ_0 on $\mathscr{E} \times \mathscr{E}$ by setting

$$\rho_0(x, y) = \inf \sum_{i=1}^{n} w(z_i - z_{i-1}),$$

where the infimum is taken over all finite sequences $\{z_0, z_1, \ldots, z_n\}$ satisfying the conditions $z_0 = x$ and $z_n = y$ (and over all positive integers n). Show that ρ_0 is an invariant pseudometric on $\mathscr{E}$, and that the associated function $\delta(x) = \rho_0(x, 0)$ satisfies the condition $\delta \leq w$. Show also that δ is a deminorm on $\mathscr{E}$ if w satisfies the conditions

(a) $w(\lambda x) \leq w(x)$ for every x in $\mathscr{E}$ and all complex numbers λ such that $|\lambda| \leq 1$, and

(b) $\alpha_n \to 0$ implies $w(\alpha_n x) \to 0$

for every null sequence $\{\alpha_n\}$ of complex numbers and every x in $\mathscr{E}$. Show finally that if w satisfies the inequality

$$w(x - y) \leq 2 \max[w(x - u), w(u - v), w(v - y)] \tag{15}$$

for all vectors u, v, x, y in $\mathscr{E}$, then δ also satisfies the condition $\delta \geq w/2$. (Hint: It is necessary to show that

$$w(z_n - z_0) \leq 2 \sum_{i=1}^{n} w(z_i - z_{i-1})$$

for an arbitrary finite sequence $\{z_0, z_1, \ldots, z_n\}$ of vectors in $\mathscr{E}$. Choose an index j such that both

$$\sum_{i<j} w(z_i - z_{i-1}) \quad \text{and} \quad \sum_{i>j} w(z_i - z_{i-1})$$

are less than or equal to $\frac{1}{2}\sum_{i=1}^{n} w(z_i - z_{i-1})$, and use induction on n.)

(ii) Let $\mathscr{E}$ be a topological vector space and let $\{V_n\}_{n=1}^{\infty}$ be a nested sequence of neighborhoods of 0 in $\mathscr{E}$:

$$V_1 \supset V_2 \supset \cdots \supset V_n \supset \cdots.$$

Set $V_0 = \mathscr{E}$, and define

$$w(x) = \inf\left\{\frac{1}{2^n} : x \in V_n\right\}, \qquad x \in \mathscr{E}.$$

Show that $w(0) = 0$ and that w satisfies condition (b) of (i). Show also that w satisfies condition (a) if the neighborhoods V_n are all balanced, and that w also possesses property (15) if the neighborhoods V_n satisfy the additional condition that

$$V_{n+1} + V_{n+1} + V_{n+1} \subset V_n$$

for every positive integer n.

(iii) Put together the results obtained in (i) and (ii) (along with other foregoing results) to prove the following fundamental theorem: If $\{V_n\}_{n=1}^{\infty}$ is an arbitrary sequence of neighborhoods of 0 in a topological linear space $\mathscr{E}$, then there exists a deminorm δ on $\mathscr{E}$ such that δ is continuous on $\mathscr{E}$ and possesses the additional property that each of the sets V_n is a neighborhood of 0 in the topology induced on $\mathscr{E}$ by δ. Conclude that every linear topology on $\mathscr{E}$ is induced by some (possibly uncountable) family of deminorms.

Y. Show that the following properties are equivalent for an arbitrary separated topological linear space $\mathscr{E}$:

(i) $\mathscr{E}$ is quasinormable (Prop. 11.32),

(ii) $\mathscr{E}$ is metrizable as a topological space,

(iii) $\mathscr{E}$ satisfies the first axiom of countability.

(Hint: Use Problem X.)

12 Bounded linear transformations

The presence of a topology on a linear space leads naturally to the extremely important classification of linear transformations into continuous and discontinuous ones. In normed spaces this distinction is facilitated by the existence of a very simple criterion for continuity.

Definition. A linear transformation T of a normed space $\mathscr{E}$ into another normed space $\mathscr{F}$ is *bounded* if there exists a scalar $M \geq 0$ such that

$$\| Tx \| \leq M \| x \|, \qquad x \in \mathscr{E}. \tag{1}$$

(It should be noted that the norms on the two sides of (1) are to be computed, in general, in two different spaces. At no time will this notational ambiguity lead to any confusion.)

If T is bounded, then, as it turns out, there is a smallest M satisfying (1) (Prob. A). This smallest bound is known as the *norm* of T and is denoted by $\| T \|$. Thus $\| T \|$ is a nonnegative real number such that $\| Tx \| \leq \| T \| \| x \|$ for every vector $\dot{x}$ in $\mathscr{E}$, and is the smallest number making this inequality valid. On the other hand, it is easy to verify that $\| T \|$ may also be described in any of the following equivalent ways (provided that T is a bounded linear transformation and that $\mathscr{E} \neq (0)$):

$$\| T \| = \sup_{\| x \| \leq 1} \| Tx \| = \sup_{\| x \| < 1} \| Tx \| = \sup_{\| x \| = 1} \| Tx \| = \sup_{x \neq 0} \frac{\| Tx \|}{\| x \|}. \tag{2}$$

Proposition 12.1. *The following conditions are equivalent for any linear transformation T of one normed space $\mathscr{E}$ into another normed space $\mathscr{F}$:*

   (i) *T is continuous,*
  (ii) *T is continuous at $x = 0$,*
 (iii) *T is bounded.*

248

PROOF. Since it is clear that (i) implies (ii), it will suffice to show that (ii) implies (iii) and that (iii) implies (i). To see that (ii) implies (iii), choose $\delta > 0$ such that $\|x\| = \|x - 0\| \leq \delta$ implies $\|Tx\| = \|Tx - T0\| \leq 1$. Then Tx belongs to the ball $\mathcal{F}_{1/\delta}$ whenever x belongs to the unit ball $\mathcal{E}_1$, and from this it follows at once that T is bounded. Finally, to see that (iii) implies (i), let x_0 be a vector in $\mathcal{E}$ and let ε be positive. Then for $\delta = \varepsilon/(\|T\| + 1)$ and $\|x - x_0\| < \delta$, we have

$$\|Tx - Tx_0\| \leq \|T\|\|x - x_0\| < \varepsilon.$$

Thus T is continuous at x_0. □

The main point of the last part of the foregoing argument is that the inequality

$$\|Tx_1 - Tx_2\| \leq \|T\|\|x_1 - x_2\|$$

is valid for an arbitrary pair of vectors x_1, x_2 in $\mathcal{E}$. Thus a continuous, and therefore bounded, linear transformation T between two normed spaces is automatically Lipschitzian. In particular, T is uniformly continuous (see Problem 4F). We deal here solely with continuous linear transformations between two normed spaces because that is the case of principal concern and because the simple criterion of boundedness is applicable in this situation. If $\mathcal{E}$ and $\mathcal{F}$ are merely quasinormed spaces, then no similarly simple criterion for continuity exists. All the more, if $\mathcal{E}$ and $\mathcal{F}$ are arbitrary topological linear spaces, then no such simple criterion for the continuity of a linear transformation of $\mathcal{E}$ into $\mathcal{F}$ can be given. Nevertheless, appropriate counterparts of many of the results developed in this chapter for bounded linear transformations can be obtained for continuous linear transformations between topological linear spaces other than normed spaces (see Problem U).

Example A. If $\mathcal{E}$ is a finite dimensional Banach space, and T denotes an arbitrary linear transformation of $\mathcal{E}$ into some normed space $\mathcal{F}$, then T is automatically bounded. Indeed, if T is one-to-one, it follows at once from Problem 11J that T is continuous, while if $\mathcal{K}(T) \neq (0)$ we may factor T as $T = \hat{T} \circ \pi$, where π denotes the natural projection of $\mathcal{E}$ onto $\mathcal{E}/\mathcal{K}(T)$ (Ex. 2L), and both $\hat{T}$ and π are continuous (Prob. 11H). Thus, in any case, T is continuous and therefore bounded. It is important to note that *only* finite dimensional Banach spaces possess this property. Indeed, if $\mathcal{E}$ is an arbitrary infinite dimensional normed space, and if X is a Hamel basis for $\mathcal{E}$ (Prob. 2A), we may select from X an infinite sequence $\{x_n\}_{n=1}^{\infty}$ of distinct vectors. Let X_0 denote the collection of vectors in X that do not appear in the sequence $\{x_n\}$, let $\mathcal{F}$ be any nontrivial normed space, and let y_0 be any nonzero vector in $\mathcal{F}$. Then

$$\varphi(x_n) = n\|x_n\|y_0, \qquad n \in \mathbb{N},$$
$$\varphi(x) = 0, \qquad x \in X_0,$$

defines a mapping of X into $\mathcal{F}$, and the linear extension T of φ to $\mathcal{E}$ (Prob. 2E) is an unbounded, and therefore discontinuous, linear transformation of $\mathcal{E}$ into $\mathcal{F}$. Taking for $\mathcal{F}$ the space $\mathbb{C}$, we note specifically that if $\mathcal{E}$ is infinite dimensional, then there always exist unbounded linear functionals on $\mathcal{E}$.

Example B. Let p be an extended real number, $1 \le p \le +\infty$. For any infinite sequence of complex numbers $x = \{\xi_n\}_{n=0}^{\infty}$ in (ℓ_p), set

$$V\{\xi_0, \xi_1, \ldots, \xi_n, \ldots\} = \{0, \xi_0, \ldots, \xi_{n-1}, \ldots\}. \tag{3}$$

Since $\|x\|_p = \|Vx\|_p$ for every x in (ℓ_p), it is clear that V is a bounded linear transformation on (ℓ_p) and that $\|V\| = 1$. The transformation V is known as the *unilateral shift* on (ℓ_p).

Example C. Let $d = \{\delta_0, \delta_1, \ldots\}$ be a fixed bounded sequence of scalars, and let p be an extended real number, $1 \le p \le +\infty$. Define M_d by writing

$$M_d x = \{\delta_0 \xi_0, \delta_1 \xi_1, \ldots\} \tag{4}$$

for an arbitrary sequence $x = \{\xi_n\}_{n=0}^{\infty}$ in (ℓ_p). Once again, it is easily seen that M_d is a bounded linear transformation on (ℓ_p), and that, in fact, $\|M_d\| = \sup_n |\delta_n| = \|d\|_\infty$. The linear transformation M_d is known as *multiplication by d*.

Example D. On each space (ℓ_p), $1 \le p \le +\infty$, the product $S = VM_d$ does the following:

$$\{\xi_0, \xi_1, \ldots\} \xrightarrow{S} \{0, \delta_0 \xi_0, \delta_1 \xi_1, \ldots\}.$$

This transformation, known as the *weighted unilateral shift* on (ℓ_p) with *weight sequence d*, is also bounded and again we have $\|S\| = \sup_n |\delta_n|$.

Example E. Let p be a real number, $p \ge 1$, and consider the linear space $(\ell_p)^\#$ of all those *two-way infinite* sequences $x = \{\xi_n\}_{n=-\infty}^{+\infty}$ with the property that $\sum_{n=-\infty}^{+\infty} |\xi_n|^p < +\infty$, and with the norm $\|x\|_p = (\sum_{n=-\infty}^{+\infty} |\xi_n|^p)^{1/p}$. (Similarly one defines $(\ell_\infty)^\#$. It is easy to see that $(\ell_p)^\#$ is a Banach space, just as is the space (ℓ_p), $1 \le p \le +\infty$. Indeed, it is possible to view this space and (ℓ_p) simply as two different versions of the same Banach space (Ex. Q).) On $(\ell_p)^\#$ let U denote the mapping defined by $U\{\xi_n\} = \{\eta_n\}$, where $\eta_n = \xi_{n-1}$ for every integer n. Then U is a bounded linear transformation of norm one called, naturally enough, the *bilateral shift* on $(\ell_p)^\#$.

In order to see more clearly how such transformations as U act on a Banach space $(\ell_p)^\#$ of two-way infinite sequences, it is desirable to introduce some special notation. We shall systematically write

$$\{\ldots, \xi_{-2}, \xi_{-1}, [\xi_0], \xi_1, \xi_2, \ldots\}$$

for such a sequence, thus putting in square brackets the term in the sequence that is indexed by zero. In this notation we have

$$\{\ldots, \xi_{-1}, [\xi_0], \xi_1, \ldots\} \xrightarrow{U} \{\ldots, \xi_{-2}, [\xi_{-1}], \xi_0, \ldots\}.$$

Example F. There are also counterparts of the transformations of Examples C and D on the spaces $(\ell_p)^\#$, $1 \le p \le +\infty$. Indeed, if $d = \{\delta_n\}_{n=-\infty}^{+\infty}$ is an arbitrary element of $(\ell_\infty)^{\#}$, then *multiplication by d* is the transformation carrying each $x = \{\xi_n\}_{n=-\infty}^{+\infty}$ in $(\ell_p)^{\#}$ to the sequence

$$M_d x = dx = \{\delta_n \xi_n\}_{n=-\infty}^{+\infty},$$

and M_d is a bounded linear transformation on $(\ell_p)^{\#}$ with $\| M_d \| = \| d \|_\infty$. Likewise, the linear transformation $T = U M_d$, where U denotes the bilateral shift of Example E, is a bounded linear transformation on $(\ell_p)^{\#}$ with $\| T \| = \| d \|_\infty$. The action of T can be better viewed using the notation introduced in that example:

$$\{\ldots, \xi_{-1}, [\xi_0], \xi_1, \ldots\} \xrightarrow{T} \{\ldots, \delta_{-2}\xi_{-2}, [\delta_{-1}\xi_{-1}], \delta_0\xi_0, \ldots\}.$$

The transformation T is called the *weighted bilateral shift* on $(\ell_p)^{\#}$ with *weight sequence d*.

Example G. Let X be a nonempty compact Hausdorff space, let f_0 be a fixed function in $\mathscr{C}(X)$ (Ex. 3D), and define M_{f_0} by

$$(M_{f_0}g)(x) = f_0(x)g(x)$$

for all x in X and all g in $\mathscr{C}(X)$. Then M_{f_0} is a bounded linear transformation on $\mathscr{C}(X)$ known as *multiplication by f_0*. Clearly $\| M_{f_0} \| = \| f_0 \|_\infty$.

Example H. Choose a fixed point x_0 in a compact Hausdorff space X, and define $\varphi(f) = f(x_0)$ for every f in $\mathscr{C}(X)$. Then φ is a bounded linear functional on $\mathscr{C}(X)$ with $\| \varphi \| = 1$. (This example has appeared before; recall Example 10E.)

The following two examples have to do with some of the basic structural concepts relating to normed spaces; they will reappear regularly in the sequel.

Example I. On any normed space $\mathscr{E}$ the dilatation $x \rightarrow \lambda x, x \in \mathscr{E}$, is a bounded linear transformation of $\mathscr{E}$ into itself. As was announced in Chapter 2, this transformation will consistently be denoted by $\lambda_{\mathscr{E}}$ or, when no confusion is possible, simply by λ. We observe that if $\lambda \neq 0$, then the dilatation λ maps $\mathscr{E}$ onto itself, and if $\mathscr{E} \neq (0)$, then $\| \lambda_{\mathscr{E}} \| = |\lambda|$.

Example J. Let $\mathscr{E}$ be a normed space and let $\mathscr{M}$ be a subspace of $\mathscr{E}$. The natural projection π of $\mathscr{E}$ onto the quotient space $\mathscr{E}/\mathscr{M}$ is a bounded linear transformation such that $\| \pi \| \leq 1$ (see Chapter 11 (8)). Since π carries the open unit ball $\mathscr{E}_1^\circ$ onto the open unit ball in $\mathscr{E}/\mathscr{M}$ (Prob. 11H), it follows that $\| \pi \| = 1$.

Notation and Terminology. If $\mathscr{E}$ and $\mathscr{F}$ are normed spaces, the set of all bounded linear transformations of $\mathscr{E}$ into $\mathscr{F}$ will be denoted by $\mathscr{L}(\mathscr{E}, \mathscr{F})$. If $\mathscr{E} = \mathscr{F}$ we employ the simpler notation $\mathscr{L}(\mathscr{E})$ for $\mathscr{L}(\mathscr{E}, \mathscr{E})$. The elements of

251

$\mathscr{L}(\mathscr{E})$ are properly known as *bounded linear operators* on $\mathscr{E}$. In this book, however, we shall not be concerned with nonlinear analysis, and we shall have only a limited interest in unbounded transformations. Accordingly, we frequently refer to the elements of $\mathscr{L}(\mathscr{E})$ simply as *operators*, or as *bounded operators*, on $\mathscr{E}$. A bounded operator T on $\mathscr{E}$ with $\|T\| \leq 1$ is called a *contraction* on $\mathscr{E}$.

It is easily seen that if $\mathscr{E}$ and $\mathscr{F}$ are arbitrary normed spaces, then $\mathscr{L}(\mathscr{E}, \mathscr{F})$ is a linear submanifold of the full space of linear transformations of $\mathscr{E}$ into $\mathscr{F}$ (Ch. 2, p. 19), and is therefore a linear space in its own right. In order to verify this, it need only be checked that if S and T are bounded, and if λ is a scalar, then $S + T$ and λS are bounded too. But for an arbitrary vector x in $\mathscr{E}$ we have

$$\|(S + T)x\| \leq \|Sx\| + \|Tx\| \leq \|S\|\|x\| + \|T\|\|x\|$$

and

$$\|(\lambda S)x\| = \|\lambda(Sx)\| \leq |\lambda|\|S\|\|x\|,$$

and from these inequalities it follows not only that $S + T$ and λS are bounded, but also that

$$\|S + T\| \leq \|S\| + \|T\|$$

and

$$\|\lambda S\| = |\lambda|\|S\|.$$

Since it is clear that $\|S\| = 0$ implies $S = 0$, we see that $\|\ \|$, as defined in (2), is, in fact, a norm on $\mathscr{L}(\mathscr{E}, \mathscr{F})$. Is the normed space $\mathscr{L}(\mathscr{E}, \mathscr{F})$ complete?

Proposition 12.2. *If $\mathscr{E}$ is a normed space and $\mathscr{F}$ is a Banach space, then $\mathscr{L}(\mathscr{E}, \mathscr{F})$ is a Banach space.*

PROOF. Let $\{T_n\}$ be a Cauchy sequence in $\mathscr{L}(\mathscr{E}, \mathscr{F})$. The inequality $\|T_n x - T_m x\| \leq \|T_n - T_m\|\|x\|$ shows that for each x in $\mathscr{E}$ the sequence $\{T_n x\}$ is a Cauchy sequence in $\mathscr{F}$. Since $\mathscr{F}$ is complete there exists a vector in $\mathscr{F}$—call it Tx—such that $T_n x \to Tx$. This defines a mapping T of $\mathscr{E}$ into $\mathscr{F}$. That T is linear is clear:

$$T(\alpha x + \beta y) = \lim_n (\alpha T_n x + \beta T_n y) = \alpha Tx + \beta Ty.$$

The proof that $\mathscr{L}(\mathscr{E}, \mathscr{F})$ is complete will be concluded by showing that T is also bounded and that $\|T - T_n\| \to 0$. To see this, let ε be a positive number, and choose N such that $\|T_n - T_m\| \leq \varepsilon$ for all $m, n \geq N$. Then $\|T_n x - T_m x\| \leq \varepsilon \|x\|$ for each x in $\mathscr{E}$ and, letting n tend to infinity, we obtain

$$\|Tx - T_m x\| \leq \varepsilon \|x\|$$

for all $m \geq N$. (Recall that closed balls in a metric space are closed sets.) Thus we see that $T - T_m$ is bounded and satisfies $\| T - T_m \| \leq \varepsilon$ for all $m \geq N$. From this it follows, in the first place, that $T = (T - T_m) + T_m$ is also bounded, and, secondly, that $\| T - T_m \| \to 0$. $\square$

> With the aid of the Hahn–Banach theorem (Th. 14.3) it can be shown, conversely, that if $\mathscr{F}$ is not complete, and if $\mathscr{E} \neq (0)$, then $\mathscr{L}(\mathscr{E}, \mathscr{F})$ is not complete. See Problem 14E.

Bounded linear transformations can be multiplied as well as added and subtracted.

Theorem 12.3. *Let $\mathscr{E}$, $\mathscr{F}$, and $\mathscr{G}$ be normed spaces, let T be a bounded linear transformation of $\mathscr{E}$ into $\mathscr{F}$, and let S be a bounded linear transformation of $\mathscr{F}$ into $\mathscr{G}$. Then the product ST is a bounded linear transformation of $\mathscr{E}$ into $\mathscr{G}$ and $\| ST \| \leq \| S \| \| T \|$.*

PROOF. We have $\| STx \| \leq \| S \| \| Tx \| \leq \| S \| \| T \| \| x \|$ for every vector x in $\mathscr{E}$. $\square$

Corollary 12.4. *If $\mathscr{E} \neq (0)$ is a normed space, then $\mathscr{L}(\mathscr{E})$ is not only a normed space but is also an algebra (Ch. 2, p. 19) in which the conditions*

$$\| ST \| \leq \| S \| \| T \|, \qquad S, T \in \mathscr{L}(\mathscr{E}), \tag{5}$$

and

$$\| 1_{\mathscr{E}} \| = 1 \tag{6}$$

hold. If $\mathscr{E}$ is complete, then $\mathscr{L}(\mathscr{E})$ is a Banach space that is also an algebra in which (5) and (6) are satisfied.

Conditions (5) and (6) are the essential ingredients in the definition of a *normed algebra*, a concept that is of the greatest importance in more advanced areas of functional analysis.

Definition. A normed space $\mathscr{A}$ that is simultaneously a linear algebra with respect to a product $(x, y) \to xy$ is a *normed algebra* if $\| xy \| \leq \| x \| \| y \|$ for every pair of elements x, y of $\mathscr{A}$. A normed algebra $\mathscr{A}$ that is complete as a normed linear space is called a *Banach algebra*. If a normed algebra [Banach algebra] $\mathscr{A}$ possesses a unit 1, and if $\| 1 \| = 1$, then $\mathscr{A}$ is said to be a *unital normed algebra* [*Banach algebra*].

Thus Corollary 12.4 says that if $\mathscr{E}$ is an arbitrary normed space other than (0), then $\mathscr{L}(\mathscr{E})$ is a unital normed algebra, and that $\mathscr{L}(\mathscr{E})$ is a unital Banach algebra if $\mathscr{E}$ is a Banach space. Other examples of Banach algebras abound. Indeed, many of the Banach spaces we have already encountered possess a natural multiplicative structure.

Example K. If $x = \{\xi_n\}_{n=0}^{\infty}$ and $y = \{\eta_n\}_{n=0}^{\infty}$ are two elements of (ℓ_p) (for some fixed value of p, $1 \le p \le +\infty$) then the sequence $xy = \{\xi_n \eta_n\}_{n=0}^{\infty}$ also belongs to (ℓ_p), and it is readily verified that $\|xy\|_p \le \|x\|_p \|y\|_p$. (Indeed, $\|xy\|_p \le \|x\|_\infty \|y\|_p$, and it is clear that $\|x\|_\infty \le \|x\|_p$.) Moreover, it is easily seen that (ℓ_p) is a linear algebra with respect to this product. Thus each of the spaces (ℓ_p) is a Banach algebra with respect to termwise or co-ordinatewise multiplication. The algebra (ℓ_∞) is unital, the algebras (ℓ_p), $1 \le p < +\infty$, are not. Similarly, the spaces $(\ell_p)^{\#}$, $1 \le p \le +\infty$, are all Banach algebras with respect to coordinatewise multiplication.

Example L. If X is a nonempty compact Hausdorff space, then $\mathscr{C}(X)$ is a unital Banach algebra with respect to pointwise multiplication. More generally, if X is an arbitrary nonempty topological space, and if $\mathscr{C}_b(X)$ denotes the linear space of bounded continuous complex-valued functions on X equipped with the sup norm (Prob. 11D), then $\mathscr{C}_b(X)$ is a unital Banach algebra with respect to pointwise multiplication. (This example contains, as a special case, the Banach algebra (ℓ_∞), which coincides with $\mathscr{C}_b(\mathbb{N}_0)$ when $\mathbb{N}_0$ is equipped with the discrete topology.)

Definition. If $\mathscr{E}$ and $\mathscr{F}$ are normed spaces, then an element T of $\mathscr{L}(\mathscr{E}, \mathscr{F})$ is *invertible* if there exists an element T^{-1} of $\mathscr{L}(\mathscr{F}, \mathscr{E})$ (called the *inverse* of T) such that $TT^{-1} = 1_{\mathscr{F}}$ and $T^{-1}T = 1_{\mathscr{E}}$. (It is clear that the inverse of an element of $\mathscr{L}(\mathscr{E}, \mathscr{F})$ is unique if it exists.) More generally, if $T \in \mathscr{L}(\mathscr{E}, \mathscr{F})$ and if there exists an element R of $\mathscr{L}(\mathscr{F}, \mathscr{E})$ with the property that $RT = 1_{\mathscr{E}}$, then R is called a *left inverse* of T and T is said to be *left invertible*. Similarly, if there exists an element S of $\mathscr{L}(\mathscr{F}, \mathscr{E})$ having the property that $TS = 1_{\mathscr{F}}$, then S is a *right inverse* of T and T is said to be *right invertible*.

It is a trivial calculation to verify that if T has both a left inverse R and a right inverse S, then T is invertible and $R = S = T^{-1}$. It is likewise clear that when $\mathscr{E} = \mathscr{F}$ the notions of invertibility and inverse introduced above reduce to the usual ones appropriate to the (normed) algebra $\mathscr{L}(\mathscr{E})$. In connection with these concepts the following notion is quite useful.

Definition. If T is a linear transformation of a normed space $\mathscr{E}$ into another normed space $\mathscr{F}$, then T is *bounded below* if there exists a positive number M such that

$$\|Tx\| \ge M\|x\|, \qquad x \in \mathscr{E}.$$

Moreover, if this inequality is satisfied, we shall say that T is bounded below *by M*.

Proposition 12.5. *Let $\mathscr{E}$ and $\mathscr{F}$ be normed spaces and let T be an element of $\mathscr{L}(\mathscr{E}, \mathscr{F})$. If T is left invertible, then T is bounded below. If T is right* *invertible, then $\mathscr{R}(T) = \mathscr{F}$. Moreover, T is invertible if and only if T is both onto and bounded below.*

PROOF. If $RT = 1_\mathscr{E}$ then for each vector x in $\mathscr{E}$ we have $x = RTx$, and therefore

$$\|x\| \leq \|R\|\|Tx\|.$$

Thus T is bounded below. (If $\|R\| = 0$ then $\mathscr{E} = (0)$, and T is trivially bounded below.) If $TS = 1_\mathscr{F}$ then for each vector y in $\mathscr{F}$ we have $y = T(Sy)$, so T maps $\mathscr{E}$ onto $\mathscr{F}$. To complete the proof it suffices to show that T is invertible if $\mathscr{R}(T) = \mathscr{F}$ and T is bounded below by some positive number M. If these hypotheses are satisfied, then, in the first place, $\mathscr{K}(T) = (0)$ so that T is one-to-one. Consequently T possesses a set-theoretic inverse S defined on $\mathscr{R}(T) = \mathscr{F}$ by $S(Tx) = x$, and S is a linear transformation of $\mathscr{F}$ onto $\mathscr{E}$. Finally, since $\|S(Tx)\| = \|x\| \leq (1/M)\|Tx\|$, we see that S is bounded and that $\|S\| \leq 1/M$, so that $S = T^{-1}$ is an element of $\mathscr{L}(\mathscr{F}, \mathscr{E})$. $\square$

The foregoing result can be improved upon somewhat when the spaces $\mathscr{E}$ and $\mathscr{F}$ are complete.

Proposition 12.6. *If $\mathscr{E}$ and $\mathscr{F}$ are Banach spaces, and if T is an element of $\mathscr{L}(\mathscr{E}, \mathscr{F})$ that is bounded below, then the range $\mathscr{R} = \mathscr{R}(T)$ is closed in $\mathscr{F}$. Hence T is invertible if and only if $\mathscr{R}$ is dense in $\mathscr{F}$.*

PROOF. Let T be bounded below by M, $M > 0$. It suffices to show that $\mathscr{R}$ is closed. Let $\{x_n\}$ be a sequence in $\mathscr{E}$ such that the sequence $\{Tx_n\}$ is convergent in $\mathscr{F}$. Since $M\|x_m - x_n\| \leq \|Tx_m - Tx_n\|$, it follows that the sequence $\{x_n\}$ is Cauchy, and therefore convergent—say to x. But then $\{Tx_n\}$ converges to Tx, which belongs to $\mathscr{R}$, and we see that $\mathscr{R}$ is closed. $\square$

Example M. Let p be an extended real number, $1 \leq p \leq +\infty$, and let S denote the weighted unilateral shift on (ℓ_p) with weight sequence $d = \{\delta_0, \delta_1, \ldots\}$ (Ex. D). Then S is bounded below on (ℓ_p) if and only if the sequence d is bounded away from zero. Moreover, if this is the case, then the range of S is readily seen to be the subspace of (ℓ_p) consisting of sequences $x = \{\xi_n\}_{n=0}^\infty$ with $\xi_0 = 0$, and a left inverse for S is given by the *backward weighted unilateral shift*

$$\{\xi_0, \xi_1, \ldots\} \xrightarrow{R} \{\xi_1/\delta_0, \xi_2/\delta_1, \ldots\} \tag{7}$$

with weight sequence $\{1/\delta_n\}_{n=0}^\infty$. (Observe that, while the inverse of an invertible linear transformation is uniquely determined, a left invertible transformation T may possess many left inverses, since the left inverse of T is determined by T only on the range of T. Thus the transformation R in (7) is the unique left inverse of the shift S that annihilates the sequence $e_0 = \{1, 0, 0, \ldots\}$. No forward unilateral shift can be right invertible since its range inevitably fails to contain e_0.)

The distinction between invertible and noninvertible linear transformations gives rise to a number of important concepts. The simplest of these is new in name only.

Definition. If $\mathscr{E}$ and $\mathscr{F}$ are normed spaces, then an invertible element of $\mathscr{L}(\mathscr{E}, \mathscr{F})$ is also known as an *equivalence* between $\mathscr{E}$ and $\mathscr{F}$, and $\mathscr{E}$ and $\mathscr{F}$ are said to be *equivalent* if there exists an equivalence between them.

It is clear that the inverse of an equivalence and the product of two equivalences are also equivalences so that this relation possesses the defining properties of an equivalence relation.

Example N. Two finite dimensional Banach spaces are equivalent if and only if they possess the same dimension (see Problem 11J).

Example O. Let $\mathscr{E}$ be a linear space and let $\| \ \|_1$ and $\| \ \|_2$ be two norms on $\mathscr{E}$. Then $\| \ \|_1$ and $\| \ \|_2$ are equivalent norms (Ex. 11B) if and only if the identity mapping on $\mathscr{E}$ is an equivalence between $\mathscr{E}$ equipped with $\| \ \|_1$ and $\mathscr{E}$ equipped with $\| \ \|_2$.

Proposition 12.7. *If $\mathscr{E}$ and $\mathscr{F}$ are equivalent normed spaces, and if either $\mathscr{E}$ or $\mathscr{F}$ is a Banach space, then both must be.*

PROOF. Suppose without loss of generality that $\mathscr{E}$ is complete, and that $T : \mathscr{E} \to \mathscr{F}$ is an equivalence. Let $\{y_n\}$ be a Cauchy sequence in $\mathscr{F}$, and let $x_n = T^{-1} y_n$, $n \in \mathbb{N}$. Then $\|x_m - x_n\| \le \|T^{-1}\| \|y_m - y_n\|$ for all m and n, so $\{x_n\}$ is a Cauchy sequence in $\mathscr{E}$. But then $\{x_n\}$ is convergent in $\mathscr{E}$, whence it follows that the sequence $\{y_n = Tx_n\}$ is also convergent. $\quad\square$

If $\mathscr{E}$ and $\mathscr{F}$ are normed spaces, then it is easily seen that a linear transformation V of $\mathscr{E}$ into $\mathscr{F}$ is an isometry of the metric space $\mathscr{E}$ into the metric space $\mathscr{F}$ if and only if V preserves norms:

$$\| Vx \| = \| x \|, \qquad x \in \mathscr{E}.$$

Such a linear transformation is accordingly known as a *(linear) isometry*, and it is in terms of this concept that we define the notion of isomorphism appropriate to the present context.

Definition. If $\mathscr{E}$ and $\mathscr{F}$ are normed spaces, then a mapping $U : \mathscr{E} \to \mathscr{F}$ of $\mathscr{E}$ onto $\mathscr{F}$ is an *isometric isomorphism* if U is a linear space isomorphism that is also an isometry, and $\mathscr{E}$ and $\mathscr{F}$ are said to be *isometrically isomorphic* if there exists such a mapping between them.

It is, once again, clear that the inverse of an isometric isomorphism and the product of two isometric isomorphisms are themselves isometric isomorphisms, so that this relation between normed spaces has the properties of an equivalence relation. It is also clear that an isometric isomorphism is an equivalence.

Example P. The bilateral shift U of Example E is an isometric isomorphism of $(\ell_p)^{\#}$ onto itself for each value of p, $1 \leq p \leq +\infty$. The unilateral shift of Example B is an isometry of (ℓ_p) into itself that is *not* an isometric isomorphism of (ℓ_p) onto itself.

Example Q. The mapping φ that assigns to each sequence $\{\xi_n\}_{n=0}^{\infty}$ in (ℓ_p) the two-way infinite sequence

$$\{\ldots \xi_4, \xi_2, [\xi_0], \xi_1, \xi_3, \ldots\}$$

if an isometric isomorphism of (ℓ_p) onto the space $(\ell_p)^{\#}$, $1 \leq p \leq +\infty$. Similarly the assignment of the sequence

$$\{\ldots \xi_n, \ldots, \xi_1, [\xi_0], \xi_{-1}, \ldots, \xi_{-n}, \ldots\}$$

to each two-way infinite sequence $\{\xi_n\}_{n=-\infty}^{+\infty}$ is an isometric isomorphism of $(\ell_p)^{\#}$ onto itself.

Example R. Let p be fixed, $1 \leq p \leq +\infty$, and let Φ denote the mapping that assigns to each sequence d in (ℓ_∞) the operator M_d on (ℓ_p). Then, as was noted in Example C, Φ is a norm preserving mapping of (ℓ_∞) onto $\mathcal{R} = \mathcal{R}(\Phi) \subset \mathcal{L}((\ell_p))$. Much more is true, however; Φ is also a linear transformation and therefore an isometric isomorphism of (ℓ_∞) onto $\mathcal{R}$. (Since (ℓ_∞) is not separable (Ex. 11H), this observation shows that $\mathcal{L}((\ell_p))$ is likewise not a separable Banach space, $1 \leq p \leq +\infty$.) Moreover, Φ also preserves the coordinatewise multiplication on (ℓ_∞) (Ex. K). Thus Φ is an isometric isomorphism of (ℓ_∞) onto $\mathcal{R}$ that is simultaneously an algebra isomorphism between (ℓ_∞) and $\mathcal{R}$. Such a mapping is a *Banach algebra isomorphism*.

There are some very important special facts concerning the invertibility of operators on a Banach space to which we now turn.

Lemma 12.8. *If $\mathcal{E}$ is a Banach space and T is a bounded operator on $\mathcal{E}$ such that $\|T\| < 1$, then $1 - T$ is invertible and $(1 - T)^{-1} = \sum_{n=0}^{\infty} T^n$.*

PROOF. If $\|T\| = r < 1$, then $\|T^n\| \leq r^n$ for every positive integer n, so the series $\sum_{n=0}^{\infty} T^n$ is absolutely convergent in $\mathcal{L}(\mathcal{E})$ (see Problem 11G). If we set

$$S = \sum_{n=0}^{\infty} T^n,$$

then it is easy to see that $ST = TS = S - 1$, and hence that $S(1 - T) = (1 - T)S = 1$. $\square$

Lemma 12.8 may also be formulated by saying that if $\mathscr{E}$ is a Banach space, then the open ball of radius one centered at $1_{\mathscr{E}}$ in $\mathscr{L}(\mathscr{E})$ consists entirely of invertible operators. The consequences of this fact are far-reaching.

Proposition 12.9. *Let $\mathscr{E}$ and $\mathscr{F}$ be Banach spaces, let T be a left [right] invertible element of $\mathscr{L}(\mathscr{E}, \mathscr{F})$, let R be a left [right] inverse of T, and let $d = \|R\|$. Then every linear transformation S in $\mathscr{L}(\mathscr{E}, \mathscr{F})$ such that $\|S - T\| < 1/d$ is also left [right] invertible. Thus the set of left [right] invertible elements of $\mathscr{L}(\mathscr{E}, \mathscr{F})$ is open. Hence the set $\mathscr{U}$ of invertible elements of $\mathscr{L}(\mathscr{E}, \mathscr{F})$ is also open. Moreover, $T \rightarrow T^{-1}$ is a continuous mapping of $\mathscr{U}$ into $\mathscr{L}(\mathscr{F}, \mathscr{E})$.*

PROOF. To verify the first part of the proposition it suffices to show that if R is a left inverse of T with $\|R\| = d$, and if $\|S - T\| < 1/d$, then S is left invertible. But in these circumstances we have

$$\|1_{\mathscr{E}} - RS\| = \|R(T - S)\| \leq \|R\| \|S - T\| < 1,$$

so that RS is invertible by the preceding lemma. Thus $(RS)^{-1}(RS) = 1$, and $(RS)^{-1}R$ is a left inverse of S. To complete the proof, let $0 < \varepsilon < \frac{1}{2}$ and suppose that $T \in \mathscr{U}$ and that $\|S - T\| < \varepsilon / \|T^{-1}\|$. Then, as we have just seen, $T^{-1}(T - S) = 1 - T^{-1}S$ and $r = \|T^{-1}(T - S)\| < \varepsilon$. Hence, by Lemma 12.8, the series $\sum_{n=0}^{\infty} [T^{-1}(T - S)]^n$ converges to

$$[1 - (1 - T^{-1}S)]^{-1} = (T^{-1}S)^{-1} = S^{-1}T$$

(Prob. J), so the inverse of S is given explicitly by the formula

$$S^{-1} = \sum_{n=0}^{\infty} [T^{-1}(T - S)]^n T^{-1}$$

$$= T^{-1} + T^{-1}(T - S)T^{-1} + T^{-1}(T - S)T^{-1}(T - S)T^{-1} + \cdots.$$

Hence

$$S^{-1} - T^{-1} = \sum_{n=1}^{\infty} [T^{-1}(T - S)]^n T^{-1},$$

and therefore

$$\|S^{-1} - T^{-1}\| \leq \frac{r\|T^{-1}\|}{1 - r} < 2\varepsilon \|T^{-1}\|$$

by Problem 11G, whence the proposition follows.   □

Note. This argument actually proves more than is asserted. The mapping $T \rightarrow T^{-1}$ is clearly *uniformly* continuous on some open neighborhood of any subset of $\mathscr{U}$ on which it is bounded.

Two further important concepts relating to inverses and invertibility are those of the *spectrum* and *resolvent* of a bounded linear operator on a Banach space.

Definition. Let $\mathscr{E}$ be a Banach space and let T be a bounded operator on $\mathscr{E}$. Then the *spectrum* $\sigma(T)$ of T is the set of those complex numbers λ such that the operator $\lambda - T$ is not invertible. Furthermore, the set of those complex numbers λ such that $\lambda - T$ is not left [right] invertible will be called the *left [right] spectrum* of T and will be denoted by $\sigma_l(T)[\sigma_r(T)]$. (Clearly the spectrum of T is the union of the left and right spectra of T. We observe that $\lambda - T$ is [left, right] invertible if and only if $T - \lambda$ is.) The complement $\mathbb{C}\backslash\sigma(T)$, consisting of the set of those scalars λ such that $\lambda - T$ *is* invertible, is called the *resolvent set* of T, and the $\mathscr{L}(\mathscr{E})$-valued mapping R_T defined by setting

$$R_T(\lambda) = (\lambda - T)^{-1}, \qquad \lambda \notin \sigma(T),$$

is called the *resolvent* of T. (If $\mathscr{E}$ is the trivial space (0), then $0_{\mathscr{E}} = 1_{\mathscr{E}}$, and careful scrutiny of the definition shows that the spectrum of this operator is the empty set. As will be shown in Chapter 15, no bounded operator on a Banach space $\mathscr{E} \neq (0)$ has empty spectrum.)

Each value $R_T(\lambda)$ of the resolvent of an operator T doubly commutes with $\lambda - T$, and therefore with T itself (Ch. 2, p. 20). In particular, the values of R_T all commute with one another. Moreover, it is an immediate consequence of Proposition 12.9 that the resolvent R_T is a continuous $\mathscr{L}(\mathscr{E})$-valued mapping on the resolvent set $\mathbb{C}\backslash\sigma(T)$.

Example S. Let $\mathscr{E}$ be a finite dimensional Banach space, and let T denote an arbitrary linear transformation of $\mathscr{E}$ into itself. Then T is automatically a bounded operator on $\mathscr{E}$ (Ex. A), and T fails to be invertible only when $\mathscr{K}(T) \neq (0)$ (Prob. 2G). Thus $\sigma(T)$ coincides with the set of eigenvalues of T, i.e., with the set of solutions of the equation $\det(\lambda - T) = 0$ (see Problem 2O). In general, it is clear that if λ is an eigenvalue of an operator T (on an arbitrary Banach space) so that $\mathscr{K}(\lambda - T) \neq (0)$, then $\lambda - T$ is certainly not invertible, and therefore $\lambda \in \sigma(T)$. Thus the set of all eigenvalues of an operator T, usually known as the *point spectrum* of T (notation: $\sigma_0(T)$), is always a subset of $\sigma(T)$. (More precisely, $\sigma_0(T) \subset \sigma_l(T)$.) On a finite dimensional Banach space the spectrum and point spectrum of an operator coincide; on infinite dimensional spaces that need not be the case.

Example T. Fix an extended real number $p, 1 \leq p \leq +\infty$, and consider the multiplication operator M_d of Example C that multiplies each sequence $\{\xi_n\}_{n=0}^{\infty}$ in (ℓ_p) by the sequence $d = \{\delta_n\}_{n=0}^{\infty}$. Each of the terms of the sequence d is clearly an eigenvalue of M_d (the eigenspace associated with δ_n contains the vector e_n), so the range W of d is contained in the point spectrum $\sigma_0(M_d)$. Moreover, if $\lambda \notin W$ then it is easily seen that λ is not an eigenvalue of M_d. Thus the point spectrum $\sigma_0(M_d)$ coincides with W. On the other hand, if $\lambda \notin W^{-}$, then d is bounded away from λ, and it is readily verified that multiplication by the (bounded) sequence $\{1/(\lambda - \delta_n)\}$ is $(\lambda - M_d)^{-1}$. Thus

259

$\sigma(M_d) \subset W^-$. Finally, if $\lambda \in W^- \setminus W$ then $\lambda - M_d$, which is just multiplication by the sequence $\lambda - d$, is not bounded below, and is therefore not invertible (Prop. 12.5). Thus $\sigma(M_d) = W^-$, and any sequence d for which W and W^- are different provides an example of an operator in $\mathscr{L}((\ell_p))$ whose spectrum is properly larger than its point spectrum. (As noted, each value $R_{M_d}(\lambda)$ of the resolvent of M_d is given by multiplication by the sequence $1/(\lambda - d)$; cf. Example R.)

The notions of spectrum and resolvent of a bounded operator T on a Banach space $\mathscr{E}$ relate only to the properties T has as an element of the Banach algebra $\mathscr{L}(\mathscr{E})$. As has already been observed, the algebras of all bounded operators on Banach spaces are special cases of a broad class of algebras known as unital Banach algebras. Many of the concepts of interest in algebras of the form $\mathscr{L}(\mathscr{E})$ are meaningful and equally important in the setting of an abstract unital Banach algebra. In particular, this is true of the notions of spectrum and resolvent.

Definition. Let $\mathscr{A}$ be a unital Banach algebra and let x be an element of $\mathscr{A}$. A complex number λ is said to belong to the *left spectrum* of x *with respect to* $\mathscr{A}$ (notation: $\lambda \in \sigma_{l,\mathscr{A}}(x)$) if $\lambda - x(= \lambda 1_{\mathscr{A}} - x)$ is not *left invertible* in $\mathscr{A}$, i.e., if there does not exist any element y of $\mathscr{A}$ such that $y(\lambda - x) = 1$. Similarly, the *right spectrum* $\sigma_{r,\mathscr{A}}(x)$ of x with respect to $\mathscr{A}$ consists of the collection of all scalars λ such that $\lambda - x$ is not *right invertible* in $\mathscr{A}$, and the *spectrum* $\sigma_{\mathscr{A}}(x)$ of x *with respect to* $\mathscr{A}$ consists of the collection of all scalars λ such that $\lambda - x$ is not invertible in $\mathscr{A}$. (Clearly $\sigma_{\mathscr{A}}(x) = \sigma_{l,\mathscr{A}}(x) \cup \sigma_{r,\mathscr{A}}(x)$ for every element x of $\mathscr{A}$.) Finally, the complement $\mathbb{C} \setminus \sigma_{\mathscr{A}}(x)$ is the *resolvent set* of x with respect to $\mathscr{A}$, and the $\mathscr{A}$-valued mapping R_x defined on the resolvent set of x with respect to $\mathscr{A}$ by

$$R_x(\lambda) = (\lambda - x)^{-1}, \qquad \lambda \notin \sigma_{\mathscr{A}}(x),$$

is the *resolvent* of x.

Just as in the case of a bounded linear operator, each value $R_x(\lambda)$ of the resolvent of an element x of a unital Banach algebra $\mathscr{A}$ doubly commutes with x (i.e., commutes with every element y of $\mathscr{A}$ that commutes with x). In particular, the values of R_x all commute with one another. Moreover, the resolvent R_x is a continuous $\mathscr{A}$-valued mapping on the resolvent set $\mathbb{C} \setminus \sigma_{\mathscr{A}}(x)$ (Prob. K). Another elementary but central fact about resolvents is set forth in the following proposition.

Proposition 12.10. *Let $\mathscr{A}$ be a unital Banach algebra, and let x be an element of $\mathscr{A}$. Then for every pair α, β of complex numbers in the resolvent set of x, the resolvent R_x satisfies the equation*

$$R_x(\alpha) - R_x(\beta) = -(\alpha - \beta)R_x(\alpha)R_x(\beta)$$

(*known as the* resolvent equation).

PROOF. For arbitrary complex numbers α and β we have

$$(\alpha - x) - (\beta - x) = \alpha - \beta.$$

If both α and β belong to the resolvent set of x, then, multiplying by $R_x(\alpha)R_x(\beta)$, we obtain

$$R_x(\beta) - R_x(\alpha) = (\alpha - \beta)R_x(\alpha))R_x(\beta). \qquad \square$$

Here, and again below, we note explicitly that whatever is true in general of spectra and resolvents of elements of an abstract Banach algebra is also true in particular for bounded linear operators. Thus if T is a bounded operator on a Banach space $\mathscr{E}$, then the resolvent R_T satisfies the resolvent equation

$$R_T(\alpha) - R_T(\beta) = -(\alpha - \beta)R_T(\alpha)R_T(\beta)$$

identically for α, β not belonging to $\sigma(T)$.

Obtaining information about the behavior of the resolvent of an element of a unital Banach algebra would be an exercise in futility if that element had an empty resolvent set. The following result shows that this can never happen. (For a more precise result see Problem M.)

Proposition 12.11. *If $\mathscr{A}$ is a unital Banach algebra and x is an arbitrary element of $\mathscr{A}$, then the series*

$$\sum_{n=0}^{\infty} \frac{x^n}{\lambda^{n+1}} \qquad (8)$$

converges to $(\lambda - x)^{-1}$ for every complex number λ such that $|\lambda| > \|x\|$. Hence the spectrum $\sigma_{\mathscr{A}}(x)$ is bounded, being a subset of the disc $\{\lambda \in \mathbb{C} : |\lambda| \le \|x\|\}$, and the resolvent set of x contains the complement of this disc.

PROOF. If $|\lambda| > \|x\|$ and if we define $r = \|x/\lambda\|$, then $r < 1$ and $\|x^n/\lambda^n\| \le r^n$ for every positive integer n. Hence the series (8) is absolutely convergent and therefore convergent (Prob. 11G). Moreover, multiplying this series (on either side) term by term by $\lambda - x$ gives rise to the difference

$$\sum_{n=0}^{\infty} \left(\frac{x}{\lambda}\right)^n - \sum_{n=0}^{\infty} \left(\frac{x}{\lambda}\right)^{n+1} = 1_{\mathscr{A}},$$

and the result follows. $\qquad \square$

The series expansion (8) may be thought of as the power series expansion of R_x about the point at infinity. A slight modification of the same argument shows that the resolvent R_x can be expanded in a power series about any point in its domain of definition. The following result is also an immediate consequence of the Banach algebra analog of Proposition 12.9; see Problem K.

Proposition 12.12. *Let $\mathscr{A}$ be a unital Banach algebra, let x be an element of $\mathscr{A}$, let λ_0 be a complex number in the resolvent set of x, and let $d = \|R_x(\lambda_0)\|$. Then every scalar λ in the disc $D = \{\lambda \in \mathbb{C} : |\lambda - \lambda_0| < 1/d\}$ belongs to the resolvent set of x, and for each λ in D,*

$$R_x(\lambda) = \sum_{n=0}^{\infty} (\lambda_0 - \lambda)^n R_x(\lambda_0)^{n+1}. \tag{9}$$

In particular, no point of $\sigma_{\mathscr{A}}(x)$ lies in D.

PROOF. If $|\lambda - \lambda_0| < 1/d$, then

$$\|(\lambda_0 - \lambda)R_x(\lambda_0)\| = r < 1$$

and

$$\|(\lambda_0 - \lambda)^n R_x(\lambda_0)^n\| \le r^n$$

for every n. Hence the series (9) is absolutely convergent and therefore convergent. Moreover, multiplying this series (on either side) term by term by $\lambda - x = (\lambda_0 - x) - (\lambda_0 - \lambda)$ gives rise to the difference

$$\sum_{n=0}^{\infty} ((\lambda_0 - \lambda)R_x(\lambda_0))^n - \sum_{n=0}^{\infty} ((\lambda_0 - \lambda)R_x(\lambda_0))^{n+1} = 1_{\mathscr{A}}. \qquad \square$$

Corollary 12.13. *For an arbitrary element x of a unital Banach algebra $\mathscr{A}$, the spectrum $\sigma_{\mathscr{A}}(x)$ is a compact set in $\mathbb{C}$.*

PROOF. Proposition 12.12 shows that the complement of $\sigma_{\mathscr{A}}(x)$ is open, and we have just seen that $\sigma_{\mathscr{A}}(x)$ is bounded (Prop. 12.11). $\qquad \square$

Considered as a (set-valued) function of the elements of a unital Banach algebra $\mathscr{A}$, the spectrum has some favorable properties and others that are not so agreeable. As an example of the former we note the obvious fact that for any complex number λ and element x of $\mathscr{A}$ we have $\sigma_{\mathscr{A}}(x + \lambda) = \sigma_{\mathscr{A}}(x) + \lambda$. (For a noteworthy generalization of this fact see Problem O.) As an example of a less desirable property of spectra, we observe that $\sigma_{\mathscr{A}}(x)$ is not, in general, a *continuous* function of x in any reasonable sense of that term.

Example U (G. Lumer). For each nonnegative number r let W_r denote the bilateral shift on $(\ell_1)^*$ having the weight sequence

$$\{\ldots, 1, \ldots, 1, [r], 1, \ldots, 1, \ldots\}$$

(cf. Example F). According to Problems E and M, when r is positive the spectrum of W_r is contained in the closed disc with center at the origin and radius $\limsup_n r^{1/n} = 1$. On the other hand, when r is positive the operator W_r is also invertible, and a very similar calculation discloses that the spectrum of W_r^{-1} is likewise contained in the closed disc with center at the origin and radius $\limsup_n (1/r)^{1/n} = 1$ (see Problem L). This, in turn, implies (Prob. O) that the spectrum $\sigma(W_r)$ lies on the unit circle, and therefore coincides with the unit circle for all $r > 0$ (Prob. R). Moreover, we have $\lim_{r \downarrow 0} W_r = W_0$; indeed, it is easy to see that $\| W_r - W_0 \| = r$, $r > 0$. But W_0 has zero for an eigenvalue ($W_0 e_0 = 0$), so $\sigma(W_0)$ contains a point at constant distance one from all of the sets $\sigma(W_r)$, $r > 0$. (We note that the sequence $\{\xi_n\}_{n=-\infty}^{+\infty}$ defined by

$$\xi_n = \begin{cases} \lambda^n, & n < 0, \\ 1, & n = 0, \\ 0, & n > 0, \end{cases}$$

belongs to $(\ell_1)^*$ for every λ such that $|\lambda| < 1$, and that if we write x_λ for this vector, then $W_0 x_\lambda = \lambda x_\lambda$. Thus $\sigma(W_0)$ contains the open unit disc, and since $\sigma(W_0)$ is closed and is contained in the closed unit disc (Prop. 12.11 and Cor. 12.13), this shows that, in fact, $\sigma(W_0) = \{\lambda \in \mathbb{C} : |\lambda| \le 1\}$. The choice $p = 1$ was made merely to fix ideas; the same calculations yield the same results on any of the spaces $(\ell_p)^*$, $1 \le p \le +\infty$.)

The notion of the spectrum of an element of a unital Banach algebra permeates all of functional analysis, and we shall return to it over and over. For the present we conclude our discussion of spectra with the following result, which shows that the mapping $x \to \sigma_{\mathscr{A}}(x)$ is at least *semicontinuous* in an appropriate sense.

Proposition 12.14. *Let $\mathscr{A}$ be a unital Banach algebra, let x_0 be an element of $\mathscr{A}$, and let ε be a positive number. Then there exists a positive number δ such that if $x \in \mathscr{A}$ and $\| x - x_0 \| < \delta$, then $\sigma_{\mathscr{A}}(x)$ lies in the open set $U_\varepsilon = \{\lambda \in \mathbb{C} : d(\lambda, \sigma_{\mathscr{A}}(x_0)) < \varepsilon\}$.*

PROOF. Let F denote the closed set of all those complex numbers λ such that $d(\lambda, \sigma_{\mathscr{A}}(x_0)) \ge \varepsilon$, and let K denote the compact set consisting of all those complex numbers λ in F such that $|\lambda| \le \| x_0 \| + 1$. The function $\| (\lambda - x_0)^{-1} \|$ is continuous along with $(\lambda - x_0)^{-1}$ on the resolvent set of x_0 (see Problem K), and therefore assumes a maximum M on the compact set K. If $\| x - x_0 \| < 1/M$, then $\| (x - x_0)(\lambda - x_0)^{-1} \| < 1$ for all λ in K, so $1 + (x - x_0)(\lambda - x_0)^{-1}$ is invertible (Prob. K), and since $\lambda - x = [1 + (x_0 - x)(\lambda - x_0)^{-1}](\lambda - x_0)$, we see that $\lambda - x$ is invertible, $\lambda \in K$, whence it follows that $\sigma_{\mathscr{A}}(x)$ is disjoint from K. If $\| x - x_0 \|$ is also less than one, then $\| x \| < \| x_0 \| + 1$, and we conclude that $\sigma_{\mathscr{A}}(x)$ is contained in the disc $\{\lambda \in \mathbb{C} : |\lambda| \le \| x_0 \| + 1\}$. Hence $\sigma_{\mathscr{A}}(x) \subset U_\varepsilon$ and we may take δ to be $1 \wedge (1/M)$. $\quad\square$

Note. Both the statement and the proof of Proposition 12.14 assume the fact that $\sigma_{\mathscr{A}}(x_0)$ is not empty. Consequently, until we prove in Chapter 15 that no element of $\mathscr{A}$ has empty spectrum, care must be taken to apply Proposition 12.14 only in circumstances where it can be shown that the spectrum of x_0 is nonempty. Since our first application of Proposition 12.14 comes in Chapter 17, this presents no difficulty. It may be noted that if we suppose (contrary to fact) that $\sigma_{\mathscr{A}}(x_0)$ *is* empty, then the proof of Proposition 12.14 (slightly modified) shows that all of the elements of some open ball centered at x_0 likewise have the property that their spectra are empty. The reader is reminded that the results developed here are valid, in particular, for bounded linear operators on a Banach space.

When for some Banach space $\mathscr{E}$ the algebra $\mathscr{L}(\mathscr{E})$ is regarded as a Banach algebra, then it is the relations between the various operators in $\mathscr{L}(\mathscr{E})$ that are of paramount importance. It should be kept in mind, however, that a Banach algebra of the form $\mathscr{L}(\mathscr{E})$ has a richer theory than does the general (unital) Banach algebra because the elements of $\mathscr{L}(\mathscr{E})$ also interact with the vectors in $\mathscr{E}$. (Thus, for example, there is no obvious counterpart of the very useful Proposition 12.6 in the general theory of Banach algebras.) We close this chapter with the first of several important applications of the Baire category theorem (Th. 4.8) to the theory of bounded linear transformations on Banach spaces. Observe that in this application the interaction between linear transformations and vectors is of central importance. In regard to terminology we note that since the elements of $\mathscr{L}(\mathscr{E}, \mathscr{F})$ are referred to individually as bounded, it has become customary to speak of a bounded subset of $\mathscr{L}(\mathscr{E}, \mathscr{F})$ as *uniformly* bounded.

Theorem 12.15 (Uniform Boundedness Theorem). *If $\mathscr{T}$ is a collection of bounded linear transformations of a Banach space $\mathscr{E}$ into a normed space $\mathscr{F}$, and if for every vector x in $\mathscr{E}$ the set $\mathscr{T}x = \{Tx : T \in \mathscr{T}\}$ is a bounded set of vectors in $\mathscr{F}$, then the collection $\mathscr{T}$ is uniformly bounded, i.e., there exists a constant K such that $\| T \| \leq K$ for every T in $\mathscr{T}$.*

Proof. For each positive integer n let S_n denote the set

$$ S_n = \{x \in \mathscr{E} : \| Tx \| \leq n, \ T \in \mathscr{T}\}. $$

Since each T in $\mathscr{T}$ is continuous, it is clear that S_n is a closed set. Moreover, it follows from the hypothesis that the sequence $\{S_n\}$ of closed sets covers the complete space $\mathscr{E}$. Hence by the Baire category theorem some S_n—say S_{n_0}—must have nonvoid interior. Thus there exists a vector x_0 and a number $\varepsilon > 0$ such that $x_0 + \mathscr{E}_\varepsilon \subset S_{n_0}$. Since every vector x in $\mathscr{E}$ can be written as $x = (x_0 + x) - x_0$, it follows that for every T in $\mathscr{T}$ and every x with $\| x \| < \varepsilon$,

$$ \| Tx \| \leq \| T(x_0 + x) \| + \| Tx_0 \| \leq 2n_0. $$

Hence $\| T \| \leq 2n_0/\varepsilon$ for all T in $\mathscr{T}$. $\qquad\qquad\square$

Example V. Let $\{T_n\}_{n=1}^{\infty}$ be an infinite sequence of bounded linear transformations of a Banach space $\mathscr{E}$ into a Banach space $\mathscr{F}$, and suppose that for every vector x in $\mathscr{E}$ the sequence $\{T_n x\}_{n=1}^{\infty}$ is convergent in $\mathscr{F}$ (briefly: $\{T_n\}$ is pointwise convergent on $\mathscr{E}$). Then for each x in $\mathscr{E}$ the set $\{T_n x : n \in \mathbb{N}\}$ is, *a fortiori*, a bounded set, and it follows that there exists a constant M such that $\|T_n\| \leq M$ for every n in $\mathbb{N}$. From this it is a simple matter to conclude that the (linear) mapping of $\mathscr{E}$ into $\mathscr{F}$ defined by setting $Tx = \lim_n T_n x$ is bounded (by M). Note that we do not assert that the sequence $\{T_n\}$ converges to T in $\mathscr{L}(\mathscr{E}, \mathscr{F})$.

PROBLEMS

A. Let T be a linear transformation of a normed space $\mathscr{E}$ into a normed space $\mathscr{F}$. Verify the fact that if the set Φ of all positive numbers M that satisfy (1) is not empty, then $\inf \Phi \in \Phi$.

B. If a linear transformation T of a normed space $\mathscr{E}$ into another normed space $\mathscr{F}$ is bounded on *any* nonempty open set, then T is continuous. Hence, in particular, if the function $\|Tx\|$ is continuous (or even upper semi-continuous; Example 3K) at a single point x_0 of $\mathscr{E}$, then T is bounded.

C. An $m \times n$ matrix $A = (\alpha_{ij})$ defines, in a natural way, a linear transformation T from $\mathbb{C}^n$ into $\mathbb{C}^m$ according to the formula

$$T(\xi_1, \ldots, \xi_n) = A \begin{pmatrix} \xi_1 \\ \vdots \\ \xi_n \end{pmatrix}$$

(Prob. 2H). Verify that T is bounded when $\mathbb{C}^n$ and $\mathbb{C}^m$ are equipped with the norms $\| \ \|_p$ and $\| \ \|_{p'}$, respectively, $1 < p, p' < +\infty$, by showing that, in fact,

$$\|T\| \leq \left(\sum_{i=1}^{m} \|r_i\|_q^{p'} \right)^{1/p'}$$

where $r_i = (\alpha_{i1}, \ldots, \alpha_{in})$ denotes the ith row of A and q is the Hölder conjugate of p. Devise analogous estimates using the norms $\| \ \|_1$ and $\| \ \|_{\infty}$.

D. Find the norm of the matrix

$$\begin{pmatrix} 1 & 1 \\ 1 & 2 \end{pmatrix} \tag{10}$$

regarded as a linear operator on $\mathbb{C}^2$ when $\mathbb{C}^2$ is given the norm $\| \ \|_1$. Do the same when $\mathbb{C}^2$ is given the norm $\| \ \|_{\infty}$. Try to calculate the norm of (10) when $\| \ \|_2$ is used.

E. Describe the action of S^2 where S denotes the weighted unilateral shift of Example D. What is $\|S^2\|$? More generally, describe the action of S^m, $m \in \mathbb{N}$, and show that

$$\|S^m\| = \sup_{n \in \mathbb{N}_0} |\delta_n \delta_{n+1} \cdots \delta_{n+m-1}|.$$

What are the analogous facts regarding the weighted bilateral shift T of Example F?

265

F. Let $\mathscr{E}$ be a normed space, let T be a linear transformation of $\mathscr{E}$ into a normed space $\mathscr{F}$, and let $\mathscr{M}$ be a subspace of $\mathscr{E}$ contained in the kernel $\mathscr{K}(T)$. Show that the result $\hat{T}$ of factoring T through $\mathscr{E}/\mathscr{K}$ (Ex. 2L) is bounded if and only if T is, and that, when T is bounded, then $\| \hat{T} \| = \| T \|$. Show likewise that $\hat{T}$ is open if and only if T is. (Hint: See Problem 11H.)

G. Let M be a subset of a normed space $\mathscr{E}$, let $\mathscr{M}$ be the subspace of $\mathscr{E}$ spanned by M, and let φ be a mapping of M into a normed space $\mathscr{F}$. Show that there exists a bounded linear transformation $T : \mathscr{M} \to \mathscr{F}$ that extends φ if and only if there exists a positive constant K such that

$$\left\| \sum_{i=1}^{n} \lambda_i \varphi(x_i) \right\| \le K \left\| \sum_{i=1}^{n} \lambda_i x_i \right\|$$

for every linear combination $\lambda_1 x_1 + \cdots + \lambda_n x_n$ of vectors in M (and every positive integer n).

H. Let $\mathscr{E}$ and $\mathscr{F}$ be normed spaces, and suppose that each is a direct sum of two normed spaces—say $\mathscr{E} = \mathscr{E}_1 \oplus_1 \mathscr{E}_2$ and $\mathscr{F} = \mathscr{F}_1 \oplus_1 \mathscr{F}_2$—so that every vector in $\mathscr{E}$ is a pair (x_1, x_2) where $x_i \in \mathscr{E}_i$, $i = 1, 2$, and similarly for $\mathscr{F}$.

(i) If $T \in \mathscr{L}(\mathscr{E}, \mathscr{F})$ and x is an arbitrary vector in $\mathscr{E}_1$, then $T(x, 0)$ can be written as

$$T(x, 0) = (T_{11}x, T_{21}x)$$

where T_{11} and T_{21} are mappings of $\mathscr{E}_1$ into $\mathscr{F}_1$ and $\mathscr{F}_2$, respectively. Show that T_{11} and T_{21} are bounded linear transformations satisfying $\| T_{i1} \| \le \| T \|$, $i = 1, 2$. Similarly, if x is an arbitrary vector in $\mathscr{E}_2$, then setting

$$T(0, x) = (T_{12}x, T_{22}x)$$

defines bounded linear transformations T_{12} and T_{22} of $\mathscr{E}_2$ into $\mathscr{F}_1$ and $\mathscr{F}_2$, respectively, such that $\| T_{i2} \| \le \| T \|$, $i = 1, 2$. In this way we associate with every T in $\mathscr{L}(\mathscr{E}, \mathscr{F})$ a unique matrix $M = M(T)$ given by

$$M = \begin{pmatrix} T_{11} & T_{12} \\ T_{21} & T_{22} \end{pmatrix}, \qquad T_{ij} \in \mathscr{L}(\mathscr{E}_j, \mathscr{F}_i), \qquad i, j = 1, 2, \tag{11}$$

such that if $T(x_1, x_2) = (y_1, y_2)$, $x_i \in \mathscr{E}_i$, $i = 1, 2$, then

$$y_1 = T_{11}x_1 + T_{12}x_2 \quad \text{and} \quad y_2 = T_{21}x_1 + T_{22}x_2.$$

In other words, if we write (x_1, x_2) and (y_1, y_2) as column vectors, then

$$\begin{pmatrix} y_1 \\ y_2 \end{pmatrix} = \begin{pmatrix} T_{11} & T_{12} \\ T_{21} & T_{22} \end{pmatrix} \begin{pmatrix} x_1 \\ x_2 \end{pmatrix}, \tag{12}$$

where the indicated formal matrix multiplication is performed according to the usual row by column rule. The matrix $M(T)$ is called the *operator matrix* for T relative to the decompositions $\mathscr{E} = \mathscr{E}_1 \oplus_1 \mathscr{E}_2$ and $\mathscr{F} = \mathscr{F}_1 \oplus_1 \mathscr{F}_2$.

(ii) Show conversely that if M is a matrix of the form (11), then (12) defines an element T of $\mathscr{L}(\mathscr{E}, \mathscr{F})$ with the property that $M(T) = M$. Show also that the collection $\mathscr{M}$ of all matrices of the form (11) constitutes a linear space under the operations $(S_{ij}) + (T_{ij}) = (S_{ij} + T_{ij})$ and $\alpha(S_{ij}) = (\alpha S_{ij})$, and that the mapping $M : \mathscr{L}(\mathscr{E}, \mathscr{F}) \to \mathscr{M}$ of (i) is a linear space isomorphism of $\mathscr{L}(\mathscr{E}, \mathscr{F})$ onto $\mathscr{M}$.

(iii) Let $\mathcal{G}$ be another normed space that is a direct sum $\mathcal{G} = \mathcal{G}_1 \oplus_1 \mathcal{G}_2$ and let $S \in \mathcal{L}(\mathcal{F}, \mathcal{G})$. Show that if the operator matrix for S relative to the decompositions $\mathcal{F} = \mathcal{F}_1 \oplus_1 \mathcal{F}_2$ and $\mathcal{G} = \mathcal{G}_1 \oplus_1 \mathcal{G}_2$ is

$$\begin{pmatrix} S_{11} & S_{12} \\ S_{21} & S_{22} \end{pmatrix}$$

and if T is as in (i), then the operator matrix for the product ST relative to the decompositions $\mathcal{E} = \mathcal{E}_1 \oplus_1 \mathcal{E}_2$ and $\mathcal{G} = \mathcal{G}_1 \oplus_1 \mathcal{G}_2$ is the matrix

$$\begin{pmatrix} S_{11}T_{11} + S_{12}T_{21} & S_{11}T_{12} + S_{12}T_{22} \\ S_{21}T_{11} + S_{22}T_{21} & S_{21}T_{12} + S_{22}T_{22} \end{pmatrix} = \begin{pmatrix} S_{11} & S_{12} \\ S_{21} & S_{22} \end{pmatrix}\begin{pmatrix} T_{11} & T_{12} \\ T_{21} & T_{22} \end{pmatrix} \quad (13)$$

obtained by forming the formal product $(S_{ij})(T_{ij})$ by the row by column rule.

(iv) If $\mathcal{E}_i = \mathcal{F}_i, i = 1, 2$, so that $\mathcal{E} = \mathcal{F}$ and $\mathcal{L}(\mathcal{E}, \mathcal{F}) = \mathcal{L}(\mathcal{E})$, show that the product defined by (13) turns the linear space $\mathcal{M}$ of (ii) into an algebra. Show also that the mapping M is an algebra isomorphism of $\mathcal{L}(\mathcal{E})$ onto $\mathcal{M}$.

(v) Show, more generally, that if $\mathcal{E} = \mathcal{E}_1 \oplus_1 \cdots \oplus_1 \mathcal{E}_n$ and $\mathcal{F} = \mathcal{F}_1 \oplus_1 \cdots \oplus_1 \mathcal{F}_m$, then the above construction may be broadened to assign to each element T of $\mathcal{L}(\mathcal{E}, \mathcal{F})$ an $m \times n$ operator matrix

$$M = (T_{ij}), T_{ij} \in \mathcal{L}(\mathcal{E}_j, \mathcal{F}_i), \quad i = 1, \ldots, m, \quad j = 1, \ldots, n,$$

and that all of the foregoing assertions generalize to this setting.

I. Let $\mathcal{E}$ be a quasinormed space and let T be a continuous linear transformation of a linear submanifold $\mathcal{L}$ of $\mathcal{E}$ into an F-space $\mathcal{F}$. Show that T is uniformly continuous on $\mathcal{E}$ and hence that there exists a unique continuous linear transformation $\tilde{T}$ of $\mathcal{L}^-$ into $\mathcal{F}$ that extends T (Prob. 4H). (The linear transformation $\tilde{T}$ is said to result from *extending* T *by continuity*.) In particular, every continuous linear transformation of a quasinormed space $\mathcal{E}$ into another quasinormed space $\mathcal{F}$ possesses a unique continuous extension mapping $\hat{\mathcal{E}}$ into $\hat{\mathcal{F}}$, where $\hat{\mathcal{E}}$ and $\hat{\mathcal{F}}$ denote the completions of $\mathcal{E}$ and $\mathcal{F}$, respectively (Th. 11.16). Show also that if $\mathcal{E}$ is a normed space and $\mathcal{F}$ is a Banach space, and if T is a bounded linear transformation of a linear submanifold $\mathcal{L}$ of $\mathcal{E}$ into $\mathcal{F}$, then $\|\tilde{T}\| = \|T\|$.

J. Let $\mathcal{E}, \mathcal{F}$, and $\mathcal{G}$ be Banach spaces, and let S and T be invertible elements of $\mathcal{L}(\mathcal{F}, \mathcal{G})$ and $\mathcal{L}(\mathcal{E}, \mathcal{F})$, respectively. Show that ST is invertible, and that, in fact, $(ST)^{-1} = T^{-1}S^{-1}$. In particular, if S and T are elements of the open set $\mathfrak{G}$ of invertible operators in $\mathcal{L}(\mathcal{E})$ (cf. the note following Proposition 12.9), then ST belongs to $\mathfrak{G}$. Likewise, if $T \in \mathfrak{G}$ then $T^{-1} \in \mathfrak{G}$, since $(T^{-1})^{-1} = T$. (These facts are expressed by saying that $\mathfrak{G}$ forms a *group*.) Show also that if T is an element of $\mathfrak{G}$, then $(T^n)^{-1} = (T^{-1})^n$ for each positive integer n. (This operator is customarily denoted by T^{-n}.) Verify, finally, that if $T_1, \ldots, T_n$ are pairwise commuting elements of $\mathcal{L}(\mathcal{E})$, then the product $T_1 \cdots T_n$ is invertible if and only if each T_i is invertible.

K. Let $\mathcal{A}$ be an arbitrary unital Banach algebra with unit 1. Show that if x and y are invertible elements of $\mathcal{A}$, then $(xy)^{-1} = y^{-1}x^{-1}$ and $(x^{-1})^{-1} = x$. (Thus the set $\mathfrak{G}_\mathcal{A}$ of invertible elements of $\mathcal{A}$ forms a group.) Show too that if $x \in \mathfrak{G}_\mathcal{A}$, then $(x^n)^{-1} = (x^{-1})^n$ for every positive integer n. (Once again, this element is denoted by x^{-n}.) Prove that if x is any element of $\mathcal{A}$ such that $\|1 - x\| < 1$, then $x \in \mathfrak{G}_\mathcal{A}$. Conclude that the set of left [right] invertible elements in $\mathcal{A}$ is open. Conclude also that the set $\mathfrak{G}_\mathcal{A}$ itself is open and that the mapping $x \to x^{-1}$ of $\mathfrak{G}_\mathcal{A}$ onto itself is continuous and is, in fact, uniformly continuous on (some open neighborhood of) any subset of $\mathfrak{G}_{\mathscr{A}}$ upon which it is bounded. (Hint: Follow the proofs of Lemma 12.8 and Proposition 12.9.) Verify, finally that if $x_1, \ldots, x_n$ are elements of $\mathscr{A}$ that commute in pairs, then the product $x_1 \cdots x_n$ is invertible in $\mathscr{A}$ if and only if each x_i is.

L. Let T be the weighted bilateral shift of Example F. Show that T is invertible on $(\ell_p)^*, 1 \leq p \leq +\infty$, if and only if T is bounded below, and that this is true, in turn, if and only if the weight sequence $d = \{\delta_n\}$ is bounded away from zero. Verify also that if d is bounded away from zero, then

$$\| T^{-m} \| = \sup_{n \in \mathbb{Z}} \frac{1}{|\delta_n \delta_{n+1} \cdots \delta_{n+m-1}|}, \qquad m \in \mathbb{N}.$$

(Hint: Construct T^{-1} explicitly, and recall Problem E. An operator such as T^{-1} is known as a *backward weighted bilateral shift*.)

M. Let x be an element of a unital Banach algebra $\mathscr{A}$, and let $s = \lim \sup_n \| x^n \|^{1/n}$. Show that $0 \leq s \leq \| x \|$. Show also that the power series

$$\sum_{n=0}^{\infty} \frac{x^n}{\lambda^n}$$

converges for $|\lambda| > s$ and diverges for $|\lambda| < s$ (cf. Example 5A). Conclude that

$$(\lambda - x)^{-1} = \sum_{n=0}^{\infty} \frac{x^n}{\lambda^{n+1}}$$

for all λ such that $|\lambda| > s$, and hence that the spectrum $\sigma_{\mathscr{A}}(x)$ is contained in the disc $\{\lambda \in \mathbb{C} : |\lambda| \leq s\}$. (Hint: Use the root test and the Cauchy criterion.) In particular, if T is a bounded operator on a Banach space $\mathscr{E}$, then $\sigma(T)$ is a compact subset of the disc $\{\lambda \in \mathbb{C} : |\lambda| \leq \lim \sup_n \| T^n \|^{1/n}\}$. These results should be contrasted with Proposition 12.11.

N. Let $\mathscr{A}$ denote a unital Banach algebra, and let x be an element of $\mathscr{A}$.

(i) Show that if p is any polynomial, then $p(\sigma_{\mathscr{A}}(x)) = \sigma_{\mathscr{A}}(p(x))$. (Hint: Write $p(\lambda) = \alpha_n \lambda^n + \cdots + \alpha_0$ as $\alpha_n(\lambda - \rho_1) \cdots (\lambda - \rho_n)$ ($\alpha_n \neq 0$), so that $p(x) = \alpha_n(x - \rho_1) \cdots (x - \rho_n)$, and use Problem K to conclude that zero belongs to $\sigma_{\mathscr{A}}(p(x))$ if and only if some one of the roots $\rho_1, \ldots, \rho_n$ of the equation $p(\lambda) = 0$ lies in $\sigma_{\mathscr{A}}(x)$.)

(ii) Let r be a rational function on $\mathbb{C}$ with the property that no pole of r lies in $\sigma_{\mathscr{A}}(x)$ (briefly: let r be a rational function with *poles off* $\sigma_{\mathscr{A}}(x)$). Then r can be expressed as p/q, where p and q are polynomials and $q(x)$ is invertible. Show that setting $r(x) = p(x)(q(x))^{-1}$ defines an algebra homomorphism of the algebra of all rational functions with poles off $\sigma_{\mathscr{A}}(x)$ into $\mathscr{A}$, and verify that $\sigma_{\mathscr{A}}(r(x)) = r(\sigma_{\mathscr{A}}(x))$ for every such rational function r. (Hint: For any one fixed complex number α, λ is a solution of the equation $\alpha = r(\lambda)$ if and only if λ is a solution of the equation $\alpha q(\lambda) - p(\lambda) = 0$; likewise, $\alpha - r(x)$ fails to be invertible if and only if $\alpha q(x) - p(x)$ fails to be.) We observe that this shows, in particular, that if T is a bounded operator on a Banach space $\mathscr{E}$, and if r is any rational function with poles off $\sigma(T)$, then $\sigma(r(T)) = r(\sigma(T))$. Thus, for example, if T is invertible, then $\sigma(T^{-1}) = \{\lambda^{-1} : \lambda \in \sigma(T)\}$.

O. A linear transformation N on a linear space is *nilpotent* if there exists a positive integer p such that $N^p = 0$. (The smallest such positive integer p is the *index of nilpotence* of N.) A bounded operator Q on a Banach space is *quasinilpotent* if $\sigma(Q) = \{0\}$.

   (i) Prove that a nilpotent operator on a Banach space is automatically quasinilpotent. (Hint: Use the preceding problem.) Show, conversely, that if Q is a quasinilpotent operator on a *finite dimensional* Banach space $\mathscr{E}$, then Q is nilpotent with index of nilpotence no greater than dim $\mathscr{E}$. (Hint: Recall Problem 2P.)

   (ii) Let p be an extended real number, $1 \le p \le +\infty$, let d denote the sequence $d = \{1/(n+1)\}_{n=0}^{\infty}$, and let Q be the weighted unilateral shift on (ℓ_p) with weight sequence d (Ex. D). Show that Q is quasinilpotent. (Hint: Use Problems E and M, and recall Proposition 12.5.) Thus on an infinite dimensional Banach space it is quite possible for an operator to be quasinilpotent but not nilpotent.

P. If T is a bounded operator on a Banach space $\mathscr{E}$, then the *spectral radius* of T is, by definition, the radius

$$r(T) = \sup_{\lambda \in \sigma(T)} |\lambda|$$

of the smallest closed disc centered at zero that contains the spectrum of T. More generally, if $\mathscr{A}$ is an arbitrary unital Banach algebra, then the *spectral radius with respect to $\mathscr{A}$* of an element x of $\mathscr{A}$ is the radius $r_{\mathscr{A}}(x)$ of the smallest closed disc in $\mathbb{C}$ that is centered at zero and contains $\sigma_{\mathscr{A}}(x)$. Problem M shows that

$$r_{\mathscr{A}}(x) \le \limsup_n \|x^n\|^{1/n} \le \|x\|.$$

Use Problem N to show that $r_{\mathscr{A}}(x^n) = (r_{\mathscr{A}}(x))^n$ and hence that

$$r_{\mathscr{A}}(x) = (r_{\mathscr{A}}(x^n))^{1/n} \le \|x^n\|^{1/n}$$

for every positive integer n. Thus $r_{\mathscr{A}}(x) \le \liminf_n \|x^n\|^{1/n}$.

   As we shall see in Chapter 15, the limit $\lim_n \|x^n\|^{1/n}$ always exists for an arbitrary element x of a normed algebra $\mathscr{A}$. Hence

$$\liminf_n \|x^n\|^{1/n} = \limsup_n \|x^n\|^{1/n},$$

so the result of Problem P is not really any stronger than that of Problem M.

Q. (i) If R and S are bounded operators on Banach spaces $\mathscr{E}$ and $\mathscr{F}$, respectively, then R and S are said to be *similar* if there exists an equivalence T of $\mathscr{E}$ onto $\mathscr{F}$ such that $R = T^{-1}ST$. Show that if R and S are similar, then $\sigma(R) = \sigma(S)$. More precisely, show that $\sigma_r(R) = \sigma_r(S)$, $\sigma_l(R) = \sigma_l(S)$, and $\sigma_0(R) = \sigma_0(S)$ (Ex. S).

   (ii) If x and y are elements of a unital Banach algebra $\mathscr{A}$, then x and y are said to be *similar* in $\mathscr{A}$ if there exists an invertible element z of $\mathscr{A}$ such that $x = z^{-1}yz$. Show that similarity is an equivalence relation on $\mathscr{A}$, and that, if x and y are similar elements in $\mathscr{A}$, then $\sigma_{\mathscr{A}}(x) = \sigma_{\mathscr{A}}(y)$. More precisely, show that $\sigma_{r,\mathscr{A}}(x) = \sigma_{r,\mathscr{A}}(y)$ and $\sigma_{l,\mathscr{A}}(x) = \sigma_{l,\mathscr{A}}(y)$.

R. Let $d = \{\delta_n\}_{n=0}^{\infty}$ and $d' = \{\delta_n'\}_{n=0}^{\infty}$ be two weight sequences in (ℓ_x) such that the sequence of ratios $\{\rho_n = (\delta_0 \cdots \delta_{n-1})/(\delta_0' \cdots \delta_{n-1}')\}_{n=1}^{\infty}$ has the property that there exist numbers m and M such that

$$0 < m \le |\rho_n| \le M < +\infty$$

for every positive integer n. Show that the weighted shifts VM_d and $VM_{d'}$ (Ex. D) are similar on (ℓ_p), $1 \leq p \leq +\infty$. (Hint: Set $r = \{1, \rho_1, \ldots, \rho_n, \ldots\}$ and compute $M_{1/r} VM_d M_r$.) Show that the same criterion is sufficient to ensure the similarity of the backward weighted unilateral shifts with weight sequences d and d' (see Example M). Show, similarly, that if $d = \{\delta_n\}$ and $d' = \{\delta'_n\}$ are bounded two-way infinite sequences such that the sliding ratios

$$\rho_{k,n} = \frac{\delta_n \delta_{n+1} \cdots \delta_{n+k-1}}{\delta'_n \delta'_{n+1} \cdots \delta'_{n+k-1}}$$

satisfy the condition that there exist numbers m and M such that

$$0 < m \leq |\rho_{k,n}| \leq M < +\infty$$

for all k and n, then the weighted bilateral shifts (both forward and backward) having these weight sequences are all similar to one another. Conclude that if S is any weighted shift, unilateral or bilateral, forward or backward, then S is similar to γS for every complex number γ of modulus one, and therefore that the spectrum of S possesses circular symmetry in the sense that if $\lambda \in \sigma(S)$, then $\gamma\lambda \in \sigma(S)$ for every γ of modulus one.

S. Let $\mathscr{R}$ denote the collection of multiplication operators on (ℓ_p) for some fixed p, $1 \leq p \leq +\infty$ (see Example R). Show that if an operator T on (ℓ_p) commutes with every operator in $\mathscr{R}$, then T must belong to $\mathscr{R}$. Conclude that there does not exist any commutative subset of $\mathscr{L}((\ell_p))$ that contains $\mathscr{R}$ as a proper subset. (This fact is usually expressed by saying that $\mathscr{R}$ is a *maximal abelian subalgebra* of $\mathscr{L}((\ell_p))$.) Conclude also that an operator on (ℓ_p) doubly commutes with a multiplication operator M_d if and only if it is itself a multiplication operator. (Hint: Among the operators in $\mathscr{R}$ are the multiplications by sequences of zeros and ones.)

T. Let $\{\mathscr{E}_\gamma\}_{\gamma \in \Gamma}$ be an indexed family of Banach spaces, and let $\mathscr{E}$ denote either the Banach space $\mathscr{B}$ of Problem 11S consisting of the bounded indexed families $\{x_\gamma\}_{\gamma \in \Gamma}$ (with norm $\|\{x_\gamma\}\|_\infty = \sup_\gamma \|x_\gamma\|$), or the Banach space $\mathscr{K}_p$ of Problem 11T for some p, $1 \leq p < +\infty$ (with norm $\|\{x_\gamma\}\|_p = [\sum_\gamma \|x_\gamma\|^p]^{1/p}$). Verify that the Banach space $\mathscr{L}(\mathscr{E})$ is not separable unless all of the spaces $\mathscr{L}(\mathscr{E}_\gamma)$ are separable and $\mathscr{E}_\gamma = (0)$ for all but a finite number of indices γ. (Hint: Recall Example R.)

U. Let $\mathscr{E}$ and $\mathscr{F}$ denote arbitrary topological linear spaces, and let T be a linear transformation of $\mathscr{E}$ into $\mathscr{F}$.

(i) Show that the following conditions are equivalent:

(1) T is continuous,
(2) T is continuous at $x = 0$,
(3) T is continuous at any one vector x_0 in $\mathscr{E}$.

Verify also that the collection $\mathscr{L}$ of all continuous linear transformations of $\mathscr{E}$ into $\mathscr{F}$ is a linear submanifold of the linear space of all linear transformations of $\mathscr{E}$ into $\mathscr{F}$, and that, when $\mathscr{E} = \mathscr{F}$, the linear manifold $\mathscr{L}$ is also an algebra with identity $1_\mathscr{E}$.

(ii) Show that T is automatically continuous when $\mathscr{E}$ is separated and finite dimensional. Show, similarly, that if $\mathscr{F}$ is separated and finite dimensional, then T is continuous if and only if $\mathscr{K}(T)$ is closed in $\mathscr{E}$. (Hint: Recall Problem 11P(ii).)

(iii) Suppose the topology on $\mathscr{E}$ is induced by some indexed family $\{v_\alpha\}_{\alpha \in A}$ of deminorms and that $\mathscr{F}$ is likewise topologized by a second indexed family $\{v_\beta\}_{\beta \in B}$ of deminorms. (Recall that every linear topology can be so obtained (Prob. 11X).) Show that T is continuous if and only if, given a positive number ε and an index β, there exist a positive number δ and indices $\alpha_1, \ldots, \alpha_n$ such that $v_{\alpha_i}(x) < \delta$, $i = 1, \ldots, n$, implies $v_\beta(Tx) < \varepsilon$.

(iv) Suppose $\mathscr{E}$ and $\mathscr{F}$ are topologized by saturated indexed families of pseudonorms $\{\sigma_\alpha\}_{\alpha \in A}$ and $\{\sigma_\beta\}_{\beta \in B}$, respectively. Show that T is continuous if and only if for every index β there exist a positive number M and an index α such that

$$\sigma_\beta(Tx) \leq M\sigma_\alpha(x), \qquad x \in \mathscr{E}.$$

V. Just as in the case of normed spaces, a continuous one-to-one linear transformation T of a quasinormed space $\mathscr{E}$ onto another quasinormed space $\mathscr{F}$ is called an *equivalence* between $\mathscr{E}$ and $\mathscr{F}$ if T is *invertible*, that is, if there exists a continuous linear transformation $S : \mathscr{F} \to \mathscr{E}$ such that $ST = 1_\mathscr{E}$ and $TS = 1_\mathscr{F}$. (Since the set-theoretic inverse of a linear transformation is automatically linear, this is equivalent to requiring that T be open.) Show that if there exists an equivalence between two quasinormed spaces, and if either of them is complete, then both are. Conclude that if two quasinorms on the same linear space $\mathscr{E}$ are equivalent (Ex. 11I), and if $\mathscr{E}$ is complete with respect to either quasinorm, then it is complete with respect to both. (In Chapter 13 we shall obtain a rather remarkable converse to the latter assertion; see Example 13A.)

W. Let $\mathscr{E}_1, \mathscr{E}_2$, and $\mathscr{F}$ be normed spaces, and let φ be a bilinear transformation of the algebraic direct sum $\mathscr{E}_1 + \mathscr{E}_2$ into $\mathscr{F}$. Let us say that φ is *bounded* if there exists a positive number M such that

$$\|\varphi(x, y)\| \leq M \|x\| \|y\| \tag{14}$$

for all vectors x in $\mathscr{E}_1$ and y in $\mathscr{E}_2$. Show that if φ is bounded, then there exists a smallest number M satisfying (14). Show likewise that, if we denote this smallest bound (called the norm of c) by $\|\varphi\|$ (and if $\mathscr{E}_1 \neq (0) \neq \mathscr{E}_2$), then

$$\|\varphi\| = \sup_{\|x\| \cdot \|y\| \leq 1} |\varphi(x, y)| = \sup_{\substack{x \neq 0 \\ y \neq 0}} \frac{|\varphi(x, y)|}{\|x\| \|y\|}.$$

Show, finally, that φ is continuous (with respect to the product topology on $\mathscr{E}_1 + \mathscr{E}_2$) if it is bounded, and that this, in turn, is true if φ is continuous at the origin $(0, 0)$.

X. Let $\mathscr{E}_1$ and $\mathscr{E}_2$ be normed spaces, let $\mathscr{L}_1$ and $\mathscr{L}_2$ be linear manifolds in $\mathscr{E}_1$ and $\mathscr{E}_2$, respectively, and let φ be a bounded bilinear transformation of $\mathscr{L}_1 + \mathscr{L}_2$ into a Banach space $\mathscr{F}$. Show that there exists a unique bounded bilinear transformation $\tilde{\varphi}$ of $\mathscr{L}_1^- + \mathscr{L}_2^-$ into $\mathscr{F}$ that extends φ.

Y. Let $\mathscr{E}_1$ and $\mathscr{E}_2$ be Banach spaces, and let φ be a bilinear transformation of $\mathscr{E}_1 + \mathscr{E}_2$ into a Banach space $\mathscr{F}$. Show that if $\varphi(x, y)$ is continuous in x for each y in $\mathscr{E}_2$ and continuous in y for each x in $\mathscr{E}_1$, then φ is necessarily continuous on $\mathscr{E}_1 + \mathscr{E}_2$. (Hint: For each vector x in $\mathscr{E}_1$, let T_x denote the linear transformation $y \to \varphi(x, y)$ of $\mathscr{E}_2$ into $\mathscr{F}$, and consider the collection $\mathscr{T} = \{T_x : \|x\| \leq 1\}$.)

13 The open mapping theorem

A distinguished role is played in mathematical analysis by the open mapping theorems, that is, by those theorems asserting that, under suitable hypotheses, a continuous mapping must be open. (For one-to-one mappings this is equivalent to the assertion that the inverse mapping is also continuous.) The reader is already familiar with one theorem of this type, viz., the theorem asserting that a continuous and one-to-one mapping of a compact topological space onto a Hausdorff space is a homeomorphism (Prop. 3.3). We shall here study an open mapping theorem peculiar to linear transformations. Although we are principally interested in linear transformations between Banach spaces, the results that follow are valid in F-spaces, and it is in this context that we present them. We single out the substance of what is to be proved in the form of two preliminary lemmas. (Recall that in an F-space $\mathscr{E}$ the closed ball with radius r centered at the origin is denoted by $\mathscr{E}_r$, and the corresponding open ball by $\mathscr{E}_r^\circ$.)

Lemma 13.1. *Let $\mathscr{E}$ and $\mathscr{F}$ be F-spaces, let T be a continuous linear transformation of $\mathscr{E}$ onto $\mathscr{F}$, and let ε be a positive number. Then $(T(\mathscr{E}_\varepsilon))^-$ is a neighborhood of 0 in $\mathscr{F}$.*

PROOF. For each x in $\mathscr{E}$ the mapping $\lambda \to \lambda x$ is continuous on $\mathbb{C}$. Hence $x/n \in \mathscr{E}_{\varepsilon/2}$ for all sufficiently large positive integers n, and it follows that the sequence of sets $\{n\mathscr{E}_{\varepsilon/2}\}_{n=1}^\infty$ covers $\mathscr{E}$. Since $\mathscr{R}(T) = \mathscr{F}$ the images $T(n\mathscr{E}_{\varepsilon/2}) = nT(\mathscr{E}_{\varepsilon/2})$ cover $\mathscr{F}$. Since $\mathscr{F}$ is complete, it follows by the Baire category theorem (Th. 4.8) that for some positive integer k the closed set $(kT(\mathscr{E}_{\varepsilon/2}))^- = k(T(\mathscr{E}_{\varepsilon/2}))^-$ has nonvoid interior. But then so does the set $(T(\mathscr{E}_{\varepsilon/2}))^-$, since multiplication by k is a homeomorphism on $\mathscr{F}$. Thus there exist a positive radius δ and a vector w_0 such that the ball $w_0 + \mathscr{F}_\delta$ is contained in $(T(\mathscr{E}_{\varepsilon/2}))^-$.

In particular, w_0 itself belongs to $(T(\mathscr{E}_{\varepsilon/2}))^-$. Moreover, a moment's reflection shows that by reducing δ and perturbing w_0 slightly we may arrange things so that w_0 lies in the image $T(\mathscr{E}_{\varepsilon/2})$, and we assume this done.

Now let D denote the set of all differences of pairs of vectors belonging to $T(\mathscr{E}_{\varepsilon/2})$:

$$D = \{Tx' - Tx'' : |x'|, |x''| \le \varepsilon/2\}.$$

Clearly D^- contains the closure $(T(\mathscr{E}_{\varepsilon/2}) - w_0)^-$ of $T(\mathscr{E}_{\varepsilon/2}) - w_0$. But translations are homeomorphisms on $\mathscr{E}$, so $(T(\mathscr{E}_{\varepsilon/2}) - w_0)^- = (T(\mathscr{E}_{\varepsilon/2}))^- - w_0$, and therefore

$$D^- \supset (T(\mathscr{E}_{\varepsilon/2}))^- - w_0 \supset \mathscr{F}_\delta.$$

Finally, since $Tx' - Tx'' = T(x' - x'')$, we have $D \subset T(\mathscr{E}_\varepsilon)$, and therefore

$$\mathscr{F}_\delta \subset (T(\mathscr{E}_\varepsilon))^-. \qquad \square$$

Lemma 13.2. *Let $\mathscr{E}, \mathscr{F}$, and T be as in Lemma 13.1, and let V be a neighborhood of 0 in $\mathscr{E}$. Then $T(V)$ is a neighborhood of 0 in $\mathscr{F}$.*

PROOF. Choose a positive number ε such that $\mathscr{E}_\varepsilon \subset V$, and write $\varepsilon_n = \varepsilon/2^n$, $n \in \mathbb{N}$. According to Lemma 13.1, for each n there exists a positive number δ_n such that

$$\mathscr{F}_{\delta_n} \subset (T(\mathscr{E}_{\varepsilon_n}))^-, \qquad (1)$$

and it is clear that we may assume that $\delta_n \to 0$. Let y_0 be a vector in $\mathscr{F}$ such that $|y_0| < \delta_1$. Then according to (1) (with $n = 1$) there exists a vector x_1 in $\mathscr{E}$ such that

$$|x_1| \le \varepsilon_1 \quad \text{and} \quad |y_0 - Tx_1| < \delta_2.$$

Let $y_1 = Tx_1$. Applying (1) once again, this time to $y_0 - y_1$ (with $n = 2$), we obtain a vector x_2 in $\mathscr{E}$ such that, setting $y_2 = Tx_2$, we have

$$|x_2| \le \varepsilon_2 \quad \text{and} \quad |y_0 - y_1 - y_2| < \delta_3.$$

Continuing in this manner, we obtain by mathematical induction a pair of sequences $\{x_n\}_{n=1}^\infty$ and $\{y_n = Tx_n\}_{n=1}^\infty$ in $\mathscr{E}$ and $\mathscr{F}$, respectively, such that

$$|x_n| \le \varepsilon_n \quad \text{and} \quad |y_0 - (y_1 + \cdots + y_n)| < \delta_{n+1}, \qquad n \in \mathbb{N}.$$

If for each n we define $z_n = x_1 + \cdots + x_n$, then $y_1 + \cdots + y_n = Tz_n$ so $Tz_n \to y_0$. Likewise

$$|z_{n+p} - z_n| \le \varepsilon_{n+1} + \cdots + \varepsilon_{n+p} < \varepsilon/2^n, \qquad p \in \mathbb{N},$$

so the sequence $\{z_n\}$ is Cauchy in $\mathscr{E}$. Since $\mathscr{E}$ is complete there exists a vector z_0 in $\mathscr{E}$ such that $z_n \to z_0$. Moreover, since T is continuous, we have $Tz_0 = \lim_n Tz_n = y_0$. Finally, since $|z_n| \le \varepsilon$ for all n, it follows that $|z_0| \le \varepsilon$. Thus y_0 belongs to $T(\mathscr{E}_\varepsilon)$, and we have shown that $T(V)$ contains the open ball $\mathscr{F}_{\delta_1}^\circ$. $\qquad \square$

Theorem 13.3 (Open Mapping Theorem). *If T is a continuous linear transformation of an F-space $\mathscr{E}$ onto an F-space $\mathscr{F}$, then T is an open mapping of $\mathscr{E}$ onto $\mathscr{F}$. In particular, if T is one-to-one, then the inverse transformation T^{-1} is continuous.*

PROOF. It suffices to prove that T is open. Let U be an open set in $\mathscr{E}$, and suppose y_0 is a vector belonging to $T(U)$. There exists a vector x_0 in U such that $Tx_0 = y_0$, and since U is open, x_0 is an interior point of U. It follows that the translate $U - x_0$ is a neighborhood of the origin in $\mathscr{E}$, and, by the lemma just established, $T(U - x_0) = T(U) - y_0$ is a neighborhood of the origin in $\mathscr{F}$. Thus $T(U) = (T(U) - y_0) + y_0$ contains y_0 as an interior point, and the theorem follows. $\qquad\square$

Example A. If $|\ |_1$ and $|\ |_2$ are two quasinorms on the same linear space $\mathscr{E}$, let us say that $|\ |_1$ *dominates* $|\ |_2$ if, whenever $|x_n|_1 \to 0$ for a sequence $\{x_n\}$ in $\mathscr{E}$, then $|x_n|_2 \to 0$. This is equivalent to saying that the topology induced on $\mathscr{E}$ by $|\ |_1$ refines the one induced by $|\ |_2$, or, what comes to the same thing, that the identity mapping on $\mathscr{E}$ is continuous from the quasi-normed space $(\mathscr{E}, |\ |_1)$ obtained by equipping $\mathscr{E}$ with $|\ |_1$ to the space $(\mathscr{E}, |\ |_2)$ obtained by using $|\ |_2$. (It is also easily seen that $|\ |_1$ dominates $|\ |_2$ if and only if the real-valued function $|\ |_2$ is continuous on $(\mathscr{E}, |\ |_1)$.) If $\mathscr{E}$ is *complete* with respect to both quasinorms, and if one of them dominates the other, then by the open mapping theorem the quasinorms must be equivalent, that is, must induce the same topology (Ex. 11I). In particular, if a linear space $\mathscr{E}$ is a Banach space with respect to two norms, and if one norm dominates the other, then the norms must be equivalent. (Note that if $\|\ \|_1$ and $\|\ \|_2$ are norms on $\mathscr{E}$, then $\|\ \|_1$ dominates $\|\ \|_2$ if and only if there exists a positive constant M such that $\|x\|_2 \le M\|x\|_1$ for every x in $\mathscr{E}$; see Proposition 12.1.)

Example B. On the Banach space $\mathscr{C}^{(1)}([a, b])$ of all continuously differentiable complex-valued functions on the nondegenerate interval $[a, b]$ (Prob. 11F), we consider the first order linear differential expression

$$Ly = y' - py,$$

where p denotes an arbitrary fixed coefficient in $\mathscr{C}([a, b])$. It is clear that L is a bounded linear transformation of $\mathscr{C}^{(1)}([a, b])$ into $\mathscr{C}([a, b])$, and hence that if c is any fixed point in $[a, b]$, then the linear transformation

$$Jy = (Ly, y(c))$$

of $\mathscr{C}^{(1)}([a, b])$ into $\mathscr{C}([a, b]) \oplus_1 \mathbb{C}$ is also bounded. The substance of the basic existence and uniqueness theorem for first order linear differential equations is that the mapping J is one-to-one and onto. Since both $\mathscr{C}^{(1)}([a,b])$ and $\mathscr{C}([a, b]) \oplus_1 \mathbb{C}$ are Banach spaces, it follows from Theorem 13.3 that J is invertible. In the language of differential equations we have arrived at the following result: There exists a positive constant M such that if $\varepsilon > 0$, if q_1 and q_2 are any two functions in $\mathscr{C}([a, b])$ such that $\| q_1 - q_2 \|_\infty \le \varepsilon/2M$, and if α_1 and α_2 are any two complex numbers such that $|\alpha_1 - \alpha_2| \le \varepsilon/2M$, then the solutions y_1 and y_2 of the initial value problems

$$y' = py + q_1, \qquad y(c) = \alpha_1,$$

and

$$y' = py + q_2, \qquad y(c) = \alpha_2,$$

respectively, satisfy the condition $\| y_1 - y_2 \|_\infty + \| y_1' - y_2' \|_\infty < \varepsilon$. This contains the following more conventional result: There exists a positive constant M such that if $\varepsilon > 0$ and if $|\alpha_1 - \alpha_2| < \varepsilon/M$, then the solutions y_1 and y_2 of the differential equation

$$y' = py + q \tag{2}$$

satisfying the initial conditions $y_1(c) = \alpha_1$ and $y_2(c) = \alpha_2$ also satisfy the condition $\| y_1 - y_2 \|_\infty \le \varepsilon$. This is customarily expressed by saying that the solutions of (2) "depend continuously on their initial values."

It should be noted, of course, that these facts may be obtained directly, without recourse to the open mapping theorem, since the general solution of (2) can be explicitly written down and examined. The method used here, however, applies with equal ease to the kth order linear equation

$$y^{(k)} = a_{k-1} y^{(k-1)} + \cdots + a_0 y + q,$$

where q and the coefficients $a_0, \ldots, a_{k-1}$ are continuous complex-valued functions on the interval $[a, b]$. Moreover, the method can also be extended so as to deal with partial differential equations.

The open mapping theorem has a surprising and extremely useful reformulation in terms of the graph of a linear transformation. In order to develop this point of view, we recall that if $\mathscr{E}$ and $\mathscr{F}$ are quasinormed spaces, then $|(x, y)| = |x| + |y|$ defines a quasinorm on the direct sum $\mathscr{E} + \mathscr{F}$, and that the topology induced on $\mathscr{E} + \mathscr{F}$ by this quasinorm coincides with the product topology. (See Example 11J; when the quasinorms on $\mathscr{E}$ and $\mathscr{F}$ are norms, the normed space given by this construction coincides with the normed space direct sum $\mathscr{E} \oplus_1 \mathscr{F}$, and it will be convenient to use this notation in the quasinormed case as well.) If T is a linear transformation of one quasinormed space $\mathscr{E}$ into a second quasinormed space $\mathscr{F}$, then by the *graph* of T is meant, of course, the set

$$\mathscr{G} = \{(x, Tx) : x \in \mathscr{E}\}.$$

Thus the graph of a linear transformation $T : \mathscr{E} \to \mathscr{F}$ is a subset of $\mathscr{E} \oplus_1 \mathscr{F}$. In fact, because of the linearity of T, the set $\mathscr{G}$ is a linear manifold in $\mathscr{E} \oplus_1 \mathscr{F}$. (Conversely, it may be noted that if the graph of any mapping of $\mathscr{E}$ into $\mathscr{F}$ is a linear manifold in $\mathscr{E} + \mathscr{F}$, then the mapping must be a linear transformation.) When is the graph of T closed in $\mathscr{E} \oplus_1 \mathscr{F}$?

Theorem 13.4 (Closed Graph Theorem). *If T is a linear transformation of an F-space $\mathscr{E}$ into an F-space $\mathscr{F}$, then the graph of T is closed in $\mathscr{E} \oplus_1 \mathscr{F}$ if and only if T is continuous.*

PROOF. The sufficiency of the condition is easily established. To see this, let T be continuous, and let (x_0, y_0) be a vector in $\mathscr{E} \oplus_1 \mathscr{F}$ that does not belong to $\mathscr{G}$, so that $y_0 \neq Tx_0$. Then there exist disjoint open sets W_1 and W_2 in $\mathscr{F}$ such that $y_0 \in W_1$ and $Tx_0 \in W_2$. Moreover, since T is continuous, there exists a neighborhood V of x_0 in $\mathscr{E}$ such that $T(V) \subset W_1$, and it is readily seen that $V \times W_2$ is a neighborhood of (x_0, y_0) in $\mathscr{E} \oplus_1 \mathscr{F}$ that is disjoint from $\mathscr{G}$. Hence $\mathscr{G}$ is closed.

To prove the necessity of the condition we note, to begin with, that if $\mathscr{G}$ is closed in $\mathscr{E} \oplus_1 \mathscr{F}$, then $\mathscr{G}$ is itself an F-space (since a closed subset of a complete metric space is complete, and $\mathscr{E} \oplus_1 \mathscr{F}$ is complete by Example 11L). Consider the mapping S of $\mathscr{G}$ into $\mathscr{E}$ defined by $S(x, Tx) = x$, $x \in \mathscr{E}$. Since S is the restriction to $\mathscr{G}$ of the projection π_1 of $\mathscr{E} \oplus_1 \mathscr{F}$ onto $\mathscr{E}$, it is clear that S is a continuous linear transformation. But also, just as clearly, S is a one-to-one mapping of $\mathscr{G}$ onto $\mathscr{E}$. Hence the open mapping theorem may be invoked to conclude that $S^{-1} : \mathscr{E} \to \mathscr{G}$ is also continuous. Since $T = \pi_2 \circ S^{-1}$, where π_2 denotes the projection $(x, y) \to y$ of $\mathscr{E} \oplus_1 \mathscr{F}$ onto $\mathscr{F}$, it follows that T is continuous too, and the proof is complete. $\square$

Example C. Let $\mathscr{E}$ and $\mathscr{F}$ be F-spaces with quasinorms $| \ |_1$ and $| \ |_2$, respectively, and let $\mathscr{T}_0$ be an arbitrary linear Hausdorff topology on $\mathscr{F}$ that is comparable with the quasinorm topology induced by $| \ |_2$. If $\mathscr{F}_0$ denotes the topological linear space obtained by equipping $\mathscr{F}$ with $\mathscr{T}_0$, and if $T : \mathscr{E} \to \mathscr{F}_0$ is an arbitrary continuous linear transformation, then $T : \mathscr{E} \to \mathscr{F}$ is also continuous when $\mathscr{F}$ is equipped with the quasinorm $| \ |_2$. Indeed, if $\mathscr{T}_0$ refines the topology induced by $| \ |_2$, this result is trivial. Suppose, on the other hand, that $T : \mathscr{E} \to \mathscr{F}_0$ is continuous where $\mathscr{T}_0$ is coarser than the topology induced by $| \ |_2$. Then the graph $\mathscr{G}$ of T is closed in $\mathscr{E} \times \mathscr{F}_0$ (this may be shown by exactly the same argument as the one given in the proof of Theorem 13.4), and the topology on $\mathscr{E} \times \mathscr{F}_0$ is clearly refined by the quasinorm topology on $\mathscr{E} \oplus_1 \mathscr{F}$. Hence $\mathscr{G}$ is also closed in $\mathscr{E} \oplus_1 \mathscr{F}$, and therefore $T : \mathscr{E} \to \mathscr{F}$ is continuous by the closed graph theorem.

It is very instructive to examine the sequential implications of the closed graph theorem. In general, in order to prove that a linear transformation T of a quasinormed space $\mathscr{E}$ into a second quasinormed space $\mathscr{F}$ is continuous, one must verify that $x_n \to 0$ implies $Tx_n \to 0$. However, when $\mathscr{E}$ is an F-space, the task is made significantly easier by the closed graph theorem.

Corollary 13.5. *Let T be a linear transformation of an F-space $\mathscr{E}$ into a quasinormed space $\mathscr{F}$, and suppose that for every sequence $\{z_n\}$ in $\mathscr{E}$ such that*

$z_n \to 0$ *and such that the sequence* $\{Tz_n\}$ *is Cauchy in* $\mathscr{F}$, *it is true that* $Tz_n \to 0$. *Then* T *is continuous.*

PROOF. Consider the graph $\mathscr{G}$ of T as a linear submanifold of $\mathscr{E} \oplus_1 \hat{\mathscr{F}}$, where $\hat{\mathscr{F}}$ denotes the completion of $\mathscr{F}$ (Th. 11.16). If $\{(x_n, Tx_n)\}$ is a sequence in $\mathscr{G}$ that converges in $\mathscr{E} \oplus_1 \hat{\mathscr{F}}$ to a limit (x_0, y_0), and if we set $z_n = x_n - x_0$, then $z_n \to 0$ and $Tz_n - Tz_m = Tx_n - Tx_m$, so $\{Tz_n\}$ is Cauchy. Hence, by hypothesis, $Tz_n \to 0$. But $\{Tz_n = Tx_n - Tx_0\}$ tends to $y_0 - Tx_0$. Thus $y_0 = Tx_0$, so $\mathscr{G}$ is closed in $\mathscr{E} \oplus_1 \hat{\mathscr{F}}$, and therefore T is continuous as a mapping of $\mathscr{E}$ into $\hat{\mathscr{F}}$. Since T takes its values in $\mathscr{F}$, it is continuous as a mapping of $\mathscr{E}$ into $\mathscr{F}$ as well. $\square$

Example D ([1; Ch. III, Lemme 1]). Suppose $\mathscr{E}_1$ and $\mathscr{E}_2$ are F-spaces, and T_1 and T_2 are continuous linear transformations of $\mathscr{E}_1$ and $\mathscr{E}_2$, respectively, into a quasinormed space $\mathscr{F}$. Suppose also that for each x in $\mathscr{E}_1$ there is exactly one y in $\mathscr{E}_2$ such that $T_1 x = T_2 y$. If we define $Sx = y$, then it is clear that S is a linear transformation of $\mathscr{E}_1$ into $\mathscr{E}_2$ such that $T_1 = T_2 S$. Moreover, if $\{z_n\}$ is a sequence in $\mathscr{E}_1$ such that $z_n \to 0$ and such that $\{Sz_n\}$ is Cauchy, then on the one hand, $\{Sz_n\}$ converges in $\mathscr{E}_2$—say to y_0, and since T_2 is continuous, we have $T_2 Sz_n \to T_2 y_0$. On the other hand, $T_2 Sz_n = T_1 z_n \to 0$ since T_1 is continuous, so $T_2 y_0 = 0$. Since $0 = T_1 0 = T_2 0$, this implies that $y_0 = 0$. Hence S is automatically continuous by Corollary 13.5. Note, in particular, that if T_2 is one-to-one, then T_1 can be factored as $T_1 = T_2 S$ with S continuous if and only if $\mathscr{R}(T_1) \subset \mathscr{R}(T_2)$.

It is appropriate to devote some further attention to what it is that makes the proof of the closed graph theorem work. The crucial fact is that, under the hypotheses of Theorem 13.4, the graph of T is a complete metric space. For this reason it is customary to give a special name to such linear transformations.

Definition. Let $\mathscr{E}$ and $\mathscr{F}$ be quasinormed linear spaces, and let T be a linear transformation of $\mathscr{E}$ into $\mathscr{F}$. Then T is said to be *closed* if its graph is complete as a metric space (in the metric of $\mathscr{E} \oplus_1 \mathscr{F}$).

When $\mathscr{E}$ is an F-space, the closed linear transformations defined on $\mathscr{E}$ coincide with the continuous ones—that is essentially the substance of the closed graph theorem proved above. When $\mathscr{E}$ is not complete, however, matters are quite different, and closed linear transformations on $\mathscr{E}$ need not be continuous.

Example E. Let $\mathscr{D}$ denote the linear space $\mathscr{C}^{(1)}([a, b])$ of continuously differentiable functions on the interval $[a, b]$, $a < b$, regarded as a linear manifold in the space $\mathscr{C}([a, b])$. (In other words, $\mathscr{D}$ is the linear space $\mathscr{C}^{(1)}([a, b])$ equipped with the sup norm rather than the norm introduced in Problem 11F.) The linear transformation L of $\mathscr{D}$ onto $\mathscr{C}([a, b])$ defined in Example B is clearly not bounded, but it is closed. (Recall from advanced calculus that if $\{f_n\}$ is a uniformly convergent sequence of continuously differentiable functions such that the sequence $\{f'_n\}$ also converges uniformly, then $\lim_n f'_n = (\lim_n f_n)'$.)

Theorem 13.6. *Let $\mathscr{E}$ and $\mathscr{F}$ be quasinormed spaces, and let T be a linear transformation of $\mathscr{E}$ into $\mathscr{F}$. Then any two of the following three conditions imply the third:*

  (i) *$\mathscr{E}$ is complete,*
  (ii) *T is continuous,*
  (iii) *T is closed.*

PROOF. That (ii) and (iii) imply one another in the presence of (i) is, once again, essentially the content of the closed graph theorem. Thus it need only be shown that if a quasinormed space $\mathscr{E}$ is the domain of a linear transformation that is both closed and continuous, then $\mathscr{E}$ must be complete. Suppose that $T: \mathscr{E} \to \mathscr{F}$ is such a transformation, and that $\{x_n\}$ is a Cauchy sequence in $\mathscr{E}$. If ε is a given positive number, if $\delta > 0$ is chosen so that $|x| < \delta$ implies $|Tx| < \varepsilon$, and if N is chosen large enough so that $|x_n - x_m| < \delta$ for all $m, n \geq N$, then $|Tx_n - Tx_m| = |T(x_n - x_m)| < \varepsilon$ for all $m, n \geq N$. Hence $\{Tx_n\}$ is also a Cauchy sequence in $\mathscr{F}$, and it follows at once that the sequence of pairs $\{(x_n, Tx_n)\}$ is Cauchy in $\mathscr{E} \oplus_1 \mathscr{F}$. Since this sequence belongs to the graph of T, and T is closed by hypothesis, we conclude that there exists a vector y in $\mathscr{E}$ such that $|y - x_n| + |Ty + Tx_n|$ tends to zero. But then, in particular, $\{x_n\}$ converges to y.   □

Example F. Let X be a set, and let $\mathscr{E}$ be a Banach space whose elements are complex-valued functions defined on X. If $\mathscr{E}$ possesses the property that all *point evaluations* $f \to f(x)$, $x \in X$, are bounded linear functionals on $\mathscr{E}$, then $\mathscr{E}$ is said to be a *Banach function space*. Important, but by no means exhaustive, examples of such spaces are $\mathscr{B}(X)$ in general, and $\mathscr{C}_b(X)$ when X is a topological space, both in the sup norm (see Problem 11D). Other examples are $\mathscr{C}^{(n)}([a, b])$ in the norm introduced in Problem 11F, and the spaces (ℓ_p) (where $X = \mathbb{N}_0$). Suppose now that $\mathscr{E}$ and $\mathscr{F}$ are Banach function spaces on the same set X, and suppose given a complex-valued function m on X with the property that the product $(mf)(x) = m(x)f(x)$ belongs to $\mathscr{F}$ whenever f belongs to $\mathscr{E}$. Then the multiplication $M: \mathscr{E} \to \mathscr{F}$ defined by setting $Mf = mf, f \in \mathscr{E}$, is automatically bounded. Indeed, if $\{f_n\}$ is a sequence in $\mathscr{E}$ such that $f_n \to f$ in $\mathscr{E}$ and $Mf_n \to g$ in $\mathscr{F}$, then $\{f_n\}$ converges pointwise to f and $\{Mf_n\}$ converges pointwise to g, since $\mathscr{E}$ and $\mathscr{F}$ are Banach function spaces. But $\{Mf_n = mf_n\}$ obviously converges pointwise to mf. Thus $g = Mf$, so M is closed, and the desired result follows at once from the preceding theorem (or from the closed graph theorem itself). Frequently in this situation it is possible to conclude that the multiplier m is also bounded. Suppose, for example, that $\{\lambda_n\}_{n=0}^{\infty}$ is a sequence of scalars with the property that for some fixed pair of real numbers p and p', $1 \leq p$, $p' < +\infty$, the sequence $Lx = \{\lambda_n \xi_n\}_{n=0}^{\infty}$ belongs to $(\ell_{p'})$ whenever $x = \{\xi_n\}_{n=0}^{\infty}$ belongs to (ℓ_p). Then L is bounded, as we have just seen, and if $\|L\| = r$, then it is easily verified that $\|\{\lambda_n\}\|_{\infty} = r$. (Indeed, if the vectors e_n are as in Example 11H, we have $|\lambda_n| = \|Le_n\|_{p'} \leq r\|e_n\|_p = r$ for every nonnegative integer n. The fact that $\|\{\lambda_n\}\|_{\infty} = \|L\|$ can also be derived quite directly, of course, without recourse to any version of the closed graph theorem.)

An important theme in Chapters 12 and 13 has been the application of the Baire category theorem (Th. 4.8) to obtain important information concerning linear transformations on F-spaces and Banach spaces. The main theorems in this context are certainly the ones already dealt with, viz., the uniform boundedness theorem, and the open mapping and closed graph theorems. There are other applications, however, and we close this section with one more such result. It is instructive to compare the following proposition with Problem 12B.

Proposition 13.7. *Let $\mathscr{E}$ be a Banach space, let T be a linear transformation of $\mathscr{E}$ into a normed space $\mathscr{F}$, and suppose the function $x \rightarrow \|Tx\|$ is lower semicontinuous on $\mathscr{E}$ (Ex. 3K). Then T is bounded.*

PROOF. The sets $M_n = \{x \in \mathscr{E} : \|Tx\| \leq n\}$ are closed for all positive integers n and cover $\mathscr{E}$. Hence, by the Baire category theorem, at least one M_n must have nonempty interior, and T is bounded on that open set. But then T is bounded. $\qquad\square$

PROBLEMS

A. Let $\mathscr{E}$ and $\mathscr{F}$ be Banach spaces, and let T be a bounded linear transformation of $\mathscr{E}$ onto $\mathscr{F}$ such that the range $\mathscr{R} = \mathscr{R}(T)$ is closed in $\mathscr{F}$. Show that there exists a constant $M > 0$ such that every y in $\mathscr{R}$ can be written as $y = Tx$ with $\|x\| \leq M\|y\|$. In particular, if T is one-to-one, then T is bounded below by $1/M$. (Hint: Factor out the null space of T.)

B. Let $\mathscr{E}$ and $\mathscr{F}$ be F-spaces, and let T be a continuous one-to-one linear transformation of $\mathscr{E}$ into $\mathscr{F}$. The set-theoretic inverse $T^{-1}:\mathscr{R}(T) \rightarrow \mathscr{E}$ is a (one-to-one) linear transformation that is continuous if and only if $\mathscr{R}(T)$ is closed in $\mathscr{F}$.

C. Let $\mathscr{E}$ be an F-space, and suppose $\mathscr{M}$ and $\mathscr{N}$ are closed linear submanifolds of $\mathscr{E}$ such that $\mathscr{M} \cap \mathscr{N} = (0)$ and $\mathscr{M} + \mathscr{N} = \mathscr{E}$, so that every vector x in $\mathscr{E}$ has a unique representation as $x = y + z$, where $y \in \mathscr{M}$ and $z \in \mathscr{N}$. (Such linear manifolds are said to be *complementary* in $\mathscr{E}$ (Prob. 2C).) Show that the mapping $(y, z) \rightarrow y + z$ of $\mathscr{M} \oplus_1 \mathscr{N}$ onto $\mathscr{E}$ is an equivalence, and conclude that the projections P and Q defined by $Px = y$ and $Qx = z$ are continuous linear transformations on $\mathscr{E}$. Show also that if $\mathscr{E}$ is a Banach space, and $\mathscr{M}$ and $\mathscr{N}$ are subspaces of $\mathscr{E}$ satisfying the stated conditions (so that P and Q are bounded operators on $\mathscr{E}$), then $\|P\|, \|Q\| \geq 1$. Show finally, by giving an example, that it is quite possible for $\|P\|$ and $\|Q\|$ both to exceed one.

D. Let $\mathscr{E}$ be an F-space, and let P be a linear transformation of $\mathscr{E}$ into itself that is *idempotent*, i.e., satisfies the equation $P^2 = P$. Show that the kernel $\mathscr{K}$ of P and the range $\mathscr{R}$ of P are complementary linear manifolds in $\mathscr{E}$, that is, satisfy the conditions $\mathscr{K} \cap \mathscr{R} = (0)$ and $\mathscr{K} + \mathscr{R} = \mathscr{E}$. Show also that $\mathscr{K}$ and $\mathscr{R}$ are closed in $\mathscr{E}$ if and only if P is continuous.

E. Let $\mathscr{E}$ and $\mathscr{F}$ be F-spaces, and let S and T be continuous linear transformations of $\mathscr{E}$ into $\mathscr{F}$ and of $\mathscr{F}$ into $\mathscr{E}$, respectively, such that $TS = 1_\mathscr{E}$ (so that T is a left inverse of S and S is a right inverse of T). Show that the range of S and the null space of T are necessarily closed complementary submanifolds of $\mathscr{F}$. (Hint: ST is idempotent.)

F. Show that the assumption that the space $\mathscr{F}$ is complete cannot be omitted from the open mapping theorem, even when $\mathscr{E}$ and $\mathscr{F}$ are both normed spaces. (Hint: Construct a bounded linear operator T on a Banach space $\mathscr{E}$ so that T is one-to-one but is *not* bounded below. If $\mathscr{R} = \mathscr{R}(T)$, then $T : \mathscr{E} \to \mathscr{R}$ is a bounded linear transformation of $\mathscr{E}$ onto $\mathscr{R}$, but this mapping is not open.)

G. Let $\mathscr{F}$ be an arbitrary infinite dimensional Banach space, and let $\{x_\gamma\}_{\gamma \in \Gamma}$ be an indexed Hamel basis for $\mathscr{F}$ consisting of unit vectors. Consider the normed space $\mathscr{L}$ consisting of all similarly indexed families of complex numbers $\{\lambda_\gamma\}_{\gamma \in \Gamma}$ with the property that $\lambda_\gamma = 0$ except for a finite number of indices γ, equipped with the norm $\| \{\lambda_\gamma\} \| = \sum_\gamma |\lambda_\gamma|$. (In connection with this construction the reader may wish to recall Problem 11T.) The linear transformation T of $\mathscr{L}$ onto $\mathscr{F}$ defined by

$$T(\{\lambda_\gamma\}) = \sum_\gamma \lambda_\gamma x_\gamma, \qquad \{\lambda_\gamma\} \in \mathscr{L},$$

is bounded but is not an open mapping. (Hint: Recall Proposition 12.7.) (This problem shows that the assumption that $\mathscr{E}$ be complete likewise cannot be delted from the statement of the open mapping theorem, even when $\mathscr{E}$ and $\mathscr{F}$ are both normed spaces.)

H. Show that the open mapping theorem remains valid if the assumptions that $\mathscr{F}$ is complete and that T maps $\mathscr{E}$ onto $\mathscr{F}$ are replaced by the single (apparently weaker) assumption that the range $\mathscr{R}$ of T is of second category in $\mathscr{F}$. (Hint: Modify appropriately the proof of Lemma 13.1.) Show also that this result is not really stronger than Theorem 13.3 by verifying that if $\mathscr{R}$ is of second category in $\mathscr{F}$, then $\mathscr{F}$ must be complete and $\mathscr{R}$ must coincide with $\mathscr{F}$.

I. Use the result of Example A to derive the open mapping theorem for a one-to-one linear transformation.

J. Show that $\| f \|_1 = \int_0^1 |f(t)| dt$ is a norm on $\mathscr{C}([0, 1])$ that is dominated by the sup norm, and conclude that $\mathscr{C}([0, 1])$ is not complete in the norm $\| \ \|_1$.

K. Let $\mathscr{A}$ denote the linear space of all holomorphic functions on the open unit disc D, and for every f in $\mathscr{A}$ and every radius $r, 0 < r < 1$, write

$$M_r(f) = \max_{|\lambda| \le r} |f(\lambda)|.$$

Show that each M_r is a norm on $\mathscr{A}$ and that, if we set $M_n = M_{r_n}$ where $r_n = n/(n + 1)$, $n \in \mathbb{N}$, then

$$|f| = \sum_{n=1}^{\infty} \frac{1}{2^n} \frac{M_n(f)}{M_n(f) + 1}, \qquad f \in \mathscr{A},$$

is a quasinorm on $\mathscr{A}$ turning $\mathscr{A}$ into an F-space. Show also that the topology induced by this quasinorm is the topology of uniform convergence on compact subsets of D (see Problem 11U). Let $\mathscr{M}_0$ denote the linear manifold consisting of all those functions f in $\mathscr{A}$ such that $f(0) = 0$. Use the open mapping theorem to show that the mapping of $\mathscr{M}_0$ onto $\mathscr{A}$ consisting of multiplication by the function $m(\lambda) = 1/\lambda$ is continuous in the topology of uniform convergence on compact subsets of D.

This last application illustrates very clearly a general principle concerning the open mapping theorem. It is, by its nature, restricted to yielding *qualitative* as opposed to *quantitative* information. As a matter of fact, if $f \in \mathscr{A}$ and $f(0) = 0$, then by the maximum modulus principle (Ex. 5M),

$$|f(\lambda)| \le \frac{M_r(f)}{r} |\lambda|, \qquad |\lambda| \le r, \tag{3}$$

for all $0 < r < 1$, and from this quantitative estimate the asserted continuity follows at once. (The inequality (3) is known as *Schwarz's lemma*.)

L. In the text the open mapping theorem was used to deduce the closed graph theorem. Show, conversely, that the closed graph theorem can be used to prove the open mapping theorem. (Hint: First treat the case of a one-to-one linear transformation.)

M. Suppose that a linear space $\mathscr{E}$ is complete with respect to each of two quasinorms $|\ |_1$ and $|\ |_2$, and suppose also that there exists some separated linear topology on $\mathscr{E}$ that is refined by both of the topologies induced by $|\ |_1$ and $|\ |_2$. Show that $|\ |_1$ and $|\ |_2$ must be equivalent. (Hint: Use Example C.)

N. Let $\mathscr{F}$ and $\mathscr{G}$ be quasinormed spaces, and suppose given a one-to-one continuous linear transformation S of $\mathscr{F}$ into $\mathscr{G}$. Show that if T is a linear transformation of an F-space $\mathscr{E}$ into $\mathscr{F}$ such that ST is continuous, then T is continuous. (Hint: Apply Corollary 13.5.)

O. ([1; Ch. III, Th. 8]) Let $\mathscr{F}$ and $\{\mathscr{G}_\gamma\}_{\gamma \in \Gamma}$ be quasinormed spaces, and suppose given a separating family $\{S_\gamma\}$ of continuous linear transformations $S_\gamma : \mathscr{F} \to \mathscr{G}_\gamma$, $\gamma \in \Gamma$ (see Example 3R). Let T be a linear transformation of an F-space $\mathscr{E}$ into $\mathscr{F}$ such that $S_\gamma T$ is continuous for each index γ. Show that T is continuous.

P. Let $\mathscr{E}$ and $\mathscr{F}$ be F-spaces, and let T be a linear transformation of $\mathscr{E}$ into $\mathscr{F}$. Show that $|x|_1 = |x| + |Tx|$ defines a quasinorm on $\mathscr{E}$, and show that $\mathscr{E}$ is complete with respect to $|\ |_1$ if and only if T is closed. Use these observations to show that the result of Example A implies the closed graph theorem.

Q. Let $\mathscr{E}$ be an arbitrary quasinormed space that is of second category in itself. Show that if there exists a closed linear transformation T of $\mathscr{E}$ into some quasinormed space $\mathscr{F}$, then $\mathscr{E}$ must be complete and T must be continuous. (Hint: Use Problem H. This shows that if in the closed graph theorem we assume outright that T is closed, then even the hypothesis that $\mathscr{E}$ is complete can be weakened somewhat.)

R. Let $\mathscr{E}$ and $\mathscr{F}$ be quasinormed spaces, and let T be a one-to-one linear transformation of $\mathscr{E}$ onto $\mathscr{F}$. Show that T is closed if and only if its set-theoretic inverse is closed.

S. A linear manifold $\mathscr{M}$ in an F-space $\mathscr{F}$ is sometimes called *paraclosed* if there exists an F-space $\mathscr{E}$ and a continuous linear transformation T of $\mathscr{E}$ into $\mathscr{F}$ such that $\mathscr{M} = \mathscr{R}(T)$. Show that a paraclosed linear submanifold of an F-space $\mathscr{F}$ that is not closed in $\mathscr{F}$ is necessarily of first category in its own closure, and therefore of first category in $\mathscr{F}$. (Hint: See Problem H.)

Problems Q and S raise the question whether it is possible for a linear submanifold of an F-space $\mathscr{F}$ to be of second category in $\mathscr{F}$ without coinciding with $\mathscr{F}$. That this is indeed always possible when $\mathscr{F}$ is infinite dimensional is shown by the following construction (Hausdorff [35]).

T. Let $\mathscr{F}$ be an F-space, and let X be a Hamel basis for $\mathscr{F}$. Let $\{x_n\}_{n=1}^{\infty}$ be an infinite sequence of distinct vectors in X, and let X_0 denote the subset of X consisting of those vectors that do not appear in the sequence $\{x_n\}$. Show that if for each positive integer n, $\mathscr{M}_n$ denotes the linear submanifold of $\mathscr{F}$ generated algebraically by $X_0 \cup \{x_1, \ldots, x_n\}$, then no $\mathscr{M}_n$ coincides with $\mathscr{F}$, but at least one $\mathscr{M}_n$ is of second category in $\mathscr{F}$.

The Hahn–Banach theorem

14

Many of the important applications of the theory of topological linear spaces concern the relations between a space $\mathscr{E}$ and the space of continuous linear functionals on $\mathscr{E}$. This space is known as the *conjugate space* or *dual space* of $\mathscr{E}$ and will be denoted by $\mathscr{E}^*$ in the sequel. (When $\mathscr{E}$ is a normed space, $\mathscr{E}^* = \mathscr{L}(\mathscr{E}, \mathbb{C})$ is a Banach space (Prop. 12.2); otherwise $\mathscr{E}^*$ will simply denote a vector space (Prob. 12U).)

Example A. The most general linear functional on the space $\mathbb{C}^n$ is given by the formula

$$f_a(\xi_1, \ldots, \xi_n) = \sum_{i=1}^{n} \alpha_i \xi_i, \tag{1}$$

where $a = (\alpha_1, \ldots, \alpha_n)$ is an arbitrary but fixed element of $\mathbb{C}^n$ (Prob. 2H(iii)). For any $x = (\xi_1, \ldots, \xi_n)$ it is clear that

$$|f_a(x)| \leq \|a\|_\infty \sum_{i=1}^{n} |\xi_i| = \|a\|_\infty \|x\|_1.$$

Hence, if $\mathbb{C}_1^n$ denotes the result of equipping $\mathbb{C}^n$ with the norm $\|\ \|_1$ (see Example 11A), the norm of f_a as a functional on $\mathbb{C}_1^n$ is no greater than $\|a\|_\infty$. On the other hand, if an index i_0 is chosen so that $|\alpha_{i_0}| = \|a\|_\infty$, and if γ is a complex number of absolute value one such that $\gamma \alpha_{i_0} = |\alpha_{i_0}|$, then the n-tuple x_0 with the i_0-th entry equal to γ and all other entries zero has the properties that $\|x_0\|_1 = 1$ and $f_a(x_0) = \|a\|_\infty$. Thus $\|f_a\| = \|a\|_\infty$. Similarly, it is easy to see that if $\mathbb{C}_\infty^n$ is the result of equipping $\mathbb{C}^n$ with the norm $\|\ \|_\infty$, then $\|f_a\|$ on $\mathbb{C}_\infty^n$ is given by $\|a\|_1$.

283

Finally, let us find the norm of f_a on the space $\mathbb{C}_p^n$ obtained by equipping $\mathbb{C}^n$ with the norm $\| \ \|_p$, $1 < p < \infty$. In the first place, letting q denote the Hölder conjugate of p, we have

$$| f_a(x) | \leq \| a \|_q \| x \|_p$$

for every n-tuple x, by virtue of the Hölder inequality (Prob. 11B). Thus $\| f_a \| \leq \| a \|_q$. We shall show that, in fact, $\| f_a \| = \| a \|_q$. To this end it suffices to treat the case $\| a \|_q = 1$. For each $i = 1, \ldots, n$, let γ_i be a scalar of absolute value one such that $\gamma_i \alpha_i = |\alpha_i|$, and let

$$\xi_i = |\alpha_i|^{q-1} \gamma_i, \qquad i = 1, \ldots, n.$$

Then the n-tuple $x = (\xi_1, \ldots, \xi_n)$ has the property that

$$\| x \|_p^p = \sum_{i=1}^n |\alpha_i|^{p(q-1)} = \sum_{i=1}^n |\alpha_i|^q = 1,$$

so that $\| x \|_p = 1$. But also

$$f_a(x) = \sum_{i=1}^n |\alpha_i|^{q-1}(\gamma_i \alpha_i) = \sum_{i=1}^n |\alpha_i|^q = 1,$$

and therefore $\| f_a \| = 1$.

Example B. For each nonnegative integer m, let $\mathscr{L}_m$ denote the linear space of all infinite sequences $\{\xi_n\}_{n=0}^\infty$ such that $\xi_n = 0$ for all $n > m$. Then $\mathscr{L}_m$ is a subspace of (ℓ_p), $1 \leq p \leq \infty$, and the correspondence

$$\{\xi_n\} \leftrightarrow (\xi_0, \ldots, \xi_m)$$

is an isometric isomorphism between $\mathscr{L}_m$ and the space $\mathbb{C}_p^{m+1}$ of Example A. Let us consider the case $p = 1$. If f is a bounded linear functional on (ℓ_1), then for each nonnegative integer m there is an $(m+1)$-tuple $(\alpha_0^{(m)}, \alpha_1^{(m)}, \ldots, \alpha_m^{(m)})$ such that $|\alpha_i^{(m)}| \leq \| f \|$, $i = 0, \ldots, m$, and such that

$$f(x) = \sum_{i=0}^m \alpha_i^{(m)} \xi_i, \qquad x = \{\xi_n\} \in \mathscr{L}_m. \tag{2}$$

Moreover, if $m < m'$, then $f | \mathscr{L}_m$ is the same as the restriction to $\mathscr{L}_m$ of the functional $f | \mathscr{L}_{m'}$, whence we conclude that there exists a single infinite sequence $\{\alpha_n\}$ with $\| \{\alpha_n\} \|_\infty \leq \| f \|$ such that $\alpha_n^{(m)} = \alpha_n$ for all nonnegative integers m and n.

Now if $x = \{\xi_n\} \in (\ell_1)$ and if we define, for each nonnegative integer m, $x_m = \{\xi_0^{(m)}, \ldots, \xi_n^{(m)}, \ldots\}$, where

$$\xi_n^{(m)} = \begin{cases} \xi_n, & n = 0, \ldots, m, \\ 0, & n > m, \end{cases}$$

then the sequence $\{x_m\}$ converges to x in (ℓ_1) and therefore $\{f(x_m)\}$ converges to $f(x)$. But also

$$f(x_m) = \sum_{i=0}^{m} \alpha_i \xi_i$$

for every m, and since the sequence $\{\alpha_n\}$ is bounded, we see that

$$f(x_m) \rightarrow \sum_{n=0}^{\infty} \alpha_n \xi_n.$$

Thus every element f of $(\ell_1)^*$ is given by an element $a = \{\alpha_n\}$ of (ℓ_∞) according to the formula

$$f(x) = \sum_{n=0}^{\infty} \alpha_n \xi_n, \quad x = \{\xi_n\} \in (\ell_1),$$

where $\|a\|_\infty \leq \|f\|$. On the other hand, it is quite obvious that if $a = \{\alpha_n\}$ is any element of (ℓ_∞), then

$$f_a(x) = \sum_{n=0}^{\infty} \alpha_n \xi_n, \quad x = \{\xi_n\} \in (\ell_1), \tag{3}$$

determines a bounded linear functional f_a on (ℓ_1) with $\|f_a\| \leq \|a\|_\infty$. Thus the correspondence $a \leftrightarrow f_a$ is an isometry, and since it is obviously linear, we have proved the following basic fact: *The mapping $a \rightarrow f_a$, where f_a is defined as in (3), is an isometric isomorphism of (ℓ_∞) onto $(\ell_1)^*$.* It is customary to identify these two Banach spaces via this canonical isomorphism, and to write simply, by a slight abuse of language, $(\ell_1)^* = (\ell_\infty)$.

Example C. The same device may be employed to study $(\ell_p)^*$, $1 < p < +\infty$. Letting q denote the Hölder conjugate of p and using the results of Example A, we find that if f is a bounded linear functional on (ℓ_p), then there exists a sequence $a = \{\alpha_n\}$ with $\|a\|_q \leq \|f\|$ such that

$$f(x) = \sum_{i=0}^{m} \alpha_i \xi_i$$

for every $x = \{\xi_n\}$ in $\mathscr{L}_m$. But then, of course, using the Hölder inequality and the technique of Example B, we conclude that

$$f(x) = \sum_{n=0}^{\infty} \alpha_n \xi_n$$

for every $x = \{\xi_n\}$ in (ℓ_p). Finally, using the Hölder inequality once again, we observe that if $a = \{\alpha_n\}$ is any element of (ℓ_q), then

$$f_a(x) = \sum_{n=0}^{\infty} \alpha_n \xi_n, \quad x = \{\xi_n\} \in (\ell_p), \tag{4}$$

defines a bounded linear functional f_a on (ℓ_p) with $\| f_a \| \leq \| a \|_q$. We have sketched the proof of the following fact: *The mapping $a \to f_a$, where f_a is defined as in (4), is an isometric isomorphism of (ℓ_q) onto $(\ell_p)^*$.* Here again it is customary to identify these two spaces, writing, by a slight abuse of language, $(\ell_p)^* = (\ell_q)$.

Since $p = 2$ is the unique positive number that is its own Hölder conjugate, this identification has special implications for the space (ℓ_2). In particular, the relation $(\ell_2)^* = (\ell_2)$ permits the definition, by a slight modification of (4), of an *inner product*

$$(\{\xi_n\}, \{\eta_n\}) = \sum_{n=0}^{\infty} \xi_n \bar{\eta}_n, \quad \{\xi_n\}, \{\eta_n\} \in (\ell_2),$$

that turns (ℓ_2) into a *Hilbert space*. The study of Hilbert spaces and operators on them is undertaken in Volume II of this treatise.

Examples A, B, and C give detailed description of some dual spaces. We shall have occasion as we go along to develop precise descriptions of the duals of other specific normed spaces. Concerning the dual $\mathscr{E}^*$ of a general normed space $\mathscr{E}$, we know not only that it is a normed space, but also that it is complete (even if $\mathscr{E}$ itself is not; see Proposition 12.2). What we do not know as yet is whether, in general, there are any linear functionals in $\mathscr{E}^*$ other than the zero functional. The main purpose of this chapter is to establish the fact that bounded linear functionals always exists in abundance on a normed space.

If f is a linear functional defined on a linear submanifold $\mathscr{M}$ of a linear space $\mathscr{E}$, and if f_0 is a linear functional defined on a linear submanifold $\mathscr{M}_0$ that contains $\mathscr{M}$, then f_0 is a *linear extension* of f if $f_0(x) = f(x)$ for every vector x in $\mathscr{M}$. We shall be concerned with the extension of bounded linear functionals. Our first step in this direction deals with the extension of a *real* linear functional on a *real* normed space.

Lemma 14.1. *Let $\mathscr{E}$ be a real normed space, and let $\mathscr{M}$ be a linear manifold in $\mathscr{E}$. Let f be a bounded linear functional on $\mathscr{M}$, and suppose that x_0 is a vector in $\mathscr{E}$ that does not belong to $\mathscr{M}$. If $\mathscr{M}_0$ denotes the linear manifold generated by $\mathscr{M}$ and x_0, then there exists a linear extension f_0 of the functional f to $\mathscr{M}_0$ satisfying the condition $\| f_0 \| = \| f \|$.*

PROOF. It is clear that we may assume, without loss of generality, that $\| f \| = 1$. The linear manifold $\mathscr{M}_0$ consists of all vectors of the form $y + tx_0$, where y is an arbitrary vector in $\mathscr{M}$ and t is a real number. If r denotes a real number to be determined, and if we define

$$f_0(y + tx_0) = f(y) + tr \tag{5}$$

for all y in $\mathscr{M}$ and all real t, then f_0 is clearly the most general linear extension of f to $\mathscr{M}_0$. Thus the problem reduces to choosing r so that $\| f_0 \| = 1$. This condition amounts to the requirement that

$$- \| y + tx_0 \| \leq f(y) + tr \leq \| y + tx_0 \|, \tag{6}$$

or, equivalently,

$$-\left\|\frac{y}{t}+x_0\right\|-f\left(\frac{y}{t}\right)\leq r\leq\left\|\frac{y}{t}+x_0\right\|-f\left(\frac{y}{t}\right),\qquad(7)$$

for all vectors y in $\mathcal{M}$ and all real numbers $t\neq0$. (The equivalence of (6) and (7) is most readily verified by separating the two cases $t\gtrless0$.) Moreover, since y/t belongs to $\mathcal{M}$ along with y, condition (7) will be satisfied if

$$-\|y+x_0\|-f(y)\leq r\leq\|y+x_0\|-f(y)\qquad(8)$$

for all y in $\mathcal{M}$. Thus if we write

$$r_1=\sup_{y\in\mathcal{M}}\{-\|y+x_0\|-f(y)\},\qquad r_2=\inf_{y\in\mathcal{M}}\{\|y+x_0\|-f(y)\},$$

it suffices to show that $r_1\leq r_2$, for the required extension will then be given by (5), where r is any number such that $r_1\leq r\leq r_2$.

Let y_1 and y_2 be any two vectors belonging to $\mathcal{M}$. Then

$$f(y_2)-f(y_1)\leq\|y_2-y_1\|=\|(y_2+x_0)-(y_1+x_0)\|$$
$$\leq\|y_2+x_0\|+\|y_1+x_0\|,$$

so that

$$-\|y_1+x_0\|-f(y_1)\leq\|y_2+x_0\|-f(y_2).$$

From this it follows that

$$r_1\leq\|y_2+x_0\|-f(y_2)$$

for every y_2 in $\mathcal{M}$, and hence that $r_1\leq r_2$, as desired. $\square$

Theorem 14.2 (Real Hahn–Banach Theorem). *Let $\mathcal{M}$ be a linear submanifold of a real normed space $\mathcal{E}$, and let f be a bounded linear functional defined on $\mathcal{M}$. Then there exists a linear extension f_0 of f to the entire space $\mathcal{E}$ satisfying the condition $\|f_0\|=\|f\|$.*

PROOF. Let $\mathcal{P}$ denote the collection of all pairs $(g,\mathcal{G})$, where $\mathcal{G}$ is a linear submanifold of $\mathcal{E}$ containing $\mathcal{M}$ and g is a linear functional on $\mathcal{G}$ that extends f and satisfies $\|g\|=\|f\|$. There is a natural partial ordering on $\mathcal{P}$ obtained by setting $(g_1,\mathcal{G}_1)\prec(g_2,\mathcal{G}_2)$ when $\mathcal{G}_1\subset\mathcal{G}_2$ and g_2 is an extension of g_1, and it is clear that linearly ordered subsets of $\mathcal{P}$ have upper bounds in $\mathcal{P}$. Hence, by Zorn's lemma, $\mathcal{P}$ contains a maximal element $(f_0,\mathcal{F}_0)$. If $\mathcal{F}_0$ were not all of $\mathcal{E}$, then Lemma 14.1 could be applied to obtain an element of $\mathcal{P}$ dominating $(f_0,\mathcal{F}_0)$, a manifest contradiction. Hence $\mathcal{F}_0=\mathcal{E}$, and f_0 is the desired extension (better: is *one* such extension). $\square$

The next step in the program is to extend Theorem 14.2 to complex spaces. To see how this goes, note first that if $\mathcal{E}$ is a (complex) normed space, then $\mathcal{E}$ can also be regarded as a real normed space (by refusing to multiply by any but real scalars), and also that if $\mathcal{M}$ is a linear manifold in $\mathcal{E}$, then $\mathcal{M}$ is also a (real) linear manifold in $\mathcal{E}$ regarded as a real space. Likewise, if f is a (complex) linear functional defined on $\mathcal{M}$, and if g and h are defined by $g(x) = \operatorname{Re} f(x)$ and $h(x) = \operatorname{Im} f(x)$ for all x in $\mathcal{M}$, so that

$$f = g + ih, \tag{9}$$

then g and h are real linear functionals on $\mathcal{M}$ regarded as a real space. But more than this is true. If, for a fixed vector x in $\mathcal{M}$, γ is a complex number of modulus one chosen so that $\gamma f(x)$ is real and nonnegative, then $|f(x)| = f(\gamma x) = g(\gamma x)$, whence it follows that f is bounded if and only if g is, and that, when both are bounded, $\|f\| = \|g\|$. Moreover, since $f(ix) = if(x)$, it follows from (9) that $g(ix) + ih(ix) = -h(x) + ig(x)$, so that $h(x) = -g(ix)$ identically on $\mathcal{M}$. Thus (9) assumes the form

$$f(x) = g(x) - ig(ix), \qquad x \in \mathcal{M}. \tag{10}$$

Suppose now that f is a given bounded, complex linear functional on $\mathcal{M}$. We split f into its real and imaginary parts as in (10), and recall that g is a bounded real linear functional on $\mathcal{M}$ with $\|g\| = \|f\|$. Hence, applying the real Hahn–Banach theorem, we obtain a bounded real linear functional g_0 on $\mathcal{E}$ that extends g and satisfies the condition $\|g_0\| = \|g\|$. Then, following (10), we use g_0 to define a complex-valued functional f_0 on $\mathcal{E}$ according to the formula

$$f_0(x) = g_0(x) - ig_0(ix), \qquad x \in \mathcal{E}. \tag{11}$$

It is easily seen that f_0 is additive on $\mathcal{E}$ and that the set of complex scalars λ such that $f_0(\lambda x) = \lambda f_0(x)$ for all x in $\mathcal{E}$ is a real linear submanifold $\mathcal{L}$ of $\mathbb{C}$ that contains the real axis. Since direct calculation shows that $f_0(ix) = if_0(x)$, it follows that $\mathcal{L}$ coincides with $\mathbb{C}$. In other words, f_0 is a complex linear functional. Finally, the same argument as above shows that $\|f_0\| = \|g_0\|$, and hence that $\|f_0\| = \|f\|$. We have proved the following result, in the statement of which the scalar field is, as usual, understood to be $\mathbb{C}$ once again.

Theorem 14.3 (Hahn–Banach Theorem). *Let $\mathcal{M}$ be a linear manifold in a normed space $\mathcal{E}$, and let f be a bounded linear functional defined on $\mathcal{M}$. Then there exists a linear extension f_0 of f to the entire space $\mathcal{E}$ satisfying the condition $\|f_0\| = \|f\|$.*

Example D. Suppose φ is a bounded linear functional on the space $\mathscr{C}([0, 1])$ of continuous complex-valued functions on $[0, 1]$. Since $\mathscr{C}([0, 1])$ is a linear submanifold of the normed space $\mathscr{B}([0, 1])$ of all bounded functions on $[0, 1]$ (cf. Problem 11E), the Hahn–Banach theorem assures us of the existence of a linear extension φ_0 of φ to this larger space such that $\|\varphi_0\| = \|\varphi\|$. Hence we may and do define a function α by setting

$$\alpha(t) = \begin{cases} 0, & t = 0, \\ \varphi_0(\chi_{[0, t]}), & 0 < t \le 1. \end{cases}$$

We shall show that α is of bounded variation on $[0, 1]$ (cf. Problem 1I). Indeed, if $0 = t_0 < t_1 < \cdots < t_N = 1$ is an arbitrary partition of $[0, 1]$, we write $I_1 = [0, t_1]$ and $I_j = (t_{j-1}, t_j]$ for $j = 2, \ldots, N$, and define $\chi_j = \chi_{I_j}$, $j = 1, \ldots, N$. Then for each $j = 1, \ldots, N$,

$$\alpha(t_j) - \alpha(t_{j-1}) = \varphi_0(\chi_j),$$

and it follows that there exist numbers γ_j of modulus one such that

$$\sum_{j=1}^{N} |\alpha(t_j) - \alpha(t_{j-1})| = \varphi_0(f),$$

where f denotes the step function $f = \sum_{j=1}^{N} \gamma_j \chi_j$. But f is a unit vector in $\mathscr{B}([0, 1])$, and therefore

$$\sum_{j=1}^{N} |\alpha(t_j) - \alpha(t_{j-1})| \le \|\varphi_0\|.$$

Thus α has total variation V no greater than $\|\varphi\|$.

Suppose now that f is a continuous function on $[0, 1]$, and let ε be a positive number. Choose a positive number δ small enough so that $|f(t) - f(t')| < \varepsilon$ for all $0 \le t, t' \le 1$ such that $|t - t'| < \delta$, and let $0 = t_0 < t_1 < \cdots < t_N = 1$ be a partition with mesh less than δ. Then the step function

$$g(t) = \begin{cases} f(t_1), & 0 \le t \le t_1, \\ f(t_j), & t_{j-1} < t \le t_j, \quad j = 2, \ldots, N, \end{cases}$$

has the property that $\|f - g\|_\infty \le \varepsilon$, so that $|\varphi_0(g) - \varphi(f)| \le \varepsilon\|\varphi\|$. On the other hand,

$$\varphi_0(g) = \sum_{j=1}^{N} f(t_j)[\alpha(t_j) - \alpha(t_{j-1})],$$

which is a Riemann–Stieltjes sum approximating the integral $\int_0^1 f(t)\, d\alpha(t)$. Indeed, according to standard estimates (see Problem 8L),

$$\left| \varphi_0(g) - \int_0^1 f(t)\, d\alpha(t) \right| \le 2\varepsilon V \le 2\varepsilon\|\varphi\|.$$

But then

$$\left| \varphi(f) - \int_0^1 f(t)\, d\alpha(t) \right| \le 3\varepsilon\|\varphi\|,$$

and since ε is arbitrary it follows that

$$\varphi(f) = \int_0^1 f(t)\, d\alpha(t).$$

Finally, we observe that for every f in $\mathscr{C}([0, 1])$ we have $|\varphi(f)| \leq \|f\|_\infty V$, and hence that $V = \|\varphi\|$. This completes the proof of the following result: *For every element φ of $\mathscr{C}([0, 1])^*$ there exists a complex-valued function α on $[0, 1]$ having total variation equal to $\|\varphi\|$ and satisfying the condition*

$$\varphi(f) = \int_0^1 f(t)\, d\alpha(t), \qquad f \in \mathscr{C}([0, 1]). \tag{12}$$

It is also true, of course, that every complex-valued function α of finite total variation V on $[0, 1]$ defines a bounded linear functional φ_α on $\mathscr{C}([0, 1])$ via formula (12), and for this functional we have $\|\varphi_\alpha\| \leq V$. Moreover, the assignment $\alpha \to \varphi_\alpha$ is clearly linear. However, it is in general possible to have $\varphi_\alpha = 0$ even when $V \neq 0$. Thus we must resist the temptation to identify $\mathscr{C}([0, 1])^*$ with the space of functions of bounded variation on $[0, 1]$. We shall return to this point in Chapter 18. See also Problems F and G.

For many applications it is important to know that the Hahn–Banach theorem is also valid for spaces equipped with pseudonorms in place of norms. In order to formulate the desired result, we define a linear functional f on a linear space $\mathscr{E}$ to be *bounded* with respect to a pseudonorm σ on $\mathscr{E}$ if there exists a number M such that $|f(x)| \leq M\sigma(x)$ for every x in $\mathscr{E}$ (cf. Problem 12U). Continuing the analogy, we likewise define the *norm* of f with respect to σ to be the smallest such number M. (That a smallest M does indeed exist whenever f is bounded with respect to σ is easily verified as before; see Problem 12A.)

Proposition 14.4. *Let $\mathscr{E}$ be a linear space equipped with a pseudonorm σ, and let f be a linear functional defined on a linear submanifold $\mathscr{M}$ of $\mathscr{E}$ such that f is bounded on $\mathscr{M}$ with respect to σ. Then there exists a linear extension f_0 of f to the entire space $\mathscr{E}$ such that f_0 has the same norm as f with respect to σ.*

PROOF. We employ Proposition 11.17. If $\mathscr{Z} = \{z \in \mathscr{E} : \sigma(z) = 0\}$ denotes the zero space of σ, then it is clear that f must vanish on $\mathscr{M} \cap \mathscr{Z}$. It follows at once that if we write $\tilde{\mathscr{M}} = \mathscr{M} + \mathscr{Z}$, then there is a unique linear functional $\dot{g}$ on the space $\dot{\mathscr{M}} = \tilde{\mathscr{M}}/\mathscr{Z}$ satisfying the condition

$$\dot{g}(x) = f(x), \qquad x \in \mathscr{M},$$

and the norm of $\dot{g}$ (with respect to the norm on the associated space $\dot{\mathscr{E}}$ obtained by factoring out $\mathscr{Z}$) coincides with the norm of f with respect to σ. Hence by the Hahn–Banach theorem there exists a linear functional $\dot{f}_0$ on $\dot{\mathscr{E}}$ that extends $\dot{g}$ and has the same norm as f. But then the product $f_0 = \dot{f}_0 \circ \pi$ (where π denotes the natural projection of $\mathscr{E}$ onto $\dot{\mathscr{E}}$) satisfies both of the desired conditions. □

Example E. Let (c) denote the linear space of convergent sequences $x = \{\xi_n\}_{n=0}^\infty$ (Ex. 2D). If we equip (c) with the sup norm, then it becomes a subspace of the Banach space (ℓ_∞), and

$$f(x) = \lim x = \lim_n \xi_n$$

defines a linear functional of norm one on (c). Hence by the Hahn–Banach theorem there exists a linear functional φ of norm one on (ℓ_∞) with the property that $\varphi(x) = \lim x$ whenever x is convergent. Clearly any such functional has some claim to be called a "generalized limit". But how well behaved is such a notion of limit? One good point turns up at once. If φ is such a functional and if c_0 is a sequence tending to zero, then $\varphi(x + c_0) = \varphi(x)$. In particular, $\varphi(x)$ is unchanged if we modify the sequence x at any finite number of terms. Thus φ reflects only the "ultimate" behavior of each sequence, just as any respectable notion of limit should. But if φ is not selected with considerably greater care, it may fail in other ways to behave respectably. Suppose, for example, that x_0 denotes the sequence $\{1, 0, 1, 0, \ldots\}$ of alternating ones and zeros and that $\mathcal{M}_0$ denotes the linear manifold in (ℓ_∞) generated by (c) and x_0. Then $\mathcal{M}_0$ consists of all sequences of the form $x = y + \lambda x_0$, where y is a convergent sequence and λ is a complex number. For any such sequence it is clear that $\|x\|_\infty \geq \max\{|\lim y|, |\lambda + \lim y|\}$. Hence for any r such that $0 \leq r \leq 1$, the equation

$$f_0(y + \lambda x_0) = \lim y + r\lambda$$

defines a linear functional f_0 of norm one on $\mathcal{M}_0$ with the property that $f_0(y) = \lim y$ for every y in (c). This calculation shows that a "generalized limit" φ as above can be constructed so that $\varphi(x_0)$ is any desired number r in the unit interval. For any such choice it is clear that the other sequence $x_1 = \{0, 1, 0, 1, \ldots\}$ of alternating zeros and ones must have "limit" $\varphi(x_1) = 1 - r$, since $x_0 + x_1$ is the constant sequence $\{1, 1, 1, \ldots\}$. But x_1 is actually a *tail* of x_0 (obtained by deleting the first term), and we should, therefore, expect to have $\varphi(x_0) = \varphi(x_1)$. Hence the "right" choice for r is $r = \frac{1}{2}$. Can a generalized limit φ be constructed so as to satisfy the extra condition

$$\varphi(\{\xi_0, \xi_1, \ldots\}) = \varphi(\{\xi_m, \xi_{m+1}, \ldots\}) \tag{13}$$

for every bounded sequence $x = \{\xi_0, \xi_1, \ldots\}$ and every positive integer m? The answer is affirmative, but the construction is a little tricky and requires the use of Proposition 14.4.

To begin with, let us write x_m for the mth tail $\{\xi_m, \xi_{m+1}, \ldots\}$ of the sequence $x = \{\xi_0, \xi_1, \ldots\}$, so that, in particular, $x_0 = x$, and then define

$$\sigma(x) = \inf \left\| \frac{1}{M} \sum_{i=1}^{M} x_{m_i} \right\|_\infty,$$

where the infimum is taken over all finite sequences of nonnegative integers $m_1, \ldots, m_M$ (and all positive integers M). Then σ is a pseudonorm on (ℓ_∞). Indeed, it is perfectly clear that $\sigma(x) \geq 0$ and that $\sigma(\lambda x) = |\lambda| \sigma(x)$ for all sequences x in (ℓ_∞) and all scalars λ. The hard work comes in verifying the triangle inequality. Note first that for any average

$$x' = \frac{1}{M} \sum_{i=1}^{M} x_{m_i}$$

of tails of a sequence x, we have $\|x'\|_\infty \le \|x\|_\infty$. But then, of course, a second average

$$x'' = \frac{1}{N}\sum_{j=1}^{N} x'_{n_j} = \frac{1}{MN}\sum_{i=1}^{M}\sum_{j=1}^{N} x_{m_i+n_j}$$

must also have the property that $\|x''\|_\infty \le \|x'\|_\infty$. It follows that for any average $x' = 1/M \sum_{i=1}^{M} x_{m_i}$ of tails of x and any average $y' = 1/N \sum_{j=1}^{N} y_{n_j}$ of tails of a second sequence y, the "double average"

$$z = \frac{1}{MN}\sum_{i=1}^{M}\sum_{j=1}^{N}(x+y)_{m_i+n_j}$$

of tails of $x + y$ has the property that $\|z\|_\infty \le \|x'\|_\infty + \|y'\|_\infty$. But then, of course, $\sigma(x + y) \le \|x'\|_\infty + \|y'\|_\infty$, and it follows that $\sigma(x + y) \le \sigma(x) + \sigma(y)$. Thus σ is a pseudonorm on (ℓ_∞), and it is readily verified that for any convergent sequence y we have $|\lim y| \le \sigma(y)$. Hence, by Proposition 14.4, there exists a linear functional φ on (ℓ_∞) having norm one with respect to σ and satisfying the condition that $\varphi(y) = \lim y$ whenever y is convergent. Finally, that φ satisfies (13) will be verified by showing that $\sigma(x - x_1) = 0$ for an arbitrary bounded sequence x, for this will guarantee that $\varphi(x_1) = \varphi(x)$, and hence that $\varphi(x) = \varphi(x_m)$ for all m. To see this, we simply note that

$$\sigma(x - x_1) \le \left\|\frac{1}{M}\sum_{i=1}^{M}(x - x_1)_i\right\|_\infty = \left\|\frac{1}{M}(x_1 - x_{M+1})\right\|_\infty \le \frac{2\|x\|_\infty}{M}$$

for every positive integer M.

A linear functional φ on (ℓ_∞) that satisfies (13), together with the condition that $\varphi(y) = \lim y$ whenever y is convergent, and that has norm one with respect to the pseudonorm σ, is clearly also of norm one with respect to the sup norm $\|\ \|_\infty$. Such a linear functional is called a *Banach generalized limit*.

Proposition 14.4 has a useful consequence bearing on those topological linear spaces whose topologies are induced by a family of pseudonorms.

Proposition 14.5. *Let $\{\sigma_\gamma\}_{\gamma\in\Gamma}$ be an indexed family of pseudonorms on a linear space $\mathscr{E}$, let $\mathscr{M}$ be a linear submanifold of $\mathscr{E}$, and let f be a linear functional on $\mathscr{M}$ that is continuous with respect to the linear topology induced on $\mathscr{M}$ by the family $\{\sigma_\gamma\}$. Then there exists a linear extension f_0 of f to the entire space $\mathscr{E}$ that is continuous with respect to the topology induced by the family $\{\sigma_\gamma\}$.*

PROOF. By Problem 12U there exist indices $\gamma_1, \ldots, \gamma_n$ and a positive number ε such that if $x \in \mathscr{M}$ and if $\sigma_{\gamma_i}(x) \le \varepsilon$, $i = 1, \ldots, n$, then $|f(x)| \le 1$. If we set $\sigma = \sigma_{\gamma_1} \vee \cdots \vee \sigma_{\gamma_n}$, then σ is a pseudonorm on $\mathscr{E}$ (Prop. 11.30) with the property that if $x \in \mathscr{M}$ and $\sigma(x) \le \varepsilon$ then $|f(x)| \le 1$, and it follows that f is bounded on $\mathscr{M}$ with respect to σ by $1/\varepsilon$. Hence by Proposition 14.4 there exists a linear extension f_0 of f to $\mathscr{E}$ that is also bounded (by $1/\varepsilon$) with respect to σ, and it is clear that f_0 is continuous on $\mathscr{E}$ with respect to the topology induced by the family $\{\sigma_y\}$. $\qquad\qquad\qquad\qquad\qquad\qquad\quad\;\square$

For real linear spaces there are other versions of the Hahn–Banach theorem that are extremely useful in certain situations. (See also Problems V and W.)

Proposition 14.6. *Let $\mathscr{E}$ be a real linear space, and suppose given a real-valued function p on $\mathscr{E}$ satisfying the following conditions:*

(i) $p(x + y) \leq p(x) + p(y)$ *for all x and y in $\mathscr{E}$,*
(ii) $p(tx) = tp(x)$ *for all x in $\mathscr{E}$ and all $t > 0$.*

Let $\mathscr{M}$ be a linear manifold in $\mathscr{E}$, and suppose given a linear functional f defined on $\mathscr{M}$ and satisfying the condition

$$f(x) \leq p(x) \tag{14}$$

for all x in $\mathscr{M}$. Then there exists a linear extension f_0 of f to the entire space $\mathscr{E}$ satisfying (14) at every vector x.

If p were nonnegative, and if $p(-x)$ coincided with $p(x)$ for every x, then p would be a real pseudonorm on $\mathscr{E}$, and this proposition would just be the real version of Proposition 14.4. It should be noted that in this formulation of the Hahn–Banach theorem it is not assumed that $\mathscr{E}$ possesses a topology.

SKETCH OF PROOF OF PROPOSITION 14.6. We may suppose that $\mathscr{M} \neq \mathscr{E}$. Let x_0 be a vector in $\mathscr{E}$ that does not belong to $\mathscr{M}$, and, as in Lemma 14.1, let $\mathscr{M}_0$ denote the linear submanifold of $\mathscr{E}$ generated by $\mathscr{M}$ and x_0. If we show that there exists an extension of f to $\mathscr{M}_0$ satisfying (14), then the rest of the argument will go exactly like the proof of Theorem 14.2. Thus, what is needed is to show the existence of a number r such that

$$f(y) + tr \leq p(y + tx_0) \tag{15}$$

for every y in $\mathscr{M}$ and every real number t. For $t > 0$, (15) reduces to the requirement

$$f\left(\frac{y}{t}\right) + r \leq p\left(\frac{y}{t} + x_0\right)$$

for all y in $\mathscr{M}$, or, since y/t belongs to $\mathscr{M}$ along with y, to the requirement

$$f(y) + r \leq p(y + x_0)$$

for all y in $\mathscr{M}$. Thus if $r_2 = \inf_{y \in \mathscr{M}} \{p(y + x_0) - f(y)\}$, then we must choose r so that $r \leq r_2$. Similarly, for $t < 0$, (15) reduces to the requirement

$$f(y) - r \leq p(y - x_0),$$

293

so that, if $r_1 = \sup_{y \in \mathcal{M}} \{-p(y - x_0) + f(y)\}$, then we must also choose r so that $r \geq r_1$. Hence it suffices to show that $r_1 \leq r_2$, and the rest of the proof goes exactly like the proof of Lemma 14.1. $\qquad\square$

Example F. Let $\mathscr{E}$ be a real linear space, and let C be a convex subset of $\mathscr{E}$ with the property that for every vector x in $\mathscr{E}$ there exists a positive number t_x such that $t_x x \in C$. (This is customarily expressed by saying that 0 is an *internal point* of C; note that it follows that 0 belongs to C.) For each vector x in $\mathscr{E}$ the set of real numbers

$$M_x = \left\{ t > 0 : \frac{1}{t} x \in C \right\}$$

is a nonempty ray to the right in $\mathbb{R}$ that is bounded below (by zero), and we may and do define $p(x) = p_C(x) = \inf M_x$. The function $p = p_C$, known as the *Minkowski function* of C, possesses properties (i) and (ii) of Proposition 14.6. Indeed, property (ii) follows immediately from the identity $tM_x = M_{tx}$, valid for all x in $\mathscr{E}$ and all $t > 0$. To verify (i), suppose that x and y are vectors in $\mathscr{E}$ and that s and t are any real numbers such that $p(x) < s$ and $p(y) < t$. Then $x' = x/s$ and $y' = y/t$ belong to C and, setting $u = s/(s + t)$, so that $1 - u = t/(s + t)$, we conclude that $ux' + (1 - u)y' = (x + y)/(s + t)$ also belongs to C. But then $p(x + y) \leq s + t$, and it follows that $p(x + y) \leq p(x) + p(y)$. Thus p possesses property (i).

The relation between C and its Minkowski function p makes it clear that if $p(x) < 1$ then $x \in C$, and likewise that if $x \in C$ then $p(x) \leq 1$. Thus

$$\{x \in \mathscr{E} : p(x) < 1\} \subset C \subset \{x \in \mathscr{E} : p(x) \leq 1\}.$$

Suppose now that x_0 is a vector in $\mathscr{E}$ that does not belong to C. If we define $f(tx_0) = t$ on the linear manifold $\mathscr{L} = \{tx_0 : t \in \mathbb{R}\}$, then f is obviously a linear functional on $\mathscr{L}$. Moreover, $f(x) \leq p(x)$ for every x in $\mathscr{L}$. Indeed, for $t > 0$ we have $p(tx_0) = tp(x_0) \geq t = f(tx_0)$, while for $t < 0$ we have $f(tx_0) = t \leq 0 \leq p(tx_0)$ because p is nonnegative. Hence, by Proposition 14.6, there exists a linear functional f_0 extending f to the entire space $\mathscr{E}$ and satisfying the condition $f_0(x) \leq p(x)$ everywhere. Note that $f_0(x_0) = 1$ and that $f_0(x) \leq 1$ at every point of C.

Thus, in summary, if C is a convex set having 0 as an internal point in a real linear space $\mathscr{E}$, and if $x_0 \notin C$, then there exists a (real) nonzero linear functional f on $\mathscr{E}$ such that $f(x) \leq f(x_0)$ for every x in C. More generally, it is easy to see that if C is a convex subset of $\mathscr{E}$ possessing an arbitrary internal point z_0 (that is, a point z_0 such that for every vector x in $\mathscr{E}$ there exists a positive number t_x such that $z_0 + t_x x$ belongs to C), and if x_0 is any vector not belonging to C, then there exists a nonzero linear functional f on $\mathscr{E}$ such that $f(x) \leq f(x_0)$ for every x in C (for if z_0 is an internal point of C, then 0 is an internal point of the convex set $C - z_0$).

Example G. The construction of the preceding example admits a generalization that is of some importance. Let C_1 and C_2 be disjoint nonempty convex sets in a real linear space $\mathscr{E}$, and suppose that one of them—say C_1—possesses an internal point. Let us denote by C the set of differences $C_1 - C_2 = \{u - v : u \in C_1, v \in C_2\}$. If x and y are any two vectors belonging to C, then there exist vectors u_1, u_2 in C_1 and v_1, v_2 in C_2 such that $x = u_1 - v_1$ and $y = u_2 - v_2$. Hence if $0 \le t \le 1$ then

$$tx + (1 - t)y = [tu_1 + (1 - t)u_2] - [tv_1 + (1 - t)v_2]$$

belongs to C. Thus C too is a convex set. Moreover, if v_0 is an arbitrary vector in C_2, and z_0 is an internal point of C_1, then $z_0 - v_0$ is an internal point of C (for if $z_0 + t_x x \in C_1$, then $(z_0 - v_0) + t_x x \in C$). Finally, $0 \notin C$ since C_1 and C_2 are disjoint. Hence, by Example F, there exists a nonzero linear functional f on $\mathscr{E}$ such that $f(x) \le f(0) = 0$ for every x in C. But this says that $f(u) \le f(v)$ for every vector u in C_1 and every vector v in C_2. Thus we have established the following important fact: If C_1 and C_2 are arbitrary disjoint nonempty convex sets in a real linear space $\mathscr{E}$, and if either of the sets C_1 or C_2 possesses a single internal point, then there exists a nonzero linear functional f on $\mathscr{E}$ such that $f(u) \le f(v)$ for all u in C_1 and v in C_2.

The construction of Example F also leads to some very important consequences for complex linear spaces. (Recall that a complex linear space is also a real linear space.) The following concepts are pertinent.

Definition. If $\mathscr{E}$ is a linear space and C is a subset of $\mathscr{E}$ that is both convex and balanced (Prob. 11M) then C is said to be *absolutely convex* (see Problem N). A subset A of $\mathscr{E}$ is said to be *absorbing* if for every vector x in $\mathscr{E}$ there is a disc $D_r = \{\lambda \in \mathbb{C} : |\lambda| < r\}$ in $\mathbb{C}$ with positive radius r such that $D_r x \subset A$.

Suppose now that C is an absorbing convex subset of a linear space $\mathscr{E}$. Then 0 is certainly an internal point of C in the sense of Example F, so C possesses a Minkowski function p_C satisfying (i) and (ii) of that example. If C is also balanced, i.e., if C is absolutely convex, then $p_C(\gamma x) = p_C(x)$ for every complex number γ of modulus one and every vector x in $\mathscr{E}$, and it follows at once that p_C is a pseudonorm. This observation leads to a very useful characterization of those topological linear spaces whose topologies are induced by pseudonorms. (It is instructive to compare the following result with Problem 11X.)

Proposition 14.7. *In any topological linear space $\mathscr{E}$ the following conditions are equivalent:*

(i) *The topology on $\mathscr{E}$ is induced by a family of pseudonorms,*
(ii) *There is a base for the topology on $\mathscr{E}$ consisting of convex sets,*
(iii) *There is a neighborhood base at the origin in $\mathscr{E}$ consisting of convex sets,*
(iv) *There is neighborhood base at the origin in $\mathscr{E}$ consisting of absolutely convex sets*

PROOF. If σ is a pseudonorm on $\mathscr{E}$ then every ball $\{x \in \mathscr{E} : \sigma(x - x_0) < r\}$ is convex, and it follows at once that (i) implies (ii). Moreover, it is clear that (ii) implies (iii). To see that (iii) implies (iv), let C be a convex neighborhood of 0 in $\mathscr{E}$, and let V be a balanced neighborhood of 0 such that $V \subset C$ (Prob. 11M). Then

$$W = \bigcap_{|\gamma| = 1} \gamma C$$

contains V, and is therefore a neighborhood of 0. Since W is an intersection of convex sets, W is also convex, and it is obvious that W is balanced. Thus W is an absolutely convex neighborhood of 0 that is contained in C.

It remains to verify that (iv) implies (i). To this end, suppose $\{V_\gamma\}_{\gamma \in \Gamma}$ is a neighborhood base at 0 in $\mathscr{E}$ such that each V_γ is absolutely convex. Since a neighborhood of 0 is absorbing, each V_γ has a Minkowski function σ_γ (Ex. F), and, as noted above, σ_γ is a pseudonorm on $\mathscr{E}$ for each index γ. Since

$$\{x \in \mathscr{E} : \sigma_\gamma(x) < 1\} \subset V_\gamma \subset \{x \in \mathscr{E} : \sigma_\gamma(x) \leq 1\}, \qquad \gamma \in \Gamma,$$

it is clear that the topology induced by the family of pseudonorms $\{\sigma_\gamma\}_{\gamma \in \Gamma}$ coincides with the given topology on $\mathscr{E}$ (see Proposition 11.26), and the proof is complete. $\qquad \square$

Note. For future purposes it is important to observe that the neighborhood bases referred to in conditions (iii) and (iv) of the foregoing proposition may be taken to consist exclusively either of open sets or of closed sets. To verify the former of these assertions, it suffices to take interiors (see Problem N). To verify the latter, it is advisable to replace each [absolutely] convex neighborhood W by an [absolutely] convex neighborhood W' such that $W' + W' \subset W$ (cf. Problem 11M).

Definition. A [real] topological linear space possessing one, and therefore all, of the properties set forth in Proposition 14.7 is said to be a [*real*] *locally convex* topological linear space or, more briefly, a [*real*] *locally convex space*.

According to Proposition 14.7 the locally convex topological linear spaces are precisely the ones to which the discussion at the end of Chapter 11 (Prop. 11.27–Prop. 11.32) applies. Thus if $\mathscr{E}$ is a locally convex space and $\mathscr{M}$ is a linear submanifold of $\mathscr{E}$, then $\mathscr{M}$ is a locally convex space in its relative topology, and $\mathscr{E}/\mathscr{M}$ is a locally convex space in the quotient topology (see Example 11O). Similarly, it is not difficult to see that the full algebraic direct sum of any indexed family $\{\mathscr{E}_\gamma\}_{\gamma \in \Gamma}$ of locally convex spaces is a locally convex space in the product topology (cf. Problem 11R). In the same vein, it is an easy consequence of Proposition 11.32 that a locally convex space is metrizable if and only if its topology is induced by a separating *sequence* of pseudonorms (cf. also Problem 11Y). The following is little more than a summary of earlier results restated in terms of local convexity.

Proposition 14.8. *In any topological linear space $\mathcal{E}$ the following conditions are equivalent:*

(i) *The topology on $\mathcal{E}$ is induced by a countably determined separating family of pseudonorms,*

(ii) *The topology on $\mathcal{E}$ is induced by a countable separating family of pseudonorms,*

(iii) *$\mathcal{E}$ is locally convex and quasinormable,*

(iv) *$\mathcal{E}$ is locally convex and metrizable.*

PROOF. If the given topology on $\mathcal{E}$ is induced by some countably determined separating family $\{\sigma_\gamma\}_{\gamma \in \Gamma}$ of pseudonorms, and if $\{\sigma_\gamma\}_{\gamma \in \Gamma_0}$ is a countable cofinal subfamily, then $\{\sigma_\gamma\}_{\gamma \in \Gamma_0}$ also induces the given topology on $\mathcal{E}$ (and is therefore also separating). Likewise, if a countable separating family $\{\sigma_\gamma\}_{\gamma \in \Gamma_0}$ of pseudonorms is arranged in a sequence $\{\sigma_n\}$, then according to the proof of Proposition 11.32,

$$|x| = \sum_{n=1}^{\infty} \frac{1}{2^n} \frac{\sigma_n(x)}{1 + \sigma_n(x)}, \qquad x \in \mathcal{E}, \tag{16}$$

defines a quasinorm that also induces the given topology on $\mathcal{E}$. Thus (i) implies (ii), (ii) implies (iii), and it is obvious that (iii) implies (iv). The proof will be completed by showing that (iv) implies (i). To this end it would suffice to invoke Propositions 11.32 and 14.7, but a direct proof is easy to give. Suppose $\mathcal{E}$ is locally convex and metrizable, and let $\{U_n\}_{n=1}^{\infty}$ be a countable neighborhood base at 0 in $\mathcal{E}$. It follows from Proposition 14.7 that for every positive integer n there exists an absolutely convex neighborhood V_n of 0 such that $V_n \subset U_n$, and it is easily seen that the countable family $\{V_n\}$ is also a neighborhood base at 0 in $\mathcal{E}$. Moreover, each V_n has a Minkowski function σ_n that is a pseudonorm on $\mathcal{E}$ (Ex. F), and, arguing as in Proposition 14.7, we see that the family $\{\sigma_n\}_{n=1}^{\infty}$ induces the given topology on $\mathcal{E}$. Finally, since every vector $x \neq 0$ in $\mathcal{E}$ lies outside some set U_n, it follows that the sequence $\{\sigma_n\}$ is separating on $\mathcal{E}$. ☐

According to Proposition 14.8 every metrizable locally convex space is quasinormable, and it is in these terms that the following definition is given. In this connection we recall (Prob. 12V) that if a topological vector space is complete with respect to any one quasinorm that induces its topology, then it is also complete with respect to every such quasinorm.

Definition. A locally convex topological linear space that is complete with respect to some one (and therefore every) quasinorm that induces its topology is a *Frechét space*.

Example H. Let $\mathscr{F}$ be a topological vector space whose topology is induced by an indexed family $\{\sigma_\gamma\}_{\gamma \in \Gamma}$ of pseudonorms, and let T be a linear transformation of a linear space $\mathcal{E}$ into $\mathscr{F}$. If for each index γ we define $\tau_\gamma = \sigma_\gamma \circ T$, then $\{\tau_\gamma\}_{\gamma \in \Gamma}$ is a family of pseudonorms on $\mathscr{E}$, and it may be seen that the topology induced on $\mathscr{E}$ by the family $\{\tau_\gamma\}$ is simply the topology inversely induced by T, i.e., the coarsest topology on $\mathscr{E}$ making $T : \mathscr{E} \to \mathscr{F}$ continuous. Indeed, if $\mathscr{T}$ denotes this latter topology, then $\mathscr{T}$ is a linear topology on $\mathscr{E}$ (Prob. 11R), and every pseudonorm τ_γ is continuous on $\mathscr{E}$ with respect to $\mathscr{T}$. Hence $\mathscr{T}$ refines the topology induced by the family $\{\tau_\gamma\}$ (Prop. 11.26). On the other hand, if V is an arbitrary neighborhood of 0 with respect to $\mathscr{T}$, then it is a straightforward consequence of Proposition 11.27 (and the definition of $\mathscr{T}$) that there exist indices $\gamma_1, \ldots, \gamma_n$ and a positive number ε such that

$$T^{-1}(\{y \in \mathscr{F} : \sigma_{\gamma_i}(y) < \varepsilon, i = 1, \ldots, n\}) = \{x \in \mathscr{E} : \tau_{\gamma_i}(x) < \varepsilon, i = 1, \ldots, n\}$$

is contained in V, which shows that $\mathscr{T}$ is, in turn, refined by the topology induced by the family $\{\tau_\gamma\}$. Thus, in particular, if $T : \mathscr{E} \to \mathscr{F}$ is a linear transformation and if $\mathscr{F}$ is a locally convex space, then $\mathscr{E}$ is also locally convex in the topology inversely induced by T.

Next suppose given an indexed family $\{T_\delta\}_{\delta \in \Delta}$ of linear transformations of a linear space $\mathscr{E}$ into a (similarly indexed) family $\{\mathscr{F}_\delta\}$ of locally convex spaces, so that $T_\delta : \mathscr{E} \to \mathscr{F}_\delta$, $\delta \in \Delta$, and for each index δ, let $\{\sigma_\gamma^{(\delta)}\}_{\gamma \in \Gamma_\delta}$ be an indexed family of pseudonorms that induces the topology on $\mathscr{F}_\delta$. Then from what has already been said it is clear that the topology inversely induced on $\mathscr{E}$ by the family $\{T_\delta\}$ is locally convex and is, in fact, induced by the doubly indexed family of pseudonorms $\{\tau_\gamma^{(\delta)} = \gamma_\gamma^{(\delta)} \circ T_\delta\}_{\gamma \in \Gamma_\delta, \delta \in \Delta}$. If each of the spaces $\mathscr{F}_\delta$ is metrizable, and if the index family Δ is countable, then the topology inversely induced by the family $\{T_\delta\}$ is also metrizable if and only if $\{T_\delta\}$ is separating.

Example I. If $\{\mathscr{E}_n\}_{n=0}^\infty$ is a sequence of metrizable locally convex spaces, then according to the preceding example the full algebraic direct sum $\mathscr{E} = \sum_{n=0}^\infty + \mathscr{E}_n$ is also a metrizable locally convex space in the product topology (Ex. 3N). Moreover, the space $\mathscr{E}$ is a Frechét space if and only if each of the factors $\mathscr{E}_n$ is (see Examples 11J and 11L). (If, for each nonnegative integer n, $\{\sigma_m^{(n)}\}_{m=1}^\infty$ is a sequence of pseudonorms that induces the topology on $\mathscr{E}_n$, and if for each element $X = \{x_n\}_{n=0}^\infty$ of $\mathscr{E}$ we write $\tau_{m,n}(X) = \sigma_m^{(n)}(x_n)$, then the doubly indexed family of pseudonorms $\{\tau_{m,n}\}$ induces the product topology on $\mathscr{E}$. Thus Proposition 14.9 below provides an alternate proof that $\mathscr{E}$ is complete.) In particular, the space $(\mathfrak{o})$ of all complex sequences is a Frechét space in the topology of termwise convergence. If $\{\mathscr{E}_\gamma\}_{\gamma \in \Gamma}$ is an uncountable indexed family of Frechét spaces, then the full algebraic direct sum $\sum_{\gamma \in \Gamma} + \mathscr{E}_\gamma$ is a locally convex topological linear space in the product topology, but this space is not metrizable, and is therefore not a Frechét space (unless all but countably many of the spaces $\mathscr{E}_\gamma$ are trivial).

While the definition of a Frechét space is stated in terms of quasinorms, it is, generally speaking, a nuisance to have to choose any one particular quasinorm to work with. Moreover, in view of Proposition 14.8, it is by and large unnecessary to do so. The following criterion is frequently useful in verifying the completeness of a Frechét space.

Proposition 14.9. *Let $\mathscr{E}$ be a linear space and let $\{\sigma_n\}_{n=1}^\infty$ be a separating sequence of pseudonorms on $\mathscr{E}$. Then $\mathscr{E}$ is a Frechét space in the topology induced by the sequence $\{\sigma_n\}$ if and only if each sequence $\{x_p\}_{p=1}^\infty$ in $\mathscr{E}$ that is Cauchy with respect to (the pseudometric defined by) every σ_n is also convergent with respect to every σ_n (to a limit that is independent of n).*

PROOF. The quasinorm $|\ |$ defined in (16) induces the same topology on $\mathscr{E}$ as does the sequence $\{\sigma_n\}$. Hence a sequence $\{x_p\}_{p=1}^\infty$ in $\mathscr{E}$ converges to a limit x with respect to $|\ |$ if and only if it converges to x with respect to each of the pseudonorms σ_n (Prop. 11.27). Consequently it suffices to verify that $\{x_p\}_{p=1}^\infty$ is Cauchy with respect to $|\ |$ if and only if it is Cauchy with respect to all of the pseudonorms σ_n. Moreover, it is clear that if $\{x_p\}$ is Cauchy with respect to $|\ |$, then $\{x_p\}$ is also Cauchy with respect to every σ_n. Suppose, on the other hand, that $\{x_p\}$ is Cauchy with respect to each σ_n. Let ε be a positive number, and choose a positive integer K such that $1/2^K = \sum_{n=K+1}^\infty 1/2^n < \varepsilon/2$. Then there exists a second positive integer P such that $\sigma_n(x_p - x_q) < \varepsilon/2K$ for all $p, q > P$ and $n = 1, \ldots, K$, whereupon it follows that $|x_p - x_q| < \varepsilon/2 + \varepsilon/2 = \varepsilon$ for all $p, q > P$. Thus the sequence $\{x_p\}$ is Cauchy with respect to $|\ |$, and the result follows. $\square$

Example J. Let X be a locally compact Hausdorff space, let $\mathscr{C}(X)$ denote, as usual, the linear space of continuous complex-valued functions on X, and suppose X is σ-compact (Ch. 10, p.193). If $\{K_n\}_{n=1}^\infty$ is a compact covering of X, and if, for each n, U_n is a topologically bounded open subset of X such that $K_n \subset U_n$ (Prob. 10A), then, setting $L_n = U_1^- \cup \cdots \cup U_n^-$ for each positive integer n, we obtain a sequence $\{L_n\}$ of compact subsets of X with the property that the sequence $\{L_n^\circ\}$ is also a nested covering of X. Hence for any nonempty compact subset K of X there exists a positive integer n such that $K \subset L_n$, and it follows that the sequence of pseudonorms $\{\sigma_{L_n}\}$ induces the topology of uniform convergence on compact subsets of X. (See Problem 11U for notation and terminology.) Moreover, it is a routine matter to show, using the criterion of Proposition 14.9, that $\mathscr{C}(X)$ is, in fact, a Frechét space in the topology of uniform convergence on compact subsets. (If X is not σ-compact, then $\mathscr{C}(X)$ is not even metrizable in the topology of uniform convergence on compact subsets. Indeed, if $\{K_n\}$ is any sequence of nonempty compact subsets of X, and if x_0 is a point of X that is not in any K_n, then for each positive integer p there is a function f_p in $\mathscr{C}(X)$ such that $f_p(x_0) = 1$ while $f_p = 0$ on K_p (Prob. 3V), and the sequence $\{f_p\}_{p=1}^\infty$ clearly tends to 0 in the topology induced on $\mathscr{C}(X)$ by the sequence of pseudonorms $\{\sigma_{K_n}\}$.)

Example K. If U is a nonempty open subset of $\mathbb{R}^n$, then the topology of D-convergence introduced in Problem 11V turns the linear space $\mathscr{C}^{(\infty)}(U)$ into a Frechét space. Indeed, the results of Problem 11V show that $\mathscr{C}^{(\infty)}(U)$ is a metrizable locally convex topological linear space in the topology of D-convergence. To see that $\mathscr{C}^{(\infty)}(U)$ is also complete with respect to this topology, we employ Proposition 14.9 and note that if a sequence $\{f_p\}_{p=1}^{\infty}$ in $\mathscr{C}^{(\infty)}(U)$ is Cauchy with respect to all of the pseudonorms $\sigma_{K,m}$ that induce the topology of D-convergence, then every one of the derived sequences $\{D^{k_1,\dots,k_n}f_p\}_{p=1}^{\infty}$ is convergent to *some* continuous limit in the topology of uniform convergence on compact subsets of U. But from this, via an easy mathematical induction, we conclude that if $\lim_p f_p = f$, then $\lim_p D^{k_1,\dots,k_n}f_p = D^{k_1,\dots,k_n}f$ for all nonnegative integers $k_1,\dots,k_n$. (The central element in this argument is the fact, established in advanced calculus, that if a sequence $\{g_n\}$ of continuously differentiable functions of a single real variable converges uniformly to a limit g on an open interval (a, b), and if the derived sequence $\{g'_n\}$ is uniformly Cauchy on (a, b), then $\{g'_n\}$ converges uniformly to g' on (a, b). Cf. Example 13E.)

Example L. Once again, let U denote a nonempty open subset of $\mathbb{R}^n$. The space $\mathscr{C}_0^{(\infty)}(U)$ of test functions on U (Prob. 11V) becomes a separated, metrizable, locally convex topological vector space if it is equipped with its relative topology as a linear submanifold of $\mathscr{C}^{(\infty)}(U)$, but $\mathscr{C}_0^{(\infty)}(U)$ is not a Frechét space in this topology. Equivalently, $\mathscr{C}_0^{(\infty)}(U)$ is not a closed linear submanifold in $\mathscr{C}^{(\infty)}(U)$. To see this, let $\{K_p\}_{p=1}^{\infty}$ be a sequence of compact subsets of U such that $K_1^{\circ} \neq \varnothing$, such that $K_p \subset K_{p+1}^{\circ}$ for every positive integer p, and such that $U = \bigcup_{p=1}^{\infty} K_p^{\circ}$; see Example 10A. According to Problem 11V(ii), there exists for each index p a test function f_p on U with support contained in $K_{p+1}^{\circ} \backslash K_p$ (with $K_0 = \varnothing$), and the sequence of partial sums $\sum_{p=1}^{N} f_p$ is D-convergent to the sum $\sum_{p=1}^{\infty} f_p$, an infinitely differentiable function on U that does *not* have compact support, and is therefore not a test function.

We close this chapter with some particularly useful consequences of the Hahn–Banach theorem.

Proposition 14.10. *Let $\mathscr{M}$ be a (closed) subspace of a normed space $\mathscr{E}$, and suppose x_0 is a vector in $\mathscr{E}$ that does not belong to $\mathscr{M}$. Then there exists a functional f in $\mathscr{E}^*$ satisfying the conditions:*

(i) $f(\mathscr{M}) = (0)$,
(ii) $f(x_0) = 1$,
(iii) $\|f\| = 1/d$,

where $d = d(x_0, \mathscr{M})$ denotes the (positive) distance from x_0 to $\mathscr{M}$.

PROOF. Let $\mathcal{M}_0$ denote the linear submanifold of $\mathcal{E}$ generated by $\mathcal{M}$ and x_0. If we define a linear functional g on $\mathcal{M}_0$ by setting

$$g(y + \lambda x_0) = \lambda$$

for all vectors y in $\mathcal{M}$ and all scalars λ, then $g(\mathcal{M}) = (0)$ and $g(x_0) = 1$. If we show that g also satisfies the condition $\|g\| = 1/d$, then the result will follow at once from the Hahn–Banach theorem. To this end we observe that the set $H = \{z \in \mathcal{M}_0 : g(z) = 1\}$ coincides with the coset $\mathcal{M} + x_0$, and hence that the distance $d(0, H)$ from the origin to H is the same as the distance $d = d(x_0, \mathcal{M})$. It follows that if $z \in \mathcal{M}_0$ and $\|z\| < d$, then $g(z) \neq 1$. Since the set of vectors $D = \{\lambda z : |\lambda| \leq 1\}$ is mapped by g onto the disc $D_0 = \{\lambda \in \mathbb{C} : |\lambda| \leq |g(z)|\}$ in the complex plane, it is clear that, in fact, $|g(z)| < 1$, and the result follows.   □

Corollary 14.11. *For each nonzero vector x_0 in a normed space $\mathcal{E}$ there exists a linear functional f in $\mathcal{E}^*$ such that $\|f\| = 1$ and $f(x_0) = \|x_0\|$. Consequently, if x and y are distinct vectors in $\mathcal{E}$, there exists a functional f in $\mathcal{E}^*$ such that $f(x) \neq f(y)$.*

Proposition 14.4 may be used to obtain analogous results for pseudo-normed spaces.

Proposition 14.12. *Let $\mathcal{E}$ be a linear space equipped with a pseudonorm σ, let $\mathcal{M}$ be a linear manifold in $\mathcal{E}$, and suppose x_0 is a vector in $\mathcal{E}$ such that, for some positive number ε, the ball $\{x \in \mathcal{E} : \sigma(x - x_0) < \varepsilon\}$ is disjoint from $\mathcal{M}$. Then there is a linear functional f on $\mathcal{E}$ that is bounded with respect to σ (by $1/\varepsilon$) and satisfies the conditions*

$$f(\mathcal{M}) = (0) \quad and \quad f(x_0) = 1.$$

In particular, if $\sigma(x_0) > 0$, then there is a linear functional f on $\mathcal{E}$ that is bounded with respect to σ such that $f(x_0) = \sigma(x_0)$. Hence if x and y are vectors in $\mathcal{E}$ such that $\sigma(x - y) > 0$, then there exists a linear functional f on $\mathcal{E}$ that is bounded with respect to σ such that $f(x) \neq f(y)$.

The following is just a summary of Propositions 14.5 and 14.12 in the language of locally convex spaces.

Proposition 14.13. *Let $\mathcal{E}$ be a locally convex space and let $\mathcal{M}$ be a linear submanifold of $\mathcal{E}$. For each linear functional f in $\mathcal{M}^*$ there exists a linear functional f_0 in $\mathcal{E}^*$ such that $f_0|\mathcal{M} = f$. Moreover, if $\mathcal{M}$ is closed and if x_0 is a vector in $\mathcal{E}$ that does not belong to $\mathcal{M}$, then there exists a linear functional f in $\mathcal{E}^*$ such that*

$$f(\mathcal{M}) = (0) \quad and \quad f(x_0) \neq 0. \tag{17}$$

Thus $\mathcal{M}$ is the intersection of the kernels of all of those linear functionals in $\mathscr{E}^*$ that vanish on $\mathcal{M}$. In particular, if $\mathscr{E}$ is separated, then $\mathscr{E}^*$ is separating on $\mathscr{E}$.

PROOF. By Proposition 14.7 the given topology on $\mathscr{E}$ is induced by a family $\{\sigma_\gamma\}$ of pseudonorms. Hence the first assertion is an immediate consequence of Proposition 14.5. Moreover, if $\mathcal{M}$ is closed and $x_0 \notin \mathcal{M}$, then there exist indices $\gamma_1, \ldots, \gamma_n$ and a positive number ε such that if we set $\sigma = \sigma_{\gamma_1} \vee \cdots \vee \sigma_{\gamma_n}$, then the ball $\{x \in \mathcal{M} : \sigma(x - x_0) < \varepsilon\}$ is disjoint from $\mathcal{M}$. Hence there exists a linear functional f on $\mathscr{E}$ such that (17) is satisfied and such that f is bounded with respect to σ. But then f is continuous on $\mathscr{E}$. Finally, if $\mathscr{E}$ is Hausdorff as a topological space, then (0) is a closed set in $\mathscr{E}$. Hence if $x_0 \neq 0$ there exists a functional f in $\mathscr{E}^*$ such that $f(x_0) \neq 0$, and the result follows. □

If the constructions of Examples F and G are used instead of Proposition 14.4, a remarkable generalization of Proposition 14.13 is obtained.

Definition. If $\mathscr{E}$ is either a real or complex topological vector space, then a *hyperplane* in $\mathscr{E}$ is any set of the form $H = \{x \in \mathscr{E} : f(x) = c\}$, where c is a real number and f is a continuous nonzero real linear functional on $\mathscr{E}$ (regarded as a real topological vector space). (If $\mathscr{E}$ is complex, then, as we have seen (cf. the proof of Theorem 14.3), a hyperplane may equally well be described as a set of the form $H = \{x \in \mathscr{E} : \operatorname{Re} g(x) = c\}$, where c is a real number and g is a continuous nonzero (complex) linear functional on $\mathscr{E}$.) The hyperplane $H = \{x \in \mathscr{E} : f(x) = c\}$ splits $\mathscr{E}$ into two *closed half-spaces*:

$$H_- = \{x \in \mathscr{E} : f(x) \leq c\} \quad \text{and} \quad H_+ = \{x \in \mathscr{E} : f(x) \geq c\}.$$

The hyperplane H is said to *separate* two sets if one is a subset of H_- and the other a subset of H_+.

Proposition 14.14. *Let $\mathscr{E}$ be a topological linear space, and let C and D be nonempty disjoint convex subsets of $\mathscr{E}$, one of which at least has nonempty interior. Then there exists a hyperplane separating C and D.*

PROOF. Since an interior point of a convex set is certainly an internal point in the sense of Example F, it follows at once from Example G that there exists a nonzero real linear functional f on $\mathscr{E}$ such that $f(u) \leq f(v)$ whenever $u \in C$ and $v \in D$. Hence $c = \sup\{f(u) : u \in C\}$ is a real number such that $f(u) \leq c$ for all u in C, while $f(v) \geq c$ for all v in D. To complete the proof, it suffices to show that f is continuous, and hence that $H = \{x \in \mathscr{E} : f(x) = c\}$ is a hyperplane. But if f were not continuous it would have to assume every real value on the interior of either C or D, whichever has nonempty interior (Prob. O), and this is manifestly impossible. Hence f is continuous, and the proposition is proved. □

Proposition 14.15. *Let $\mathscr{E}$ be a locally convex topological linear space, let C be a closed convex set in $\mathscr{E}$, and let x_0 be a vector in $\mathscr{E}$ that does not belong to C. Then there exists a linear functional g in $\mathscr{E}^*$ and a real number c such that $\operatorname{Re} g(x) \le c$ for all x in C, while $\operatorname{Re} g(x_0) > c$. Thus C is the intersection of all of the closed half-spaces of $\mathscr{E}$ that contain C.*

PROOF. Let V be a convex open neighborhood of x_0 that does not meet C. Then, as was seen in the proof of Proposition 14.14, there exists a continuous nonzero real linear functional f on $\mathscr{E}$ such that $f(u) \le f(v)$ whenever $u \in C$ and $v \in V$. If we set $c = \sup\{f(u) : u \in C\}$, and if g is the complex linear functional on $\mathscr{E}$ such that $\operatorname{Re} g = f$ (see the proof of Theorem 14.3), then g is continuous along with f, and the inequality $\operatorname{Re} g(x) \le c$ holds for all vectors x in C. To complete the proof we note that $f(V)$ is an open set in $\mathbb{R}$. (This can be seen in many ways. The simplest way is to fix a vector z_0 such that $f(z_0) \ne 0$, and to choose for each vector v in V a positive number $\delta = \delta_v$ such that $v + tz_0 \in V$ for all $-\delta < t < +\delta$. Then $f(V)$ contains the interval $\{f(v) + tf(z_0) : -\delta < t < +\delta\}$ about $f(v)$.) Hence if V contained any vector v such that $f(v) = c$, then V would also have to contain vectors v such that $f(v) < c$, a contradiction. Thus $f(v) > c$ for every v in V; in particular, $f(x_0) > c$. $\qquad\square$

The foregoing results suggest that local convexity is closely connected with the abundance of continuous linear functionals on a topological linear space. The following two examples shed further light on this connection.

Example M. Let μ denote Lebesgue–Borel measure on the unit interval $[0, 1]$, and let $\mathscr{M}$ denote the linear space of equivalence classes of Borel measurable functions on $[0, 1]$ (with equivalence defined as equality a.e. $[\mu]$) equipped with the quasinorm

$$\| f \| = \int_0^1 \frac{|f|}{1 + |f|} \, d\mu$$

(cf. Example 11K). Suppose φ is a continuous linear functional on $\mathscr{M}$. Then there exists a positive number δ such that $\| f \| < \delta$ implies $|\varphi([f])| < 1$. Choose a positive integer N such that $N > 1/\delta$, and let χ_k denote the characteristic function of the interval $((k - 1)/N, k/N), k = 1, \dots, N$. If f is an arbitrary Borel measurable function on $[0, 1]$, and if we write $f_k = Nf\chi_k, k = 1, \dots, N$, then $\| f_k \| < 1/N < \delta$, and therefore $|\varphi([f_k])| < 1$ for every k. Since $f = (f_1 + \cdots + f_N)/N$ a.e. $[\mu]$, it is also the case that $|\varphi([f])| < 1$, and since f is arbitrary, it follows that $\varphi = 0$. Thus *the zero functional is the only continuous linear functional on $\mathscr{M}$.*

(There is another way of viewing the situation in the space $\mathscr{M}$ that ties in with the notion of local convexity. Essentially the same argument as that presented above shows that if C is any convex subset of $\mathscr{M}$ such that $C^\circ \ne \varnothing$, then C is *dense* in $\mathscr{M}$.)

Example N. If $0 < p < 1$ the linear space (ℓ_p) (Ex. 11D) is not (in any natural way) a normed space (cf. Example 11C), but

$$|x| = \sum_{n=0}^{\infty} |\xi_n|^p, \qquad x = \{\xi_n\}_{n=0}^{\infty} \in (\ell_p), \tag{18}$$

defines a useful quasinorm on (ℓ_p) (see Problem U). If ε is a positive number, then the ball $(\ell_p)_\varepsilon = \{x \in (\ell_p) : |x| \leq \varepsilon\}$ contains all of the vectors εe_m, $m \in \mathbb{N}_0$ (where e_m denotes the sequence $\{\delta_{mn}\}_{n=0}^{\infty}$ as usual; cf. Example 11H). Since $x_n = \varepsilon(e_0 + \cdots + e_{n-1})/n$ is a convex combination of the vectors $\varepsilon e_0, \ldots, \varepsilon e_{n-1}$, and since $|x_n| = \varepsilon^p/n^{p-1}$ for each nonnegative integer n, we see that the convex hull of $(\ell_p)_\varepsilon$ is not contained in any ball $(\ell_p)_r$, $r > 0$. Hence no bounded neighborhood of the origin 0 in (ℓ_p) contains a convex neighborhood of 0, so the space (ℓ_p) equipped with the quasinorm (18) is certainly not locally convex. Nevertheless the linear functionals $\{\xi_n\}_{n=0}^{\infty} \to \xi_m$ are all continuous on (ℓ_p) with respect to the quasinorm (18), and this family of linear functionals is separating on (ℓ_p). Thus the collection $\mathscr{E}^*$ of continuous linear functionals on a topological linear space $\mathscr{E}$ may be separating even when $\mathscr{E}$ is *not* locally convex.

PROBLEMS

A. Let $\mathscr{M}$ denote the linear submanifold of $\mathbb{C}^n$ consisting of n-tuples of the form $(\xi, 0, \ldots, 0)$, and let f be the linear functional whose value at $(\xi, 0, \ldots, 0)$ is ξ. If $\mathbb{C}^n$ is given the norm $\| \ \|_1$ (Ex. 11B), what are the linear extensions f_0 of f to $\mathbb{C}^n$ satisfying the condition $\|f_0\| = \|f\|$? Discuss the same problem for the norms $\| \ \|_p$, $1 < p \leq +\infty$.

B. (i) The technique of Examples B and C may be applied to the study of the bounded linear functionals on (ℓ_∞), but the procedure fails to identify $(\ell_\infty)^*$. Explain what goes right and what goes wrong.
   (ii) Find the duals of the spaces (c_0) and (c) (Ex. 2D) equipped with the sup norm.

C. If $\mathscr{E}$ is a normed space and $\hat{\mathscr{E}}$ is its completion, then every functional in $\mathscr{E}^*$ extends to a unique functional in $(\hat{\mathscr{E}})^*$, and this correspondence between $\mathscr{E}^*$ and $(\hat{\mathscr{E}})^*$ is an isometric isomorphism. (It is customary simply to identify $\mathscr{E}^*$ with $(\hat{\mathscr{E}})^*$ via this canonical isomorphism.)

D. (i) Let $\mathscr{E}_1, \ldots, \mathscr{E}_n$ be Banach spaces, and let $\mathscr{E} = \mathscr{E}_1 \oplus_1 \cdots \oplus_1 \mathscr{E}_n$ be their direct sum (Ch. 11, p. 222). Show that $\mathscr{E}^*$ can be identified in a natural way via an isometric isomorphism with the Banach space consisting of the linear space $\mathscr{E}_1^* + \cdots + \mathscr{E}_n^*$ equipped with the norm $\|(f_1, \ldots, f_n)\|_\infty = \|f_1\| \vee \cdots \vee \|f_n\|$ (cf. Problem 11S). Show similarly that if $1 < p < +\infty$ and $\mathscr{E} = \mathscr{E}_1 \oplus_p \cdots \oplus_p \mathscr{E}_n$ (Prob. 11T), then $\mathscr{E}^*$ may be identified in a natural way with the Banach space $\mathscr{E}_1^* \oplus_q \cdots \oplus_q \mathscr{E}_n^*$, where q denotes the Hölder conjugate of p.
   (ii) Let $\{\mathscr{E}_\gamma\}_{\gamma \in \Gamma}$ be an indexed family of Banach spaces, and let p be a number such that $1 \leq p < +\infty$. Denote by $\mathscr{K}_p$ the Banach space consisting of those elements

$\{x_\gamma\}$ of the full algebraic direct sum of the family $\{\mathscr{E}_\gamma\}$ such that $\sum_\gamma \|x_\gamma\|^p < +\infty$, equipped with the norm

$$\|\{x_\gamma\}\|_p = \left[\sum_\gamma \|x_\gamma\|^p\right]^{1/p}$$

(see Problem 11T). Find the dual space $\mathscr{H}_p^*$.

E. Prove the converse of Proposition 12.2 by showing that if $\mathscr{E}$ and $\mathscr{F}$ are normed spaces with $\mathscr{E} \neq (0)$, and if $\mathscr{L}(\mathscr{E}, \mathscr{F})$ is complete, then $\mathscr{F}$ is also complete.

F. Let $[a, b]$ be a real interval, $a \leq b$, and let $\mathscr{V} = \mathscr{V}([a, b])$ denote the collection of all complex-valued functions α of bounded variation on $[a, b]$. Then $\mathscr{V}$ is a linear space and $V(\alpha) = V(\alpha; a, b)$ (the total variation of α over $[a, b]$) defines a pseudo-norm on $\mathscr{V}$. Show that the zero space $\mathscr{Z}$ of V is the one-dimensional space consisting of the constant functions, and verify that the associated space $\dot{\mathscr{V}}$ is complete in the norm obtained by factoring out $\mathscr{Z}$ (Prop. 11.17). (Thus the elements of the Banach space $\dot{\mathscr{V}}$ are really equivalence classes of functions differing by additive constants. Nevertheless it is customary to refer to them as functions, to write $\mathscr{V}([a, b])$ for $\dot{\mathscr{V}}$, and also to refer to V as the *total variation norm* on $\mathscr{V}$. It should be noted that the ambiguity thus introduced places on the reader the rather trivial burden of determining, in some situations, just which meaning to assign to the symbol $\mathscr{V}$.) Show also that the mapping that assigns to each α in $\dot{\mathscr{V}}$ the linear functional φ_α defined in (12) is a linear transformation of $\dot{\mathscr{V}}$ onto $\mathscr{C}([a, b])^*$ having norm one.

G. (i) Let α be defined on $[0, 1]$ as follows:

$$\alpha(t) = \begin{cases} 0, & t \neq \tfrac{1}{2}, \\ 1, & t = \tfrac{1}{2}. \end{cases}$$

Show that α has total variation two, but that the linear functional φ_α defined by α according to (12) is the zero functional.

(ii) Let $\{\delta_n\}$ be a sequence of positive numbers such that $\sum_n \delta_n < +\infty$, and let $\{s_n\}$ be any similarly indexed sequence of distinct points in the open unit interval. Define α on $[0, 1]$ as follows:

$$\alpha = \sum_n \delta_n \chi_{\{s_n\}}.$$

Show that the total variation of α is given by $2\sum_n \delta_n$, but that the linear functional φ_α defined by α is again the zero functional. (Hint: First do the case of a finite sum; then, for $\varepsilon > 0$, choose N so that $\sum_{n>N} \delta_n < \varepsilon$.)

(iii) Show that if α is a *continuous* function of total variation V on $[0, 1]$, then the linear functional φ_α defined by α has norm V.

A careful scrutiny of the arguments used in the preceding problem shows that a necessary and sufficient condition for $\|\varphi_\alpha\|$ in (12) to equal the total variation of the complex-valued integrator α is that, at every point t_0 of discontinuity of α in $(0, 1)$, the value of α should lie in the rectangle in $\mathbb{C}$ having $\alpha(t_0 +)$ and $\alpha(t_0 -)$ as two diagonally opposite vertices. Thus it *is* possible to use formula (12) to identify $\mathscr{C}([0, 1])^*$ with various subspaces of $\mathscr{V}([0, 1])$—for instance, with the subspace of $\mathscr{V}([0, 1])$ consisting of all functions that are right-continuous on $(0, 1)$; see Example 18A.

H. Suppose $x = \{\xi_n\}_{n=0}^{\infty}$ is an *ultimately periodic* sequence of complex numbers, i.e., suppose there exist complex numbers $\alpha_1, \ldots, \alpha_p$ and a positive integer N such that $\xi_n = \alpha_i$ whenever n is congruent to i modulo p and $n > N$, and let φ be an arbitrary Banach generalized limit (Ex. E). Show that $\varphi(x) = (\alpha_1 + \cdots + \alpha_p)/p$.

I. For an arbitrary infinite sequence $x = \{\xi_n\}_{n=0}^{\infty}$ of complex numbers let us write $Cx = \{\eta_n\}_{n=0}^{\infty}$, where

$$\eta_n = \frac{\xi_0 + \cdots + \xi_n}{n+1}, \qquad n \in \mathbb{N}_0.$$

The sequence x is said to be *limitable (in the sense of) Cesàro* to λ_0 (notation: C-$\lim x = \lambda_0$) if the sequence Cx is convergent and $\lim Cx = \lambda_0$. Show that the collection $\mathscr{L}$ of all bounded sequences x that are limitable Cesàro forms a subspace of (ℓ_∞), and that $\varphi(x) = C$-$\lim x$ is a bounded linear functional on $\mathscr{L}$. Show also that φ may be extended to a Banach generalized limit on (ℓ_∞). (Hint: Show first that C acts as a contraction on (ℓ_∞), and hence that φ has norm one. Verify that if $x = \{\xi_n\}$ is limitable Cesàro, and if x_m denotes the tail $\{\xi_m, \xi_{m+1}, \ldots\}$ of x, then C-$\lim x_m = C$-$\lim x$. Conclude that $(c) \subset \mathscr{L}$, that φ extends the functional $\lim$ on (c), and that φ has norm one with respect to the pseudonorm σ introduced in Example E.)

J. If $x = \{\xi_n\}_{n=0}^{\infty}$ is a bounded sequence, then the value at x of the pseudonorm σ in Example E is given by

$$\sigma(x) = \inf \sup_n \left| \frac{1}{M} \sum_{i=1}^{M} \xi_{m_i + n} \right|,$$

where the infimum is taken over all finite sequences of nonnegative integers $m_1, \ldots, m_M$. Suppose we restrict attention to *real* sequences only, and modify this definition, writing

$$p_0(x) = \inf \sup_n \left\{ \frac{1}{M} \sum_{i=1}^{M} \xi_{m_i + n} \right\},$$

where the infimum is taken just as before, but where we have omitted the absolute value bars. Show that p_0 is also given by

$$p_0(x) = \inf \limsup_n \left\{ \frac{1}{M} \sum_{i=1}^{M} \xi_{m_i + n} \right\}.$$

Show further that p_0 possesses the properties (i) and (ii) of Proposition 14.6. Conclude that there exists a linear functional φ_0 on the space $(\ell_\infty)_{\mathbb{R}}$ of all bounded real sequences that satisfies the condition

$$\varphi_0(x) \leq p_0(x), \qquad x \in (\ell_\infty)_{\mathbb{R}}. \tag{19}$$

Show that any such functional φ_0 satisfies condition (13), and possesses the additional property that if $x \geq 0$, i.e., if every term of x is nonnegative, then $\varphi_0(x) \geq 0$. (In other words, φ_0 is a positive linear functional on $(\ell_\infty)_{\mathbb{R}}$ (Ch. 10, p. 197)). Show, finally, that for any bounded real sequence x we have

$$\liminf x \leq \varphi_0(x) \leq \limsup x,$$

and conclude that $\varphi_0(x) = \lim x$ whenever x is convergent.

K. (i) Let φ be an arbitrary linear functional on (ℓ_∞) that takes the value one at the constant sequence $e = \{1, 1, \ldots\}$. Show that φ is a positive linear functional when and only when $\| \varphi \| = 1$. (Hint: Show that if $\| \varphi \| = 1$ and if x is a sequence in (ℓ_∞) all of whose terms lie in some closed disc D^- in $\mathbb{C}$, then $\varphi(x) \in D^-$; use this fact to conclude that $\mathrm{Re}\, \varphi(x) \geq 0$ whenever $\mathrm{Re}\, x \geq 0$.)

(ii) Let φ be a Banach generalized limit on (ℓ_∞) as defined in Example E. Show that φ is the complexification of its restriction to $(\ell_\infty)_\mathbb{R}$ (Ex. 2M), and that if φ_0 denotes the restriction of φ to $(\ell_\infty)_\mathbb{R}$, then φ_0 satisfies condition (13). Show conversely that if φ_0 is any linear functional on $(\ell_\infty)_\mathbb{R}$ satisfying (13), then the complexification of φ_0 has norm one with respect to the pseudonorm σ of Example E, and is therefore a Banach generalized limit. (In other words, the real generalized limits generated by the procedure set forth in Problem J are precisely the restrictions to $(\ell_\infty)_\mathbb{R}$ of the Banach generalized limits of Example E.)

L. The function p_0 introduced in Problem J has a number of interesting properties.

(i) Verify first that if $x = \{\xi_n\}$, then

$$-p_0(-x) = \sup_n \inf\left\{\frac{1}{M}\sum_{i=1}^M \xi_{m_i+n}\right\} = \sup \lim_n \inf\left\{\frac{1}{M}\sum_{i=1}^M \xi_{m_i+n}\right\},$$

where the supremum is taken, as before, over all finite sequences $m_1, \ldots, m_M$ of nonnegative integers.

(ii) Let φ_0 be a real Banach generalized limit, that is, a linear functional on $(\ell_\infty)_\mathbb{R}$ satisfying condition (19) of Problem J, and suppose $x \in (\ell_\infty)_\mathbb{R}$. Show that

$$\liminf x \leq -p_0(-x) \leq \varphi_0(x) \leq p_0(x) \leq \limsup x,$$

and conclude that $p_0(x) = -p_0(-x) = \lim x (= \varphi_0(x))$ whenever x is convergent.

(iii) Show that if $x_0 \in (\ell_\infty)_\mathbb{R}$ and y denotes a convergent real sequence, then $p_0(x_0 + y) = p_0(x_0) + p_0(y) (= p_0(x_0) + \lim y)$, and conclude that if $x_0 \notin (c)$, then

$$p_0(x_0) = \inf_{y \in (c)}\{p_0(x_0 + y) - \lim y\}$$

and

$$-p_0(-x_0) = \sup_{y \in (c)}\{-p_0(y - x_0) + \lim y\}.$$

Use this fact to show that there exist Banach generalized limits φ' and φ'' such that $\varphi'(x_0) \neq \varphi''(x_0)$ if and only if $-p_0(-x_0) < p_0(x_0)$. (Hint: The condition is clearly necessary by virtue of (ii). To see that it is sufficient, consult the sketch of the proof of Proposition 14.6.)

It is customary to say of a bounded sequence x that it is *almost convergent* if $\varphi(x)$ is the same for every Banach generalized limit φ. Thus, for example, every ultimately periodic sequence is almost convergent (Prob. H). The foregoing result may be paraphrased by saying that a bounded real sequence is almost convergent if and only if $p_0(-x) = -p_0(x)$. Readers wishing to learn more about this phenomenon may consult [29] or [4], or the original paper [44].

M. Show that there exist linear functionals φ on the real Banach space $\mathscr{B}_{\mathbb{R}}([0, +\infty))$ of all bounded real functions on the half-line $[0, +\infty)$ satisfying the following conditions for every f in $\mathscr{B}_{\mathbb{R}}([0, +\infty))$:

(i) If $\lim_{t \to +\infty} f(t)$ exists, then $\varphi(f) = \lim_{t \to +\infty} f(t)$,

(ii) If $f \geq 0$, then $\varphi(f) \geq 0$,

(iii) If we write $f_s(t) = f(s + t)$, then $\varphi(f_s) = \varphi(f)$ for all $s \geq 0$.

Show also that if φ is any such functional, then

(iv) $\liminf_{t \to +\infty} f(t) \leq \varphi(f) \leq \limsup_{t \to +\infty} f(t)$ for every f in $\mathscr{B}_{\mathbb{R}}([0, +\infty))$, and

(v) $\|\varphi\| = 1$.

N. (i) Show that if a subset A of a topological linear space $\mathscr{E}$ is [absolutely] convex, then A^- and A° are also [absolutely] convex. (Hint: If x, $y \in \mathscr{E}$ and V is a neighborhood of 0, and if $0 \leq t \leq 1$, then $t(x + V) + (1 - t)(y + V) = (tx + (1 - t)y) + (tV + (1 - t)V)$; recall Problem 11M.) Show also that a subset A of $\mathscr{E}$ is absolutely convex if and only if x, $y \in A$ and $|\alpha| + |\beta| \leq 1$ imply that $\alpha x + \beta y \in A$. (It is this characterization that led to the term "absolutely convex".) Verify that an arbitrary subset M of $\mathscr{E}$ is contained in a smallest absolutely convex set A, and that this set, known as the *absolutely convex hull* of M, consists of the collection of all linear combinations $\alpha_1 x_1 + \cdots + \alpha_n x_n$ of vectors $x_1, \ldots, x_n$ in M such that $|\alpha_1| + \cdots + |\alpha_n| \leq 1$. Show finally that an arbitrary subset M of $\mathscr{E}$ is contained in a smallest closed [absolutely] convex subset of $\mathscr{E}$, and that this set is, in fact, the closure of the [absolutely] convex hull of M. (The smallest closed [absolutely] convex set containing M is called the *closed [absolutely] convex hull of M.*)

(ii) (Mazur) Let $\mathscr{E}$ be an F-space, and let K be a compact subset of $\mathscr{E}$. Show that the closed convex hull of K is also compact. (Hint: If $N_\varepsilon = \{x_1, \ldots, x_n\}$ is an ε-net in K (Prop. 4.3), then the convex hull C_0 of N_ε is an ε-net in the convex hull C of K. The set C_0 is a bounded subset of a finite dimensional subspace of $\mathscr{E}$, and is therefore totally bounded. If $M_\varepsilon = \{y_1, \ldots, y_m\}$ is an ε-net in C_0, then M_ε is also a 2ε-net in C. Hence C is totally bounded. Use Problem 4Q.)

It is also true, and sometimes useful to know, that the closed convex hull of any *weakly* compact subset of a Banach space (as defined in Chapter 15) is also *weakly* compact (cf. [23], p. 434).

O. Let $\mathscr{E}$ be a (real or complex) topological linear space, and let f be a linear functional on $\mathscr{E}$. As we have already seen (Prob. 12U), a necessary and sufficient condition for f to be continuous is that its null space $\mathscr{K}(f)$ be closed. Show that, in fact, if f is discontinuous, then $\mathscr{K}(f)$ is dense in $\mathscr{E}$. Conclude, more generally, that if f is discontinuous, then the set on which f assumes the value α is dense in $\mathscr{E}$ for every scalar α. (Hint: Show first that the linear submanifold of $\mathscr{E}$ generated by $\mathscr{K}(f)$ and any vector not in $\mathscr{K}(f)$ coincides with $\mathscr{E}$, and use the fact that if $f(x_0) = \alpha$, then $\{x \in \mathscr{E} : f(x) = \alpha\} = x_0 + \mathscr{K}(f).$)

P. Let $\mathscr{E}$ and $\mathscr{F}$ be separated locally convex topological linear spaces, let $x_1, \ldots, x_k$ be linearly independent vectors in $\mathscr{E}$, and let $y_1, \ldots, y_k$ be arbitrary vectors in $\mathscr{F}$. Show that there exists a continuous linear transformation T of $\mathscr{E}$ into $\mathscr{F}$ such that $Tx_i = y_i$, $i = 1, \ldots, k$.

Q. Let Δ be a domain in the complex plane, and let $\mathscr{A}$ denote the space of all analytic functions on Δ. Show that $\mathscr{A}$ is a Frechét space in the topology of uniform convergence on compact subsets of Δ. (Hint: See Problem 11U.)

R. Show that if $\mathscr{E}$ is a normed space such that $\mathscr{E}^*$ is separable, then $\mathscr{E}$ itself is separable. (Hint: Let $\{f_n\}$ be a countable dense subset of $\mathscr{E}^*$ and for each n let x_n be a unit vector in $\mathscr{E}$ such that $f_n(x_n) > \| f_n \|/2$. Use Proposition 14.10 to show that the countable set $\{x_n\}$ spans $\mathscr{E}$; then use Problem 11K.) Show by an example that the separability of $\mathscr{E}$ does not imply that of $\mathscr{E}^*$.

S. Let C be a convex subset of a real linear space $\mathscr{E}$. A subset D of C is called an *extreme subset* of C if, for every line segment σ in C such that $\sigma \cap D \neq \varnothing$, either $\sigma \subset D$ or else $\sigma \cap D$ is a singleton consisting of an endpoint of σ. A point x_0 such that the singleton $\{x_0\}$ is an extreme subset of C is called an *extreme point* of C.

 (i) Let C be a convex set in a real linear space $\mathscr{E}$, let f be a (real) linear functional on $\mathscr{E}$, and suppose f is bounded above on C. Show that the set

$$D = \left\{ x \in C : f(x) = \sup_{y \in C} f(y) \right\}$$

  is an extreme subset of C.
 (ii) Show that an extreme subset D of a convex set C has the property that both D and $C \backslash D$ are convex sets. Show also that if D is an extreme subset of a convex set C, and if E is an extreme subset of D, then E is also an extreme subset of C.
 (iii) Let C be a convex set in a real linear space $\mathscr{E}$, and let $\mathscr{D}$ be an arbitrary nonempty collection of extreme subsets of C that is simply ordered with respect to set inclusion. Show that the intersection $\bigcap \mathscr{D}$ is also an extreme subset of C.

T. (Krein-Mil'man Theorem) Let $\mathscr{E}$ be a real locally convex space, and let C be a nonempty compact convex subset of $\mathscr{E}$.

 (i) Show that every closed nonempty extreme subset of C contains a *minimal* closed nonempty extreme subset of C (in the inclusion ordering). (Hint: Use Problem S(iii) and Zorn's lemma.)
 (ii) Show that a minimal closed nonempty extreme subset of C is necessarily a singleton, and conclude that C possesses an extreme point. (Hint: The set of continuous linear functionals is separating on $\mathscr{E}$ by the real version of Proposition 4.13.)
 (iii) Show that, in fact, C coincides with the closed convex hull of the set of all extreme points of C. (Hint: Let C' denote the closed convex hull of the set of all extreme points of C. The set C' is a compact convex subset of C, so if $C' \neq C$, then there exists a continuous (real) linear functional f on $\mathscr{E}$ and a real number c such that $f(x) \leq c$ for every $x \in C'$, while $f(x_0) > c$ for at least one point x_0 of C. Set $c'' = \sup_{x \in C} f(x)$ and $C'' = \{x \in C : f(x) = c''\}$, and verify that every extreme point of C'' is also an extreme point of C.)

U. Let a, b, and p be positive numbers. Show that

$$(a + b)^p \gtreqless a^p + b^p \quad \text{when} \quad p \gtreqless 1,$$

and conclude that the function $\| \ \|_p$ introduced in Example M is indeed a quasinorm on (ℓ_p) for $0 < p < 1$. Show also that the linear manifold (φ) of finitely nonzero sequences (Ex. 11H) is dense in (ℓ_p) with respect to this quasinorm, and use this fact to verify that every continuous linear functional on (ℓ_p), $0 < p < 1$, is of the form

$$f_a(x) = \sum_{n=0}^{\infty} \alpha_n \xi_n, \qquad x = \{\xi_n\}_{n=0}^{\infty} \in (\ell_p),$$

where $a = \{\alpha_n\}_{n=0}^{\infty}$ is a bounded sequence of complex numbers. Show, finally, in the converse direction, that if $\{\alpha_n\} \in (\ell_\infty)$, then $f(x) = \sum_{n=0}^{\infty} \alpha_n \xi_n$ defines a continuous linear functional on (ℓ_p), $0 < p < 1$. (Hint: If $|\,x\,|_p \le 1$, then $\|x\|_1 \le |\,x\,|_p$.)

> The following two problems are concerned with yet another version of the Hahn–Banach theorem, valid in the context of partially ordered linear spaces.

V. If $\mathscr{E}$ is a real linear space that is equipped with a partial ordering $\le$, then $\mathscr{E}$ is said to be a *partially ordered linear space* if the following conditions are satisfied: (1) if $x \le y$ in $\mathscr{E}$ then $x + z \le y + z$ for every z in $\mathscr{E}$, and (2) if $x \le y$ and if t is a nonnegative real number, then $tx \le ty$.

(i) Show that if $\mathscr{E}$ is a partially ordered linear space, then the set P of *positive elements* of $\mathscr{E}$, that is, elements x of $\mathscr{E}$ such that $x \ge 0$, is a convex subset of $\mathscr{E}$ with the property that if $x \in P$ and t is a nonnegative real number then $tx \in P$. (Such a convex set is called a *convex cone*.) Show conversely that if P is a convex cone in an arbitrary real linear space $\mathscr{E}$, and if we define $x \le y$ to mean that $y - x \in P$, then $\le$ turns $\mathscr{E}$ into a partially ordered linear space. A linear functional on a partially ordered linear space $\mathscr{E}$ is *positive* if it assumes only nonnegative values on the convex cone P of positive elements of $\mathscr{E}$. Show that a linear functional on $\mathscr{E}$ is positive if and only if it is a monotone increasing mapping of $\mathscr{E}$ into $\mathbb{R}$. Show also that if $\mathscr{F}$ is both a linear space and a function lattice of real-valued functions on a set X, then $\mathscr{F}$ is a partially ordered linear space in its natural ordering, and a linear functional on $\mathscr{F}$ is positive in the present sense if and only if it is a positive linear functional in the sense introduced in Chapter 10. (Thus the present notion of positivity is a natural generalization of the former one.)

(ii) Let $\mathscr{E}$ be a partially ordered linear space, and let $\mathscr{M}$ be a cofinal linear submanifold of $\mathscr{E}$, that is, let $\mathscr{M}$ be a linear submanifold of $\mathscr{E}$ such that for every vector x in $\mathscr{E}$ there exists a vector y in $\mathscr{M}$ such that $x \le y$ (Prob. 1P). Show that it is also the case that for any vector x in $\mathscr{E}$ there exist vectors y in $\mathscr{M}$ such that $y \le x$.

W. (i) Let $\mathscr{M}$ be a cofinal linear manifold in a partially ordered linear space $\mathscr{E}$, and let f be a positive linear functional defined on $\mathscr{M}$. Show that there exists a positive linear extension of f to the entire space $\mathscr{E}$. (Hint: Suppose first that x_0 is a single vector in $\mathscr{E}$ not belonging to $\mathscr{M}$, and write $\mathscr{M}_0$ for the linear manifold generated by $\mathscr{M}$ and x_0. In order to extend f to a positive linear functional f_0 on $\mathscr{M}_0$ it is necessary to choose a real number r such that $y + tx_0 \ge 0$ implies $f(y) + tr \ge 0$ for every vector y in $\mathscr{M}$ and every real number t. Follow the pattern of the proofs of Lemma 14.1, Proposition 14.6, and Theorem 14.2.)

(ii) Use the preceding fact to give new proofs of the results of Problems J and M.

Local convexity and weak topologies 15

Along with its norm topology every normed space possesses a second topology of great importance known as its *weak* topology.

Definition. If $\mathscr{E}$ is a normed space, then the coarsest topology on $\mathscr{E}$ making all of the linear functionals in $\mathscr{E}^*$ continuous is the *weak topology* on $\mathscr{E}$. (Equivalently, the weak topology on $\mathscr{E}$ is the topology inversely induced on $\mathscr{E}$ by the family of linear functionals in $\mathscr{E}^*$; see Chapter 3, page 43.)

This definition has the merit of brevity, and serves to show at once how the weak topology on $\mathscr{E}$ compares with the norm topology. Since the functionals in $\mathscr{E}^*$ are all norm continuous by definition, the weak topology is clearly refined by the norm topology. (That is, in fact, what the term "weak" is intended to suggest.) To see more clearly what the weak topology on $\mathscr{E}$ looks like, let us write

$$\sigma_f(x) = |f(x)|, \qquad x \in \mathscr{E}, f \in \mathscr{E}^*. \tag{1}$$

Lemma 15.1. *For any normed space $\mathscr{E}$ and for each f in $\mathscr{E}^*$ the function σ_f defined in (1) is a pseudonorm on $\mathscr{E}$. Moreover, if $\mathscr{T}$ denotes an arbitrary linear topology on $\mathscr{E}$, then f is continuous with respect to $\mathscr{T}$ if and only if σ_f is. Hence f is continuous with respect to $\mathscr{T}$ if and only if $\mathscr{T}$ refines the pseudometric topology induced by σ_f, and the weak topology on $\mathscr{E}$ coincides with the topology induced by the family of pseudonorms $\{\sigma_f\}_{f \in \mathscr{E}^*}$.*

PROOF. The verification that σ_f is a pseudonorm is routine and is left to the reader (cf. Example 11M). Suppose that f is continuous with respect to some topology $\mathscr{T}$ on $\mathscr{E}$ and let x_0 be a vector in $\mathscr{E}$. Then for any given $\varepsilon > 0$

there exists a neighborhood V of x_0 with respect to $\mathcal{T}$ such that $x \in V$ implies $|f(x) - f(x_0)| < \varepsilon$. But then

$$|\sigma_f(x) - \sigma_f(x_0)| = ||f(x)| - |f(x_0)|| \le |f(x) - f(x_0)| < \varepsilon,$$

and it follows that σ_f is also continuous with respect to $\mathcal{T}$. (Observe that this part of the argument does not even require that the topology $\mathcal{T}$ be linear.)

Suppose next that σ_f is continuous with respect to a topology $\mathcal{T}$ on $\mathcal{E}$. Then for any given $\varepsilon > 0$ there exists a neighborhood V of the origin in $\mathcal{E}$ with respect to $\mathcal{T}$ such that $|\sigma_f(x) - \sigma_f(0)| = |f(x)| < \varepsilon$ for every x in V. Thus f is continuous at the origin, and if $\mathcal{T}$ is a linear topology on $\mathcal{E}$, this implies that f is continuous on $\mathcal{E}$ with respect to $\mathcal{T}$ (Prob. 12U). Hence the weak topology on $\mathcal{E}$ coincides with the supremum of the linear topologies induced by the various pseudonorms σ_f, $f \in \mathcal{E}^*$, or, in other words, with the topology induced by the family $\{\sigma_f\}_{f \in \mathcal{E}^*}$ (Ch. 11, p. 234). ☐

Proposition 15.2. *The weak topology on a normed space $\mathcal{E}$ is a linear topology on $\mathcal{E}$ turning $\mathcal{E}$ into a separated locally convex space. The weak topology coincides with the norm topology when and only when $\mathcal{E}$ is finite dimensional.*

PROOF. The weak topology on $\mathcal{E}$ coincides with the (linear) topology induced on $\mathcal{E}$ by the family of pseudonorms $\{\sigma_f\}_{f \in \mathcal{E}^*}$, and is therefore locally convex (Prop. 14.7). To see that the weak topology is also Hausdorff, note that $\mathcal{E}^*$ is separating on $\mathcal{E}$ (Cor. 14.11) and that $\{\sigma_f\}_{f \in \mathcal{E}^*}$ is therefore also separating on $\mathcal{E}$. The weak topology coincides with the norm topology when $\mathcal{E}$ is finite dimensional because a finite dimensional linear space possesses only one separated linear topology (Prob. 11P). That the weak topology is strictly coarser than the norm topology whenever $\mathcal{E}$ is infinite dimensional is shown in Problem H. ☐

All questions concerning the weak topology on a normed space $\mathcal{E}$ can be settled, at least in principle, by Lemma 15.1 and Proposition 15.2. Nevertheless it is frequently useful to know an explicit base for the weak topology. In order to describe one let us agree to write

$$U(f_1, \ldots, f_n ; \varepsilon) = \{x \in \mathcal{E} : |f_i(x)| < \varepsilon, i = 1, \ldots, n\} \tag{2}$$

for every finite subset $\{f_1, \ldots, f_n\}$ of $\mathcal{E}^*$ and every positive number ε.

Proposition 15.3. *In any normed space $\mathcal{E}$ the collection $\mathcal{V}$ of all sets of the form (2) constitutes a base of open neighborhoods of 0 in the weak topology on $\mathcal{E}$; the collection of all translates*

$$x_0 + U(f_1, \ldots, f_n ; \varepsilon) = \{x \in \mathcal{E} : |f_i(x - x_0)| < \varepsilon, i = 1, \ldots, n\}, \quad x_0 \in \mathcal{E},$$

of the sets in $\mathcal{V}$ is a base for the weak topology on $\mathcal{E}$.

PROOF. It suffices to show that $\mathcal{V}$ is a base of open neighborhoods of 0 in the weak topology (Prob. 11L(ii)). To see this we note that $U(f_1, \ldots, f_n ; \varepsilon) = $

$U(f_1; \varepsilon) \cap \cdots \cap U(f_n; \varepsilon)$ is the intersection of the open balls with center 0 and radius ε with respect to the pseudometrics defined by the pseudonorms $\sigma_{f_1}, \ldots, \sigma_{f_n}$. Hence each set in $\mathcal{V}$ is a weakly open neighborhood of 0. On the other hand, if U is any weakly open subset of $\mathcal{E}$ such that $0 \in U$, then there exist functionals $f_1, \ldots, f_n$ in $\mathcal{E}^*$ and positive radii $\varepsilon_1, \ldots, \varepsilon_n$ such that

$$U(f_1; \varepsilon_1) \cap \cdots \cap U(f_n; \varepsilon_n) \subset U$$

(Ex. 3M) and hence such that

$$U(f_1, \ldots, f_n; \varepsilon_1 \wedge \cdots \wedge \varepsilon_n) \subset U. \qquad \square$$

Example A. Suppose $\| \; \|_1$ and $\| \; \|_2$ are two equivalent norms on a linear space $\mathcal{E}$, and let $\mathcal{E}_1$ and $\mathcal{E}_2$ denote the spaces obtained by equipping $\mathcal{E}$ with $\| \; \|_1$ and $\| \; \|_2$, respectively. Then the norm topologies on $\mathcal{E}_1$ and $\mathcal{E}_2$ are identical, and consequently $\mathcal{E}_1$ and $\mathcal{E}_2$ have exactly the same linear functionals in their dual spaces (though the norm of a functional in $\mathcal{E}_1^*$ need not be the same as its norm in $\mathcal{E}_2^*$). Hence the weak topology on $\mathcal{E}_1$ is identical with the weak topology on $\mathcal{E}_2$. If the norm $\| \; \|_1$ merely dominates $\| \; \|_2$ (Ex. 13A), then the norm topology on $\mathcal{E}_1$ refines the norm topology on $\mathcal{E}_2$, so every linear functional in $\mathcal{E}_2^*$ belongs to $\mathcal{E}_1^*$ (though, once again, not necessarily with the same norm), so the weak topology on $\mathcal{E}_1$ refines the weak topology on $\mathcal{E}_2$.

Proposition 15.4. *A net $\{x_\lambda\}$ in a normed space $\mathcal{E}$ converges to a limit x_0 in the weak topology on $\mathcal{E}$ if and only if the net $\{f(x_\lambda)\}$ converges to $f(x_0)$ for every f in $\mathcal{E}^*$.*

PROOF. According to Proposition 11.27, a net $\{x_\lambda\}$ in $\mathcal{E}$ converges to x_0 with respect to the weak topology on $\mathcal{E}$ if and only if $\lim \sigma_f(x_0 - x_\lambda) = 0$ for every f in $\mathcal{E}^*$, i.e., if and only if $\lim_\lambda |f(x_0) - f(x_\lambda)| = 0$ for every f in $\mathcal{E}^*$. $\qquad \square$

Corollary 15.5. *Let $\mathcal{F}$ be a normed space and let $\mathcal{F}_w$ denote the topological linear space obtained by equipping $\mathcal{F}$ with its weak topology. Let $\mathcal{E}$ be an arbitrary topological linear space, and let T be a linear transformation of $\mathcal{E}$ into $\mathcal{F}$. Then $T: \mathcal{E} \to \mathcal{F}_w$ is continuous if and only if $f \circ T$ is continuous for every f in $\mathcal{F}^*$.*

PROOF. It is enough to show that the condition is sufficient. Let $\{x_\lambda\}$ be a net in $\mathcal{E}$ such that $\lim_\lambda x_\lambda = 0$. Then the net $\{f(Tx_\lambda)\}$ converges to zero in $\mathbb{C}$ for each f in $\mathcal{F}^*$, and it follows by Proposition 15.4 that the net $\{Tx_\lambda\}$ converges to 0 in $\mathcal{F}_w$. Hence T is continuous at the origin, and the result follows (Prob. 12U). $\qquad \square$

Example B. Let $\{x_n\}_{n=1}^{\infty}$ be a sequence of vectors in (ℓ_1), and let

$$x_n = \{\xi_0^{(n)}, \xi_1^{(n)}, \ldots\}, \qquad n \in \mathbb{N}.$$

Since $(\ell_1)^* = (\ell_\infty)$ by Example 14B, the sequence $\{x_n\}$ converges weakly to zero if and only if

$$\lim_n \sum_{m=0}^\infty \zeta_m^{(n)} \alpha_m = 0$$

for every element $a = \{\alpha_0, \alpha_1, \ldots\}$ of (ℓ_∞). In particular, setting $a = e_m$ (cf. Example 11H), we see that if $\{x_n\}$ converges weakly to 0 in (ℓ_1), then $\lim_n \zeta_m^{(n)} = 0$ for every m in $\mathbb{N}_0$. Hence if $\{x_n\}$ converges weakly to a limit x in (ℓ_1), then $\{x_n\}$ also converges termwise to x. That this necessary condition is not sufficient can be seen by considering the sequence $\{e_n\}_{n=0}^\infty$ itself, which converges termwise to zero, but does not converge weakly in (ℓ_1) to anything. On the other hand, $\{e_n\}_{n=0}^\infty$ does converge weakly to zero in (ℓ_p), $1 < p < +\infty$ (cf. Example 14C).

In obvious analogy with the notion of weak convergence, it is customary to say that a set M in a normed space $\mathscr{E}$ is *weakly bounded* if $f(M)$ is a bounded set of scalars for every f in $\mathscr{E}^*$. As it turns out, however, this definition does not lead to anything new. This is a fact having many important consequences, and we pause to study the matter more closely.

The central idea is that if x is a vector in a normed space $\mathscr{E}$, then x defines in a natural way a functional F_x on the dual space $\mathscr{E}^*$ according to the formula

$$F_x(f) = f(x), \qquad f \in \mathscr{E}^*. \tag{4}$$

Since $|F_x(f)| = |f(x)| \le \|f\|\,\|x\|$, we see that F_x is, in fact, a bounded linear functional on $\mathscr{E}^*$ and that $\|F_x\| \le \|x\|$. Thus if we write

$$j(x) = F_x, \qquad x \in \mathscr{E},$$

then j is a mapping of $\mathscr{E}$ into the dual space $\mathscr{E}^{**} = (\mathscr{E}^*)^*$ of $\mathscr{E}^*$—the *second dual* of $\mathscr{E}$. It is obvious that j is a linear transformation:

$$j(\alpha x + \beta y)(f) = f(\alpha x + \beta y) = \alpha f(x) + \beta f(y) = \alpha j(x)(f) + \beta j(y)(f).$$

Since $\|j(x)\| = \|F_x\| \le \|x\|$, the transformation j is a contraction. In particular, j is bounded. More precise information is supplied by the following proposition.

Proposition 15.6. *If $\mathscr{E}$ is an arbitrary normed space, then the linear transformation j of $\mathscr{E}$ into $\mathscr{E}^{**}$ defined by setting*

$$j(x)(f) = f(x), \qquad f \in \mathscr{E}^*, \qquad x \in \mathscr{E},$$

is an isometry.

PROOF. Let x be a vector in $\mathscr{E}$. By Corollary 14.11 there exists a linear functional f_0 in $\mathscr{E}^*$ such that $\|f_0\| = 1$ and such that $j(x)(f_0) = f_0(x) = \|x\|$. Thus $\|j(x)\| \ge \|x\|$, and the result follows. $\qquad\square$

Definition. The linear isometry j of Proposition 15.6 is known as the *natural embedding* of the normed space $\mathcal{E}$ in the second dual $\mathcal{E}^{**}$ of $\mathcal{E}$. (This mapping will play a central role in Chapter 16.)

Using the natural embedding of a normed space $\mathcal{E}$ in $\mathcal{E}^{**}$ we readily characterize the weakly bounded subsets of $\mathcal{E}$.

Proposition 15.7. *A set M in a normed space $\mathcal{E}$ is weakly bounded if and only if it is bounded* (*in norm*).

PROOF. It is clear that if M is bounded, then M is weakly bounded. On the other hand, if M is weakly bounded, and if j denotes the natural embedding of $\mathcal{E}$ in $\mathcal{E}^{**}$, then $j(M)$ is, by hypothesis, pointwise bounded on $\mathcal{E}^*$, and is therefore bounded in $\mathcal{E}^{**}$ by the uniform boundedness theorem (Th. 12.15). Since j is an isometry, the proposition follows. (Note that we here use the fact that $\mathcal{E}^*$ is always complete, whether $\mathcal{E}$ itself is complete or not; see Proposition 12.2.) □

Example C. Let $\{x_n\}_{n=1}^{\infty}$ be a weakly convergent sequence in a normed space $\mathcal{E}$. Since convergent sequences in $\mathbb{C}$ are bounded, it follows that $\{x_n\}$ is weakly bounded, and therefore bounded. It is worth noting that this argument fails for more general nets, and that, in fact, it is possible for an unbounded net to be weakly convergent (cf. Problem A or Problem 3L).

Example D. Let $\mathcal{E}$ and $\mathcal{F}$ be normed spaces, and let $\mathcal{E}_w$ and $\mathcal{F}_w$ denote the topological linear spaces obtained by equipping $\mathcal{E}$ and $\mathcal{F}$, respectively, with their weak topologies. Suppose T is a bounded linear transformation of $\mathcal{E}$ into $\mathcal{F}$. If $f \in \mathcal{F}^*$ then $f \circ T \in \mathcal{E}^*$, so $f \circ T$ is continuous on $\mathcal{E}_w$. Hence $T : \mathcal{E}_w \to \mathcal{F}_w$ is also continuous by Corollary 15.5. But then $T : \mathcal{E} \to \mathcal{F}_w$ is certainly continuous too, since the norm topology on $\mathcal{E}$ is finer than the weak topology. Finally, to close the circle, suppose $T : \mathcal{E} \to \mathcal{F}_w$ is continuous. Then for each f in $\mathcal{F}^*$ the functional $f \circ T$ is bounded on the unit ball $\mathcal{E}_1$. In other words, $T(\mathcal{E}_1)$ is weakly bounded in $\mathcal{F}$. But then $T(\mathcal{E}_1)$ is norm bounded as well, and it follows that T is bounded. Thus we see that the following three conditions are equivalent for an arbitrary linear transformation T of $\mathcal{E}$ into $\mathcal{F}$:

(i) $T : \mathcal{E} \to \mathcal{F}$ is bounded,
(ii) $T : \mathcal{E}_w \to \mathcal{F}_w$ is continuous,
(iii) $T : \mathcal{E} \to \mathcal{F}_w$ is continuous.

Example E. Let Δ be a domain in the complex plane, and let $\Phi : \Delta \to \mathcal{E}$ be a mapping of Δ into a Banach space $\mathcal{E}$ with the property that $f \circ \Phi$ is analytic

315

on Δ for every f in $\mathscr{E}^*$. (Such an $\mathscr{E}$-valued mapping is said to be *weakly analytic* on Δ.) If we write $Q(\Phi; \zeta, \xi)$ for the difference quotient

$$Q(\Phi; \zeta, \xi) = \frac{\Phi(\zeta + \xi) - \Phi(\zeta)}{\xi}$$

whenever it is defined (that is, for all ζ in Δ and $\xi \neq 0$ such that $\zeta + \xi$ is also in Δ), then it is clear from linearity that

$$f(Q(\Phi; \zeta, \xi)) = Q(f \circ \Phi: \zeta, \xi)$$

for every f in $\mathscr{E}^*$. Hence if K denotes a compact subset of Δ, then there exists a positive number ε and, for each f in $\mathscr{E}^*$, a positive number M_f, such that

$$\frac{1}{|\xi - \eta|} |Q(f \circ \Phi: \zeta, \xi) - Q(f \circ \Phi: \zeta, \eta)| \leq M_f$$

for all ζ in K and all ξ, η such that $0 < |\xi|, |\eta| < \varepsilon$ and $\xi \neq \eta$ (see Example 5I). In other words the set

$$\left\{ \frac{1}{\xi - \eta} [Q(\Phi; \zeta, \xi) - Q(\Phi; \zeta, \eta)] : \zeta \in K, 0 < |\xi|, |\eta| < \varepsilon, \xi \neq \eta \right\}$$

is weakly bounded in $\mathscr{E}$, and is therefore norm bounded. Hence there exists a single positive number M such that

$$\| Q(\Phi; \zeta, \xi) - Q(\Phi; \zeta, \eta) \| \leq M |\xi - \eta|, \zeta \in K, 0 < |\xi|, |\eta| < \varepsilon. \quad (5)$$

From this it follows at once that, if ζ is any point of K and $\{\zeta_n\}$ is an arbitrary sequence in $\Delta \backslash \{\zeta\}$ that converges to ζ, then the sequence $\{Q(\Phi; \zeta, \zeta_n - \zeta)\}$ is Cauchy, and therefore convergent, in $\mathscr{E}$. Moreover, it is readily seen that

$$\lim_n Q(\Phi; \zeta, \zeta_n - \zeta)$$

depends only on Φ and ζ, and not on the sequence $\{\zeta_n\}$. Since K can be taken, in particular, to be any singleton $\{\zeta\}$ in Δ, this shows that the limit

$$\lim_{\xi \to 0} Q(\Phi, \zeta, \xi) = \lim_{\xi \to 0} \frac{\Phi(\zeta + \xi) - \Phi(\zeta)}{\xi} \quad (6)$$

actually exists in the norm sense at every point of Δ. (This also shows that Φ is norm continuous.) The limit (6) is called the *derivative* of Φ at ζ and is denoted by $\Phi'(\zeta)$ just as in elementary calculus. If we agree to call an $\mathscr{E}$-valued mapping *analytic* on Δ if it is *differentiable*, that is, if it possesses a derivative at every point of Δ, then this argument shows that a weakly analytic Banach space valued mapping is automatically analytic. (In the converse direction, it is obvious that an analytic $\mathscr{E}$-valued mapping Φ is weakly analytic, and that, in fact, if $f \in \mathscr{E}^*$, then $(f \circ \Phi)' = f \circ \Phi'$; cf. Problem E.) Finally, returning to (5) and letting η tend to zero, we learn that the difference quotient of Φ tends to Φ' uniformly on compact subsets of Δ, whence it follows at once that Φ' *is also (weakly) analytic on* Δ.

316

Example F. Let $\mathscr{E}$ and $\mathscr{F}$ be two Banach spaces, and let Ψ be a mapping of a domain Δ in $\mathbb{C}$ into $\mathscr{L}(\mathscr{E}, \mathscr{F})$. An argument entirely similar to the one presented in the preceding example shows that if Ψ possesses the property that the mapping $\Psi_x(\lambda) = \Psi(\lambda)(x)$ is an analytic $\mathscr{F}$-valued mapping for every vector x in $\mathscr{E}$, then the mapping Ψ is automatically analytic as a mapping of Δ into $\mathscr{L}(\mathscr{E}, \mathscr{F})$. The details are left to the interested reader.

Example G. Analytic mappings taking their values in Banach spaces turn up frequently in functional analysis. We reconsider here one very important instance of such a mapping that has already appeared. Let $\mathscr{A}$ be a unital Banach algebra and let x be an element of $\mathscr{A}$. We recall from Chapter 12 that the spectrum $\sigma_{\mathscr{A}}(x)$ is, by definition, the set of complex numbers λ such that $\lambda - x$ is not invertible in $\mathscr{A}$, and that if $\lambda_0 \notin \sigma_{\mathscr{A}}(x)$, so that $\lambda_0 - x$ is invertible in $\mathscr{A}$, then

$$(\lambda - x)^{-1} = \sum_{n=0}^{\infty} (\lambda_0 - \lambda)^n (\lambda_0 - x)^{-(n+1)}$$

for every complex number λ such that $|\lambda - \lambda_0| < 1/\|(\lambda_0 - x)^{-1}\|$ (Prop. 12.12). As was noted in Chapter 12, this shows that the spectrum $\sigma_{\mathscr{A}}(x)$ is closed, but it also shows that the resolvent $R_x(\lambda) = (\lambda - x)^{-1}$ is a *locally analytic* $\mathscr{A}$-valued mapping on the resolvent set $\mathbb{C}\backslash\sigma_{\mathscr{A}}(x)$, i.e., $R_x(\lambda)$ is an analytic $\mathscr{A}$-valued mapping on each component of the complement of $\sigma_{\mathscr{A}}(x)$ (see Problem F). In this connection we also recall that

$$R_x(\lambda) = (\lambda - x)^{-1} = \sum_{n=0}^{\infty} \frac{x^n}{\lambda^{n+1}}$$

for all complex numbers λ such that $|\lambda| > \|x\|$, and hence that $\sigma_{\mathscr{A}}(x)$ is bounded and therefore compact (Prop. 12.11). Moreover, this last expression for the resolvent shows that $\lim_{|\lambda| \to +\infty} R_x(\lambda) = 0$. In particular, R_x is bounded outside of some sufficiently large disc centered at $\lambda = 0$. (Following the standard usage of complex analysis, this is frequently expressed by saying that the resolvent R_x has a *removable singularity at* ∞.) Employing these observations we establish the following fact: *If $\mathscr{A}$ is a unital Banach algebra and x belongs to $\mathscr{A}$, then the spectrum $\sigma_{\mathscr{A}}(x)$ is nonempty.*

Indeed, suppose $\sigma_{\mathscr{A}}(x) = \varnothing$. Then for each bounded linear functional f on $\mathscr{A}$ the complex-valued function $\varphi_f = f \circ R_x$ is an entire function on $\mathbb{C}$ with the property that $\lim_{|\lambda| \to +\infty} \varphi_f(\lambda) = 0$. In particular φ_f is bounded, and is therefore a constant by Liouville's theorem (Ex. 5K). But then, since φ_f tends to zero at infinity, we must have $\varphi_f(\lambda) = f(R_x(\lambda)) \equiv 0$. Finally, since this holds for every bounded linear functional f on $\mathscr{A}$, we conclude (Cor. 14.11) that $R_x(\lambda) \equiv 0$, a manifest impossibility since $R_x(\lambda) = (\lambda - x)^{-1}$ is an invertible element of $\mathscr{A}$ for all λ not belonging to $\sigma_{\mathscr{A}}(x)$.

Finally, we note that this shows, in particular, that if T is a bounded linear operator on a Banach space $\mathscr{E} \neq (0)$, then $\sigma(T) \neq \varnothing$ and the resolvent R_T is a locally analytic $\mathscr{L}(\mathscr{E})$-valued mapping on the complement $\mathbb{C}\backslash\sigma(T)$.

There is a natural and far-reaching generalization of the notion of the weak topology on a normed space. Suppose $\mathscr{E}$ is an arbitrary linear space and $\mathscr{M}$ is an arbitrary linear manifold of linear functionals defined on $\mathscr{E}$. The coarsest topology on $\mathscr{E}$ making all of the functionals in $\mathscr{M}$ continuous, that is, the topology (inversely) induced on $\mathscr{E}$ by the family of functionals in $\mathscr{M}$ (see Chapter 3) will be called simply the topology *induced* on $\mathscr{E}$ by $\mathscr{M}$. (There is no loss of generality in assuming $\mathscr{M}$ to be a linear manifold; if M is an arbitrary set of linear functionals on $\mathscr{E}$, and if $\mathscr{M}$ denotes the linear submanifold of the full algebraic dual of $\mathscr{E}$ generated algebraically by M, then every functional in $\mathscr{M}$ is continuous with respect to the topology inversely induced by M (Prob. 12U), so the latter topology coincides with the topology induced by $\mathscr{M}$.) If we continue to use the notation (1), writing σ_f for the pseudonorm

$$\sigma_f(x) = |f(x)|, \qquad x \in \mathscr{E},$$

for each f in $\mathscr{M}$, then by Lemma 15.1 the topology induced on $\mathscr{E}$ by $\mathscr{M}$ coincides with the (linear) topology induced on $\mathscr{E}$ by the indexed family $\{\sigma_f\}_{f \in \mathscr{M}}$. Thus we obtain the following result, whose proof is virtually a repetition of the proofs of Propositions 15.2 and 15.3 and Corollary 15.4, and is therefore omitted.

Proposition 15.8. *The topology induced on a linear space $\mathscr{E}$ by a linear manifold $\mathscr{M}$ of linear functionals on $\mathscr{E}$ is a linear topology on $\mathscr{E}$ that turns $\mathscr{E}$ into a locally convex topological linear space. This topology is separated if and only if $\mathscr{M}$ is separating on $\mathscr{E}$. A net $\{x_\lambda\}_{\lambda \in \Lambda}$ in $\mathscr{E}$ converges to a limit x_0 in this topology if and only if the net $\{f(x_\lambda)\}$ converges in $\mathbb{C}$ to $f(x_0)$ for each functional f in $\mathscr{M}$. A linear transformation of a topological linear space $\mathscr{F}$ into $\mathscr{E}$ is continuous with respect to this topology if and only if $f \circ T$ is continuous on $\mathscr{F}$ for every f in $\mathscr{M}$. Finally, the collection of all sets of the form (2), where $f_1, \ldots, f_n$ are selected from $\mathscr{M}$ and ε denotes an arbitrary positive number, constitutes a base of open neighborhoods at 0, and the translates of these neighborhoods form a base for this topology.*

An important property of the topology induced by a linear manifold of functionals is expressed in the following result.

Proposition 15.9. *Let $\mathscr{E}$ be a linear space, let $\mathscr{M}$ be a linear manifold of linear functionals on $\mathscr{E}$, and let $\mathscr{F}$ be a linear submanifold of $\mathscr{E}$ that is closed with respect to the topology induced on $\mathscr{E}$ by $\mathscr{M}$. Then for any vector x_0 in $\mathscr{E}$ that does not belong to $\mathscr{F}$ there are functionals f in $\mathscr{M}$ such that $f(\mathscr{F}) = (0)$ while $f(x_0) \neq 0$.*

PROOF. As noted above, the topology induced on $\mathscr{E}$ by $\mathscr{M}$ is the same as that induced by the family of pseudonorms $\{\sigma_f\}_{f \in \mathscr{M}}$, and is therefore locally convex. Hence Proposition 14.13 applies, and since every linear functional on $\mathscr{E}$ that is continuous with respect to the topology induced by $\mathscr{M}$ actually belongs to $\mathscr{M}$ (Prob. J), the result follows.   □

Example H. Let $\mathscr{E}$ be a linear space and take for $\mathscr{M}$ the full algebraic dual of $\mathscr{E}$. The topology induced on $\mathscr{E}$ by $\mathscr{M}$ in this case is a separated locally convex topology on $\mathscr{E}$ with respect to which *every* linear functional is continuous.

Example I. If $\mathscr{E}$ is an arbitrary separated locally convex space, then the dual $\mathscr{E}^*$ (the collection of all continuous linear functionals on $\mathscr{E}$) induces a separated locally convex linear topology on $\mathscr{E}$. It will be convenient to continue to call this the *weak topology* on $\mathscr{E}$, just as when $\mathscr{E}$ is a normed space.

Other examples of topologies induced by linear functionals will occupy us from time to time as we go along. For the present, however, the weak topology on a normed space and the following *weak* topology* on the dual of a normed space are the ones of greatest interest.

Definition. If $\mathscr{E}$ is a normed space, then the topology induced on the dual space $\mathscr{E}^*$ by the subspace $j(\mathscr{E})$ of $\mathscr{E}^{**}$ is the *weak* topology* on $\mathscr{E}^*$.

The following is nothing more than a summary of Proposition 15.8 as it applies to the weak* topology.

Proposition 15.10. *For any normed space $\mathscr{E}$ the weak* topology on $\mathscr{E}^*$ turns $\mathscr{E}^*$ into a separated, locally convex, topological linear space. A net $\{f_\lambda\}$ in $\mathscr{E}^*$ converges to a limit f in the weak* topology if and only if $\lim_\lambda f_\lambda(x) = f(x)$ in $\mathbb{C}$ for every vector x in $\mathscr{E}$. A linear transformation T of a topological linear space $\mathscr{F}$ into $\mathscr{E}^*$ is continuous with respect to the weak* topology if and only if the mapping $y \to (Ty)(x)$ is a continuous linear functional on $\mathscr{F}$ for every x in $\mathscr{E}$. Finally, the collection of all sets of the form*

$$\{f \in \mathscr{E}^* : |f(x_i)| < \varepsilon, \ i = 1, \ldots, n\},$$

where $x_1, \ldots, x_n$ are selected from $\mathscr{E}$ and ε denotes an arbitrary positive number, constitutes a base of open neighborhoods of 0 in the weak* topology, and the translates of these neighborhoods form a base for that topology.

PROOF. It is obvious that $j(\mathscr{E})$ is separating on $\mathscr{E}^*$. Everything else follows from Proposition 15.8.   □

The most important single property of weak* topologies is stated in the following theorem.

Theorem 15.11 (Alaoglu's Theorem). *If $\mathscr{E}$ is a normed linear space, then the closed unit ball $(\mathscr{E}^*)_1$ in the dual space $\mathscr{E}^*$ is compact in the weak* topology.*

Proof. For each vector x in $\mathscr{E}$ set $D_x^- = \{\lambda \in \mathbb{C} : |\lambda| \leq \|x\|\}$, and form the topological product

$$\Pi = \prod_{x \in \mathscr{E}} D_x^-.$$

Each D_x^- is a compact disc in $\mathbb{C}$, and it follows by Tihonov's theorem (Theorem 3.15) that Π is compact. Moreover, if we define

$$\Phi(f) = \{f(x)\}_{x \in \mathscr{E}}, \qquad f \in (\mathscr{E}^*)_1,$$

then it is clear that Φ is a continuous one-to-one mapping of $(\mathscr{E}^*)_1$ into Π. The proof will be completed by showing that the range $R = \Phi((\mathscr{E}^*)_1)$ of Φ is closed in Π (which will show that R is compact; cf. Proposition 3.2) and that the mapping $\Phi^{-1} : R \to (\mathscr{E}^*)_1$ is continuous when $\mathscr{E}^*$ is given the weak* topology (which will show that Φ is a homeomorphism between $(\mathscr{E}^*)_1$ and R). To that end, let $\{f_\lambda\}_{\lambda \in \Lambda}$ be a net in $(\mathscr{E}^*)_1$ with the property that the net $\{\Phi(f_\lambda)\}_{\lambda \in \Lambda}$ is convergent in Π. We must show that there exists a linear functional f_0 in $(\mathscr{E}^*)_1$ such that $\Phi(f_\lambda) \to \Phi(f_0)$ (this will show that R is closed in Π) and such that $f_\lambda \to f_0$ in the weak* topology (this will show that Φ^{-1} is continuous on R). But to say that the net $\{\Phi(f_\lambda)\}$ converges in Π is the same as saying that all of the coordinate nets $\{\Phi(f_\lambda)_x = f_\lambda(x)\}$ converge in $\mathbb{C}$ (Prop. 3.20). Accordingly we may and do define f_0 by setting

$$f_0(x) = \lim_\lambda f_\lambda(x), \qquad x \in \mathscr{E}.$$

If we knew f_0 to be a bounded linear functional with $\|f_0\| \leq 1$, then from this definition (and Proposition 15.10) it would be clear at once that $f_\lambda \to f_0$ in the weak* topology on $\mathscr{E}^*$ and likewise that $\Phi(f_\lambda) \to \Phi(f_0)$ in Π. Thus it only remains to verify that f_0 is an element of $(\mathscr{E}^*)_1$. Suppose now that x and y are vectors in $\mathscr{E}$. Then

$$f_0(x + y) = \lim_\lambda f_\lambda(x + y) = \lim_\lambda f_\lambda(x) + \lim_\lambda f_\lambda(y) = f_0(x) + f_0(y),$$

which shows that f_0 is additive. Similarly, if α is a scalar and $x \in \mathscr{E}$, then $f_0(\alpha x) = \alpha f_0(x)$, and we see that f_0 is a linear functional. Finally, since $f_0(x) \in D_x^-$ for every x in $\mathscr{E}$, it is clear that f_0 is bounded and that, in fact, $\|f_0\| \leq 1$. (The reader may wish to compare this construction with that employed in Example 3R.) $\qquad\square$

Let $\{x_n\}_{n=0}^{\infty}$ be a sequence in the unit ball $(\ell_p)_1$ for some p, $1 < p < +\infty$, and let $x_n = \{\xi_m^{(n)}\}_{m=0}^{\infty}$ for each n. The sequence $\{\xi_0^{(n)}\}_{n=0}^{\infty}$ is bounded in $\mathbb{C}$ and therefore possesses a convergent subsequence. Hence there is a subsequence $\{x_{k_0(n)}\}_{n=0}^{\infty}$ of $\{x_n\}$ that is convergent along the 0th coordinate to some complex number β_0 with $|\beta_0| \leq 1$. That sequence in turn possesses a subsequence $\{x_{k_1(n)}\}_{n=0}^{\infty}$ converging along its 0th coordinate to β_0 and also converging along its 1st coordinate to a complex number β_1 with $|\beta_1| \leq 1$. Continuing in this manner, we obtain by mathematical induction an infinite sequence $\{x_{k_m(n)}\}$ of sequences, $m = 1, 2, \ldots$, each of which is a subsequence of the foregoing one, and with the property that $\{x_{k_m(n)}\}$ converges along the first $m + 1$ coordinates to the first $m + 1$ terms of a sequence $b = \{\beta_m\}_{m=0}^{\infty}$ belonging to $(\ell_r)_1$. But then the *diagonal sequence* $\{x'_n = x_{k_n(n)}\}_{n=0}^{\infty}$ is eventually a subsequence of every one of the sequences $\{x_{k_m(n)}\}$ and therefore converges coordinatewise to b. Hence, by Problem C, b belongs to $(\ell_p)_1$ and $\{x'_n\}$ converges weakly to b. We have shown that every sequence in $(\ell_p)_1$ possesses weakly convergent subsequences, and since $(\ell_p)_1$ is weakly metrizable (Prob. M), this implies that $(\ell_p)_1$ is weakly compact. Thus, by the device of extracting a whole sequence of successively finer subsequences, and then extracting the diagonal from the resulting infinite tableau—that is, by employing the *diagonal process*; cf. Problem 4P—we obtain an independent proof of the weak* compactness of the unit ball in the space (ℓ_p), $1 < p < +\infty$. Historically it was this construction and others like it that led to the discovery of Alaoglu's theorem.

The topologies induced on a linear space by various families of linear functionals are of interest by virtue of being the *coarsest* topologies satisfying certain conditions. There are also many situations in which a linear topology is interesting because it is the *finest* topology satisfying certain specified conditions. To see how this goes we need the following result.

Proposition 15.12. *If $\{\mathcal{T}_\gamma\}_{\gamma \in \Gamma}$ is an arbitrary indexed family of locally convex linear topologies on a linear space $\mathcal{E}$, then $\sup_\gamma \mathcal{T}_\gamma$ is also a locally convex linear topology on $\mathcal{E}$.*

PROOF. It suffices to show that the topology $\sup_\gamma \mathcal{T}_\gamma$ is locally convex (Prop. 11.25). If $V_i = V_{\gamma_i}$ is an absolutely convex neighborhood of 0 in $\mathcal{E}$ with respect to $\mathcal{T}_{\gamma_i}$, $i = 1, \ldots, n$, then $V_1 \cap \cdots \cap V_n$ is also absolutely convex. Since sets of this form constitute a neighborhood base at 0 in the topology $\sup_\gamma \mathcal{T}_\gamma$, the result follows, at least when $\Gamma \neq \varnothing$; the supremum of the empty family of topologies on $\mathcal{E}$ is the indiscrete topology, clearly locally convex and linear. $\qquad\qquad\square$

Example J. The supremum of the collection of *all* locally convex linear topologies on a linear space $\mathcal{E}$ is a locally convex linear topology on $\mathcal{E}$. If $\mathcal{E}$ is equipped with this topology, then every pseudonorm on $\mathcal{E}$ becomes continuous; equivalently, every absorbing absolutely convex subset of $\mathcal{E}$ is a neighborhood of the origin in $\mathcal{E}$. Likewise, an arbitrary linear transformation of $\mathcal{E}$ into a locally convex topological linear space is continuous on $\mathcal{E}$ with respect to this topology. In particular, the dual $\mathcal{E}^*$ of $\mathcal{E}$ coincides with the full algebraic dual of $\mathcal{E}$.

Example K. Let $\mathcal{E}$ be a locally convex topological linear space, and let $\mathcal{E}^*$ denote, as usual, the dual of $\mathcal{E}$. In the collection of all those locally convex linear topologies on $\mathcal{E}$ with the property that, when $\mathcal{E}$ is equipped with that topology, the dual space of $\mathcal{E}$ is precisely the set $\mathcal{E}^*$ (of which the given topology on $\mathcal{E}$ is one by hypothesis) we already know that there is a coarsest topology—namely, the weak topology on $\mathscr{E}$. According to Proposition 15.12 there is also a *finest* locally convex linear topology on $\mathscr{E}$ admitting $\mathscr{E}^*$ as the dual space. This topology is known as the *Mackey topology* on $\mathscr{E}$.

Another application of Proposition 15.12 arises in connection with the notion of *inductive limit*.

Definition. Suppose given a linear space $\mathscr{E}$, a directed set Λ of indices, and a monotone increasing net $\{\mathscr{M}_\lambda\}_{\lambda \in \Lambda}$ of linear submanifolds of $\mathscr{E}$ (in the inclusion ordering) such that $\mathscr{E} = \bigcup_\lambda \mathscr{M}_\lambda = \sum_\lambda \mathscr{M}_\lambda$. Suppose also that each $\mathscr{M}_\lambda$ is equipped with a locally convex linear topology, and that these topologies are all coherent in the sense that $\mathscr{M}_\lambda$ is topologically a subspace of $\mathscr{M}_{\lambda'}$ as well as a linear submanifold of $\mathscr{M}_{\lambda'}$ whenever $\lambda \leq \lambda'$ in Λ (cf. Example 3L). Then the finest locally convex linear topology on $\mathscr{E}$ with respect to which all of the inclusion mappings $i_\lambda : \mathscr{M}_\lambda \to \mathscr{E}$ are continuous (the indiscrete topology on $\mathscr{E}$ is one such) is called the *inductive limit topology* on $\mathscr{E}$, and $\mathscr{E}$ equipped with this topology is known as the *inductive limit* of the net $\{\mathscr{M}_\lambda\}$.

The notion of inductive limit introduced here is more restrictive than the usual definition, and corresponds to what many authors refer to as the *strict* inductive limit.

It should be remarked that this inductive limit topology may equally well be described as the finest locally convex linear topology on $\mathscr{E}$ with the property that the relative topology induced by it on each of the submanifolds $\mathscr{M}_\lambda$ is refined by the given topology on $\mathscr{M}_\lambda$. Concerning inductive limits we have the following basic result.

Proposition 15.13. *Suppose $\mathscr{E}$ is the inductive limit of a net of linear submanifolds $\{\mathscr{M}_\lambda\}$. Then a pseudonorm σ on $\mathscr{E}$ is continuous if and only if the restriction of σ to each $\mathscr{M}_\lambda$ is continuous with respect to the given topology on $\mathscr{M}_\lambda$. Similarly, an absorbing absolutely convex subset A of $\mathscr{E}$ is a neighborhood of 0 in $\mathscr{E}$ if and only if $A \cap \mathscr{M}_\lambda$ is a neighborhood of 0 in $\mathscr{M}_\lambda$ for every λ. Finally, if $\mathscr{F}$ is any locally convex topological linear space and T is a linear transformation of $\mathscr{E}$ into $\mathscr{F}$, then T is continuous if and only if each restriction $T \,|\, \mathscr{M}_\lambda$ is continuous.*

PROOF. A pseudonorm σ on $\mathscr{E}$ is continuous with respect to the inductive limit topology if and only if the topology $\mathscr{T}_\sigma$ induced by σ is refined by that topology, i.e., if and only if the inclusion mappings $i_\lambda : \mathscr{M}_\lambda \to \mathscr{E}$ all become continuous when $\mathscr{E}$ is equipped with the topology $\mathscr{T}_\sigma$. But this, in turn, is true if and only if each restriction $\sigma \,|\, \mathscr{M}_\lambda$ is continuous. (We have here used the results of Proposition 11.26.) If A is an absorbing absolutely convex set in $\mathscr{E}$, and if p_A denotes its Minkowski function (Ex. 14F), then A is a neighborhood of 0 if and only if the pseudonorm p_A is continuous, and it is readily seen that this is the case if and only if $A \cap \mathcal{M}_\lambda$ is a neighborhood of 0 in $\mathcal{M}_\lambda$ for every index λ. Suppose, finally, that T is a linear transformation of $\mathcal{E}$ into some locally convex topological vector space $\mathcal{F}$. If T is continuous with respect to the inductive limit topology on $\mathcal{E}$, then it is clear that $T|\mathcal{M}_\lambda$ is continuous on $\mathcal{M}_\lambda$ for every λ. On the other hand, if the latter condition is satisfied, and if V is any absolutely convex neighborhood of 0 in $\mathcal{F}$, then $A = T^{-1}(V)$ is an absolutely convex subset of $\mathcal{E}$, and, as is readily seen, A is also absorbing. Moreover, each intersection $A \cap \mathcal{M}_\lambda$ is a neighborhood of 0 in $\mathcal{M}_\lambda$ by virtue of the hypothesis. Hence A is a neighborhood of 0 in $\mathcal{E}$, so T is continuous. $\square$

Example L. The (self-indexed) collection of all finite dimensional linear submanifolds of an arbitrary linear space $\mathcal{E}$ satisfies the requirements of the above definition if each finite dimensional submanifold is equipped with its unique separated (locally convex) linear topology (Prob. 11P). Since every pseudonorm σ on $\mathcal{E}$ becomes continuous when restricted to a finite dimensional linear manifold (see Problem 11P again), it follows from Proposition 15.13 that every pseudonorm on $\mathcal{E}$ is continuous with respect to the inductive limit topology on $\mathcal{E}$. Hence the inductive limit topology in this instance coincides with the finest locally convex linear topology on $\mathcal{E}$ (Ex. J).

Example M. Suppose given an indexed family $\{\mathcal{E}_\gamma\}_{\gamma \in \Gamma}$ of locally convex topological linear spaces. For each subset Γ' of Γ it is clear that the full algebraic direct sum $\sum_{\gamma \in \Gamma'} \dotplus \mathcal{E}_\gamma$ can be identified in a natural way with a linear manifold in $\sum_{\gamma \in \Gamma} \dotplus \mathcal{E}_\gamma$ (by identifying each element $\{x_\gamma\}_{\gamma \in \Gamma'}$ with the element of $\sum_{\gamma \in \Gamma} \dotplus \mathcal{E}_\gamma$ that agrees with it on Γ' and is equal to 0 elsewhere). In particular, if this is done for each finite subset D of Γ, then the net of finite direct sums $\{\sum_{\gamma \in D} \dotplus \mathcal{E}_\gamma\}$ is identified with a net of linear manifolds in $\sum_{\gamma \in \Gamma} \dotplus \mathcal{E}_\gamma$ indexed by the directed set of all finite subsets of Γ. The sum (or union) of this net consists of the collection $\mathcal{S}$ of those elements of the full algebraic direct sum $\sum_{\gamma \in \Gamma} \dotplus \mathcal{E}_\gamma$ that vanish off some finite subset of Γ, and is known as the *algebraic direct sum* of the family $\{\mathcal{E}_\gamma\}_{\gamma \in \Gamma}$.

Suppose now that for each finite subset D of Γ we equip the finite direct sum $\sum_{\gamma \in D} \dotplus \mathcal{E}_\gamma$ with its product topology. Then the conditions in the definition of inductive limit are satisfied so there exists an inductive limit topology on the algebraic direct sum $\mathcal{S}$ of the family $\{\mathcal{E}_\gamma\}_{\gamma \in \Gamma}$ of topological linear spaces. (The topology here defined on $\mathcal{S}$ is known as the *direct sum topology*. If Γ is infinite it is definitely finer than the topology $\mathcal{S}$ acquires as a linear manifold in the full algebraic direct sum of the family $\{\mathcal{E}_\gamma\}_\gamma$ in the product topology (Prop. 11.24).)

Example N. Let U be a nonempty open subset of Euclidean space $\mathbb{R}^n$, and write $\mathscr{C}_0^{(\infty)}(U)$ for the linear space of test functions on U (that is, the space of all those infinitely differentiable functions on U that have compact support;

see Problem 11V). Let us write $\mathscr{K}$ for the collection of all nonempty compact subsets of U, and for each K in $\mathscr{K}$ let us write $\mathscr{D}_K(U)$ for the set of all those test functions on U that are *supported* on K, i.e., vanish on $U\backslash K$. Then $\mathscr{D}_K(U)$ is a linear manifold in $\mathscr{C}_0^{(\infty)}(U)$. Moreover, if, as before, we write σ_K for the pseudonorm

$$\sigma_K(f) = \sup_{x \in K} |f(x)|, \qquad f \in \mathscr{C}^{(\infty)}(U),$$

and $D^{k_1, \ldots, k_n}f$ for the partial derivative of f obtained by differentiating k_i times with respect to the variable x_i, $i = 1, \ldots, n$, then

$$\sigma_{K,m}(f) = \sup_{k_1 + \cdots + k_n \leq m} \sigma_K(D^{k_1, \ldots, k_n}f), \qquad m \in \mathbb{N}_0,$$

defines a sequence of pseudonorms on $\mathscr{D}_K(U)$ that can be seen (by considering the convergence of sequences, for example) to induce on $\mathscr{D}_K(U)$ the same locally convex topology as the relative topology $\mathscr{D}_K(U)$ acquires as a linear manifold in the Frechét space $\mathscr{C}^{(\infty)}(U)$ of all infinitely differentiable functions on U in the topology of D-convergence (cf. Example 14K). Moreover, $\mathscr{D}_K(U)$ is a closed linear manifold in $\mathscr{C}^{(\infty)}(U)$, and is therefore a Frechét space in its own right. The system $\{\mathscr{D}_K(U)\}$ is a net of linear submanifolds of $\mathscr{C}^{(\infty)}(U)$ indexed by $\mathscr{K}$ (directed in the inclusion ordering) and the union (or sum) of this net is the linear space $\mathscr{C}_0^{(\infty)}(U)$ of all test functions on U. Moreover, if K and K' are nonempty compact subsets of U such that $K \subset K'$, then the various pseudonorms $\sigma_{K',m}$ on $\mathscr{D}_{K'}(U)$ agree with the corresponding pseudonorms $\sigma_{K,m}$ on $\mathscr{D}_K(U)$ for each nonnegative integer m. Hence the space $\mathscr{D}_K(U)$ is topologically embedded in $\mathscr{D}_{K'}(U)$ whenever $K \subset K'$. Thus the conditions set forth in the definition of the inductive limit topology are all satisfied, and it results that there exists an inductive limit topology on the space $\mathscr{C}_0^{(\infty)}(U)$. A linear functional on $\mathscr{C}_0^{(\infty)}(U)$ that is continuous with respect to the inductive limit topology, i.e., an element of $\mathscr{C}_0^{(\infty)}(U)^*$, is called a *distribution* on U. Thus, according to Proposition 15.13, a linear functional on $\mathscr{C}_0^{(\infty)}(U)$ is a distribution on U if and only if its restriction to each space $\mathscr{D}_K(U)$ is continuous there in the topology of D-convergence. It should be noted that for each fixed K in $\mathscr{K}$ the sequence $\{\sigma_{K,m}\}_{m=0}^{\infty}$ is saturated on $\mathscr{D}_K(U)$, and hence that a linear functional T on $\mathscr{C}_0^{(\infty)}(U)$ is continuous, and therefore a distribution on U, if and only if for each K in $\mathscr{K}$ there exists a nonnegative integer m such that T is bounded (and therefore continuous) with respect to $\sigma_{K,m}$ (see Problem 12U). In regard to this criterion it is important to note that the integer m will depend not only on T but also, in general, on K.

That the inductive limit topology is definitely finer than the topology of D-convergence on $\mathscr{C}_0^{(\infty)}(U)$ may be seen as follows: If $\{\varphi_n\}$ is an infinite sequence in $\mathscr{C}_0^{(\infty)}(U)$ that converges to 0 in the inductive limit topology, then there exists a single compact subset K of U such that φ_n vanishes on $U\backslash K$ for every n (cf. Example 14L). Indeed, if this condition is not satisfied, then it is a simple matter to construct by mathematical induction an increasing sequence $\{K_n\}$ of compact subsets of U, a sequence $\{x_n\}$ of points, and a subsequence $\{\varphi_{k_n}\}$ of the given sequence of test functions such that the following conditions are satisfied:

(i) $U = \bigcup_{n=1}^{\infty} K_n^{\circ}$,

(ii) $x_n \in L_n = K_{n+1} \setminus K_n$, $\qquad n \in \mathbb{N}$,

(iii) $\varphi_{k_n}(x_n) \neq 0$, $\qquad n \in \mathbb{N}$

(cf. Example 10A). Let $a_n = |\varphi_{k_n}(x_n)|$, and define

$$\sigma(\varphi) = \sum_n \sup_{x \in L_n} \frac{|\varphi(x)|}{a_n}, \qquad \varphi \in \mathscr{C}_0^{(\infty)}(U).$$

(By (i) only a finite number of terms in this series are nonzero for any one test function φ.) It is clear that σ is a pseudonorm on $\mathscr{C}_0^{(\infty)}(U)$ and it is easily seen that the restriction of σ to each space $\mathscr{D}_K(U)$ is continuous there. Hence σ is continuous on $\mathscr{C}_0^{(\infty)}(U)$ in the inductive limit topology by Proposition 15.13. But for each positive integer n, $1 \leq \sup_{x \in L_n} |\varphi_{k_n}(x)|/a_n \leq \sigma(\varphi_{k_n})$. Thus the sequence $\{\varphi_n\}$ does not converge to 0 in that topology.

Example O. Let U be as in the preceding example. Most ordinary complex-valued functions on U can be interpreted as distributions on U. To see how this goes, consider a Borel measurable function f on U that is integrable (with respect to n-dimensional Lebesgue measure μ_n) over every compact subset of U. (Such a function is customarily said to be *locally integrable* on U. Clearly every continuous function on U is locally integrable on U.) For every test function φ on U the integral $\int_U \varphi f \, d\mu_n$ exists, and setting

$$T_f(\varphi) = \int_U \varphi f \, d\mu_n, \qquad \varphi \in \mathscr{C}_0^{(\infty)}(U),$$

defines a linear functional T_f on $\mathscr{C}_0^{(\infty)}(U)$. Moreover, this functional is readily seen to be a distribution on U. (Indeed, T_f is bounded with respect to the pseudonorm $\sigma_K = \sigma_{K,0}$ on every $\mathscr{D}_K(U)$, $K \in \mathscr{K}$; this fact is expressed by saying that the distribution T_f is of *order zero* on U.) Thus every locally integrable function on U defines in a simple and unique fashion a distribution on U. Moreover, the mapping that assigns to each such function f on U the associated distribution T_f is clearly a linear transformation of the linear space of locally integrable functions on U into the space $\mathscr{C}_0^{(\infty)}(U)^*$ of distributions on U. This mapping is not one-to-one, however. Indeed, it is obvious that T_f and T_g coincide whenever $f = g$ a.e. $[\mu_n]$. It is an important fact in the theory of distributions that the converse of this latter assertion is also valid, that is, if $T_f = T_g$ for some pair of locally integrable functions f and g on U, then $f = g$ a.e. $[\mu_n]$. To verify this it suffices to show that $T_f = 0$ implies $f = 0$ a.e. $[\mu_n]$. Suppose, accordingly, that f gives rise to the trivial distribution. Recall (Prob. 11V) that for each nondegenerate closed cell Z contained in U there exists a test function φ having support Z that is strictly positive on Z° and has the property that φ^r is also a test function for every positive power r. Moreover, we may also suppose that φ is bounded by one.

But then the sequence $\{\varphi^{1/p}\}_{p=1}^{\infty}$ converges pointwise on U to the characteristic function of Z° and is uniformly bounded (by one). Hence

$$\lim_{p} \int_{Z} \varphi^{1/p} f \, d\mu_n = \int_{Z} f \, d\mu_n$$

by the bounded convergence theorem (Th. 7.13). But also

$$\int_{Z} \varphi^{1/p} f \, d\mu_n = \int_{U} \varphi^{1/p} f \, d\mu_n = T_f(\varphi^{1/p}) \equiv 0.$$

Thus $\int_{Z} f \, d\mu_n = 0$ for every closed cell Z contained in U.

Suppose now that V is an arbitrary open set such that V^- is compact and contained in U. If we write

$$v_f(E) = \int_{E} f \, d\mu_n$$

for every Borel set E contained in V, then v_f is a complex measure on the σ-ring of Borel subsets of V such that $v_f(Z) = 0$ for every closed cell Z contained in V, whence it follows at once that v_f is the zero measure on V (see Problem 8A). But then f must vanish a.e. $[\mu_n]$ on V. Finally, since U can be expressed as the union of a countable collection of open sets such as V, this implies that f vanishes a.e. $[\mu_n]$ on U.

This discussion shows that the ordinary (locally integrable) functions on U may be thought of as forming a linear submanifold of the space $\mathscr{C}_0^{(\infty)}(U)^*$ of distributions on U (provided we agree to identify two functions that are equal almost everywhere $[\mu_n]$). Because of this fact, distributions on U are also frequently called *generalized functions* on U. The distributions of the form T_f, that is, those that come from locally integrable functions on U, are known as the *regular* distributions on U.

Example P. Let U and V be nonempty open subsets of $\mathbb{R}^n$ such that $V \subset U$. A test function φ on V may be viewed in a natural way as a test function on U (just extend φ to be zero on $U \backslash V$), and in this manner $\mathscr{C}_0^{(\infty)}(V)$ may be identified with a linear submanifold of $\mathscr{C}_0^{(\infty)}(U)$. Consequently, if T is a distribution on U and φ is a test function on V, then it makes sense to evaluate T at φ, and the linear functional thus defined on $\mathscr{C}_0^{(\infty)}(V)$ is called the *restriction* of T to V, denoted by $T|V$. It is obvious that T is a distribution on V, and that if T_f is the regular distribution associated with a locally integrable function f on U as in Example O, then $T_f|V = T_{f|V}$.

Consider now a distribution T on U and let U_0 be the union of the collection of all those nonempty open subsets V of U such that $T|V = 0$. (If there are no such open sets, the union U_0 is empty, and the following discussion is rendered trivial.) We shall show that $T|U_0 = 0$. Indeed, suppose K is an arbitrary compact subset of U_0 and φ is a test function on U_0 that vanishes on $U_0 \backslash K$. We must show that $T\varphi = 0$. If K is empty there is nothing to prove. If not, then, by the definition of U_0, there exist open subsets $V_1, \ldots, V_m$ of U_0 that cover K and have the property that $T | V_i = 0$, $i = 1, \ldots, m$. But then there also exist test functions $\varphi_1, \ldots, \varphi_m$ constituting a partition of unity subordinate to the covering $\{V_1, \ldots, V_m\}$ (Prob. U). Since

$$\varphi = \varphi\varphi_1 + \cdots + \varphi\varphi_m$$

and since $\varphi\varphi_i$ is supported on some compact subset of V_i, $i = 1, \ldots, m$, it follows that $T\varphi = \sum_{i=1}^m T\varphi\varphi_i = 0$. Thus $T | U_0 = 0$, as was asserted.

This argument shows that for each distribution T on U there is a *smallest relatively closed subset* S of U such that $T | (U \backslash S) = 0$. The set S is called the *support* of T.

Example Q. Let us take for U an open interval (a, b) in $\mathbb{R}$ $(a < b)$, and suppose that f is a continuously differentiable function on U. If φ is a test function on U, then φ vanishes outside some closed subinterval $[c, d] \subset (a, b)$, and, integrating by parts, we obtain

$$T_{f'}(\varphi) = \int_c^d \varphi(t) f'(t) dt = [\varphi(d)f(d) - \varphi(c)f(c)] - \int_c^d \varphi'(t) f(t) dt$$

$$= - \int_c^d \varphi'(t) f(t) dt.$$

Thus, in brief, $T_{f'}(\varphi) = - T_f(\varphi')$. Inspired by this observation one defines the *distributional derivative* T' of an arbitrary distribution T on U by setting

$$T'(\varphi) = - T(\varphi'), \qquad \varphi \in \mathscr{C}_0^{(\infty)}(U).$$

It is easily verified that T', thus defined, is again a distribution on U. (If T is bounded on some nonempty compact subset K of U with respect to $\sigma_{K, m}$, then T' is bounded on K with respect to $\sigma_{K, m+1}$.) Hence *every distribution on U is infinitely differentiable* in that it may be differentiated as often as desired in the distributional sense. A formula for the nth distributional derivative of a distribution T on U is readily computed:

$$T^{(n)}(\varphi) = (-1)^n T(\varphi^{(n)}), \qquad \varphi \in \mathscr{C}_0^{(\infty)}(U).$$

Moreover, according to the discussion that motivated the definition, if T_f is a regular distribution associated with a function f in $\mathscr{C}^{(1)}(U)$, then

$$(T_f)' = T_{f'}.$$

An elementary induction argument shows that, in fact, if $f \in \mathscr{C}^{(m)}(U)$, then $(T_f)^{(k)} = T_{f^{(k)}}$ for every $k = 1, \ldots, m$.

What can be said about the derivative of a regular distribution T_f when f is *not* continuously differentiable on U? To shed some light on this question, let us suppose that α is a right-continuous function on U that is of bounded variation (Prob. 1I) on every closed subinterval of U (such a function is said to be locally of bounded variation on U; see Problem 10R). We shall show that if φ is a test function on U that is supported on the closed subinterval $[c, d]$, then the value of the distributional derivative $(T_\alpha)'$ at φ is given by

$$\int_c^d \varphi \, d\zeta_\alpha = \int_a^b \varphi \, d\zeta_\alpha$$

where ζ_α denotes the Stieltjes–Borel measure associated with α on the interval $[c, d]$ (see Example 8K). To this end we use the Fubini theorem (Prob. 9F). Let Q denote the square $[c, d] \times [c, d]$ equipped with the Borel measure $\mu \times \zeta_\alpha$, where μ denotes Lebesgue–Borel measure on $\mathbb{R}$. Consider the set $D = \{(t, u) \in Q : t < u\}$ and the integral $J = \int_D \varphi'(t) d(\mu \times \zeta_\alpha)(t, u)$. Clearly the hypotheses of the Fubini theorem are all satisfied here, so this double integral may be computed as an iterated integral in either order. If we integrate first with respect to u, we obtain

$$J = \int_c^d \left[\varphi'(t) \int_{(t, d)} d\zeta_\alpha(u) \right] dt = \int_c^d \varphi'(t)[\alpha(d) - \alpha(t)] dt,$$

and since $\int_c^d \varphi'(t) dt = 0$, this simplifies to

$$J = -\int_c^d \varphi'(t)\alpha(t) dt = -T_\alpha(\varphi') = (T_\alpha)'(\varphi).$$

If, on the other hand, we integrate first with respect to t, we obtain

$$J = \int_c^d \left[\int_c^u \varphi'(t) dt \right] d\zeta_\alpha(u) = \int_c^d \varphi(u) d\zeta_\alpha(u),$$

and our assertion is proved.

Thus the distributional derivative $(T_\alpha)'$ is given by integration with respect to a local Borel measure on U (see Problem 10R), and it follows by the Radon–Nikodym theorem (Th. 9.6) that $(T_\alpha)'$ is again a regular distribution on U if and only if the measure ζ_α is absolutely continuous with respect to μ on every closed subinterval of U.

Finally, let us consider the special case of the *Heaviside* function H defined on $\mathbb{R}$ as follows:

$$H(t) = \begin{cases} 0, & t < 0, \\ 1, & t \geq 0. \end{cases}$$

If φ is a test function on $\mathbb{R}$, then

$$-\int_{-\alpha}^{+\infty} \varphi'(t) H(t) dt = -\int_0^{+\infty} \varphi'(t) dt = \varphi(0).$$

Thus the distributional derivative of T_H is integration with respect to the Dirac mass concentrated at the origin (cf. Example 7I), which is, of course, the Borel measure on $\mathbb{R}$ associated with H.

PROBLEMS

A. If $\{s_n\}_{n=1}^{\infty}$ is an arbitrary sequence of positive real numbers, then $\|s_n e_n\|_p = s_n$ in (ℓ_p) for every value of p, $1 \le p < +\infty$ (see Example 11H). Hence if $\lim_n s_n = +\infty$, then no subsequence of the sequence $\{s_n e_n\}$ can be weakly convergent in (ℓ_p) for any value of p. Nevertheless, the countable set $S = \{s_n e_n : n \in \mathbb{N}\}$ may have weak accumulation points. Show, for example, that 0 belongs to the weak closure of S in (ℓ_p) if we set $s_n = n^{1/q}$, $n \in \mathbb{N}$, where $p > 1$ and q denotes the Hölder conjugate of p (Ex. 11B). (Hint: Show first that for each weak neighborhood V of 0 in (ℓ_p) there exists a single vector $a = \{\alpha_n\}_{n=0}^{\infty}$ in (ℓ_q) and a positive number ε such that V contains the set

$$\left\{ x = \{\xi_n\} \in (\ell_p) : \left| \sum_{n=0}^{\infty} \alpha_n \xi_n \right| < \varepsilon \right\}.$$

If V contained no point of S it would follow that $|\alpha_n| s_n \ge \varepsilon > 0$, and hence that $|\alpha_n| \ge \varepsilon/n^{1/q}$ for every n in $\mathbb{N}_0$.)

> The preceding exercise shows very clearly that the weak topology on (ℓ_p), $1 < p < +\infty$, does not satisfy the first axiom of countability, and is therefore not metrizable. As a matter of fact, the weak topology is never metrizable on an infinite dimensional Banach space (cf. Problems K and M below), and it follows that the convergence of infinite sequences cannot be used to characterize weak topologies (see Chapter 3). Nonetheless the study of weak convergence of sequences is frequently interesting and occasionally of critical importance. The following three problems concern this topic.

B. Let $\{x_n\}$ be a sequence of elements of the Banach space (c_0) of null sequences (sup norm). Prove that $\{x_n\}$ converges weakly to an element x of (c_0) if and only if it converges termwise to x and the numerical sequence $\{\|x_n\|_\alpha\}$ is bounded. Is the same true in the space (c) of all convergent sequences? (Hint: See Problem 14B.)

C. Let $\{x_n\}$ be a sequence of elements of (ℓ_p), $1 < p < +\infty$, such that the numerical sequence $\{\|x_n\|_p\}$ is bounded and such that the sequence $\{x_n\}$ converges termwise to some limit x. Show that $x \in (\ell_p)$ and that $\{x_n\}$ converges weakly to x.

D. Conditions for the weak convergence of a sequence in (ℓ_1) are dramatically different from those found in Problem C for a sequence in (ℓ_p), $1 < p < +\infty$. Show that if $\{x_n\}_{n=1}^{\infty}$ is a sequence in (ℓ_1) such that $\|x_n\| = 1$ for every n, then there exists a linear functional f in $(\ell_1)^*$ such that the sequence $\{f(x_n)\}$ does not converge to zero in $\mathbb{C}$. Use this fact to prove that a sequence $\{x_n\}$ converges weakly in (ℓ_1) if and only if it converges in norm. (Hint: Recall (Ex. 14B) that $(\ell_1)^* = (\ell_\infty)$. Let $\{x_n\}$ be a sequence of unit vectors in (ℓ_1), and let $x_n = \{\xi_m^{(n)}\}_{m=0}^{\infty}$ for each n. One may clearly assume that $\lim_n \xi_m^{(n)} = 0$ for every nonnegative integer m. Construct by induction a disjoint sequence $\{D_n\}$ of finite subsets of $\mathbb{N}_0$ and a corresponding strictly increasing sequence $\{k_n\}$ of positive integers such that

$$\sum_{m \in D_n} |\xi_m^{(k_n)}| > \frac{2}{3}, \qquad n \in \mathbb{N}.$$

For each n in $\mathbb{N}$ and for every m in D_n let ω_m be a complex number of modulus one such that $\omega_m \zeta_m^{(k_n)} = |\zeta_m^{(k_n)}|$, and set $\omega_m = 0$ if m does not belong to any of the sets D_n. Then $\{\omega_m\}_{m=0}^{\infty}$ is a unit vector in (ℓ_{∞}) such that

$$\left| \sum_{m=0}^{\infty} \omega_m \zeta_m^{(k_n)} \right| \geq \frac{1}{3}$$

for every positive integer n.)

E. Let $\mathscr{E}$ and $\mathscr{F}$ be Banach spaces, and let T be a bounded linear transformation of $\mathscr{E}$ into $\mathscr{F}$. Show that if Φ is an analytic $\mathscr{E}$-valued mapping on a domain Δ in $\mathbb{C}$, then $T \circ \Phi$ is an analytic $\mathscr{F}$-valued mapping on Δ, and verify that $(T \circ \Phi)' = T \circ \Phi'$.

F. Let $\mathscr{E}$ be a Banach space, and let $\{a_n\}_{n=0}^{\infty}$ be a sequence of "coefficients" in $\mathscr{E}$. Show that if the sequence $\{r^n a_n\}$ is bounded in $\mathscr{E}$ for some positive number r, then for an arbitrary complex number λ_0 the series

$$\sum_{n=0}^{\infty} (\lambda - \lambda_0)^n a_n \qquad (7)$$

is absolutely convergent in $\mathscr{E}$ for all $|\lambda - \lambda_0| < r$ (Prob. 11G). Conclude that, just as in the case of complex-valued functions, every power series of the form (7) possesses a *radius of convergence* R, $0 \leq R \leq +\infty$, such that (7) converges absolutely for $|\lambda - \lambda_0| < R$ and defines an $\mathscr{E}$-valued mapping on the disc $D_R(\lambda_0) = \{\lambda \in \mathbb{C} : |\lambda - \lambda_0| < R\}$. (If $R = +\infty$, then (7) converges everywhere on $\mathbb{C}$ and defines an *entire* $\mathscr{E}$-valued mapping.) Show that if a series of the form (7) has a positive radius of convergence R, then the $\mathscr{E}$-valued mapping Φ defined by

$$\Phi(\lambda) = \sum_{n=0}^{\infty} (\lambda - \lambda_0)^n a_n$$

on the disc $D_R(\lambda_0)$ is analytic on $D_R(\lambda_0)$. (Hint: If $f \in \mathscr{E}^*$, then $(f \circ \Phi)(\lambda) = \sum_{n=0}^{\infty} f(a_n)(\lambda - \lambda_0)^n$; use Example E.) Suppose, conversely, that Φ is a given analytic $\mathscr{E}$-valued mapping on some domain Δ in $\mathbb{C}$, and let λ_0 be a point of Δ. According to Example E, Φ is infinitely differentiable on Δ, so we may construct (formally) the Taylor series

$$\sum_{n=0}^{\infty} (\lambda - \lambda_0)^n \frac{\Phi^{(n)}(\lambda_0)}{n!}. \qquad (8)$$

of Φ at λ_0 (Ex. 5A). Show that this series actually converges to Φ on the largest open disc about λ_0 contained in Δ. Show, finally, that if Φ can be continued analytically onto some larger domain containing a disc $D_R(\lambda_0)$, then the radius of convergence of the Taylor series (8) is at least R. (Hint: If $f \in \mathscr{E}^*$, then $(f \circ \Phi)^{(n)} = f \circ \Phi^{(n)}$ for each nonnegative integer n, so $\sum_{n=0}^{\infty} f(\Phi^{(n)}(\lambda_0))(\lambda - \lambda_0)^n / n!$ is the Taylor series of the analytic function $f \circ \Phi$.)

G. Let x be an element of a unital Banach algebra $\mathscr{A}$. Show that if

$$s = \limsup_n \| x^n \|^{1/n} > 0,$$

then the spectrum $\sigma_{\mathscr{A}}(x)$ is not only contained in the closed disc $\{\lambda \in \mathbb{C} : |\lambda| \leq s\}$, but actually contains a point of the circle $\{\lambda \in \mathbb{C} : |\lambda| = s\}$. Conclude that the sequence $\{\|x^n\|^{1/n}\}_{n=1}^{\infty}$ is always convergent and that the spectral radius $r_{\mathscr{A}}(x)$ is given by the formula

$$r_{\mathscr{A}}(x) = \lim_n \|x^n\|^{1/n}.$$

Thus, in particular, if $\mathscr{E}$ is a Banach space and T is a bounded operator on $\mathscr{E}$, then

$$r(T) = \lim_n \|T^n\|^{1/n}.$$

(Hint: Review Problems 12M and 12P, and use Problem F.)

H. Let $\mathscr{E}$ be an infinite dimensional linear space, and let $\mathscr{L}$ be a linear manifold of linear functionals on $\mathscr{E}$. Show that in the topology induced on $\mathscr{E}$ by $\mathscr{L}$ every neighborhood of the origin contains an infinite dimensional linear submanifold of $\mathscr{E}$. Conclude that every neighborhood of an arbitrary vector x_0 in this topology contains a line through x_0. Conclude also that the topology induced by $\mathscr{L}$ is not induced by any norm on $\mathscr{E}$. (This is expressed by saying that the space $\mathscr{E}$ equipped with the topology induced by $\mathscr{L}$ is not *normable*.) (Hint: If $\{f_1, \ldots, f_p\}$ is any finite set of linear functionals on $\mathscr{E}$, then the linear manifold $\mathscr{N} = \{x \in \mathscr{E} : f_i(x) = 0, i = 1, \ldots, p\}$ is infinite dimensional.)

I. Let $\mathscr{E}$ be an infinite dimensional normed space, and let C be a closed bounded convex subset of $\mathscr{E}$. Show that C is the weak closure of the boundary ∂C (taken in the norm topology). Thus, for example, the closed unit ball $\mathscr{E}_1$ is the weak closure of the unit sphere $\{x \in \mathscr{E} : \|x\| = 1\}$. (Hint: Use Problem H.)

J. Let $f_1, \ldots, f_p$ be linear functionals on a linear space $\mathscr{E}$, and let

$$\mathscr{N} = \{x \in \mathscr{E} : f_i(x) = 0, i = 1, \ldots, p\}.$$

Prove that a linear functional g on $\mathscr{E}$ is a linear combination of $f_1, \ldots, f_p$ if and only if $g(\mathscr{N}) = (0)$. (Hint: $\mathscr{E}/\mathscr{N}$ is isomorphic to a linear submanifold of $\mathbb{C}^p$.) Use this fact to show that if M is a set of linear functionals on $\mathscr{E}$, then a linear functional g on $\mathscr{E}$ is continuous with respect to the topology inversely induced on $\mathscr{E}$ by M if and only if g is a linear combination of functionals in M. Thus, for example, if $\mathscr{E}$ is a normed space, then the weak* continuous linear functionals on $\mathscr{E}^*$ are precisely the ones in $j(\mathscr{E})$.

K. (i) Let $\mathscr{E}$ be a linear space, and let g and $f_1, \ldots, f_p$ be linear functionals on $\mathscr{E}$. Show that a necessary condition for the existence of two positive numbers δ and ε such that

$$U(f_1, \ldots, f_p; \delta) \subset U(g; \varepsilon) \qquad (9)$$

(see (2)) is that g be a linear combination of $f_1, \ldots, f_p$. Show also that if g is a linear combination of $f_1, \ldots, f_p$, then for each $\varepsilon > 0$ there is a $\delta > 0$ such that (9) holds.

(ii) Let $\mathscr{E}$ be a linear space, let $\mathscr{L}$ be a linear manifold of linear functionals on $\mathscr{E}$, and suppose given a neighborhood base at 0 in the topology induced by $\mathscr{L}$ consisting of sets of the form $U(f_1, \ldots, f_p; \varepsilon)$, $f_i \in \mathscr{L}$, $i = 1, \ldots, p$. Prove that the union of the finite sets $\{f_1, \ldots, f_p\}$ used in forming the neighborhood base must generate $\mathscr{L}$ algebraically. Conclude that the weight of $\mathscr{E}$ at each vector x with respect to the topology induced by $\mathscr{L}$ (Prob. 3A) is at least as great as the Hamel dimension of $\mathscr{L}$.

331

(iii) Conclude that if $\mathscr{E}$ is an infinite dimensional normed space, then the weak topology on $\mathscr{E}$ fails to satisfy the first axiom of countability, and hence is not metrizable. Show further that if $\mathscr{E}$ is complete, then the weak* topology on $\mathscr{E}*$ also fails to satisfy the first axiom of countability. (Hint: Recall Problem 11Q.)

> It is a surprising but useless consequence of Problems J and K that there is no weak neighborhood base at the origin of an infinite dimensional normed space $\mathscr{E}$ (and therefore none at any point of $\mathscr{E}$) that is simply ordered under inclusion. Indeed, suppose $\mathscr{V}$ is such a neighborhood base. Then each set V in $\mathscr{V}$ contains a special neighborhood $U = U(f_1, \ldots, f_p; \varepsilon)$. Using Problem J and the fact that $\mathscr{V}$ is a neighborhood base, it is easy to construct inductively a sequence $\{V_n\}$ of neighborhoods belonging to $\mathscr{V}$ and a sequence $\{U_n\}$ of special neighborhoods (2) such that
>
> $$V_1 \supset U_1 \supset V_2 \supset U_2 \supset \cdots \supset V_n \supset U_n \supset \cdots$$
>
> and such that, if $\mathscr{F}_n$ denotes the finite dimensional subspace of $\mathscr{E}*$ generated by the functionals used in defining U_n, then the sequence $\{\mathscr{F}_n\}$ is strictly increasing. It follows that $\bigcup_n \mathscr{F}_n$ is infinite dimensional, and therefore that $\bigcap_n V_n$ contains no neighborhood of the origin. But then, since $\mathscr{V}$ is nested, every V in $\mathscr{V}$ must include some V_n, which implies that the sequence $\{V_n\}$ is itself a neighborhood base, in contradiction of Problem K.

L. Let $\mathscr{E}$ be a separated locally convex space. Show that all hyperplanes and all closed half-spaces of $\mathscr{E}$ are also weakly closed (see Example I). Use this fact to show that every convex set in $\mathscr{E}$ that is closed in the given topology is also weakly closed, and hence that the weak closure of every convex subset of $\mathscr{E}$ coincides with its closure in the given topology. (Thus, in particular, if $\mathscr{E}$ is a normed space and C is a convex set in $\mathscr{E}$, then the weak closure and the norm closure of C are identical.) Is it true that all norm closed convex subsets of the dual $\mathscr{E}*$ of a normed space $\mathscr{E}$ are weak* closed?

M. Problem K shows that if $\mathscr{E}$ is an infinite dimensional normed space, then $\mathscr{E}$ is not metrizable in its weak topology, but it is possible for subsets of $\mathscr{E}$ to be (relatively) weakly metrizable. Let $\mathscr{E}$ be a normed space with the property that its dual space $\mathscr{E}*$ is separable, and let $\{f_n\}_{n=1}^\infty$ be a sequence dense in $\mathscr{E}*$. Then the sequence $\{\sigma_n = \sigma_{f_n}\}$ of pseudonorms induces a metrizable topology on $\mathscr{E}$ (Prop. 11.32). Show that this metrizable topology coincides with the weak topology on every closed ball $\mathscr{E}_r, r > 0$, and conclude that the (relative) weak topology on any bounded subset of $\mathscr{E}$ is metrizable. Show also that the (relative) weak topology satisfies the second axiom of countability on bounded subsets of $\mathscr{E}$. (Hint: The space $\mathscr{E}$ is separable too (Prob. 14R) and therefore satisfies the second axiom of countability, so there is a sequence $\{x_n\}_{n=1}^\infty$ in $\mathscr{E}_r$ that is norm dense in $\mathscr{E}_r$. Show that sets of the form

$$\left\{ x \in \mathscr{E}_r : |f_k(x - x_n)| < \frac{1}{m}, \ k = 1, \ldots, N \right\}$$

form a base for the relative weak topology on $\mathscr{E}_r$, where m, n, and N range independently over all positive integers.)

N. Let $\mathscr{E}$ be a separable Banach space, let $\{x_n\}_{n=1}^\infty$ be a sequence dense in $\mathscr{E}$, and let

$$\sigma_n(f) = |f(x_n)|, \qquad f \in \mathscr{E}*, \qquad n \in \mathbb{N}.$$

Show that the sequence $\{\sigma_n\}_{n=1}^{\infty}$ of pseudonorms on $\mathscr{E}^*$ induces the weak* topology on every ball $(\mathscr{E}^*)_r$, $r > 0$, and use this fact to prove that the (relative) weak* topology on $(\mathscr{E}^*)_r$ is metrizable. Conclude that the (relative) weak* topology is metrizable and satisfies the second axiom of countability on bounded subsets of $\mathscr{E}^*$. (Hint: The closed ball $(\mathscr{E}^*)_r$, $r > 0$, is compact in the weak* topology by Alaoglu's theorem (Th. 15.11).)

It is easy to give an explicit formula for a metric that induces the relative weak* topology on the ball $(\mathscr{E}^*)_r$ of Problem N. Indeed,

$$\rho(f, g) = \sum_{n=1}^{\infty} \frac{1}{2^n} \frac{|f(x_n) - g(x_n)|}{1 + |f(x_n) - g(x_n)|}, \qquad f, g \in (\mathscr{E}^*)_r,$$

defines such a metric (see Proposition 11.32). Similarly the formula

$$\rho(x, y) = \sum_{n=1}^{\infty} \frac{1}{2^n} \frac{|f_n(x - y)|}{1 + |f_n(x - y)|}, \qquad x, y \in \mathscr{E}_r,$$

defines explicitly a metric on the ball $\mathscr{E}_r$ of Problem M that induces the relative weak topology there.

O. Let $\mathscr{E}$ be a Banach space with the property that the dual $\mathscr{E}^*$ is separable. Show that a subset of $\mathscr{E}$ is weakly compact if and only if it is weakly sequentially compact (Prob. 3U). (Hint: If K is a subset of $\mathscr{E}$ that is either weakly compact or weakly sequentially compact, then K is (weakly) bounded, and the weak topology on K is therefore metrizable by Problem M. Use Problem 4Q.) Verify similarly that if $\mathscr{E}$ is separable, then a subset of $\mathscr{E}^*$ is weak* compact if and only if it is weak* sequentially compact.

There is a theorem more general and considerably deeper than the first assertion of Problem O that holds without any hypothesis of separability in an arbitrary Banach space. The following fact is known as the Eberlein–Šmul'yan theorem: *Let K be a subset of a Banach space $\mathscr{E}$. Then K is weakly compact if and only if it is weakly closed and weakly sequentially compact* [23; Part I, p. 430].

P. Let $\mathscr{E}$ be a Banach space and let M be a subset of $\mathscr{E}^*$ that spans it as a Banach space. Show that if a net $\{x_\lambda\}_{\lambda \in \Lambda}$ in $\mathscr{E}$ is bounded, and if $f(x_\lambda) \to f(x)$ in $\mathbb{C}$ for every f in M, then $\{x_\lambda\}$ converges weakly to x. Show, similarly, that if M is a subset of $\mathscr{E}$ that spans $\mathscr{E}$, and if $\{f_\lambda\}_{\lambda \in \Lambda}$ is a bounded net in $\mathscr{E}^*$ such that $f_\lambda(x) \to f(x)$ for every x in M, then $\lim_\lambda f_\lambda = f$ in the weak* topology on $\mathscr{E}^*$.

Q. Let us call a subset M of the dual $\mathscr{E}^*$ of a normed space $\mathscr{E}$ *weak* bounded* if $\{f(x): f \in M\}$ is a bounded subset of $\mathbb{C}$ for every vector x in $\mathscr{E}$. Prove that if $\mathscr{E}$ is complete, then every weak* bounded subset of $\mathscr{E}^*$ is bounded. Conclude, in particular, that if $\mathscr{E}$ is complete, then every weak* convergent sequence $\{f_n\}$ in $\mathscr{E}^*$ is bounded. Show also, by example, that the requirement of completeness cannot be omitted in this result.

R. It is customary to say that a net $\{x_\lambda\}_{\lambda \in \Lambda}$ in a normed space $\mathscr{E}$ is *weakly Cauchy* if the net $\{f(x_\lambda)\}$ is Cauchy in $\mathbb{C}$ for every f in $\mathscr{E}^*$. (Note that this is the same as saying that $\{x_\lambda\}$ is Cauchy with respect to all of the pseudonorms σ_f, $f \in \mathscr{E}^*$.) Similarly, a net $\{f_\lambda\}_{\lambda\in\Lambda}$ in $\mathscr{E}^*$ is *weak* Cauchy* if the net $\{f_\lambda(x)\}$ is Cauchy in $\mathbb{C}$ for every x in $\mathscr{E}$. As usual it is obvious that every weakly [weak*] convergent net is weakly [weak*] Cauchy. Following established custom, we define $\mathscr{E}$ to be *weakly complete* if every weakly Cauchy net in $\mathscr{E}$ is weakly convergent. Similarly, we define $\mathscr{E}^*$ to be *weak* complete* if every weak* Cauchy net in $\mathscr{E}^*$ is weak* convergent. As it turns out, however, these concepts are not very interesting.

  (i) Show that if F is *any* linear functional on $\mathscr{E}$, then there exists a net $\{f_\lambda\}$ in $\mathscr{E}^*$ such that $f_\lambda(x) \to F(x)$ for every x in $\mathscr{E}$, and verify that the net $\{f_\lambda\}$ is weak* Cauchy. Conclude that $\mathscr{E}^*$ is weak* complete when and only when $\mathscr{E}$ (and therefore $\mathscr{E}^*$) is finite dimensional. (Hint: The restriction of F to each finite dimensional subspace $\mathscr{F}$ of $\mathscr{E}$ is bounded on $\mathscr{F}$, and therefore possesses an extension $f_\mathscr{F}$ belonging to $\mathscr{E}^*$. Use the fact that the finite dimensional subspaces of $\mathscr{E}$ form a directed set in the inclusion ordering, and also the fact (Ex. 12A) that there exist unbounded linear functionals on $\mathscr{E}$ if $\mathscr{E}$ is infinite dimensional.)
  (ii) Follow the line of argument used in (i) to show that if φ is an *arbitrary* linear functional on $\mathscr{E}^*$, then there exists a net $\{x_\lambda\}$ in $\mathscr{E}$ such that $\lim_\lambda f(x_\lambda) = \varphi(f)$ for every f in $\mathscr{E}^*$. Conclude that $j(\mathscr{E})$ is weak* dense in the second dual $\mathscr{E}^{**}$, and also that $\mathscr{E}$ is weakly complete if and only if it is finite dimensional. (Hint: Recall Proposition 14.5.)
  (iii) Other conditions weaker than completeness are sometimes of interest. Show, for example, that for any normed space $\mathscr{E}$, every bounded weak* Cauchy net in $\mathscr{E}^*$ is weak* convergent. (Hint: Use Alaoglu's theorem and recall Problem 3R.) This fact is usually expressed by saying that $\mathscr{E}^*$ is *weak* boundedly complete*. Frame a definition of the concept of a *weakly boundedly complete* Banach space, and show that (ℓ_p) is weakly boundedly complete for all $1 < p < +\infty$.
  (iv) If $\mathscr{E}$ is any Banach space, then every weak* Cauchy *sequence* in $\mathscr{E}^*$ is weak* convergent. This fact is usually expressed by saying that $\mathscr{E}^*$ is *weak* sequentially complete*. Frame a definition of the concept of a *weakly sequentially complete* Banach space, and show that (c_0) does not satisfy the conditions of the definition.

S. Let $\mathscr{E}$ be the inductive limit of a net $\{\mathscr{M}_\lambda\}_{\lambda\in\Lambda}$ of locally convex topological linear spaces, and suppose the net $\{\mathscr{M}_\lambda\}$ is countably determined (Prob. 1P). Let $\{\lambda_n\}_{n=1}^\infty$ be a cofinal increasing sequence in Λ, and let U be an open absolutely convex neighborhood of 0 in $\mathscr{M}_{\lambda_1}$. According to the definition of inductive limits, there exists a sequence $\{U_n\}_{n=1}^\infty$ of sets in $\mathscr{E}$ such that $U_1 = U$ and such that U_n is an open absolutely convex neighborhood of 0 in $\mathscr{M}_{\lambda_n}$ with the property that $U_{n+1} \cap \mathscr{M}_{\lambda_n} \subset U_n$, $n \in \mathbb{N}$. Prove that the union $\bigcup_{n=1}^\infty U_n$ is a balanced absorbing set in $\mathscr{E}$, and deduce that the absolutely convex hull V of this union is simply the sum $\sum_{n\in\mathbb{N}} U_n$. Show also that $V \cap \mathscr{M}_{\lambda_1} = U$, and hence that $\mathscr{M}_{\lambda_1}$ is topologically embedded in $\mathscr{E}$, that is, is topologically a subspace of $\mathscr{E}$ as well as a linear submanifold of $\mathscr{E}$. Conclude that the submanifolds $\mathscr{M}_\lambda$ are all topologically embedded in $\mathscr{E}$. (Hint: Each vector $x \neq 0$ in V can be written uniquely in the form $x = x_1 + \cdots + x_n$ for some n, where $x_1 \in U$, $x_i \in \mathscr{M}_{\lambda_i} \backslash \mathscr{M}_{\lambda_{i-1}}$, $i = 2, \ldots, n$, and $x_n \neq 0$.)

T. Let $\{\mathscr{F}_\gamma\}_{\gamma\in\Gamma}$ be an indexed family of locally convex topological linear spaces, let $\mathscr{E}$ be a linear space, and suppose given for each index γ a linear transformation $T_\gamma : \mathscr{E} \to \mathscr{F}_\gamma$. The topology inversely induced on $\mathscr{E}$ by the family $\{T_\gamma\}$ is called a *projective limit* topology on $\mathscr{E}$, and $\mathscr{E}$ equipped with this topology is called the *projective limit* of the family $\{\mathscr{F}_\gamma\}$ via the linear transformations T_γ. Verify that the projective limit topology is a locally convex linear topology on $\mathscr{E}$ that is separated if and only if the family $\{T_\gamma\}$ is separating on $\mathscr{E}$. Verify also that if T is a linear transformation of some locally convex topological linear space $\mathscr{E}_0$ into $\mathscr{E}$, then T is continuous with respect to the projective limit topology on $\mathscr{E}$ if and only if every composition $T_\gamma \circ T$ is continuous.

The following problem introduces some technical equipment that is very useful in the theory of distributions. In particular, it is needed for the program set forth in Example P.

U. If φ is a continuously differentiable function on $\mathbb{R}$ that is supported on some closed interval $[c, d]$, then φ' is bounded on $\mathbb{R}$. Use this fact to show that the difference quotient $(\varphi(s + h) - \varphi(s))/h$ tends boundedly to $\varphi'(s)$ on $\mathbb{R}$ as h tends to zero. (Show, that is, that there exists a positive constant M such that $|(\varphi(s + h) - \varphi(s))/h| \leq M$ for all $h \neq 0$.) Conclude that if f is a locally integrable function on $\mathbb{R}$ (Ex. O), then the function

$$g(s) = \int_{\mathbb{R}} f(t)\varphi(s - t)dt$$

has derivative

$$g'(s) = \int_{\mathbb{R}} f(t)\varphi'(s - t)dt$$

at every point s of $\mathbb{R}$. (Hint: See Problem 7W.))

(i) Let r be a positive number and let φ_r be a nonnegative test function on $\mathbb{R}^n$ that is supported on the ball $(\mathbb{R}^n)_r = \{x \in \mathbb{R}^n : \|x\|_2 \leq r\}$ and satisfies the condition $\int_{\mathbb{R}^n} \varphi_r \, d\mu_n = 1$ (see Problem 11V). Show that if f is an arbitrary locally integrable complex-valued function on $\mathbb{R}^n$, then the function

$$g(x) = \int_{\mathbb{R}^n} f(y)\varphi_r(x - y)d\mu_n(y)$$

belongs to $\mathscr{C}^{(\infty)}(\mathbb{R}^n)$. (The function g is called the *convolution* of f and φ_r, and is denoted by $f * \varphi_r$.) Verify that g is real-valued if f is, and that if $c \leq f(y) \leq d$ a.e. $[\mu_n]$ on some ball $x + (\mathbb{R}^n)_r$, then $c \leq g(x) \leq d$. In particular, if f is constant and equal to c a.e. on $x + (\mathbb{R}^n)_r$, then $g(x) = c$. (More generally, if $f(y)$ lies in some disc $\{\lambda \in \mathbb{C} : |\lambda - \lambda_0| \leq a\}$ for almost every y in $x + (\mathbb{R}^n)_r$, then $|g(x) - \lambda_0| \leq a$.)

(ii) Let K be a nonempty compact subset of $\mathbb{R}^n$, and let ε be a positive number. Show that if F denotes the compact set $F = \{x \in \mathbb{R}^n : d(x, \mathbb{R}^n \backslash K) \geq \varepsilon\}$ and U the open set $U = \{x \in \mathbb{R}^n : d(x, K) < \varepsilon\}$, then there exists a test function φ on $\mathbb{R}^n$ such that

$$\chi_F \leq \varphi \leq \chi_U.$$

(iii) Let K be a nonempty compact subset of $\mathbb{R}^n$, and let $U_1, \ldots, U_m$ be an open covering of K. Show that there exist nonnegative test functions $\varphi_1, \ldots, \varphi_m$ on $\mathbb{R}^n$ such that

(1) $\varphi_1 + \cdots + \varphi_m \leq 1$ on $\mathbb{R}^n$,
(2) $\varphi_1(x) + \cdots + \varphi_m(x) = 1$ for every x in K,
(3) $\varphi_i(x) = 0$ for $x \notin U_i$, $i = 1, \ldots, m$.

(Hint: This is just an infinitely differentiable version of Example 3G, qv. One may assume the open sets $U_1, \ldots, U_m$ to be bounded, and it is convenient to do so. There exists a closed covering of $\mathbb{R}^n$ consisting of a closed set F_0 that is disjoint from K and compact sets $K_1, \ldots, K_m$ such that $K_i \subset U_i, i = 1, \ldots, m$ (see Example 3G). Let $\eta_0 = d(K, F_0)$, let $\eta_i = d(K_i, \mathbb{R}^n \backslash U_i), i = 1, \ldots, m$, and set $\varepsilon = \eta_0 \wedge \eta_1 \wedge \cdots \wedge \eta_m$. (If any set K_i is empty, we delete it from the list and set the corresponding φ_i equal to zero.) If $L_i = \{x \in \mathbb{R}^n : d(x, K_i) \le \varepsilon/2\}$, $E_0 = \{x \in \mathbb{R}^n : d(x, F_0) \le \varepsilon/2\}$, and if φ_r is chosen as in (i) with $r = \varepsilon/2$, then the convolutions $\psi_0 = \chi_{E_0} * \varphi_r$ and $\psi_i = \chi_{L_i} * \varphi_r, i = 1, \ldots, m$, are nonnegative test functions on $\mathbb{R}^n$ such that $\psi = \psi_0 + \psi_1 + \cdots + \psi_m \ge 1$ everywhere on $\mathbb{R}^n$, and it follows at once that the functions $\varphi_i = \psi_i/\psi, i = 1, \ldots, m$, satisfy all the stated requirements.)

V. Let U be a nonempty open subset of $\mathbb{R}^n$. For each index $i = 1, \ldots, n$ and for each distribution T on U we define the *distributional derivative* $D_i T$ by setting

$$D_i T(\varphi) = -T\left(\frac{\partial \varphi}{\partial x_i}\right)$$

for each test function φ on U (cf. Example Q). Show that $D_i T$ is again a distribution on U, and that D_i is a linear transformation of $\mathscr{C}_0^{(\infty)}(U)^*$ into itself, $i = 1, \ldots, n$. Show also that the linear transformations $D_1, \ldots, D_n$ commute in pairs, and hence that it makes sense to define the differential operator $D^{k_1, \ldots, k_n} = D_1^{k_1} \cdots D_n^{k_n}$ on $\mathscr{C}_0^{(\infty)}(U)^*$. (If $k_i = 0$ then $D_i^{k_i}$ is taken, as usual, to be the identity operator.) Verify that

$$D^{k_1, \ldots, k_n}T(\varphi) = (-1)^{k_1 + \cdots + k_n}T(D^{k_1, \ldots, k_n}\varphi), \qquad \varphi \in \mathscr{C}_0^{(\infty)}(U).$$

Verify finally that if $f \in \mathscr{C}^{(m)}(U)$, then, for any nonnegative integers $k_1, \ldots, k_n$ such that $k_1 + \cdots + k_n \le m$, the distributional derivative $D^{k_1, \ldots, k_n}T_f$ coincides with the regular distribution associated with the (continuous) function $D^{k_1, \ldots, k_n}f$.

W. Let U be a nonempty open subset of $\mathbb{R}^n$, and let f be a locally integrable function on U (see Example O). Prove that there exists a largest open subset V of U such that $f = 0$ a.e. $[\mu_n]$ on V, and show that $U \backslash V$ coincides with the support of the regular distribution T_f on U (cf. Example P).

X. Let U be a nonempty open subset of $\mathbb{R}^n$. If ν is a local complex Borel measure on U (see Problem 10R), then

$$T_\nu(\varphi) = \int_U \varphi \, d\nu, \qquad \varphi \in \mathscr{C}_0^{(\infty)}(U),$$

defines a distribution T_ν on U. If ν is *absolutely continuous* with respect to μ_n, that is, if it vanishes on every topologically bounded Borel set in U that has Lebesgue measure zero, then ν is the indefinite integral with respect to μ_n of some locally integrable function f on U (see Theorem 9.6). In this event the distribution T_ν coincides with the regular distribution T_f. If ν is not absolutely continuous $[\mu_n]$, then T_ν is not a regular distribution. Thus integration with respect to the Dirac mass δ concentrated at the origin is a distribution T_δ on $\mathbb{R}$ that is *singular*, i.e., not regular (see Example Q). Show that $(T_\delta)'$ (the distributional derivative of T_δ) is a distribution on $\mathbb{R}$ that is neither regular nor of the form T_ν for any measure ν. (Hint: Consider the support of $(T_\delta)'$; see Example P.)

Duality 16

In this chapter we continue our investigation of the interplay between a separated locally convex topological linear space $\mathscr{E}$ and its dual $\mathscr{E}^*$. The matters to be discussed here, however, are most useful and most easily understood in the context of Banach spaces (where a natural topology, viz., the norm topology, exists on $\mathscr{E}^*$); accordingly, we shall focus our attention principally on that important special case. However, in order not to restrict unnecessarily our treatment of duality theory, we shall formulate both theorems and definitions in the general context of locally convex spaces whenever that is convenient. The following concept is of basic importance.

Definition. Let $\mathscr{E}$ be a normed space with dual space $\mathscr{E}^*$, and let j denote the natural embedding of $\mathscr{E}$ in the second dual $\mathscr{E}^{**}$ (Prop. 15.6). The space $\mathscr{E}$ is said to be *reflexive* if j maps $\mathscr{E}$ onto $\mathscr{E}^{**}$, i.e., if the only bounded linear functionals on $\mathscr{E}^*$ are those defined by elements x of $\mathscr{E}$ according to the formula

$$f \to f(x), \qquad f \in \mathscr{E}^*.$$

If $\mathscr{E}$ is reflexive then j is an isometric isomorphism between $\mathscr{E}$ and $\mathscr{E}^{**}$ (Prop. 15.6), and it follows that $\mathscr{E}$ is necessarily complete (Prop. 12.7). A reflexive Banach space $\mathscr{E}$ may be viewed as being identical with its own second dual (that is, $\mathscr{E}$ and $\mathscr{E}^{**}$ may be identified via the natural embedding j) so that $\mathscr{E}$ and $\mathscr{E}^*$ become the dual spaces of one another. Reflexive Banach spaces constitute a special, particularly well-behaved, class of Banach spaces.

Example A. Let $\mathscr{E}$ be a finite dimensional Banach space, and let $\{x_j\}_{j=1}^n$ be a basis for $\mathscr{E}$. For each $i = 1, \ldots, n$, there exists a unique (bounded)

337

linear functional f_i on $\mathscr{E}$ such that $f_i(x_j) = \delta_{ij}, j = 1, \ldots, n$ (Ex. 2B), and it is easy to see that $\{f_i\}_{i=1}^n$ is a basis for $\mathscr{E}^*$. (If $f \in \mathscr{E}^*$ then $f(x_i) = \alpha_i, i = 1, \ldots, n$, if and only if $f = \sum_{i=1}^n \alpha_i f_i$. The basis $\{f_i\}_{i=1}^n$ is called the basis *dual to* the given basis $\{x_j\}_{j=1}^n$.) It follows that dim $\mathscr{E}^* = n = $ dim $\mathscr{E}$. Hence, by the same token, dim $\mathscr{E}^{**} = n$ also, and since the natural embedding j of $\mathscr{E}$ into $\mathscr{E}^{**}$ has rank n (Prob. 2G), it follows that j is onto. Thus every finite dimensional Banach space is reflexive.

Example B. If $1 < p < +\infty$, and if we use the formula set forth in Example 14C to identify $(\ell_p)^*$ with (ℓ_q), where q denotes the Hölder conjugate of p, then we must simultaneously identify $(\ell_q)^* = (\ell_p)^{**}$ with (ℓ_p). It is not very difficult to check that under these identifications the natural embedding of (ℓ_p) in itself becomes the identity mapping on (ℓ_p). (We shall return to this point in Example H.) Thus (ℓ_p) is reflexive for $1 < p < +\infty$. On the other hand, the dual of (ℓ_1) may be identified with (ℓ_∞) (Ex. 14B), and the latter is not a separable space (Ex. 11H), whence it follows that $(\ell_1)^{**} = (\ell_\infty)^*$ is not separable either (Prob. 14R). Since (ℓ_1) itself is separable, this shows that (ℓ_1) is *irreflexive*, i.e., not reflexive.

Example C. The dual of (c_0) may be identified in a natural way with (ℓ_1) (see Problem 14B), so there is also a natural way of identifying $(c_0)^{**}$ with $(\ell_1)^* = (\ell_\infty)$. Once again it is not difficult to verify that under these identifications the natural embedding of (c_0) in (ℓ_∞) becomes the inclusion mapping. Thus (c_0) is likewise irreflexive.

We have already employed the notion of the weak topology on an arbitrary separated locally convex topological linear space $\mathscr{E}$ (cf. Example 15I). It is also convenient to introduce the appropriate generalization of the weak* topology on the dual space $\mathscr{E}^*$.

Definition. Let $\mathscr{E}$ be a separated locally convex topological linear space, and let $\mathscr{E}^*$ be its dual. For each vector x in $\mathscr{E}$ the mapping $F_x(f) = f(x), f \in \mathscr{E}^*$, is a linear functional on $\mathscr{E}^*$, and the topology induced on $\mathscr{E}^*$ by the collection of all linear functionals of the form $F_x, x \in \mathscr{E}$, is known as the *weak* topology* on $\mathscr{E}^*$ (or as the topology *induced on $\mathscr{E}^*$ by $\mathscr{E}$*).

It is obvious that this topology on $\mathscr{E}^*$ coincides with the weak* topology introduced in Chapter 15 when $\mathscr{E}$ is a normed space, and that the mapping assigning the functional F_x to each vector x in $\mathscr{E}$ (the natural embedding j in the case of a normed space) is, in general, a linear space isomorphism of $\mathscr{E}$ into the full algebraic dual of $\mathscr{E}^*$ (Prop. 14.13). It should also be noted that the weak* topology on $\mathscr{E}^*$ enjoys all of the properties set forth in Proposition 15.10.

Definition. If M is any nonempty subset of a locally convex topological linear space $\mathscr{E}$, then the *annihilator* of M is the set

$$M^a = \{f \in \mathscr{E}^* : f(M) = (0)\}$$

consisting of all the continuous linear functionals on $\mathscr{E}$ that vanish identically on M. Dually, if M is a nonempty subset of $\mathscr{E}^*$, then the *preannihilator* of M is the set ${}^aM$ of all those vectors x in $\mathscr{E}$ such that $f(x) = 0$ for every f in M. (By convention the empty subset of $\mathscr{E}$ is assigned $\mathscr{E}^*$ as its annihilator; similarly, the preannihilator of the empty subset of $\mathscr{E}^*$ is taken to be the entire space $\mathscr{E}$.)

The following two propositions and accompanying corollary give a summary of the elementary properties of annihilators and preannihilators. Proposition 16.1 is self-evident and requires no proof; it is recorded here solely for convenience of reference. The lack of symmetry in the formulation of Proposition 16.2 stems from the fact that every closed convex subset of a locally convex space $\mathscr{E}$ is also weakly closed; see Problem 15L.

Proposition 16.1. *Let $\mathscr{E}$ be a locally convex topological linear space. If M and N are subsets of $\mathscr{E}$ such that $M \subset N$, then $M^a \supset N^a$. Likewise, if $\{M_\gamma\}_{\gamma \in \Gamma}$ is a nonempty indexed family of subsets of $\mathscr{E}$, then*

$$\left(\bigcup_\gamma M_\gamma \right)^a = \bigcap_\gamma M_\gamma^a.$$

Dually, if M and N are subsets of $\mathscr{E}^*$ such that $M \subset N$, then ${}^aM \supset {}^aN$, and if $\{M_\gamma\}_{\gamma \in \Gamma}$ is a nonempty indexed family of subsets of $\mathscr{E}^*$, then

$$^a\left(\bigcup_\gamma M_\gamma \right) = \bigcap_\gamma {}^aM_\gamma.$$

Proposition 16.2. *If M is a subset of a locally convex topological linear space $\mathscr{E}$, then the annihilator M^a is a weak* closed linear submanifold of $\mathscr{E}^*$, and the preannihilator ${}^a(M^a)$ of M^a is the smallest closed linear manifold in $\mathscr{E}$ containing M. If M is a subset of $\mathscr{E}^*$, then ${}^aM$ is a closed linear manifold in $\mathscr{E}$, and the annihilator $({}^aM)^a$ of ${}^aM$ is the smallest weak* closed linear manifold in $\mathscr{E}^*$ containing M.*

PROOF. All parts of the proposition are trivially valid for $M = \varnothing$; accordingly, we assume in what follows that M is nonempty. For each single vector x in $\mathscr{E}$, the annihilator $\{x\}^a = \{f \in \mathscr{E}^* : f(x) = 0\}$ coincides with the null space of the functional F_x defined on $\mathscr{E}^*$ by x. Thus $\{x\}^a$ is a weak* closed linear manifold, and so therefore is

$$M^a = \bigcap_{x \in M} \{x\}^a,$$

whenever $M \subset \mathscr{E}$. Similarly, if $f \in \mathscr{E}^*$ then $^a\{f\}$ is just the kernel of f, whence it is clear that

$$^aM = \bigcap_{f \in M} {}^a\{f\}$$

is a closed linear submanifold of $\mathscr{E}$ whenever $M \subset \mathscr{E}^*$.

Suppose now that M is a subset of $\mathscr{E}$ and that $\mathscr{M}$ is the smallest closed linear manifold in $\mathscr{E}$ that contains M. From what has already been shown it is clear that $\mathscr{M} \subset {}^a(M^a)$. On the other hand, if x_0 is any vector in $\mathscr{E}$ that does not belong to $\mathscr{M}$, then there is an element f_0 of M^a such that $f_0(x_0) \neq 0$ (Prop. 14.13). But then $x_0 \notin {}^a(M^a)$, and it follows that $\mathscr{M}$ and $^a(M^a)$ coincide.

Similarly, if $M \subset \mathscr{E}^*$ and $\mathscr{M}$ denotes the smallest weak* closed linear submanifold of $\mathscr{E}^*$ that contains M, then it is clear that $\mathscr{M} \subset ({}^aM)^a$. But again, if f_0 is any element of $\mathscr{E}^*$ that does not belong to $\mathscr{M}$, then there exists a vector x_0 in ^aM such that $f_0(x_0) \neq 0$. Thus $f_0 \notin ({}^aM)^a$, and it follows that $\mathscr{M} = ({}^aM)^a$. (Recall (Prob. 15J) that the weak* continuous linear functionals on $\mathscr{E}^*$ are precisely the ones of the form $F_x, x \in \mathscr{E}$.) $\qquad \square$

Corollary 16.3. *If M is an arbitrary subset of a locally convex space $\mathscr{E}$, then $(({}^a(M^a))^a = M^a$. Furthermore if $\mathscr{L}$ is a linear manifold in $\mathscr{E}$, then $^a(\mathscr{L}^a)$ coincides with the closure of $\mathscr{L}$. In particular, if $\mathscr{M}$ is a closed linear manifold in $\mathscr{E}$, then $^a(\mathscr{M}^a) = \mathscr{M}$. Similarly, if M is an arbitrary subset of $\mathscr{E}^*$, then $^a(({}^aM)^a) = {}^aM$. Likewise if $\mathscr{L}$ is a linear manifold in $\mathscr{E}^*$, then $({}^a\mathscr{L})^a$ coincides with the weak* closure of $\mathscr{L}$, and $({}^a\mathscr{L})^a = \mathscr{L}$ if and only if $\mathscr{L}$ is itself weak* closed.*

The following assertions are nothing more than paraphrases of a portion of the foregoing results in the important special case in which $\mathscr{E}$ is a Banach space.

Proposition 16.4. *If M is a subset of a Banach space $\mathscr{E}$, then the annihilator M^a is a weak* closed subspace of $\mathscr{E}^*$, and $^a(M^a)$ is the subspace of $\mathscr{E}$ spanned by M. If M is a subset of $\mathscr{E}^*$, then ^aM is a subspace of $\mathscr{E}$, and $({}^aM)^a$ is the smallest weak* closed subspace of $\mathscr{E}^*$ containing M.*

Corollary 16.5. *If $\mathscr{L}$ is a linear submanifold of a Banach space $\mathscr{E}$, then $^a(\mathscr{L}^a) = \mathscr{L}^-$. In particular, if $\mathscr{M}$ is a subspace of $\mathscr{E}$, then $^a(\mathscr{M}^a) = \mathscr{M}$. Similarly, if $\mathscr{L}$ is a linear submanifold of $\mathscr{E}^*$, then $({}^a\mathscr{L})^a$ coincides with the weak* closure of $\mathscr{L}$, and $({}^a\mathscr{L})^a = \mathscr{L}$ if and only if $\mathscr{L}$ is itself a weak* closed subspace of $\mathscr{E}^*$.*

Example D. Let Δ be a domain in the complex plane, let $\mathscr{E}$ be a Banach space, and let Φ be an analytic $\mathscr{E}$-valued mapping defined on Δ (see Example 15E). Suppose $\mathscr{M}$ is a subspace of $\mathscr{E}$ and $\Phi(\lambda) \in \mathscr{M}$ for all scalars λ in some subset

M of Δ. Then $f \circ \Phi = 0$ on M for every f in $\mathcal{M}^a$. If M has any points of accumulation in Δ, it follows that $f \circ \Phi = 0$ identically on Δ for every f in $\mathcal{M}^a$ (Th. 5.2), and hence that the entire range $\Phi(\Delta)$ belongs to $^a(\mathcal{M}^a) = \mathcal{M}$. Thus if $\Phi(\lambda) \in \mathcal{M}$ for every λ in some subset of Δ possessing a single point of accumulation in Δ, then the entire range of Φ is contained in $\mathcal{M}$.

Suppose now, once again, that $\mathcal{E}$ is a separated locally convex space, and let $\mathcal{M}$ denote a closed linear submanifold of $\mathcal{E}$. As is well known (cf. Example 2L), a linear functional f in $\mathcal{E}^*$ can be factored through $\mathcal{E}/\mathcal{M}$ if and only if $\mathcal{M}$ is contained in the kernel of f, that is, if and only if $f \in \mathcal{M}^a$. Moreover, if f does belong to $\mathcal{M}^a$, and if $\hat{f}$ denotes the unique linear functional on $\mathcal{E}/\mathcal{M}$ such that $f = \hat{f} \circ \pi$, where π denotes the natural projection of $\mathcal{E}$ onto $\mathcal{E}/\mathcal{M}$, then $\hat{f}$ is clearly a continuous linear functional on $\mathcal{E}/\mathcal{M}$ (since π is an open mapping; recall Proposition 11.22). Finally, if $\hat{f}$ denotes an arbitrary element of $(\mathcal{E}/\mathcal{M})^*$, then $f = \hat{f} \circ \pi$ belongs to $\mathcal{M}^a$. Thus if we write $\alpha(\hat{f}) = \hat{f} \circ \pi$, $\hat{f} \in (\mathcal{E}/\mathcal{M})^*$, then α maps $(\mathcal{E}/\mathcal{M})^*$ onto $\mathcal{M}^a$, and it is a triviality to verify that α is, in fact, an isomorphism between these two spaces.

Consider now the situation when $\mathcal{E}$ is a Banach space. What more can then be said about α? The answer is immediate and is contained, in fact, in an earlier exercise (cf. Problem 12F). The spaces $(\mathcal{E}/\mathcal{M})^*$ and $\mathcal{M}^a$ are themselves Banach spaces, and $\|\alpha(\hat{f})\| = \|\hat{f}\|$ for every $\hat{f}$ in $(\mathcal{E}/\mathcal{M})^*$. Briefly, the mapping α is an *isometric isomorphism*—a Banach space isomorphism—between $(\mathcal{E}/\mathcal{M})^*$ and $\mathcal{M}^a$. This result, formulated below as Proposition 16.6, is typical of those theorems in the duality theory of Banach spaces that do not have any obvious counterparts in the general context of locally convex spaces.

Proposition 16.6. *Let $\mathcal{E}$ be a Banach space, let $\mathcal{M}$ be a subspace of $\mathcal{E}$, and let π denote the natural projection of $\mathcal{E}$ onto $\mathcal{E}/\mathcal{M}$. Then the mapping α of $(\mathcal{E}/\mathcal{M})^*$ onto $\mathcal{M}^a$ defined by setting $\alpha(\hat{f}) = \hat{f} \circ \pi$, $\hat{f} \in (\mathcal{E}/\mathcal{M})^*$, is an isometric isomorphism of $(\mathcal{E}/\mathcal{M})^*$ onto $\mathcal{M}^a$.*

The isomorphism α between $(\mathcal{E}/\mathcal{M})^*$ and $\mathcal{M}^a$ defined in Proposition 16.6 is known as the *natural* isomorphism between these two spaces. Similarly, the mapping $\hat{\beta}$ of the following theorem is the *natural isomorphism* of $\mathcal{E}^*/\mathcal{M}^a$ onto $\mathcal{M}^*$.

Proposition 16.7. *Let $\mathcal{E}$ be a locally convex space, and let $\mathcal{M}$ be a closed linear submanifold of $\mathcal{E}$. Then the mapping $\beta(f) = f|\mathcal{M}$ is a linear transformation of $\mathcal{E}^*$ onto $\mathcal{M}^*$ having the annihilator $\mathcal{M}^a$ for its kernel. Hence there exists a unique one-to-one linear transformation $\hat{\beta}$ of $\mathcal{E}^*/\mathcal{M}^a$ onto $\mathcal{M}^*$ with the property that $\hat{\beta}(f + \mathcal{M}^a) = f|\mathcal{M}$ for every f in $\mathcal{E}^*$. When $\mathcal{E}$ is a Banach space, this mapping $\hat{\beta}$ is an isometric isomorphism.*

PROOF. It is obvious that β is a linear transformation of $\mathcal{E}^*$ into $\mathcal{M}^*$, and that the null space of β coincides with $\mathcal{M}^a$. Moreover, it is an immediate consequence of Proposition 14.13 that β maps $\mathcal{E}^*$ onto $\mathcal{M}^*$. Thus setting

$\hat{\beta}(f + \mathcal{M}^a) = f \mid \mathcal{M}$ defines a linear isomorphism of $\mathscr{E}^*/\mathcal{M}^a$ onto $\mathcal{M}^*$. Suppose next that $\mathscr{E}$ is a Banach space. Then, according to Problem 12F, $\hat{\beta}$ is bounded with $\| \hat{\beta} \| \leq 1$. On the other hand, if f is any element of $\mathcal{M}^*$, then by the Hahn–Banach theorem (Th. 14.3) there exists an extension f_0 of f to the entire space $\mathscr{E}$ such that $\| f_0 \| = \| f \|$, and for this functional we clearly have $\hat{\beta}(f_0 + \mathcal{M}^a) = f$. But then $\| f \| = \| f_0 \| \geq \| f_0 + \mathcal{M}^a \| \geq \| f \|$ since $\| \hat{\beta} \| \leq 1$, which shows that $\hat{\beta}$ is an isometry.   □

Propositions 16.6 and 16.7 show that subspaces and quotient spaces of Banach spaces are dual to one another. There is also a duality for linear transformations, and, here again, the basic definitions make sense in the general context of locally convex spaces. To see how this goes, let $\mathscr{E}$ and $\mathscr{F}$ be locally convex topological linear spaces, and let T be a continuous linear transformation of $\mathscr{E}$ into $\mathscr{F}$. Then for each functional f in the dual $\mathscr{F}^*$ the composition $f \circ T$ is an element of $\mathscr{E}^*$. Let us write f^* for this functional, so that

$$f^*(x) = f(Tx), \qquad x \in \mathscr{E}, \qquad f \in \mathscr{F}^*.$$

Thus, setting $T^*f = f^*$, $f \in \mathscr{F}^*$, defines a mapping T^* of $\mathscr{F}^*$ into $\mathscr{E}^*$, and it is a triviality to verify that T^* is a linear transformation.

Definition. With $\mathscr{E}$, $\mathscr{F}$ and T as above, the linear transformation T^* is called the *adjoint* of T.

Proposition 16.8. *Let $\mathscr{E}$ and $\mathscr{F}$ be separated locally convex topological linear spaces, and let $\mathscr{L}(\mathscr{E}, \mathscr{F})$ denote the linear space of all continuous linear transformations of $\mathscr{E}$ into $\mathscr{F}$. Then the mapping $T \to T^*$ is a linear space isomorphism of $\mathscr{L}(\mathscr{E}, \mathscr{F})$ into the linear space of all linear transformations of $\mathscr{F}^*$ into $\mathscr{E}^*$. Moreover, if T is a continuous linear transformation of $\mathscr{E}$ into $\mathscr{F}$ and S is a continuous linear transformation of $\mathscr{F}$ into some third separated locally convex space $\mathscr{G}$, then $(ST)^* = T^*S^*$. Finally, the adjoint $(1_\mathscr{E})^*$ of the identity mapping on $\mathscr{E}$ is the identity mapping $1_{\mathscr{E}^*}$.*

PROOF. The verification that the mapping $T \to T^*$ is a linear transformation is routine, and will be left to the reader. To show that it is an isomorphism of $\mathscr{L}(\mathscr{E}, \mathscr{F})$ into the space of all linear transformations of $\mathscr{F}^*$ into $\mathscr{E}^*$ it suffices to show that it has trivial kernel. Suppose, accordingly, that $T^* = 0$. Then for every f in $\mathscr{F}^*$ and x in $\mathscr{E}$ we have $(T^*f)(x) = f(Tx) = 0$. Since f is arbitrary, this implies that $Tx = 0$ (Prop. 14.13); since x is also arbitrary, it follows that $T = 0$.

Next let S and T be as in the statement of the proposition. If x is an arbitrary vector in $\mathscr{E}$ and f an arbitrary functional in $\mathscr{G}^*$, then

$$(T^*S^*f)(x) = (S^*f)(Tx) = f(STx) = ((ST)^*f)(x).$$

Thus $T^*S^* = (ST)^*$. Finally, if f is an element of $\mathscr{E}^*$, then $((1_\mathscr{E})^*f)(x) = f(x)$ for every x in $\mathscr{E}$, so $(1_\mathscr{E})^*$ is the identity mapping on $\mathscr{E}^*$.   □

Example E. The most general (continuous) linear functional on $\mathbb{C}^n$ is of the form f_y, where $y = (\eta_1, \ldots, \eta_n)$ is itself an element of $\mathbb{C}^n$ and $f_y(x) = \sum_{i=1}^n \xi_i \eta_i$ for each $x = (\xi_1, \ldots, \xi_n)$. Moreover, in this notation, the basis for $(\mathbb{C}^n)^*$ dual to the natural basis $\{e_1, \ldots, e_n\}$ (Ex. 2B) is the basis $\{f_i = f_{e_i}\}_{i=1}^n$ (cf. Example A).

Suppose now that A is an $m \times n$ complex matrix, and let T denote the (continuous) linear transformation of $\mathbb{C}^n$ into $\mathbb{C}^m$ defined by A, so that $T(\xi_1, \ldots, \xi_n) = (\zeta_1, \ldots, \zeta_m)$, where

$$\begin{pmatrix} \zeta_1 \\ \vdots \\ \zeta_m \end{pmatrix} = A \begin{pmatrix} \xi_1 \\ \vdots \\ \xi_n \end{pmatrix}$$

for each $(\xi_1, \ldots, \xi_n)$ in $\mathbb{C}^n$ (Prob. 2H). Then A is the matrix of T with respect to the natural bases for $\mathbb{C}^m$ and $\mathbb{C}^n$. Moreover, if $y = (\eta_1, \ldots, \eta_m)$ belongs to $\mathbb{C}^m$ and $x = (\xi_1, \ldots, \xi_n)$ to $\mathbb{C}^n$, then

$$f_y(Tx) = (T^*f_y)(x) = \sum_{i=1}^m \sum_{j=1}^n \eta_i \alpha_{ij} \xi_j.$$

A straightforward calculation based on this formula shows that if $B = (\beta_{ij})$ denotes the matrix of T^* with respect to the bases dual to the natural bases for $\mathbb{C}^m$ and $\mathbb{C}^n$, respectively, then $\beta_{ij} = \alpha_{ji}$. In other words, $B = A^t$, the transpose of A (Ch. 2, p. 21).

> This observation suggests that it might be more appropriate to call T^* the *transpose* of T, and this is, in fact, sometimes done, most notably in [12]. The terminology employed here, while somewhat unfortunate, is nevertheless quite standard in the literature.

As usual, things go much better when there are norms around.

Proposition 16.9. *If $\mathscr{E}$ and $\mathscr{F}$ are Banach spaces, then the mapping that assigns to each bounded linear transformation T in $\mathscr{L}(\mathscr{E}, \mathscr{F})$ its adjoint T^* is an isometric isomorphism of $\mathscr{L}(\mathscr{E}, \mathscr{F})$ into $\mathscr{L}(\mathscr{F}^*, \mathscr{E}^*)$. Hence, when $\mathscr{E} = \mathscr{F}$, the mapping $T \to T^*$ is an isometric linear space isomorphism of the Banach algebra $\mathscr{L}(\mathscr{E})$ into $\mathscr{L}(\mathscr{E}^*)$ such that $(1_\mathscr{E})^* = 1_{\mathscr{E}^*}$ and $(ST)^* = T^*S^*$ for all S, T in $\mathscr{L}(\mathscr{E})$. (Such a mapping may be called a* unital isometric Banach algebra anti-isomorphism.)*

PROOF. All that is needed is to prove that T^* is bounded and $\|T^*\| = \|T\|$ for every T in $\mathscr{L}(\mathscr{E}, \mathscr{F})$, for everything else has already been established. Moreover, since $\|(T^*f)(x)\| \leq \|f\| \|T\| \|x\|$ for every f in $\mathscr{F}^*$ and x in $\mathscr{E}$, it is clear that T^* is bounded and that $\|T^*\|$ cannot exceed $\|T\|$. On the other hand, for any given T in $\mathscr{L}(\mathscr{E}, \mathscr{F})$ and $\varepsilon > 0$ there is a vector x_0 in $\mathscr{E}$ with $\|x_0\| = 1$ such that $\|Tx_0\| > \|T\| - \varepsilon$, and there exists a functional f_0 in $\mathscr{F}^*$ with $\|f_0\| = 1$ such that $f_0(Tx_0) = \|Tx_0\|$ (Cor. 14.11). Thus $(T^*f_0)(x_0) > \|T\| - \varepsilon$, whence it follows that $\|T^*\| > \|T\| - \varepsilon$, and, since ε is an arbitrary positive number, this proves that $\|T^*\| = \|T\|$. $\square$

Corollary 16.10. *If T is an invertible element of $\mathscr{L}(\mathscr{E}, \mathscr{F})$, then T^* is invertible in $\mathscr{L}(\mathscr{F}^*, \mathscr{E}^*)$ and $(T^*)^{-1} = (T^{-1})^*$.*

Corollary 16.11. *If T is an isometric isomorphism of $\mathscr{E}$ onto $\mathscr{F}$, then T^* is an isometric isomorphism of $\mathscr{F}^*$ onto $\mathscr{E}^*$.*

PROOF. Since T is an isometric isomorphism we have $\|T\| = \|T^{-1}\| = 1$, from which it follows, according to Proposition 16.9 and Corollary 16.10, that $\|T^*\| = \|(T^*)^{-1}\| = 1$. Hence $\|f\| \le \|T^*f\| \le \|f\|$ for every f in $\mathscr{F}^*$, which shows that T^* is an isometry. □

Example F. If $1 < p < +\infty$, and if V denotes the unilateral shift on (ℓ_p) (see Example 12B), then V^* is the backward shift

$$V^*\{\xi_0, \xi_1, \ldots\} = \{\xi_1, \xi_2, \ldots\}$$

on (ℓ_q), where q denotes the Hölder conjugate of p (cf. Example 12M). Similarly, if U denotes the bilateral shift on $(\ell_p)^*$, $1 < p < +\infty$ (see Example 12E), then U^* is the backward bilateral shift on $(\ell_q)^*$. That is, if $\{\xi_n\}_{n=-\infty}^{+\infty}$ is an arbitrary element of $(\ell_q)^*$, then

$$U^*\{\cdots \xi_{-1}, [\xi_0], \xi_1, \cdots\} = \{\cdots \xi_0, [\xi_1], \xi_2, \cdots\}$$

(cf. Problem 12L). Likewise, the adjoint of the unilateral shift on (ℓ_1) is the backward shift on (ℓ_∞), and the adjoint of the bilateral shift on the space $(\ell_1)^*$ is the backward bilateral shift on $(\ell_\infty)^*$.

Example G. It follows from Proposition 16.8 and Corollary 16.10 that if $T \in \mathscr{L}(\mathscr{E})$, where $\mathscr{E}$ is a Banach space, then $\lambda - T^*$ is invertible on $\mathscr{E}^*$ if and only if $\lambda - T$ is invertible on $\mathscr{E}$. In other words, the spectra $\sigma(T)$ and $\sigma(T^*)$ coincide.

Example H. Let p and q be Hölder conjugates, $1 < p, q < +\infty$, and let Ψ_p denote the isometric isomorphism of $(\ell_p)^*$ onto (ℓ_q) via which we identify these two spaces (cf. Example 14C), so that $\Psi_p f_a = a$ for each sequence a in (ℓ_q). (In other words, $\Psi_p f = \{f(e_n)\}_{n=0}^{\infty}$, $f \in (\ell_p)^*$.) Then the adjoint Ψ_p^* is an isometric isomorphism of $(\ell_q)^*$ onto $(\ell_p)^{**}$ via which these two spaces are likewise to be identified. Let $a = \{\alpha_n\}$ denote an arbitrary sequence in (ℓ_p), and let $f_a = \Psi_q^{-1}a$ be the corresponding element of $(\ell_q)^*$. Then for each element g of $(\ell_p)^*$ we have

$$(\Psi_p^* f_a)(g) = f_a(\Psi_p g) = \sum_{n=0}^{\infty} \alpha_n g(e_n) = j(a)(g).$$

Thus the composition $\Psi_p^* \circ \Psi_q^{-1}$, via which we identify (ℓ_p) with $(\ell_p)^{**}$, coincides with the natural embedding, as asserted in Example B.

Similarly, if Φ denotes the isometric isomorphism of $(c_0)^*$ onto (ℓ_1) via which one may identify these two spaces (Prob. 14B), and if Ψ_1 denotes the isometric isomorphism of $(\ell_1)^*$ onto (ℓ_∞) used to identify the latter spaces (Ex. 14B), then a simple calculation shows that the mapping $\Phi^* \circ \Psi_1^{-1}$ via which we identify $(c_0)^{**}$ with (ℓ_∞) has the property that $\Phi^* \circ \Psi_1^{-1}(a) = j(a)$ whenever $a \in (c_0)$, as asserted in Example C.

Example I. The mapping β of Proposition 16.7 is nothing other than the adjoint i^*, where i denotes the inclusion mapping of $\mathscr{M}$ into $\mathscr{E}$. Similarly, if π denotes the natural projection of $\mathscr{E}$ onto $\mathscr{E}/\mathscr{M}$, then the mapping α of Proposition 16.6 agrees with π^* at every element of $(\mathscr{E}/\mathscr{M})^*$, so α is simply π^* regarded as a mapping of $(\mathscr{E}/\mathscr{M})^*$ onto the range $\mathscr{R}(\pi^*) = \mathscr{M}^a$.

By piecing together the natural isomorphisms of Propositions 16.6 and 16.7 along with their adjoints, we can construct other natural isomorphisms. For example, if $\mathscr{M}$ is a subspace of a Banach space $\mathscr{E}$ and if $\hat{\beta}$ denotes the natural isomorphism of $\mathscr{E}^*/\mathscr{M}^a$ onto $\mathscr{M}^*$, then $\hat{\beta}^*$ provides a natural isomorphism of $\mathscr{M}^{**}$ onto $(\mathscr{E}^*/\mathscr{M}^a)^*$. But, according to Proposition 16.6, there is also a natural isomorphism α of $(\mathscr{E}^*/\mathscr{M}^a)^*$ onto $\mathscr{M}^{aa}$ (the annihilator of $\mathscr{M}^a$ in $\mathscr{E}^{**}$). Thus the composition $\gamma = \alpha \circ \hat{\beta}^*$ is an isometric isomorphism of $\mathscr{M}^{**}$ onto $\mathscr{M}^{aa}$ that also deserves to be called *natural*. Here is a diagram that elucidates this construction:

$$\overset{\gamma}{\overbrace{\mathscr{M}^{**} \xrightarrow{\hat{\beta}^*} (\mathscr{E}^*/\mathscr{M}^a)^* \xrightarrow{\alpha} \mathscr{M}^{aa}}}$$

$$\mathscr{M}^* \xleftarrow{\hat{\beta}} \mathscr{E}^*/\mathscr{M}^a.$$

Care must be taken in computing the precise action of the isomorphism γ. If φ is an element of $\mathscr{M}^{**}$ and f an element of $\mathscr{E}^*$, then $(\hat{\beta}^*\varphi)(f + \mathscr{M}^a) = \varphi(\hat{\beta}(f + \mathscr{M}^a)) = \varphi(f\,|\,\mathscr{M})$. But then $(\gamma\varphi)(f) = (\hat{\beta}^*\varphi)(f + \mathscr{M}^a) = \varphi(f\,|\,\mathscr{M})$. Thus we have proved one half of the following result. The other half is similar and its proof is left to the reader (see Problem G).

Proposition 16.12. *If $\mathscr{E}$ is a Banach space and $\mathscr{M}$ is a subspace of $\mathscr{E}$, then the mapping γ of $\mathscr{M}^{**}$ into $\mathscr{E}^{**}$ obtained by setting*

$$(\gamma\varphi)(f) = \varphi(f\,|\,\mathscr{M}), \qquad \varphi \in \mathscr{M}^{**}, \qquad f \in \mathscr{E}^*,$$

is an isometric isomorphism of $\mathscr{M}^{**}$ onto $\mathscr{M}^{aa}$. Similarly, the linear transformation δ of $\mathscr{E}^{**}$ onto $(\mathscr{E}/\mathscr{M})^{**}$ obtained by setting

$$(\delta\psi)(\hat{f}) = \psi(\hat{f} \circ \pi), \qquad \psi \in \mathscr{E}^{**}, \qquad \hat{f} \in (\mathscr{E}/\mathscr{M})^*,$$

has null space $\mathscr{M}^{aa}$ and the property that the mapping $\hat{\delta}$ obtained by factoring out $\mathscr{M}^{aa}$, i.e., the mapping $\hat{\delta}(\psi + \mathscr{M}^{aa}) = \delta(\psi)$, is an isometric isomorphism of $\mathscr{E}^{**}/\mathscr{M}^{aa}$ onto $(\mathscr{E}/\mathscr{M})^{**}$. (In this statement π denotes, as always, the natural projection onto the appropriate quotient space.)

It is only when $\mathscr{E}$ is a reflexive Banach space that $\mathscr{E}$ and $\mathscr{E}^*$ may reasonably be thought of as duals of one another, so that even to use the terms "dual" and "duality" when $\mathscr{E}$ is not reflexive is a mild (but entirely standard) abuse of language. It is to be expected that the foregoing results should simplify somewhat when $\mathscr{E}$ is reflexive, and that is indeed the case. For example, when $\mathscr{E}$ is reflexive, every (norm closed) subspace of $\mathscr{E}^*$ is weak* closed as well, so that Corollary 16.5 says in this case that $(^a\mathscr{L})^a = \mathscr{L}$ for every subspace $\mathscr{L}$ of $\mathscr{E}^*$. Similarly, when $\mathscr{E}$ is reflexive and $\mathscr{M}$ is a subspace of $\mathscr{E}$, then $\mathscr{M}$ and $\mathscr{E}/\mathscr{M}$ are reflexive too (Prob. H), and if we use the natural embeddings to identify $\mathscr{E}$ with $\mathscr{E}^{**}$, $\mathscr{M}$ with $\mathscr{M}^{**}$, and $\mathscr{E}/\mathscr{M}$ with $(\mathscr{E}/\mathscr{M})^{**}$, then the mappings γ and δ of Proposition 16.12 become the inclusion mapping of $\mathscr{M}$ into $\mathscr{E}$ and the natural projection of $\mathscr{E}$ onto $\mathscr{E}/\mathscr{M}$, respectively (see Problems F and G).

Reflexivity is clearly an important property for a Banach space to have, and it is desirable to have some criteria for determining if a given space possesses it. One such criterion is easily established. As we know (Prob. 15J), two linear spaces of linear functionals on a linear space $\mathscr{E}$ coincide if and only if they induce the same topology on $\mathscr{E}$. Applying this observation to the dual space of a normed space $\mathscr{E}$ we conclude that $\mathscr{E}$ is *reflexive if and only if the weak and weak* topologies on $\mathscr{E}^*$ coincide*. This test, while easy to state and to derive, is awkward to apply in practice, and it is highly desirable to be able to check whether a space is reflexive without having to have recourse to its dual. Such a test—the *Šmul'yan criterion*—will be obtained later as a by-product of our further study of locally convex spaces. (An alternate proof using only Banach space theory is outlined in the problem set; see Problems C and D.)

The foregoing discussion constitutes a brief synopsis of the bare essentials of what is generally known as *duality theory* for Banach spaces. Let us turn now, at least briefly, to the problem of generalizing this useful and fairly compact circle of ideas to more general locally convex spaces. (This project is far from simple, and has generated an extensive literature. All we can hope to do here is to convey some feeling for the texture of the theory, along with, perhaps, a notion of how it is related to other areas of the study of topological linear spaces; readers wishing to pursue further the study of duality theory are advised to consult either [55] or [12].)

The first difficulty confronting us in the program upon which we now embark is immediately apparent and, as it turns out, this obvious first difficulty is also a most serious one. If $\mathscr{E}$ is a normed space, then $\mathscr{E}^*$ is also a normed space with respect to a norm the usefulness of which is not in doubt. If $\mathscr{E}$ is merely a separated locally convex topological linear space, then $\mathscr{E}^*$ is nothing more than a linear space. It is true that we know of a linear topology on $\mathscr{E}^*$ that is both interesting and useful, viz., the weak* topology, but the weak* topology never agrees with the norm topology on the dual $\mathscr{E}^*$ of a normed space $\mathscr{E}$ except in the finite dimensional case (Prob. 15H), a trivial case that we may safely ignore. Consequently it is pointless to try to base a generalized duality theory on the weak* topology alone, although we should expect it to play a significant role, to be sure. Not surprisingly, then, the generalization of duality theory requires yet one more consideration in depth of the whole idea of the topologization of locally convex spaces. The fundamental tool in this reconsideration is the following notion.

Definition. Let $\mathscr{E}$ and $\mathscr{F}$ be linear spaces, and let ψ be a bilinear functional on $\mathscr{E} + \mathscr{F}$. The triple $(\mathscr{E}, \mathscr{F}, \psi)$ is a *dual pair* if the following two conditions are satisfied:

(i) For each nonzero x in $\mathscr{E}$ there is a vector y in $\mathscr{F}$ such that $\psi(x, y) \neq 0$,
(ii) For each nonzero y in $\mathscr{F}$ there is a vector x in $\mathscr{E}$ such that $\psi(x, y) \neq 0$.

When dealing with a dual pair $(\mathscr{E}, \mathscr{F}, \psi)$, it is customary to supress the symbol for the distinguished bilinear functional ψ, writing $\psi(x, y) = \langle x, y \rangle$ instead. Likewise it will be convenient to write $\langle \mathscr{E}, \mathscr{F} \rangle$ for the dual pair consisting of $\mathscr{E}$ and $\mathscr{F}$ equipped with the bilinear functional $\langle , \rangle$.

If $\langle \mathscr{E}, \mathscr{F} \rangle$ is a dual pair then for each y in $\mathscr{F}$ the function f_y defined on $\mathscr{E}$ by setting $f_y(x) = \langle x, y \rangle$ is a linear functional, and it is the substance of condition (ii) in the above definition that the linear transformation $y \to f_y$ is an isomorphism of $\mathscr{F}$ into the full algebraic dual $\mathscr{E}'$ of $\mathscr{E}$. It is both customary and convenient to identify $\mathscr{F}$ with this collection of linear functionals $f_y, y \in \mathscr{F}$, and we shall frequently make this identification without further comment. Thus the topology induced on $\mathscr{E}$ by the set of all functionals f_y is called the topology *induced on $\mathscr{E}$ by $\mathscr{F}$*. The topology induced on $\mathscr{E}$ by $\mathscr{F}$ is locally convex and is separated by condition (i).

In the same vein, of course, the mapping $x \to f^x$, $x \in \mathscr{E}$, where $f^x(y) = \langle x, y \rangle$, $y \in \mathscr{F}$, is a linear space isomorphism of $\mathscr{E}$ into the full algebraic dual $\mathscr{F}'$ of $\mathscr{F}$ (which we frequently use to identify $\mathscr{E}$ with a linear submanifold of $\mathscr{F}'$), and the topology induced on $\mathscr{F}$ by this collection of linear functionals (called the topology *induced on $\mathscr{F}$ by $\mathscr{E}$*) is locally convex and separated. Indeed, it is clear from the symmetry of the definition that if $\langle \mathscr{E}, \mathscr{F} \rangle$ is a dual pair, then $\mathscr{F}$ and $\mathscr{E}$ form another dual pair with respect to the bilinear functional $\langle y, x \rangle = \langle x, y \rangle$. (We shall consistently employ the notation $\langle \mathscr{F}, \mathscr{E} \rangle$ for this dual pair.) Thus *every* theorem concerning the relation between the spaces of a dual pair has a valid counterpart with the roles of the two spaces reversed. In the following discussion we shall ordinarily not even bother to state this dual result (unless, as above, useful notation or terminology is thereby introduced), but rather leave it to the reader to bear in mind at all times that an appropriate reformulation of each theorem is valid.

Example J. Let $\mathscr{E}$ be a separated locally convex space with dual space $\mathscr{E}^*$. Then $\langle \mathscr{E}, \mathscr{E}^* \rangle$ is a dual pair with respect to the bilinear functional $\langle x, f \rangle = f(x)$. The topology induced on $\mathscr{E}$ by $\mathscr{E}^*$ is the weak topology; the topology induced on $\mathscr{E}^*$ by $\mathscr{E}$ is the weak* topology. In the corresponding dual pair $\langle \mathscr{E}^*, \mathscr{E} \rangle$ these topologies are reversed. (The notation $\langle \mathscr{E}, \mathscr{E}^* \rangle$ and $\langle \mathscr{E}^*, \mathscr{E} \rangle$ will be used in this manner consistently in the sequel.) Dual pairs of the form $\langle \mathscr{E}, \mathscr{E}^* \rangle$ are by far the most important ones we shall consider, and it is convenient to borrow some of the terminology developed in this special case. Thus for an arbitrary dual pair $\langle \mathscr{E}, \mathscr{F} \rangle$ we shall sometimes refer to the topology induced on $\mathscr{E}$ by $\mathscr{F}$ as the *weak topology* on $\mathscr{E}$.

Example K. Let $\mathscr{E}$ be an arbitrary linear space and let $\mathscr{E}'$ be its full algebraic dual. Then it is clear that $\langle \mathscr{E}, \mathscr{E}' \rangle$ is a dual pair with respect to the bilinear functional $\langle x, f \rangle = f(x)$ (see Problem 2D).

An important tool in the study of duality and dual pairs is the following concept.

Definition. Let $\langle \mathscr{E}, \mathscr{F} \rangle$ be a dual pair and let M be a nonempty subset of $\mathscr{E}$. Then the *polar* of M is the set

$$M^0 = \{y \in \mathscr{F} : \sup_{x \in M} |\langle x, y \rangle| \le 1\}.$$

Dually, if M is a nonempty subset of $\mathscr{F}$, then the *prepolar* of M is the set ^0M of all those vectors x in $\mathscr{E}$ such that $|\langle x, y \rangle| \le 1$ for every y in M. (By convention the empty subset of $\mathscr{E}$ is assigned $\mathscr{F}$ as its polar; similarly the prepolar of the empty subset of $\mathscr{F}$ is taken to be the entire space $\mathscr{E}$. The prepolar of a subset of $\mathscr{F}$ is clearly just its polar in the reversed dual pair $\langle \mathscr{F}, \mathscr{E} \rangle$.)

Example L. If $\langle \mathscr{E}, \mathscr{F} \rangle$ is a dual pair, and $\mathscr{L}$ is a linear submanifold of $\mathscr{E}$, then the polar of $\mathscr{L}$ coincides with the *annihilator*

$$\mathscr{L}^a = \{y \in \mathscr{F} : \langle x, y \rangle = 0, x \in \mathscr{L}\}.$$

Similarly, if $\mathscr{L}$ is a linear submanifold of $\mathscr{F}$, then the prepolar $^0\mathscr{L}$ is also the *preannihilator*

$$^a\mathscr{L} = \{x \in \mathscr{E} : \langle x, y \rangle = 0, y \in \mathscr{L}\}.$$

Example M. Let $\mathscr{E}$ be a normed space. Then the polar $(\mathscr{E}_1)^0$ of the closed unit ball in the dual pair $\langle \mathscr{E}, \mathscr{E}^* \rangle$ is the closed unit ball $(\mathscr{E}^*)_1$. More generally, $(\mathscr{E}_r)^0 = (\mathscr{E}^*)_{1/r}$ for each positive radius r.

Example N. Let $\mathscr{E}$ be a separated locally convex topological linear space and let $\mathscr{E}^*$ be its dual. A subset E of $\mathscr{E}^*$ is said to be *equicontinuous* on $\mathscr{E}$ if for each $\varepsilon > 0$ there exists a neighborhood V_ε of the origin in $\mathscr{E}$ such that $|f(x)| \le \varepsilon$ for all x in V_ε and all f in E. (Note that this implies that if x_0 is an arbitrary point in $\mathscr{E}$, then $|f(x) - f(x_0)| \le \varepsilon$ for every x in $x_0 + V$ and every f in E; cf. Problem 4R.) If E is an equicontinuous subset of $\mathscr{E}^*$, then it is clear that there is a neighborhood V of 0 in $\mathscr{E}$ such that E is contained in the polar of V in the dual pair $\langle \mathscr{E}, \mathscr{E}^* \rangle$ (set $\varepsilon = 1$). On the other hand, if $E \subset V^0$ for some neighborhood V of 0 in $\mathscr{E}$, then $|f(x)| \leq \varepsilon$ for every f in E provided $x \in \varepsilon V$. Thus a subset E of $\mathscr{E}^*$ is equicontinuous if and only if there exists a neighborhood V of 0 in $\mathscr{E}$ such that $E \subset V^0$.

Example M shows that the closed balls centered at 0 in the dual of a normed space $\mathscr{E}$ are polars, and thus that polars, in a sense, generalize closed balls centered at 0. Example L shows that polars are simultaneously analogs of the annihilators introduced earlier in our study of normed spaces. It is on the basis of this latter analogy that the following summary of the elementary facts about polars is formulated (cf. Propositions 16.1 and 16.2, and Corollary 16.3). Proposition 16.13 is self-evident and requires no proof; it is recorded here solely for convenience of reference. The reader is reminded that the corresponding assertions concerning prepolars are also valid; their statements are simply suppressed as being redundant.

Proposition 16.13. *Let $\langle \mathscr{E}, \mathscr{F} \rangle$ be a dual pair. If M and N are subsets of $\mathscr{E}$ such that $M \subset N$, then $M^0 \supset N^0$. Likewise if $\{M_\gamma\}_{\gamma \in \Gamma}$ is a nonempty indexed family of subsets of $\mathscr{E}$, then*

$$\left(\bigcup_{\gamma \in \Gamma} M_\gamma \right)^0 = \bigcap_{\gamma \in \Gamma} M_\gamma^0.$$

Proposition 16.14. *If $\langle \mathscr{E}, \mathscr{F} \rangle$ is a dual pair and if M is a subset of $\mathscr{E}$, then the polar M^0 is an absolutely convex subset of $\mathscr{F}$ that is closed in the topology induced on $\mathscr{F}$ by $\mathscr{E}$, and the prepolar $^0(M^0)$ of M^0 is the closed absolutely convex hull of M in the topology induced on $\mathscr{E}$ by $\mathscr{F}$.*

The proof of Proposition 16.14 is so like that of Proposition 16.2 in every particular that we omit it. The trick is to recall Problem 15J and use the following lemma in place of Proposition 14.13.

Lemma 16.15. *Let $\mathscr{E}$ be a separated locally convex topological linear space. If A is a nonempty, closed, absolutely convex subset of $\mathscr{E}$ and if x_0 is a vector in $\mathscr{E}$ not belonging to A, then there exists a functional f_0 in $\mathscr{E}^*$ such that $|f_0(x)| \leq 1$ for all x in A and such that $f_0(x_0) > 1$.*

PROOF. Since x_0 is a vector not belonging to A, then by Proposition 14.15 there exists a functional g in $\mathscr{E}^*$ and a real number c such that

$$\operatorname{Re} g(x) \leq c, x \in A, \quad \text{while} \quad \operatorname{Re} g(x_0) > c. \qquad (1)$$

But A is balanced as well as convex, whence it is clear that $g(A)$ is a disc in $\mathbb{C}$ with center 0 and, according to (1), radius no greater than c (which must be nonnegative). Thus, replacing g by a functional f of the form γg, where γ denotes a suitable complex number of modulus one, we obtain

$$|f(x)| \leq c, x \in A, \quad \text{while} \quad f(x_0) > c.$$

If $c > 0$ we have but to define $f_0 = f/c$; if $c = 0$ we may set $f_0 = 2f/f(x_0)$. $\square$

Corollary 16.16. *If $\langle \mathscr{E}, \mathscr{F} \rangle$ is a dual pair and M is an arbitrary subset of $\mathscr{E}$, then $({}^0(M^0))^0 = M^0$. If A is an absolutely convex subset of $\mathscr{E}$, then ${}^0(A^0)$ coincides with the closure of A in the weak topology. In particular, if A is an absolutely convex set in $\mathscr{E}$ that is closed in the weak topology, then ${}^0(A^0) = A$.*

If $\langle \mathscr{E}, \mathscr{F} \rangle$ is a dual pair, then the polar M^0 of an arbitrary subset M of $\mathscr{E}$ is absolutely convex. It is natural to enquire when M^0 is also absorbing, or, what comes to the same thing, when the Minkowski function of M^0 is a pseudonorm on $\mathscr{F}$ (cf. Example 14F and the discussion preceding Proposition 14.7). The answer to this question involves a very important concept in the general theory of locally convex spaces.

Definition. If M and N are subsets of a linear space $\mathscr{E}$, then M is said to *absorb* N if there exists a positive number k large enough so that $N \subset \lambda M$ for every λ such that $|\lambda| \geq k$. (Thus a subset A of $\mathscr{E}$ is absorbing in the sense defined in Chapter 14 if and only if it absorbs every singleton.) A subset B of a topological linear space $\mathscr{E}$ is said to be *bounded* if every neighborhood of 0 in $\mathscr{E}$ absorbs B.

Example O. If $\mathscr{E}$ is a normed space, then a subset B of $\mathscr{E}$ is bounded if and only if B is contained in some ball $\mathscr{E}_r, r > 0$. Thus B is bounded in $\mathscr{E}$ regarded as a topological linear space if and only if B is bounded in $\mathscr{E}$ regarded as a metric space.

Example P. Let $\mathscr{E}$ be a separated locally convex space and suppose there exists a bounded neighborhood U_0 of the origin in $\mathscr{E}$. If W_0 denotes an absolutely convex neighborhood of 0 contained in U_0 (Prop. 14.7), then W_0 is also bounded. Hence, if V is an arbitrary neighborhood of 0 in $\mathscr{E}$, there exists a positive number ε such that $\varepsilon W_0 \subset V$. In other words, if σ denotes the Minkowski function of W_0, then the balls $D_\varepsilon = \{x \in \mathscr{E} : \sigma(x) < \varepsilon\}$ constitute a neighborhood base at 0. Moreover, since $\mathscr{E}$ is separated, $\bigcap_{\varepsilon > 0} D_\varepsilon = (0)$, so σ is actually a *norm* on $\mathscr{E}$. Moreover, it is clear that the norm σ induces the given locally convex topology on $\mathscr{E}$. A topological linear space whose topology is induced by a norm is said to be *normable* (cf. Problem 15H). Thus, putting this example together with the preceding one, we see that *a necessary and sufficient condition for a separated locally convex space to be normable is that it possess a bounded neighborhood of the origin.*

Example Q. Every compact subset of a separated locally convex space $\mathscr{E}$ is bounded. Indeed, if V is an absolutely convex neighborhood of 0 in $\mathscr{E}$, then the sequence $\{nV^\circ\}_{n=1}^\infty$ is a nested open covering of $\mathscr{E}$. Hence if K is a compact set in $\mathscr{E}$, then some one set nV contains K. In particular, every convergent sequence in $\mathscr{E}$ is bounded.

Example R. If $\mathscr{E}$ is an arbitrary separated locally convex topological linear space, then a subset B of $\mathscr{E}$ is bounded with respect to the weak topology on $\mathscr{E}$ if and only if it is *weakly bounded*, i.e., if and only if each functional f in $\mathscr{E}^*$ is bounded on B. Indeed it is obvious that if B is bounded with respect to the weak topology on $\mathscr{E}$, then this condition must be satisfied since the weak neighborhood $U(f; 1)$ must absorb B for each f in $\mathscr{E}^*$ (cf. Chapter 15(2)). Suppose, on the other hand, that B is weakly bounded. Every weak neighborhood of 0 in $\mathscr{E}$ contains a neighborhood of the form $U = (f_1, \ldots, f_n; \varepsilon)$, and if K is chosen so that $|f_i(x)| \le K$ for all x in B and $i = 1, \ldots, n$, then $B \subset \lambda U$ whenever $|\lambda| \ge K/\varepsilon$, which shows that U absorbs B. It should be noted that this example shows, in general, that if $\langle \mathscr{E}, \mathscr{F} \rangle$ is a dual pair then a nonempty subset M of $\mathscr{E}$ is bounded in the weak topology if and only if $\sup_{x \in M} |\langle x, y \rangle| < +\infty$ for each y in $\mathscr{F}$.) .

The notion of boundedness leads at once to a simple answer to a question raised earlier.

Proposition 16.17. *Let $\langle \mathscr{E}, \mathscr{F} \rangle$ be a dual pair. If M is a subset of $\mathscr{E}$ that is bounded in the topology induced on $\mathscr{E}$ by $\mathscr{F}$, then the polar M^0 is absorbing. Dually, if A is an absorbing subset of $\mathscr{E}$, then A^0 is bounded with respect to the topology induced on $\mathscr{F}$ by $\mathscr{E}$.*

PROOF. If $\sup_{x \in M} |\langle x, y \rangle| = K < +\infty$, then $\lambda y \in M^0$ whenever $|\lambda| \le 1/K$. Thus M^0 is absorbing whenever M is bounded in the topology induced by $\mathscr{F}$. Similarly, if A is absorbing and $x \in \mathscr{E}$, and if $\varepsilon x \in A$ for some $\varepsilon > 0$, then $|\langle x, y \rangle| \le 1/\varepsilon$ for every y in A^0, which shows that A^0 is bounded in the topology induced by $\mathscr{E}$. $\square$

Corollary 16.18. *If $\langle \mathscr{E}, \mathscr{F} \rangle$ is a dual pair, then a subset M of $\mathscr{E}$ is bounded in the topology induced on $\mathscr{E}$ by $\mathscr{F}$ if and only if M^0 is absorbing. Likewise if A is a subset of $\mathscr{E}$ that is absolutely convex and closed in the topology induced by $\mathscr{F}$, then A is absorbing if and only if A^0 is bounded in the topology induced on $\mathscr{F}$ by $\mathscr{E}$.*

PROOF. It suffices to give the half of the proof that has not already been given. If M is a subset of $\mathscr{E}$ such that M^0 is absorbing, then $^0(M^0)$ is bounded in the topology induced by $\mathscr{F}$ (simply apply the foregoing result to the reversed pair $\langle \mathscr{F}, \mathscr{E} \rangle$), and since $M \subset {}^0(M^0)$, we see that M is also bounded. Similarly, if A^0 is bounded in the topology induced on $\mathscr{F}$ by $\mathscr{E}$, then $^0(A^0)$ is absorbing. But if A is as stated, then $A = {}^0(A^0)$ by Corollary 16.16. $\square$

The foregoing discussion shows that if $\langle \mathscr{E}, \mathscr{F} \rangle$ is a dual pair and M is a subset of $\mathscr{E}$, then M^0 is an absorbing absolutely convex subset of $\mathscr{F}$ if and only if M is weakly bounded. If M is a weakly bounded subset of $\mathscr{E}$, then the Minkowski function σ of M^0 is a pseudonorm on $\mathscr{F}$ called the pseudonorm *associated with* M. A trivial calculation shows that if $M \neq \varnothing$, then σ is given by the formula

$$\sigma(y) = \sup_{x \in M} |\langle x, y \rangle|, \qquad y \in \mathscr{F}.$$

(The pseudonorm associated with the empty subset of $\mathscr{E}$ is identically zero.) Moreover, it is clear that M^0 is the closed unit ball with respect to σ, and that a net $\{y_\lambda\}$ in $\mathscr{F}$ is convergent to a limit y with respect to σ if and only if $\{y_\lambda\}$ converges uniformly to y on M (with the elements of $\mathscr{F}$ identified with functionals on $\mathscr{E}$ in the standard way). Thus we are lead to the following terminology.

Definition. Let $\langle \mathscr{E}, \mathscr{F} \rangle$ be a dual pair and let M be a weakly bounded subset of $\mathscr{E}$. Then the topology induced on $\mathscr{F}$ by the pseudonorm associated with M is called the *topology of uniform convergence on* M. More generally, if $\mathscr{C}$ is a nonempty collection of weakly bounded subsets of $\mathscr{E}$, then the (locally convex, linear) topology $\mathscr{T}$ induced on $\mathscr{F}$ by the family of pseudonorms associated with the various sets in $\mathscr{C}$ is characterized by the property that a net $\{y_\lambda\}$ in $\mathscr{F}$ converges to a limit y with respect to $\mathscr{T}$ if and only if $\{y_\lambda\}$ converges to y uniformly on each set M in $\mathscr{C}$, and is called the *topology of uniform convergence on the sets in* $\mathscr{C}$. All such topologies are known as *polar topologies* on $\mathscr{F}$.

Example S. If $\mathscr{E}$ is a normed space, then the norm topology on $\mathscr{E}^*$ is the topology of uniform convergence on the unit ball $\mathscr{E}_1$ in the dual pair $\langle \mathscr{E}, \mathscr{E}^* \rangle$. This topology may also be described as the topology of uniform convergence on the collection of all balls $\mathscr{E}_r$, $r > 0$, or as the topology of uniform convergence on all of the bounded subsets of $\mathscr{E}$ (see Example O).

Example T. For an arbitrary dual pair $\langle \mathscr{E}, \mathscr{F} \rangle$ the topology of uniform convergence on the finite subsets of $\mathscr{E}$ (which coincides with the topology of simple pointwise convergence on $\mathscr{E}$, i.e., the topology of uniform convergence on the singletons in $\mathscr{E}$) is the topology induced on $\mathscr{F}$ by $\mathscr{E}$. In particular, if $\mathscr{E}$ is a given separated locally convex space, then the topology of uniform convergence on the finite subsets of $\mathscr{E}$ in the dual pair $\langle \mathscr{E}, \mathscr{E}^* \rangle$ is the weak* topology on $\mathscr{E}^*$.

As these examples show, it is possible for the same polar topology to be induced by quite disparate collections of weakly bounded sets. In this connection the following observations are to the point.

Example U. Let $\langle \mathscr{E}, \mathscr{F} \rangle$ be a dual pair and let $\mathscr{C}_1$ and $\mathscr{C}_2$ be two collections of weakly bounded subsets of $\mathscr{E}$. Then a sufficient (not a necessary) condition for the two to induce the same polar topology on $\mathscr{F}$ is that each set in $\mathscr{C}_1$ be contained in the union of a finite number of sets belonging to $\mathscr{C}_2$, and conversely. In the same vein we note that if $\mathscr{C}$ is a collection of weakly bounded subsets of $\mathscr{E}$ and if $\mathscr{C}_0$ denotes the collection $\{^0(M^0) : M \in \mathscr{C}\}$, then $\mathscr{C}$ and $\mathscr{C}_0$ induce the same polar topology on $\mathscr{F}$ by Corollary 16.16.

For our purposes it is desirable to impose a modest restriction on the collections of sets used to define polar topologies.

Definition. If $\langle \mathscr{E}, \mathscr{F} \rangle$ is a dual pair, then a collection $\mathscr{A}$ of weakly bounded subsets of $\mathscr{E}$ is *admissible* if the following conditions are satisfied:

(i) If M_1 and M_2 belong to $\mathscr{A}$, then there is a set N in $\mathscr{A}$ such that $M_1 \cup M_2 \subset N$,
(ii) If $M \in \mathscr{A}$ and $\lambda \neq 0$, then $\lambda M \in \mathscr{A}$,
(iii) $\bigcup \mathscr{A}$ generates $\mathscr{E}$ algebraically.

A polar topology is said to be *admissible* if there exists an admissible collection $\mathscr{A}$ of weakly bounded subsets of $\mathscr{E}$ that induces it.

Proposition 16.19. *Let $\langle \mathscr{E}, \mathscr{F} \rangle$ be a dual pair, and let $\mathscr{A}$ be an admissible collection of weakly bounded subsets of $\mathscr{E}$. Then the topology $\mathscr{T}$ of uniform convergence on the sets in $\mathscr{A}$ is a separated locally convex linear topology on $\mathscr{F}$ in which a neighborhood base at 0 is provided by the collection of all polars M^0, $M \in \mathscr{A}$, and with respect to which all of the vectors in $\mathscr{E}$ are continuous when regarded as linear functionals on $\mathscr{F}$.*

PROOF. The topology $\mathscr{T}$ is automatically locally convex, being induced by a family of pseudonorms. That $\mathscr{T}$ is separated follows at once from condition (iii) in the foregoing definition. Moreover, condition (i) implies that the family of pseudonorms inducing $\mathscr{T}$ is saturated, so a neighborhood base at 0 is indeed provided by the sets M^0, $M \in \mathscr{A}$ (Prop. 11.31). Finally, if x is a vector belonging to some M in $\mathscr{A}$, then x is bounded on the polar M^0 and is therefore continuous in the topology induced on $\mathscr{F}$ by $\mathscr{A}$. Since the continuous linear functionals on $\mathscr{F}$ form a linear submanifold of the full algebraic dual of $\mathscr{F}$, the result follows by condition (iii) in the definition of admissibility. □

If $\langle \mathscr{E}, \mathscr{F} \rangle$ is a dual pair and $\mathscr{C}$ is an arbitrary nonempty collection of weakly bounded subsets of $\mathscr{E}$, then it is always a simple matter to replace $\mathscr{C}$ by another collection $\mathscr{C}'$ that induces the same polar topology on $\mathscr{F}$ as $\mathscr{C}$ does and that also satisfies conditions (i) and (ii) in the definition of admissibility. Moreover, it is clear that (i) and (ii) imply that the polars M^0, $M \in \mathscr{C}'$, form a neighborhood base at the origin in $\mathscr{F}$ in the polar topology induced by $\mathscr{C}$. Likewise, it may easily be seen that (iii) is satisfied in the presence of (i) and (ii) if and only if the sets $^0(M^0)$, $M \in \mathscr{C}'$, cover $\mathscr{E}$, and therefore if and only if the vectors in $\mathscr{E}$ are all continuous as functionals on $\mathscr{F}$ in the topology of uniform convergence on the sets in $\mathscr{C}$. Hence a polar topology on $\mathscr{F}$ is admissible if and only if this latter criterion is satisfied.

If $\langle \mathscr{E}, \mathscr{F} \rangle$ is a dual pair, then, according to Proposition 16.19, there is a coarsest admissible polar topology on $\mathscr{F}$, viz., the topology induced by $\mathscr{E}$. There is likewise a finest admissible polar topology on $\mathscr{F}$, namely the topology of uniform convergence on *all* the weakly bounded subsets of $\mathscr{E}$. (The collection of all weakly bounded subsets of $\mathscr{E}$ is readily perceived to be admissible.) This topology is frequently called the *strong topology* on $\mathscr{F}$. Every admissible polar topology on $\mathscr{F}$ with respect to the dual pair $\langle \mathscr{E}, \mathscr{F} \rangle$ lies between the topology induced by $\mathscr{E}$ and the strong topology.

Example V. Let $\mathscr{E}$ be a Banach space with dual $\mathscr{E}^*$, and consider the dual pair $\langle \mathscr{E}^*, \mathscr{E} \rangle$. According to Problem 15Q a subset of $\mathscr{E}^*$ is bounded in the topology induced by $\mathscr{E}$ (i.e., is weak* bounded) if and only if it is bounded in norm. Hence the strong topology on $\mathscr{E}$ is identified as the topology of uniform convergence on the (norm) bounded subsets of $\mathscr{E}^*$. Thus the strong topology on $\mathscr{E}$ (with respect to the dual pair $\langle \mathscr{E}^*, \mathscr{E} \rangle$) is simply the norm topology.

Definition. If $\langle \mathscr{E}, \mathscr{F} \rangle$ is a dual pair, then any locally convex linear topology on $\mathscr{E}$ with respect to which the dual space $\mathscr{E}^*$ coincides with $\mathscr{F}$ (as a linear manifold in the full algebraic dual of $\mathscr{E}$) is called a *topology of the dual pair* $\langle \mathscr{E}, \mathscr{F} \rangle$.

It is clear that such topologies always exist; the topology induced by $\mathscr{F}$ itself is one example of a topology of the dual pair $\langle \mathscr{E}, \mathscr{F} \rangle$ (Prob. 15J). As it turns out, every topology of a dual pair $\langle \mathscr{E}, \mathscr{F} \rangle$ is an admissible polar topology (with respect to the *reversed* pair $\langle \mathscr{F}, \mathscr{E} \rangle$).

Proposition 16.20. *If $\mathscr{T}$ is a topology of a dual pair $\langle \mathscr{E}, \mathscr{F} \rangle$, then the collection $\mathscr{A}$ of all those subsets of $\mathscr{F}$ that are equicontinuous with respect to $\mathscr{T}$ is an admissible collection of weakly bounded subsets of $\mathscr{F}$ in the dual pair $\langle \mathscr{F}, \mathscr{E} \rangle$, and $\mathscr{T}$ is the topology of uniform convergence on the sets in $\mathscr{A}$. Likewise, $\mathscr{T}$ may be characterized as the topology of uniform convergence on the sets of the form V^0, where V runs through an arbitrary neighborhood base at 0 in $\mathscr{E}$. In particular, $\mathscr{T}$ is an admissible polar topology.*

PROOF. It is obvious that the topologies of uniform convergence on the polars of any two neighborhood bases at 0 in $\mathscr{E}$ coincide (Ex. U). Let $\mathscr{V}$ be a neighborhood base at 0 in $\mathscr{E}$ with respect to $\mathscr{T}$ that consists exclusively of closed absolutely convex sets (cf. Proposition 14.7) and let $\mathscr{B}$ denote the collection

$$\mathscr{B} = \{tV : V \in \mathscr{V}, t > 0\}.$$

If $B \in \mathscr{B}$, then B^0 is bounded in the topology induced on $\mathscr{F}$ by $\mathscr{E}$ (Prop. 16.17), and $B = {}^0(B^0)$. From this observation it will follow directly (Prop. 16.19) that $\mathscr{T}$ is the polar topology induced by $\mathscr{B}_0 = \{B^0 : B \in \mathscr{B}\}$, and hence by $\mathscr{V}_0 = \{V^0 : V \in \mathscr{V}\}$, once we have proved that $\mathscr{B}_0$ is admissible. To this end we note first that if $B_1 = t_1 V_1$ and $B_2 = t_2 V_2$, with $t_1, t_2 > 0$

and V_1, V_2 in $\mathscr{V}$, and if V is a neighborhood in $\mathscr{V}$ such that $V \subset V_1 \cap V_2$, then $B = (t_1 \wedge t_2) V \subset B_1 \cap B_2$, and $B^0 \supset B_1^0 \cup B_2^0$. Thus condition (i) in the definition of admissibility holds. As for the other two conditions, (ii) is clearly satisfied, while (iii) follows at once from the fact that $\mathscr{F} = \bigcup_{V \in \mathscr{V}} V^0$ since $\mathscr{T}$ is a topology of the dual pair $\langle \mathscr{E}, \mathscr{F} \rangle$. Finally, to verify that $\mathscr{V}_0$ may be replaced by the collection $\mathscr{A}$ of all equicontinuous subsets of $\mathscr{F}$ with respect to $\mathscr{T}$, we refer to Example N. $\square$

The converse of Proposition 16.20 is false.

Example W. Consider the normed space (φ), consisting of all sequences $\{\xi_n\}_{n=0}^{\infty}$ that are eventually zero, equipped with the sup norm (see Example 11H). The space (φ) is a dense linear manifold in (c_0) and therefore has dual (ℓ_1) according to the standard identification (Probs. 14B and 14C). It follows from Example R that, in the dual pair $\langle (\ell_1), (\varphi) \rangle$, a subset M of (ℓ_1) is bounded in the topology induced by (φ) if and only if it is termwise bounded. It follows that for an *absolutely arbitrary* sequence $A = \{a_m\}_{m=0}^{\infty}$ of non-negative numbers, setting

$$\sigma_A(x) = \sum_{n=0}^{\infty} a_n |\xi_n|, \qquad x = \{\xi_n\} \in (\varphi),$$

defines a pseudonorm on (φ) which is continuous in the strong topology on (φ) (in the dual pair $\langle (\ell_1), (\varphi) \rangle$). Thus the strong topology on (φ) is decidedly finer than the sup norm topology, and (ℓ_1) is *not* the dual of (ϕ) in the strong topology. Indeed, the dual of (φ) in the strong topology is readily seen to be the space (∂) of *all* complex sequences (the full algebraic dual of (φ)).

It is an important problem to determine just which admissible polar topologies with respect to a dual pair $\langle \mathscr{F}, \mathscr{E} \rangle$ are topologies of the reversed pair $\langle \mathscr{E}, \mathscr{F} \rangle$. The solution to this problem turns out to hinge on a couple of compactness arguments.

Lemma 16.21. *Let $\mathscr{E}$ be a separated locally convex space with dual $\mathscr{E}^*$, and let A be an absolutely convex compact subset of $\mathscr{E}$. Then every linear functional on $\mathscr{E}^*$ that is bounded on the polar A^0 is determined by a vector x in $\mathscr{E}$ (that is absorbed by A).*

PROOF. Let $\mathscr{E}^{*\prime}$ denote the full algebraic dual of $\mathscr{E}^*$, and let k denote the natural mapping of $\mathscr{E}$ into $\mathscr{E}^{*\prime}$. Then k is a linear space isomorphism of $\mathscr{E}$ into $\mathscr{E}^{*\prime}$ and is also clearly a homeomorphism of $\mathscr{E}$ onto $k(\mathscr{E})$ if $\mathscr{E}$ and $\mathscr{E}^{*\prime}$ are both given the topologies induced by $\mathscr{E}^*$ in the dual pairs $\langle \mathscr{E}, \mathscr{E}^* \rangle$ and $\langle \mathscr{E}^*, \mathscr{E}^{*\prime} \rangle$, respectively. (The former of these topologies is just the weak topology on $\mathscr{E}$, of course; cf. Example 15I.) Since A is compact in $\mathscr{E}$ it is also weakly compact, so $k(A)$ is a compact, and therefore a closed, subset of $\mathscr{E}^{*\prime}$

in the topology induced by $\mathscr{E}^*$. Thus $k(A)$ is an absolutely convex subset of $\mathscr{E}^{*\prime}$ that is closed in the topology induced by $\mathscr{E}^*$. Moreover, the prepolar of $k(A)$ in $\mathscr{E}^*$ coincides with the polar A^0. Hence if φ is an element of $\mathscr{E}^{*\prime}$ and if, say, $\sup_{f \in A^0} |\varphi(f)| = K < +\infty$, then $\varphi/K \in ({}^0k(A))^0 = k(A)$. Consequently there exists a vector x in A such that $\varphi/K = k(x)$, and hence such that $\varphi = Kk(x)$. $\qquad\square$

Theorem 16.22 (Mackey–Arens Theorem). *A locally convex linear topology $\mathscr{T}$ on the space $\mathscr{E}$ of a dual pair $\langle \mathscr{E}, \mathscr{F} \rangle$ is a topology of that dual pair if and only if it is the topology of uniform convergence on the sets of some admissible collection of absolutely convex subsets of $\mathscr{F}$ each of which is compact in the topology induced on $\mathscr{F}$ by $\mathscr{E}$.*

PROOF. Suppose first that the topology $\mathscr{T}$ is induced by such a collection of sets, and let φ be a linear functional on $\mathscr{E}$ that is continuous with respect to $\mathscr{T}$. Then, since $\mathscr{A}$ is admissible, there exists a set A in $\mathscr{A}$ such that φ is bounded on ${}^0A$, and, by the preceding lemma (applied to the space $\mathscr{F}$ in the topology induced by $\mathscr{E}$), the functional φ is one of those defined on $\mathscr{E}$ by some vector in $\mathscr{F}$. Thus $\mathscr{T}$ is a topology of the dual pair $\langle \mathscr{E}, \mathscr{F} \rangle$, and one half of the theorem is proved. In view of Proposition 16.20, to prove the other half it suffices to establish the following result, a fact of interest in its own right. $\qquad\square$

Theorem 16.23 (Alaoglu–Bourbaki Theorem). *If $\mathscr{E}$ is a separated locally convex space and if V is an arbitrary neighborhood of the origin in $\mathscr{E}$, then the polar V^0 is weak* compact.*

PROOF. Since V is a neighborhood of 0 it is absorbing. For each vector x in $\mathscr{E}$ let $t(x)$ be a positive number large enough so that $x \in t(x)V$ (in particular, set $t(x) = 1$ for each x in V), and form the product

$$\Pi = \prod_{x \in \mathscr{E}} D_{t(x)}^-,$$

where $D_{t(x)}^- = \{\lambda \in \mathbb{C} : |\lambda| \le t(x)\}$, $x \in \mathscr{E}$. Then Π is a compact Hausdorff space (Th. 3.15) and setting $\Psi(f) = \{f(x)\}_{x \in \mathscr{E}}$, $f \in V^0$, defines a one-to-one mapping of V^0 into Π. The balance of the proof is a word for word repetition of the argument used to prove Theorem 15.11 and is omitted. $\qquad\square$

The facts discussed thus far do not by any means exhaust the theory of duality; indeed, it would be more accurate to say that they constitute its beginnings. Nevertheless, to pursue these matters at greater length would only take us further from the topics of central interest in this book, which primarily pertain to normed spaces. Consequently we close this chapter here with a second important application of Lemma 16.21 that shows that generalized duality theory can lead to results that are of great interest even in the context of normed spaces.

Theorem 16.24 (Šmul'yan Criterion). *A normed space $\mathscr{E}$ is reflexive if and only if its closed unit ball $\mathscr{E}_1$ is weakly compact.*

PROOF. As was noted in the proof of Proposition 16.2 in a somewhat more general context, it is obvious that the natural embedding j of $\mathscr{E}$ in $\mathscr{E}^{**}$ is a homeomorphism between $\mathscr{E}$ in its weak topology and $j(\mathscr{E})$ in the (relative) weak* topology (see also Problem B). If $\mathscr{E}$ is reflexive, so that j maps $\mathscr{E}$ onto $\mathscr{E}^{**}$, then, since j is an isometry, it must also map $\mathscr{E}_1$ onto $(\mathscr{E}^{**})_1$, and the latter set is weak* compact by Alaoglu's theorem (Th. 15.11). Hence $\mathscr{E}_1$ is compact in the weak topology on $\mathscr{E}$. Suppose, on the other hand, that $\mathscr{E}_1$ is weakly compact, and consider the space $\mathscr{E}_w$ consisting of $\mathscr{E}$ equipped with its weak topology. The set $\mathscr{E}_1$ is compact and absolutely convex and has polar $(\mathscr{E}^*)_1$. Hence by Lemma 16.21, a linear functional on $\mathscr{E}^*$ is bounded on $(\mathscr{E}^*)_1$ if and only if it is of the form $j(x)$ for some vector x in $\mathscr{E}$. In other words, j maps $\mathscr{E}$ onto $\mathscr{E}^{**}$. $\qquad\square$

PROBLEMS

A. (i) Let $\mathscr{E}_1,\ldots,\mathscr{E}_n$ be Banach spaces, and let $\mathscr{E} = \mathscr{E}_1 \oplus_p \cdots \oplus_p \mathscr{E}_n$ for some number p, $1 < p < +\infty$ (Prob. 11T). Show that $\mathscr{E}$ is reflexive if and only if each $\mathscr{E}_i$, $i = 1,\ldots,n$, is reflexive. (Hint: See Problem 14D(i).) Is the same result valid for $p = 1$?

(ii) Let $\{\mathscr{E}_\gamma\}_{\gamma\in\Gamma}$ be an indexed family of Banach spaces, let p be a number such that $1 < p < +\infty$, and let $\mathscr{K}_p$ denote the Banach space of all those elements $x = \{x_\gamma\}$ of the full algebraic direct sum of the family $\{\mathscr{E}_\gamma\}$ such that $\|x\| = [\sum_\gamma \|x_\gamma\|^p]^{1/p}$ is finite (see Problem 14D(ii)). Show that $\mathscr{K}_p$ is reflexive if and only if each $\mathscr{E}_\gamma$ is.

B. Let $\mathscr{E}$ be a normed space and let j denote the natural embedding of $\mathscr{E}$ in $\mathscr{E}^{**}$. Show that a net $\{j(x_\lambda)\}_{\lambda\in\Lambda}$ in $j(\mathscr{E})$ converges to a functional $j(x)$ in the weak* topology on $\mathscr{E}^{**}$ if and only if $f(x_\lambda) \to f(x)$ for every f in $\mathscr{E}^*$, and conclude that j is a homeomorphism between $\mathscr{E}$ in its weak topology and $j(\mathscr{E})$ in its (relative) weak* topology. Conclude that the restriction of j to the unit ball $\mathscr{E}_1$ is also a homeomorphism between $\mathscr{E}_1$ in the weak topology and $j(\mathscr{E}) \cap (\mathscr{E}^{**})_1$ in the weak* topology on $\mathscr{E}^{**}$. Show also that $j(\mathscr{E})$ is always weak* dense in $\mathscr{E}^{**}$. (Hint: Use Corollary 16.3.)

C. Let $\mathscr{E}$ be a normed space and let $\mathscr{F}$ denote a finite dimensional subspace of $\mathscr{E}^*$. Show that for a given linear functional φ on $\mathscr{F}$ and a given $\varepsilon > 0$ there exists a vector x in $\mathscr{E}$ such that $\|x\| < \|\varphi\| + \varepsilon$ and such that $\varphi(f) = f(x)$ for every f in $\mathscr{F}$. (Hint: If $\mathscr{N} = {}^a\mathscr{F}$, then $\mathscr{F} = \mathscr{N}^a$. Hence the mapping α of Proposition 16.6 is an isometric isomorphism of $(\mathscr{E}/\mathscr{N})^*$ onto $\mathscr{F}$. If we set $\hat{\phi}(\hat{f}) = \varphi(\alpha(\hat{f}))$ for each linear functional $\hat{f}$ on $\mathscr{E}/\mathscr{N}$, then $\hat{\phi}$ is an element of $(\mathscr{E}/\mathscr{N})^{**}$ with $\|\hat{\phi}\| = \|\varphi\|$, and $\mathscr{E}/\mathscr{N}$ is reflexive by Example A.) Conclude that the image under the natural embedding j of the open unit ball $\mathscr{E}_1^\circ$ is weak* dense in the open unit ball $(\mathscr{E}^{**})_1^\circ$.

D. (Šmul'yan Criterion) Prove that a normed space $\mathscr{E}$ is reflexive if and only if the closed unit ball $\mathscr{E}_1$ is weakly compact. (Hint: To show the necessity of the criterion use Problem B and the theorem of Alaoglu (Th. 15.11); to show its sufficiency use Problems B and C. For an alternate argument see Theorem 16.24.)

E. Let $\mathscr{E}$ be a reflexive Banach space and let T be a bounded linear transformation of $\mathscr{E}$ into a normed space $\mathscr{F}$. Show that T carries an arbitrary closed, bounded, convex subset of $\mathscr{E}$ onto a closed, bounded, convex subset of $\mathscr{F}$. (Hint: Recall Example 15D and the fact (Prob. 15L) that a closed convex set is also weakly closed. Use the Šmul'yan criterion.)

F. Let $\mathscr{E}$ and $\mathscr{F}$ be normed spaces, let T be an element of $\mathscr{L}(\mathscr{E}, \mathscr{F})$, and let $j_{\mathscr{E}}$ and $j_{\mathscr{F}}$ denote the natural embeddings of $\mathscr{E}$ and $\mathscr{F}$, respectively, in their second duals. Show that commutativity holds in the diagram

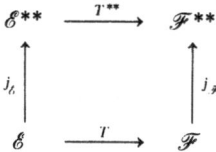

(Thus if $\mathscr{E}$ and $\mathscr{F}$ are reflexive, and if their natural embeddings are used to identify them with their respective second duals, then T^{**} is identified with T.)

G. Let $\mathscr{M}$ be a subspace of a Banach space $\mathscr{E}$.

(i) Verify that the mapping δ described in Proposition 16.12 is given by the composition $\alpha^* \circ \beta$ as shown in the diagram

$$\overset{\delta}{\overbrace{\mathscr{E}^{**} \overset{\beta}{\to} (\mathscr{M}^a)^* \overset{\alpha^*}{\to} (\mathscr{E}/\mathscr{M})^{**}}}$$
$$\mathscr{M}^a \overset{\alpha}{\leftarrow} (\mathscr{E}/\mathscr{M})^*,$$

and complete the proof of the proposition.

(ii) Let i and π denote, respectively, the inclusion mapping of $\mathscr{M}$ into $\mathscr{E}$ and the natural projection of $\mathscr{E}$ onto $\mathscr{E}/\mathscr{M}$. Show that the mapping δ of Proposition 16.12 also coincides with π^{**}. Show similarly that the mapping γ of Proposition 16.12 agrees with i^{**} at every element of $\mathscr{M}^{**}$, so that γ is simply i^{**} regarded as a mapping of $\mathscr{M}^{**}$ onto the range $\mathscr{R}(i^{**}) = \mathscr{M}^{aa}$. Conclude that, in the sequence

$$\mathscr{M}^{**} \overset{i^{**}}{\to} \mathscr{E}^{**} \overset{\pi^{**}}{\to} (\mathscr{E}/\mathscr{M})^{**},$$

i^{**} is one-to-one, π^{**} is onto, and $\mathscr{R}(i^{**}) = \mathscr{K}(\pi^{**})(= \mathscr{M}^{aa})$.

H. Let $\mathscr{E}$ be a Banach space and let $\mathscr{M}$ be a subspace of $\mathscr{E}$.

(i) Show that if $\mathscr{E}$ is reflexive then $\mathscr{M}$ and $\mathscr{E}/\mathscr{M}$ are also reflexive. (Hint: This can be proved using either Problems F and G or the Šmul'yan criterion (Prob. D).)
(ii) Show conversely that if $\mathscr{M}$ and $\mathscr{E}/\mathscr{M}$ are both reflexive, then $\mathscr{E}$ is reflexive too. (Hint: Use Problems F and G.)

In the language of homological algebra Problem G says that

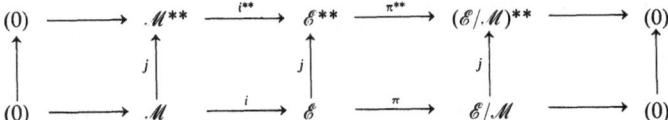

$$(0) \xrightarrow{} \mathcal{M}^{**} \xrightarrow{i^{**}} \mathcal{E}^{**} \xrightarrow{\pi^{**}} (\mathcal{E}/\mathcal{M})^{**} \xrightarrow{} (0)$$

is an exact sequence. Likewise, Problem F says commutativity holds in the diagram

$$\begin{array}{ccccccccc}
(0) & \longrightarrow & \mathcal{M}^{**} & \xrightarrow{i^{**}} & \mathcal{E}^{**} & \xrightarrow{\pi^{**}} & (\mathcal{E}/\mathcal{M})^{**} & \longrightarrow & (0) \\
\uparrow & & \uparrow{\scriptstyle j} & & \uparrow{\scriptstyle j} & & \uparrow{\scriptstyle j} & & \uparrow \\
(0) & \longrightarrow & \mathcal{M} & \xrightarrow{i} & \mathcal{E} & \xrightarrow{\pi} & \mathcal{E}/\mathcal{M} & \longrightarrow & (0)
\end{array}$$

where the j's denote the appropriate natural embeddings. Thus the calculations involved in Problem H all reduce to routine diagram chases. Indeed, we are looking at just one more version of the protean "five lemma" [47; p. 14]; see [67].

I. Let $\mathcal{E}$ and $\mathcal{F}$ be normed spaces and let T be an element of $\mathscr{L}(\mathcal{E}, \mathcal{F})$ having kernel $\mathcal{K} = \mathcal{K}(T)$ and range $\mathcal{R} = \mathcal{R}(T)$. Show that the kernel of T^* is $\mathcal{R}^a$ and that $\mathcal{K} = {}^a(\mathcal{R}(T^*))$. Show also that T^* is continuous when $\mathcal{F}^*$ and $\mathcal{E}^*$ are both equipped with their weak* topologies.

J. Let $\mathcal{E}$ and $\mathcal{F}$ be Banach spaces, and let T be a bounded linear transformation of $\mathcal{E}$ into $\mathcal{F}$ such that $\mathcal{R} = \mathcal{R}(T)$ is closed in $\mathcal{F}$. Prove that the range of T^* coincides with $\mathcal{K}^a$ where $\mathcal{K} = \mathcal{K}(T)$. Thus if T has closed range, then the range of T^* is weak* closed (and therefore norm closed as well). (Hint: If $\hat{T}$ denotes the mapping of $\mathcal{E}/\mathcal{K}$ onto $\mathcal{R}$ obtained by factoring T through $\mathcal{E}/\mathcal{K}$, then T factors into $T = i \circ \hat{T} \circ \pi$, where π is the natural projection of $\mathcal{E}$ onto $\mathcal{E}/\mathcal{K}$ and i is the inclusion mapping of $\mathcal{R}$ into $\mathcal{F}$. Recall Example I and Corollary 16.10.)

K. Let $\mathcal{E}$ and $\mathcal{F}$ be Banach spaces, and let T be an element of $\mathscr{L}(\mathcal{E}, \mathcal{F})$ such that T^* is invertible. Show that T must also be invertible. (Hint: According to Problem I and Corollary 16.5, T is one-to-one and has dense range. Let S denote the linear transformation of $\mathcal{R}(T)$ onto $\mathcal{E}$ that is the set-theoretic inverse of T, and show that S is both closed and bounded. Use Theorem 13.6 to conclude that $\mathcal{R}(T) = \mathcal{F}$.)

It is tempting at this point to conclude that the converse of Problem J also holds, i.e., that if T^* has (norm) closed range, then T must have closed range too, and as a matter of fact, this is true. (This implies that, if an adjoint transformation T^* has closed range, then its range is automatically weak* closed.) But the simple strategy of mimicking the technique used in Problem J fails, and a new approach must be sought; see, for example, [13].

L. Let $\mathcal{E}$ and $\mathcal{F}$ be Banach spaces, and let S and T be elements of $\mathscr{L}(\mathcal{E}, \mathcal{F})$. Show that the range of S^* is included in the range of T^* if and only if there exists a positive constant k such that

$$\|Sx\| \le k\|Tx\|, \qquad x \in \mathcal{E}.$$

(Hint: Let M denote the set $\{x \in \mathcal{E} : \|Tx\| \le 1\}$. If $\mathcal{R}(S^*) \subset \mathcal{R}(T^*)$ and if $f \in \mathcal{F}^*$, then there exists a functional g in $\mathcal{F}^*$ such that $g(Tx) = f(Sx), x \in \mathcal{E}$. Use this fact to show that $S(M)$ is (weakly) bounded in $\mathcal{F}$, and hence that there exists a positive constant k such that $\|Tx\| \le 1$ implies $\|Sx\| \le k$. To go the other way, show that if the stated criterion is satisfied and if $f \in \mathcal{F}^*$, then $Tx \to f(Sx)$ defines a bounded linear functional on $\mathcal{R}(T)$, and use the Hahn–Banach theorem. See also Example 13D.)

M. Let $\mathscr{E}$ be a normed space and let $j_\mathscr{E}$ and $j_{\mathscr{E}^*}$ denote, respectively, the natural embeddings of $\mathscr{E}$ in $\mathscr{E}^{**}$ and $\mathscr{E}^*$ in $\mathscr{E}^{***}$. Verify that $(j_\mathscr{E})^*$ is a left inverse of $j_{\mathscr{E}^*}$, and use this fact to show that $(j_\mathscr{E}(\mathscr{E}))^a$ and $j_{\mathscr{E}^*}(\mathscr{E}^*)$ are complementary subspaces in $\mathscr{E}^{***}$. (Hint: See Problem 13E.)

N. Show that a Banach space $\mathscr{E}$ is reflexive if and only if $\mathscr{E}^*$ is reflexive. (Hint: This fact is an immediate consequence of the preceding problem, but it may also be derived from the Šmul'yan criterion.)

O. Let (φ) denote the vector space consisting of all those complex sequences that are eventually zero, and let (σ) denote the vector space of all complex sequences. Show that $\langle(\varphi),(\sigma)\rangle$ is a dual pair with respect to the bilinear functional defined on $(\varphi) \dotplus (\sigma)$ by

$$\langle x, y \rangle = \sum_{n=0}^{\infty} \xi_n \eta_n,$$

where $x = \{\xi_n\}_{n=0}^{\infty}$ and $y = \{\eta_n\}_{n=0}^{\infty}$. Show also that if $\{x_m\}_{m=1}^{\infty}$ is a sequence in (φ) and if $x_m = \{\xi_n^{(m)}\}_{n=0}^{\infty}$ for each index m, then the sequence $\{x_m\}$ is weakly convergent in (φ) if and only if the sequence $\{\xi_n^{(m)}\}_{m=1}^{\infty}$ is convergent in $\mathbb{C}$ for every nonnegative integer n and there exists a fixed nonnegative integer N such that $\xi_n^{(m)} = 0$ for all $n \geq N$ and all sufficiently large m.

P. Let $\langle \mathscr{E}, \mathscr{F} \rangle$ be a dual pair, let $\mathscr{L}$ be a linear manifold in $\mathscr{E}$, and let $\mathscr{L}^a$ be the annihilator of $\mathscr{L}$. (See Example L for notation and terminology.)

   (i) Show that $\langle \mathscr{L}, \mathscr{F}/\mathscr{L}^a \rangle$ is a dual pair with respect to the bilinear functional obtained by setting $\langle x, y + \mathscr{L}^a \rangle = \langle x, y \rangle$, $x \in \mathscr{L}$, $y \in \mathscr{F}$, and show that the topology induced on $\mathscr{L}$ by $\mathscr{F}/\mathscr{L}^a$ in this dual pair coincides with the relative topology induced on $\mathscr{L}$ by the weak topology on $\mathscr{E}$.

   (ii) Show, similarly, that $\langle \mathscr{E}/\mathscr{L}, \mathscr{L}^a \rangle$ is a dual pair with respect to the bilinear functional $\langle x + \mathscr{L}, y \rangle = \langle x, y \rangle$, $x \in \mathscr{E}$, $y \in \mathscr{L}^a$, if and only if $\mathscr{L}$ is weakly closed in $\mathscr{E}$. Verify that, when this latter condition is satisfied, the topology induced on $\mathscr{E}/\mathscr{L}$ by $\mathscr{L}^a$ coincides with the quotient topology on $\mathscr{E}/\mathscr{L}$ obtained from the weak topology on $\mathscr{E}$. (Hint: Suppose $\mathscr{L}$ is weakly closed in $\mathscr{E}$, and let $\mathscr{T}$ denote the quotient topology on $\mathscr{E}/\mathscr{L}$ obtained from the weak topology on $\mathscr{E}$. The functionals defined on $\mathscr{E}/\mathscr{L}$ by $\mathscr{L}^a$ in the dual pair $\langle \mathscr{E}/\mathscr{L}, \mathscr{L}^a \rangle$ are clearly continuous with respect to $\mathscr{T}$, so $\mathscr{T}$ refines the topology induced on $\mathscr{E}/\mathscr{L}$ by $\mathscr{L}^a$. Thus the trick is to show that the latter refines $\mathscr{T}$. Let $U = U(y_1, \ldots, y_p; \varepsilon)$ be a typical weak neighborhood of 0 in $\mathscr{E}$, let $\mathscr{N}$ denote the linear submanifold of $\mathscr{F}$ generated algebraically by $\mathscr{L}^a$ and the vectors $y_1, \ldots, y_p$, and let $\mathscr{M}$ be an algebraic complement of $\mathscr{L}^a$ in $\mathscr{N}$, so that $\mathscr{L}^a \cap \mathscr{M} = (0)$ and $\mathscr{L}^a + \mathscr{M} = \mathscr{N}$. Note that $\mathscr{M}$ is finite dimensional (dim $\mathscr{M} \leq p$), and use this fact and the fact that $\mathscr{M} \cap \mathscr{L}^a = (0)$ to show that the linear functionals defined on $\mathscr{M}$ by the vectors in $\mathscr{L}$ fill the entire (algebraic) dual of $\mathscr{M}$. Hence to each vector x in $\mathscr{E}$ there corresponds a vector v_x in $\mathscr{L}$ such that $\langle x, z \rangle = \langle v_x, z \rangle$, $z \in \mathscr{M}$. Write $y_i = w_i + z_i$, where $w_i \in \mathscr{L}^a$ and $z_i \in \mathscr{M}$, $i = 1, \ldots, p$. Show that $\langle x - v_x, y_i \rangle = \langle x, w_i \rangle$ for $i = 1, \ldots, p$, and every vector x in $\mathscr{E}$, and conclude that $\pi(U(w_1, \ldots, w_p; \varepsilon))$ is a weak neighborhood of 0 in $\mathscr{E}/\mathscr{L}$ for the dual pair $\langle \mathscr{E}/\mathscr{L}, \mathscr{L}^a \rangle$ that is contained in $\pi(U)$.

(iii) Show, dually, that if $\mathcal{M}$ is an arbitrary linear manifold in $\mathcal{F}$, and if $^a\mathcal{M}(=\,^0\mathcal{M})$ is the preannihilator of $\mathcal{M}$ in $\mathcal{E}$, then $\langle \mathcal{E}/^a\mathcal{M}, \mathcal{M} \rangle$ is a dual pair with respect to the bilinear functional $\langle x + ^a\mathcal{M}, y \rangle = \langle x, y \rangle$, $x \in \mathcal{E}$, $y \in \mathcal{M}$. Verify also that the topology induced on $\mathcal{E}/^a\mathcal{M}$ by $\mathcal{M}$ coincides with the quotient topology on $\mathcal{E}/^a\mathcal{M}$ obtained from the weak topology on $\mathcal{E}$ when and only when $\mathcal{M}$ is closed in the topology induced on $\mathcal{F}$ by $\mathcal{E}$. (Hint: According to (ii) the quotient topology on $\mathcal{E}/^a\mathcal{M}$ is induced by $(^a\mathcal{M})^a$.)

Q. Let $\langle \mathcal{E}, \mathcal{F} \rangle$ and $\langle \mathcal{E}', \mathcal{F}' \rangle$ be dual pairs, and let T be a linear transformation of $\mathcal{E}$ into $\mathcal{E}'$ Show that there exists a mapping $T^*: \mathcal{F}' \to \mathcal{F}$ satisfying the condition

$$\langle Tx, y' \rangle = \langle x, T^*y' \rangle, \qquad x \in \mathcal{E}, \qquad y' \in \mathcal{F}', \tag{2}$$

if and only if T is *weakly continuous*, i.e., continuous when $\mathcal{E}$ and $\mathcal{E}'$ are given the topologies induced by $\mathcal{F}$ and $\mathcal{F}'$, respectively. Show also that if T is weakly continuous then T^* is a uniquely determined linear transformation. (The mapping T^* is called the *adjoint* of T.) In the remainder of this problem we suppose T to be weakly continuous.

(i) Verify that the adjoint T^* is in turn weakly continuous with respect to the reversed dual pairs $\langle \mathcal{F}', \mathcal{E}' \rangle$ and $\langle \mathcal{F}, \mathcal{E} \rangle$, and that the adjoint of T^* coincides with T. State and prove analogs of Propositions 16.8 and 16.9 and Corollary 16.10 for this "weak" notion of adjoint.

(ii) Show that $\mathcal{K}(T^*) = (\mathcal{R}(T))^a$, and likewise that $\mathcal{K}(T) = {}^a(\mathcal{R}(T^*))$.

(iii) Show that if $\mathcal{K}(T) = (0)$, then $\langle \mathcal{E}, \mathcal{R}(T^*) \rangle$ is a dual pair, and that T is, in fact, a homeomorphism of $\mathcal{E}$ onto $T(\mathcal{E})$ if $\mathcal{E}$ is given the topology induced by $\mathcal{R}(T^*)$ and $T(\mathcal{E})$ is given the relative weak topology. (Hint: Use nets and the adjoint relation (2); see Corollary 3.19.)

(iv) Prove that T is an open mapping of $\mathcal{E}$ (in the weak topology) onto $T(\mathcal{E})$ (in the relative weak topology) if and only if $\mathcal{R}(T^*)$ is closed in $\mathcal{F}$ in the topology induced by $\mathcal{E}$. (Hint: Set $\mathcal{R} = \mathcal{R}(T^*)$ and let $T = \hat{T} \circ \pi$ be the factorization of T through $\mathcal{E}/\mathcal{K}(T) = \mathcal{E}/^a\mathcal{R}$. According to (iii), $\hat{T}$ is a homeomorphism of $\mathcal{E}/^a\mathcal{R}$ onto $T(\mathcal{E})$ if $\mathcal{E}/^a\mathcal{R}$ is given the topology induced by $\mathcal{R}$. On the other hand, T is clearly open when and only when $\hat{T}$ is open in the quotient topology on $\mathcal{E}/^a\mathcal{R}$. Recall (Prob. P) that these two topologies on $\mathcal{E}/^a\mathcal{R}$ agree if and only if $\mathcal{R}$ is closed in the topology induced on $\mathcal{F}$ by $\mathcal{E}$. The reader should contrast this result with Problem J.)

R. Let $\langle \mathcal{E}, \mathcal{F} \rangle$ be a dual pair and let C be a convex subset of $\mathcal{E}$. Show that if C is weakly closed, then C is closed in all of the topologies of the dual pair $\langle \mathcal{E}, \mathcal{F} \rangle$. Conclude that the closure of C is the same for every topology of the dual pair $\langle \mathcal{E}, \mathcal{F} \rangle$. (Hint: Use Proposition 14.15.)

S. Let $\langle \mathcal{E}, \mathcal{F} \rangle$ be a dual pair. Show that the collection of all weakly compact [absolutely] convex subsets of $\mathcal{E}$ is admissible. (Hint: Show first that if $C_1, \ldots, C_p$ are arbitrary convex subsets of a linear space, then the convex hull of the union $C_1 \cup \cdots \cup C_p$ is just the set of all convex combinations $t_1 x_1 + \cdots + t_p x_p$, where $x_i \in C_i$, $i = 1, \ldots, p$. Show next that if $\mathcal{E}$ is equipped with a linear topology with respect to which the sets C_i are all compact, then the convex hull of the union $C_1 \cup \cdots \cup C_p$ is also compact (Prop. 3.3). To complete the argument, show that the convex hull of a finite union of absolutely convex sets is balanced.) Conclude that there is always a *finest* topology of a dual pair $\langle \mathcal{E}, \mathcal{F} \rangle$, viz., the topology of uniform convergence on the collection of all those absolutely convex subsets of $\mathscr{F}$ that are compact in the topology induced by $\mathscr{E}$. This topology is called the *Mackey topology* on $\mathscr{E}$ with respect to the dual pair $\langle \mathscr{E}, \mathscr{F} \rangle$. Verify that if $\mathscr{E}$ is a separated locally convex space, then the Mackey topology on $\mathscr{E}$ with respect to the dual pair $\langle \mathscr{E}, \mathscr{E}^* \rangle$ agrees with the Mackey topology as defined in Example 15K.

T. (i) A subset B of a separated locally convex space $\mathscr{E}$ is called a *barrel* in $\mathscr{E}$ if B is closed, absolutely convex, and absorbing. Verify that a subset B of $\mathscr{E}$ is a barrel if and only if there exists a weak* bounded subset A of $\mathscr{E}^*$ such that $B = {}^0 A$ in the dual pair $\langle \mathscr{E}, \mathscr{E}^* \rangle$. (Hint: If B is a barrel, then $B = {}^0(B^0)$; use Proposition 16.17.)

(ii) A separated locally convex space $\mathscr{E}$ is said to be *barreled* if every barrel in $\mathscr{E}$ is a neighborhood of 0 in $\mathscr{E}$. Show that $\mathscr{E}$ is barreled if and only if every subset of $\mathscr{E}^*$ that is weak* bounded is also equicontinuous on $\mathscr{E}$. Conclude that the given locally convex topology on a barreled space $\mathscr{E}$ coincides with both the Mackey topology and the strong topology (in the dual pair $\langle \mathscr{E}, \mathscr{E}^* \rangle$). (Hint: Recall Example N.)

U. Verify that every Frechét space is barreled. (Hint: If B is a barrel in a Frechét space $\mathscr{E}$, then $\{nB\}_{n=1}^{\infty}$ is a countable closed covering of the complete metric space $\mathscr{E}$. Use the Baire category theorem (Th. 4.8) and the fact that the interior A° of an absolutely convex set A contains 0 whenever A° is not empty (Prob. 14N).

V. Let $\mathscr{E}$ be a separated locally convex space, and let B be a barrel in $\mathscr{E}$.

(i) Let C be a bounded convex subset of $\mathscr{E}$ that contains 0. Show that if there exists a neighborhood V of 0 such that $C \cap V$ is absorbed by B, then C itself is absorbed by B. (Hint: C is absorbed by V, and if $C \subset kV$ for some $k \geq 1$, then $C \subset k(C \cap V)$.)

(ii) Let K be a compact convex subset of $\mathscr{E}$. Show that there exists a point x_0 in K and a neighborhood V of 0 in $\mathscr{E}$ such that $K \cap (x_0 + V)$ is absorbed by B. (Hint: The sequence $\{K \cap nB\}_{n=1}^{\infty}$ is a closed covering of K, and K is a compact Hausdorff space; recall Problem 4W.) Use this and (i) to deduce that, in fact, B absorbs K. (Hint: The set $K - x_0$ is compact, convex, and contains 0. Moreover, if $K \cap (x_0 + V) \subset nB$, then $(K - x_0) \cap V \subset nB - x_0$.)

W. (i) (Mackey [46]) Show that every weakly bounded subset of a separated locally convex space $\mathscr{E}$ is bounded with respect to the given topology on $\mathscr{E}$. Conclude that exactly the same sets are bounded with respect to all of the topologies of an arbitrarily given dual pair $\langle \mathscr{E}, \mathscr{F} \rangle$. (Hint: Let A be a weakly bounded subset of $\mathscr{E}$ and let V be a closed absolutely convex neighborhood of 0 in the given topology on $\mathscr{E}$. Then (Prop. 16.17) A^0 is a barrel in $\mathscr{E}^*$, while V^0 is compact and absolutely convex (Th. 16.23), both in the weak* topology. Hence A^0 absorbs V^0 (Prob. V), whence it follows at once that ${}^0(V^0) = V$ absorbs ${}^0(A^0) \supset A$.)

(ii) If the topology on a separated locally convex space $\mathscr{E}$ is metrizable (that is, if $\mathscr{E}$ satisfies the first axiom of countability; cf. Proposition 14.8), then the given topology on $\mathscr{E}$ coincides with the Mackey topology on $\mathscr{E}$ (Ex. 15K). (Hint: Let $\{V_n\}_{n=1}^{\infty}$ be a nested neighborhood base at 0 in $\mathscr{E}$, let W be an arbitrary neighborhood of 0 in the Mackey topology on $\mathscr{E}$, and suppose W does not contain any neighborhood V_n. Conclude that for each positive integer n, nW does not contain V_n, and let $x_n \in V_n \backslash nW$, $n \in \mathbb{N}$. The sequence $\{x_n\}$ converges to 0 in the given topology on $\mathscr{E}$ but is unbounded with respect to the Mackey topology, a contradiction.)

X. Let $\mathscr{E}$ be a separated locally convex space.

(i) Verify that the collection of all compact subsets of $\mathscr{E}$ is admissible, and hence that the topology $\mathscr{T}$ of uniform convergence on compact subsets of $\mathscr{E}$ is an admissible polar topology on $\mathscr{E}^*$. Verify also that if $\{f_\lambda\}_{\lambda\in\Lambda}$ is a net in $\mathscr{E}^*$ and if there exists a neighborhood V of 0 in $\mathscr{E}$ such that $f_\lambda \in V^0$, $\lambda \in \Lambda$, then $\{f_\lambda\}$ converges to a (continuous linear) functional f in the topology $\mathscr{T}$ if and only if $\{f_\lambda\}$ converges to f in the weak* topology, that is, pointwise. Conclude that $\mathscr{T}$ induces the same topology on equicontinuous subsets of $\mathscr{E}^*$ as does the weak* topology. (Hint: If $\{f_\lambda\}$ converges pointwise to f, then $f \in V^0$. Hence, for a given positive ε, f and f_λ differ by no more than $|f_\lambda(x) - f(x)| + 2\varepsilon$ on the set $x + \varepsilon V$, and sets of this form cover $\mathscr{E}$. Recall Example N.)

(ii) Let U and V be neighborhoods of 0 in $\mathscr{E}$ such that $U \subset V$, let W' be a subset of U^0 that is open in the relative weak* topology on U^0, and suppose given a subset M of $\mathscr{E}$ such that $(M \cup V)^0 \subset W'$. Show that there exists a finite subset D of V such that $(D \cup M)^0 \cap (U^0 \backslash W') = \varnothing$, and hence such that $(D \cup M \cup U)^0 \subset W'$. (Hint: Let $\mathscr{D}$ denote the collection of all sets of the form $(D \cup M)^0 \cap (U^0 \backslash W')$, where D ranges over the finite subsets of V. The intersection of $\mathscr{D}$ is $(V \cup M)^0 \cap (U^0 \backslash W') = \varnothing$, but, if the desired conclusion were false, $\mathscr{D}$ would have the finite intersection property (Prob. 3Q).) Use this fact to prove that if $\mathscr{E}$ satisfies the first axiom of countability, and if W is an arbitrary subset of $\mathscr{E}^*$ with the property that $W \cap V^0$ is open in the relative weak* topology on V^0 for every neighborhood V of 0 in $\mathscr{E}$, then W is open in the topology $\mathscr{T}$ of uniform convergence on the compact subsets of $\mathscr{E}$. (Thus the topology of uniform convergence on the compact subsets of a metrizable locally convex space $\mathscr{E}$ is characterized as the collection of all those subsets W of $\mathscr{E}^*$ such that $W \cap V^0$ is relatively weak* open in V^0 for every neighborhood V of 0 in $\mathscr{E}$.) (Hint: Let $\{V_n\}_{n=1}^\infty$ be a nested neighborhood base at 0 in $\mathscr{E}$, set $V_0 = \mathscr{E}$, and let W be a subset of $\mathscr{E}^*$ having the stated property and containing the origin 0 in $\mathscr{E}^*$. Use mathematical induction to construct finite subsets D_n of V_n such that $(D_0 \cup \cdots \cup D_n \cup V_{n+1})^0 \subset W$, $n \in \mathbb{N}_0$, set $K = \{0\} \cup \bigcup_{n=1}^\infty D_n$, and verify that $K^0 \subset W$.)

(iii) Show that if $\mathscr{E}$ is a Frechét space and if C is a convex subset of $\mathscr{E}^*$, then C is weak* closed if and only if $C \cap V_\rho^0$ is weak* closed for every neighborhood V of 0 in $\mathscr{E}$ (belonging to some specified neighborhood base at 0). (Hint: Recall Example U and the fact (Prob. 14N) that if K is a compact set in $\mathscr{E}$, then ${}^0(K^0)$ is also compact. Conclude that the topology of uniform convergence on the compact subsets of $\mathscr{E}$ is a topology of the dual pair $\langle \mathscr{E}^*, \mathscr{E} \rangle$, and use Problem R.)

(iv) Verify, in particular, that if $\mathscr{E}$ is a Banach space and if $\mathscr{L}$ is a linear manifold in $\mathscr{E}^*$, then $\mathscr{L}$ is weak* closed if and only if $\mathscr{L} \cap (\mathscr{E}^*)_1$ is.

17

Banach spaces and integration theory

In this chapter we introduce some topics that combine in different, but not unrelated, ways the theory of normed spaces discussed thus far in Part II and the theories of integration discussed in Chapters 6–10 of Part I. Our first observation is that there are natural and important generalizations of the (ℓ_p) spaces obtained by employing Lebesgue integration in place of summation.

Proposition 17.1. *If* $(X, \mathbf{S}, \mu)$ *is a measure space and if* p *is an arbitrary positive number, then the collection* $\mathscr{L}_p(X) = \mathscr{L}_p(X, \mathbf{S}, \mu)$ *of all those measurable complex-valued functions* f *on* $(X, \mathbf{S})$ *with the property that* $|f|^p$ *is integrable* $[\mu]$ *forms a linear space (with respect to pointwise linear operations). Moreover, if* $p \geq 1$ *(and, for nontrivial* X, *only if* $p \geq 1$),

$$\| f \|_p = \left[\int_X |f|^p \, d\mu \right]^{1/p} \tag{1}$$

defines a pseudonorm on $\mathscr{L}_p(X)$.

PROOF. That $\mathscr{L}_p(X)$ is a linear space for every value of p follows from the elementary inequality $(u + v)^p \leq 2^p(u^p + v^p)$, valid for $u, v \geq 0$ and all positive p. To show that the function defined in (1) is a pseudonorm on $\mathscr{L}_p(X)$ it suffices to verify the triangle inequality, and this argument is exactly analogous to the one given in the case of the (ℓ_p) spaces; cf. Problem A. Finally, if $0 < p < 1$ and if the measure space X contains any two disjoint measurable sets of positive finite measure, then it is a simple matter to verify that the function defined in (1) is not a pseudonorm on $\mathscr{L}_p(X)$; cf. Example 11C. □

364

The counterpart of Proposition 17.1 for $p = +\infty$ is also trivially valid; verification of the details is left to the interested reader. (The notions of essential boundedness and essential supremum are defined in Chapter 7.)

Proposition 17.2. *If $(X, \mathbf{S}, \mu)$ is a measure space, then the collection $\mathscr{L}_\infty(X) = \mathscr{L}_\infty(X, \mathbf{S}, \mu)$ of all essentially bounded, measurable, complex-valued functions on X is a linear space (with respect to pointwise linear operations), and setting*

$$\| f \|_\alpha = \operatorname*{ess\,sup}_X | f | \tag{2}$$

defines a pseudonorm on $\mathscr{L}_\infty(X)$.

Note. The spaces $\mathscr{L}_p(X)$ really are generalizations of the spaces (ℓ_p), $1 \le p \le +\infty$. Indeed, $\mathscr{L}_p(X, \mathbf{S}, \mu)$ coincides with (ℓ_p) if we take $\mathbb{N}_0$ for X and let μ be the counting measure on $\mathbb{N}_0$ (Ex. 7J).

If $(X, \mathbf{S}, \mu)$ is a measure space, and if $\mathscr{Z}$ denotes the linear space of all those measurable complex-valued functions on X that vanish outside of some set of measure zero $[\mu]$, then $\mathscr{Z}$ is a linear manifold in $\mathscr{L}_p(X)$ for $1 \le p \le +\infty$; indeed, $\mathscr{Z}$ is the zero space of the pseudonorm $\| \; \|_p$ for all $1 \le p \le +\infty$. Thus the elements of the associated space $\dot{\mathscr{L}}_p(X)$ (Prop. 11.17) are, in all cases, equivalence classes of functions equal a.e. $[\mu]$, and we confront a dilemma. The elements of $\mathscr{L}_p$ are familiar objects, viz., functions, but $\| \; \|_p$ is only a pseudonorm; the associated function $\| \; \|_p$ on $\dot{\mathscr{L}}_p(X)$ is a norm, all right, but the elements of $\dot{\mathscr{L}}_p(X)$ are not functions but equivalence classes of functions. What we shall do about this distinction is, briefly, ignore it, and let the symbol $\mathscr{L}_p(X)$ stand for either the pseudonormed space or the associated normed space as the occasion demands. This practice, while it is the usual one, is more convenient than logical, and in all candor it must be admitted that it brings with it a certain amount of ambiguity. There are many places in the sequel where $\mathscr{L}_p(X)$ must be thought of as a vector space of complex-valued functions equipped with the pseudonorm $\| \; \|_p$, but there are also many others where $\mathscr{L}_p(X)$ must be regarded as the associated space of equivalence classes, even though the elements of the space are spoken of just as if they were functions. (A standard way of describing this state of affairs is to say that the functions constituting $\mathscr{L}_p(X)$ are "only defined up to sets of measure zero".) In defense of our convention it should be said that at no time will this ambiguity give rise to any genuine mathematical embarrassment, and also that the reader will soon learn to spot easily and unerringly which interpretation is required in any given case.

(In actual practice matters are even a little more confused than the preceding explanation admits to, owing to the universal practice of regarding a function on a measure space as "defined" if its domain of definition contains the complement of some set of measure zero. Thus if $(X, \mathbf{S}, \mu)$ is a measure space, then a function f that is defined on a subset A of X will be regarded as an element of $\mathscr{L}_p(X, \mathbf{S}, \mu)$, $1 \leq p < +\infty$, if there exists a set $D \subset A$ with $\mu(X \backslash D) = 0$ such that f is measurable on D and such that $\int_D |f|^p \, d\mu < +\infty$. Likewise, if $\tilde{f}$ is a function defined on a subset $\tilde{A}$ of X, and if $\tilde{f}$ is measurable on a set $\tilde{D} \subset \tilde{A}$ such that $\mu(X \backslash \tilde{D}) = 0$, then f and $\tilde{f}$ will be regarded as the same element of $\mathscr{L}_p(X)$ if $f = \tilde{f}$ a.e. $[\mu]$ on $D \cap \tilde{D}$. Similar conventions are also observed concerning $\mathscr{L}_\alpha(X, \mathbf{S}, \mu)$.)

Definition. For any measure space $(X, \mathbf{S}, \mu)$ the normed spaces $\mathscr{L}_p(X, \mathbf{S}, \mu)$, $1 \leq p \leq +\infty$, are known as the *Lebesgue spaces* on X.

Example A. Let $(X, \mathbf{S}, \mu)$ be a finite measure space, and suppose $1 \leq p \leq p' < +\infty$. If f is an arbitrary element of $\mathscr{L}_{p'}(X)$, and if we write

$$E = \{x \in X : |f(x)| < 1\},$$

then

$$\int_X |f|^p \, d\mu = \left(\int_E + \int_{X \backslash E} \right) |f|^p \, d\mu \leq \mu(X) + \int_X |f|^{p'} \, d\mu < +\infty.$$

Thus $\mathscr{L}_{p'}(X) \subset \mathscr{L}_p(X)$ when $1 \leq p \leq p' < +\infty$ and $\mu(X) < +\infty$. Moreover, it is clear that $\mathscr{L}_\infty(X) \subset \mathscr{L}_p(X)$ for all $p \geq 1$ when $\mu(X) < +\infty$. (This behavior contrasts sharply with the relation between the various spaces (ℓ_p) where, as we know, $(\ell_1) \subset (\ell_p) \subset (\ell_{p'}) \subset (\ell_\infty)$ whenever $1 \leq p \leq p' \leq +\infty$ (Prob. 11C).

Example B. Let $(X, \mathbf{S}, \mu)$ be the direct sum $\sum_\gamma \oplus (X_\gamma, \mathbf{S}_\gamma, \mu_\gamma)$ of an indexed family $\{(X_\gamma, \mathbf{S}_\gamma, \mu_\gamma)\}_{\gamma \in \Gamma}$ of pairwise disjoint measure spaces (see Proposition 8.3), and let p be a real number such that $1 \leq p < +\infty$. Then the normed space $\mathscr{L}_p(X, \mathbf{S}, \mu)$ is readily seen to coincide with the space $\mathscr{K}_p$ formed as in Problem 11T from the family of spaces $\{\mathscr{L}_p(X_\gamma, \mathbf{S}_\gamma, \mu_\gamma)\}$. (In particular, if $\Gamma = \{1, \ldots, n\}$ then $\mathscr{L}_p(X) = \mathscr{L}_p(X_1) \oplus_p \cdots \oplus_p \mathscr{L}_p(X_n)$.) Likewise $\mathscr{L}_\alpha(X, \mathbf{S}, \mu)$ coincides with the space $\mathscr{B}$ formed as in Problem 11S from the family $\{\mathscr{L}_\infty(X_\gamma)\}_{\gamma \in \Gamma}$. (Note that this implies that $\mathscr{L}(\mathscr{L}_p(X))$ is never separable for any p, $1 \leq p \leq +\infty$, except when X is a finite union of atoms; see Problem 12T.)

Examples A and B illustrate nicely the notational ambiguity alluded to above. In Example A the spaces $\mathscr{L}_p(X)$ and $\mathscr{L}_{p'}(X)$ are really the pseudo-normed spaces whose elements are complex-valued functions, while in Example B the symbols $\mathscr{L}_p(X, \mathbf{S}, \mu)$ and $\mathscr{L}_p(X_\gamma, \mathbf{S}_\gamma, \mu_\gamma)$ must be interpreted as standing for the associated normed spaces $\dot{\mathscr{L}}_p(X, \mathbf{S}, \mu)$ and $\dot{\mathscr{L}}_p(X_\gamma, \mathbf{S}_\gamma, \mu_\gamma)$.

Example C. Suppose we take for $(X, \mathbf{S}, \mu)$ the real interval $[a, b]$ equipped with Lebesgue measure. Then for any p, $1 \leq p \leq +\infty$, the formula

$$(Jf)(x) = \int_a^x f(t) \, dt, \qquad a \leq x \leq b,$$

defines a bounded linear operator J on $\mathscr{L}_p([a, b])$. Indeed, for $p = +\infty$ it is
clear that J is bounded and that, in fact, $\|J\| = b - a$. To take care of the
case $1 \le p < +\infty$ we define the function

$$V(x, t) = \begin{cases} 1, & a \le t \le x \le b, \\ 0, & a \le x < t \le b, \end{cases}$$

on the square $[a, b] \times [a, b]$. Then

$$(Jf)(x) = \int_a^b V(x, t)f(t)dt, \qquad a \le x \le b, \qquad f \in \mathscr{L}_p([a, b]).$$

Hence for $p = 1$ we may use the Fubini and Tonelli theorems (Ths. 9.2 and
9.3) to write

$$\|Jf\|_1 \le \int_a^b \left[\int_a^b V(x, t)|f(t)|dx \right] dt = \int_a^b |f(t)|(b - t)dt \le (b - a)\|f\|_1$$

for an arbitrary element f of $\mathscr{L}_1([a, b])$, which shows that $\|J\| \le b - a$.
(Here, once again, it is easy to verify that $\|J\| = b - a$.) Finally, for
$1 < p < +\infty$ we employ the Hölder inequality (Prob. A). If q denotes the
Hölder conjugate of p, then

$$|(Jf)(x)| \le \left[\int_a^b V(x, t)^q dt \right]^{1/q} \|f\|_p = (x - a)^{1/q} \|f\|_p, \qquad a \le x \le b.$$

Hence $\|Jf\|_p \le [\int_a^b (x - a)^{p/q} dx]^{1/p} \|f\|_p$, which shows that

$$\|J\| \le (b - a)^{1/p + 1/q} = b - a.$$

(For $1 < p < +\infty$ it is not true that $\|J\| = b - a$.)

It is time to verify that the Lebesgue spaces are complete. The following
lemma is of some interest in its own right.

Lemma 17.3. *Let (X, S, μ) be a measure space and, for an arbitrary p,
$1 \le p \le +\infty$, let $\{g_n\}$ be a sequence of functions in $\mathscr{L}_p(X)$ with the property
that*

$$\sum_{n=1}^\infty \|g_n\|_p < +\infty.$$

*Then the series of functions $\sum_{n=1}^\infty g_n$ converges a.e. $[\mu]$ to a function g in
$\mathscr{L}_p(X)$ and*

$$\|g\|_p \le \sum_{n=1}^\infty \|g_n\|_p.$$

PROOF. The case $p = +\infty$ is trivial and will be omitted. For $1 \le p < +\infty$,
we note first that it suffices to treat the case in which the functions g_n are all

nonnegative (for we may simply replace g_n by $|g_n|$). Set $h_n = g_1 + \cdots + g_n$. Then $\|h_n\|_p \leq \|g_1\|_p + \cdots + \|g_n\|_p$ by the Minkowski inequality (Prob. A), and consequently

$$\int_X h_n^p \, d\mu = \|h_n\|_p^p \leq \left[\sum_{k=1}^n \|g_k\|_p\right]^p \leq \left[\sum_{n=1}^\infty \|g_n\|_p\right]^p < +\infty. \qquad (3)$$

Since $\{h_n^p\}_{n=1}^\infty$ is a monotone increasing sequence, it follows from the monotone convergence theorem (Th. 7.9) that the sequence $\{h_n^p\}$ converges a.e. $[\mu]$. Hence $\{h_n\}$ also converges a.e. $[\mu]$, and if we write $g = \lim_n h_n$, then $h_n^p \to g^p$ a.e. as well. But then by (3) we have $\int_X g^p \, d\mu \leq ([\sum_{n=1}^\infty \|g_n\|_p]^p)$, and the lemma follows. (The function g is, in general, undefined on a set of measure zero; nevertheless, g represents a unique element of $\mathscr{L}_p(X)$ according to the convention announced above.) $\square$

Theorem 17.4. *For any measure space* $(X, \mathbf{S}, \mu)$ *the Lebesgue spaces* $\mathscr{L}_p(X, \mathbf{S}, \mu)$, $1 \leq p \leq +\infty$, *are complete, and are therefore Banach spaces.*

PROOF. It suffices as always (see Problem 4E) to show that if a Cauchy sequence $\{f_n\}$ satisfies the added condition

$$\sum_{n=1}^\infty \|f_{n+1} - f_n\|_p < +\infty, \qquad (4)$$

then there exists a function f in $\mathscr{L}_p(X)$ such that $\lim_n \|f - f_n\|_p = 0$. But if (4) holds, then, applying the preceding lemma to the series

$$f_1 + \sum_{n=1}^\infty (f_{n+1} - f_n),$$

we ascertain that the sequence $\{f_n\}$ does indeed converge a.e. $[\mu]$ to a limit f in $\mathscr{L}_p(X)$. Moreover, if ε is a positive number, and if N is chosen large enough so that

$$\sum_{n=N}^\infty \|f_{n+1} - f_n\|_p < \varepsilon,$$

then, applying the lemma once again to the residual series

$$\sum_{n=m}^\infty (f_{n+1} - f_n),$$

we see that $\|f - f_m\|_p < \varepsilon$ for all $m \geq N$, so that the sequence $\{f_n\}$ actually converges to f in $\mathscr{L}_p(X)$. $\square$

Example D. Let n be a fixed positive integer. Let **B** denote the σ-ring of Borel subsets of Euclidean space $\mathbb{R}^n$, and let μ_n denote Lebesgue–Borel measure on $\mathbb{R}^n$. Then the spaces $\mathscr{L}_p(\mathbb{R}^n, \mathbf{B}, \mu_n)$, $1 \leq p \leq +\infty$, are important instances of Lebesgue spaces. Throughout this book the notation $\mathscr{L}_p(\mathbb{R}^n)$, used without any further explanation, will denote the space $\mathscr{L}_p(\mathbb{R}^n, \mathbf{B}, \mu_n)$. (More generally, if E is any Borel set in $\mathbb{R}^n$, then $\mathscr{L}_p(E)$ will be understood to mean the Lebesgue space $\mathscr{L}_p(E, \mathbf{B}_E, \mu_n | E)$, where $\mathbf{B}_E$ denotes the σ-ring of Borel subsets of E.) For each complex-valued function f on $\mathbb{R}^n$ and each fixed element a of $\mathbb{R}^n$ let us write $f_{(a)}$ for the *translate* of f defined by

$$f_{(a)}(x) = f(x + a), \quad x \in \mathbb{R}^n.$$

Since μ_n is translation invariant (Prob. 8F), it is clear that if $f \in \mathscr{L}_p(\mathbb{R}^n)$ for some p, $1 \le p \le +\infty$, then $f_{(a)} \in \mathscr{L}_p(\mathbb{R}^n)$ too, and $\| f_{(a)} \|_p = \| f \|_p$. Thus if we write T_a for the mapping defined by $T_a f = f_{(a)}, f \in \mathscr{L}_p(\mathbb{R}^n)$, $a \in \mathbb{R}^n$, then T_a is an isometric isomorphism of $\mathscr{L}_p(\mathbb{R}^n)$ onto itself (with $(T_a)^{-1} = T_{-a}$) for each point a of $\mathbb{R}^n$ and each p, $1 \le p \le +\infty$. Moreover, if Φ denotes the mapping $a \xrightarrow{\Phi} T_a$ of $\mathbb{R}^n$ into $\mathscr{L}(\mathscr{L}_p(\mathbb{R}^n))$, then $\Phi(a + b) = T_{a+b} = T_a T_b = \Phi(a)\Phi(b)$ for all a, b in $\mathbb{R}^n$, since $T_{a+b} f = f_{(a+b)} = (f_{(b)})_{(a)} = T_a(T_b f)$ for each f in $\mathscr{L}_p(\mathbb{R}^n)$. (This fact is customarily expressed by saying that Φ is a *representation* of the *additive group* $\mathbb{R}^n$ on $\mathscr{L}_p(\mathbb{R}^n)$.)

Another, less obvious, but also useful, fact is that if $f \in \mathscr{L}_p(\mathbb{R}^n)$ and if $1 \le p < +\infty$, then

$$\lim_{h \to 0} \| f_{(h)} - f \|_p = 0.$$

To see that this is so, note first that it is true when $f = \chi_E$ is a characteristic function (see Example 9H), and hence when f is any simple function. Since the simple functions in $\mathscr{L}_p(\mathbb{R}^n)$ are dense for $1 \le p < +\infty$ (Prob. D), the result follows. Moreover, since $\| f_{(a+h)} - f_{(a)} \|_p = \| T_a(f_{(h)} - f) \|_p = \| f_{(h)} - f \|_p$ for each f in $\mathscr{L}_p(\mathbb{R}^n)$ and all a and h in $\mathbb{R}^n$, this shows that the mapping $a \to f_{(a)} = T_a f$ of $\mathbb{R}^n$ into $\mathscr{L}_p(\mathbb{R}^n)$ is continuous for each p, $1 \le p < +\infty$, and for each fixed f. (The mapping $a \to f_{(a)}$ need not be continuous from $\mathbb{R}^n$ into $\mathscr{L}_\infty(\mathbb{R}^n)$. The mapping Φ of $\mathbb{R}^n$ into $\mathscr{L}(\mathscr{L}_p(\mathbb{R}^n))$ is not continuous for any value of p.)

Example E. A second important set of examples of Lebesgue spaces is obtained by taking for $(X, \mathbf{S}, \mu)$ the unit circle Z equipped with its σ-ring $\mathbf{B}$ of Borel sets and arc-length measure θ (Ex. 8F). Throughout this book the notation $\mathscr{L}_p(Z)$, used without any further explanation, will denote the space $\mathscr{L}_p(Z, \mathbf{B}, \theta)$. If f is a function on Z and $\gamma \in Z$, we write $f_{(\gamma)}$ for the *rotation* of f

$$f_{(\gamma)}(\lambda) = f(\gamma\lambda), \quad \lambda \in Z.$$

Since θ is rotation invariant, it follows that if $f \in \mathscr{L}_p(Z)$ for some p such that $1 \le p \le +\infty$, then $f_{(\gamma)} \in \mathscr{L}_p(Z)$ too, and $\| f_{(\gamma)} \|_p = \| f \|_p$ for all γ in Z. Thus if we write R_γ for the mapping defined by $R_\gamma f = f_{(\gamma)}, f \in \mathscr{L}_p(Z), \gamma \in Z$, then R_γ is an isometric isomorphism of $\mathscr{L}_p(Z)$ onto itself (with $(R_\gamma)^{-1} = R_{\bar{\gamma}}$) for each γ in Z and each p, $1 \le p \le +\infty$. Moreover, if Ψ denotes the mapping $\gamma \xrightarrow{\Psi} R_\gamma$ of Z into $\mathscr{L}(\mathscr{L}_p(Z))$, then $\Psi(\gamma\gamma') = R_{\gamma\gamma'} = R_\gamma R_{\gamma'} = \Psi(\gamma)\Psi(\gamma')$ for all γ, γ' in Z, since $R_{\gamma\gamma'} f = f_{(\gamma\gamma')} = (f_{(\gamma')})_{(\gamma)}$ for all f in $\mathscr{L}_p(Z)$. (Here again it is customary to express this fact by saying that Ψ is a *representation* of the *multiplicative group* Z on the space $\mathscr{L}_p(Z)$.) Just as in the preceding example it is readily verified that for each fixed f in $\mathscr{L}_p(Z)$, the mapping $\gamma \to f_{(\gamma)} = R_\gamma f$ of Z into $\mathscr{L}_p(Z)$ is continuous on Z whenever $1 \leq p < +\infty$. (The mapping $\gamma \to R_\gamma f$ need not be continuous as a mapping of Z into $\mathscr{L}_\infty(Z)$. The mapping Ψ of Z into $\mathscr{L}(\mathscr{L}_p(Z))$ is not continuous for any value of p.)

Example F. Let $\mathscr{P}$ denote the linear space of all complex-valued Borel measurable functions f on the real line $\mathbb{R}$ with the property that $f_{(2\pi)} = f$, that is, the space of all measurable periodic functions with period 2π. If f is such a measurable periodic function, and if, for some real number a, f is integrable with respect to Lebesgue measure over the period $[a, a + 2\pi]$, then

$$\int_a^{a+2\pi} f(t)dt = \int_b^{b+2\pi} f(t)dt$$

for any real number b. In this case we shall say that f is *integrable over a period* and write

$$\oint f(t)dt = \int_a^{a+2\pi} f(t)dt$$

(where, as noted, a can be taken to be any real number). Using this terminology, we define the space $\mathscr{P}_p$, $1 \leq p < +\infty$, to consist of the collection of all those functions f in $\mathscr{P}$ such that $|f|^p$ is integrable over a period, and we write

$$\|f\|_p = \left[\oint |f(t)|^p \, dt\right]^{1/p}$$

for all f in $\mathscr{P}_p$. Likewise, by $\mathscr{P}_\infty$ we shall mean the linear space of essentially bounded functions in $\mathscr{P}$, and we write

$$\|f\|_\infty = \operatorname*{ess\,sup}_{\mathbb{R}} |f|$$

for f in $\mathscr{P}_\infty$. (Note that it is a consequence of these definitions that $\mathscr{P}_\infty$ is a subspace of $\mathscr{L}_\infty(\mathbb{R})$, while $\mathscr{P}_p$ and $\mathscr{L}_p(\mathbb{R})$ have only the origin in common for $1 \leq p < +\infty$.) The spaces $\mathscr{P}_p$ are Banach spaces closely related to those of Example E. Indeed, if for each function f on Z we define $\hat{f}(t) = f(e^{it})$, $t \in \mathbb{R}$, then $\hat{f} \in \mathscr{P}$, and the mapping $f \to \hat{f}$ is readily seen to be an isometric isomorphism of $\mathscr{L}_p(Z)$ onto $\mathscr{P}_p$, $1 \leq p \leq +\infty$. Using the results of either Example D or Example E, it is easy to see that the mapping $t \to f_{(t)}$ of $\mathbb{R}$ into $\mathscr{P}_p$ is continuous on $\mathbb{R}$ for each fixed function f in $\mathscr{P}_p$ provided $p < +\infty$.

Example G. Let us take for $(X, \mathbf{S}, \mu)$ the real interval $[a, b]$ equipped with its σ-ring $\mathbf{S}$ of Lebesgue measurable sets and linear Lebesgue measure. For each fixed p, $1 \leq p < +\infty$, let $\mathscr{W}_{1,p} = \mathscr{W}_{1,p}([a,b])$ denote the linear space of all those absolutely continuous functions f on $[a, b]$ with the property that the derivative f' belongs to $\mathscr{L}_p([a, b])$ (see Problem 9N). It is easily seen that $\mathscr{W}_{1,p}$ is a dense linear submanifold of $\mathscr{L}_p([a, b])$. (Indeed, the polynomials are dense in $\mathscr{L}_p([a, b])$ by the Weierstrass theorem; cf. Example 18C.) Moreover, while the linear transformation $D: \mathscr{W}_{1,p} \to \mathscr{L}_p([a, b])$ defined by $D(f) = f', f \in \mathscr{W}_{1,p}$, is not bounded, it is *closed* (see Chapter 13). To see why this is so, note that the null space of D is the space $\mathscr{N}$ of constant functions, and let $\hat{D}: \mathscr{W}_{1,2}/\mathscr{N} \to \mathscr{L}_p([a, b])$ be the linear transformation that results from factoring out $\mathscr{N}$. If J denotes the (bounded) integral operator

$$(Jf)(x) = \int_a^x f(t)dt, \quad a \le x \le b, \quad f \in \mathscr{L}_p([a, b])$$

(Ex. C), and if π denotes the natural projection of $\mathscr{L}_p([a, b])$ onto $\mathscr{L}([a, b])/\mathscr{N}$, then $\hat{D}$ is the set-theoretic inverse of $\pi \circ J$. Since the graph of the bounded transformation $\pi \circ J$ is certainly complete, it is clear that the graph of $\hat{D}$ is also (Prob. 13R). But from this it follows readily that the graph of D is closed in $\mathscr{L}_p([a, b]) \oplus_1 \mathscr{L}_p([a, b])$. Observe that what we have shown implies that $\mathscr{W}_{1,2}$ is complete (and is therefore a Banach space) in the new norm

$$\| f \| = \left[\int_a^b |f(t)|^p \, dt + \int_a^b |f'(t)|^p \, dt \right]^{1/p}.$$

(See Problem 13O for a similar construction.)

Like considerations show that if $\mathscr{W}_{n,p}$ denotes the linear space of those $(n-1)$ times continuously differentiable functions f on $[a, b]$ possessing the properties that $f^{(n-1)}$ is absolutely continuous and $f^{(n)}$ belongs to $\mathscr{L}_p([a, b])$, then the linear transformation $D^n: \mathscr{W}_{n,p} \to \mathscr{L}_p([a, b])$ defined by $D^n(f) = f^{(n)}$ is closed, and hence that

$$\| f \| = \left[\int_a^b |f(t)|^p \, dt + \int_a^b |f'(t)|^p \, dt + \cdots + \int_a^b |f^{(n)}(t)|^p \, dt \right]^{1/p}$$

is a norm on $\mathscr{W}_{n,p}$ with respect to which $\mathscr{W}_{n,p}$ is complete. (These are the Lebesgue space counterparts of the constructions of Problem 11F. Spaces of this kind have come to be known as *Sobolev spaces*.)

Duality relations among Lebesgue spaces are substantially the same as in the special case of the spaces (ℓ_p), except that $p = 1$ plays a somewhat surprising special role.

Theorem 17.5. *If $1 < p < +\infty$ and if q denotes the Hölder conjugate of p, then for any measure space $(X, \mathbf{S}, \mu)$ the space, $\mathscr{L}_p(X, \mathbf{S}, \mu)^*$ is isometrically isomorphic to $\mathscr{L}_q(X, \mathbf{S}, \mu)$ under the correspondence $\varphi_g \to g$, where*

$$\varphi_g(f) = \int_X fg \, d\mu, \quad f \in \mathscr{L}_p(X), \quad g \in \mathscr{L}_q(X). \tag{5}$$

PROOF. It follows immediately from the Hölder inequality (see Problem A) that for any function g in $\mathscr{L}_q(X)$ the functional φ_g defined by (5) is bounded and satisfies the inequality $\|\varphi_g\| \leq \|g\|_q$. The hard part is to show that if φ is an arbitrary element of $\mathscr{L}_p(X)^*$, then there exists a function g in $\mathscr{L}_q(X)$ such that $\varphi = \varphi_g$ and $\|\varphi\| = \|g\|_q$.

We treat first the case in which $\mu(X) < +\infty$. The set function $v(E) = \varphi(\chi_E)$, $E \in \mathbf{S}$, is a complex measure on the measurable space $(X, \mathbf{S})$. Indeed, the finite additivity of v follows at once from the additivity of φ, and the countable additivity of v is then a consequence of the continuity of φ (see Problem D). Moreover, it is obvious that v is absolutely continuous $[\mu]$. Hence, by the Radon–Nikodym theorem (Prob. 9L), there exists a function g such that $v(E) = \int_E g \, d\mu$ for every measurable set E. Thus we have

$$\varphi(\chi_E) = \int_X \chi_E g \, d\mu$$

for characteristic functions of measurable sets, and therefore, by linearity,

$$\varphi(s) = \int_X sg \, d\mu$$

for every measurable simple function s on X.

Suppose now that g belongs to $\mathscr{L}_q(X)$ and that φ_g is the functional defined in (5). By what has just been shown, we have $\varphi = \varphi_g$ on the dense linear manifold of simple functions in $\mathscr{L}_p(X)$ (Prob. D), and therefore $\varphi = \varphi_g$ by continuity. Thus to complete the proof (in the case $\mu(X) < +\infty$) it suffices to verify that $g \in L_q(X)$ and that $\|g\|_q \leq \|\varphi\|$. Let M be a positive number, and let $E_0 = \{x \in X : |g(x)| \leq M\}$. If h is any bounded, measurable function that vanishes outside E_0, then there exists a sequence $\{s_n\}$ of measurable simple functions likewise vanishing outside E_0 and converging uniformly to h (Prob. 6S), so $\lim_n \|h - s_n\|_p = 0$ and therefore $\lim_n \varphi(s_n) = \varphi(h)$. On the other hand, the sequence $\{s_n g\}$ tends pointwise to hg and is uniformly bounded, so

$$\lim_n \int_X s_n g \, d\mu = \int_X hg \, d\mu$$

by the bounded convergence theorem (Th. 7.13). Hence we have

$$\varphi(h) = \int_X hg \, d\mu$$

for every such function h. In particular, setting

$$h_0(x) = \begin{cases} |g(x)|^q/g(x), & g(x) \neq 0, \\ 0, & g(x) = 0, \end{cases}$$

we obtain

$$\varphi(h_0 \chi_{E_0}) = \int_{E_0} h_0 g \, d\mu = \int_{E_0} |g|^q \, d\mu.$$

But $|h_0| = |g|^{q-1}$, so $|h_0|^p = |g|^{pq-p} = |g|^q$. Consequently

$$\| h_0 \chi_{E_0} \|_p = \left[\int_{E_0} |g|^q \, d\mu \right]^{1/p},$$

and therefore

$$\int_{E_0} |g|^q \, d\mu \le \| \varphi \| \left[\int_{E_0} |g|^q \, d\mu \right]^{1/p}.$$

From this it follows at once that $\| g \chi_{E_0} \|_q \le \| \varphi \|$ (recall that $1 - (1/p) = 1/q$), and since this inequality holds independently of the choice of M, we conclude that $\| g \|_q \le \| \varphi \|$. Thus the theorem is proved in the case $\mu(X) < +\infty$.

Suppose now that $(X, \mathbf{S}, \mu)$ is an arbitrary measure space. If E is a measurable subset of X, then the subspace $\mathscr{L}_E$ of $\mathscr{L}_p(X, \mathbf{S}, \mu)$ consisting of the functions f in $\mathscr{L}_p(X)$ that vanish outside E may be identified with the Lebesgue space $\mathscr{L}_p(E, \mathbf{S}_E, \mu|E)$ (Prob. E). Consequently, if φ is a bounded linear functional on $\mathscr{L}_p(X, \mathbf{S}, \mu)$ and if E is any measurable subset of X having finite measure, then by the case already treated there exists a function g_E in $\mathscr{L}_q(E, \mathbf{S}_E, \mu|E)$ such that

$$\varphi(f) = \int_E f g_E \, d\mu$$

for all f in $\mathscr{L}_E$. Moreover, as has been shown, the function g_E is unique up to a set of measure zero, so that if E and F are two sets of finite measure, then g_E and g_F are equal a.e. $[\mu]$ on $E \cap F$. In addition, for every set E of finite measure we have $\| g_E \|_q = \| \varphi | \mathscr{L}_E \| \le \| \varphi \|$.

Set $a = \sup_E \| g_E \|_q$, where the supremum is taken over all sets E of finite measure, and note that $a \le \| \varphi \|$. There exist sets of finite measure $E_n, n = 1, 2, \ldots$, such that $E_1 \subset \cdots \subset E_n \subset \cdots$ and such that $\lim_n \| g_{E_n} \|_q = a$. Let $E_0 = \bigcup_n E_n$, and define

$$g_n(x) = \begin{cases} g_{E_n}(x), & x \in E_n, \\ 0, & x \in X \backslash E_n. \end{cases}$$

Then for $m < n$ we have

$$\int_X |g_n - g_m|^q \, d\mu = \int_X |g_n|^q \, d\mu - \int_X |g_m|^q \, d\mu,$$

from which it follows that $\{g_n\}$ is Cauchy, and therefore convergent, in $\mathscr{L}_q(X)$. Let g denote the function in $\mathscr{L}_q(X)$ to which $\{g_n\}$ converges, and suppose f is a function in $\mathscr{L}_p(X)$ that vanishes outside the set E_m. Then $fg_n = fg_m$ a.e. $[\mu]$ for $n \geq m$, while

$$\int_X fg_n \, d\mu \to \int_X fg \, d\mu.$$

It follows that $\varphi_g(f) = \varphi(f)$ provided only that f vanishes outside some one of the sets E_m. Hence, by continuity, φ_g and φ agree on $\mathscr{L}_{E_0}$ (Prob. D). On the other hand, it is clear that g vanishes a.e. on the complement $X \backslash E_0$ (Prob. C). Thus the proof will be complete if we show that φ annihilates the subspace $\mathscr{L}_{X \backslash E_0}$.

Suppose, on the contrary, that there is some function f in $\mathscr{L}_p(X)$ such that f vanishes on E_0 and $\varphi(f) \neq 0$. Since the support of f is σ-finite (Prob. 7J), it is easily seen that there is a set F of finite measure such that F is disjoint from E_0 and such that $\varphi(f\chi_F) \neq 0$, from which it follows that $\|g_F\|_q > 0$. Set $G_n = E_n \cup F$. Then for sufficiently large n we must have

$$\|g_{G_n}\|_q^q = \int_{E_n} |g_{E_n}|^q \, d\mu + \int_F |g_F|^q \, d\mu > a^q,$$

a manifest contradiction, and the proof is complete. □

Example H. Let $(X, \mathbf{S}, \mu)$ be a measure space, and let p and q be Hölder conjugates ($1 < p, q < +\infty$). If we write Ψ_p for the natural isomorphism of $\mathscr{L}_p(X)^*$ onto $\mathscr{L}_q(X)$ just established, so that (in the notation of Theorem 17.5)

$$\Psi_p(\varphi_g) = g, \quad g \in \mathscr{L}_q(X),$$

then the adjoint Ψ_p^* is likewise an isometric isomorphism of $\mathscr{L}_q(X)^*$ onto $\mathscr{L}_p(X)^{**}$, and the composition $\Psi_p^* \circ \Psi_q^{-1}$ is a natural isometric isomorphism of $\mathscr{L}_p(X)$ onto $\mathscr{L}_p(X)^{**}$ which is readily identified as the natural embedding j (cf. Example 16H). Thus the Lebesgue space $\mathscr{L}_p(X)$ is reflexive.

It is customary to identify $\mathscr{L}_p(X)^*$ with $\mathscr{L}_q(X)$ via the isomorphism Ψ_p. Thus (in the terminology of Chapter 16), in the dual pair $\langle \mathscr{L}_p(X), \mathscr{L}_q(X) \rangle$, equipped with the distinguished bilinear functional

$$\langle f, g \rangle = \int_X fg \, d\mu, \quad f \in \mathscr{L}_p(X), \quad g \in \mathscr{L}_q(X),$$

each Banach space is the dual of the other. Since $p = 2$ is the unique positive number that is its own Hölder conjugate, among all of the Lebesgue spaces $\mathscr{L}_p(X)$, $1 \leq p \leq +\infty$, the space $\mathscr{L}_2(X)$ is special. In particular, the above formula defines a distinguished bilinear functional on $\mathscr{L}_2(X)$, and the existence of this bilinear functional is of profound importance in the study of such spaces.

A variation of the above formula obtained by writing

$$(f,g) = \langle f. \bar{g} \rangle = \int_X f\bar{g}\, d\mu, \quad f,g \in \mathscr{L}_2(X),$$

defines a sesquilinear functional on $\mathscr{L}_2(X)$ (Ch. 2, p. 22), called an *inner product*, that turns $\mathscr{L}_2(X)$ into a *Hilbert space*. The study of Hilbert spaces and operators on them is undertaken in Volume II of this treatise.

A brief examination of the argument employed in Theorem 17.5 shows that the proof goes through without substantial change for $p = 1$ when $\mu(X) < +\infty$, so that we may identify $\mathscr{L}_1(X, \mathbf{S}, \mu)^*$ with $\mathscr{L}_\infty(X, \mathbf{S}, \mu)$ in the usual way in this case. A standard argument permits us to improve on this result somewhat.

Theorem 17.6. *If $(X, \mathbf{S}, \mu)$ is a σ-finite measure space, then $\mathscr{L}_1(X, \mathbf{S}, \mu)^*$ is isometrically isomorphic to $\mathscr{L}_\infty(X, \mathbf{S}, \mu)$ under the correspondence $\varphi_g \to g$, where*

$$\varphi_g(f) = \int_X fg\, d\mu, \quad f \in \mathscr{L}_1(X), \quad g \in \mathscr{L}_\infty(X).$$

PROOF. Clearly if g is essentially bounded and f is integrable, then

$$\left| \int_X fg\, d\mu \right| \le \|g\|_\infty \|f\|_1,$$

so that $\|\varphi_g\| \le \|g\|_\infty$. Hence, as before, it suffices to show that for an arbitrary φ in $\mathscr{L}_1(X)^*$ there exists a function g in $\mathscr{L}_\infty(X)$ with $\|g\|_\infty \le \|\varphi\|$ such that $\varphi = \varphi_g$. Let $\{E_n\}$ be a partition of X into a countable sequence of disjoint sets of finite measure, and for each n let g_n be the essentially bounded measurable function ($\|g_n\|_\infty \le \|\varphi\|$) on E_n such that

$$\varphi(f) = \int_{E_n} fg_n\, d\mu$$

for all functions f in the subspace $\mathscr{L}_{E_n}$ consisting of those functions in $\mathscr{L}_1(X)$ that vanish outside E_n. If we define $g(x) = g_n(x)$ for all x in E_n, $n = 1, 2, \ldots$, then g is measurable, $\|g\|_\infty \le \|\varphi\|$, and $\varphi(f) = \varphi_g(f)$ whenever f vanishes outside some one of the sets E_n. But then $\varphi = \varphi_g$ by continuity; see Problem D. □

The above proof shows that if $(X, \mathbf{S}, \mu)$ is an arbitrary measure space, and φ is a bounded linear functional on $\mathscr{L}_1(X)$, then for every set E of finite measure in X there is a function g_E on E that gives rise to φ on $\mathscr{L}_E$. Moreover, it is clear that the family $\{g_E\}$ is uniformly essentially bounded, and "almost coherent" in the sense that $g_E = g_F$ a.e. on $E \cap F$ for all E and F. What is needed to complete the proof of Theorem 17.6 is to patch these functions together, and σ-finiteness is one simple hypothesis that makes such patching possible. As it happens, for locally finite measure spaces (see Proposition 8.1) it is known that such patching is possible (and consequently Theorem 17.6 holds) precisely when the Radon–Nikodym theorem is valid in X (see [41] or [61]). For a simple example in which $\mathscr{L}_{\infty}(X)$ does *not* yield all of $\mathscr{L}_1(X)^*$, the reader may consult [60]. (A different example is given in [30; Problem 31 (9)].)

It turns out that the Lebesgue spaces on a measure space $(X, \mathbf{S}, \mu)$ are separable if and only if the measure space itself is separable (i.e., if and only if the metric space $\dot{\mathbf{S}}_{\mathscr{F}}$ associated with the measure space is separable; see Chapter 9, page 174).

Theorem 17.7. *For any measure space* $(X, \mathbf{S}, \mu)$ *the following conditions are equivalent*:

 (i) *$\mathscr{L}_p(X)$ is separable for some one p, $1 \le p < +\infty$,*
 (ii) *$(X, \mathbf{S}, \mu)$ is separable,*
 (iii) *$\mathscr{L}_p(X)$ is separable for every p, $1 \le p < +\infty$.*

PROOF. The collection S of all characteristic functions of sets of finite measure is a subset of $\mathscr{L}_p(X)$. Consequently, if $\mathscr{L}_p(X)$ is separable for some one value of p, then S is also separable in the relative metric. (Recall (Prop. 4.1) that a metric space is separable if and only if it satisfies the second axiom of countability.) Let S_0 be a countable subset of S that is dense in S, and let $\mathscr{C}_0$ denote the corresponding collection of measurable sets. For any set F having finite measure there is a sequence $\{E_n\}$ in $\mathscr{C}_0$ such that $\| \chi_{E_n} - \chi_F \|_p = [\mu(E_n \triangledown F)]^{1/p} \to 0$. But then $\mu(E_n \triangledown F) \to 0$ too, and it follows that the subset of the metric space $\dot{\mathbf{S}}_{\mathscr{F}}$ consisting of the equivalence classes of the sets in $\mathscr{C}_0$ is dense in $\dot{\mathbf{S}}_{\mathscr{F}}$. Thus $\dot{\mathbf{S}}_{\mathscr{F}}$ is separable, so (i) implies (ii). Suppose, on the other hand, that $(X, \mathbf{S}, \mu)$ is separable and $\mathscr{C}_0$ is a countable collection of measurable sets whose equivalence classes form a dense subset of the metric space $\dot{\mathbf{S}}_{\mathscr{F}}$. Let S_0 denote the (countable) collection of characteristic functions of the sets in $\mathscr{C}_0$. Then the subspace of $\mathscr{L}_p(X)$ spanned by S_0 contains all characteristic functions of sets of finite measure, and therefore all integrable simple functions. But this subspace coincides with $\mathscr{L}_p(X)$ by Problem D. Thus $\mathscr{L}_p(X)$ is separable for all p, $1 \le p < +\infty$, so (ii) implies (iii), and since it is obvious that (iii) implies (i), this completes the proof of the theorem. $\square$

It is unrealistic to expect $\mathscr{L}_{\infty}(X)$ to be separable for any interesting measure spaces, since even $(l_{\infty}) = \mathscr{L}_{\infty}(\mathbb{N}_0)$ is nonseparable (Ex. 11H). Indeed, it is not difficult to verify that the only circumstance in which $\mathscr{L}_{\infty}(X)$ is separable is when $(X, \mathbf{S}, \mu)$ is purely atomic and consists of a finite number of atoms.

Thus far in Chapter 17 we have considered spaces consisting of (equivalence classes of) scalar-valued functions on some measure space. It is of interest to note that these constructions generalize without difficulty to spaces of Banach space valued mappings. In this connection we recall (cf. Chapter 6) that if $(X, \mathbf{S})$ is a measurable space and $\mathscr{E}$ a normed space, then an $\mathscr{E}$-valued mapping Φ defined on X is measurable if $\Phi^{-1}(M) \in \mathbf{S}$ for every Borel set M in $\mathscr{E}$.

Lemma 17.8. *Let $(X, \mathbf{S})$ be a measurable space, let $\mathscr{E}$ be a normed space, and let Φ be a measurable $\mathscr{E}$-valued mapping on X. Then the function*

$$N_\Phi(x) = \| \Phi(x) \|, \qquad x \in X, \tag{6}$$

is also measurable on $(X, \mathbf{S})$.

PROOF. The set $\{x \in X : \| \Phi(x) \| \leq r\}$ can be written as $\{x \in X : \Phi(x) \in \mathscr{E}_r\}$, $r > 0$, and the closed balls $\mathscr{E}_r$ are Borel sets in $\mathscr{E}$. (Recall Problem 6C.)   □

Proposition 17.9. *Let $(X, \mathbf{S}, \mu)$ be a measure space, let $\mathscr{E}$ be a normed space, and let p be a positive real number. Then the collection $\mathscr{L}_p(X; \mathscr{E}) = \mathscr{L}_p(X, \mathbf{S}, \mu; \mathscr{E})$ of all those measurable $\mathscr{E}$-valued mappings Φ on X such that the function N_Φ defined in (6) satisfies the condition*

$$\int_X (N_\Phi)^p \, d\mu < +\infty \tag{7}$$

is a linear space with respect to pointwise linear operations. (The integral indicated in (7) makes sense by virtue of the foregoing lemma.) Moreover, for $p \geq 1$ (and, for nontrivial X and $\mathscr{E}$, in this case only),

$$\| \| \Phi \| \|_p = \left[\int_X \| \Phi(x) \|^p \, d\mu(x) \right]^{1/p} = \| N_\Phi \|_p$$

defines a pseudonorm on $\mathscr{L}_p(X; \mathscr{E})$.

PROOF. If Φ and Ψ are arbitrary $\mathscr{E}$-valued mappings, then $N_{\Phi+\Psi} \leq N_\Phi + N_\Psi$ by the triangle inequality in $\mathscr{E}$, and since $\mathscr{L}_p(X)$ is a linear space, it follows at once that $\mathscr{L}_p(X; \mathscr{E})$ is also a linear space. Furthermore, to show that $\| \| \|_p$ is a pseudonorm for $p \geq 1$, it suffices to verify the triangle inequality, and this follows at once from the triangle inequalities in $\mathscr{E}$ and $\mathscr{L}_p(X)$:

$$\| \| \Phi + \Psi \| \|_p = \| N_{\Phi+\Psi} \|_p \leq \| N_\Phi + N_\Psi \|_p \leq \| N_\Phi \|_p + \| N_\Psi \|_p$$
$$= \| \| \Phi \| \|_p + \| \| \Psi \| \|_p.$$

Finally, if X contains two disjoint measurable sets, both having positive finite measure, and if $\mathscr{E}$ contains any vector $v \neq 0$, then it is a triviality to verify that $\| \| \|_p$ is not a pseudonorm for $0 < p < 1$ (cf. Example 11C).   □

The counterpart of Proposition 17.9 is also valid for the case $p = +\infty$; the details of the proof of the following result are left to the interested reader.

Proposition 17.10. *Let* $(X, \mathbf{S}, \mu)$ *be a measure space and let* $\mathscr{E}$ *be a normed space. The collection* $\mathscr{L}_\infty(X; \mathscr{E}) = \mathscr{L}_\infty(X, \mathbf{S}, \mu; \mathscr{E})$ *of all those measurable mappings* Φ *of* X *into* $\mathscr{E}$ *such that*

$$\|\|\Phi\|\|_\infty = \|N_\Phi\|_\infty = \operatorname*{ess\,sup}_X \|\Phi(x)\|.< +\infty$$

is a linear space with respect to pointwise linear operations, and $\|\| \ \|\|_\infty$ *is a pseudonorm on* $\mathscr{L}_\infty(X; \mathscr{E})$.

For any measure space $(X, \mathbf{S}, \mu)$ and normed space $\mathscr{E}$, the zero spaces of the various pseudonorms $\|\| \ \|\|_p$, $1 \leq p \leq +\infty$, are all the same, each coinciding with the linear manifold $\mathscr{Z}(X; \mathscr{E})$ of those measurable $\mathscr{E}$-valued mappings on X that are equal to 0 a.e. $[\mu]$. Just as before, we shall use the symbol $\mathscr{L}_p(X; \mathscr{E})$ to stand for both the linear space of $\mathscr{E}$-valued mappings equipped with the pseudonorm $\|\| \ \|\|_p$ and for the associated normed space $\dot{\mathscr{L}}_p(X; \mathscr{E})$ (whose elements are cosets modulo $\mathscr{Z}(X; \mathscr{E})$) equipped with the norm $\|\| \ \|\|_p$. In this same spirit we also continue to admit as an element of $\mathscr{L}_p(X; \mathscr{E})$ any $\mathscr{E}$-valued mapping Φ whose domain is a subset of X that contains the complement of some set Z such that $\mu(Z) = 0$, provided the restriction of Φ to $X\backslash Z$ belongs to $\mathscr{L}_p(X\backslash Z; \mathscr{E})$. With these conventions in effect, the following result is readily established.

Theorem 17.11. *For any measure space* $(X, \mathbf{S}, \mu)$ *and normed space* $\mathscr{E}$, *the Lebesgue spaces* $\mathscr{L}_p(X; \mathscr{E})$, $1 \leq p \leq +\infty$, *are complete, and are therefore Banach spaces, provided* $\mathscr{E}$ *is complete. Conversely, if the measure space* X *is nontrivial, and if* $\mathscr{L}_p(X; \mathscr{E})$ *is complete for any one value of* p $(p \geq 1)$, *then* $\mathscr{E}$ *must be complete.*

PROOF. To prove that $\mathscr{L}_p(X; \mathscr{E})$ is complete when $\mathscr{E}$ is, it suffices, as always, to show that any sequence $\{\Phi_n\}_{n=1}^\infty$ in $\mathscr{L}_p(X; \mathscr{E})$ such that

$$\sum_{n=1}^\infty \|\|\Phi_{n+1} - \Phi_n\|\|_p < +\infty$$

is convergent (Prob. 4E). For each positive integer n and each x in X, set $g_n(x) = \|\Phi_{n+1}(x) - \Phi_n(x)\|$. Then by Lemma 17.3 we have $\sum_{n=1}^\infty g_n(x) < +\infty$ for all points x in the complement of some set Z such that $\mu(Z) = 0$. Hence the series

$$\Phi_1(x) + \sum_{n=1}^\infty (\Phi_{n+1}(x) - \Phi_n(x)),$$

and with it the sequence $\{\Phi_n(x)\}_{n=1}^\infty$, is pointwise convergent in $\mathscr{E}$ on $X\backslash Z$. Moreover, if we define

$$\Phi(x) = \lim_n \Phi_n(x), \quad x \in X\backslash Z,$$

then Φ is measurable on $X\backslash Z$ (Prob. 6R), and $\|\Phi(x) - \Phi_N(x)\| \leq \sum_{n=N}^{\infty} g_n(x)$ for all x in $X\backslash Z$ and all positive integers N. Hence, appealing to Lemma 17.3 once again, we obtain

$$\| \Phi - \Phi_k \|_p \leq \left\| \sum_{n=N}^{\infty} g_n \right\|_p \leq \sum_{n=N}^{\infty} \| g_n \|_p$$

for all $k \geq N$, and the completeness of $\mathscr{L}_p(X;\mathscr{E})$ follows.

Suppose, conversely, that $\mathscr{L}_p(X;\mathscr{E})$ is complete for some value of p ($p \geq 1$), and that X contains at least one measurable set E such that $0 < \mu(E) < +\infty$. If $\{v_n\}$ is a sequence in $\mathscr{E}$ such that $\sum_n \| v_{n+1} - v_n \| < +\infty$, then the sequence $\{\chi_E v_n\}_{n=1}^{\infty}$ satisfies the condition

$$\sum_n \| \chi_E v_{n+1} - \chi_E v_n \|_p < +\infty$$

in $\mathscr{L}_p(X;\mathscr{E})$, whence it follows, just as in the first part of the proof, that the sequence $\{\chi_E v_n\}_{n=1}^{\infty}$ is pointwise convergent in $\mathscr{E}$ almost everywhere on X. If x_0 is any one point of E at which this series converges, say to a limit w, then it is clear that the sequence $\{v_n\}$ converges to w, and it follows that $\mathscr{E}$ is complete. $\square$

The existence of the Lebesgue space $\mathscr{L}_1(X;\mathscr{E})$ permits an interesting and useful extension of the theory of integration to Banach space valued mappings when $\mathscr{E}$ is a separable Banach space. The following lemma provides a substitute in the context of Banach space valued mappings for the result of Proposition 6.6 (and Problem 6S). To facilitate the discussion it will be convenient to call an $\mathscr{E}$-valued mapping on a set X *simple* if it assumes only a finite number of values, and to call a measurable simple $\mathscr{E}$-valued mapping on a measure space X *integrable* if it vanishes outside some set of finite measure (or, in other words, if it belongs to $\mathscr{L}_1(X;\mathscr{E})$).

Lemma 17.12. *Let (X, S, μ) be a measure space, let $\mathscr{E}$ be a separable Banach space, and let Φ be a measurable $\mathscr{E}$-valued mapping on X that vanishes outside some subset of X having σ-finite measure. Then there exists a sequence $\{\Sigma_n\}_{n=1}^{\infty}$ of integrable simple $\mathscr{E}$-valued mappings converging to Φ a.e. $[\mu]$ and satisfying the condition $\| \Sigma_n(x) \| \leq 2 \| \Phi(x) \|$ at every point x of X and for every positive integer n.*

PROOF. Let $\{v_k\}_{k=1}^{\infty}$ be a dense sequence of vectors in $\mathscr{E}$, and let $\{F_n\}_{n=1}^{\infty}$ be an increasing sequence of sets of finite measure in X such that $\Phi(x) = 0$ for every x outside $\bigcup_n F_n$. For each positive integer n, set

$$E_n = \left\{ x \in F_n : \| \Phi(x) \| \geq \frac{1}{n} \right\}.$$

Then it is easily seen that $\{E_n\}_{n=1}^{\infty}$ is also an increasing sequence of measurable sets of finite measure with the properties that $\| \Phi(x) \| \geq 1/n$ at every point of

E_n and $\Phi(x) = 0$ outside of $E = \bigcup_n E_n$. For each pair of positive integers k and n we define $G_{k,n} = \{x \in X : \|\Phi(x) - v_k\| \leq 1/n\}$. Then, since the vectors v_k are dense in $\mathscr{E}$, for each fixed n the sequence $\{G_{k,n}\}_{k=1}^{\infty}$ covers X, whence it follows that for each positive integer n there exists a positive integer $m = m_n$ with the property that

$$\mu(E_n \backslash (G_{1,n} \cup G_{2,n} \cup \cdots \cup G_{m_n,n})) \leq 1/2^n$$

(Prop. 7.4). We write $G_n = G_{1,n} \cup \cdots \cup G_{m_n,n}$, $n \in \mathbb{N}$, and for each $k = 1, \ldots, m_n$, we define

$$H_{k,n} = E_n \cap \left(G_{k,n} \backslash \bigcup_{i>k} G_{i,n} \right),$$

with the result that the sets $H_{1,n}, \ldots, H_{m_n,n}$ are pairwise disjoint, each $H_{k,n}$ is contained in the corresponding set $G_{k,n}$, and $H_{1,n} \cup \cdots \cup H_{m_n,n} = E_n \cap G_n$. Next we define Σ_n by setting $\Sigma_n(x) = v_k$ for all x in $H_{k,n}$ and defining $\Sigma_n(x) = 0$ outside $E_n \cap G_n$. Then from the various definitions it is obvious that Σ_n is an integrable simple $\mathscr{E}$-valued mapping on X for each n. Likewise, if $x \in E_n \cap G_n$ then $x \in H_{k,n}$ for some k, and we have $\Sigma_n(x) = v_k$, so

$$\|\Phi(x) - \Sigma_n(x)\| = \|\Phi(x) - v_k\| \leq \frac{1}{n}.$$

Hence if, for some positive integer N, x belongs to $\bigcap_{n=N}^{\infty} (E_n \cap G_n)$, then the sequence $\{\Sigma_n(x)\}$ converges to $\Phi(x)$. Let Z denote the set of those points z in X at which $\{\Sigma_n(z)\}$ fails to converge to $\Phi(z)$. Clearly each point z of Z is contained in all but a finite number of the sets E_n (since Φ and all of the mappings Σ_n vanish outside of E), and must therefore, by what has just been said, fail to be contained in some infinite number of the sets G_n. Thus for every positive integer N we have

$$Z \subset \bigcup_{n=N}^{\infty} (E_n \backslash G_n),$$

and therefore $\mu(Z) \leq \sum_{n=N}^{\infty} 1/2^n = 1/2^{N-1}$, from which it follows at once that $\mu(Z) = 0$. Hence $\{\Sigma_n\}$ converges to Φ a.e. $[\mu]$. Finally, if $\Sigma_n(x) \neq 0$ for some x and n, then $x \in H_{i,n}$ for some $i = 1, \ldots, m_n$, and therefore $x \in E_n \cap G_n$. But then, as we have seen, $\|\Phi(x) - \Sigma_n(x)\| \leq 1/n$, and since $\|\Phi(x)\| \geq 1/n$ on E_n, it follows that $\|\Sigma_n(x)\| \leq \|\Phi(x)\| + 1/n \leq 2\|\Phi(x)\|$, so the proof is complete. $\square$

Proposition 17.13. *If $(X, \mathbf{S}, \mu)$ is a measure space and $\mathscr{E}$ is a separable Banach space, then the integrable simple $\mathscr{E}$-valued mappings on X constitute a dense linear manifold in $\mathscr{L}_p(X; \mathscr{E})$ for $1 \leq p < +\infty$. Furthermore, if $(X, \mathbf{S}, \mu)$ is separable then $\mathscr{L}_p(X; \mathscr{E})$ is also separable for $1 \leq p < +\infty$.*

PROOF. That the collection $\mathscr{S}$ of integrable simple $\mathscr{E}$-valued mappings constitutes a linear submanifold of $\mathscr{L}_p(X; \mathscr{E})$ for all values of p is obvious. If Φ belongs to $\mathscr{L}_p(X; \mathscr{E})$ then the function N_Φ in (6) vanishes outside some set having σ-finite measure, and so therefore does Φ. If $\{\Sigma_n\}$ is a sequence of elements of $\mathscr{S}$ tending to Φ as in the preceding lemma, then the sequence of functions $h_n(x) = \|\Phi(x) - \Sigma_n(x)\|$ tends to zero a.e. $[\mu]$ and is dominated by the function $3\|\Phi(x)\|$. Hence $\lim_n \int_X h_n^p \, d\mu = \lim_n \|\Phi - \Sigma_n\|_p^p = 0$ by the dominated convergence theorem (Th. 7.12). Thus $\mathscr{S}$ is dense in $\mathscr{L}_p(X; \mathscr{E})$. If $\mathscr{C}_0$ is any countable collection of measurable sets of finite measure whose corresponding equivalence classes form a dense subset of the measure ring $(\dot{\mathbf{S}}, \mu)$ of $(X, \mathbf{S}, \mu)$ and V is a countable dense subset of $\mathscr{E}$, then the simple mappings of the form $\sum_{i=1}^N \chi_{E_i} v_i$, where the sets E_i are selected from $\mathscr{C}_0$ and the vectors v_i from V, constitute a countable dense set in the linear manifold $\mathscr{S}$ and therefore a dense subset of $\mathscr{L}_p(X; \mathscr{E})$, $1 \le p < +\infty$, by what has already been shown. $\qquad \square$

Theorem 17.14. *Let $(X, \mathbf{S}, \mu)$ be a measure space, and let $\mathscr{E}$ be a separable Banach space. Then there exists a unique bounded linear transformation L of $\mathscr{L}_1(X; \mathscr{E})$ into $\mathscr{E}$ satisfying the condition*

$$L(\chi_E v) = \mu(E)v, \qquad v \in \mathscr{E}, \qquad E \in \mathbf{S}, \qquad \mu(E) < +\infty. \qquad (8)$$

The value $L(\Phi)$ of L at a mapping Φ belonging to $\mathscr{L}_1(X; \mathscr{E})$ is called the Lebesgue integral of Φ with respect to μ and will be denoted by $\int_X \Phi \, d\mu$ or $\int_X \Phi(x) d\mu(x)$. The Lebesgue integral on $\mathscr{L}_1(X; \mathscr{E})$ satisfies the inequality

$$\left\| \int_X \Phi \, d\mu \right\| \le \|\Phi\|_1.$$

(In other words, L is a contraction.)

PROOF. Suppose first that T is an arbitrary bounded linear transformation of $\mathscr{L}_1(X; \mathscr{E})$ into $\mathscr{E}$ satisfying (8). If Σ is an integrable simple $\mathscr{E}$-valued mapping on X and

$$\Sigma = \sum_{i=1}^N \chi_{E_i} v_i, \qquad (9)$$

where $\mu(E_i) < +\infty$, $i = 1, \ldots, N$, then (8) and the linearity of T ensure that $T(\Sigma) = \sum_{i=1}^N \mu(E_i)v_i$. Since the set $\mathscr{S}$ of integrable simple mappings is dense in $\mathscr{L}_1(X; \mathscr{E})$, it follows that there can be at most one bounded linear transformation L satisfying (8). On the other hand, if for each integrable simple $\mathscr{E}$-valued mapping Σ on X represented as in (9), we define

$$L_0(\Sigma) = \sum_{i=1}^N \mu(E_i)v_i,$$

then it is a simple matter to verify that L_0 is a well-defined linear transformation of the linear space $\mathscr{S}$ into $\mathscr{E}$ (see Problem 1L) and that (8) is satisfied. To complete the proof we note that if the sets E_i in (9) are chosen to be pairwise disjoint, as they always can be, then the function N_Σ defined as in (6) is the simple function $\sum_{i=1}^N \| v_i \| \chi_{E_i}$, whence it follows that

$$\| L_0(\Sigma) \| \leq \sum_{i=1}^N \| v_i \| \mu(E_i) = \| N_\Sigma \|_1 = \|| \Sigma \||_1.$$

Thus L_0 is bounded (by one) and therefore possesses a unique bounded linear extension to $\mathcal{L}_1(X; \mathscr{E})$ (that is also bounded by one; see Problem 12I).   □

Example I. If $\mathscr{E}$ is a separable Banach space and if, for some fixed n, Φ is a mapping of $\mathbb{R}^n$ into $\mathscr{E}$, then, just as when $\mathscr{E} = \mathbb{C}$, for each point a of $\mathbb{R}^n$ we write $\Phi_{(a)}$ for the *translate*

$$\Phi_{(a)}(x) = \Phi(x + a), \qquad x \in \mathbb{R}^n$$

(cf. Example D). If $\mathscr{E}$ is a separable Banach space and E is a Borel set in $\mathbb{R}^n$ having finite Lebesgue measure, then the translation invariance of Lebesgue measure μ_n ensures that

$$\int_{\mathbb{R}^n} \chi_E v \, d\mu_n = \int_{\mathbb{R}^n} (\chi_E v)_{(a)} \, d\mu_n \quad (= \mu_n(E)v)$$

for each point a of $\mathbb{R}^n$ and every vector v in $\mathscr{E}$. Hence if Σ is an arbitrary integrable simple $\mathscr{E}$-valued mapping on $\mathbb{R}^n$, then

$$\int_{\mathbb{R}^n} \Sigma \, d\mu_n = \int_{\mathbb{R}^n} \Sigma_{(a)} \, d\mu_n,$$

whence it follows at once that

$$\int_{\mathbb{R}^n} \Phi \, d\mu_n = \int_{\mathbb{R}^n} \Phi_{(a)} \, d\mu_n$$

for every mapping Φ in $\mathcal{L}_1(\mathbb{R}^n; \mathscr{E})$ and every point a of $\mathbb{R}^n$. Similarly, if Z denotes the unit circle and if for each γ in Z we denote by $\Phi_{(\gamma)}$ the *rotation* $\Phi_{(\gamma)}(\zeta) = \Phi(\gamma\zeta)$, $\gamma \in Z$, of an $\mathscr{E}$-valued mapping Φ defined on Z, then

$$\int_Z \Phi \, d\theta = \int_Z \Phi_{(\gamma)} \, d\theta$$

for every mapping Φ in $\mathcal{L}_1(Z; \mathscr{E})$ and every γ in Z (where, as always, θ denotes arc-length measure on Z; cf. Examples E and 8F).

Proposition 17.15. *Let (X, S, μ) be a measure space, let $\mathscr{E}$ be a separable Banach space, and suppose T is a bounded linear transformation of $\mathscr{E}$ into a second separable Banach space $\mathscr{F}$. Then for any mapping Φ in $\mathcal{L}_1(X; \mathscr{E})$,*

$$T\left(\int_X \Phi \, d\mu \right) = \int_X (T \circ \Phi) d\mu.$$

PROOF. The composition $(T \circ \Phi)(x) = T\Phi(x)$ is measurable along with Φ, and since $\|T\Phi(x)\| \leq \|T\| \|\Phi(x)\|$, $x \in X$, it follows that $T \circ \Phi$ belongs to $\mathscr{L}_1(X; \mathscr{F})$ whenever Φ belongs to $\mathscr{L}_1(X; \mathscr{E})$. A direct calculation shows that if Σ is a simple $\mathscr{E}$-valued mapping on X, say $\Sigma = \sum_{i=1}^N \chi_{E_i} v_i$, then $T \circ \Sigma$ is the simple $\mathscr{F}$-valued mapping $\sum_{i=1}^N \chi_{E_i} Tv_i$. Hence if Σ is integrable and $\mu(E_i) < +\infty$, $i = 1, \ldots, N$, then

$$\int_X (T \circ \Sigma) d\mu = \sum_{i=1}^N \mu(E_i) Tv_i = T\left(\sum_{i=1}^N \mu(E_i) v_i\right) = T\left(\int_X \Sigma \, d\mu\right),$$

so the proposition is valid for integrable simple $\mathscr{E}$-valued mappings.

Finally, suppose Φ is an arbitrary element of $\mathscr{L}_1(X; \mathscr{E})$, and let $\{\Sigma_n\}$ be a sequence of integrable simple mappings converging to Φ as in Lemma 17.12. Then, in the first place, $\{\Sigma_n\}$ tends to Φ in $\mathscr{L}_1(X; \mathscr{E})$ by the dominated convergence theorem (see the proof of Proposition 17.13), and since $\|\int_X \Phi \, d\mu - \int_X \Sigma_n \, d\mu\| = \|\int_X (\Phi - \Sigma_n) d\mu\| \leq \|\Phi - \Sigma_n\|_1$, this implies that the sequence $\{\int_X \Sigma_n \, d\mu\}$ converges to $\int_X \Phi \, d\mu$ in $\mathscr{E}$. Secondly, the sequence $\{T \circ \Sigma_n\}$ converges a.e. $[\mu]$ to $T \circ \Phi$ and

$$\|(T \circ \Phi)(x) - (T \circ \Sigma_n)(x)\| \leq \|T\| \|\Phi(x) - \Sigma_n(x)\| \leq 3\|T\| \|\Phi(x)\|$$

at every point of X. Applying the dominated convergence theorem again, we conclude that the sequence $\{\int_X (T \circ \Sigma_n) d\mu\}$ tends to the integral $\int_X (T \circ \Phi) d\mu$ in $\mathscr{F}$, and it follows at once that $\int_X (T \circ \Phi) d\mu = T(\int_X \Phi \, d\mu)$. □

Example J. Suppose $(X, \mathbf{S}, \mu)$ is a measure space and $\mathscr{E}$ is itself a Banach space of the form $\mathscr{L}_p(Y, \mathbf{T}, \nu)$ for some p, $1 \leq p < +\infty$, where $(Y, \mathbf{T}, \nu)$ is a separable finite measure space. Then integration with respect to ν is a bounded linear functional on $\mathscr{E}$ (Prob. J). Hence, according to the foregoing proposition, if Φ is a mapping belonging to $\mathscr{L}_1(X; \mathscr{L}_p(Y))$, then

$$\int_Y \left[\int_X \Phi \, d\mu\right] d\nu = \int_X \left[\int_Y \Phi(x) d\nu\right] d\mu(x).$$

Example K. If $f \in \mathscr{L}_p(Z)$ for some fixed p, $1 \leq p < +\infty$, then, as noted in Example E, setting

$$\Psi_f(\gamma) = f_{(\gamma)}, \qquad \gamma \in Z,$$

defines a continuous mapping Ψ_f of the unit circle Z into $\mathscr{L}_p(Z)$. It follows at once that Ψ_f belongs to $\mathscr{L}_1(Z; \mathscr{L}_p(Z))$, and hence that

$$\Omega = \int_Z \Psi_f \, d\theta$$

exists (and belongs to $\mathscr{L}_p(Z)$). In order to compute Ω, we first note that if R_γ denotes the rotation operator $R_\gamma f = f_{(\gamma)}$ introduced in Example E, then $R_\gamma \Omega = \int_Z R_\gamma \circ \Psi_f \, d\theta$ for each γ in Z by Proposition 17.15. But for each point ζ in Z, $R_\gamma(\Psi_f(\zeta)) = f_{(\gamma\zeta)} = (\Psi_f)_{(\gamma)}(\zeta)$, so $R_\gamma\Omega = \int_Z (\Psi_f)_{(\gamma)} \, d\theta = \Omega$ by Example

I. Thus Ω is invariant under all rotations R_γ, $\gamma \in Z$, whence it follows that Ω is constant a.e. $[\theta]$. That is, there exists a complex number ω such that $\Omega = \omega$ a.e. $[\theta]$. (This may be seen in various ways. Suppose, for example, that the real function $g = \text{Re}(\Omega)$ is not essentially constant on Z. Then there exist real numbers m and m' where $m < m'$ such that $E = \{\zeta \in Z : g(\zeta) \leq m\}$ and $E' = \{\zeta \in Z : g(\zeta) \geq m'\}$ both have positive arc-length measure. Thus there exist real numbers t and t' and a positive number h such that the intersection of E with the arc $\{e^{iu} : t \leq u \leq t + h\}$ has arc-length measure greater than $h/2$, while the intersection of E' with the congruent arc $\{e^{iu} : t' \leq u \leq t' + h\}$ likewise has arc-length measure greater than $h/2$ (see Problem 8F). But then it is impossible for Ω and $R_\gamma \Omega$ to be equal a.e. if we take for γ the number $e^{i(t'-t)}$.) Finally, to determine the value of ω, we employ Example J to write

$$\int_Z \Omega \, d\theta = 2\pi\omega = \int_Z \left[\int_Z f_{(\gamma)} \, d\theta \right] d\theta(\gamma) = 2\pi \int_Z f \, d\theta.$$

Thus $\omega = \int_Z f \, d\theta$, and, putting everything together, we see that

$$\int_Z f_{(\gamma)} \, d\theta(\gamma) = \int_Z f \, d\theta$$

for almost every γ in Z and for every f in $\mathscr{L}_p(Z)$.

If $\gamma \to R_\gamma$ denotes the representation of Z on $\mathscr{L}_p(Z)$ introduced in Example E, then for each f in $\mathscr{L}_p(Z)$ the mapping Ψ_f of Example K may also be described as the mapping $\gamma \to R_\gamma f$, $\gamma \in Z$. Hence, on the basis of the observations of Example K, it is tempting to define $\int_Z R_\gamma \, d\theta(\gamma)$ to be the rank-one operator

$$Tf = \left(\int_Z f \, d\theta \right) u_0,$$

where u_0 denotes the function identically equal to one on Z. However, the notion of an integral of such a mapping as R_γ transcends the scope of the present treatment, since the space $\mathscr{L}(\mathscr{L}_p(Z))$ is nonseparable. Readers wishing to pursue further the study of Banach space valued integrals are referred to [52] or the treatise [18].

We shall also have need of the analog of the Riemann–Stieltjes integral for Banach space valued mappings. This theory is developed almost exactly as in the classical case, and the following treatment will therefore be rather brief. (The reader is referred to Problems P, Q and R of this chapter, as well as Problems 8L and 5D, for omitted details.)

Definition. Let $[a, b]$ be a real interval, let Φ be a bounded mapping of $[a, b]$ into a normed space $\mathscr{E}$, and let α be a bounded complex-valued function defined on $[a, b]$. Then for any partition

$$\mathscr{P} = \{a = t_0 < \cdots < t_N = b\}$$

of $[a, b]$ and any numbers τ_i such that $t_{i-1} \leq \tau_i \leq t_i$, $i = 1, \ldots, N$, the sum

$$s = \sum_{i=1}^{N} [\alpha(t_i) - \alpha(t_{i-1})]\Phi(\tau_i)$$

is a *Riemann–Stieltjes sum for* Φ *with respect to* α *based on* $\mathscr{P}$. Further, a vector r in $\mathscr{E}$ is the *Riemann–Stieltjes integral of* Φ *over* $[a, b]$ *with respect to* α (notation: $r = \int_a^b \Phi(t)d\alpha(t)$) if for each $\varepsilon > 0$ there exists $\delta > 0$ such that $\|r - s\| < \varepsilon$ for every Riemann–Stieltjes sum s for Φ with respect to α based on an arbitrary partition $\mathscr{P}$ of $[a, b]$ such that mesh $\mathscr{P} < \delta$. (See Problem 1G for basic definitions.)

It is obvious that the Riemann–Stieltjes integral of Φ over $[a, b]$ with respect to α is unique if it exists, in which case we say that Φ is (*Riemann–Stieltjes*) *integrable over* $[a, b]$ *with respect to* α. Just as in the case of scalar-valued functions, the mapping Φ in the integral $\int_a^b \Phi(t)d\alpha(t)$ is called the *integrand*, the function α the *integrator*.

Example L. Let $\alpha(t) \equiv t$, $a \leq t \leq b$. Then the integral $\int_a^b \Phi(t)d\alpha(t)$ is called the *Riemann integral of* Φ *over* $[a, b]$, and is denoted by

$$\int_a^b \Phi(t)dt,$$

whenever it exists.

Example M. If either the integrand Φ or the integrator α is constant on the closed interval $[a, b]$, then the Riemann–Stieltjes integral $\int_a^b \Phi(t)d\alpha(t)$ exists; indeed, $\int_a^b v_0 \, d\alpha(t) = [\alpha(b) - \alpha(a)]v_0$ for any integrator α, while $\int_a^b \Phi(t)d\alpha(t) = 0$ if the integrator α is constant. (In particular, if c is a point in the interval where Φ and α are defined, then $\int_c^c \Phi(t)d\alpha(t) = 0$.)

Example N. Let $\mathscr{E}$ be a normed space, and suppose Φ and Ψ are bounded $\mathscr{E}$-valued mappings defined on an interval $[a, b]$ such that $\Phi(t) = \Psi(t)$ for every t in $[a, b]$ except at a single point t_0, $a \leq t_0 \leq b$. Suppose further that t_0 is a point of continuity of a bounded complex-valued function α defined on $[a, b]$. Let M be chosen so that $\|\Phi(t)\|$, $\|\Psi(t)\| \leq M$, $a \leq t \leq b$, and let ε be a positive number. If δ is chosen so that $|\alpha(t) - \alpha(t_0)| < \varepsilon$ whenever $|t - t_0| < \delta$, and if $\mathscr{P}$ is an arbitrary partition of $[a, b]$ with mesh $\mathscr{P} < \delta$, then it is easy to see that for any Riemann–Stieltjes sum s for Φ with respect to α based on $\mathscr{P}$ there exists a Riemann–Stieltjes sum s' for Ψ with respect to α based on the same partition $\mathscr{P}$ such that $\|s - s'\| \leq 4M\varepsilon$. Hence if Φ is integrable over $[a, b]$ with respect to α, then Ψ is too, and $\int_a^b \Psi(t)d\alpha(t) = \int_a^b \Phi(t)d\alpha(t)$. It follows at once that the Riemann–Stieltjes integral of a bounded $\mathscr{E}$-valued mapping Φ over $[a, b]$ is unaffected by changing the value of Φ at any finite number of points of the interval $[a, b]$, provided the changes are made at points of continuity of the integrator α. In particular, if

$$\mathscr{P} = \{a = t_0 < \cdots < t_N = b\}$$

is a partition of $[a, b]$ such that the function α is continuous at each point $t_i, i = 1, \ldots, N$, and Σ is a mapping of $[a, b]$ into $\mathscr{E}$ such that Σ is constantly equal to some vector v_i in each interval $(t_{i-1}, t_i), i = 1, \ldots, N$, then (irrespective of the values of Σ at the points of $\mathscr{P}$) Σ is Riemann–Stieltjes integrable with respect to α and

$$\int_a^b \Sigma(t) d\alpha(t) = \sum_{i=1}^N [\alpha(t_i) - \alpha(t_{i-1})] v_i.$$

The following fact is easily verified (see Problem P).

Proposition 17.16. *Let $\mathscr{E}$ and $\mathscr{F}$ be Banach spaces and let T be an element of $\mathscr{L}(\mathscr{E}, \mathscr{F})$. If an $\mathscr{E}$-valued mapping Φ is integrable over an interval $[a, b]$ with respect to an integrator α, then the $\mathscr{F}$-valued mapping $T \circ \Phi$ is also integrable over $[a, b]$ with respect to α, and*

$$T\left(\int_a^b \Phi(t) d\alpha(t)\right) = \int_a^b T(\Phi(t)) d\alpha(t).$$

Considered as a function of the interval of integration, the Riemann–Stieltjes integral of a Banach space valued mapping is additive; the proof of the following formal assertion of that fact is elementary, and is likewise left to the reader (Prob. P).

Proposition 17.17. *Let $\mathscr{E}$ be a Banach space, let Φ be a norm-bounded $\mathscr{E}$-valued mapping defined on a real interval $[a, b]$, and let α be a bounded complex-valued function defined on $[a, b]$. If Φ is integrable with respect to the integrator α over $[a, b]$, then Φ is also integrable with respect to α over every closed subinterval $[c, d]$ of $[a, b]$. Conversely, if c is a point of $[a, b]$ at which α is continuous and Φ is integrable over both $[a, c]$ and $[c, b]$ with respect to α, then Φ is integrable over $[a, b]$ with respect to α, and*

$$\int_a^b \Phi(t) d\alpha(t) = \left(\int_a^c + \int_c^b\right) \Phi(t) d\alpha(t).$$

The entire theory of Riemann–Stieltjes integration carries over without any difficulty to Banach space valued mappings. In particular, the following result may be established in exactly the same manner as in the case $\mathscr{E} = \mathbb{C}$ (cf. Problem 8L).

Proposition 17.18. *Let Φ be a continuous mapping of a real interval $[a, b]$ into a Banach space $\mathscr{E}$, and let α be a function of bounded variation on $[a, b]$ (see*

Problem 1I). Then Φ *is integrable over* $[a, b]$ *with respect to* α, *and*

$$\left\| \int_a^b \Phi(t)d\alpha(t) \right\| \leq \int_a^b \| \Phi(t) \| \, dL(t) \leq \max_{a \leq t \leq b} \| \Phi(t) \| \, L(b)$$

where $L(t)$ *denotes the total variation of* α *on the interval* $[a, t]$, $a \leq t \leq b$.

The following basic result is in turn an immediate consequence of Proposition 17.18 and the definition of the Riemann–Stieltjes integral.

Proposition 17.19. *Let* $\mathcal{E}$ *be a Banach space, let* $[a, b]$ *be a real interval, and let* α *be a complex-valued function of bounded variation on* $[a, b]$. *Then the mapping*

$$\Phi \rightarrow \int_a^b \Phi(t)d\alpha(t)$$

is a bounded linear transformation of $\mathcal{C}([a, b]; \mathcal{E})$ *into* $\mathcal{E}$ *with norm no greater than the total variation of* α *over* $[a, b]$.

The principal role played in this book by Banach space valued Riemann–Stieltjes integrals will be in the form of the Banach space analogs of the line integrals used in complex analysis (cf. Chapter 5).

Definition. Let $\mathcal{E}$ be a Banach space, let α be a rectifiable arc in $\mathbb{C}$ defined on a real interval $[a, b]$, and let Φ be an $\mathcal{E}$-valued mapping that is defined and continuous on a subset M of the complex plane $\mathbb{C}$ containing the range of α. Then we define

$$\int_\alpha \Phi(\zeta)d\zeta = \int_a^b \Phi(\alpha(t))d\alpha(t). \tag{10}$$

(Just as in complex analysis, this integral is called the *line integral* of Φ *along* the arc α. The Riemann–Stieltjes integral in the right member of (10) exists by Proposition 17.18, since $\Phi \circ \alpha$ is continuous and α is of bounded variation. To see that this concept really is a direct generalization of the line integrals employed in Chapter 5, consult Example 8K.) More generally, if $\alpha = \alpha_1 + \cdots + \alpha_n$ is any formal sum of rectifiable arcs (Prob. 5H) each having range contained in M, then we define (unambiguously) the *line integral* of Φ along α by setting

$$\int_\alpha \Phi(\zeta)d\zeta = \sum_{i=1}^\infty \int_{\alpha_i} \Phi(\zeta)d\zeta$$

for each Φ in $\mathcal{C}(M; \mathcal{E})$.

The following is nothing more than a summary of Propositions 17.16—17.19 in terms of line integrals, and no proof is required.

Proposition 17.20. *Let $\mathscr{E}$ be a Banach space and let K be a compact subset of $\mathbb{C}$. The mapping*

$$\Phi \to \int_\alpha \Phi(\zeta)d\zeta$$

is a bounded linear transformation of $\mathscr{C}(K; \mathscr{E})$ into $\mathscr{E}$ for each fixed finite formal sum α of rectifiable arcs lying in K, the norm of the transformation being less than or equal to the length of α. If α and β are two such formal sums, then

$$\int_{\alpha+\beta} \Phi(\zeta)d\zeta = \left(\int_\alpha + \int_\beta\right)\Phi(\zeta)d\zeta, \quad \Phi \in \mathscr{C}(K; \mathscr{E}),$$

and if T is any bounded linear transformation of $\mathscr{E}$ into a second Banach space $\mathscr{F}$, then

$$T\left(\int_\alpha \Phi(\zeta)d\zeta\right) = \int_\alpha T(\Phi(\zeta))d\zeta, \quad \Phi \in \mathscr{C}(K; \mathscr{E}).$$

The fundamental theorem of all complex analysis is the Cauchy–Goursat theorem (see Theorem 5.7 and Problem 5O). It is, consequently, gratifying to learn that the Cauchy–Goursat theorem is likewise valid for Banach space valued mappings.

Theorem 17.21. *Let U be an open set of complex numbers, let $\mathscr{E}$ be a Banach space, and let Φ be a locally analytic $\mathscr{E}$-valued mapping defined on U (Ex. 15G). If γ is an arbitrary finite formal sum of closed rectifiable arcs in U such that $\gamma \sim 0$ in U (that is, such that the winding number w_γ vanishes at every point of $\mathbb{C}\backslash U$; see Problems 5H and 5O), then*

$$\int_\gamma \Phi(\zeta)d\zeta = 0.$$

PROOF. For any bounded linear functional f on $\mathscr{E}$ the complex-valued function $f \circ \Phi$ is locally analytic on U, so $\int_\gamma f(\Phi(\zeta))d\zeta = 0$ by the Cauchy–Goursat theorem for locally analytic scalar-valued functions. By Proposition 17.20 this implies that $f(\int_\gamma \Phi(\zeta)d\zeta) = 0$, and the theorem follows by Corollary 14.11, a consequence of the Hahn–Banach theorem. $\square$

The potential applications of this result are numerous, but the only use we shall make of Theorem 17.21 at present is to obtain the following corollary, in effect a uniqueness theorem. To facilitate its formulation we employ the notion of an *oriented envelope*; see Problem 5K.

Proposition 17.22. *Let U be an open subset of $\mathbb{C}$, let K be a compact subset of U, and suppose given two oriented envelopes γ_1 and γ_2 of K in U. Then*

$$\int_{\gamma_1} \Phi(\zeta)d\zeta = \int_{\gamma_2} \Phi(\zeta)d\zeta$$

for an arbitrary locally analytic Banach space valued mapping Φ defined on $U \backslash K$.

PROOF. The winding number of $\gamma_1 - \gamma_2$ on $\mathbb{C} \backslash U$ is zero since $w_{\gamma_1} = w_{\gamma_2} = 0$ there. The winding number of $\gamma_1 - \gamma_2$ on K is likewise $w_{\gamma_1} - w_{\gamma_2} = 0$. Hence $\gamma_1 - \gamma_2 \sim 0$ in $U \backslash K$, so

$$\int_{\gamma_1 - \gamma_2} \Phi(\zeta)d\zeta = 0$$

by Theorem 17.21. $\qquad\qquad\qquad\qquad\qquad\qquad\qquad\qquad\qquad\qquad\qquad$ □

We are now ready to define the concept of an analytic function of an operator T on a Banach space $\mathscr{E}$. Just as in Chapter 12, however, where a similar idea was introduced for rational functions (cf. Problem 12N), it turns out that the appropriate context in which to give this definition is that of a unital Banach algebra. In this connection we note that if a is an element of a Banach algebra $\mathscr{A}$, then multiplication by a (on either right or left) is a bounded linear transformation on $\mathscr{A}$ (cf. Problem U). Hence if $\int_\alpha \Phi(\zeta)d\zeta$ is a line integral of some $\mathscr{A}$-valued mapping Φ, then, according to Proposition 17.20,

$$a\left[\int_\alpha \Phi(\zeta)d\zeta\right] = \int_\alpha a\Phi(\zeta)d\zeta \quad \text{and} \quad \left[\int_\alpha \Phi(\zeta)d\zeta\right]a = \int_\alpha \Phi(\zeta)ad\zeta$$

for every a in $\mathscr{A}$.

Definition. Let $\mathscr{A}$ be a unital Banach algebra, let f be a complex-valued function defined and locally analytic on an open subset U of $\mathbb{C}$, and let x be an element of $\mathscr{A}$ with the property that its spectrum $\sigma_{\mathscr{A}}(x)$ is contained in U. Then we define

$$f(x) = \frac{1}{2\pi i} \int_\gamma f(\zeta)R_x(\zeta)d\zeta,$$

where γ denotes an arbitrary oriented envelope of $\sigma_{\mathscr{A}}(x)$ in U, and R_x denotes, as usual, the resolvent of x—$R_x(\lambda) = (\lambda - x)^{-1}$.

Note. Since R_x is locally analytic on the resolvent set $\mathbb{C} \backslash \sigma_{\mathscr{A}}(x)$ (Ex. 15G) and f is locally analytic on U, it is a matter of elementary calculus to verify that the product $f R_x$ is locally analytic on $U \backslash \sigma_{\mathscr{A}}(x)$, and it results from the preceding proposition that this definition really depends only on x and f, and not on the choice of the oriented envelope γ. By the same token, if f_1

and f_2 are locally analytic functions defined on open sets U_1 and U_2, respectively, and if $f_1 = f_2$ on some open subset V of $U_1 \cap U_2$ that contains $\sigma_{\mathscr{A}}(x)$, then $f_1(x) = f_2(x)$, since γ can be chosen to be an oriented envelope of $\sigma_{\mathscr{A}}(x)$ in V. In particular, if $\sigma_{\mathscr{A}}(x) \subset U_1 \subset U_2$ and if $f_1 = f_2 | U_1$, then $f_1(x) = f_2(x)$.

> The rationale of this definition is best seen by considering the special case $\mathscr{A} = \mathbb{C}$. If $\alpha \in \mathbb{C}$ then $\sigma_{\mathbb{C}}(\alpha) = \{\alpha\}$ and $R_{\alpha}(\lambda) = 1/(\lambda - \alpha)$, $\lambda \neq \alpha$. Hence
>
> $$f(\alpha) = \frac{1}{2\pi i} \int_{\gamma} f(\zeta) R_{\alpha}(\zeta) d\zeta$$
>
> for any function f that is analytic on some open neighborhood U of α and any oriented envelope γ of $\{\alpha\}$ in U by the Cauchy integral formula. Thus what we are doing in the present context is, in effect, turning the Cauchy integral formula into a definition. The use of line integration to define analytic functions of matrices dates back to the nineteenth century; the bold extension of the construction presented here is due, in substance, to F. Riesz [53], N. Dunford [21] and I. Gelfand [28].

An arrangement for defining various functions of the elements of a Banach algebra $\mathscr{A}$ in a systematic and useful manner is usually referred to as a *functional calculus* on $\mathscr{A}$. The functional calculus introduced above is customarily known as the *Riesz-Dunford functional calculus*. Our first result concerning this functional calculus is a useful technicality.

Proposition 17.23. *Let x be an element of a unital Banach algebra $\mathscr{A}$, and let f be a locally analytic function on an open neighborhood U of $\sigma_{\mathscr{A}}(x)$. Then $f(x)$ doubly commutes with x, that is, $f(x)$ commutes with every element y of $\mathscr{A}$ that commutes with x. In particular, any two analytic functions of x commute with each other.*

PROOF. If $\lambda \notin \sigma_{\mathscr{A}}(x)$ then it is clear that $R_x(\lambda)$ doubly commutes with $\lambda - x$, and therefore with x. Hence if γ is an oriented envelope of $\sigma_{\mathscr{A}}(x)$ in U, and y is an element of $\mathscr{A}$ that commutes with x, then

$$yf(x) = \frac{1}{2\pi i} \int_{\gamma} f(\zeta) y R_x(\zeta) d\zeta = \frac{1}{2\pi i} \int_{\gamma} f(\zeta) R_x(\zeta) y \, d\zeta = f(x)y,$$

so y also commutes with $f(x)$. $\qquad\qquad\square$

The following important result concerns the dependence of $f(x)$ on the locally analytic function f.

Theorem 17.24. *Suppose $\mathscr{A}$ is a unital Banach algebra and x is an element of $\mathscr{A}$. Let f and g be locally analytic functions on open neighborhoods U_1 and U_2 of $\sigma_{\mathscr{A}}(x)$, respectively, and let α and β be complex numbers. Then $\alpha f + \beta g$ and fg are both locally analytic on an open neighborhood of $\sigma_{\mathscr{A}}(x)$, and we have*

$$(\alpha f + \beta g)(x) = \alpha f(x) + \beta g(x) \quad and \quad (fg)(x) = f(x)g(x). \tag{11}$$

PROOF. Clearly $\alpha f + \beta g$ and fg are locally analytic on $U_1 \cap U_2$, and if γ is an oriented envelope of $\sigma_{\mathscr{A}}(x)$ in $U_1 \cap U_2$, then

$$\int_\gamma [\alpha f(\zeta) + \beta g(\zeta)] R_x(\zeta) d\zeta = \alpha \int_\gamma f(\zeta) R_x(\zeta) d\zeta + \beta \int_\gamma g(\zeta) R_x(\zeta) d\zeta$$

by Proposition 17.20. Hence $(\alpha f + \beta g)(x) = \alpha f(x) + \beta g(x)$. To verify the second equation in (11) is more difficult.

Let γ be an oriented envelope of $\sigma_{\mathscr{A}}(x)$ in $U_1 \cap U_2$ as above, let V denote the (open) set of those points λ in $\mathbb{C}$ such that the winding number $w_\gamma(\lambda)$ is one, and let γ' be an oriented envelope of V^- in $U_1 \cap U_2$. (Since V^- is a compact subset of $U_1 \cap U_2$, such an oriented envelope exists. Observe that if λ lies in the range W_γ of γ then the winding number of γ' at λ is one, while if λ lies in the range $W_{\gamma'}$ of γ', then the winding number of γ at λ is zero.) Then, as is easily seen, γ' is also an oriented envelope of $\sigma_{\mathscr{A}}(x)$ in $U_1 \cap U_2$, so

$$f(x) = \frac{1}{2\pi i} \int_\gamma f(\zeta) R_x(\zeta) d\zeta \quad \text{and} \quad g(x) = \frac{1}{2\pi i} \int_{\gamma'} g(\lambda) R_x(\lambda) d\lambda.$$

Hence

$$f(x) g(x) = \frac{-1}{4\pi^2} \left[\int_\gamma f(\zeta) R_x(\zeta) d\zeta \right] \left[\int_{\gamma'} g(\lambda) R_x(\lambda) d\lambda \right],$$

whence we conclude that $-4\pi^2 f(x) g(x)$ is given by the iterated integral, with respect to ζ along γ and with respect to λ along γ', of the $\mathscr{A}$-valued mapping

$$\Phi(\zeta, \lambda) = f(\zeta) g(\lambda) R_x(\zeta) R_x(\lambda).$$

Since W_γ and $W_{\gamma'}$ are disjoint by construction, the resolvent equation (Prop. 12.10) permits us to transform Φ as follows:

$$\Phi(\zeta, \lambda) = \frac{f(\zeta) g(\lambda)}{\zeta - \lambda} [R_x(\lambda) - R_x(\zeta)], \qquad \zeta \in W_\gamma, \qquad \lambda \in W_{\gamma'}.$$

Moreover, a trivial calculation shows that both of the mappings

$$\frac{f(\zeta) g(\lambda)}{\zeta - \lambda} R_x(\zeta) \quad \text{and} \quad \frac{f(\zeta) g(\lambda)}{\zeta - \lambda} R_x(\lambda)$$

are continuous as functions of two variables on the set $W_\gamma \times W_{\gamma'}$ in $\mathbb{C}^2$. Hence (Prob. R) the integrations of these mappings needed to calculate $f(x) g(x)$ may be performed in either order. Thus

$$(-4\pi^2) f(x) g(x) = \int_{\gamma'} \left[g(\lambda) R_x(\lambda) \int_\gamma \frac{f(\zeta)}{\zeta - \lambda} d\zeta \right] d\lambda$$

$$+ \int_\gamma \left[f(\zeta) R_x(\zeta) \int_{\gamma'} \frac{g(\lambda)}{\lambda - \zeta} d\lambda \right] d\zeta.$$

(12)

Furthermore the inner integrals in both summands in (12) are readily evaluated by the Cauchy integral formula. Indeed,

$$\int_\gamma \frac{f(\zeta)}{\zeta - \lambda}\, d\zeta = (2\pi i) w_\gamma(\lambda) f(\lambda) = 0,$$

while

$$\int_{\gamma'} \frac{g(\lambda)}{\lambda - \zeta}\, d\lambda = (2\pi i) w_{\gamma'}(\zeta) g(\zeta) = 2\pi i g(\zeta).$$

(Recall that γ' surrounds W_γ, but not vice versa.) Thus, putting everything together, we obtain

$$f(x)g(x) = \frac{-1}{4\pi^2} \int_\gamma 2\pi i f(\zeta) g(\zeta) R_x(\zeta) d\zeta$$

$$= \frac{1}{2\pi i} \int_\gamma f(\zeta) g(\zeta) R_x(\zeta) d\zeta,$$

and the theorem is proved. $\qquad\square$

Theorem 17.25. *Let U be an open subset of $\mathbb{C}$, and let $\mathscr{F}_U$ denote the unital algebra of all locally analytic complex-valued functions defined on U. If $\mathscr{A}$ is a unital Banach algebra and x is an element of $\mathscr{A}$ such that $\sigma_{\mathscr{A}}(x) \subset U$, then $f \xrightarrow{\rho} f(x)$ is an algebra homomorphism of $\mathscr{F}_U$ into $\mathscr{A}$ with the property that $\rho(1) = 1_{\mathscr{A}}$. Moreover, if K denotes an arbitrary compact neighborhood of $\sigma_{\mathscr{A}}(x)$ contained in U, then ρ is bounded with respect to the pseudonorm $\sigma(f) = \max_{\lambda \in K} |f(\lambda)|$, $f \in \mathscr{F}_U$. In particular, if a sequence $\{f_n\}$ in $\mathscr{F}_U$ converges to a limit f uniformly on compact subsets of U (Prob. 11U), then $\{f_n(x)\}$ converges to $f(x)$ in $\mathscr{A}$.*

PROOF. That ρ is an algebra homomorphism was established in Theorem 17.24. To see that ρ is bounded with respect to σ, note that there exists an oriented envelope γ of $\sigma_{\mathscr{A}}(x)$ in K°. If L denotes the length of γ and $f \in \mathscr{F}_U$, then $\| f(x) \| \leq L\sigma(f)$ by Proposition 17.20. The proof that $\rho(1) = 1_{\mathscr{A}}$ is given in the following example. $\qquad\square$

Example O. Let us take $U = \mathbb{C}$, so that the algebra $\mathscr{F}_U = \mathscr{F}_{\mathbb{C}}$ is the algebra of all entire functions. Then $f(x)$ is defined for every f in $\mathscr{F}_{\mathbb{C}}$ and for every element x of every unital Banach algebra. Moreover, in computing $f(x)$ we may take for the oriented envelope γ a circle C_r with center at 0 and with arbitrary (sufficiently large) radius r parametrized in the standard manner; see Example 5D. Consider, in particular, the functions $f_k(\lambda) = \lambda^k$, where k is a nonnegative integer. If $\mathscr{A}$ is any unital Banach algebra and $x \in \mathscr{A}$, the resolvent R_x has a power series expansion

$$R_x(\lambda) = \sum_{n=0}^\infty \frac{x^n}{\lambda^{n+1}}.$$

about $\lambda = \infty$ that is uniformly and absolutely convergent on any circle C_r with sufficiently large radius r (e.g., for $r > \|x\|$, see Proposition 12.11). Hence, choosing any one such circle C_r, we have

$$f_k(x) = \frac{1}{2\pi i} \int_{C_r} \lambda^k \left(\sum_{n=0}^{\infty} \frac{x^n}{\lambda^{n+1}} \right) d\lambda$$

$$= \sum_{n=0}^{\infty} \left[\frac{1}{2\pi i} \int_{C_r} \frac{d\lambda}{\lambda^{n-k+1}} \right] x^n = x^k.$$

In particular, if $f(\lambda) \equiv 1$, then $f(x) = 1_{\mathscr{A}}$, and if $f(\lambda) \equiv \lambda$, then $f(x) = x$. Hence if $p(\lambda) = \alpha_0 + \cdots + \alpha_n \lambda^n$ is an arbitrary complex polynomial, then $p(x)$ as defined by the Riesz–Dunford functional calculus coincides with the customary sense of $p(x)$ (see Problem 2L).

Example P. Suppose U denotes the open disc $D_R = \{\lambda \in \mathbb{C} : |\lambda| < R\}$. If $\mathscr{A}$ is a unital Banach algebra and x is an element of $\mathscr{A}$, then $\sigma_{\mathscr{A}}(x) \subset U$ if and only if the spectral radius $r_{\mathscr{A}}(x)$ is smaller than R (Prob. 12P). Moreover, if this is the case and if $f \in \mathscr{F}_U$, then f possesses a power series expansion $f(\lambda) = \sum_{n=0}^{\infty} \alpha_n \lambda^n$ that converges to f uniformly on compact subsets of U (Th. 5.1). Hence, writing $f_N(\lambda) = \sum_{n=0}^{N} \alpha_n \lambda^n$, we see that

$$f(x) = \lim_{N \to \infty} f_N(x) = \lim_{N \to \infty} \sum_{n=0}^{N} \alpha_n x^n = \sum_{n=0}^{\infty} \alpha_n x^n.$$

(This power series expansion of $f(x)$ is readily seen to be absolutely convergent, since $r_{\mathscr{A}}(x) = \lim_{n \to \infty} \|x^n\|^{1/n} < R$.)

Example Q. An element x of a unital Banach algebra $\mathscr{A}$ is, by definition, invertible in $\mathscr{A}$ if and only if $0 \notin \sigma_{\mathscr{A}}(x)$, which is to say, if and only if $\sigma_{\mathscr{A}}(x)$ is contained in the domain of the function $1/\lambda$. Moreover, if $0 \notin \sigma_{\mathscr{A}}(x)$ and $f(\lambda) = 1/\lambda$, then $f(x)$ has the property that $xf(x) = f(x)x = 1_{\mathscr{A}}$. Thus x is invertible in $\mathscr{A}$ and $f(x) = x^{-1}$ is the inverse of x. It follows that if $p(\lambda)$ and $q(\lambda)$ are polynomials, and if the zeros of q all lie outside $\sigma_{\mathscr{A}}(x)$, then

$$\left(\frac{p}{q} \right)(x) = p(x)(q(x))^{-1}.$$

Hence the Riesz–Dunford functional calculus also extends the functional calculus introduced in Problem 12N.

Proposition 17.24 tells us all we need to know about the dependence of $f(x)$ on the function f. It is also important to know something of how $f(x)$ depends on x.

Proposition 17.26. *Let $\mathscr{A}$ be a unital Banach algebra, and let x_0 be an element of $\mathscr{A}$ such that $\sigma_{\mathscr{A}}(x_0)$ is contained in an open set U in the complex plane. Suppose given a sequence $\{f_n\}$ of locally analytic functions on U converging* *to a limit f_0 uniformly on compact subsets of U, and likewise a sequence $\{x_n\}$ in $\mathscr{A}$ converging to x_0 (in norm). Then $f_n(x_n)$ is defined for all sufficiently large n, and the sequence $\{f_n(x_n)\}$ converges (in norm) to $f_0(x_0)$.*

PROOF. Let γ be a fixed oriented envelope of $\sigma_{\mathscr{A}}(x_0)$ in U, let W_γ denote the range of γ, and let $V = \{\lambda \in U : w_\gamma(\lambda) = 1\}$. Then V is an open neighborhood of $\sigma_{\mathscr{A}}(x_0)$ so, by Proposition 12.14, $\sigma_{\mathscr{A}}(x_n) \subset V$ for all sufficiently large n, and $f_n(x_n)$ is defined for all such positive integers n (since γ is also an oriented envelope of $\sigma_{\mathscr{A}}(x_n)$ in U). Moreover, the (continuous) function $\|R_{x_0}(\lambda)\|$ is bounded on W_γ, whence it follows (Prob. 12K) that for each positive number ε there exists a positive δ such that if $\|y - (\lambda - x_0)\| < \delta$ with λ in W_γ, then $\|y^{-1} - R_{x_0}(\lambda)\| < \varepsilon$. If n_0 is chosen so that $\|x_n - x_0\| < \delta$ for $n > n_0$, then $\|(\lambda - x_n) - (\lambda - x_0)\| = \|x_n - x_0\| < \delta$ for all λ, and therefore $\|R_{x_n}(\lambda) - R_{x_0}(\lambda)\| < \varepsilon$ uniformly on W_γ. Hence the sequence $\{f_n R_{x_n}\}$ converges uniformly to $f_0 R_{x_0}$ on W_γ, and therefore

$$f_0(x_0) - f_n(x_n) = \frac{1}{2\pi i} \int_\gamma [f_0(\zeta)R_{x_0}(\zeta) - f_n(\zeta)R_{x_n}(\zeta)]d\zeta$$

tends to 0 as n tends to infinity by Proposition 17.20. $\square$

The Riesz–Dunford functional calculus also transforms spectra correctly.

Proposition 17.27. *If $\mathscr{A}$ is a unital Banach algebra and f is a locally analytic function on some open neighborhood U of the spectrum of an element x of $\mathscr{A}$, then $\sigma_{\mathscr{A}}(f(x)) = f(\sigma_{\mathscr{A}}(x))$.*

PROOF. Since $(\alpha - f)(x) = \alpha - f(x)$ it suffices to show that $f(x)$ is invertible in $\mathscr{A}$ when and only when the function f has no zero in $\sigma_{\mathscr{A}}(x)$. One way is easy enough. If $f(\lambda)$ fails to vanish on $\sigma_{\mathscr{A}}(x)$, then $\{\lambda \in U : f(\lambda) \neq 0\}$ is an open neighborhood of $\sigma_{\mathscr{A}}(x)$ on which $g(\lambda) = 1/f(\lambda)$ is locally analytic, and we have $g(x)f(x) = f(x)g(x) = 1_{\mathscr{A}}$, so $g(x)$ is the inverse of $f(x)$ in $\mathscr{A}$. To go the other way, suppose $f(\alpha) = 0$ for some α in $\sigma_{\mathscr{A}}(x)$. Then f can be factored as $f(\lambda) = (\lambda - \alpha)g(\lambda)$, where g is also locally analytic on U. Hence $f(x) = (x - \alpha)g(x)$ by Theorem 17.24, and if $f(x)$ were invertible in $\mathscr{A}$, then $x - \alpha$ would also be invertible (Prob. 12K), contrary to hypothesis. $\square$

Our next result adds greatly to the flexibility of the Riesz–Dunford functional calculus and provides a suitable culmination of its discussion.

Proposition 17.28. *Let x be an element of a unital Banach algebra $\mathscr{A}$, let g be a locally analytic function defined on some open neighborhood U of $\sigma_{\mathscr{A}}(x)$, and suppose given a second locally analytic function f defined on some open set $\tilde{U}$ such that $g(\sigma_{\mathscr{A}}(x)) \subset \tilde{U}$. Then $f(g(x)) = (f \circ g)(x)$.*

PROOF. That the function $f(g(x))$ is defined follows from the preceding result. Moreover, the set $\{\lambda \in U : g(\lambda) \in \tilde{U}\}$ is an open neighborhood of $\sigma_{\mathscr{A}}(x)$ on which $f \circ g$ is defined and locally analytic, so $(f \circ g)(x)$ is also defined. Finally, to establish the desired equality, we first choose an oriented envelope γ_1 of $g(\sigma_{\mathscr{A}}(x))$ in $\tilde{U}$, and denote by V the (open) set $\{\lambda \in \tilde{U} : w_{\gamma_1}(\lambda) = 1\}$. Next we choose an oriented envelope γ_2 of $\sigma_{\mathscr{A}}(x)$ in the open set $\{\lambda \in U : g(\lambda) \in V\}$. Then $\sigma_{\mathscr{A}}(g(x))$ and the range of γ_1 are disjoint, so

$$
\begin{aligned}
f(g(x)) &= \frac{1}{2\pi i} \int_{\gamma_1} f(\zeta) R_{g(x)}(\zeta) d\zeta \\
&= \frac{1}{2\pi i} \int_{\gamma_1} \left[f(\zeta) \frac{1}{2\pi i} \int_{\gamma_2} \frac{R_x(\lambda)}{\zeta - g(\lambda)} d\lambda \right] d\zeta \\
&= \frac{1}{2\pi i} \int_{\gamma_2} \left[\frac{1}{2\pi i} \int_{\gamma_1} \frac{f(\zeta)}{\zeta - g(\lambda)} d\zeta \right] R_x(\lambda) d\lambda \\
&= \frac{1}{2\pi i} \int_{\gamma_2} w_{\gamma_1}(g(\lambda)) f(g(\lambda)) R_x(\lambda) d\lambda \\
&= (f \circ g)(x)
\end{aligned}
$$

since $w_{\gamma_1}(g(\lambda)) = 1$ for all λ in the range of γ_2. (The interchange of the order of integration is justified in Problem R.)   □

The Riesz–Dunford functional calculus had its origin in the study of the algebra $\mathbb{C}_{n,n}$ of complex matrices, and was pursued further by Riesz, Dunford, and others, principally as a tool in the study of operators on Banach spaces. We close this discussion (and with it Chapter 17) with the consideration of a situation in which vectors once again play an indispensable role.

Example R. Let $\mathscr{E}$ be a Banach space, let T be an element of $\mathscr{L}(\mathscr{E})$, and suppose the spectrum of T is disconnected. Let K and L be any two nonempty disjoint compact sets in $\mathbb{C}$ such that $K \cup L = \sigma(T)$, and let U and V be disjoint open neighborhoods of K and L, respectively. Then $U \cup V$ is an open neighborhood of $\sigma(T)$ and the function

$$
f(\lambda) = \begin{cases} 1, & \lambda \in U, \\ 0, & \lambda \in V, \end{cases}
$$

is locally analytic on $U \cup V$. Let us write $P = f(T)$. Then P is a bounded operator on $\mathscr{E}$ such that $P^2 = P$ (since $f^2 = f$), and the idempotent P splits $\mathscr{E}$ into the sum of two subspaces $\mathscr{M} = P(\mathscr{E})$ and $\mathscr{N} = (1 - P)(\mathscr{E})$, so that each vector x in $\mathscr{E}$ has a unique representation as $x = y + z$ where $y \in \mathscr{M}$ and $z \in \mathscr{N}$ (see Problem 13D). Since $\sigma(P) = \{0, 1\}$ (Prop. 17.27), we have $0_{\mathscr{E}} \neq P \neq 1_{\mathscr{E}}$, so neither $\mathscr{M}$ nor $\mathscr{N}$ is trivial. Moreover, as an analytic function of T, the operator P doubly commutes with T. Hence if S is any bounded operator on $\mathscr{E}$ that commutes with T and if $y \in \mathscr{M}$, then $Sy = SPy = PSy \in \mathscr{M}$ also. This shows that S carries the subspace $\mathscr{M}$ into itself

$(S(\mathcal{M}) \subset \mathcal{M})$, a fact expressed by saying that $\mathcal{M}$ is *invariant* under S, or is an *invariant subspace* of S. Thus $\mathcal{M}$ is invariant under every bounded operator on $\mathscr{E}$ that commutes with T, a fact expressed by saying that $\mathcal{M}$ is *hyperinvariant* for T, or is a *hyperinvariant subspace* of T. Similarly, of course, the subspace $\mathcal{N}$ is hyperinvariant for T. Thus, for example, if S is any analytic function of T, then $\mathcal{M}$ and $\mathcal{N}$ are both invariant under S. In particular, $\mathcal{M}$ and $\mathcal{N}$ are invariant under T itself, and we may and do define operators A and B in $\mathscr{L}(\mathcal{M})$ and $\mathscr{L}(\mathcal{N})$, respectively, by setting

$$A = T|\mathcal{M} \quad \text{and} \quad B = T|\mathcal{N}.$$

(Thus $T(y + z) = Ay + Bz$ for all y in $\mathcal{M}$ and z in $\mathcal{N}$.)

Suppose now that α is a complex number such that $\alpha \notin \sigma(A) \cup \sigma(B)$. Then setting

$$S(y + z) = (\alpha - A)^{-1}y + (\alpha - B)^{-1}z, \qquad y \in \mathcal{M}, \qquad z \in \mathcal{N},$$

defines a bounded operator on $\mathscr{E}$ (indeed,

$$\| S(y + z) \| \leq \| (\alpha - A)^{-1} \| \| y \| + \| (\alpha - B)^{-1} \| \| z \|,$$

so

$$\| S \| \leq \| (\alpha - A)^{-1} \| \| P \| + \| (\alpha - B)^{-1} \| \| 1 - P \|),$$

and it is obvious that $S = (\alpha - T)^{-1}$. Thus

$$\sigma(T) \subset \sigma(A) \cup \sigma(B).$$

On the other hand, if α is a complex number such that $\alpha \notin K$, then the function

$$g_\alpha(\lambda) = \begin{cases} 1/(\alpha - \lambda), & \lambda \in U, \quad \lambda \neq \alpha, \\ 0, & \lambda \in V, \end{cases}$$

is locally analytic on $(U \backslash \{\alpha\}) \cup V$ and satisfies the identity $(\alpha - \lambda)g_\alpha(\lambda) = f(\lambda)$ there. Hence $(\alpha - T)g_\alpha(T) = g_\alpha(T)(\alpha - T) = P$, from which it follows at once that

$$g_\alpha(T)|\mathcal{M} = (\alpha - A)^{-1},$$

and therefore that $\sigma(A) \subset K$. Similarly, of course, $\sigma(B) \subset L$, and since $K \cap L = \varnothing$, it follows that $\sigma(A) = K$ and $\sigma(B) = L$. Thus the nontrivial splitting of the spectrum $\sigma(T)$ into the union of two relatively open subsets K and L actually gives rise to a nontrivial spatial splitting of the operator T, and permits one to construct a nontrivial hyperinvariant subspace for T.

PROBLEMS

A. Let p be a real number such that $1 < p < +\infty$, let q be the Hölder conjugate of p, and let $(X, \mathbf{S}, \mu)$ be a measure space.

(i) Use the basic version of the Hölder inequality derived in Problem 11A to establish the following inequality:

$$\left| \int_X fg \, d\mu \right| \leq \| f \|_p \| g \|_q, \qquad f \in \mathscr{L}_p(X), \qquad g \in \mathscr{L}_q(X).$$

(This inequality is also known as the *Hölder inequality*.) (Hint: Review the proof of the Hölder inequality in Example 11B.)

(ii) Use the preceding result to establish the triangle inequality for the pseudonorm $\|\ \|_p$ on $\mathscr{L}_p(X)$. In other words, verify that

$$\|f + g\|_p \leq \|f\|_p + \|g\|_p, \qquad f, g \in \mathscr{L}_p(X).$$

(This inequality is also known as the *Minkowski inequality*.)

B. Let $(X, \mathbf{S}, \mu)$ be a measure space, and suppose p, q, and r are positive real numbers such that $1/p + 1/q + 1/r = 1$. Show that if f, g, and h are elements of $\mathscr{L}_p(X)$, $\mathscr{L}_q(X)$, and $\mathscr{L}_r(X)$, respectively, then

$$\|fgh\|_1 \leq \|f\|_p \|g\|_q \|h\|_r.$$

(Hint: If $s = 1 - 1/r$, then ps and qs are Hölder conjugates, as are r and $1/s$, and $|f|^{1/s}$ and $|g|^{1/s}$ belong to $\mathscr{L}_{ps}(X)$ and $\mathscr{L}_{qs}(X)$, respectively.) Show, in general, that if $f_i \in \mathscr{L}_{p_i}(X)$, $i = 1, \ldots, k$, where $1/p_1 + \cdots + 1/p_k = 1$, then $\|f_1 \cdots f_k\|_1 \leq \|f_1\|_{p_1} \cdots \|f_k\|_{p_k}$.

C. Let $(X, \mathbf{S}, \mu)$ be a measure space, let p be an extended real number such that $1 \leq p \leq +\infty$, and let $\{f_n\}_{n=1}^{\infty}$ and f be functions in $\mathscr{L}_p(X)$.

(i) Show that $\lim_n \|f - f_n\|_p = 0$ implies that $\{f_n\}$ converges to f in measure (see Problem 7X), but does not imply that $\{f_n\}$ converges to f a.e. $[\mu]$ when $p < +\infty$.

(ii) Show, on the other hand, that if $p < +\infty$ and if $\{f_n\}$ converges to f in the norm $\|\ \|_p$, then some *subsequence* of $\{f_n\}$ converges to f a.e. $[\mu]$. (Hint: Select $\{f_{n_k}\}$ so that

$$\sum_k \|f_{n_{k+1}} - f_{n_k}\|_p < +\infty.)$$

(iii) Show that, in any case, if $\{f_n\}$ converges to f in the norm $\|\ \|_p$, and if all of the functions f_n vanish a.e. $[\mu]$ on some measurable set E, then $f = 0$ a.e. $[\mu]$ on E.

D. Let $(X, \mathbf{S}, \mu)$ be a measure space, let p be a real number, $1 \leq p < +\infty$, and let f be a function belonging to $\mathscr{L}_p(X)$. Prove that if $\{f_n\}_{n=1}^{\infty}$ is an arbitrary sequence of measurable functions on X that converges to f a.e. and satisfies the condition $|f_n| \leq |f|$ a.e., $n \in \mathbb{N}$, then $\{f_n\}$ converges to f in norm in $\mathscr{L}_p(X)$. Use this fact to show that if $f \in \mathscr{L}_p(X)$ for some p, $1 \leq p < +\infty$, and if $\{E_n\}$ is an increasing sequence of measurable sets in X such that f vanishes outside $\bigcup_n E_n$, then the sequence $\{f\chi_{E_n}\}$ converges to f in $\mathscr{L}_p(X)$. Conclude also that the linear manifold $\mathscr{S}$ of integrable simple functions on X is dense in $\mathscr{L}_p(X)$ for every p, $1 \leq p < +\infty$. (Hint: Recall Problem 6S, and use the fact that every function f in $\mathscr{L}_p(X)$ has σ-finite support (Prob. 7J).)

E. Let $(X, \mathbf{S}, \mu)$ be a measure space, and suppose $1 \leq p \leq +\infty$.

(i) Let E be a measurable subset of X, and let $\mathscr{L}_E$ denote the set of all those functions in $\mathscr{L}_p(X)$ that vanish a.e. $[\mu]$ on $X \backslash E$. Show that $\mathscr{L}_E$ is a subspace of $\mathscr{L}_p(X)$, and that $f \to f|E$ defines an isometric isomorphism of $\mathscr{L}_E$ onto $\mathscr{L}_p(E, \mathbf{S}_E, \mu|E)$. (It is frequently convenient to use this isomorphism to identify $\mathscr{L}_E$ with $\mathscr{L}_p(E)$.)

(ii) Let $\{E_1, \ldots, E_m\}$ be a partition of X into disjoint measurable sets, and let the subspace $\mathscr{L}_{E_i}$ of $\mathscr{L}_p(X)$ be identified with $\mathscr{L}_p(E_i)$, $i = 1, \ldots, m$, as in (i). Then

$\mathscr{L}_p(X)$ is naturally isometrically isomorphic to the direct sum

$$\mathscr{L}_p(E_1) \oplus_p \cdots \oplus_p \mathscr{L}_p(E_m)$$

(see Problem 11T). State and prove the analogous result for a countable partition of X.

F. Show that if $\mathscr{C}$ is any countable covering of a measure space $(X, \mathbf{S}, \mu)$ by measurable sets, then $\bigcup_{E \in \mathscr{C}} \mathscr{L}_E$ spans $\mathscr{L}_p(X)$ as a Banach space for all $1 \le p < +\infty$. Show also that this result is not true in general for $p = +\infty$.

G. Let $(X, \mathbf{S}, \mu)$ be a finite measure space and suppose that $1 \le p \le +\infty$. Let g be a measurable function on X with the property that the product fg is integrable $[\mu]$ whenever $f \in \mathscr{L}_p(X)$. Show that g must then be an element of $\mathscr{L}_q(X)$, where, as usual, $1/p + 1/q = 1$ (if $p = 1$, then $q = +\infty$; if $p = +\infty$, then $q = 1$). (Hint: If the function g is bounded, and therefore does belong to $\mathscr{L}_q(X)$, then the norm of the functional $f \to \int fg\,d\mu$ on $\mathscr{L}_p(X)$ is given by $\|g\|_q$ (cf. Theorem 17.5 for the case $1 < p < +\infty$). Let $E_n = \{x \in X : |g(x)| \le n\}$, set $g_n = g\chi_{E_n}$, and recall Example 12V.) Extend this result to a σ-finite measure space.

> As a matter of fact the result of Problem G holds for any locally finite measure space $(X, \mathbf{S}, \mu)$. To see this one may verify first that if a measurable function g on a locally finite measure space does not have σ-finite support, then there exist a positive number ε_0 and a disjoint sequence $\{E_n\}$ of measurable sets of finite measure such that $1 \le \mu(E_n) < +\infty$ for all n and such that $\bigcup_n E_n$ is contained in the set $\{x \in X : |g(x)| \ge \varepsilon_0\}$. From this point it is easy to see that a function having the property stated in Problem G for some $p > 1$ must have σ-finite support, and the result then follows from Problem G itself. To take care of the case $p = 1$, one may verify directly that if g is an unbounded measurable function on a locally finite measure space, and if $\{A_n\}$ is an arbitrary monotone increasing sequence of positive real numbers, then there exists a corresponding disjoint sequence $\{E_n\}$ of sets of positive finite measure such that $|g(x)| \ge A_n$ for all x in E_n, and then use this observation to show that there exists an integrable function f on $\bigcup_n E_n$ such that fg is *not* integrable.

H. Let $(X, \mathbf{S}, \mu)$ be a measure space, and let p and q be an arbitrary pair of positive numbers. Show that if $1/r = 1/p + 1/q$, and if f and g belong to $\mathscr{L}_p(X)$ and $\mathscr{L}_q(X)$, respectively, then fg belongs to $\mathscr{L}_r(X)$ and

$$\left[\int |fg|^r\,d\mu\right]^{1/r} \le \left[\int |f|^p\,d\mu\right]^{1/p}\left[\int |g|^q\,d\mu\right]^{1/q}.$$

In particular, if $p, q > 1$ and if $1/p + 1/q \le 1$, so that $r \ge 1$, then setting

$$\Phi(f, g) = fg$$

defines a bounded bilinear to transformation of the Banach space $\mathscr{L}_p(X) \oplus_1 \mathscr{L}_q(X)$ into $\mathscr{L}_r(X)$ (see Problem 12W for definitions). What is the norm of Φ? (Hint: p/r and q/r are Hölder conjugates.)

I. Let $(X, \mathbf{S}, \mu)$ and $(Y, \mathbf{T}, \nu)$ be σ-finite measure spaces, and let f be a measurable function on the product $(X \times Y, \mathbf{S} \times \mathbf{T})$ with the property that the function

$$\|f\|_{p,y} = \left[\int_X |f(x, y)|^p\,d\mu(x)\right]^{1/p}$$

is integrable $[v]$ over Y. (Note that $\| f \|_{p,y}$ is a measurable function on Y by Lemma 9.1.) Show that the function

$$\left| \int_Y f(x,y) dv(y) \right|^p$$

is integrable $[\mu]$ over X, and that

$$\left[\int_X \left| \int_Y f(x,y) dv(y) \right|^p d\mu(x) \right]^{1/p} \le \int_Y \| f \|_{p,y} \, dv(y).$$

(Hint: Show that it suffices to treat the case of a nonnegative simple function of the form

$$\sum_{i,j} t_{ij} \chi_{E_i \times F_j}$$

where $\{E_1, \ldots, E_m\}$ and $\{F_1, \ldots, F_n\}$ are disjoint collections of measurable subsets of X and Y, respectively.) Show also that this version of the Minkowski inequality implies Lemma 17.3.

J. If $(X, \mathbf{S}, \mu)$ is a finite measure space, and if p and p' are extended real numbers such that $1 \le p < p' \le +\infty$, then $\mathscr{L}_{p'}(X)$ is contained (as a subset) in $\mathscr{L}_p(X)$ (Ex. A). Show that the inclusion mapping i of $\mathscr{L}_{p'}(X)$ into $\mathscr{L}_p(X)$ is bounded. Conclude that $\mathscr{L}_{p'}(X)$ is a dense, first category, linear submanifold of $\mathscr{L}_p(X)$ (except when X consists entirely of a finite number of atoms, in which case $\mathscr{L}_p(X)$ and $\mathscr{L}_{p'}(X)$ are simply $\mathbb{C}^n$ in two different norms). (Hint: Use the Hölder inequality to show that $\| i \| = \mu(X)^{(p'-p)/pp'}$ and recall Problem 8S.) Show, finally, that integration with respect to μ ($f \to \int_X f \, d\mu$) is a bounded linear functional on $\mathscr{L}_p(X)$ for all p, $1 \le p \le +\infty$.

K. Let $(X, \mathbf{S}, \mu)$ be a measure space, let p and q be Hölder conjugates, and let E be a measurable subset of X. If we identify $\mathscr{L}_p(E)$ with the subspace $\mathscr{L}_E$ of $\mathscr{L}_p(X)$ as in Problem E, and also identify $\mathscr{L}_p(X)^*$ with $\mathscr{L}_q(X)$ as in Example H, what subspace of $\mathscr{L}_q(X)$ becomes the annihilator $\mathscr{L}_p(E)^a$?

L. Let $(X, \mathbf{S}, \mu)$ be a measure space, and let p be a real number, $1 \le p < +\infty$.

(i) Suppose $\mathbf{R}$ is a ring of sets of finite measure with respect to μ such that $\mathbf{R}$ generates $\mathbf{S}$ as a σ-ring, and let $\mathscr{E}$ be a separable Banach space. Demonstrate that the collection of simple mappings of the form

$$\Sigma = \sum_{i=1}^n \chi_{E_i} v_i, \qquad E_i \in \mathbf{R}, \qquad v_i \in \mathscr{E}, \qquad i = 1, \ldots, n,$$

constitutes a dense linear manifold in $\mathscr{L}_p(X; \mathscr{E})$. In particular, then, this is true for $\mathscr{E} = \mathbb{C}$. (Hint: Recall Problem 8M.)

(ii) Suppose that X is a locally compact Hausdorff space (Prob. 3V) and that μ is a regular Borel measure on X (see Chapter 10), and let $\mathscr{E}$ be a separable Banach space. Prove that the continuous mappings in $\mathscr{L}_p(X; \mathscr{E})$ likewise form a dense linear manifold in $\mathscr{L}_p(X; \mathscr{E})$. (In particular then, this is true when $\mathscr{E} = \mathbb{C}$; cf. Proposition 10.7.)

M. A mapping $\Phi : X \to \mathscr{E}$ of a measurable space $(X, \mathbf{S})$ into a Banach space $\mathscr{E}$ is said to be *weakly measurable* if for each linear functional f in $\mathscr{E}^*$ the function $x \to f(\Phi(x))$

is measurable. It is obvious that a measurable mapping is weakly measurable. Show, conversely, that if $\mathscr{E}$ is separable, then every weakly measurable $\mathscr{E}$-valued mapping is actually measurable. (Hint: Let $\{v_n\}$ be a sequence of vectors that is dense in $\mathscr{E}$, and for each n let f_n be a linear functional on $\mathscr{E}$ with $\|f_n\| = 1$ and $f_n(v_n) = \|v_n\|$ (Cor. 14.11). Show that for any vector v in $\mathscr{E}$ we have $\|v\| = \sup_n |f_n(v)|$, and note that if Φ is weakly measurable, then for any vector u_0 in $\mathscr{E}$ the function $\sup_n |f_n(\Phi(x) - u_0)|$ is measurable.)

N. Let $(X, \mathbf{S}, \mu)$ be a measure space, and let $\mathscr{E}$ be a Banach space. Prove that if Φ is a measurable mapping of X into $\mathscr{E}$ and Ψ is a measurable mapping of X into the dual space $\mathscr{E}^*$, then $x \to \Psi(x)(\Phi(x))$ is a measurable complex-valued function on X. Show also that if p and q are Hölder conjugates, and if Φ and Ψ belong to $\mathscr{L}_p(X; \mathscr{E})$ and $\mathscr{L}_q(X; \mathscr{E}^*)$, respectively, then

$$\int_X |\Psi(x)(\Phi(x))| d\mu(x) \leq |||\Phi|||_p |||\Psi|||_q.$$

Hence, setting $\Omega(\Phi) = \int_X \Psi(x)(\Phi(x)) d\mu(x)$ defines a bounded linear functional on $\mathscr{L}_p(X; \mathscr{E})$ satisfying $\|\Omega\| \leq |||\Psi|||_q$. Show that if $\mathscr{E}^*$ is separable, then $\|\Omega\| = |||\Psi|||_q$, and verify the analogous facts for $p = 1$ and $q = +\infty$. (Hint: Recall Problem 14R.)

> The foregoing result, along with the absence of any obvious counter-examples, inspires the hope that it may be possible to identify $\mathscr{L}_p(X; \mathscr{E})^*$ with $\mathscr{L}_q(X; \mathscr{E}^*)$ just as is done in the special case $\mathscr{E} = \mathscr{E}^* = \mathbb{C}$. And with suitable restrictions on the measure space $(X, \mathbf{S}, \mu)$ and the Banach space $\mathscr{E}$, this is indeed the case, but the proof is surprisingly difficult. The interested reader may consult [37] or [15, 16, 17] for a full account of this matter. (There are, to be sure, special cases in which the verification is simpler.)

O. Suppose $(X, \mathbf{S}, \mu)$ and $(Y, \mathbf{T}, \nu)$ are finite measure spaces (or σ-finite at worst; the Fubini theorem will be needed), and consider the space $\mathscr{L}_p(X \times Y, \mathbf{S} \times \mathbf{T}, \mu \times \nu)$. If for each f in $\mathscr{L}_p(X \times Y)$ and each x in X we write $f_x(y) = f(x, y)$, then $f_x \in \mathscr{L}_p(Y)$ for almost every x. Show that setting $\Phi_f(x) = f_x$ defines a measurable mapping Φ_f of X into $\mathscr{L}_p(Y)$ that satisfies $|||\Phi_f|||_p = \|f\|_p$. Show also that the correspondence $\Phi_f \to f$ is an isometric isomorphism between $\mathscr{L}_p(X; \mathscr{L}_p(Y))$ and $\mathscr{L}_p(X \times Y)$, and conclude that it is possible to identify $\mathscr{L}_p(X; \mathscr{L}_p(Y))^*$ with $\mathscr{L}_q(X; \mathscr{L}_p(Y)^*)$, where q denotes the Hölder conjugate of p (if $p = 1$, then $q = +\infty$). (Hint: Φ_f is clearly measurable if f is of the special form

$$f = \sum_{i,j} \lambda_{ij} \chi_{E_i \times F_j},$$

where $\{E_1, \ldots, E_m\}$ and $\{F_1, \ldots, F_n\}$ are disjoint collections of measurable subsets of X and Y, respectively.)

P. Let $\mathscr{E}$ be a Banach space, let $[a, b]$ be a real interval ($a \leq b$), and let Φ and α be, respectively, a bounded $\mathscr{E}$-valued mapping and a bounded complex-valued function defined on $[a, b]$.

(i) Show that the Riemann–Stieltjes integral $\int_a^b \Phi(t) d\alpha(t)$ exists if and only if the following Cauchy criterion is satisfied: Given $\varepsilon > 0$ there exists $\delta > 0$ such that if $\mathscr{P}$ and $\mathscr{P}'$ are arbitrary partitions of $[a, b]$ with mesh $\mathscr{P}, \mathscr{P}' < \delta$ and if Σ and Σ' are arbitrary Riemann–Stieltjes sums for Φ with respect to α based on $\mathscr{P}$ and $\mathscr{P}'$, respectively, then $\|\Sigma - \Sigma'\| < \varepsilon$. Verify also that if the integral $\int_a^b \Phi(t)d\alpha(t)$ does exist, and if ε and δ are as stated, then

$$\left\| \Sigma - \int_a^b \Phi(t)d\alpha(t) \right\| \leq \varepsilon$$

for any Riemann–Stieltjes sum Σ for Φ with respect to α based on a partition $\mathscr{P}$ of $[a, b]$ such that mesh $\mathscr{P} < \delta$.

(ii) Use the foregoing Cauchy criterion to prove Propositions 17.16 and 17.17.

(iii) Suppose α is rectifiable, let η be a nonnegative number, and let $\tilde{\Phi}$ be a second $\mathscr{E}$-valued mapping defined on $[a, b]$ such that $\|\Phi(t) - \tilde{\Phi}(t)\| \leq \eta$ for all $a \leq t \leq b$. Suppose also that for a given positive number ε there exists a positive δ with the property that $\|\Phi(t) - \Phi(t')\| < \varepsilon$ for all t, t' in $[a, b]$ such that $|t - t'| < \delta$. Show that if $\mathscr{P}$ and $\mathscr{P}'$ are partitions of $[a, b]$ such that mesh $\mathscr{P}, \mathscr{P}' < \delta$ and Σ and Σ' are Riemann–Stieltjes sums for Φ and $\tilde{\Phi}$ with respect to α based on $\mathscr{P}$ and $\mathscr{P}'$, respectively, then $\|\Sigma - \Sigma'\| < L(2\varepsilon + \eta)$, where L denotes the length of α. (Hint: See Problem 5D.) In particular, if Σ and Σ' are both Riemann–Stieltjes sums for Φ with respect to α based on $\mathscr{P}$ and $\mathscr{P}'$, respectively, then $\|\Sigma - \Sigma'\| < 2L\varepsilon$. Complete the proof of Proposition 17.18, and verify that if Φ is continuous on $[a, b]$, and if ε and δ are related as above, then

$$\left\| \Sigma - \int_a^b \Phi(t)d\alpha(t) \right\| \leq 2L\varepsilon \qquad \left[\left\| \tilde{\Sigma} - \int_a^b \Phi(t)d\alpha(t) \right\| \leq L(2\varepsilon + \eta) \right]$$

for any Riemann–Stieltjes sum Σ for Φ [$\tilde{\Sigma}$ for $\tilde{\Phi}$] with respect to α based on any partition $\mathscr{P}$ of $[a, b]$ with mesh $\mathscr{P} < \delta$.

Q. Let $\mathscr{E}$ be a separable Banach space, let $[a, b]$ be a real interval $(a \leq b)$, and let Φ and α be, respectively, a continuous $\mathscr{E}$-valued mapping and a rectifiable complex-valued function defined on $[a, b]$. Show that the Riemann–Stieltjes integral $\int_a^b \Phi(t)d\alpha(t)$ is equal to the Lebesgue integral $\int_{[a, b]} \Phi\, d\zeta_\alpha$, where ζ_α denotes the Stieltjes–Borel measure associated with the integrator α (Ex. 8K). (Hint: If α is right-continuous, then each Riemann–Stieltjes sum for Φ with respect to α is the integral [ζ_α] of a step function that approximates Φ.)

R. Let α and β be functions of bounded variation on real intervals $[a, b]$ and $[c, d]$, respectively, and suppose Ψ is a continuous Banach space valued mapping defined on the rectangle $[a, b] \times [c, d]$. Show that

$$\int_a^b \left[\int_c^d \Psi(s, t)d\beta(t) \right] d\alpha(s) = \int_c^d \left[\int_a^b \Psi(s, t)d\alpha(s) \right] d\beta(t).$$

(Hint: See Problem 5L.)

S. Let $\mathscr{E}$ be a Banach space, let T be an element of $\mathscr{L}(\mathscr{E})$, and let f be a locally analytic complex-valued function defined on some open neighborhood of $\sigma(T)$. Show that $(f(T))^* = f(T^*)$. (Hint: Recall Example 16G.)

T. If $\mathscr{A}$ and $\mathscr{B}$ are Banach algebras and if ρ is a bounded linear transformation of $\mathscr{A}$ into $\mathscr{B}$ that is also an algebra homomorphism (Prob. 2L), then ρ is called a *Banach* *algebra homomorphism.* Likewise, if $\mathscr{A}$ and $\mathscr{B}$ are both unital Banach algebras, then an algebra homomorphism ρ of $\mathscr{A}$ into $\mathscr{B}$ is called *unital* if $\rho(1_{\mathscr{A}}) = 1_{\mathscr{B}}$. Verify that if ρ is a unital Banach algebra homomorphism of $\mathscr{A}$ into $\mathscr{B}$ and if x is an arbitrary element of $\mathscr{A}$, then $\sigma_{\mathscr{B}}(\rho(x)) \subset \sigma_{\mathscr{A}}(x)$. Show also that if a complex-valued function f is defined and locally analytic on some open neighborhood of $\sigma_{\mathscr{A}}(x)$, then $f(\rho(x)) = \rho(f(x))$.

U. Let $\mathscr{A}$ be a Banach algebra, and for each element a of $\mathscr{A}$ let us write $L_{(a)}x = ax$ and $R_{(a)}x = xa$, $x \in \mathscr{A}$. (The operators $L_{(a)}$ and $R_{(a)}$ are called *left* and *right* multiplication by a, respectively.)

   (i) Show that $L_{(a)}$ and $R_{(a)}$ both belong to $\mathscr{L}(\mathscr{A})$ for every a in $\mathscr{A}$. Demonstrate also that if $\mathscr{A}$ is unital, then the collection of all right multiplications by elements of $\mathscr{A}$ constitutes the *commutant* in $\mathscr{L}(\mathscr{A})$ of the collection of all left multiplications by elements of $\mathscr{A}$. Show, that is, that a bounded operator T on $\mathscr{A}$ is of the form $R_{(b)}$ for some (uniquely determined) element b of $\mathscr{A}$ if and only if T commutes with every operator of the form $L_{(a)}$. Show that the analogous assertions with "left" and "right" interchanged are also valid. (Hint: Consider right multiplication by the element $T1_{\mathscr{A}}$.)

   (ii) Show that if $\mathscr{A}$ is a unital Banach algebra, then setting

$$\rho_l(a) = L_{(a)}, \qquad a \in \mathscr{A},$$

defines an isometric unital Banach algebra isomorphism of $\mathscr{A}$ into $\mathscr{L}(\mathscr{A})$ that preserves spectra, and conclude that if f is locally analytic on some open neighborhood of an element a of $\mathscr{A}$, then $L_{(f(a))} = f(L_{(a)})$. What are the analogous facts concerning right multiplications?

V. If a and b are elements of a Banach algebra $\mathscr{A}$, then an element x of $\mathscr{A}$ is said to *intertwine* a and b if $ax = xb$. Show that if x intertwines elements a and b of a unital Banach algebra $\mathscr{A}$, and if a complex-valued function f is locally analytic on some open neighborhood of the union $\sigma_{\mathscr{A}}(a) \cup \sigma_{\mathscr{A}}(b)$, then x also intertwines $f(a)$ and $f(b)$. Conclude that if x is invertible in $\mathscr{A}$, so that $b = x^{-1}ax$, and if f is locally analytic on an open neighborhood of $\sigma_{\mathscr{A}}(a)$, then $f(b) = x^{-1}f(a)x$. (In particular, if a and b are similar in $\mathscr{A}$, then so are $f(a)$ and $f(b)$; cf. Problem 12Q.) (Hint: Show first that x intertwines the resolvents $R_a(\lambda)$ and $R_b(\lambda)$ whenever $\lambda \notin \sigma_{\mathscr{A}}(a) \cup \sigma_{\mathscr{A}}(b)$.) Conclude, finally, that if the spectra of a and b with respect to $\mathscr{A}$ are disjoint, then a and b are intertwined only by $x = 0$. (Hint: Adapt the construction in Example R to show that if x intertwines two elements of $\mathscr{A}$ with disjoint spectra, then x also intertwines $1_{\mathscr{A}}$ and $0_{\mathscr{A}}$.)

W. (Rosenblum [56]) Suppose that $\mathscr{A}$ is a unital Banach algebra and that a ánd b are elements of $\mathscr{A}$ such that $\sigma_{\mathscr{A}}(a) \cap \sigma_{\mathscr{A}}(b) = \varnothing$. Show that for each element q of $\mathscr{A}$ the equation $ax - xb = q$ has a unique solution in $\mathscr{A}$, and that this solution is given by

$$x = \frac{1}{2\pi i} \int_{\gamma} R_a(\zeta)q R_b(\zeta)d\zeta,$$

where γ denotes an oriented envelope of $\sigma_{\mathscr{A}}(b)$ in $\mathbb{C} \backslash \sigma_{\mathscr{A}}(a)$. (Hint: $a = \zeta - (\zeta - a)$ for every complex number ζ.) Conclude that the linear transformation $x \to ax - xb$ is an invertible operator on $\mathscr{A}$ when a and b have disjoint spectra. (Hence, in particular, if A and B are bounded operators on a Banach space $\mathscr{E}$, and if $\sigma(A) \cap \sigma(B) = \varnothing$, then $T \to AT - TB$ is an invertible operator on $\mathscr{L}(\mathscr{E})$.)

The notion of an invariant subspace for an operator on a Banach space (Ex. R) has played a major role in the development of operator theory for reasons that may be briefly stated. In the classically familiar setting of operators on finite dimensional spaces (where no serious topological considerations come into play) invariant subspaces always exist in adequate supply and are used extensively in studying the structure of such operators. (See, for example, Problem 2P.) It is natural to ask how much of the classical programs can be carried over to the infinite dimensional case, and, in that connection, to inquire into the existence (and abundance) of invariant subspaces for operators on infinite dimensional spaces. Concerning this question little is known in the general case. In particular, the *general invariant subspace problem*, that is, the question *whether an arbitrary (bounded, linear) operator on an arbitrary infinite dimensional Banach space $\mathscr{E}$ necessarily possesses a nontrivial invariant subspace*, remains open as of this writing. (The subspaces $\mathscr{E}$ and (0) are trivially invariant under any operator T on $\mathscr{E}$; any other invariant subspace is said to be *nontrivial*.) Moreover, there is no infinite dimensional Banach space of which it is known that every operator on that space has a nontrivial invariant subspace (though impressive progress has been made in the case of operators on Hilbert space). But while the invariant subspace problem in general has proved to be intractable, there are many important classes of operators for which it has been solved, in the sense that it has been shown that every operator of that class possesses nontrivial invariant subspaces. Thus Example R itself may be viewed as a solution of the invariant subspace problem for the class of operators having disconnected spectra.

As was noted at the time, the subspaces constructed in Example R, by integrating the resolvent of an operator T around some proper subset of its spectrum, are not just invariant for T—they are hyperinvariant. Moreover, many other techniques employed to construct invariant subspaces actually produce hyperinvariant ones. For this reason interest has arisen in recent years in the *hyperinvariant subspace problem*: *Does every nonscalar (bounded, linear) operator on an arbitrary infinite dimensional Banach space possess a nontrivial hyperinvariant subspace?* (A scalar operator, that is, an operator of the form α_ℓ on a Banach space $\mathscr{E}$, is readily seen to have no hyperinvariant subspaces other than $\mathscr{E}$ and (0); every nonscalar operator on a finite dimensional space possesses nontrivial hyperinvariant subspaces. Cf. Problem X below.) The purpose of the following two problems is to permit the reader to employ a number of the facts and techniques he has been studying while acquainting himself with a solution of the hyperinvariant subspace problem for an important class of operators.

X. Let $\mathscr{E}$ be a Banach space. If E is any subset of $\mathscr{E}$ and $\mathscr{T}$ any subset of $\mathscr{L}(\mathscr{E})$, then E is said to be *invariant under $\mathscr{T}$* (or *for $\mathscr{T}$*), and $\mathscr{T}$ is said to *leave E invariant*, if $T(E) \subset E$ for every T in $\mathscr{T}$. Show that if $\mathscr{L}$ is a linear submanifold of $\mathscr{E}$, then the collection of all elements of $\mathscr{L}(\mathscr{E})$ leaving $\mathscr{L}$ invariant forms a linear subalgebra of $\mathscr{L}(\mathscr{E})$ containing the identity $1_\mathscr{E}$, and also that $\mathscr{L}^-$ is invariant under every operator leaving $\mathscr{L}$ invariant. Show, in the converse direction, that if $\mathscr{A}$ is an arbitrary subalgebra of $\mathscr{L}(\mathscr{E})$, then the collection of all of the subspaces of $\mathscr{E}$ left invariant by $\mathscr{A}$ is a complete sublattice of the lattice of all subspaces of $\mathscr{E}$ (Cor. 11.8). (This lattice is the *lattice of invariant subspaces* of $\mathscr{A}$.) In particular, if $1_\mathscr{E} \in \mathscr{A}$, then the smallest invariant subspace for $\mathscr{A}$ containing a vector x_0 is the closure of $\mathscr{A}x_0 = \{Tx_0 : T \in \mathscr{A}\}$. A subalgebra of $\mathscr{L}(\mathscr{E})$ is said to be *transitive* if its lattice of invariant subspaces is trivial, i.e., consists only of $\mathscr{E}$ and (0).

(i) Show that a subalgebra $\mathscr{A}$ of $\mathscr{L}(\mathscr{E})$ is transitive if and only if $\mathscr{A}x$ is dense in $\mathscr{E}$ for every $x \neq 0$ in $\mathscr{E}$. Show also that an element T of $\mathscr{L}(\mathscr{E})$ fails to possess any nontrivial invariant subspaces if and only if the algebra $\mathscr{P}(T)$ of all polynomials in T is transitive. (Such an operator T is also said to be *transitive*. Whether transitive operators exist or not is, of course, the substance of the general invariant subspace problem.)

(ii) Show that for any subset $\mathscr{T}$ of $\mathscr{L}(\mathscr{E})$ the commutant $\mathscr{T}'$ (Prob. U) is a subalgebra of $\mathscr{L}(\mathscr{E})$ containing the identity $1_{\mathscr{E}}$, and that an element T of $\mathscr{L}(\mathscr{E})$ fails to possess any nontrivial hyperinvariant subspace if and only if the commutant $\{T\}'$ is transitive.

(iii) Prove that $\mathscr{L}(\mathscr{E})$ and the subalgebra $\mathbb{C}1_{\mathscr{E}} = \{\alpha_{\mathscr{E}} : \alpha \in \mathbb{C}\}$ are commutants of one another. (Thus $\mathbb{C}1_{\mathscr{E}}$ is the center of $\mathscr{L}(\mathscr{E})$.) Prove also that $\mathscr{L}(\mathscr{E})$ is transitive, and conclude that the scalar operators $\alpha_{\mathscr{E}}$, $\alpha \in \mathbb{C}$, do not possess any nontrivial hyperinvariant subspaces.

(iv) Show that for each element T of $\mathscr{L}(\mathscr{E})$ both the kernel $\mathscr{N}(T)$ and the range $\mathscr{R}(T)$ are invariant under the commutant $\{T\}'$, and use this fact to show that if T is an element of $\mathscr{L}(\mathscr{E})$ that does not possess any nontrivial hyperinvariant subspaces, and if T is not a scalar operator, then both T and T^* must have empty point spectrum (Ex. 12S). (Hint: $\{\alpha - T\}' = \{T\}'$ for every complex number α; recall Problem 16I.)

Y. (Lomonosov) A linear transformation K of a Banach space $\mathscr{E}$ into a Banach space $\mathscr{F}$ is *compact* if $K(\mathscr{E}_1)$ is totally bounded in $\mathscr{F}$, i.e., if $K(\mathscr{E}_1)^-$ is a compact subset of $\mathscr{F}$ (in the norm topology). (See Problem 4Q.) Prove that every compact linear transformation of $\mathscr{E}$ into $\mathscr{F}$ is automatically bounded, and possesses the property that it carries every bounded subset of $\mathscr{E}$ onto a totally bounded subset of $\mathscr{F}$. Show also that a linear transformation K of $\mathscr{E}$ into $\mathscr{F}$ is compact if and only if it possesses the property that for every bounded sequence $\{x_n\}$ in $\mathscr{E}$ the sequence $\{Kx_n\}$ has a convergent subsequence.

(i) Let K be a compact linear operator on an infinite dimensional Banach space $\mathscr{E}$. Show that if α is a nonzero complex number such that $\alpha - K$ is not bounded below, then α is an eigenvalue of K. (Hint: If $\{x_n\}$ is a sequence of unit vectors in $\mathscr{E}$ such that $\|(\alpha - K)x_n\| \to 0$, and if $Kx_n \to x_0$, then $\|x_0\| = \lim_n \|Kx_n\| = |\alpha| > 0$, and $Kx_0 = \alpha x_0$.) Use this fact to prove that every $\alpha \neq 0$ in $\sigma(K)$ is an eigenvalue either of K or of K^*, and conclude that a compact operator on $\mathscr{E}$ that does not have any nontrivial hyperinvariant subspace is necessarily quasinilpotent. (Hint: No scalar operator $\alpha_{\mathscr{E}} \neq 0$ is compact on $\mathscr{E}$ since $\mathscr{E}$ is infinite dimensional (Prob. 11P); recall Problem 12O and Problem X.)

(ii) Let K be a compact operator on an infinite dimensional Banach space $\mathscr{E}$ such that $\|K\| = 1$, and suppose K does not have any nontrivial hyperinvariant subspace. Show first that there is a vector x_0 in $\mathscr{E}$ such that $\|x_0\| > 1$ and $\|Kx_0\| > 1$, and hence such that neither the open ball $U = x_0 + \mathscr{E}_1^{\circ}$ nor the compact set $D = K(U)^-$ contains the origin 0. Then prove that there exists a finite set $\{T^{(1)}, \ldots, T^{(p)}\}$ of operators belonging to the commutant $\{K\}'$ such that the sets

$$W_{T^{(i)}} = \{x \in \mathscr{E} : T^{(i)}x \in U\}, \qquad i = 1, \ldots, p,$$

cover D. (Hint: According to Problem X the sets $W_T = \{x \in \mathscr{E} : Tx \in U\}$, $T \in \{K\}'$, cover $\mathscr{E} \setminus \{0\}$, and $0 \notin D$.) Next construct an infinite sequence of operators $\{T_n\}_{n=1}^{\infty}$ such that each T_n is one of the special operators $T^{(1)}, \ldots, T^{(p)}$, and such that if we write

$$P_n = (T_n K)(T_{n-1} K) \cdots (T_1 K),$$

then $P_n x_0 \in U$ for every positive integer n. Finally, use this sequence to show that there exist positive numbers ε_0 and M such that $M^n \| K^n \| \| x_0 \| \geq \varepsilon_0 > 0$ for all n, and conclude that $\liminf_n \| K^n \|^{1/n}$ is necessarily strictly positive. (Hint: Set $\varepsilon_0 = \| x_0 \| - 1$, and use the fact that the sequence $\{T_n\}$ is uniformly bounded.)

(iii) Combine the results of (i) and (ii) to show that every nonzero compact operator on an infinite dimensional Banach space possesses a nontrivial hyperinvariant subspace. (Hint: Recall the formula for the spectral radius obtained in Problem 15G.)

18 The spaces $\mathscr{C}(X)$

Banach spaces of the form $\mathscr{C}(X)$, where X denotes a compact Hausdorff space, have already appeared on numerous occasions (see, in particular, Example 11F). In this chapter we develop some special information regarding these spaces that constitutes an essential part of the general theory of normed spaces. Our first project is to find the dual of such a space, and to this end we employ the material of Chapter 10.

Definition. Let X be a locally compact Hausdorff space, and let $\mathbf{B}$ denote the σ-ring of Borel subsets of X. If ξ and ζ are complex Borel measures on X the sum $\xi + \zeta$ is defined setwise:

$$(\xi + \zeta)(E) = \xi(E) + \zeta(E), \quad E \in \mathbf{B}.$$

Similarly, if α is a complex number, then $\alpha\xi$ is defined setwise:

$$(\alpha\xi)(E) = \alpha\xi(E), \quad E \in \mathbf{B}.$$

The collection of all complex Borel measures on X, which is clearly a linear space with respect to these setwise linear operations, will be denoted by $\mathscr{M}(X)$.

There is a natural and useful norm on the space $\mathscr{M}(X)$. (In this connection we recall that the total variation of a complex measure ξ is denoted by $|\xi|$. For a discussion of the total variation of a complex measure see Chapter 8.)

Proposition 18.1. *The function* $\|\xi\| = |\xi|(X)$ *is a norm on* $\mathscr{M}(X)$ (*called the* total variation *norm*) *with respect to which* $\mathscr{M}(X)$ *is a Banach space.*

PROOF. It is clear that $\|\xi\| = 0$ if and only if ξ is the zero measure, and that, if ξ is any complex Borel measure on X and α is any complex number, then

$\|\alpha\xi\| = |\alpha|\,\|\xi\|$. Moreover, if ξ and ζ are two elements of $\mathscr{M}(X)$, and if $\{E_1, \ldots, E_p\}$ is any partition of X into Borel sets, then

$$\sum_{i=1}^{p} |\xi(E_i) + \zeta(E_i)| \le \sum_{i=1}^{p} |\xi(E_i)| + \sum_{i=1}^{p} |\zeta(E_i)| \le \|\xi\| + \|\zeta\|$$

and therefore $\|\xi + \zeta\| \le \|\xi\| + \|\zeta\|$. Thus we have proved that $\|\ \|$ is a norm on $\mathscr{M}(X)$. All that remains is to show that $\mathscr{M}(X)$ is complete in this norm.

Let $\{\xi_n\}$ be a sequence of complex Borel measures that is Cauchy in the total variation norm. Then for any one Borel set E we have $|\xi_m(E) - \xi_n(E)| \le \|\xi_m - \xi_n\|$, so $\{\xi_n\}$ converges setwise to a set function ξ on the σ-ring $\mathbf{B}$ of Borel subsets of X, and it is obvious that ξ is finitely additive. Hence, to prove that ξ is a complex Borel measure, it suffices to show that ξ is semicontinuous (see Problem 7E). Let $\{E_n\}$ be an increasing sequence of Borel sets in X with union E, and let ε be a positive number. Choose N such that $\|\xi_m - \xi_n\| < \varepsilon/3$ for $m, n \ge N$, and then choose K such that $|\xi_N(E_k) - \xi_N(E)| < \varepsilon/3$ for all $k \ge K$. Then $\lim_n |\xi_N(F) - \xi_n(F)| = |\xi_N(F) - \xi(F)| \le \varepsilon/3$ for every Borel set F, so we have $|\xi_N(E_k) - \xi(E_k)| \le \varepsilon/3$ and likewise $|\xi_N(E) - \xi(E)| \le \varepsilon/3$. Thus altogether $|\xi(E_k) - \xi(E)| < \varepsilon$, and since this holds for all $k \ge K$, the set function ξ is a complex Borel measure.

Finally, for any $\varepsilon > 0$, if $\|\xi_m - \xi_n\| < \varepsilon$ for $m, n \ge N$, as above, and if $\{E_1, \ldots, E_p\}$ is any partition of X into Borel sets, then

$$\sum_{i=1}^{p} |\xi_m(E_i) - \xi_n(E_i)| < \varepsilon$$

for all $m, n \ge N$. Letting n tend to infinity, we obtain

$$\sum_{i=1}^{p} |\xi_m(E_i) - \xi(E_i)| \le \varepsilon$$

for all $m \ge N$. Thus $\|\xi_m - \xi\| \le \varepsilon$ for all $m \ge N$, and the proposition is proved. $\qquad\square$

Among the complex Borel measures on a given compact Hausdorff space a special role is played by the *regular* ones, as we have seen (Ch. 10, p. 194).

Proposition 18.2. *On any compact Hausdorff space X the collection $\mathscr{M}_0(X)$ of all regular complex Borel measures is a (closed) subspace of $\mathscr{M}(X)$. If X is metrizable then $\mathscr{M}_0(X) = \mathscr{M}(X)$.*

PROOF. A complex Borel measure ξ on X is regular if and only if for each Borel set E in X and positive number ε there exist a compact set K and an open set U in X such that $K \subset E \subset U$ and such that $|\xi|(U\backslash K) < \varepsilon$. In view of this criterion it is obvious that if ξ is a regular complex Borel measure on X and α is a complex number, then $\alpha\xi$ is also regular. To see that $\mathscr{M}_0(X)$ is closed with respect to the formation of sums, let ξ_1 and ξ_2 be regular complex Borel measures on X, let E be a Borel set, and let ε be a positive number. Then there exist compact sets K_1 and K_2 and open sets U_1 and U_2 such that $K_i \subset E \subset U_i$ and $|\xi_i|(U_i \backslash K_i) < \varepsilon/2$, $i = 1, 2$. But then $K = K_1 \cup K_2 \subset E$, $E \subset U = U_1 \cap U_2$, and we have

$$|\xi_1 + \xi_2|(U \backslash K) \leq |\xi_1|(U \backslash K) + |\xi_2|(U \backslash K) < \varepsilon.$$

Thus $\mathscr{M}_0(X)$ is a linear manifold in $\mathscr{M}(X)$. We show next that $\mathscr{M}_0(X)$ is closed.

Let $\{\xi_n\}_{n=1}^\infty$ be a sequence in $\mathscr{M}_0(X)$ that converges to some limit ξ in $\mathscr{M}(X)$. We must show that ξ is regular. To this end let E be a Borel set in X, let ε be a positive number, and let N be a positive integer such that $\|\xi_n - \xi\| < \varepsilon/3$ for all $n \geq N$, so that $\|\xi_n - \xi_m\| < 2\varepsilon/3$ for all $m, n \geq N$. Choose a compact set K and an open set U such that $K \subset E \subset U$ and such that $|\xi_N|(U \backslash K) < \varepsilon/3$. Then $|(|\xi_n| - |\xi_N|)(U \backslash K)| < 2\varepsilon/3$ for $n \geq N$ (see Problem A) and, letting n tend to infinity, we find that $|(|\xi| - |\xi_N|)(U \backslash K)| \leq 2\varepsilon/3$. Hence

$$|\xi|(U \backslash K) < \frac{2\varepsilon}{3} + \frac{\varepsilon}{3} = \varepsilon,$$

which shows that $\mathscr{M}_0(X)$ is closed. Finally, the fact that every complex Borel measure on a compact metrizable space is regular is a consequence of Corollary 10.6. $\qquad\square$

The following theorem is one of the two main results of this chapter.

Theorem 18.3 (Riesz Representation Theorem). *Let X be a compact Hausdorff space. Then for each functional φ in $\mathscr{C}(X)^*$ there is a unique, regular, complex Borel measure ξ in $\mathscr{M}_0(X)$ such that*

$$\varphi(f) = \varphi_\xi(f) = \int_X f \, d\xi, \qquad f \in \mathscr{C}(X). \tag{1}$$

Moreover, the correspondence $\xi \to \varphi_\xi$ in (1) is an isometric isomorphism between $\mathscr{C}(X)^*$ and the Banach space $\mathscr{M}_0(X)$.

PROOF. It has already been seen (Prop. 10.16) that for any complex Borel measure ξ on X, regular or not, the functional φ_ξ defined in (1) is linear and satisfies the inequality $\|\varphi_\xi\| \leq |\xi|(X) = \|\xi\|$. Moreover (Prob. 10I), we also know that $\|\varphi_\xi\| = \|\xi\|$ when ξ is regular (this establishes the uniqueness of ξ, since it is clear that the correspondence $\xi \to \varphi_\xi$ is linear). Thus the proof will be complete if we show that for each φ in $\mathscr{C}(X)^*$ there exists a regular complex Borel measure ξ on X such that (1) holds. To see that such a ξ exists we note that $\varphi_1(f) = \operatorname{Re} \varphi(f)$ and $\varphi_2(f) = \operatorname{Im} \varphi(f)$ are bounded real linear functionals on the space $\mathscr{C}_{\mathbb{R}}(X)$ of continuous real-valued functions on X. Moreover, if we assume that μ_1 and μ_2 are regular signed Borel measures on X such that

$$\operatorname{Re} \varphi(f) = \int_X f \, d\mu_1 \quad \text{and} \quad \operatorname{Im} \varphi(f) = \int_X f \, d\mu_2 \qquad (2)$$

for every f in $\mathscr{C}_{\mathbb{R}}(X)$, then $\xi = \mu_1 + i\mu_2$ is a regular complex Borel measure, and φ and φ_ξ agree on the real functions in $\mathscr{C}(X)$ by (2). But then, of course, $\varphi = \varphi_\xi$ on all of $\mathscr{C}(X)$. (It should be noted, here and below, that since the space X is compact by hypothesis, all (signed) Borel measures on X are finite-valued, and are therefore complex Borel measures.) Thus it suffices to prove the following real version of the theorem. $\qquad \square$

Theorem 18.4. *Let X be a compact Hausdorff space. Then for each bounded linear functional φ on the real space $\mathscr{C}_{\mathbb{R}}(X)$ of continuous real-valued functions on X there is a unique signed measure μ in $\mathscr{M}_0(X)$ such that*

$$\varphi(f) = \varphi_\mu(f) = \int_X f \, d\mu, \qquad f \in \mathscr{C}_{\mathbb{R}}(X). \qquad (3)$$

Moreover, the correspondence $\mu \to \varphi_\mu$ in (3) is an isometric isomorphism of the real Banach space $\mathscr{C}_{\mathbb{R}}(X)^*$ onto the real Banach space $\mathscr{M}_{\mathbb{R},0}(X)$ of all regular signed Borel measures on X.

PROOF. As was noted above, all that is really necessary is to show that there exists a regular signed Borel measure μ such that (3) holds. To this end we define a new functional φ' on the collection $\mathscr{P}$ of nonnegative functions in $\mathscr{C}_{\mathbb{R}}(X)$ by setting

$$\varphi'(f) = \sup_{0 \le g \le f} \varphi(g), \qquad f \in \mathscr{P}, \qquad (4)$$

where, as indicated, the supremum is taken over all those continuous real functions g on X such that $0 \le g \le f$. Since $g = 0$ and $g = f$ are two such functions, we see that $\varphi'(f) \ge 0$ and $\varphi'(f) \ge \varphi(f)$ for every f in $\mathscr{P}$. Moreover, we have $\varphi'(f) \le \|\varphi\| \|f\|$ for every f in $\mathscr{P}$.

It is clear from the definition (4) that if f belongs to $\mathscr{P}$ and if t is a nonnegative number, then $\varphi'(tf) = t\varphi'(f)$. We show next that φ' is also additive. Indeed, if $0 \le g_i \le f_i$, where $f_i, g_i \in \mathscr{P}$, $i = 1, 2$, then $0 \le g_1 + g_2 \le f_1 + f_2$, from which it follows that $\varphi(g_1) + \varphi(g_2) \le \varphi'(f_1 + f_2)$, and hence that $\varphi'(f_1) + \varphi'(f_2) \le \varphi'(f_1 + f_2)$. On the other hand, if $0 \le g \le f_1 + f_2$ where $g \in \mathscr{P}$, and if we write

$$g' = g \wedge f_1 \quad \text{and} \quad g'' = g - g',$$

then it is readily verified that $0 \le g' \le f_1$ and $0 \le g'' \le f_2$. Hence $\varphi(g) = \varphi(g') + \varphi(g'') \le \varphi'(f_1) + \varphi'(f_2)$, and it follows that

$$\varphi'(f_1 + f_2) \le \varphi'(f_1) + \varphi'(f_2).$$

Thus $\varphi'(f_1 + f_2) = \varphi'(f_1) + \varphi'(f_2)$.

We next extend φ' to $\mathscr{C}_{\mathbb{R}}(X)$ by setting

$$\varphi'(f) = \varphi'(f^+) - \varphi'(f^-), \qquad f \in \mathscr{C}_{\mathbb{R}}(X),$$

and verify that φ' thus extended is a linear functional. Indeed, if $t \geq 0$, then $(tf)^+ = tf^+$ and $(tf)^- = tf^-$, so we have

$$\varphi'(tf) = t\varphi'(f^+) - t\varphi'(f^-) = t\varphi'(f).$$

On the other hand, if $t \leq 0$, then $(tf)^+ = -tf^-$ and $(tf)^- = -tf^+$, so

$$\varphi'(tf) = \varphi'(-tf^-) - \varphi'(-tf^+) = t\varphi'(f).$$

Thus φ' is homogeneous. In order to verify additivity, let f_1 and f_2 be two functions in $\mathscr{C}_{\mathbb{R}}(X)$ and let us write $h = f_1^+ + f_2^+ - (f_1 + f_2)^+$. Then $h \geq 0$, so

$$\varphi'(f_1^+) + \varphi'(f_2^+) = \varphi'(f_1^+ + f_2^+) = \varphi'((f_1 + f_2)^+) + \varphi'(h).$$

On the other hand, we also have $h = f_1^- + f_2^- - (f_1 + f_2)^-$, and therefore, as above,

$$\varphi'(f_1^-) + \varphi'(f_2^-) = \varphi'((f_1 + f_2)^-) + \varphi'(h),$$

whence it follows at once that $\varphi'(f_1 + f_2) = \varphi'(f_1) + \varphi'(f_2)$.

Thus, summarizing, we see that φ' is a positive linear functional on $\mathscr{C}_{\mathbb{R}}(X)$. Hence, by Theorem 10.9 there exists a regular Borel measure μ' on X such that

$$\varphi'(f) = \int_X f \, d\mu', \qquad f \in \mathscr{C}_{\mathbb{R}}(X).$$

Moreover, if $\varphi'' = \varphi' - \varphi$, then φ'' is also a positive linear functional on $\mathscr{C}_{\mathbb{R}}(X)$, so there is a second regular Borel measure μ'' on X such that

$$\varphi''(f) = \int_X f \, d\mu'', \qquad f \in \mathscr{C}_{\mathbb{R}}(X).$$

But then, of course, since $\varphi = \varphi' - \varphi''$, we have $\varphi(f) = \int_X f \, d(\mu' - \mu'')$ for every f in $\mathscr{C}_{\mathbb{R}}(X)$, and $\mu = \mu' - \mu''$ is the desired regular signed Borel measure. $\qquad \square$

Example A. It was noted in Chapter 14 (see Example 14D) that, in the special case $X = [a, b]$, the functionals in $\mathscr{C}(X)^*$ can be realized as integrals with respect to functions of bounded variation on $[a, b]$. In fact, the assignment $\alpha \overset{\Psi}{\rightarrow} \varphi_\alpha$, where

$$\varphi_\alpha(f) = \int_a^b f(t) d\alpha(t), \qquad f \in \mathscr{C}([a, b]), \tag{5}$$

is a contractive linear transformation of the Banach space $\mathscr{V}([a, b])$ of all functions of bounded variation on $[a, b]$ in the total variation norm onto

$\mathscr{C}([a, b])^*$ (see Problem 14F). Theorem 18.3 permits us to shed further light on this relationship. In the first place, the interval $[a, b]$ is a metric space, so $\mathscr{M}([a, b])$ and $\mathscr{M}_0([a, b])$ coincide. Moreover, if for each element ξ of $\mathscr{M}([a, b])$ we define

$$\alpha_\xi(t) = \begin{cases} 0, & t = a, \\ \xi([a, t]), & a < t \le b, \end{cases}$$

then α_ξ is a complex-valued function of bounded variation on $[a, b]$, and it is easy to see that α_ξ is right-continuous on (a, b). From this it follows that ξ is, in turn, the (Stieltjes–) Borel measure associated with α_ξ (see Example 8K), that

$$\int_{[a, b]} f\, d\xi = \int_a^b f(t) d\alpha_\xi(t), \qquad f \in \mathscr{C}([a, b]) \qquad (6)$$

(Prob. 8X), and also that if we define

$$\Phi(\xi) = \alpha_\xi, \qquad \xi \in \mathscr{M}([a, b]),$$

then Φ, which is obviously a linear transformation of $\mathscr{M}([a, b])$ into $\mathscr{V}([a, b])$, is in fact an *isometry* of $\mathscr{M}([a, b])$ onto a subspace $\mathscr{R}$ of $\mathscr{V}([a, b])$ (see Problem 8W). Finally, we note that $\mathscr{R}$ consists precisely of those functions in $\mathscr{V}([a, b])$ that are right-continuous on (a, b) and vanish at a. (If β is a function of bounded variation on $[a, b]$ that is right-continuous on (a, b), and if $\zeta = \zeta_\beta$ is the Stieltjes–Borel measure associated with β, then β and α_ζ differ by an additive constant; hence if β also vanishes at a, then $\beta = \alpha_\zeta$.) Thus we see that it is possible to identify the dual $\mathscr{C}([a, b])^*$ with the Banach space $\mathscr{R}$ of all those functions of bounded variation on $[a, b]$ that vanish at a and are right-continuous on (a, b). Indeed, *the mapping* $\Psi_0 = \Psi | \mathscr{R}$ *is an isometric isomorphism of* $\mathscr{R}$ *onto* $\mathscr{C}([a, b])^*$. (For another view of the relationship between $\mathscr{M}([a, b])$ and $\mathscr{V}([a, b])$ see Problem C.)

Example B. Let $\hat{W}$ denote the ordinal number segment consisting of all ordinal numbers less than or equal to Ω (see Problem 1W), and let μ_a denote the nonregular Borel measure on $W(\Omega)$ defined in Example 10C. Then $\hat{W}$ is a compact Hausdorff space (Prob. 3D), and it is easy to see that

$$\varphi(f) = \int_{W(\Omega)} f\, d\mu_a, \qquad f \in \mathscr{C}(\hat{W}),$$

defines a bounded linear functional on $\mathscr{C}(\hat{W})$. Hence there must exist a *regular* Borel measure on $\hat{W}$ that yields this same functional, and it is easy to see that this measure is nothing other than the Dirac mass at Ω multiplied by a. (See Problem 3D(ii) and recall Example 10E.)

As was noted earlier (Ex. 12L), if X is a compact Hausdorff space, then $\mathscr{C}(X)$ is not only a Banach space but is also a Banach algebra, and we shall frequently have occasion to use properties of $\mathscr{C}(X)$ that pertain to it by virtue of its being an algebra. Among other things, we shall need some information concerning the subalgebras of $\mathscr{C}(X)$. In this connection we recall (Ex. 3R) that a subset M of $\mathscr{C}(X)$ is said to be *separating* on X if for every two points $x_1 \neq x_2$ in X there exists a function f in M such that $f(x_1) \neq f(x_2)$. We shall also say that M is *strongly separating* on X if M is separating and if for each x in X there is a function f in M such that $f(x) \neq 0$.

The central fact that we wish to verify is that if $\mathscr{A}$ is a self-conjugate, strongly separating subalgebra of $\mathscr{C}(X)$, then $\mathscr{A}$ is dense in $\mathscr{C}(X)$. (Recall (Ex. 2H) that a linear space of complex-valued functions on a set X is said to be self-conjugate if $\bar{f}$ belongs to the set whenever f does.) Since the closure of a self-conjugate subalgebra is again a self-conjugate subalgebra, this assertion is equivalent to the following theorem.

Theorem 18.5 (Stone–Weierstrass Theorem). *Let X be a compact Hausdorff space. Then the only (sup norm) closed, self-conjugate, strongly separating subalgebra of $\mathscr{C}(X)$ is the algebra $\mathscr{C}(X)$ itself. Similarly, the only closed, strongly separating subalgebra of the real algebra $\mathscr{C}_{\mathbb{R}}(X)$ is $\mathscr{C}_{\mathbb{R}}(X)$ itself.*

The proof of this theorem is fairly lengthy, and will be given in part as a sequence of preparatory propositions. Our first step is to elucidate the role of the hypothesis of self-conjugacy.

Lemma 18.6. *A self-conjugate subalgebra $\mathscr{A}$ of $\mathscr{C}(X)$ coincides with $\mathscr{C}(X)$ if and only if the real algebra $\mathscr{A}_{\mathbb{R}}$ of real-valued functions in $\mathscr{A}$ coincides with the real algebra $\mathscr{C}_{\mathbb{R}}(X)$ of continuous real-valued functions on X. Also, $\mathscr{A}$ is [strongly] separating on X if and only if $\mathscr{A}_{\mathbb{R}}$ is.*

PROOF. All three assertions are immediate consequences of the fact that a function f belongs to $\mathscr{A}$ if and only if $\operatorname{Re} f$ and $\operatorname{Im} f$ do (see Example 2H). $\qquad\square$

The effect of Lemma 18.6 is to reduce the proof of Theorem 18.5 to the real case, and there the most important fact we shall need is the following one.

Lemma 18.7. *There exists a sequence $\{p_n(t)\}$ of real polynomials that converges uniformly to the function $\sqrt{1-t}$ on the closed interval $[0,1]$.*

PROOF. The power series

$$\sum_{n=0}^{\infty} \binom{\frac{1}{2}}{n} \lambda^n, \tag{7}$$

where $\binom{1/2}{n}$ denotes, as usual, the binomial coefficient

$$\frac{\frac{1}{2}(\frac{1}{2}-1)\cdots(\frac{1}{2}-n+1)}{n!},$$

is the Taylor series of the function $\sqrt{1 + \lambda}$ and therefore converges to $\sqrt{1 + \lambda}$ for $|\lambda| < 1$ (see Example 5J and Problem 5R). Since all of the factors $(\frac{1}{2} - k)$ in the numerator of $\binom{1/2}{n}$ are negative except the first, it is clear that if we define

$$A_n = \left| \binom{\frac{1}{2}}{n} \right|,$$

then

$$\binom{\frac{1}{2}}{n} = (-1)^{n+1} A_n$$

for all n except $n = 0$. Hence, if s_n denotes the nth partial sum of the series (7), and if we define $p_n(\lambda) = s_n(-\lambda)$, then

$$p_n(\lambda) \equiv 1 - \sum_{k=1}^{n} A_k \lambda^k$$

and the sequence $\{p_n(\lambda)\}$ converges to $\sqrt{1 - \lambda}$ for all $|\lambda| < 1$. In particular, if t is real and $0 \le t < 1$, then

$$1 - \sum_{n=1}^{\infty} A_n t^n = \sqrt{1 - t}.$$

It follows that $p_n(t) \ge 0$ for all n and all $0 \le t < 1$, and hence, by continuity, that $p_n(1) \ge 0$ for all n. Suppose now that ε is a given positive number. We first choose a number t_0, $0 < t_0 < 1$, such that $\sqrt{1 - t_0} < \varepsilon/2$, and then a positive integer n_0 such that $p_{n_0}(t_0) < \varepsilon$. Then for $t_0 \le t \le 1$ and for all $n \ge n_0$, we have

$$0 \le \sqrt{1 - t} \le p_n(t) \le p_{n_0}(t) \le p_{n_0}(t_0) < \varepsilon,$$

since $p_{n_0}(t)$ is clearly a decreasing function of t and since, for fixed t, the numerical sequence $\{p_n(t)\}$ is decreasing. In addition

$$p_{n_0}(t) - \sqrt{1 - t} = \sum_{k=n_0+1}^{\infty} A_k t^k, \qquad 0 \le t \le t_0,$$

and the right member of this equation is a monotone increasing function of t. Hence

$$0 \le p_n(t) - \sqrt{1 - t} < \varepsilon, \qquad n \ge n_0, \qquad 0 \le t \le t_0.$$

Thus $\{p_n(t)\}$ converges uniformly to $\sqrt{1 - t}$ on the closed interval $[0, 1]$. $\square$

This lemma allows us to prove the following important result. (See Problem 1D for the notion of a function lattice.)

Proposition 18.8 (Kakutani). *Every closed subalgebra $\mathscr{A}$ of the algebra $\mathscr{C}_{\mathbb{R}}(X)$ is also a sublattice of $\mathscr{C}_{\mathbb{R}}(X)$.*

PROOF. According to Problem 1D it suffices to verify that if $f \in \mathscr{A}$, then $|f| \in \mathscr{A}$ also. To establish the latter assertion observe that if we set $q_n(t) = p_n(1 - t)$, where $\{p_n(t)\}$ is a sequence of real polynomials converging uniformly to $\sqrt{1 - t}$ on the interval $[0, 1]$, then the sequence $\{q_n(t)\}$ converges uniformly to the function $\sqrt{t}$ on $[0, 1]$. In particular, if $a_n = q_n(0)$, then $a_n \to 0$. Hence if $r_n(t) = q_n(t) - a_n$, then the sequence of polynomials $\{r_n(t)\}$ also converges uniformly to $\sqrt{t}$ on $[0, 1]$, and each r_n has zero constant term. Note that $r_n \circ f \in \mathscr{A}$ whenever $f \in \mathscr{A}$ by virtue of the fact that $r_n(0) = 0$.

Suppose now that f is a nonzero element of $\mathscr{A}$. Set $g(x) = (f(x)/\|f\|_\infty)^2$ for all x in X, and observe that g belongs to $\mathscr{A}$ and that the sequence $\{r_n \circ g\}$ does so too. But this latter sequence converges uniformly on X to the limit $\sqrt{g} = |f|/\|f\|_\alpha$, whence it follows at once that $|f|$ belongs to $\mathscr{A}$. □

Lemma 18.9. *Suppose that $\mathscr{A}$ is a strongly separating subalgebra of $\mathscr{C}_{\mathbb{R}}(X)$ and that x_1 and x_2 are distinct points of X. Then for any two real numbers t_1 and t_2 there is a function f in $\mathscr{A}$ such that $f(x_1) = t_1$ and $f(x_2) = t_2$.*

PROOF. To verify this, let $\mathscr{L} = \{(f(x_1), f(x_2)) : f \in \mathscr{A}\}$. Then $\mathscr{L}$ is a non-trivial subalgebra of $\mathbb{R}^2$, and since $\mathscr{A}$ is strongly separating, it follows from Problem M that this subalgebra is all of $\mathbb{R}^2$. □

With this preparation we are ready to prove the Stone–Weierstrass theorem.

PROOF OF THEOREM 18.5. By virtue of Lemma 18.6 it suffices to show that a closed strongly separating *real* subalgebra $\mathscr{A}$ of the *real* algebra $\mathscr{C}_{\mathbb{R}}(X)$ necessarily coincides with $\mathscr{C}_{\mathbb{R}}(X)$. To this end, let g belong to $\mathscr{C}_{\mathbb{R}}(X)$, let y_0 be any fixed point in X, and let ε be a given positive number. Then, by Lemma 18.9, for every x in X there exists a function f_x in $\mathscr{A}$ such that $f_x(y_0) = g(y_0)$ and such that $f_x(x) = g(x)$. Since both f and g are continuous, there exists an open neighborhood U_x of x such that $f_x(y) > g(y) - \varepsilon$ for all y in U_x. Thus we obtain an open covering $\{U_x\}_{x \in X}$ of X, and since X is compact, there exists a finite subset $\{x_1, \ldots, x_n\}$ of X such that the sets $U_{x_1}, \ldots, U_{x_n}$ cover X. Furthermore, the functions f_{x_i} all belong to $\mathscr{A}$, and $\mathscr{A}$ is a lattice by Proposition 18.8. Hence the function

$$F_{y_0} = f_{x_1} \vee f_{x_2} \vee \cdots \vee f_{x_n}$$

also belongs to $\mathscr{A}$. Clearly $F_{y_0}(y_0) = g(y_0)$ and $F_{y_0}(x) > g(x) - \varepsilon$ for all x in X (since every x in X belongs to some U_{x_i}). Thus we have shown that for each y in X and each given $\varepsilon > 0$ there exists a function F_y in $\mathscr{A}$ such that $F_y(y) = g(y)$ and such that $F_y > g - \varepsilon$. By continuity, there is an open neighborhood V_y of y such that $F_y(z) < g(z) + \varepsilon$ for all z in V_y. Then, just as in the foregoing argument, there exist points $y_1, \ldots, y_m$ in X such that the open sets $V_{y_1}, \ldots, V_{y_m}$ cover X, and if we define

$$G = F_{y_1} \wedge F_{y_2} \wedge \cdots \wedge F_{y_m},$$

then $G \in \mathscr{A}$, $G > g - \varepsilon$ (since each F_{y_i} has this property), and also $G < g + \varepsilon$ (since every point of X belongs to some V_{y_i}). Thus $\|G - g\| \le \varepsilon$, and the theorem is proved. $\square$

Corollary 18.10. *If X is a compact Hausdorff space, and if $\mathscr{F}$ is any strongly separating set of functions in $\mathscr{C}(X)$, then the linear submanifold of $\mathscr{C}(X)$ generated algebraically by the set of all finite products of elements of $\mathscr{F}$ and their complex conjugates is dense in $\mathscr{C}(X)$. Likewise, if $\mathscr{F}$ is a strongly separating subset of $\mathscr{C}_{\mathbb{R}}(X)$, the linear submanifold of $\mathscr{C}_{\mathbb{R}}(X)$ generated algebraically by the set of all finite products of elements of $\mathscr{F}$ is dense in $\mathscr{C}_{\mathbb{R}}(X)$.*

PROOF. In both cases the system described is a self-conjugate strongly separating subalgebra. $\square$

Example C. The pair $\{t, 1\}$ is strongly separating on any interval $[a, b]$. Hence the complex polynomial functions $p(t)$ are dense in $\mathscr{C}([a, b])$ and the real polynomial functions are dense in $\mathscr{C}_{\mathbb{R}}([a, b])$. Note that this may be paraphrased by saying that *for any continuous function f on $[a, b]$ there is a sequence of polynomials $\{p_n\}$ converging uniformly to f on $[a, b]$.* This is the original *theorem of Weierstrass.*

Example D. The projections $(x, y) \to x$ and $(x, y) \to y$ are real-valued and separating on the plane $\mathbb{R}^2$. Hence if K is any compact subset of $\mathbb{R}^2$, the complex polynomial functions $p(x, y)$ form a dense subalgebra of $\mathscr{C}(K)$. Similarly, for substantially the same reason, if K is a compact subset of the complex plane $\mathbb{C}$, then the functions of the form $p(\lambda, \bar{\lambda})$, where p is a complex polynomial in two indeterminates, constitute a dense subalgebra of $\mathscr{C}(K)$.

If K is a compact subset of $\mathbb{C}$ then the complex polynomials $p(\lambda)$ (restricted to K) form a strongly separating subalgebra of $\mathscr{C}(K)$, but this subalgebra is not (ordinarily) self-conjugate, and need not be dense in $\mathscr{C}(K)$. In connection with this observation the following notions are pertinent.

Definition. For any compact subset K of the complex plane $\mathbb{C}$ we shall write $\mathscr{P}(K)$ for the closure in $\mathscr{C}(K)$ of the algebra of (the restrictions to K of) all polynomials $p(\lambda)$. (In other words, $\mathscr{P}(K)$ is the subspace of $\mathscr{C}(K)$ consisting of those functions on K that can be uniformly approximated by polynomial functions.) We shall also write $\mathscr{A}(K)$ for the algebra of functions f in $\mathscr{C}(K)$ such that f is locally analytic on K°. (If K° is empty this requirement becomes vacuous and $\mathscr{A}(K) = \mathscr{C}(K)$.)

The relations between $\mathscr{C}(K)$, $\mathscr{P}(K)$ and $\mathscr{A}(K)$ are quite difficult to determine in full generality, but certain elementary facts are easily established.

Proposition 18.11. *For every compact subset K of $\mathbb{C}$, $\mathscr{P}(K)$ and $\mathscr{A}(K)$ are closed subalgebras of $\mathscr{C}(K)$ such that $\mathscr{P}(K) \subset \mathscr{A}(K)$.*

PROOF. Since every polynomial function clearly belongs to $\mathscr{A}(K)$, it suffices to prove that $\mathscr{P}(K)$ is a subalgebra of $\mathscr{C}(K)$ and that $\mathscr{A}(K)$ is closed. To show the former, let f and g be functions in $\mathscr{P}(K)$, and let $\{p_n\}$ and $\{q_n\}$ be sequences of polynomials such that $p_n \to f$ and $q_n \to g$ uniformly on K. Then $p_n q_n \to fg$ uniformly on K, so $fg \in \mathscr{P}(K)$. To show the latter, let $\{f_n\}$ be a sequence in $\mathscr{A}(K)$ that converges in $\mathscr{C}(K)$ to a limit g. If $D = D_r(\alpha)$ is an open disc contained in K°, then $\{f_n\}$ also converges uniformly to g on D, and since each of the functions f_n is analytic on D, it follows that g is analytic on D (see Problem 11U). But this shows that g is locally analytic on K°, and hence that $g \in \mathscr{A}(K)$. $\square$

Corollary 18.12. *If K is an arbitrary compact subset of $\mathbb{C}$, a necessary and sufficient condition for $\mathscr{P}(K)$ to coincide with $\mathscr{C}(K)$ is that the function $\bar{\lambda}$ belong to $\mathscr{P}(K)$. Similarly, $\mathscr{A}(K) = \mathscr{C}(K)$ if and only if $\bar{\lambda} \in \mathscr{A}(K)$. Consequently $\mathscr{A}(K) \neq \mathscr{C}(K)$ when $K^\circ \neq \varnothing$.*

PROOF. If $\bar{\lambda}$ belongs to either $\mathscr{P}(K)$ or $\mathscr{A}(K)$, then all polynomials $p(\lambda, \bar{\lambda})$ in λ and $\bar{\lambda}$ do so too; cf. Example D. If $K^\circ \neq \varnothing$, then $\bar{\lambda} \notin \mathscr{A}(K)$ (Prob. 5B).

Example E. If K is a compact subset of $\mathbb{C}$ that is mapped onto $\tilde{K}$ by some affine mapping of the form $\lambda \xrightarrow{A} \alpha\lambda + \beta$ ($\alpha \neq 0$), then $f \to f \circ A$ maps $\mathscr{C}(\tilde{K})$ onto $\mathscr{C}(K)$ in such a way that $\mathscr{A}(\tilde{K})$ is carried onto $\mathscr{A}(K)$ and $\mathscr{P}(\tilde{K})$ is carried onto $\mathscr{P}(K)$. Thus if any two of $\mathscr{C}(K)$, $\mathscr{A}(K)$ and $\mathscr{P}(K)$ coincide, then the corresponding relation holds on $\tilde{K}$ too. For example, since $\mathscr{P}(K) = \mathscr{C}(K)$ when K is a compact subset of the real axis (Ex. C), the same is true for any compact subset of any line in $\mathbb{C}$.

Example F (The disc algebra). Let $D = \{\lambda \in \mathbb{C} : |\lambda| < 1\}$, and set $K = D^-$. Then $\mathscr{A}(K)$ (known as the *disc algebra*), may also be described as the algebra of all those analytic functions on D that possess continuous extensions to D^-. Moreover, every function f in $\mathscr{A}(K)$ is uniformly continuous on K (Prop. 4.6), from which it follows easily that, if we write $f_s(\lambda) = f(s\lambda)$ for each s such that $0 < s < 1$, then the net $\{f_s(\lambda)\}_{0 < s < 1}$ converges uniformly in λ to f on K as $s \uparrow 1$. On the other hand, if $f(\lambda) = \sum_{n=0}^{\infty} \alpha_n \lambda^n$ is the power series expansion of f on D, then the series

$$\sum_{n=0}^{\infty} \alpha_n s^n \lambda^n$$

converges uniformly in λ to f_s on $D^- = K$ for every s, $0 < s < 1$, and it follows at once that f belongs to $\mathscr{P}(K)$. Thus $\mathscr{A}(K) = \mathscr{P}(K)$ when K is a closed disc.

The following idea is also useful in dealing with polynomial approximation.

Definition. If K is a compact subset of $\mathbb{C}$ and if U_∞ denotes the unbounded component of $\mathbb{C}\backslash K$, then we shall write $\hat{K}$ for the compact set $\mathbb{C}\backslash U_\infty$, so that $\hat{K}$ consists of the union of K with all of the various holes in K. (See Chapter 3 for a discussion of this terminology.) The boundary of $\hat{K}$ (which is the same as the boundary of U_∞) is called the *outer boundary* of K.

Proposition 18.13. *If K is a compact subset of $\mathbb{C}$ then the algebra $\mathscr{P}(K)$ coincides with the restriction to K of the subalgebra $\mathscr{P}(\hat{K})$ of $\mathscr{C}(\hat{K})$. Thus the functions in $\mathscr{P}(K)$ are not only locally analytic on K°, but can be extended so as to be locally analytic on the interior of $\hat{K}$.*

PROOF. A sequence $\{p_n\}$ of polynomials converging uniformly on K automatically converges uniformly on the outer boundary of K. But then, by the maximum modulus principle (Ex. 5M), $\{p_n\}$ converges uniformly on $\hat{K}$, and the limit is therefore differentiable at each point of $\hat{K}^\circ$ (see Problem 11U).   □

The foregoing discussion shows that for a compact subset K of $\mathbb{C}$, the algebra $\mathscr{P}(K)$ may fail to be equal to $\mathscr{C}(K)$ for two quite different reasons, either because $K^\circ \neq \varnothing$, or because K has one or more holes.

Example G. Let us take for K the unit circle Z. Then $\hat{K}$ is just the closed unit disc D^- of Example F, so $\mathscr{P}(Z)$ consists of the restrictions to Z of all of the functions in $\mathscr{P}(D^-) = \mathscr{A}(D^-)$. We observe that the mapping $f \to f\,|\,Z$ is an isometric Banach algebra isomorphism of the disc algebra $\mathscr{P}(D^-)$ onto $\mathscr{P}(Z)$ by the maximum modulus principle (Ex. 5M). (The algebra $\mathscr{P}(Z)$ is also frequently called the disc algebra.)

Example H. Let K be any compact subset of $\mathbb{C}$ such that $K^\circ \neq \varnothing$. Then, just as in the preceding example, the mapping $f \to f\,|\,(\partial K)$ is an isometric Banach algebra isomorphism of $\mathscr{A}(K)$ onto a subalgebra of $\mathscr{C}(\partial K)$, so any bounded linear functional φ on $\mathscr{A}(K)$ may be regarded as a functional on a subspace of $\mathscr{C}(\partial K)$, which may then be extended to a bounded linear functional on $\mathscr{C}(\partial K)$ by the Hahn–Banach theorem (Th. 14.3). Hence, by the Riesz representation theorem (Th. 18.3), there exists a complex Borel measure ξ on ∂K such that

$$\varphi(f) = \int_{\partial K} f \, d\xi, \qquad f \in \mathscr{A}(K).$$

(If $\mathscr{A}(K)$ is replaced by $\mathscr{P}(K)$, the boundary ∂K may be replaced by the *outer* boundary of K.)

Suppose, in particular, we take for K the closed unit disc D^-, so that ∂K is the unit circle Z. Then for each point α of D^- evaluation at α is a bounded

417

linear functional on $\mathscr{A}(D^-)$, so there exists a complex Borel measure ξ_α on Z such that

$$f(\alpha) = \int_Z f\, d\xi_\alpha, \qquad f \in \mathscr{A}(D^-).$$

Thus, for example, for $\alpha = 0$ we have

$$f(0) = \frac{1}{2\pi i} \int_Z \frac{f(\xi)}{\xi}\, d\xi = \frac{1}{2\pi}\int_0^{2\pi} f(e^{it})\,dt = \frac{1}{2\pi}\int_Z f\,d\theta$$

(where θ denotes arc-length measure; see Example 8F), so we may take ξ_0 to be $\theta/2\pi$. (This use of the Cauchy integral formula is readily justified because of the uniform continuity of f on D^-.)

Example I. Let r and R be radii such that $0 < r < R$, let A denote the annular domain $\{\lambda \in \mathbb{C} : r < |\lambda| < R\}$, and let $K = A^-$. Then the circle $C_R = \{\lambda \in \mathbb{C} : |\lambda| = R\}$ is the outer boundary of K and $\hat{K}$ is the closed disc $D_R^- = \{\lambda \in \mathbb{C} : |\lambda| \le R\}$. Thus every function f in $\mathscr{P}(K)$ is the restriction to K of a function g in the algebra $\mathscr{A}(D_R^-) = \mathscr{P}(D_R^-)$.

Next let us consider the algebra $\mathscr{A}(K)$. If $f \in \mathscr{A}(K)$ then f possesses a Laurent expansion

$$f(\lambda) = \sum_{n=-\infty}^{+\infty} \alpha_n \lambda^n, \qquad \lambda \in A$$

(Prop. 5.9), and we may write $f = f_1 + f_2$ where

$$f_1(\lambda) = \sum_{n=0}^{\infty} \alpha_n \lambda^n \quad \text{and} \quad f_2(\lambda) = \sum_{n=-1}^{-\infty} \alpha_n \lambda^n.$$

The function f_1 can be extended to the closed disc D_R^- by continuity (since f can be and f_2 is already continuous in a neighborhood of C_R), and there exists a sequence $\{p_n\}$ of polynomials $p_n(\lambda) = \beta_0^{(n)} + \cdots + \beta_n^{(n)}\lambda^n$ converging uniformly to f_1 on D_R^-, and therefore on K (Ex. F). Similarly, f_2 can be extended by continuity to the closed region $\mathbb{C}\backslash D_r = \{\lambda \in \mathbb{C} : |\lambda| \ge r\}$, and an argument exactly like the one in Example F shows that there exists a sequence $\{q_n\}$ of functions of the form

$$q_n(\lambda) = \frac{\beta_{-1}^{(n)}}{\lambda} + \cdots + \frac{\beta_{-n}^{(n)}}{\lambda^n}$$

converging uniformly to f_2 on $\mathbb{C}\backslash D_r$, and therefore on K. Thus, summarizing, we see that there exists a sequence of rational functions of the form

$$r_n(\lambda) = \frac{\beta_{-n}^{(n)}}{\lambda^n} + \cdots + \beta_0^{(n)} + \beta_1^{(n)}\lambda + \cdots + \beta_n^{(n)}\lambda^n$$

converging uniformly to f on K.

The foregoing example suggests the introduction of yet one more Banach algebra.

Definition. If K is a compact subset of $\mathbb{C}$ we shall denote by $\mathscr{R}(K)$ the closure in $\mathscr{C}(K)$ of the algebra of all rational functions with poles off K.

It is clear that $\mathscr{R}(K)$ is a closed subalgebra of $\mathscr{C}(K)$ and that

$$\mathscr{P}(K) \subset \mathscr{R}(K) \subset \mathscr{A}(K)$$

for any compact set K of complex numbers. Example I shows that $\mathscr{R}(K)$ may coincide with $\mathscr{A}(K)$ when K has a hole. But also in Example I every function in $\mathscr{A}(K)$ is approximable by means of special rational functions with poles only at 0 (and ∞). As a matter of fact, a similar restriction on the poles of approximating rational functions can be effected whenever, as in Example I, the functions in $\mathscr{R}(K)$ can be uniformly approximated on K by functions that are locally analytic on some neighborhood of K.

Proposition 18.14. *Let K be a nonempty compact subset of $\mathbb{C}$, let P be a subset of the Riemann sphere $\hat{\mathbb{C}}$ (Prob. 3W) with the property that P contains at least one point of each connected component of $\hat{\mathbb{C}} \backslash K$, and let f be a locally analytic function on some open neighborhood U of K. Then there exists a sequence $\{r_n\}$ of rational functions converging uniformly to f on K such that each r_n has poles only in P.*

PROOF. Let us denote by $\mathscr{L}$ the linear submanifold of $\mathscr{C}(K)$ consisting of (the restrictions to K of) those rational functions having their poles in P. We are to show that f belongs to $\mathscr{L}^-$, and by Proposition 14.10 (a consequence of the Hahn–Banach theorem) it suffices to show that if a bounded linear functional φ on $\mathscr{C}(K)$ annihilates $\mathscr{L}$, then $\varphi(f) = 0$. By virtue of Theorem 18.3 this amounts to showing that if ξ is a complex Borel measure on K such that $\int_K r \, d\xi = 0$ for every rational function r with poles in P, then $\int_K f \, d\xi = 0$. Suppose, accordingly, that $\int_K r \, d\xi = 0$ for every rational function r with poles in P. Let V be a component of $\hat{\mathbb{C}} \backslash K$, suppose α is a point of $V \cap P$ such that $\alpha \neq \infty$, and let $d = d(\alpha, K)$. Clearly $d > 0$, and for each λ such that $|\lambda - \alpha| < d$ the power series

$$\sum_{n=0}^{\infty} \frac{(\lambda - \alpha)^n}{(\zeta - \alpha)^{n+1}} \tag{8}$$

converges to $1/(\zeta - \lambda)$ uniformly (in ζ) on K. Hence the Cauchy transform

$$h(\lambda) = \int_K \frac{d\xi(\zeta)}{\zeta - \lambda}$$

(see Example 10F) may be computed by integrating the series (8) term by term, and since

$$\int_K \frac{d\xi(\zeta)}{(\zeta - \alpha)^{n+1}} = 0, \qquad n \in \mathbb{N}_0,$$

by assumption, this shows that h vanishes identically on the disc $D_d(\alpha)$, and hence on all of V by the identity theorem (Th. 5.2). Thus $h \equiv 0$ on $\mathbb{C} \backslash K$. (If the only point of P in the unbounded component of $\hat{\mathbb{C}} \backslash K$ is the point at infinity, then the argument must be modified on that component in that the series (8) is replaced by

$$-\sum_{n=0}^{\infty} \frac{\zeta^n}{\lambda^{n+1}}, \qquad |\lambda| > R,$$

where R is chosen large enough so that $K \subset D_{\bar{R}}$.)

Now let γ be an oriented envelope of K in U (Prob. 5K), so that

$$f(\lambda) = \frac{1}{2\pi i} \int_{\gamma} \frac{f(\zeta)}{\zeta - \lambda} d\zeta$$

for all λ in some neighborhood of K by the Cauchy integral formula (Prob. 5O). Then

$$\int_K f \, d\xi = \frac{1}{2\pi i} \int_K \left[\int_{\gamma} \frac{f(\zeta)}{\zeta - \lambda} d\zeta \right] d\xi(\lambda)$$

$$= \frac{1}{2\pi i} \int_{\gamma} f(\zeta) \left[\int_K \frac{d\xi(\lambda)}{\zeta - \lambda} \right] d\zeta$$

$$= \frac{-1}{2\pi i} \int_{\gamma} f(\zeta) h(\zeta) d\zeta = 0$$

by the Fubini theorem for complex measures (see Problem 9F). $\qquad \square$

Example J. Let K be a compact subset of the unit circle Z such that $K \neq Z$. Then the function $\bar{\lambda}$ possesses an analytic extension (namely $1/\lambda$) to a neighborhood of K, and $\mathbb{C} \backslash K$ is connected. Hence we may set $P = \{\infty\}$ in Proposition 18.14, and it follows that $\bar{\lambda} \in \mathscr{P}(K)$. But then, by Corollary 18.12, we see that $\mathscr{P}(K) = \mathscr{C}(K)$.

It is a theorem due to Lavrentiev [43] that the two obvious necessary conditions in order that $\mathscr{P}(K)$ coincide with $\mathscr{C}(K)$, viz., that K° be empty and K have no holes, are actually sufficient. In other words, for any compact subset K of the plane $\mathbb{C}$, *the algebras $\mathscr{P}(K)$ and $\mathscr{C}(K)$ coincide if and only if K° is empty and the complement $\mathbb{C} \backslash K$ is connected.*

Corollary 18.15. *For any compact subset K of $\mathbb{C}$ the Banach algebra $\mathscr{R}(K)$ contains every function in $\mathscr{A}(K)$ that can be continued analytically onto some open neighborhood of K.*

As a matter of fact, Corollary 18.15 can be proved much more simply and directly. Indeed, if f is locally analytic on an open set U containing K, and if

γ is an oriented envelope of K in U, then any suitably chosen Riemann sum approximating the Cauchy integral

$$\frac{1}{2\pi i}\int_\gamma \frac{f(\zeta)}{\zeta - \lambda}d\zeta$$

will approximate f as closely as desired in $\mathscr{C}(K)$. The fact that Proposition 18.14 permits us to place the poles of the approximating rational functions in a specified set (provided this set contains at least one point of each component of $\hat{\mathbb{C}}\backslash K$) is particularly useful.

Proposition 18.16 (Runge's Theorem). *Let U be an open subset of $\mathbb{C}$ and let P be a subset of $\hat{\mathbb{C}}\backslash U$ with the property that P contains at least one point of each connected component of $\hat{\mathbb{C}}\backslash U$. Then for any locally analytic function f on U there exists a sequence $\{r_n\}$ of rational functions, each with poles confined to P, such that $\{r_n\}$ converges to f uniformly on compact subsets of U.*

PROOF. If $U = \mathbb{C}$ then $P = \{\infty\}$, f is entire, and the partial sums of any power series expansion of f may be used. Otherwise, for each positive integer n let D_n^- denote the closed disc $\{\lambda \in \mathbb{C} : |\lambda| \leq n\}$, and set

$$K_n = \left\{\lambda \in D_n^- : d(\lambda, \mathbb{C}\backslash U) \geq \frac{1}{n}\right\}.$$

Then $\{K_n\}$ is an increasing sequence of compact subsets of U with the properties that $K_n \subset K_{n+1}^\circ, n \in \mathbb{N}$, and $U = \bigcup_{n=1}^\infty K_n$. Hence for any compact subset K of U there exists a positive integer n such that $K \subset K_n^\circ$, so it suffices to show that for each n there exists a rational function r_n with poles in P such that $|r_n(\lambda) - f(\lambda)| \leq 1/n$ for all λ in K_n. Hence by Proposition 18.14 it suffices to verify that each connected component of $\hat{\mathbb{C}}\backslash K_n$ contains a point of $\hat{\mathbb{C}}\backslash U$, for it will then contain an entire component of $\hat{\mathbb{C}}\backslash U$, and therefore a point of P. But each component of $\hat{\mathbb{C}}\backslash K_n$ is the union of some subcollection of the collection consisting of the (connected) open set

$$\{\infty\} \cup \{\lambda \in \mathbb{C} : |\lambda| > n\}$$

and the various open discs $D_{1/n}(\alpha), \alpha \notin U$, and every set in this collection meets $\hat{\mathbb{C}}\backslash U$. (Recall that $U \subset \mathbb{C}$, so that $\infty \notin U$.) $\square$

Example K. Suppose Δ is a simply connected domain in $\mathbb{C}$, so that the complement of Δ in the Riemann sphere $\hat{\mathbb{C}}$ is connected (see Problem 5P). In this situation we may select $P = \{\infty\}$, and we arrive at the following conclusion: *If f is an analytic function on a simply connected domain Δ in $\mathbb{C}$, then there exists a sequence of polynomials that converges to f uniformly on compact* *subsets of* Δ. (This special case of Proposition 18.16 is also sometimes referred to as Runge's theorem.)

Precise (but quite complicated) necessary and sufficient conditions on the compact set K in order that $\mathscr{R}(K)$ coincide with $\mathscr{A}(K)$ are known. The relevant theorems are due to Vituškin [64]. A nice account of them may be found in [27]. (Sample result: If K is a compact set in $\mathbb{C}$ and if every point of K has a neighborhood that meets only a finite number of components of $\mathbb{C}\setminus K$, then $\mathscr{R}(K) = \mathscr{A}(K)$.) On the other hand, a simple necessary and sufficient condition that $\mathscr{P}(K)$ coincide with $\mathscr{A}(K)$ may be stated at once: If K is a compact set in $\mathbb{C}$, then $\mathscr{P}(K) = \mathscr{A}(K)$ if and only if K does not separate the plane, that is, if and only if there are no holes in K. This is *Mergelyan's theorem* [48]. More readable accounts may be found in [27] or [57].

PROBLEMS

A. Let ξ and ζ be regular complex Borel measures on a locally compact Hausdorff space X. Verify that

$$|\,|\xi|(E) - |\zeta|(E)\,| \le \|\xi - \zeta\|$$

for every Borel set E in X.

B. Show that if $[a, b]$ is a closed interval of real numbers, then $\mathscr{C}([a, b])^*$ may not only be identified with the subspace $\mathscr{R}$ of $\mathscr{V}([a, b])$ consisting of those functions that are right-continuous on (a, b) and vanish at a, as in Example A, but may also be identified with a subspace of $\mathscr{V}([a, b])$ consisting of *left*-continuous functions.

C. Show that if, for an arbitrary function α of bounded variation on an interval $[a, b]$, we denote by ζ_α the complex Borel measure associated with α (Ex. 8K), then setting $\sigma(\alpha) = \|\zeta_\alpha\|$ defines a pseudonorm on the space $\mathscr{V}([a, b])$ (Prob. 14F), and find the zero space of this pseudonorm. Use this to identify $\mathscr{C}([a, b])^*$ as a quotient space of $\mathscr{V}([a, b])$. (Hint: See Problem 14G.)

D. Show that if X is a compact Hausdorff space, then $\mathscr{C}(X)$ is reflexive if and only if X is finite. (Hint: If E is any fixed Borel set, then $\mu \to \mu(E)$ defines a bounded linear functional on $\mathscr{M}_0(X)$. Recall Example 10E.)

E. If X is a compact Hausdorff space, then a sequence $\{f_n\}$ in $\mathscr{C}(X)$ converges weakly to a limit f if and only if $\{f_n\}$ is uniformly bounded and converges pointwise to f. Similarly, a sequence $\{\mu_n\}$ in $\mathscr{M}_0(X)$ converges weak* to a regular complex Borel measure μ if and only if $\{\mu_n\}$ is uniformly bounded (in the total variation norm) and converges setwise to μ.

F. Show that if X is a compact Hausdorff space, then $\mathscr{C}(X)$ is separable if and only if X satisfies the second axiom of countability, or, in other words, if and only if X is metrizable (see Corollary 4.4). (Hint: If $\mathscr{U}_0$ is a countable base of open sets in X, and if, for each pair U, V of sets in $\mathscr{U}_0$ such that $U^- \subset V$, $f_{U,V}$ denotes some one continuous function such that $\chi_U \le f_{U,V} \le \chi_V$ (Prob. 3V), then the system of all finite suprema of the functions $f_{U,V}$ spans $\mathscr{C}(X)$ as a Banach space. On the other hand, if $\{f_n\}$ is any sequence dense in $\mathscr{C}(X)$, and if for each n we set $U_n = \{x \in X : |f_n(x)| > \frac{1}{2}\}$, then $\{U_n\}$ is a base of open sets in X.)

G. Let X and Y be compact Hausdorff spaces, and let $F : X \to Y$ be a continuous mapping. If we write $\Phi_F(f) = f \circ F$ for every f in $\mathscr{C}(Y)$, then Φ_F maps $\mathscr{C}(Y)$ into $\mathscr{C}(X)$. Show that Φ_F is a contraction. Show also that if F is onto, then Φ_F is an isometry, and that if f is one-to-one, then Φ_F is onto. Consequently, if F is a homeomorphism between X and Y, then Φ_F is a unital Banach algebra isomorphism between $\mathscr{C}(Y)$ and $\mathscr{C}(X)$. (Hint: Use the Tietze extension theorem; see Example 11G.)

H. Show that, for any Banach space $\mathscr{E}$ whatever, there exists a compact Hausdorff space $X = X_{\mathscr{E}}$ and an isometric isomorphism of $\mathscr{E}$ onto a subspace of $\mathscr{C}(X)$. Show also that X can be taken to be metrizable if and only if $\mathscr{E}$ is separable. (Hint: For each x in $\mathscr{E}$ consider $j(x)|(\mathscr{E}^*)_1$.)

> The results of the preceding problem are frequently expressed by saying that $\mathscr{C}(X)$ is "universal" for Banach spaces—meaning that every Banach space may be viewed as a subspace of some $\mathscr{C}(X)$— and that $\mathscr{C}(X)$, with X a compact metric space, is universal for separable Banach spaces. It is interesting to pursue these ideas a bit further. According to a standard theorem in topology (see Problem 4T), if X is any compact metric space, then there exists a continuous mapping of the Cantor set C onto X. Hence, according to Problem G, $\mathscr{C}(X)$ is isomorphic as a Banach space to some subspace of $\mathscr{C}(C)$, so $\mathscr{C}(C)$ all by itself is, in fact, universal for separable Banach spaces. Moreover, if for each continuous function f on the Cantor set C we define $\hat{f}$ to be the uniquely determined continuous function on the unit interval $I = [0, 1]$ that agrees with f on C and is linear on each of the intervals complementary to C, then $f \to \hat{f}$ is readily seen to be a linear isometry of $\mathscr{C}(C)$ into $\mathscr{C}(I)$. Thus $\mathscr{C}(I)$ is also universal for separable Banach spaces.

I. Let X be a *locally* compact Hausdorff space, and let $\mathscr{C}_0(X)$ denote the linear space of all continuous complex-valued functions on X with the property that for every $\varepsilon > 0$ there is a compact subset K of X such that $|f(x)| < \varepsilon$ for all x in $X \backslash K$ (such functions are said to *vanish at infinity*). Show that $\mathscr{C}_0(X)$ is a (closed) subspace of $\mathscr{C}_b(X)$ in the sup norm (Prob. 11D). What is the dual of $\mathscr{C}_0(X)$? Let us say, more generally, that a complex-valued function f on X has *limit L at infinity* if for every $\varepsilon > 0$ there is a compact subset K of X such that $|L - f(x)| < \varepsilon$ for all x in $X \backslash K$. Show similarly that the space $\mathscr{C}_l(X)$ of all continuous functions on X that have limits at infinity is a Banach space in the sup norm, and find the dual of $\mathscr{C}_l(X)$. (Hint: Use the one-point compactification (Prob. 3W).)

J. Consider the Banach space $\mathscr{C}^{(1)} = \mathscr{C}^{(1)}([a, b])$ of continuously differentiable functions on the interval $[a, b]$ introduced in Problem 11F. Among the bounded linear functionals on $\mathscr{C}^{(1)}$ there are the point evaluations $f \to f(t_0)$ and also integrations, such as

$$f \to \int_a^b f' \, d\xi$$

for various complex Borel measures ξ. Show that, in fact, the most general bounded linear functional φ on $\mathscr{C}^{(1)}$ is of the form

$$\varphi(f) = \alpha f(t_0) + \int_a^b f'(t) d\xi(t), \tag{9}$$

where ξ is a complex Borel measure on $[a, b]$, t_0 denotes a fixed point of $[a, b]$, and α is a complex number. Show also that this representation is unique in the sense that, once t_0 has been chosen, the measure ξ and the coefficient α are completely determined by φ. Likewise, find the most general bounded linear functional on the space $\mathscr{C}^{(k)}$ of k times continuously differentiable functions. (Hint: If $D: \mathscr{C}^{(1)} \to \mathscr{C}^{(0)}$ is defined by $Df = f'$, and $V: \mathscr{C}^{(0)} \to \mathscr{C}^{(1)}$ is the transformation given by

$$(Vf)(x) = \int_a^x f(t)dt, \quad a \le x \le b,$$

then D is a left inverse of V; use Problem 13E.)

K. If ζ is a complex Borel measure on the interval $[a, b]$, then

$$f \to \int_a^b f \, d\zeta$$

is a bounded linear functional on the space $\mathscr{C}^{(1)}$ of the preceding problem, and must therefore be of the form (9) for some complex Borel measure ξ. Find ξ. (Hint: It is convenient to replace Lebesgue integration with respect to Stieltjes–Borel measures by Riemann–Stieltjes integration with respect to (suitably normalized) functions of bounded variation. It is also convenient to set $t_0 = a$.)

L. Let X be a compact Hausdorff space, let ξ be a regular complex Borel measure on X, and let φ denote the bounded linear functional on $\mathscr{C}(X)$ determined by ξ.

(i) Verify that φ is self-conjugate (Ex. 2H) if and only if ξ is real-valued, and likewise that φ is positive if and only if ξ is a (nonnegative) measure.

(ii) Verify, in the same vein, that φ is an algebra homomorphism of $\mathscr{C}(X)$ onto $\mathbb{C}$ if and only if φ is a point evaluation at some point of X, i.e., if and only if ξ is the Dirac mass concentrated at some point of X (Ex. 7I). (Hint: Prove that if there exist any two disjoint closed sets E and F in X such that $\xi(E) \ne 0 \ne \xi(F)$, then φ does not preserve multiplication, and use this fact to show that the support of $|\xi|$ (Prob. 10P) is a singleton.)

M. Prove the Stone–Weierstrass theorem (Th. 18.5) directly for the case of a two point space by showing that the only strongly separating subalgebra of $\mathbb{C}^2[\mathbb{R}^2]$ (with respect to termwise multiplication) is $\mathbb{C}^2[\mathbb{R}^2]$ itself. Conclude that every subalgebra of $\mathbb{C}^2$ is self-conjugate. (Hint: Aside from the whole two-dimensional algebra and the trivial subalgebra (0) there are but three (one-dimensional) subalgebras; find them.)

N. Let X be a compact Hausdorff space, let $\mathscr{A}$ be a subalgebra of $\mathscr{C}(X)$, and let $\mathscr{A}'$ denote the result of adjoining the constant functions to $\mathscr{A}$ (that is, $\mathscr{A}' = \{f + \lambda : f \in \mathscr{A}, \lambda \in \mathbb{C}\}$). Show that $\mathscr{A}'$ is a subalgebra of $\mathscr{C}(X)$ that is self-conjugate if $\mathscr{A}$ is, closed if $\mathscr{A}$ is, and strongly separating if $\mathscr{A}$ is separating. Use these facts to describe the most general closed, separating, self-conjugate subalgebra of $\mathscr{C}(X)$.

O. Verify that the *trigonometric polynomials*

$$T(\lambda) = \sum_{n=-N}^{+N} \alpha_n \lambda^n$$

are dense in $\mathscr{C}(Z)$ where Z denotes the unit circle.

P. Improve upon the theorem of Weierstrass (Ex. C) by showing that the polynomial functions $p(t)$ are actually dense in the Banach space $\mathscr{C}^{(k)}([a, b])$ of Problem 11F. (As a matter of historical fact, Weierstrass himself proved this version of the theorem.)

Q. Let us say that a function f on a set X is *constant on* a nonempty subset E of X if there exists a scalar λ_0 such that $f(x) = \lambda_0$ for all x in E.

  (i) Show that if X is a compact Hausdorff space, and if $\{E_\gamma\}_{\gamma \in \Gamma}$ is an arbitrary disjoint collection of nonempty subsets of X, then the collection $\mathscr{A}$ of all those continuous complex-valued functions on X that are constant on each E_γ is a closed self-conjugate subalgebra of $\mathscr{C}(X)$. Show too that the subset of $\mathscr{A}$ consisting of those functions in $\mathscr{A}$ that vanish on some one designated set E_{γ_0} is also a closed self-conjugate subalgebra of $\mathscr{C}(X)$.

  (ii) The subalgebras described in (i) are the most general closed self-conjugate subalgebras of $\mathscr{C}(X)$. To see this let $\mathscr{A}$ be an arbitrary closed self-conjugate subalgebra of $\mathscr{C}(X)$. We may use $\mathscr{A}$ to define an equivalence relation $\sim$ on X, setting $x_1 \sim x_2$ if $f(x_1) = f(x_2)$ for every f in $\mathscr{A}$. Let us write $[x]$ to designate the equivalence class of a point x with respect to this equivalence relation. There may or may not be an equivalence class such that all the functions in $\mathscr{A}$ vanish on it; if there is, let us call this distinguished class the *zero class* and denote it by $[x]_0$. Prove that $\mathscr{A}$ coincides with the collection of all those continuous functions on X that are constant on each equivalence class and that vanish on the zero class if there is one. (Hint: Start with the case in which there is no zero class, verify the counterpart of Lemma 18.9 directly for $x_1 \not\sim x_2$, and follow the proof in the text. Alternatively, one may use the quotient topology (Prob. 3H) to reduce this result to Theorem 18.5.)

R. Let X be a compact Hausdorff space, and let M be an arbitrary nonempty subset of $\mathscr{C}(X)$. Define $x_1 \sim x_2$ to mean, as above, that $f(x_1) = f(x_2)$ for every f in M. Show that the closed, self-conjugate, subalgebra of $\mathscr{C}(X)$ generated by M, i.e., the smallest one containing M, is the algebra of all those continuous functions on X that are constant on all of the equivalence classes of the relation $\sim$ and that vanish on the zero class if there is one. Show likewise that if $M \subset \mathscr{C}_\mathbb{R}(X)$, then the closed subalgebra of $\mathscr{C}_\mathbb{R}(X)$ generated by M coincides with the collection of all those continuous real functions on X answering the same description.

S. What is the closed self-conjugate subalgebra of $\mathscr{C}(\mathbb{R})$ generated by the pair of functions $\{\sin x, \cos x\}$? What is the closed, self-conjugate, subalgebra generated by the function $\sin x$ alone? (Hint: Introduce a quotient topology.)

T. It was shown in Problem M that every subalgebra of $\mathbb{C}^2$ is self-conjugate. Show that this result holds for the algebra $\mathbb{C}^n$ for every n, and find the most general subalgebra of $\mathbb{C}^n$. Use your results to improve upon Lemma 18.9 by showing that if $\mathscr{A}$ is any strongly separating subalgebra of $\mathscr{C}(X)$, and if $x_1, x_2, \ldots, x_n$ are distinct points of X, then for an arbitrary n-tuple of scalars $(\lambda_1, \lambda_2, \ldots, \lambda_n)$ there is a function f in $\mathscr{A}$ such that $f(x_i) = \lambda_i, i = 1, 2, \ldots, n$. (Hint: Use polynomial interpolation.)

U. Let X be a locally compact Hausdorff space, and let $\mathscr{C}_0(X)$ denote the space of continuous functions on X that vanish at infinity (see Problem I). Describe the most general closed, self-conjugate, subalgebra of $\mathscr{C}_0(X)$. Similarly, describe the most general closed subalgebra of the real algebra $\mathscr{C}_{0,\mathbb{R}}(X)$. (Hint: Use the one-point compactification (Prob. 3W).)

V. Let K be a compact subset of the complex plane, and let V denote the set of all those complex numbers α such that the function $1/(\lambda - \alpha)$ belongs to the algebra $\mathscr{P}(K)$.

   (i) Show that each hole in K is either disjoint from V or wholly contained in V. Show also that the unbounded component U_∞ of $\mathbb{C}\backslash K$ is always contained in V, and conclude that $\mathscr{P}(\hat{K}) = \mathscr{R}(\hat{K})$. (Hint: Show that V is both open and closed relative to $\mathbb{C}\backslash K$. To verify that $U_\infty \subset V$ use the fact that the series $\sum_{n=0}^{\infty} \lambda^n/\alpha^{n+1}$ converges uniformly in λ on K for $|\alpha|$ sufficiently large. Recall that $\hat{K} = \mathbb{C}\backslash U_\infty$.)

   (ii) Use the fact that $U_\infty \subset V$ to give a different proof of Runge's theorem on a simply connected domain (see Example K). (Hint: Recall the remark following Corollary 18.15.)

   The following problem outlines an alternate proof of the Stone–Weierstrass theorem (Th. 18.5) based on Theorem 18.4 and the Krein–Mil'man theorem (Prob. 14T). This proof is due to L. de Branges [14].

W. Let X be a compact Hausdorff space, let $\mathscr{L}$ be a linear submanifold of the real Banach algebra $\mathscr{C}_\mathbb{R}(X)$, and let K denote the subset of the annihilator $\mathscr{L}^a$ (in the dual space $\mathscr{M}_{\mathbb{R},0}(X)$ of all regular signed Borel measures on X) consisting of the measures μ in $\mathscr{L}^a$ such that $\|\mu\| \le 1$. (In other words, $K = \mathscr{L}^a \cap (\mathscr{M}_{\mathbb{R},0}(X))_1$.)

   (i) Prove that K is a weak* compact convex subset of $\mathscr{M}_{\mathbb{R},0}(X)$, and verify that if $\mathscr{L}$ is not dense in $\mathscr{C}_\mathbb{R}(X)$, then K contains measures μ such that $\|\mu\| = 1$.

   (ii) Let μ be a measure belonging to K such that $\|\mu\| = 1$, and let g be a bounded Borel measurable function on X with the property that the indefinite integral v of g (with respect to μ; Problem 9M) belongs to $\mathscr{L}^a$. Show that if $g' = Ag + B$ for any two real numbers A and B, then the indefinite integral v' of g' (with respect to μ) belongs to $\mathscr{L}^a$, and also that A and B can be chosen (with $A > 0$) so that g' is nonnegative-valued on X and satisfies the equation

$$\int_X g' \, d|\mu| = 1. \tag{10}$$

   Verify also that for any such choice of g' we have $\|v'\| \le 1$, and therefore $v' \in K$.

   (iii) Let g, g', and v' be as in (ii), and let M denote the essential supremum of g' with respect to $|\mu|$. Show, on the one hand, that if $M \le 1$, then g is constant almost everywhere $[|\mu|]$. Show also, on the other hand, that if $M > 1$, and if we define $t = 1/M$, so that $0 < t < 1$, then the indefinite integral v'' of the function

$$g'' = \frac{1 - tg'}{1 - t}$$

   (with respect to μ) also belongs to K, and

$$tv' + (1 - t)v'' = \mu.$$

   (Hint: If $M \le 1$ then $|1 - g'| = 1 - g'$ a.e. $[|\mu|]$, and (10) implies that $g' = 1$ a.e. $[|\mu|]$; if $M > 1$ then $1 - tg' \ge 0$ and $tg' + (1 - t)g'' = 1$ a.e. $[|\mu|]$.)

   (iv) Suppose next that $\mathscr{L}$ is not dense in $\mathscr{C}_\mathbb{R}(X)$ and that μ is an extreme point of K (see Problem 14S). Show that if g is a continuous function on X whose indefinite integral with respect to μ belongs to $\mathscr{L}^a$, then g is constant on the support of $|\mu|$ (Prob. 10P). (Hint: Show first that $\|\mu\| = 1$; then use the continuity of g.)

(v) Suppose, finally, that $\mathscr{L} = \mathscr{A}$ is a sub*algebra* of $\mathscr{C}_{\mathbb{R}}(X)$. Conclude that if μ is an extreme point of K, then every function in $\mathscr{A}$ is constant on the support of $|\mu|$, and use the Krein–Mil'man theorem to give a new proof of the real version of the Stone–Weierstrass theorem.

19 Vector sums and bases

We conclude this introduction to the theory of normed spaces with a discussion of two related topics. First we treat some notions of the *sum* of an infinite family of vectors in a normed space, and, secondly, we give a brief account of the corresponding notions of a *basis* for a Banach space.

The simplest and most familiar concept of an infinite sum stems directly from classical analysis and has already been introduced (see Problem 11G). If $\{x_n\}$ is a *sequence* in a normed space $\mathscr{E}$, indexed, say, by $\mathbb{N}_0$, then we write

$$x = \sum_{n=0}^{\infty} x_n$$

to indicate that the sequence $\{\sum_{n=0}^{N} x_n\}_{N=0}^{\infty}$ converges in $\mathscr{E}$ to x (in the norm topology). In this case the *infinite series* $\sum_{n=0}^{\infty} x_n$ is said to *converge* and x is called its *sum*. If $\{x_n\}$ is indexed by $\mathbb{N}$, this definition is, of course, modified in an appropriate fashion. If $\{x_n\}$ is a two-way infinite sequence, i.e., indexed by $\mathbb{Z}$, then we write

$$x = \sum_{n=-\infty}^{+\infty} x_n$$

to indicate either that the two infinite series $\sum_{n=1}^{\infty} x_{-n} = \sum_{n=-\infty}^{-1} x_n$ and $\sum_{n=0}^{\infty} x_n$ converge, and that $x = \sum_{n=-\infty}^{-1} x_n + \sum_{n=0}^{\infty} x_n$, in which case we say that $\sum_{n=-\infty}^{+\infty} x_n$ converges to x in the *primary sense*, or else that

$$x = \lim_{N} \sum_{n=-N}^{+N} x_n,$$

in which case we say that $\sum_{n=-\infty}^{+\infty} x_n$ converges to x in the *secondary* or *Cauchy sense*, or that x is the *principal value* of the series. (It is clear that a

428

two-way infinite series $\sum_{n=-\infty}^{+\infty} x_n$ that converges to a sum x in the primary sense does so also in the Cauchy sense, but the converse is false. Thus, to avoid ambiguity, it is necessary to make clear in any given case just which sense of sum is intended.) The following observations are trivial consequences of the definitions; they are formally stated here mainly for the sake of symmetry.

Proposition 19.1. *If* $\{x_n\}_{n=0}^{\infty}$ *and* $\{y_n\}_{n=0}^{\infty}$ *are two sequences of vectors in a normed space* $\mathscr{E}$, *and if*

$$x = \sum_{n=0}^{\infty} x_n \quad \text{and} \quad y = \sum_{n=0}^{\infty} y_n,$$

then for any pair of complex numbers α *and* β,

$$\sum_{n=0}^{\infty} (\alpha x_n + \beta y_n) = \alpha x + \beta y.$$

Likewise, if T *is any bounded linear transformation of* $\mathscr{E}$ *into some normed space* $\mathscr{F}$, *then* $Tx = \sum_{n=0}^{\infty} Tx_n$. *Moreover, the counterparts of these assertions are all valid, with appropriate notational changes, for sequences indexed by other systems of the same order type as* $\mathbb{N}_0$, *such as* $\mathbb{N}$, *and also for sequences indexed by* $\mathbb{Z}$ *(in either sense of convergence).*

Proposition 19.2 (Cauchy Criterion for Infinite Series). *If* $\{x_n\}_{n=0}^{\infty}$ *is a sequence of vectors in a Banach space* $\mathscr{E}$, *then the infinite series* $\sum_{n=0}^{\infty} x_n$ *is convergent in* $\mathscr{E}$ *if and only if, for each given positive number* ε, *there exists a positive integer* K *such that* $\| \sum_{n=K}^{K+p} x_n \| < \varepsilon$ *for every positive integer* p. *(Suitably worded versions of the Cauchy criterion are also valid, of course, for infinite series of the form* $\sum_{n=1}^{\infty} x_n$ *and* $\sum_{n=-\infty}^{+\infty} x_n$.)

Another notion of the sum of an infinite system of vectors is obtained simply by extending to indexed families of vectors the idea of the sum of an indexed family of scalars (Ex. 3Q).

Definition. Let $\mathscr{E}$ be a normed space, and let $\{x_\gamma\}_{\gamma \in \Gamma}$ be an indexed family of vectors in $\mathscr{E}$. If for each finite set D of indices we write

$$s_D = \sum_{\gamma \in D} x_\gamma,$$

then $\{s_D\}_{D \in \mathscr{D}}$ is a net indexed by the directed set $\mathscr{D}$ of all finite subsets of the index family Γ. The family $\{x_\gamma\}_{\gamma \in \Gamma}$ is said to be *summable*, and to have *sum* x, if the net $\{s_D\}$ converges to x in the norm topology on $\mathscr{E}$. In this case, we write

$$x = \sum_{\gamma \in \Gamma} x_\gamma.$$

Note that this definition can be paraphrased as follows: The family $\{x_\gamma\}_{\gamma \in \Gamma}$ has sum x if for every positive number ε there exists a finite subset D_ε of Γ such that $\| s_D - x \| < \varepsilon$ for every finite subset D of Γ such that $D \supset D_\varepsilon$.

Proposition 19.3. *If $\{x_\gamma\}_{\gamma \in \Gamma}$ and $\{y_\gamma\}_{\gamma \in \Gamma}$ are two summable families of vectors in a normed space $\mathscr{E}$, both indexed by the same index set Γ, and if*

$$x = \sum_{\gamma \in \Gamma} x_\gamma \quad and \quad y = \sum_{\gamma \in \Gamma} y_\gamma,$$

then, for any pair of scalars α and β,

$$\sum_{\gamma \in \Gamma} (\alpha x_\gamma + \beta y_\gamma) = \alpha x + \beta y.$$

Furthermore, if T is any bounded linear transformation of $\mathscr{E}$ into some normed space $\mathscr{F}$, then

$$Tx = \sum_{\gamma \in \Gamma} Tx_\gamma.$$

PROOF. These facts follow immediately from the definition of summability.   ☐

Proposition 19.4 (Cauchy Criterion). *An indexed family $\{x_\gamma\}_{\gamma \in \Gamma}$ of vectors in a Banach space $\mathscr{E}$ is summable if and only if for every $\varepsilon > 0$ there exists a finite set of indices D_ε such that $\| s_D = \sum_{\gamma \in D} x_\gamma \| < \varepsilon$ for every finite set of indices D such that D and D_ε are disjoint.*

PROOF. The proof of the necessity of the condition is trivial, as usual, and will be omitted. Suppose then that $\mathscr{E}$ is a Banach space and that $\{x_\gamma\}_{\gamma \in \Gamma}$ is an indexed family in $\mathscr{E}$ satisfying the stated condition. Let ε be a positive number, let D_ε be chosen so as to have the property stated in the condition, and let D' and D'' be any two finite sets of indices, both of which contain D_ε. Then $\| s_{D'} - s_{D''} \| = \| s_{D' \nabla D''} \| < \varepsilon$ since $D' \nabla D''$ is disjoint from D_ε. Thus the net $\{s_D\}$ is Cauchy, and therefore convergent (see Problem 4M).   ☐

Corollary 19.5. *If $\{x_\gamma\}_{\gamma \in \Gamma}$ is a summable family of vectors in a normed space $\mathscr{E}$, then $\{\gamma \in \Gamma : x_\gamma \neq 0\}$ is countable.*

PROOF. For each positive integer n, let D_n be a finite set of indices such that $\| s_D \| < 1/n$ whenever $D \cap D_n = \varnothing$, and let

$$J = \bigcup_{n=1}^{\infty} D_n.$$

If $\gamma \notin J$, then $\{\gamma\} \cap D_n = \varnothing$, and therefore $\| x_\gamma \| < 1/n$, for every positive integer n. Thus $x_\gamma = 0$ except when $\gamma \in J$.   ☐

Thus, just as in the case of indexed families of scalars (Prob. 4N), an uncountable indexed family of vectors may be summable, but such a family can possess only a countable number of nonzero terms.

Corollary 19.6. *Let $\mathscr{E}$ be a Banach space and let $\{x_\gamma\}_{\gamma\in\Gamma}$ be an indexed family of vectors in $\mathscr{E}$. A sufficient condition for the family $\{x_\gamma\}$ to be summable is for it to be absolutely summable, i.e., for the family of norms $\{\|x_\gamma\|\}$ to be a summable family of real numbers.*

PROOF. For any finite set of indices $D \subset \Gamma$ we have

$$\|s_D\| \le \sum_{\gamma\in D} \|x_\gamma\|$$

and the result follows by Proposition 19.4. $\qquad\qquad\qquad\square$

Example A. Let x_n be the sequence whose nth term is $1/n$ and the rest of whose terms are all zero. Then the family $\{x_n\}_{n\in\mathbb{N}}$ is summable in (ℓ_p) for $1 < p < +\infty$, but is not summable in (ℓ_1). On the other hand, $\{x_n\}$ is not absolutely summable in any of the spaces (ℓ_p).

Note. In dealing with vector families $\{x_n\}$ indexed by $\mathbb{N}$ it is important to distinguish between the notion of the sum of the indexed family $\{x_n\}_{n\in\mathbb{N}}$ and the sum of the infinite series $\sum_{n=1}^{\infty} x_n$. If the family $\{x_n\}_{n\in\mathbb{N}}$ is summable, then the series $\sum_{n=1}^{\infty} x_n$ is convergent, but the converse is false even for sequences of scalars (recall Problem 4N). Similar remarks are valid, of course, for the index systems $\mathbb{N}_0$ and $\mathbb{Z}$.

As has already been noted several times, the sequence of unit vectors $\{e_n\}_{n=0}^{\infty}$ plays a quite special role in the spaces (ℓ_p), at least for finite values of p. One aspect of this role is the fact that if $x = \{\xi_n\}_{n=0}^{\infty}$ belongs to (ℓ_p) for some $p, 1 \le p < +\infty$, then

$$x = \sum_{n=0}^{\infty} \xi_n e_n \qquad\qquad (1)$$

in (ℓ_p). For this reason the sequence $\{e_n\}_{n=0}^{\infty}$ is called a *basis* for (ℓ_p). We conclude this chapter with a brief account of some of the facts concerning such bases.

The essential feature of a basis of any kind in a vector space is that every vector in the space should be uniquely expressible as a sum of scalar multiples of the basis vectors. If only ordinary (finite) algebraic sums are employed, the basis is simply a Hamel basis, a purely algebraic concept that does not reflect in any way the topological structure that a normed space possesses. Hence it is more appropriate to allow infinite sums, and, in principle at least, any notion of infinite sum gives rise to a corresponding notion of basis. In fact, however, for reasons both substantive and historical, by far the greatest attention has been paid to ordinary (norm) convergent series, and we shall therefore be principally concerned with bases that are sequences, indexed, for the most part, either by $\mathbb{N}_0$ or by $\mathbb{Z}$.

431

Definition. A sequence $X = \{x_n\}_{n=0}^{\infty}$ in a Banach space $\mathscr{E}$ is a *(Schauder) basis* for $\mathscr{E}$ if for each vector x in $\mathscr{E}$ there exists a unique sequence $\{\alpha_n\}_{n=0}^{\infty}$ of complex numbers such that

$$x = \sum_{n=0}^{\infty} \alpha_n x_n. \tag{2}$$

If X is a basis for $\mathscr{E}$ then the scalars α_n appearing in (2) are the *coefficients* of x with respect to the basis X, the sequence $\{\alpha_n\}_{n=0}^{\infty}$ is the *coefficient sequence* of x with respect to X, and the series (2) is the *expansion* of x with respect to the basis X. A basis for $\mathscr{E}$ indexed by $\mathbb{N}$ (or any other index system with the same order type as $\mathbb{N}_0$) is defined in exactly the same fashion with the obvious changes in notation. A two-way infinite sequence $X = \{x_n\}_{n=-\infty}^{+\infty}$ in $\mathscr{E}$ is a *(Schauder) basis* for $\mathscr{E}$ in the *primary sense* [*Cauchy sense*] if for each vector x in $\mathscr{E}$ there exists a unique sequence $\{\alpha_n\}_{n=-\infty}^{+\infty}$ of complex numbers such that

$$x = \sum_{n=-\infty}^{+\infty} \alpha_n x_n$$

in the primary sense [Cauchy sense], and the terms *coefficient* and *expansion* are employed accordingly.

Example B. As noted above, the sequence $\{e_n\}_{n=0}^{\infty}$ is a basis for (ℓ_p), $1 \le p < +\infty$. This basis has the special property that the coefficient sequence of each element x of (ℓ_p) with respect to it is the sequence x itself. Similarly, for each p, $1 \le p < +\infty$, the two-way infinite sequence $\{e_n\}_{n=-\infty}^{+\infty}$ is a basis for $(\ell_p)^{\#}$ (in the primary sense) having the property that the coefficient sequence of each vector in $(\ell_p)^{\#}$ with respect to it is that vector itself.

Example C. Let $\mathscr{E}$ be a Banach space and let $\{x_n\}_{n=0}^{\infty}$ be a Schauder basis for $\mathscr{E}$. Then the set of all vectors of the form $\sum_{n=0}^{\infty} \alpha_n x_n$, where only finitely many α_n are nonzero and all α_n have rational real and imaginary parts, is a countable dense subset of $\mathscr{E}$, so $\mathscr{E}$ is separable. Similar considerations apply to spaces with Schauder bases indexed by $\mathbb{N}$ or $\mathbb{Z}$. It follows that no nonseparable Banach space possesses a Schauder basis.

Example D. Let the (piecewise linear, continuous) function g be defined on $\mathbb{R}$ as follows:

$$g(x) = \begin{cases} 2t, & 0 \le t \le \frac{1}{2}, \\ 2(1-t), & \frac{1}{2} \le t \le 1, \\ 0, & t < 0, t > 1. \end{cases}$$

For reasons that will be clear shortly we denote by f_2 the restriction of g to $[0, 1]$. Similarly, we denote by f_3 the restriction to $[0, 1]$ of $g(2t)$, and by f_4 the restriction to $[0, 1]$ of $g(2t - 1)$. In general, for $k = 1, \ldots, 2^n$, $f_{2^n + k}$

is defined to be the restriction to $[0, 1]$ of the function $g(2^n t - k + 1)$, and in this way we obtain an infinite sequence $\{f_2, f_3, \ldots\}$ of piecewise linear elements of $\mathscr{C}([0, 1])$ all of which vanish at both $t = 0$ and $t = 1$. Moreover, an examination of the details of the construction discloses that for each nonnegative integer n, and for each dyadic number $p/2^{n+1}$, where $p = 2q - 1$, $q = 1, \ldots, 2^n$ (so that p is an odd integer between 1 and 2^{n+1}), there exists exactly one index $m > 2^n$ (viz., $m = 2^n + q$) such that $f_m(p/2^{n+1}) = 1$, while $f_r(p/2^{n+1}) = 0$ for all other indices $r > 2^n$. It follows, of course, that if f is an element of the Banach space $\mathscr{C}_0([0, 1])$ consisting of those functions in $\mathscr{C}([0, 1])$ that vanish at 0 and 1, and if $\{\alpha_n\}_{n=2}^\infty$ is a sequence of complex numbers such that $\sum_{n=2}^\infty \alpha_n f_n$ converges uniformly to a function f, then the coefficients α_m, $m = 2^n + 1, \ldots, 2^n + 2^n$, must be chosen so that the partial sum

$$\sum_{n=2}^{2^n+q} \alpha_n f_n$$

agrees with f at $(2q - 1)/2^{n+1}$. Thus α_2 must equal $f(\frac{1}{2})$, α_3 and α_4 must equal $f(\frac{1}{4}) - \frac{1}{2}f(\frac{1}{2})$ and $f(\frac{3}{4}) - \frac{1}{2}f(\frac{1}{2})$, etc.

This shows that the sequence $\{\alpha_n\}_{n=2}^\infty$ possessing the stated property is uniquely determined by f if it exists. On the other hand, if the sequence $\{\alpha_n\}$ is chosen in the manner just described, then each partial sum $s_m = \sum_{n=2}^m \alpha_n f_n$ is a piecewise linear function in $\mathscr{C}_0([0, 1])$ obtained by interpolating linearly between various values of the given function f, and if $m > 2^n$, then the set of numbers in the interval $[0, 1]$ at which s_m equals f contains all of the dyadic numbers $p/2^n$, $p = 0, 1, \ldots, 2^n$. Since every function f in $\mathscr{C}_0([0, 1])$ is uniformly continuous on $[0, 1]$, $\{s_m\}$ converges uniformly to f, and it follows that the sequence $\{f_n\}_{n=2}^\infty$ is a Schauder basis for $\mathscr{C}_0([0, 1])$.

Finally, the quotient space $\mathscr{C}([0, 1])/\mathscr{C}_0([0, 1])$ is generated by the cosets of the functions $f_0(t) = t$ and $f_1(t) = 1 - t$. Hence a Schauder basis for $\mathscr{C}([0, 1])$ is given by the sequence $\{f_n\}_{n=0}^\infty$, where $f_0(t) = t$, $f_1(t) = 1 - t$, and $\{f_n\}_{n=2}^\infty$ is as above. This is a basis for $\mathscr{C}([0, 1])$ of the type originally defined by Schauder [59].

Since a basis for a Banach space $\mathscr{E}$ is certainly linearly independent, the following result, while stated in somewhat more general terms than required for present purposes, is clearly pertinent to a discussion of bases.

Proposition 19.7. *Let $\mathscr{E}$ be a Banach space, let $\{x_n\}_{n=0}^\infty$ be a linearly independent sequence of vectors in $\mathscr{E}$, and let $\mathscr{L}$ denote the collection of all those sequences $a = \{\alpha_n\}_{n=0}^\infty$ of complex numbers such that the series $\sum_{n=0}^\infty \alpha_n x_n$ is convergent in $\mathscr{E}$. Then $\mathscr{L}$ is a linear submanifold of (a) (Ex. 2D), and setting*

$$\|a\| = \sup_{N \in \mathbb{N}_0} \left\| \sum_{n=0}^N \alpha_n x_n \right\| \tag{3}$$

defines a norm on $\mathscr{L}$ with respect to which $\mathscr{L}$ is complete. (Analogous results hold with appropriate notational changes when other index systems such as $\mathbb{N}$ or $\mathbb{Z}$ are used in place of $\mathbb{N}_0$.)

PROOF. We treat only the case of sequences indexed by $\mathbb{N}_0$; see Problem M. It is an immediate consequence of Proposition 19.1 that $\mathscr{L}$ is a linear space with respect to termwise linear operations. Likewise the function defined in (3) is finite-valued on $\mathscr{L}$ (since a convergent sequence is necessarily bounded; cf. Chapter 4) and is easily seen to be a pseudonorm. Moreover, if $a = \{\alpha_n\} \neq 0$, and if α_N is the first nonzero term in a, then $\sum_{n=0}^{N} \alpha_n x_n \neq 0$, so $\|a\| > 0$. Thus $\| \ \|$ as defined in (3) turns $\mathscr{L}$ into a normed space. It remains to show that $\mathscr{L}$ is complete in this norm.

We begin with the observation that if $a = \{\alpha_n\}$ belongs to $\mathscr{L}$, and if M and N are arbitrary nonnegative integers such that $M < N$, then

$$\sum_{n=M+1}^{N} \alpha_n x_n = \sum_{n=0}^{N} \alpha_n x_n - \sum_{n=0}^{M} \alpha_n x_n,$$

and therefore $\|\sum_{n=M+1}^{N} \alpha_n x_n\| \leq 2\|a\|$. In particular, $|\alpha_N| \|x_N\| \leq 2\|a\|$ for each N in $\mathbb{N}_0$, and since $x_N \neq 0$, this shows that each coordinate function $a \to \alpha_N(a)$ is a bounded linear functional on $\mathscr{L}$. Suppose now that $\{a^{(q)}\}_{q=1}^{\infty}$ is a Cauchy sequence in $\mathscr{L}$, so that $\lim_{p,q} \|a^{(p)} - a^{(q)}\| = 0$, and let $a^{(q)} = \{\alpha_n^{(q)}\}_{n=0}^{\infty}$ for each q. Then, by what has just been shown, each numerical sequence $\{\alpha_n^{(q)}\}_{q=1}^{\infty}$ is convergent, so $\{a^{(q)}\}$ converges termwise to a sequence $\tilde{a} = \{\tilde{\alpha}_n\}_{n=0}^{\infty}$. We must show that $\tilde{a}$ belongs to $\mathscr{L}$ and that $\lim_q \|\tilde{a} - a^{(q)}\| = 0$. Let ε be a positive number and let Q be a positive integer such that $\|a^{(p)} - a^{(q)}\| < \varepsilon/2$ for all $p, q \geq Q$. Then for every pair of nonnegative integers K and r,

$$\left\| \sum_{n=K}^{K+r} (\alpha_n^{(p)} - \alpha_n^{(q)}) x_n \right\| < \varepsilon, \qquad p, q \geq Q.$$

Letting p tend to infinity in this inequality we find that

$$\left\| \sum_{n=K}^{K+r} (\tilde{\alpha}_n - \alpha_n^{(q)}) x_n \right\| \leq \varepsilon \qquad (4)$$

for all K and r and all $q \geq Q$. Thus the infinite series $\sum_{n=0}^{\infty} (\tilde{\alpha}_n - \alpha_n^{(q)}) x_n$ satisfies the Cauchy criterion, and is therefore convergent, since $\mathscr{E}$ is complete (Prop. 19.2). Hence $\tilde{a} - a^{(q)}$ belongs to $\mathscr{L}$, and so therefore does $\tilde{a}$. Moreover, setting $K = 0$ in (4) we see that $\|\tilde{a} - a^{(q)}\| \leq \varepsilon$ for all $q \geq Q$. Thus the proof is complete. $\square$

Example E. Let $\mathscr{E} = (\ell_\infty)$ and let $x_n = e_n$, $n \in \mathbb{N}_0$. Then the space $\mathscr{L}$ of those complex sequences $\{\alpha_n\}_{n=0}^{\infty}$ with the property that $\sum_{n=0}^{\infty} \alpha_n \varepsilon_n$ converges in (ℓ_∞) is just the subspace (c_0), and the norm (3) coincides with $\| \ \|_\infty$ on that subspace. It follows at once that in (ℓ_∞) the sequence $\{e_n\}_{n=0}^{\infty}$ is a basis for the subspace (c_0). Similarly the two-way infinite sequence $\{e_n\}_{n=-\infty}^{+\infty}$

is a basis in $(\ell_\infty)^\#$ (in the primary sense) for the subspace of $(\ell_\infty)^\#$ consisting of those sequences $\{\xi_n\}_{n=-\infty}^{+\infty}$ in $(\ell_\infty)^\#$ such that $\lim_{|n| \to +\infty} |\xi_n| = 0$. (Since (ℓ_∞) and $(\ell_\infty)^\#$ are not separable, neither can possess a Schauder basis (Ex. C).)

Proposition 19.7 permits us to establish a very important property of bases.

Theorem 19.8. *Let $\mathscr{E}$ be a (separable) Banach space, let $\{x_n\}_{n=0}^\infty$ be a basis for $\mathscr{E}$, and for each nonnegative integer n and vector x in $\mathscr{E}$ let us write $\alpha_n(x)$ for the nth coefficient of x with respect to the basis $\{x_n\}$, so that*

$$x = \sum_{n=0}^\infty \alpha_n(x)x_n, \qquad x \in \mathscr{E}. \tag{5}$$

Then α_n is a bounded linear functional on $\mathscr{E}$. (Analogous results hold, with appropriate notational changes, when other systems such as $\mathbb{N}$ or $\mathbb{Z}$ are used in place of $\mathbb{N}_0$; see Problem M.)

PROOF. As has been noted, the uniqueness requirement in the definition of a basis ensures that $\{x_n\}$ is linearly independent, so Proposition 19.7 applies. Hence the linear space $\mathscr{L}$ of all the coordinate sequences of vectors in $\mathscr{E}$ (with respect to the given basis $\{x_n\}$) is a Banach space in the norm (3). Moreover, it is clear that if we write $\Phi(a) = \sum_{n=0}^\infty \alpha_n x_n$ for each $a = \{\alpha_n\}$ in $\mathscr{L}$, then $\|\Phi(a)\| \leq \sup_N \|\sum_{n=0}^N \alpha_n x_n\| = \|a\|$. Thus Φ is a bounded one-to-one linear transformation of the Banach space $\mathscr{L}$ onto the Banach space $\mathscr{E}$, and it follows (Th. 13.3) that Φ is an equivalence between $\mathscr{L}$ and $\mathscr{E}$. Hence it suffices to show that each coordinate map $a \to \alpha_n(a)$ is a bounded linear functional on $\mathscr{L}$ in the norm (3), and this was demonstrated in the proof of Proposition 19.7. □

The foregoing result shows that whenever a Schauder basis $X = \{x_n\}$ is given in a Banach space $\mathscr{E}$ there corresponds to that basis a similarly indexed sequence $A = \{\alpha_n\}$ of *coordinate functionals* in $\mathscr{E}^*$ such that (5) is satisfied. Moreover, from the uniqueness requirement satisfied by X it is readily deduced that the two sequences $\{x_n\}$ and $\{\alpha_n\}$ satisfy the condition

$$\alpha_m(x_n) = \delta_{mn}$$

for all values of m and n. This observation leads naturally to the following idea.

Definition. Let $\mathscr{E}$ be a Banach space. If $X = \{x_n\}$ and $A = \{\alpha_n\}$ are similarly indexed sequences in $\mathscr{E}$ and $\mathscr{E}^*$, respectively, then the pair (X, A) is a *biorthogonal system* for $\mathscr{E}$ if $\alpha_m(x_n) = \delta_{mn}$ for all values of m and n. If (X, A) is a biorthogonal system for $\mathscr{E}$ and $x \in \mathscr{E}$, then the (formal) series $\sum_n \alpha_n(x)x_n$ is called the *series expansion* of x in the system (X, A). It is customary to write $x \sim \sum_n \alpha_n(x)x_n$ to indicate this relationship.

Example F. The functions $f_n(\lambda) = \lambda^n$ are continuous on the unit circle Z for every n in $\mathbb{Z}$ and therefore belong to $\mathscr{L}_p(Z)$ for all p, $1 \le p \le +\infty$. Moreover,

$$\int_Z f_m f_{-n} \, d\theta = \int_Z \lambda^{m-n} \, d\theta(\lambda) = 2\pi\delta_{mn}, \qquad m, n \in \mathbb{Z}.$$

Hence the two-way infinite sequence $\{u_n = f_n/\sqrt{2\pi}\}_{n=-\infty}^{+\infty}$ and the companion sequence $\{\alpha_n = f_{-n}/\sqrt{2\pi} = u_{-n}\}_{n=-\infty}^{+\infty}$ (regarded as a sequence in $\mathscr{L}_q(Z)$, where q denotes the Hölder conjugate of p) constitute a biorthogonal system for $\mathscr{L}_p(Z)$, $1 \le p < +\infty$ (for $p = 1$ set $q = +\infty$). This *trigonometric biorthogonal system* is of the greatest importance both intrinsically and for historical reasons. The (formal) series expansion of an integrable function f in this system is the *Fourier series* of f, frequently denoted by $S(f)$. The generic term of $S(f)$ is $\alpha_n(f)u_n(\lambda) = (1/2\pi)\lambda^n \int_Z f(\zeta)\zeta^{-n} \, d\theta(\zeta)$, and it is customary to write c_n (or $c_n(f)$ when necessary) for the nth *Fourier coefficient* of f:

$$c_n = \frac{1}{2\pi} \int_Z f(\zeta)\zeta^{-n} \, d\theta(\zeta).$$

Thus

$$S(f) = \sum_{n=-\infty}^{+\infty} c_n \lambda^n = \frac{1}{2\pi} \sum_{n=-\infty}^{+\infty} \lambda^n \int_Z f(\zeta)\zeta^{-n} \, d\theta(\zeta)$$

for each f in $\mathscr{L}_1(Z)$ (this calculation is the same for all values of p since $\mathscr{L}_p(Z) \subset \mathscr{L}_1(Z)$; see Example 17A). Finally, for each N in $\mathbb{N}_0$ the Nth *partial sum* of the Fourier series $S(f)$ is the symmetric sum

$$s_N(f)(\lambda) = \sum_{n=-N}^{+N} c_n \lambda^n. \tag{6}$$

Those who prefer to associate Fourier series with periodic functions on the real line may simply transplant this entire discussion into the Banach space $\mathscr{P}_1$ (consisting of those measurable functions on $\mathbb{R}$ that are periodic with period 2π and integrable over a period) via the standard isomorphism $f \to \hat{f}$, where $\hat{f}(t) = f(e^{it})$ (see Example 17F). In that notation we have

$$S(\hat{f}) = \sum_{n=-\infty}^{+\infty} c_n e^{int} = \frac{1}{2\pi} \sum_{n=-\infty}^{+\infty} e^{int} \oint \hat{f}(u) e^{-inu} \, du$$

for the *Fourier series* of $\hat{f}$, where the nth *Fourier coefficient* of $\hat{f}$ is given by

$$c_n = \frac{1}{2\pi} \oint \hat{f}(u) e^{-inu} \, du,$$

and the Nth partial sum of $S(\hat{f})$ is given by

$$s_N(\hat{f})(t) = \sum_{n=-N}^{+N} c_n e^{int}.$$

It turns out that if f is a function in $\mathscr{L}_p(Z)$ for any p such that $1 < p < +\infty$, then the sequence $\{s_N(f)\}_{N=0}^\infty$ converges to f in the norm $\|\ \|_p$, but for $p = 1$ the situation is quite different; there exist functions f in $\mathscr{L}_1(Z)$ such that the partial sums $s_N(f)$ do *not* converge to f in the norm $\|\ \|_1$. In other words, the sequence $U = \{u_n\}_{n=-\infty}^{+\infty}$ is a basis (in the Cauchy sense) for $\mathscr{L}_p(Z)$, $1 < p < +\infty$, but is *not* a basis (even in the Cauchy sense) for $\mathscr{L}_1(Z)$. The proofs of these assertions are quite deep, and well beyond the scope of this discussion; the interested reader may consult [3; Ch. VIII, §§20, 22].

Example G. Let $U = \{u_n\}_{n=-\infty}^{+\infty}$ and $A = \{\alpha_n\}_{n=-\infty}^{+\infty}$ be as in the preceding example, and for each integer n write ρ_n for the Borel measure on Z that is the indefinite integral of α_n with respect to θ. If we set $R = \{\rho_n\}_{n=-\infty}^{+\infty}$, then the pair (U, R) is a biorthogonal system for $\mathscr{C}(Z)$ (see Proposition 9.5 and Theorem 18.3), but the sequence U is not a basis (even in the Cauchy sense) for $\mathscr{C}(Z)$. This is equivalent to saying that there exist continuous functions f on Z such that the partial sums $s_N(f)$ given by (6) do not converge uniformly to f, a fact that is well known, though a concrete counterexample is not easily constructed. (One may be found in [3; Chap. I, §44].)

As the foregoing examples show, given a biorthogonal system (X, A) for a Banach space $\mathscr{E}$, so that X and A are similarly indexed sequences in $\mathscr{E}$ and $\mathscr{E}^*$, respectively, the question whether X is a basis for $\mathscr{E}$ may turn out to be quite delicate. One obvious *necessary* condition is that X must span $\mathscr{E}$, but this condition is satisfied in every case by the trigonometric system discussed in Examples F and G (recall Problem 18O), so this necessary condition is clearly not sufficient. The following definition makes it possible to formulate a criterion (Prop. 19.10) that is sometimes useful.

Definition. Let $\mathscr{E}$ be a Banach space and let $X = \{x_n\}_{n=0}^\infty$ and $A = \{\alpha_n\}_{n=0}^\infty$ be (similarly indexed) sequences in $\mathscr{E}$ and $\mathscr{E}^*$, respectively, such that (X, A) is a biorthogonal system. Then for each nonnegative integer N the equation

$$E_N x = \sum_{n=0}^N \alpha_n(x)x_n, \qquad x \in \mathscr{E},$$

defines a bounded linear transformation E_N on $\mathscr{E}$ called an *expansion operator* of the system (X, A). (If X and A are indexed by some other index set, such as $\mathbb{N}$, with the same order type as $\mathbb{N}_0$, this definition is to be modified by making the obvious changes in notation. For biorthogonal systems consisting of sequences indexed by $\mathbb{Z}$ there are two kinds of expansion operators corresponding to the two types of convergence of two-way infinite series; see Problem N.)

Lemma 19.9. *Let* (X, A) *be a biorthogonal system for a Banach space $\mathscr{E}$ as in the foregoing definition. Then the expansion operators E_N of the system (X, A) have the property that $E_M E_N = E_N E_M = E_{M \wedge N}$ for every pair of nonnegative integers M and N. In particular, each E_N is idempotent.*

PROOF. These assertions are immediate consequences of the biorthogonality relations.

Proposition 19.10. *Let $\mathscr{E}$ be a Banach space, let $X = \{x_n\}_{n=0}^\infty$ and $A = \{\alpha_n\}_{n=0}^\infty$ be sequences in $\mathscr{E}$ and $\mathscr{E}^*$, respectively, such that (X, A) is a biorthogonal system for $\mathscr{E}$, and let $\mathscr{M}$ denote the subspace of $\mathscr{E}$ spanned by X. Then X is a basis for $\mathscr{M}$ if and only if the sequence $\{E_N\}_{N=0}^\infty$ of expansion operators is uniformly bounded on $\mathscr{M}$.*

PROOF. If X is a basis for $\mathscr{M}$, then for each vector x in $\mathscr{M}$ we have $x = \lim_N E_N x$, from which it follows at once (by the uniform boundedness theorem; Theorem 12.15) that the sequence $\{E_N\}$ is uniformly bounded on $\mathscr{M}$ (cf. Example 12V). Suppose, on the other hand, that there is a positive constant M such that $\| E_N y \| \leq M \| y \|$ for all N in $\mathbb{N}_0$ and all y in $\mathscr{M}$, and let x be a vector in $\mathscr{M}$. Then for any positive number ε there exists a linear combination $x' = \lambda_0 x_0 + \cdots + \lambda_K x_K$ of vectors belonging to X such that $\| x - x' \| < \varepsilon$, and, by the preceding lemma, $E_N x' = x'$ for all $N \geq K$. Thus

$$\| x - E_N x \| \leq \| x - x' \| + \| x' - E_N x' \| + \| E_N x' - E_N x \| \leq (1 + M)\varepsilon$$

for all $N \geq K$, and the proposition follows. $\qquad\square$

Corollary 19.11. *If (X, A) is a biorthogonal system for a Banach space $\mathscr{E}$, then necessary and sufficient conditions for X to be a basis for $\mathscr{E}$ are that X span $\mathscr{E}$ and the system of expansion operators be uniformly bounded.*

Note. For biorthogonal systems (X, A) composed of sequences indexed by $\mathbb{Z}$ there are two versions of this criterion, one for X to be a basis in the Cauchy sense, the other for X to be a basis in the primary sense; see Problem N.

Example H. Corollary 19.11 may be employed to give a simple proof that the trigonometric sequence $U = \{u_n\}_{n=-\infty}^{+\infty}$ introduced in Example F is a basis for $\mathscr{L}_p(\mathbb{Z})$ in the (very special) case $p = 2$. To this end let us define

$$(s_D(f))(\lambda) = \sum_{n \in D} c_n(f)\lambda^n = \frac{1}{2\pi} \sum_{n \in D} \lambda^n \int_Z f(\zeta)\zeta^{-n} \, d\theta(\zeta)$$

for each f in $\mathscr{L}_1(\mathbb{Z})$ and finite subset D of $\mathbb{Z}$. Then a straightforward calculation, using the fact that $1/\lambda = \bar{\lambda}$ on Z, shows that

$$\int_Z f\bar{s}_D(f)d\theta = \sum_{n \in D} \bar{c}_n(f) \int_Z f(\lambda)\lambda^{-n} \, d\theta(\lambda) = 2\pi \sum_{n \in D} |c_n(f)|^2,$$

and similarly that

$$\int_Z \bar{f} s_D(f) d\theta = 2\pi \sum_{n \in D} |c_n(f)|^2.$$

In particular, setting $f = s_D(f)$, we obtain $\| s_D(f) \|_2^2 = 2\pi \sum_{n \in D} |c_n(f)|^2$. (We here use the fact that s_D, like all expansion operators, is idempotent; recall that $\| g \|_2^2 = \int_Z g \bar{g} \, d\theta$ for every function g in $\mathcal{L}_2(Z)$.) Hence, summarizing, we see that if $f \in \mathcal{L}_2(Z)$, then

$$\| f - s_D(f) \|_2^2 = \int_Z (f - s_D(f))(\bar{f} - \bar{s}_D(f)) d\theta = \| f \|_2^2 - 2\pi \sum_{n \in D} |c_n(f)|^2$$

$$= \| f \|_2^2 - \| s_D(f) \|_2^2. \tag{7}$$

Thus s_D (regarded as an operator on $\mathcal{L}_2(Z)$) is a contraction for every choice of D. In particular, this is true for $D = \{-M, -M + 1, \ldots, N - 1, N\}$, where M and N are arbitrary nonnegative integers, and it follows (Prob. N) that U is actually a basis for $\mathcal{L}_2(Z)$ in the primary sense. (The sequence U is also an *indexed* basis for $\mathcal{L}_2(Z)$, as defined in Problem R; these observations are not of great importance since the sum of a Fourier series is invariably taken to be its principal value.) The inequality

$$2\pi \sum_{n=-N}^{+N} |c_n(f)|^2 \leq \| f \|_2^2,$$

valid for all nonnegative integers N and every f in $\mathcal{L}_2(Z)$ according to (7), is known as *Bessel's inequality*. On the basis of this inequality we may also conclude that

$$2\pi \sum_{n \in Z} |c_n(f)|^2 \leq \| f \|_2^2$$

for every f in $\mathcal{L}_2(Z)$, an inequality also sometimes called Bessel's inequality. Finally, setting $D = \{-N, \ldots, +N\}$ in (7), we find that

$$\| f - s_N(f) \|_2^2 = \| f \|_2^2 - \| s_N(f) \|_2^2, \qquad N \in \mathbb{N}_0,$$

and hence that

$$\| f \|^2 = 2\pi \sum_{n \in Z} |c_n(f)|^2 = 2\pi \sum_{n=-\infty}^{+\infty} |c_n(f)|^2$$

for every function f in $\mathcal{L}_2(Z)$, an equation known as *Parseval's identity*.

Note. In the Lebesgue spaces of the special form $\mathcal{L}_2(X, \mathbf{S}, \mu)$ it is possible to introduce an *inner product* by writing

$$(f, g) = \int_X f \bar{g} \, d\mu, \qquad f, g \in \mathcal{L}_2(X).$$

(The function $f\bar{g}$ is integrable $[\mu]$ by the Hölder inequality; we are here using the fact that $p = 2$ is its own Hölder conjugate.) This inner product is sesquilinear (that is, linear in the first variable f and conjugate linear in the second variable g; see Chapter 2) and is not to be confused with the bilinear pairing of $\mathscr{L}_2(X)$ with itself introduced in Example 17H, from which the inner product may be derived, to be sure, by writing $(f, g) = \langle f, \bar{g} \rangle$, $f, g \in \mathscr{L}_2(X)$. In terms of the inner product $(\ ,\ )$ the norm $\|\ \|_2$ is given by

$$\| f \|_2 = (f, f)^{1/2}, \qquad f \in \mathscr{L}_2(X).$$

A Banach space in which the norm is given in this way by a sesquilinear functional is called a *Hilbert space*. Thus, in particular, $\mathscr{L}_2(Z)$ is a Hilbert space, and in terms of inner products it is the substance of the calculations in Example H that

$$(f, s_D(f)) = (s_D(f), f) = \| s_D(f) \|_2^2$$

and therefore

$$\| f - s_D(f) \|_2^2 = (f - s_D(f), f - s_D(f)) = \| f \|_2^2 - \| s_D(f) \|_2^2$$

for every function f in $\mathscr{L}_2(Z)$. The theory of Hilbert spaces and operators thereon will form the principal subject matter of Volume II.

The explicit construction of a basis for a given Banach space may present serious difficulties, and (as Example D shows) require considerable ingenuity. Nevertheless, within a few years of the publication of Schauder's original memoir [59], bases had been constructed for all of the classical separable Banach spaces, and as new separable Banach spaces were introduced and studied over the intervening years, bases for them were eventually found. Accordingly, the *basis problem* was formulated: *Is it the case that every separable Banach space possesses a basis?* This problem proved to be very deep. So deep, in fact, that for nearly fifty years every effort to solve it failed. Then in 1973, Per Enflo presented to the world [25] a very complicated example of a separable Banach space that does not possess *any* basis at all, and the basis problem was thereby finally solved in the negative. (It should be emphasized that the *basis problem* here referred to has to do with *Schauder* bases. Similar problems for other more restricted types of bases had been solved earlier; see Problem V and the note preceding it.)

PROBLEMS

A. Let (X, S, μ) be a measure space, and let $\{E_\gamma\}_{\gamma \in \Gamma}$ be a disjoint indexed family of measurable subsets of X, each of which has finite measure. If $f_\gamma = \chi_{E_\gamma}$, determine, for each value of p, $1 \le p \le +\infty$, when $\{f_\gamma\}_{\gamma \in \Gamma}$ is a summable family in $\mathscr{L}_p(X)$, and find the sum when it exists.

B. Let Γ be an infinite index family, let $\{x_\gamma\}_{\gamma \in \Gamma}$ be a summable family of vectors indexed by Γ in a normed space $\mathscr{E}$, and let $\{\gamma_n\}_{n=1}^\infty$ be an enumeration of any subset Γ_0 of indices such that $x_\gamma = 0$ for all γ not in Γ_0. (That such an enumeration exists is assured by Corollary 19.5.) Show that

$$\sum_{\gamma \in \Gamma} x_\gamma = \sum_{n=1}^\infty x_{\gamma_n}.$$

C. If $\{x_\gamma\}_{\gamma\in\Gamma}$ is a summable family of vectors in a Banach space $\mathscr{E}$, and if Γ' is any subset of Γ, then $\{x_\gamma\}_{\gamma\in\Gamma'}$ is also summable. Show also that, if we set $\Gamma'' = \Gamma\backslash\Gamma'$, then

$$\sum_{\gamma\in\Gamma} x_\gamma = \sum_{\gamma\in\Gamma'} x_\gamma + \sum_{\gamma\in\Gamma''} x_\gamma.$$

Show, more generally, that if $\{\Gamma_\delta\}_{\delta\in\Delta}$ is an arbitrary indexed partition of the index set Γ into disjoint subsets Γ_δ, then

$$\sum_{\gamma\in\Gamma} x_\gamma = \sum_{\delta\in\Delta} \left(\sum_{\gamma\in\Gamma_\delta} x_\gamma\right).$$

D. Show that if $\mathscr{E}$ is a finite dimensional Banach space, and if $\{x_\gamma\}_{\gamma\in\Gamma}$ is an arbitrary indexed family of vectors in $\mathscr{E}$, then $\{x_\gamma\}$ is summable if and only if it is absolutely summable. (Hint: See Problem 4N for the one dimensional case. Recall (Prob. 11J) that if $\mathscr{E}$ is finite dimensional with basis $\{x_1, \ldots, x_r\}$, then $\|\lambda_1 x_1 + \cdots + \lambda_r x_r\|_0 = |\lambda_1| + \cdots + |\lambda_r|$ defines a norm on $\mathscr{E}$ that is equivalent with the given one.)

> Our experience to date with the differences between the finite and infinite dimensional cases would lead us to expect the converse of Problem D to be valid, i.e., to guess that in an infinite dimensional normed space there should exist summable families of vectors that are not absolutely summable. That this is indeed the case is the substance of the (surprisingly deep) *Dvoretsky–Rogers theorem* (see [24]).

E. Let $\mathscr{E}$ be a Banach space and let $\{x_\gamma\}_{\gamma\in\Gamma}$ be an indexed family of vectors in $\mathscr{E}$.

(i) Show that $\{x_\gamma\}$ *fails* to satisfy the Cauchy criterion for indexed families (Prop. 19.4) if and only if there exists a number $\varepsilon_0 > 0$ and a corresponding disjoint infinite sequence $\{D_n\}$ of finite subsets of Γ such that

$$\left\|\sum_{\gamma\in D_n} x_\gamma\right\| \geq \varepsilon_0$$

for every n. Conclude that the family $\{x_\gamma\}$ is summable in $\mathscr{E}$ if and only if $\lim_n \|\sum_{\gamma\in D_n} x_\gamma\| = 0$ for every disjoint infinite sequence $\{D_n\}$ of finite subsets of Γ.

(ii) Suppose $\mathscr{E}$ is finite dimensional. Show in this case that $\{x_\gamma\}$ fails to satisfy the Cauchy criterion for indexed families if and only if there exists a sequence $\{D_n\}$ of finite subsets of Γ such that

$$\lim_n \left\|\sum_{\gamma\in D_n} x_\gamma\right\| = +\infty.$$

Conclude that the family $\{x_\gamma\}$ is summable if and only if

$$\sup_D \left\|\sum_{\gamma\in D} x_\gamma\right\| < +\infty,$$

where the supremum is taken over all finite subsets D of Γ. (In particular, then, this is true for $\mathscr{E} = \mathbb{C}$.)

F. Let $\mathscr{E}$ be a Banach space, and suppose given a sequence $\{x_n\}_{n=0}^{\infty}$ of vectors in $\mathscr{E}$.

(i) The series $\sum_{n=0}^{\infty} x_n$ is said to be *unconditionally convergent* if it is convergent to a sum x and if, for every permutation σ of $\mathbb{N}_0$, the series $\sum_{n=0}^{\infty} x_{\sigma(n)}$ also converges to x. (A permutation of an infinite set is defined, just as for finite sets, as a one-to-one mapping of the set onto itself. As a matter of fact, it is easy to see that if all of the permuted series $\sum_{n=0}^{\infty} x_{\sigma(n)}$ converge, then they all must converge to the same vector x.) Prove that $\sum_{n=0}^{\infty} x_n$ is unconditionally convergent if and only if the indexed family $\{x_n\}_{n \in \mathbb{N}_0}$ is summable, and that, in this event,

$$\sum_{n=0}^{\infty} x_n = \sum_{n \in \mathbb{N}_0} x_n.$$

(Hint: Use Problem E(i).)

(ii) A *subseries* of the series $\sum_{n=0}^{\infty} x_n$ is any series of the form $\sum_{k=0}^{\infty} x_{n_k}$ where $\{x_{n_k}\}_{k=0}^{\infty}$ is a subsequence of $\{x_n\}$. The series $\sum_{n=0}^{\infty} x_n$ is said to be *subseries convergent* if every subseries of the given series is convergent. Prove that $\sum_{n=0}^{\infty} x_n$ is subseries convergent if and only if the indexed family $\{x_n\}_{n \in \mathbb{N}_0}$ is summable. (Hint: Use Problem E(i).)

> Problem F, together with Problem 4N, yields a proof of the classical result that an infinite series of scalars is unconditionally convergent if and only if it is absolutely convergent.

G. Let $\mathscr{E}$ be a Banach space and let $\{x_\gamma\}_{\gamma \in \Gamma}$ be a summable family of vectors in $\mathscr{E}$. Then, according to Proposition 19.4, for any given $\varepsilon > 0$ there is a finite set D_ε of indices such that $\|\sum_{\gamma \in D} x_\gamma\| < \varepsilon$ whenever D is a finite set of indices such that $D \cap D_\varepsilon = \varnothing$, and hence such that

$$\left\| \sum_{\gamma \in D} \varepsilon_\gamma x_\gamma \right\| < \varepsilon$$

for any such D and any family $\{\varepsilon_\gamma\}$ of zeros and ones. Show that, in fact,

$$\left\| \sum_{\gamma \in D} t_\gamma x_\gamma \right\| < \varepsilon$$

whenever $D \cap D_\varepsilon = \varnothing$ and all of the scalars t_γ belong to the unit interval. (Hint: Each t_γ has a binary expansion

$$t_\gamma = \sum_{n=1}^{\infty} \frac{\varepsilon_{n,\gamma}}{2^n},$$

where every $\varepsilon_{n,\gamma}$ equals 0 or 1. Hence

$$\sum_{\gamma \in D} t_\gamma x_\gamma = \sum_{n=1}^{\infty} \frac{1}{2^n} \sum_{\gamma \in D} \varepsilon_{n,\gamma} x_\gamma.)$$

Use this observation to prove that for an *arbitrary bounded* family of scalars $\{\lambda_\gamma\}_{\gamma \in \Gamma}$, the family of vectors $\{\lambda_\gamma x_\gamma\}$ is summable along with $\{x_\gamma\}$. Show also, in the converse direction, that if an infinite series

$$\sum_{n=1}^{\infty} x_n$$

in $\mathscr{E}$ has the property that $\sum_{n=0}^{\infty} \lambda_n x_n$ is convergent for every bounded sequence of scalars $\{\lambda_n\}$, then $\{x_n\}_{n \in \mathbb{N}_0}$ is summable as an indexed family.

H. Let us equip an index family Γ with the counting measure κ. (See Example 7J; it doesn't matter which σ-ring of subsets of Γ we declare to be measurable so long as it contains all the singletons $\{\gamma\}$.) Show that for any separable Banach space $\mathscr{E}$ the collection of all absolutely summable vector families $\{x_\gamma\}_{\gamma \in \Gamma}$ in $\mathscr{E}$ coincides with the Banach space $\mathscr{L}_1(\Gamma; \mathscr{E})$ of Theorem 17.11, and that for any such family we have

$$\sum_{\gamma \in \Gamma} x_\gamma = \int_\Gamma x_\gamma \, d\kappa(\gamma)$$

(cf. Theorem 17.14 and Problem 7Q).

I. An indexed family $\{x_\gamma\}_{\gamma \in \Gamma}$ of vectors in a Banach space $\mathscr{E}$ is said to be *weakly summable* if the family of scalars $\{f(x_\gamma)\}_{\gamma \in \Gamma}$ is summable in $\mathbb{C}$ for every f in $\mathscr{E}^*$. Likewise, $\{x_\gamma\}$ is *weakly summable to sum* x if (for some x in $\mathscr{E}$)

$$\sum_\gamma f(x_\gamma) = f(x)$$

for every f in $\mathscr{E}^*$.

(i) Show that $\{x_\gamma\}$ is weakly summable if and only if the net $\{s_D\}_{D \in \mathscr{D}}$ of finite sums $s_D = \sum_{\gamma \in D} x_\gamma$ is weakly Cauchy in $\mathscr{E}$. (See Problem 15R.)

(ii) Show that the indexed family $\{e_n\}_{n \in \mathbb{N}_0}$ (Ex. 11H) is weakly summable in the space (c_0), but is not weakly summable to any sum in that space. Is the family $\{e_n\}$ weakly summable (or weakly summable to a sum x) in any of the spaces (ℓ_p)? (Hint: The space (ℓ_∞) contains (c_0) as a (weakly closed) subspace and the sequence $\{e_n\}$ lies in (c_0).)

(iii) If $\mathscr{E}$ is finite dimensional, then every weakly summable indexed family of vectors in $\mathscr{E}$ is (absolutely) summable.

J. Let $\{x_\gamma\}_{\gamma \in \Gamma}$ be a weakly summable indexed family of vectors in a Banach space $\mathscr{E}$. Prove that

$$Tf = \{f(x_\gamma)\}_{\gamma \in \Gamma}, \qquad f \in \mathscr{E}^*,$$

defines a bounded linear transformation T of $\mathscr{E}^*$ into the normed space $(\ell_1; \Gamma)$ of all summable indexed families $\{\lambda_\gamma\}_{\gamma \in \Gamma}$ of complex numbers. (Hint: The space $(\ell_1; \Gamma)$ is complete by Problem H. Show that T is closed.)

K. Let $\mathscr{E}$ be a Banach space. An indexed family $\{f_\gamma\}_{\gamma \in \Gamma}$ of functionals in $\mathscr{E}^*$ is said to be *weak* summable* if the family of scalars $\{f_\gamma(x)\}_{\gamma \in \Gamma}$ is summable in $\mathbb{C}$ for every x in $\mathscr{E}$, and to be *weak* summable to sum* f if (for some f in $\mathscr{E}^*$)

$$\sum_\gamma f_\gamma(x) = f(x)$$

for every x in $\mathscr{E}$. Show, for an arbitrary Banach space $\mathscr{E}$, that an indexed family $\{f_\gamma\}$ in $\mathscr{E}^*$ that is weak* summable is weak* summable to some f in $\mathscr{E}^*$. Conclude that if $\mathscr{E}$ is a reflexive Banach space, then every weakly summable indexed family in $\mathscr{E}$ is weakly summable to a sum in $\mathscr{E}$.

It is a remarkable result due to Bessaga and Pelczynski [9] that if $\mathscr{E}$ is a Banach space in which there exists a *single* indexed family that is weakly summable but not summable (in norm), then $\mathscr{E}$ has a subspace that is equivalent to the space (c_0). (The converse result is obvious in view of Problem I(ii).) Thus in the presence of any condition on $\mathscr{E}$ that prevents it from having such a subspace, every weakly summable indexed family in $\mathscr{E}$ is necessarily summable. Reflexivity is one such condition, to be sure (Ex. 16C; Prob. 16H), but there are many others. Thus, for example (Orlicz [51]): *If $\mathscr{E}$ is weakly sequentially complete, then every weakly summable indexed family of vectors in $\mathscr{E}$ is summable.*

In Problems F and G it was seen that several senses of summability are equivalent for families of vectors indexed by $\mathbb{N}_0$ (or any other index system with the same order type). Since the Cauchy criterion is the critical ingredient in these arguments, and since the Cauchy criterion may fail for weak summability, as noted in Problem I, the fact developed in the following problem comes as a distinctly pleasant surprise.

L. (Orlicz–Pettis Theorem) Let $\mathscr{E}$ be a Banach space, let $X = \{x_n\}_{n=0}^\infty$ be a sequence in $\mathscr{E}$ such that the series $\sum_{n=0}^\infty x_n$ is *weakly convergent* to a sum x, that is, such that $\sum_{n=0}^\infty f(x_n) = f(x)$ for every f in $\mathscr{E}^*$, and let $\mathscr{M}$ be the subspace of $\mathscr{E}$ spanned by X.

(i) Let $\{f_p\}_{p=1}^\infty$ be a sequence in the unit ball $(\mathscr{E}^*)_1$ with the property that there exists a bounded linear functional f_0 on $\mathscr{E}$ such that $\lim_p f_p(z) = f_0(z)$ for every z in $\mathscr{M}$. Show that

$$\lim_p \sum_{n=0}^\infty f_p(x_n) = \sum_{n=0}^\infty f_0(x_n).$$

(Hint: The vector x belongs to $\mathscr{M}$.)

(ii) Suppose further that the series $\sum_{n=0}^\infty x_n$ is *weakly subseries convergent* in the sense that, for every infinite subset J of $\mathbb{N}_0$, there exists a vector x_J in $\mathscr{E}$ such that $\sum_{n=0}^\infty \chi_J(n) f(x_n) = f(x_J)$ for every f in $\mathscr{E}^*$. Show that the indexed family $\{x_n\}_{n \in \mathbb{N}_0}$ is weakly summable as defined in Problem I.

(iii) Let $\sum_{n=0}^\infty x_n$ be weakly subseries convergent as in (ii), and let T denote the bounded linear transformation of $\mathscr{E}^*$ into (ℓ_1) introduced in Problem J (existent because of (ii)). For each infinite subset J of $\mathbb{N}_0$ write F_J for the bounded linear functional on (ℓ_1) defined by χ_J (regarded as an element of (ℓ_∞)), and let the sequence $\{f_p\}$ in $\mathscr{E}^*$ be as in (i). Show that

$$\lim_p F_J(Tf_p) = F_J(Tf_0) \qquad (8)$$

for every J, and conclude that the sequence $\{Tf_p\}_{p=1}^\infty$ converges to Tf_0 in (ℓ_1). (Hint: To verify (8) one may simply replace the given series $\sum_{n=0}^\infty x_n$ in (i) by the series $\sum_{n=0}^\infty \chi_J(n) x_n$, weakly convergent because $\sum_{n=0}^\infty x_n$ is weakly subseries convergent. Since the sequence $\{Tf_p\}_{p=1}^\infty$ is bounded in (ℓ_1) (by $\|T\|$) and the functions χ_J span (ℓ_∞) as a Banach space (Ex. 11H), one may conclude that the sequence $\{f_p\}$ converges weakly to f_0 (Prob. 15P). Recall Problem 15D.)

(iv) Use (iii) to prove that if $\sum_{n=0}^{\infty} x_n$ is weakly subseries convergent and if $\{f_p\}_{p=1}^{\infty}$ is an arbitrary infinite sequence in $(\mathscr{E}^*)_1$, then $\{f_p\}$ has a subsequence $\{f_{p_k}\}$ such that $\{Tf_{p_k}\}$ is convergent in (ℓ_1), and conclude that $T((\mathscr{E}^*)_1)^-$ is compact in (ℓ_1). (Hint: Since $\mathscr{M}$ is separable, the sequence $\{f_p|\mathscr{M}\}_{p=1}^{\infty}$ has a weak* convergent subsequence $\{f_{p_k}|\mathscr{M}\}$ (Prob. 15P). Use the Hahn–Banach theorem to extend $\lim_k (f_{p_k}|\mathscr{M})$ to a bounded linear functional f_0 in $(\mathscr{E}^*)_1$, and invoke (iii) to conclude that $T((\mathscr{E}^*)_1)^-$ is sequentially compact. Finally, recall Prob. 4Q.) Thus T is a compact linear transformation (Prob. 17Y).

(v) (Pettis [52]) Prove that a weakly subseries convergent infinite series in a Banach space is actually subseries convergent and therefore summable (in norm) as an indexed family. (Hint: If for each nonnegative integer k we write P_k for the projection

$$P_k\{\xi_0,\ldots,\xi_k,\xi_{k+1},\ldots\} = \{0,\ldots,0,\xi_k,\xi_{k+1},\ldots\},$$

then $\|P_k x\| \to 0$ for each x in (ℓ_1), and therefore, by a standard compactness argument, $\{P_k\}_{k=1}^{\infty}$ tends to zero uniformly on $T((\mathscr{E}^*)_1)$; in other words, $\lim_k \|P_k T\| = 0$. Use the fact that

$$\left\| x_J - \sum_{n=0}^{N} \chi_J(n)x_n \right\| \le \sup\left\{ \sum_{n=N+1}^{\infty} |f(x_n)| : f \in (\mathscr{E}^*)_1 \right\}$$

$$= \|P_{N+1}T\|.)$$

M. (i) Let $\{x_n\}_{n=-\infty}^{+\infty}$ be a two-way infinite linearly independent sequence in a Banach space $\mathscr{E}$, and let $\mathscr{L}$ denote the collection of all complex sequences $\{\alpha_n\}_{n=-\infty}^{+\infty}$ such that the series $\sum_{n=-\infty}^{+\infty} \alpha_n x_n$ is convergent in $\mathscr{E}$ in the primary sense. Verify that

$$\|a\| = \sup_{M,N\in\mathbb{N}_0} \left\| \sum_{n=-M}^{N} \alpha_n x_n \right\|$$

defines a norm on $\mathscr{L}$ with respect to which $\mathscr{L}$ is a Banach space, and conclude that if $\{x_n\}$ is a basis for $\mathscr{E}$ in the primary sense, then each α_n is a bounded linear functional on $\mathscr{E}$. Show also that if $\mathscr{L}_C$ denotes the (generally larger) collection of all complex sequences $a = \{\alpha_n\}_{n=-\infty}^{+\infty}$ such that the series $\sum_{n=-\infty}^{+\infty} \alpha_n x_n$ is convergent in the Cauchy sense, then

$$\|a\| = \sup_{N\in\mathbb{N}_0} \left\| \sum_{n=-N}^{+N} \alpha_n x_n \right\|$$

defines a norm on $\mathscr{L}_C$ with respect to which $\mathscr{L}_C$ becomes a Banach space, and conclude that if $\{x_n\}$ is a basis for $\mathscr{E}$ in the Cauchy sense, then the functional $f_n(x) = \alpha_n(x) + \alpha_{-n}(x)$ is bounded for each nonnegative integer n.

(ii) Use the results of (i) to derive appropriate versions of Theorem 19.8 for bases indexed by $\mathbb{Z}$.

N. Let $\mathscr{E}$ be a Banach space, let (X, A) be a biorthogonal system for $\mathscr{E}$ such that $X = \{x_n\}$ and $A = \{\alpha_n\}$ are both indexed by $\mathbb{Z}$, and for each nonnegative integer N set

$$E_N x = \sum_{n=-N}^{+N} \alpha_n(x)x_n, \qquad x \in \mathscr{E}.$$

Show that X is a basis in the Cauchy sense for the subspace $\mathcal{M}$ of $\mathcal{E}$ that it spans if and only if the sequence $\{E_N\}_{N=0}^{\infty}$ is uniformly bounded on $\mathcal{M}$. (Hint: Show first that $E_M E_N = E_N E_M = E_{M \wedge N}$, and then adapt the proof of Proposition 19.10.) Show also that X is a basis in the primary sense for $\mathcal{M}$ if and only if the double sequence $\{E_{M,N}\}_{M,N=0}^{\infty}$ is uniformly bounded on $\mathcal{M}$, where

$$E_{M,N}x = \sum_{n=-M}^{+N} \alpha_n(x)x_n, \qquad x \in \mathcal{E},$$

for all nonnegative integers M and N.

O. A sequence $X = \{x_n\}_{n=0}^{\infty}$ in a Banach space $\mathcal{E}$ is said to be a *weak basis* for $\mathcal{E}$ if for every vector x in $\mathcal{E}$ there exists a unique sequence $\{\alpha_n\}_{n=0}^{\infty}$ of complex numbers such that the series $\sum_{n=0}^{\infty} \alpha_n x_n$ converges weakly to x (see Problem L). Prove that a weak basis for a Banach space $\mathcal{E}$ is actually a Schauder basis for $\mathcal{E}$. (Hint: Use Proposition 19.7 to show that X and the corresponding sequence A of coefficient functionals form a biorthogonal system; then use the uniform boundedness theorem to prove that the sequence of expansion operators is uniformly bounded.) Develop the analogous results for a weak basis indexed by $\mathbb{Z}$.

P. If $X = \{x_n\}_{n=0}^{\infty}$ and $A = \{\alpha_n\}_{n=0}^{\infty}$ form a biorthogonal system for a Banach space $\mathcal{E}$, then $(A, j(X))$ is a biorthogonal system for the dual space $\mathcal{E}^*$ (where j denotes, as usual, the natural embedding of $\mathcal{E}$ in $\mathcal{E}^{**}$; see Chapter 15). Moreover, if X is a basis for $\mathcal{E}$, then A is a basis for the subspace of $\mathcal{E}^*$ spanned by A. (Hint: The expansion operators $\tilde{E}_N$ for the system $(A, j(X))$ are the adjoints of the expansion operators E_N for (X, A).) Develop analogous results for biorthogonal systems indexed by $\mathbb{Z}$.

Q. A sequence $\{f_n\}_{n=0}^{\infty}$ in the dual $\mathcal{E}^*$ of a Banach space $\mathcal{E}$ is a *weak* basis* for $\mathcal{E}^*$ if for every element f of $\mathcal{E}^*$ there exists a unique sequence $\{\alpha_n\}_{n=0}^{\infty}$ of complex numbers such that the sequence $\{\sum_{n=0}^{N} \alpha_n f_n\}_{N=0}^{\infty}$ converges to f in the weak* topology. Show that if $X = \{x_n\}_{n=0}^{\infty}$ is a Schauder basis for $\mathcal{E}$ and $A = \{\alpha_n\}_{n=0}^{\infty}$ denotes the corresponding sequence of coefficient functionals, then A is a weak* basis for $\mathcal{E}^*$. Develop analogous results for bases indexed by $\mathbb{Z}$.

R. An indexed family $X = \{x_\gamma\}_{\gamma \in \Gamma}$ of vectors in a Banach space $\mathcal{E}$ is an *indexed basis* for $\mathcal{E}$ if for each vector x in $\mathcal{E}$ there exists a unique indexed family $\{\alpha_\gamma\}_{\gamma \in \Gamma}$ of complex numbers such that

$$x = \sum_{\gamma \in \Gamma} \alpha_\gamma x_\gamma. \tag{9}$$

If X is an indexed basis for $\mathcal{E}$ then the scalars $\{\alpha_\gamma\}$ appearing in (9) are the *coefficients* of x with respect to the indexed basis X, the indexed family $\{\alpha_\gamma\}_{\gamma \in \Gamma}$ is the *coefficient family* of x with respect to X, and the expression (9) is the *expansion* of x with respect to the indexed basis X.

(i) Let $\{x_\gamma\}_{\gamma \in \Gamma}$ be an indexed basis for $\mathcal{E}$, and for each index γ write α_γ for the linear functional on $\mathcal{E}$ that assigns to each vector x its coefficient with respect to X having index γ, so that

$$x = \sum_{\gamma \in \Gamma} \alpha_\gamma(x)x_\gamma, \qquad x \in \mathcal{E}.$$

Prove that α_γ is bounded for each index γ, so that X and the indexed family $A = \{\alpha_\gamma\}_{\gamma \in \Gamma}$ of *coefficient functionals* constitute an *indexed biorthogonal system* for $\mathscr{E}$, that is, a pair of similarly indexed families $X = \{x_\gamma\}$ and $A = \{\alpha_\gamma\}$ in $\mathscr{E}$ and $\mathscr{E}^*$, respectively, satisfying the biorthogonality relation $\alpha_{\gamma'}(x_{\gamma''}) = \delta_{\gamma'\gamma''}$ (the Kronecker delta) for all γ', γ'' in Γ. (Hint: Let $\mathscr{L}$ denote the linear space consisting of all those families of scalars $\{\alpha_\gamma\}$ indexed by Γ with the property that the family $\{\alpha_\gamma x_\gamma\}_{\gamma\in\Gamma}$ is summable in $\mathscr{E}$. Use Problem 4N to show that the sum $\sum_{\gamma\in\Gamma}|\alpha_\gamma f(x_\gamma)|$ is finite for each f in $\mathscr{E}^*$ and $\{\alpha_\gamma\}$ in $\mathscr{L}$. Then use this fact to show that

$$\|a\| = \sup_D \left\| \sum_{\gamma\in D} \alpha_\gamma x_\gamma \right\|,$$

where the indicated supremum is taken over all finite subsets D of Γ, is a norm on $\mathscr{L}$ turning it into a Banach space; follow the proofs of Proposition 19.7 and Theorem 19.8.)

(ii) Let $(X = \{x_\gamma\}_{\gamma\in\Gamma}, \ A = \{\alpha_\gamma\}_{\gamma\in\Gamma})$ be an indexed biorthogonal system for a Banach space $\mathscr{E}$, and for each finite subset D of Γ set

$$E_D x = \sum_{\gamma\in D} \alpha_\gamma(x)x_\gamma, \qquad x \in \mathscr{E}.$$

Show that X is an indexed basis for $\mathscr{E}$ if and only if X spans $\mathscr{E}$ and the family of *expansion operators* $\{E_D\}$ is uniformly bounded.

S. Let $\{x_\gamma\}_{\gamma\in\Gamma}$ be an indexed basis for a Banach space $\mathscr{E}$, and let $\{\alpha_\gamma\}$ be the corresponding family of coefficient functionals. Let $\Gamma = \bigcup_{\delta\in\Delta} \Gamma_\delta$ be an arbitrary indexed partition of Γ, and for each index δ let $\mathscr{M}_\delta$ denote the subspace of $\mathscr{E}$ spanned by $\{x_\gamma : \gamma\in\Gamma_\delta\}$. Prove that Γ_δ is an indexed basis for $\mathscr{M}_\delta$, $\delta\in\Delta$, and further that for each x in $\mathscr{E}$ there exists a unique indexed family $\{x_\delta\}_{\delta\in\Delta}$ with x_δ in $\mathscr{M}_\delta$, $\delta\in\Delta$, such that $x = \sum_{\delta\in\Delta} x_\delta$. Show too that for each index δ in Δ

$$\mathscr{M}_\delta = \bigcap_{\gamma\notin\Gamma_\delta} \mathscr{K}(\alpha_\gamma),$$

where $\mathscr{K}(\alpha_\gamma)$ denotes the null space of α_γ. What are the corresponding results for an ordinary Schauder basis?

T. (i) Give an example of an uncountable indexed basis in a nonseparable Banach space.

(ii) Let $\{x_\gamma\}_{\gamma\in\Gamma}$ be an indexed basis for a Banach space $\mathscr{E}$, let M be a subset of $\mathscr{E}$ of $\mathscr{E}$ that spans $\mathscr{E}$ as a Banach space, and for each vector y in M set $N_y = \{\gamma\in\Gamma : \alpha_\gamma(y) \neq 0\}$. Show that

$$\Gamma = \bigcup_{y\in M} N_y.$$

Use this fact to verify that any two indexed bases for $\mathscr{E}$ have the same cardinal number. (Hint: Use Corollary 19.5, Problem 1T, and the Cantor–Bernstein theorem (Prob. 1R).) Conclude also that if $\mathscr{E}$ is separable, then an indexed basis for $\mathscr{E}$ must be countable.

This last result shows that an indexed basis for a separable infinite dimensional Banach space can always be taken to be indexed by $\mathbb{N}_0$, in which case the requirement that the indexed family $\{\alpha_n(x)x_n\}$ be summable is equivalent to the requirement that the series $\sum_{n=0}^{\infty} \alpha_n(x)x_n$ converge unconditionally (Prob. F). For this reason the notion we have called an *indexed basis* is also frequently called an *unconditional basis*.

U. Let $\mathscr{E}$ be a Banach space, let $X = \{x_n\}_{n \in \mathbb{N}_0}$ be an indexed basis for $\mathscr{E}$, and let $A = \{\alpha_n\}_{n \in \mathbb{N}_0}$ be the corresponding sequence of coefficient functionals.

(i) Show that $(A, j(X))$ is an indexed biorthogonal system for $\mathscr{E}^*$ (Prob. R), and that the family of expansion operators

$$\tilde{E}_D f = \sum_{n \in D} f(x_n) \alpha_n$$

of this system is uniformly bounded. Conclude that if f and φ denote elements of $\mathscr{E}^*$ and $\mathscr{E}^{**}$, respectively, then

$$\sum_n |f(x_n)\varphi(\alpha_n)| < +\infty. \tag{10}$$

(Hint: Recall Problem P.)

(ii) Suppose now, in addition, that $\mathscr{E}^*$ is weakly sequentially complete (see Problem 15R). Show that for each f in $\mathscr{E}^*$ the series

$$\sum_{n=0}^{\infty} f(x_n)\alpha_n$$

is unconditionally convergent to f, and conclude that the sequence A is an indexed basis for $\mathscr{E}^*$. (Hint: If $\{N_k\}$ is an arbitrary strictly increasing sequence of nonnegative integers, then the subseries

$$\sum_{k=0}^{\infty} f(x_{N_k})\alpha_{N_k}$$

is weakly summable by (10). Since $\mathscr{E}^*$ is weakly sequentially complete, it follows that $\sum_{n=0}^{\infty} f(x_n)\alpha_n$ is weakly subseries convergent in $\mathscr{E}^*$. Use the Orlicz–Pettis theorem (Prob. L).)

(iii) (Karlin [39]) Conclude, in summary, that if $\mathscr{E}$ is a separable Banach space that admits an indexed basis, and if $\mathscr{E}^*$ is weakly sequentially complete, then $\mathscr{E}^*$ is also a separable Banach space that admits an indexed basis.

It is an immediate consequence of this last result that if a separable Banach space $\mathscr{E}$ has a dual space $\mathscr{E}^*$ that is both nonseparable and weakly sequentially complete, then $\mathscr{E}$ *does not admit* an indexed basis. As it turns out, the dual $\mathscr{M}([0, 1])$ of the space $\mathscr{C}([0,1])$ is weakly sequentially complete according to a result of Banach and Mazur [2; Satz 2]. Since $\mathscr{M}([0, 1])$ is visibly nonseparable, this shows that the familiar space $\mathscr{C}([0, 1])$ is an example of a separable Banach space that does not admit an indexed, or unconditional, basis. (That $\mathscr{C}([0, 1])$ possesses an ordinary Schauder basis was shown in Example D.)

V. An indexed basis $\{x_\gamma\}_{\gamma \in \Gamma}$ for a Banach space $\mathscr{E}$ is called an *absolute basis* for $\mathscr{E}$ if, in the expansion

$$x = \sum_{\gamma \in \Gamma} \alpha_\gamma(x) x_\gamma$$

of an arbitrary vector x in $\mathscr{E}$, the indexed family $\{\alpha_\gamma(x)x_\gamma\}_{\gamma \in \Gamma}$ is absolutely summable.

Let $\{x_\gamma\}$ be such an absolute basis, let $(\ell_1; \Gamma)$ denote the Banach space of all absolutely summable families of scalars indexed by Γ (cf. Problem J), and define

$$Tx = \{\alpha_\gamma(x)\|x_\gamma\|\}_{\gamma \in \Gamma}, \qquad x \in \mathcal{E}.$$

Show that T is a linear transformation of $\mathcal{E}$ onto $(\ell_1; \Gamma)$. Show further that T is bounded, and is therefore an equivalence between $\mathcal{E}$ and $(\ell_1; \Gamma)$. Thus, in particular, a separable Banach space admits an absolute basis if and only if it is equivalent to (ℓ_1). (Hint: Use the continuity of the coefficient functionals α_γ to prove that T is closed.)

Bibliography

1. Banach, S., *Théorie des opérations linéaires*, Warsaw, 1932.
2. Banach, S. and S. Mazur, Zur Theorie der linearen Dimension, Studia Math., *4* (1933), 100–112.
3. Bary, N. K., *Trigonometric series*, New York, 1964.
4. Bennett, G. and N. J. Kalton, Consistency theorems for almost convergence, Trans. Amer. Math. Soc., *198* (1974), 23–43.
5. Berberian, S. K., *Measure and integration*, New York, 1965.
6. Berberian, S. K., The product of two measures, Amer. Math. Monthly, *69* (1962), 961–968.
7. Berberian, S. K., Counterexamples in Haar measure, Amer. Math. Monthly, *73* (1966), 135–140.
8. Berberian, S. K. and J. F. Jakobsen, A note on Borel sets, Amer. Math. Monthly, *70* (1963), 55.
9. Bessaga, C. and A. Pelczynski, On bases and unconditional convergence of series in Banach spaces, Studia Math., *17* (1958), 151–164.
10. Bourbaki, N., *Éléments de mathématique: Livre* I, *Théorie des ensembles*, Paris, 1966.
11. Bourbaki, N., *Éléments de mathématique: Livre* III, *Topologie générale*, Paris, 1948.
12. Bourbaki, N., *Éléments de mathématique: Livre* V, *Espaces vectoriels topologiques*, Paris, 1967.
13. Brown, A., On the adjoint of a closed transformation, Proc. Amer. Math. Soc., *15* (1964), 239–240.
14. de Branges, L., The Stone–Weierstrass theorem, Proc. Amer. Math. Soc., *10* (1959), 822–824.
15. Dieudonné, J., Sur le théorème de Lebesgue–Nikodym (III), Ann. de l'Université de Grenoble, *23* (1947-8), 25–53.

16. Dieudonné, J., Sur le théorème de Lebesgue–Nikodym (IV), J. Indian Math. Soc., N. S. *15* (1951), 77–86.

17. Dieudonné, J., Sur le théorème de Lebesgue–Nikodym (V), Canad. J. Math., *3* (1951), 129–139.

18. Dinculeanu, N., *Vector measures*, Berlin, 1966.

19. Dixon, J. D., A brief proof of Cauchy's integral theorem, Proc. Amer. Math. Soc., *29* (1971), 625–626.

20. Dugundji, J., *Topology*, Boston, 1966.

21. Dunford, N., Spectral theory I. Convergence to projections, Trans. Amer. Math. Soc., *54* (1943), 185–217.

22. Dunford, N. and B. J. Pettis, Linear operations on summable functions, Trans. Amer. Math. Soc., *47* (1940), 323–392.

23. Dunford, N. and J. Schwartz, *Linear operators Part* I: *General theory*, New York, 1958.

24. Dvoretzky, A. and C. A. Rogers, Absolute and unconditional convergence in normed linear spaces, Proc. Nat. Acad. Sci., *36* (1950), 192–197.

25. Enflo, P., A counterexample to the approximation problem in Banach spaces, Acta Math., *130* (1973), 309–317.

26. Federer, H., *Geometric measure theory*, New York, 1969.

27. Gamelin, T. W., *Uniform algebras*, Englewood Cliffs, N.J., 1969.

28. Gelfand, I., Normierte Ringe, Mat. Sbornik, (N.S.) *9* (1941), 3–24.

29. Goffman, C. and G. Pedrick, *First course in functional analysis*, Englewood Cliffs, N.J., 1965.

30. Halmos, P. R., *Measure theory*, New York, 1950.

31. Halmos, P. R., *Naive set theory*, New York, 1974.

32. Halmos, P. R., *Finite dimensional vector spaces*, New York, 1974.

33. Halmos, P. R. and J. von Neumann, Operator methods in classical mechanics II, Ann. Math., *43* (1942), 332–350.

34. Hausdorff, F., *Mengenlehre* (dr. Aufl.), New York, 1944.

35. Hausdorff, F., Zur Theorie der linearen metrischen Räume, J. für d. reine u. angew. Math., *167* (1932), 294–311.

36. Hewitt, E. and K. Stromberg, *Real and abstract analysis*, New York, 1965.

37. Ionescu Tulcea, A. and C. Ionescu Tulcea, *Topics in the theory of lifting*, New York, 1969.

38. Jacobson, N., *Lectures in abstract algebra*: *Vol.* II, *Linear algebra*, New York, 1953.

39. Karlin, S., Bases in Banach spaces, Duke J. Math., *15* (1948), 971–985.

40. Kelley, J. L., *General topology*, Princeton, 1953.

41. Kelley, J. L., Decomposition and representation theorems in measure theory, Math. Ann., *163* (1966), 89–94.

42. Kuratowski, K., *Topology*, New York, 1966.

43. Lavrentiev, M., *Sur les fonctions d'une variable complex representables par des séries de polynomes*, Paris, 1936.

44. Lorentz, G. G., A contribution to the theory of divergent sequences, Acta Math., *80* (1948), 167–190.

45. Luther, N. Y., Unique extension and product measures, Canad. J. Math., *19* (1967), 757-763.

46. Mackey, G. W., On convex topological linear spaces, Trans. Amer. Math. Soc., *60* (1946), 519-537.

47. MacLane, S., *Homology*, New York, 1975.

48. Mergel'yan, S. N., Uniform approximation to functions of a complex variable, Uspehi Mat. Nauk, 7 vyp. 2 (1952), 31-122 (A.M.S. Transl., Ser. 1., Vol. 3).

49. Mukherjea, A., A remark on Tonelli's theorem on integration in product spaces, Pac. J. Math., *42* (1972), 177-185.

50. Mukherjea, A., Remark on Tonelli's theorem on integration in product spaces—II, Ind. U. Math. J., *23* (1974), 679-684.

51. Orlicz, W., Beiträge zur Theorie der Orthogonalentwicklungen II, Studia Math., *1* (1929), 241-255.

52. Pettis, B. J., On integration in vector spaces, Trans. Amer. Math. Soc., *44* (1938), 277-304.

53. Riesz, F., *Les systèmes d'équations linéaires à une infinité d'inconnues*, Paris, 1913.

54. Riesz, F., Sur l'existence de la dérivée des fonctions monotones et sur quelques problèmes qui s'y rattachent, Acta Szeged, *5* (1930-2), 208-221.

55. Robertson, A. P. and W. Robertson, *Topological vector spaces*, Cambridge, 1964.

56. Rosenblum, M., On the operator equation $BX - XA = Q$, Duke J. Math., *23* (1956), 263-269.

57. Rudin, W., *Real and complex analysis*, 2nd Ed., New York, 1974.

58. Saks, S., *Theory of the integral*, New York, 1937.

59. Schauder, J., Zur Theorie stetiger Abbildungen in Funktionalräumen, Math. Zeit., *26* (1927), 47-65.

60. Schwartz, J., A note on the space L_p^*, Proc. Amer. Math. Soc., *2* (1951), 270-275.

61. Segal, I. E., Equivalences of measure spaces, Amer. J. Math., *73* (1951), 275-313.

62. Singer, I., *Bases in Banach spaces*, New York, 1970.

63. Stone, M. H., The theory of representations for Boolean algebras, Trans. Amer. Math. Soc., *40* (1936), 37-111.

64. Vituškin, A. G., Analytic capacity of sets in problems in approximation theory, Uspehi Math. Nauk, *22* vyp. 6 (1967), 141-199 (Russian Math. Surveys, *22* (1967), 139-200).

65. Weil, A., *Sur les espaces à structure uniforme*, Paris, 1938.

66. Whyburn, G. T., *Analytic topology*, Mem. Amer. Math. Soc., No. 28, New York, 1942.

67. Yang, K. -W., A note on reflexive Banach spaces, Proc. Amer. Math. Soc., *18* (1967), 859-861.

Index

Index

References
to the
examples, propositions, and problems

Ex.1A	8	Prop.2.2	24, 25, 31
Ex.1B	33, 45	Prop.2.3	25
Ex.1C	10	Prob.2A	13, 31, 249
Prob.1A	9, 10	Prop.2B	14
Prob.1C	6, 9	Prop.2C	26, 279
Prob.1D	74, 106, 116, 197, 413, 414	Prop.2D	30, 348
Prob.1F	69, 174, 175	Prop.2E	23, 31, 249
Prob.1G	8, 60, 154, 158, 385	Prop.2F	14
Prob.1H	158	Prop.2G	20, 28, 259, 338
Prob.1I	52, 60, 74, 147, 149, 154,	Prop.2H	20, 29, 106, 265, 283, 343
	206, 289, 327, 387	Prob.2J	29
Prob.1J	8, 147, 162	Prob.2L	20, 21, 29, 393, 401
Prob.1K	60, 146, 147, 154, 162	Prob.2M	20, 21
Prob.1L	106, 108, 179, 181, 381	Prob.2O	259
Prob.1M	45	Prob.2P	269, 403
Prob.1P	11, 49, 135, 237, 243, 310,	Prob.2Q	24
	334	Ex.3A	38, 39, 110, 137
Prob.1R	26, 103, 447	Ex.3B	42, 44, 50, 110, 182, 195, 208
Prob.1S	157	Ex.3C	45, 62, 104
Prob.1T	26, 49, 447	Ex.3D	67, 217, 251
Prob.1U	109	Ex.3E	87, 244
Prob.1V	109, 207	Ex.3F	81, 110, 137
Prob.1W	50, 109, 110, 182, 195, 411	Ex.3G	81, 198, 336
Ex.2A	15, 22	Ex.3H	37
Ex.2B	20, 28, 221, 338, 343	Ex.3I	38
Ex.2C	25	Ex.3J	50, 51, 67, 157
Ex.2D	28, 215, 221, 225, 290, 304,	Ex.3K	265, 279
	433	Ex.3L	322
Ex.2F	17, 19	Ex.3M	48, 233, 234, 235, 313
Ex.2H	106, 116, 117, 130, 197, 412,	Ex.3N	48, 233, 298
	424	Ex.3O	45
Ex.2I	17	Ex.3P	52, 139
Ex.2J	73, 116, 239	Ex.3Q	65, 118, 119, 123, 126, 133,
Ex.2L	23, 29, 124, 249, 266, 341		243, 429
Ex.2M	117, 130, 307	Ex.3R	59, 235, 281, 320, 412
Ex.2N	21, 29	Prop.3.2	35, 50, 320
Prop.2.1	23, 25	Prop.3.3	40, 217, 272, 361

Graduate Texts in Mathematics

Soft and hard cover editions are available for each volume up to vol. 14, hard cover only from Vol. 15